THE WORLD ENVIRONMENT 1972–1992

*The problems that overwhelm us today
are precisely those we failed to solve decades ago.*
M. K. Tolba (1982)

The United Nations Environment Programme

The United Nations Environment Programme (UNEP) was established in 1972 by the General Assembly of the United Nations with a mandate to 'safeguard and enhance the environment for the benefit of present and future generations'. It does so with the help of a small Secretariat in Nairobi, Kenya, making it the first United Nations body with headquarters in a developing country.

Working with other members of the international community, UNEP helps to monitor, assess and manage the global environment. Earthwatch facilities use satellite data and aerial photography, while a world-wide network of collection-points gathers, collates and distributes the data. UNEP experts translate the data into workable information for today's decision-makers.

UNEP provides a range of services for protecting the world's environment: they include legal instruments to protect the ozone layer; digitalized maps for use by national planners; an international regime to manage shared water resources; advice to industry, and agreements to protect the world's seas from land-based sources of pollution.

THE WORLD ENVIRONMENT 1972–1992

Two decades of challenge

Edited by
Mostafa K. Tolba, United Nations Environment Programme,
Nairobi, Kenya and
Osama A. El-Kholy, Cairo University, Egypt

in association with
E. El-Hinnawi, National Research Centre, Egypt,
M. W. Holdgate, IUCN–The World Conservation Union, Switzerland,
D. F. McMichael, Environment and Heritage Consultants, Australia, and
R. E. Munn, University of Toronto, Canada.

Published by Chapman & Hall
on behalf of
The United Nations Environment Programme

UNEP

CHAPMAN & HALL

London · Glasgow · New York · Tokyo · Melbourne · Madras

Published by Chapman & Hall, 2–6 Boundary Row, London SE1 8HN

Chapman & Hall, 2-6 Boundary Row, London SE1 8HN, UK

Blackie Academic & Professional, Wester Cleddens Road, Bishopbriggs, Glasgow G64 2NZ, UK

Chapman & Hall, 29 West 35th Street, New York, NY 10001, USA

Chapman & Hall Japan, Thomson Publishing Japan, Hirakawacho Nemoto Building, 6F, 1-7-11 Hirahawa-cho, Chiyoda-ku, Tokyo 102, Japan

Chapman & Hall Australia, Thomas Nelson Australia, 102 Dodds Street, South Melbourne, Victoria 3205, Australia

Chapman & Hall India, R. Seshadri, 32 Second Main Road, CIT East, Madras 600 035, India

First edition 1992

© 1992 United Nations Environment Programme
    P.O. Box 30552, Nairobi, Kenya

ISBN 0 412 46990 1 (Hardback)
ISBN 0 412 47000 4 (Paperback)

Prepared for publication by Anagram Editorial Service, Guildford, UK, and Hardlines Illustration and Design, Charlbury, Oxford, UK.
Index compiled by Paul Nash.

Printed in Great Britain at the University Press, Cambridge.

A catalogue record for this book is available from the British Library.
Library of Congress Cataloging-in-Publication data available.

Contents

Foreword

Our planet is under siege. Assaults on the atmosphere - the greenhouse effect, the depletion of the ozone layer and increasing air pollution - pose a still-unquantified threat to human life. The dumping of hazardous wastes, and land-based sources of pollution, present a similar threat to the oceans. On land we are destroying a tropical forest the size of Austria every year, and more than a hundred species of wild plants and animals are lost forever each day.

When the General Assembly of the United Nations established UNEP it charged us with reporting on the changing state of the world's environment, tracking the underlying causes of change, and working with governments to develop responses to those changes.

Every year since 1974, UNEP has produced a *State of the Environment* report, focusing on one or more emerging environmental issues and always stressing the human factor - the impact of environmental quality on people and society. Three times since its inception (1982, 1987 and 1992) UNEP has undertaken a more wide-ranging study. The results of the present study are the most disturbing of the three.

However, not all the signs are negative. Throughout the 1970s and into the 1980s, UNEP was able to report progress in some important areas. Environmental monitoring capacity was being rapidly improved in many parts of the world, and Ministries of the Environment were being set up in an attempt to deal with environmental threats in a more coherent way.

Now I am obliged to report to governments and the public that progress has slowed. The commitment to set up ministries and to enter into international agreements has not always led to an equal commitment to action. Environment Ministries exist, but their role in national decision-making is frequently marginal. Agreements have been entered into freely, but the will to enforce them has often been lacking.

There is a paradox here. On the one hand public concern has been growing steadily, as manifested by the growing power and influence of 'green consumers'. The narrow concern over pollution seen in the late 1960s and early 1970s has blossomed into a wider debate about the root causes of environmental degradation and its ultimate consequences. The popular media quickly and correctly connected the great African famines of the last decade not only with drought but also with the deterioration of the natural resource base. Discussions about global warming now focus not only on the immediate problem but also on the decision-making processes that encourage industry and consumers to burn energy now and ask questions later.

On the other hand, the pace of government action has faltered. Even at a time of enormous upheaval, when governments are being ousted for failing to be responsive enough to their constituencies, the slow-down is evident. A regional agreement on sulphur dioxide and a global agreement on ozone-depleting chemicals led to a general optimism that the world was developing a more coherent response to a range of environmental problems. Some are saying that the optimism was misplaced. These agreements may be nothing more than isolated successes, reached in the face of immediate and overwhelming public pressure, rather than a sign of things to come. For while the biosphere continues to take a beating, the apathy persists. The goal of marrying economic development and environmental quality remains remote. None the less, the two treaties signed at the Earth Summit on climate change and biodiversity offer a ray of hope.

UNEP takes very seriously its obligation to report on the state of the environment, and it is not with any satisfaction that I am forced to report this failure of political will. But that is the conclusion we have reached in UNEP, on the basis not only of our own work but also on inputs from over thirty of the world's leading scholars.

Yet, in the face of all this I do retain a certain optimism. I have seen what can be accomplished when there is a will, and I have seen the speed and cost-effectiveness of these isolated responses. This report documents those accomplishments. It is my belief that when this new resolve is manifested we will learn the full benefit of living in harmony with our environment. *The World Environment 1972-1992* records not only our descent into apathy but also the positive examples that offer a path towards a more rational future. That path must lead us towards a global partnership: a partnership not just of words, but also of actions.

Mostafa K. Tolba
Executive Director, UNEP
Nairobi

Editors' preface

The World Environment 1972-1992 looks back over the past 20 years and reviews some of the trends: progress in some areas offset by accelerating destruction in others. It also looks into the future, not so much in an attempt to predict how governments and communities will act, but more as an effort to present a spectrum of issues that must be addressed if a decent environment and sustainable development are to become realities.

Along with advances in the way in which information is gathered, there has also been a shift in the issues themselves. A number of issues that attracted little or no attention in 1972 have now taken on a new significance. At the time of writing, for example, nations are being brought back to the negotiating table to discuss again the impact on stratospheric ozone of chlorine and bromine compounds. Even after a largely successful international campaign against chlorofluorocarbons and halons there is evidence that the ozone layer is not out of danger.

In spite of a new sophistication in monitoring, and in spite of the new issues now presenting themselves to the international community, a number of things have remained unchanged since the first report eighteen years ago, and even since Stockholm.

One concern is the flow of information from the Third World. Advances in remote sensing techniques have led to some progress, but the critical mass of knowledge necessary for coherent analysis and intelligent policy-making is often not available in an ordered and up-to-date form. Despite a recent flood of information, the same is still true of Central and Eastern Europe.

On the management side, a number of the problems that presented themselves at Stockholm are still with us. Among the most obvious is the way in which we approach the environment. Despite all the advances and all the changes, human concern for the environment is still essentially reactive rather than precautionary, although there is some evidence, particularly in the period since 1987, that governments are being edged by their constituencies into taking more preventive action. The basic pattern, however, is one of crisis-and-response. It is an approach that can - and must - be changed.

The World Environment 1972-1992 offers a look at the environment from three perspectives. Ten introductory chapters detail the range of environmental threats and examine how they have unfolded over the past two decades. The threats include those that were well known two decades ago, such as toxic chemicals, air and water pollution, and land degradation (desertification, deforestation and soil loss). They also include those that have emerged over the period under review - stratospheric ozone depletion, loss of biodiversity and climate change.

The second section reviews the different sectors of the economy, analysing how each of them has impacted on the human environment. Here, patterns of development in both developed and developing countries are addressed, and a closer look is taken at the way in which poverty degrades the environment through the proliferation of 'dirty' technology, the growth of human populations, and the pressure to destroy renewable resources.

The final major section of the book analyses the range of responses to the changing environment. On the scientific front, advances in the natural and social sciences are discussed: space monitoring, sophisticated computer modelling, evaluation of natural resources and improved techniques of cost-benefit analysis. New economic tools for valuing natural resources and for improving cost-benefit analysis are explained. A range of national responses is examined, and new international treaties are presented and their effectiveness evaluated. The impacts of 'green consumerism' and 'green politics' are reviewed.

The body of knowledge has continued to expand rapidly, and for that reason this report cannot give the last word on the subject. The report was finalized early in 1992: thus, developments since that time are not covered. Similarly, technical, economic, legal and political developments that have taken place since the third quarter of 1991 have, by necessity, been dealt with only cursorily.

It has been our wish to make this publication as accessible as possible to as wide a readership as possible. Most of the text can be readily understood without a strong background in the natural sciences, and it is hoped that it will provide a useful reference for specialists and non-specialists alike.

M. K. Tolba
O. A. El-Kholy
E. El-Hinnawi
M. W. Holdgate
D. F. McMichael
R. E. Munn

Acknowledgements

A project such as *The World Environment 1972-1992* could not be undertaken by a single author, and I am deeply grateful to the scholars and officials who have given so much of their time to help me see it through to completion. First, to my co-editors, it would be unfair simply to offer thanks. Together we made it happen: they are partners in this enterprise, and share the pride of producing such a volume. Thanks must also go to the many contributors - from the academic community and from international organizations - who produced drafts on specific topics; and with them we must also thank the technical reviewers: whether they joined us for review meetings or undertook their work at long range by fax and by phone, their work was critical to us. Thanks also to the UNEP staff members who contributed in so many ways. And lastly, we wish to thank those people who made it possible to produce this report almost a year in advance of its target date - the production editors, graphic artists, publishers and many others. To all of those who participated in this project we are deeply indebted, though we, the editors, alone are responsible for any errors and omissions.

M. K. Tolba

PART ONE

THE ISSUES

An introductory overview

The Earth's environment is continually being transformed through natural processes and human interventions. Change is inevitable, and while some changes are for the better, others are for the worse. Occasionally, of course, society must sacrifice long-term environmental benefits in order to meet basic human needs. In other cases, damage to the environment is inadvertent because the harmful consequences of particular human actions are sometimes not recognized.

There are several contemporary views of the ability of the global environment to withstand changes and shocks. One is the view of '*Nature forgiving*'. Witness, for example, the case of some early Central American settlements which drastically transformed the landscape in those days. In many areas the land has now returned to its original state - dense tropical jungle. A second example is an acidified lake in Northern Ontario which recovered once a nearby smelter was closed down (Gunn and Keller, 1990). Another view is that of '*Nature unforgiving*'. In this paradigm the world is finite, and the environmental impacts of society are cumulative and difficult to reverse: witness the problems of desertification and urbanization. A third view is that of '*Nature resilient*'. In this case, ecological systems, including human ones, are strengthened by environmental stresses so that they can withstand even greater shocks should the need arise (although there is, of course, a limit beyond which a system can no longer survive). Kano, for example, a large city in northern Nigeria, has been able to withstand intermittent droughts over many centuries. Another example is that of forest fires, which release essential nutrients from the forest and so contribute to rapid regeneration.

Many ecologists accept the third paradigm, believing that ecosystems should be managed in ways that permit them to adapt to continually changing conditions. In fact, the goal of enhancing an ecosystem's ability to evolve rather than collapse is at the heart of current efforts to promote sustainable development. (Historically, of course, ecosystems were managed with the goal of maintaining the *status quo* no matter how much the environment changed, and this tended to stress the systems increasingly.) Of course, whenever rates of environmental change become too rapid, successful management may in any case be impossible.

The initial step in developing sound environmental policies for the coming decades is to have a good idea of the recent changes that have occurred. Thus, the first ten chapters of this book will focus on recent environmental trends, particularly those that have occurred since the Stockholm Conference, and the associated key issues that have emerged. Where appropriate, chronologies are given for the most important events of the last 20 years. The chapters are structured by medium (atmosphere, oceans, etc.) but this is for convenience only, and linkages between media must always be borne in mind. In this regard, the idea of global and regional biogeochemical cycles of trace substances has provided a unifying theme for the study of environmental issues in recent years. These cycles have remained in balance over many centuries, but recently they have become seriously disrupted. For example, the quantities of carbon passing into and out of the atmosphere remained approximately in balance until recently. Now, the annual net gain of carbon by the atmosphere is three gigatonnes (10^9 metric tonnes; an enormous amount) - and rising.

Given a series of measurements of an environmental indicator, trend detection is often exceedingly difficult because of the great variability encountered, and signal cannot easily be separated from noise. Nevertheless, some worrisome trends have been reported in the last 20 years, including increases in global and urban populations, ozone depletion in the Antarctic stratosphere, increasing concentrations of some of the atmospheric trace gases (carbon dioxide, methane and the chlorofluorocarbons) and a reduction in biodiversity.

Not all of the trends are for the worse, however. In many developed countries, for example, atmospheric lead levels have declined with a switch to lead-free gasoline, and pesticide levels have also declined. In developing countries encouraging environmental achievements have been seen in the increasing numbers of protected areas, and in some of the UNEP Regional Seas programmes leading, for example, to an increase in the number of waste water treatment facilities in the Mediterranean.

In this connection it must be emphasized that the biosphere is very complex, even in the absence of major human impacts. Taken alone, therefore, historical trends of the behaviour of particular environmental indicators are not very reliable indicators of the future. Instead, a broad understanding of the environment is required, particularly of the linkages between the different parts of the global system. On the ecological side, the biogeochemical cycling of nutrients and wastes is a useful conceptual framework, but because the causes of disruptions in the natural cycles are often socio-economic, the first ten chapters of this book can be evaluated only within the context of some of the material contained in later chapters.

References

Holdgate, M. W. (1979) *A Perspective of Environmental Pollution*, Cambridge University Press, Cambridge, UK.

Gunn, J. M. and Keller, W. (1990) Biological recovery of an acid lake after reduction in industrial emissions of sulphur, *Nature*, **345**, 431-33.

CHAPTER 1

Air pollution

The issue

Air pollution has been recognized for centuries as an undesirable by-product of civilization. The first book on the subject (*Fumifugium*) was written by John Evelyn in the seventeenth century; and in several industrialized countries in the nineteenth century, concern over urban air pollution, especially that from furnaces, smelters and chemical works, led to sporadic anti-pollution campaigns, and to control legislation (Ashby and Anderson, 1981). In more recent times, several 'episodes' causing illness and death occurred in the 1930s to 1950s, and air pollution became a major public issue in a number of industrialized countries (Box 1). In particular, the 1952 London episode led to the passage of the UK Clean Air Act of 1956 and the establishment of smoke control areas. Other legislation followed in the UK (The Clean Air Acts of 1964 and 1968) with similar actions in many other countries in North America and Western Europe and in Japan. This in turn led to the establishment of supporting regulatory, monitoring and assessment bodies.

BOX 1

Significant air pollution episodes occurring pre-1970.

	Meuse Valley, Belgium (1930)	Donora, Pennsylvania (1948)	Poza Rica, Mexico (1950)	London (1952)
Mortality and morbidity				
Deaths	60	15	22	4000
Illnesses	6000	5900	320	>20000
Age groups affected				
	Elderly	Elderly	All ages	Elderly at first
Weather				
	Anticyclonic inversion and fog	Anticyclonic inversion and fog	Nocturnal inversion; low winds	Anticyclonic inversion and fog
Geographical setting				
	River valley	River valley	Coastal valley	River plain
Sources				
	Steel and zinc manufacture	Steel and zinc manufacture	Sulphur recovery-accident	Domestic coal burning
Pollutants				
	SO_2 and smoke	SO_2 and smoke	H_2S	SO_2 and smoke

Source: Brimblecombe (1986).

The first international inquiry into air pollution damage was undertaken in the 1930s, in the Columbia River Valley of the Rocky Mountains, where a smelter in Canada was identified as the cause of vegetation damage in the United States. The problem was ultimately resolved by the US-Canada International Joint Commission. In a number of developed countries up until the mid-1970s, local air quality standards were achieved by building tall chimneys (the so-called 'tall-stack policy') in order to disperse pollutants over a wide area and so reduce their local impacts. Various studies have since demonstrated that although this dispersion improves local air quality, it may cause long-range problems. Sometimes even the global atmosphere is over-burdened. Acidification of Scandinavian lakes by deposited sulphur and nitrogen oxides was established in the 1960s (Odén, 1968), and in Europe and North America acid rain subsequently became widely accepted as an important regional-scale problem, contributing to damage to lakes, rivers and forests up to 1,000 kilometres from the emitting sources of pollution. In the 1960s too, trace amounts of pollutants were discovered in areas thousands of kilometres from pollution sources.

Often, of course, the atmosphere is merely a carrier of unwanted material, and accumulation takes place in other media, including fresh waters, soils, biota and food chains.

BOX 2

Air pollution issues and research directions of the last 20 years.

Issues causing public concern:

- Indoor air pollution.

- Urban air pollution.

- Control of emissions of lead (Pb).

- Tropospheric ozone.

- Acid rain and long-range transport of air pollutants.

- 'Toxics',

- Accidental releases of hazardous materials.

Additional issues of current interest and future policy relevance:

- Background air pollution

- Economic assessments of costs and benefits of pollution control.

- Control of greenhouse gas emissions.

- Control of ozone-depleting chemicals.

Air pollution problems highlighted in the 1982 UNEP *State of the Environment* report (Holdgate *et al.*, 1982) were smoke and sulphur oxides in urban air, photochemical oxidants, sulphate and nitrate hazes, acid rain, airborne toxic substances, accumulation of chlorofluorocarbons (CFCs) in the stratosphere, and the increasing concentrations of gases that might affect climate. These issues have remained of concern in the 1980s, and some have become a focus of international attention. Climate change and ozone depletion have gained such prominence that they have been treated separately in Chapters 2 and 3. An overview of contemporary air pollution issues and research directions is given in Box 2.

The adverse effects of atmospheric pollution are not restricted to human beings, but can also result in serious damage to vegetation and the pollution of water resources. The major health effects are shown in Box 3. Impacts on vegetation are discussed below (see sections on Tropospheric ozone and Acid rain).

Trends in air pollution levels and impacts over the past 20 years

Indoor air pollution

Indoor air pollution is the 'original' air pollution problem, and it still exists (WHO/UNEP, 1990). Initial concern focused on the working environment, where exposure to dust, fibres (such as asbestos), and toxic and carcinogenic fumes was directly related to occupational illnesses such as silicosis, pneumoconiosis, mesothelioma, acute and chronic mercury poisoning, cancers and other conditions. The critical analysis of dose/effect relationships and the establishment of 'criteria' by WHO and other bodies owes much to the analysis of occupational exposure.

In recent decades, other indoor environments have also become a focus of concern (see Chapters 17 and 18). In industrialized countries, as cleaner and more efficient forms of energy such as natural gas and electricity replaced domestic coal burning, many new indoor air pollution problems have arisen (Box 4). Such problems are exacerbated by modern airtight buildings, designed with a view to conserving energy. This, for example, has been a major reason why radon emitted from the ground has become a concern. When enclosed in airtight indoor environments or mines, radon decays into isotopes that can be a health hazard. In parts of Europe and North America, formaldehyde emitted from insulating foam, and asbestos from building materials, also became public issues in the late 1970s, and were ultimately withdrawn from the market. In North America and some other parts of the world, the wood stove has become increasingly popular, causing indoor pollution problems as well as adding to

BOX 3

Health effects of major air pollutants, and associated WHO health effects criteria

Sulphur dioxide
- Respiratory irritation, shortness of breath, impaired pulmonary function, increased susceptibility to infection, illness in the lower respiratory tract (particularly in children), chronic lung disease and pulmonary fibrosis.
- Increased toxicity in combination with other pollutants.

($500\mu g/m^3$ for 10 min; $350\mu g/m^3$ for 1 hour).

Respirable particulate matter
- Irritation, altered immune defence, systemic toxicity, decreased pulmonary function and stress on the heart.
- Acts in combination with SO_2; effects depend on the chemical and biological properties of the individual particles.

(no health effects criteria).

Oxides of nitrogen
- Eye and nasal irritation, respiratory tract disease, lung damage, decreased pulmonary function and right heart stress.

($400\mu g/m^3$ for 1 hour; $150\mu g/m^3$ for 24 hours).

Carbon monoxide
- Interferes with oxygen uptake into the blood (chronic anoxia).
- Can result in heart and brain damage, impaired perception, asphyxiation; or in lower doses, weakness, fatigue, headaches and nausea.

($100mg/m^3$ for 15 min; $60mg/m^3$ for 30 min; $30mg/m^3$ for 1 hour; $10mg/m^3$ for 8 hours).

Lead
- Kidney disease and neurological impairments.
- Primarily affects children.

(0.5-$1.0\mu g/m^3$ for 1 year).

Photochemical oxidants (e.g. ozone)
- Decreased pulmonary function, heart stress or failure, emphysema, fibrosis, and aging of lung and respiratory tissue.

(150-$200\mu g/m^3$ for 1 hour; 100-$120\mu g/m^3$ for 8 hours).

Sources: WHO (1987); Brimblecombe (1986); Kapchella and Hyland (1986).

outdoor ambient loadings and depleting forest wood supplies. Another public issue of the last decade has been tobacco smoke. The issue first arose in North America but the concern is spreading. Many restaurants and offices are now required to have smoke-free areas, and some airlines have banned smoking. Finally, the phrase *sick building syndrome* has become current, following several outbreaks of Legionnaire's disease (a serious bacterial infection named after an outbreak at a meeting of the American Legion in 1976) and a growing perception by office workers that eye, nose and throat irritation, mental fatigue and headaches are associated with particular buildings.

BOX 4

Indoor air pollutants.

Pollutant	Source	Possible effects
Formaldehyde	Plywood, particle board, foam insulation, cigarettes	Eye and respiratory problems, headaches and nausea, at high concentrations.
Nitrogen dioxide	Gas stoves, kerosene heaters	Eye and nasal irritation, respiratory and pulmonary damage.
Particles	Tobacco smoke, cooking, aerosol sprays, rubber backed carpets, wood stoves.	Effects depend on particle properties, but include respiratory irritation, altered immune defence, and pulmonary problems.
Radon	Stone, concrete, water, soil.	Lung cancer a possible result of over-exposure.
Carbon monoxide	Vehicles in garages, kerosene heaters, cigarettes, wood stoves.	Over-exposure may result in headaches, nausea, impaired perception, heart or brain damage.
Polyaromatic hydrocarbons (PAHs) (e.g. benzopyrene)	Burning of wood, coal or dung, industrial solvents, wood stoves	Sinusitis, lung damage, including lung cancer.
Sulphur dioxide	Kerosene heaters	Respiratory and pulmonary irritation or disease.
Molds, fungi, viruses	Dampness	Respiratory effects.

Source: Brimblecombe (1986); WHO/UNEP (1990); Kapchella and Hyland (1986).

In many developing countries, the combustion of biomass fuels (wood, crop residues and dung) in primitive unvented stoves for cooking and heating can lead to high exposures to pollutants such as carbon monoxide and polycyclic aromatic hydrocarbons (PAHs) (Brimblecombe, 1986; Cleary and Blackburn, 1968; Master, 1974). Women and children, who spend much time indoors, are particularly threatened (Chen *et al.*, 1990). Biomass burning has been identified by the WHO as the major indoor air pollution health problem in the world today.

Urban air pollution

The five main pollutants of concern in urban environments are sulphur dioxide, the nitrogen oxides, carbon monoxide, ozone and suspended particulate matter (including lead). In 1990 it was estimated that, world-wide, 99 million tonnes of sulphur oxides (SO_x), 68 million tonnes of nitrogen oxides (NO_x), 57 million tonnes of suspended particulate matter (SPM) and 177 million tonnes of carbon monoxide (CO) were released into the atmosphere as a result of human activities (OECD, 1991). The OECD countries accounted for about 40 per cent of SO_x, 52 per cent of NO_x, 71 per cent of CO and 23 per cent of SPM (Figure 1).

Figure 1: *Man-made emissions of common air pollutants.*

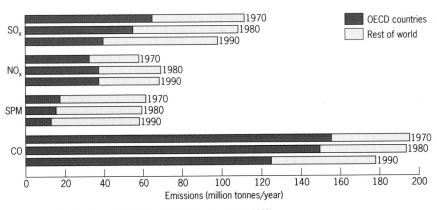

Source: Based on data by OECD (1985, 1991); Dignon and Hameed (1989).

A more recent estimate (Hameed and Dignon, 1992) is that during the period 1970-86, global emission rates of NO_x increased by almost one-third, while emissions of SO_x increased by approximately 18 per cent (Figures 2 to 7).

In most industrialized countries, combustion of fossil fuels (coal and oil) at stationary sources is responsible for the majority of man-made sulphur emissions, with contributions from metal smelting and other industrial processes. Nitrogen oxides are primarily emitted by fossil fuel combustion, with the transport sector typically accounting for between one-third and one-half of national emissions. In the last 20 years, many industrialized countries have made considerable progress in reducing emissions of sulphur dioxide and suspended particulate matter, in part by imposing progressively stricter emission and air quality standards but also through improved energy efficiency and economic incentives.

For the nitrogen oxides, however, trends in national emissions during the past 20 years have been mixed. In the majority of countries with the exception of Africa and South America, emissions appear to have remained relatively steady or have increased slightly since the early 1970s (Dignon and Hameed, 1989). In Japan, concentrations of NO_x declined from the mid-1970s until 1985 but then began to increase again (Swinbanks, 1991). These trends broadly mirror those observed in ambient nitrogen oxide concentrations in many cities of the industrialized world (UNEP, 1991). Increases in vehicle numbers are generally considered to be the main cause of such trends. Strict control of the number of motor vehicles in downtown Singapore provides a notable exception. Oxides of nitrogen are more difficult to control than oxides of sulphur. (They come from multiple sources and they depend *inter alia* on combustion temperature).

In Eastern Europe and the developing world, urban air pollution is, in contrast, a continuing and indeed an increasingly severe problem. There are indications that emissions of the oxides of sulphur and nitrogen are increasing in many such areas, although the data base is limited. In the former USSR, emissions of air pollutants began to decline from 1986 to 1990 although 68 industrial centres still have intolerably high air pollution indices (Danilov-Danilyan and Arski, 1991).

Figure 2: *NO_x emissions in million tons of equivalent N in the period 1970-86. Results for Asia, Europe, North America and the former USSR.*

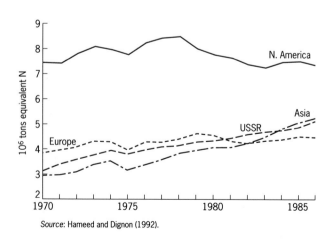

Source: Hameed and Dignon (1992).

In most cases, technologies are available to control emissions: the main obstacles are the lack of the economic ability and political will to do so. At present, increased emphasis is being placed on the use of flue gas de-sulphurization and new types of burners which reduce SO_x and NO_x emissions.

Figure 3: *NO_x emissions in million tons of equivalent N in the period 1970-86. Results for Africa, Oceania and South America.*

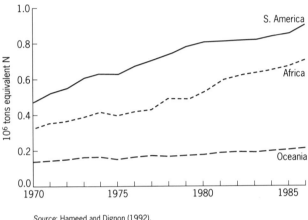

Source: Hameed and Dignon (1992).

Figure 4: *Comparison of the five highest NO_x emitters estimated for 1970, 1975, 1980 and 1986 in million tons of equivalent N.*

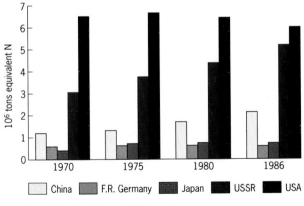

Source: Hameed and Dignon (1992).

Figure 5: *SO$_x$ emissions in million tons of equivalent S in the period 1970-86. Results for Asia, Europe, North America and the former USSR.*

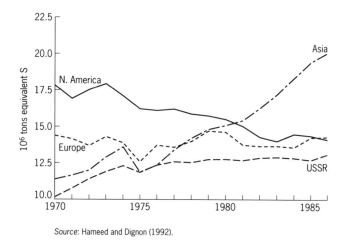

Source: Hameed and Dignon (1992).

Figure 6: *SO$_x$ emissions in million tons of equivalent S in the period 1970-86. Results for Africa, Oceania and South America.*

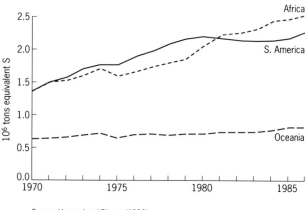

Source: Hameed and Dignon (1992).

A few countries have made significant reductions in national emissions of carbon monoxide. In the USA, for example, mandatory introduction of catalytic convertors on motor vehicles, the dominant source of CO emissions since the 1970s, has contributed to the 40 per cent reduction that has occurred between 1970 and today, in spite of the increasing numbers of vehicle-kilometres travelled (EPA, 1991).

Figure 7: *Comparison of the five highest SO$_x$ emitters estimated for 1970, 1975, 1980 and 1986 in million tons of equivalent S.*

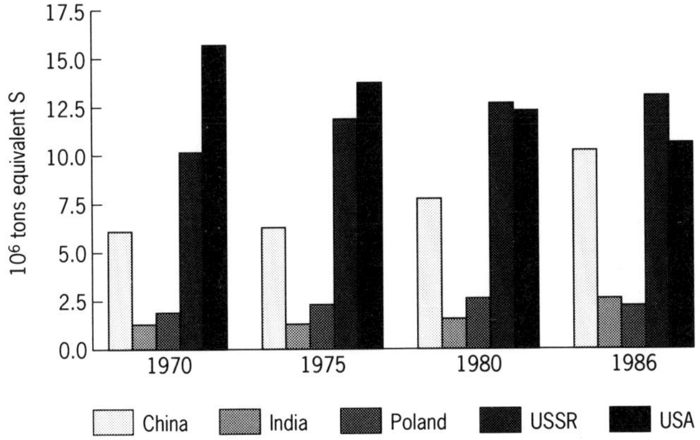

Source: Hameed and Dignon (1992).

One of the striking successes of the last two decades has been the steep decline in national lead emissions and concomitant decreases in urban lead levels in those countries that have reduced the content of lead in gasoline (see Box 5 and Figure 8).

BOX 5

Declines in ambient lead levels.

Most lead emissions to the atmosphere come from the use of leaded gasoline. Since the 1970s, lead additives have been progressively phased out in North America, the European Community and Japan. In these areas there have been dramatic declines in ambient levels. For example:

- A 21-site network in the UK reported a 50% decline between the first quarters of 1985 and 1986. The lead content of gasoline was reduced from 0.4 to 0.15 g per litre on 1 January, 1986. (McInnes, 1988).

- In the United States, 189 monitoring sites reported an 87% decrease in ambient lead values between 1980 and 1989. Lead emissions declined by 96% between 1970 and 1986 (and mostly between 1980 and 1989) (EPA, 1991).

- Sampling stations in 53 Canadian cities reported a 55% decline in annual mean lead levels between 1975 and 1983 (Hilborn and Still, 1990).

Figure 8: *Trends in lead levels in ambient air, trends in lead emissions, and blood lead levels in Canada.*

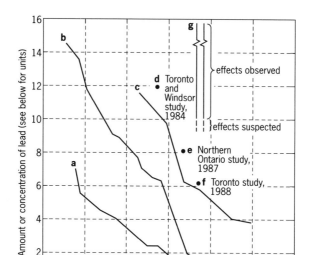

a – Lead concentration in ambient air (μg/m³ x 10).

b – Lead emissions from gasoline combustion (thousands of tonnes).

c – Lead emissions from all sources (thousands of tonnes).

d – Average blood lead levels in children in two southern Ontario urban areas (Toronto and Windsor) in 1984 (μg/dL).

e – Average blood lead levels in children in a northern Ontario urban area in 1987 (μg/dL).

f – Average blood lead levels in children in a southern Ontario urban area in 1988 (same Toronto neighbourhood as in the 1984 study) (μg/dL).

g – Current lowest range of blood lead levels at which deleterious effects have been observed in some population groups (μg/dL).

Source: Hilborn and Still (1990). Based on information from Environment Canada;the Department of National Heath and Welfare; Ontario Ministry of Health 1985, 1990; and Smith 1990.

That these reductions in the concentrations of atmospheric lead are not just a local phenomenon is demonstrated in Figure 9, which shows a sharp rise in lead levels in Greenland snow cores in the first half of this century, followed by a decline beginning about 1970 (Boutron *et al.*, 1991). During these last 20 years, there has been very little decrease in cadmium, zinc and copper, measured at the same site.

Despite the effort that has been put into air pollution abatement in many industrialized countries, two additional issues continue to daunt pollution control authorities - *ozone episodes* and *brown haze*, which sometimes shrouds metropolitan areas during fine weather, even when air quality criteria are not exceeded.

Figure 9: *Changes in lead concentrations in Greenland ice and snow from 5,500 BP to the present. The decline in the last 20 years is mainly a consequence of the decreasing use of lead additives in gasoline.*

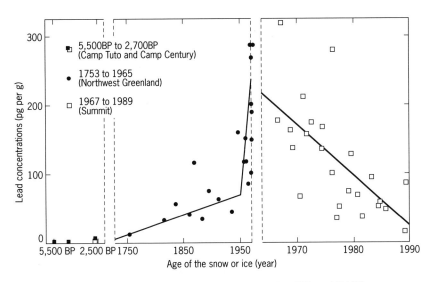

Source: (Boutron *et al.* 1991), reprinted by permission from *Nature*, volume 353, copyright 1992, Macmillan Magazines Ltd.

Tropospheric ozone

Photochemical smog is formed from chemical reactions among nitrogen oxides and reactive hydrocarbons in the presence of sunlight. The main photochemical oxidants are ozone, the hydroxyl radical, hydrogen peroxide and peroxyacetyl nitrate. These species are very reactive chemically and cause damage to vegetation, materials and human health. They have been implicated as a factor in the widespread decline of European forests during the last 20 years.

Photochemical pollution is an evolving issue. Tropospheric oxidant levels have increased over the last two decades as a result of increased nitrogen oxide emissions and fairly constant hydrocarbon emissions; see, for example, Figure 10 (Volz and Kley, 1988). In fact, urban ozone concentrations regularly exceed known thresholds for damage in countries where ozone is monitored. But high ozone concentrations are not limited to urban areas. Long-range transport of precursors causes ozone episodes far from source regions. The precursors originate not only in urban and industrial areas but also in tropical regions where there is extensive biomass burning. Thus tropospheric ozone is an issue of world-wide significance.

One of the dilemmas facing pollution control agencies is the relative degree of control to be placed on sources of nitrogen oxides and of hydrocarbons (which are likely to be the limiting factor in the photochemical reactions creating oxidants). Nevertheless, control programmes are either in place or being

Figure 10: *Comparison of a unique series of tropospheric ozone measurements from Montsouris near Paris (1876-1910) with more recent data from Arkona on the Baltic coast (1956-1983).*

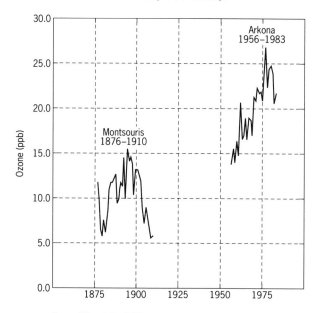

Source: Volz and Kley (1988), reprinted by permission from *Nature*, volume 332, copyright 1988 Macmillan Magazines Ltd.

negotiated in North America, Europe and Japan. At the same time, the problem of photochemical oxidant smog is becoming more acute in cities and downwind areas in parts of the developing world where road traffic is increasing rapidly (e.g. Mexico City, Santiago).

Acid rain and long-range transport of air pollutants

A concern about acidification damage (e.g. declines in fish populations and ultimate extinctions in rivers and lakes) was first voiced by Svante Odén in Sweden in the 1960s (Odén, 1968). Odén argued that because distant sources contribute to local acidification, intergovernmental action is required to solve the problem. Subsequently, a Swedish Case Study on Acid Rain was prepared for the 1972 Stockholm Conference on the Human Environment (Sweden, 1972).

The principal acid deposition precursors are the oxides of sulphur and nitrogen, which can be transported by winds for distances of up to 1,000 kilometres. They can damage vegetation, fabrics and structures either directly (dry deposition), or when dissolved in mist or rain to form dilute sulphuric and nitric acids (wet deposition). Lichens and some coniferous trees are especially sensitive to dry deposition, while wet deposition affects freshwater ecosystems, soils and forests (see Chapters 4 and 7). Acidification has been seen most noticeably in Europe and northeastern North America.

Man-made emissions of sulphur dioxide, derived largely from fossil-fuel combustion sources, are now similar in magnitude to those from natural sources. On the global scale, emissions of the nitrogen oxides are also roughly equally divided between man-made sources (fossil fuel combustion and biomass burning) and natural sources (microbial action, lightning).

Considerable work has been directed towards estimating global and regional atmospheric budgets of sulphur and nitrogen. Figure 11 gives one example, a regional sulphur budget for southeast Europe, where forest decline became a major public issue in the early 1980s, and where acidic deposition was implicated as one of several causative factors. A second example, of wet deposition of sulphur averaged over six years in eastern North America, is given in Figure 12. Evidence that changes in regional sulphur emissions result in corresponding changes in deposition has taken some years to collect. However, intensive monitoring and analysis have shown that the downturn in emissions over the last ten years is indeed paralleled by the acid deposition record (Figure 13). Nevertheless, the deposition levels over large portions of both Europe and eastern North America still exceed those at which damage occurs to materials and to sensitive portions of aquatic and terrestrial ecosystems.

Figure 11: *A regional budget of atmospheric sulphur fluxes for southeast Europe, in million tonnes of sulphur per year. The region of southeast Europe includes Albania, Bulgaria, Greece, Italy, Romania, western Turkey and Yugoslavia.*

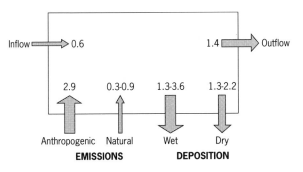

Source: Katsoulis and Whelpdale (1990).

Knowledge of acid deposition is quite advanced in terms of overall understanding and an appreciation of the remedial actions required. Emission controls in Europe and North America have reduced deposition rates significantly, and recovery of aquatic ecosystems in some areas has occurred: see, for example, Gunn and Keller (1990). However, acidification damage to soils, forests and lakes continues in some countries, and even further sulphur reductions will be needed. International agreements are in place but it will be some time before

Figure 12: *Six-year mean SO$_4$ wet deposition in eastern North America, 1982-87, in kilograms SO$_4$ per hectare per year.*

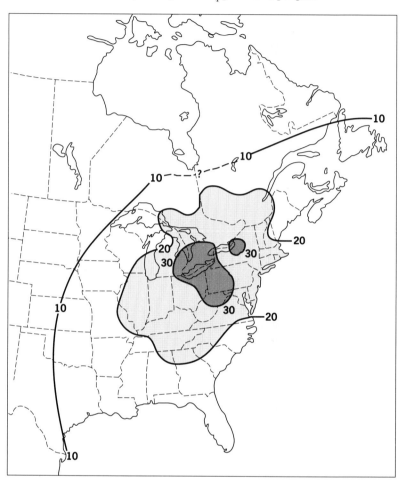

Source: RMCC (1990).

the overall impact of current reductions becomes known. In this connection, the simulation modelling projections carried out at the International Institute for Applied Systems Analysis (IIASA) indicate that even with currently expected reduction schedules, additional damage will occur in Europe over the next three or four decades (Alcamo *et al.*, 1987).

The acidification phenomenon is potentially also a threat to some parts of the developing world. A combination of sensitive soils (i.e. those with poor buffering capacities) and increasing industrial emissions may result in acidification of soils and surface waters in regions indicated in Figure 14. Figure 15 shows sulphur emissions for the year 2020, for assumed growth rates of population and per capita consumption of energy (Galloway, 1989).

Figure 13: *Long-term changes in the concentration of excess (non-seasalt) sulphate in precipitation in Scandinavia show a close correlation with changes in sulphur dioxide emissions in Europe.*

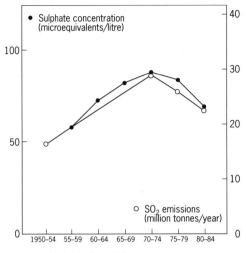

Source: Leck and Rodhe (1989).

Figure 14: *Schematic map showing regions that currently have acidification problems, and regions where, based on soil sensitivity and expected future industrialization, acidification might become severe in the future.*

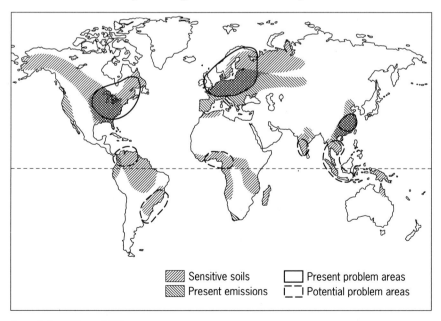

Source: Rodhe and Herrera (1988).

21

Figure 15: *Estimated sulphur emissions for the year 2020 for assumed growth rates of population and per capita consumption of energy.*

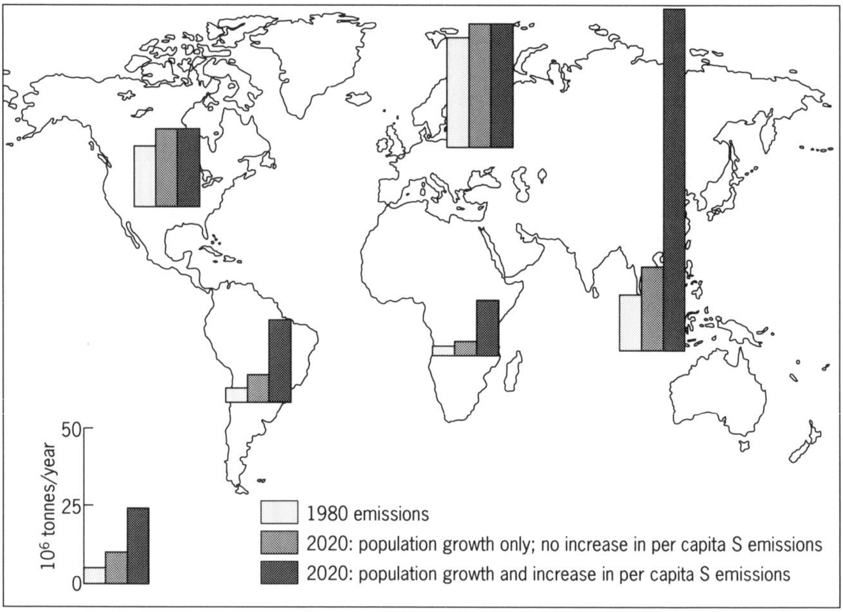

50

25

0

10^6 tonnes/year

1980 emissions

2020: population growth only; no increase in per capita S emissions

2020: population growth and increase in per capita S emissions

Source: Galloway (1989).

Toxic metals and organic compounds ('toxics')

The atmosphere serves as an effective distribution system for a range of potentially toxic substances including trace metals and synthetic organic compounds. Several of these substances have long environmental lifetimes, relatively high toxicities and industrial sources that greatly exceed natural ones. Public concern over their importance has grown tremendously during the past two decades, but little routine monitoring has been carried out world-wide.

Recent estimates show that emissions of selected trace metals, including lead, mercury and cadmium, from human sources exceed natural ones (Hutchinson and Meema, 1987). A measure of their significance as global pollutants is their fluxes to the world oceans. Atmospheric inputs of their dissolved components are comparable to (for copper, nickel, arsenic) or greater than (for zinc, cadmium, lead) their inputs from the world's rivers (GESAMP, 1987; Chapter 5).

Synthetic organic chemicals are becoming ubiquitous, and in the last 20 years chlorinated hydrocarbon pesticides and a number of industrial organic compounds have been found in Arctic food chains, in the Antarctic, and in deep-sea organisms. In inland and enclosed marine water bodies such as the Great Lakes and the Baltic Sea, these substances are a major health and environmental concern. The atmosphere is their primary pathway for entry into the global environment. For example, 80 per cent of the PCBs and 98 per cent of DDT in the world oceans have come from the atmosphere (Atlas and Giam, 1986; Duce *et al.*, 1991).

Accidental releases of hazardous materials into the atmosphere

Accidents such as Seveso, Bhopal and Chernobyl have heightened awareness of the need for strict management and safety regulations at all stages of production, storage, transport and use of hazardous materials (see Chapter 9). On the atmospheric side, these accidents have stimulated the development of better mathematical dispersion models and associated mesoscale monitoring systems.

Responses

Responses to air pollution problems have traditionally been in the form of increased monitoring and control programmes. In recent decades more and more agencies have taken holistic approaches at all stages of production, storage, transport, use and disposal of hazardous chemicals (see Chapters 9 and 10). On the atmospheric side, programmes are based on the idea of critical loads (see Box 6) and rather complex long-range transport and deposition models (Alcamo *et al.*, 1987). Some of the milestones in this high visibility issue are given in Box 7.

In 1973, WHO set up a global programme to assist countries in operational air pollution monitoring. In 1976, this became part of UNEP's Global Environmental Monitoring System (GEMS). At present, some 50 countries participate in the GEMS/AIR monitoring project. Data is gathered from about 175 sites in 75 cities, of which one-third are in developing countries. The GEMS/AIR assessment has concluded that, world-wide, nearly 900 million people living in urban areas are exposed to unhealthy levels of sulphur oxides, and more than one billion are exposed to high levels of particulates which could be a health hazard.

The control of air pollution in the working environment has been the concern of factory inspectors and health officers for many decades, and standards for permissible exposures have been tightened progressively (see Chapter 18). Control of air pollution in other indoor environments is less well advanced. In most cases technological solutions are available, but their application is not always feasible (although action has been taken against sources of asbestos and formaldehyde in homes, schools and offices in many developed countries, and a number of houses with high radon exposures have been identified and modified). The reduction of indoor air pollution is highly dependent on public realization of the problems, and willingness to take or demand the necessary

BOX 6

The relationships between critical loads and target loads, and their effects on ecosystem sensitivity.

The internationally accepted definition of a Critical Load/Level is:
'a quantitative estimate of and exposure to one or more pollutants below which significant harmful effects on specified sensitive elements do not occur according to present knowledge.'

This definition implies that pollution above the Critical Load/Level will lead to permanent damage to a given ecosystem.

Research activities have set the limits of Critical Loads. These loads are related to ecosystem sensitivity. The highest politically acceptable load of a pollutant is the 'Target Load'. The Target Load can be set at, above or below the Critical Load. If the Target Load is set above the Critical Load, one accepts that the most sensitive ecosystems will be negatively affected. If it is set below, a safety margin is included. This relationship is illustrated below.

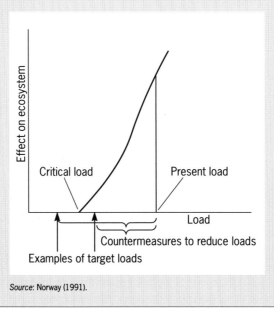

Source: Norway (1991).

action. Concerted efforts to monitor human exposure in non-industrial indoor environments began in many countries in the early 1980s, largely through the initiatives of WHO.

The adverse health effects of atmospheric pollution have provided the basis for most air quality guidelines and standards (see Box 3) while damage to vegetation has led to the formulation of so-called secondary air quality standards.

BOX 7

Acid precipitation milestones.

1972   Sweden's case study for the Stockholm Conference documented the occurrence and effects of transboundary pollution and acid precipitation, and provided the impetus for subsequent international action.

1972-1977 The OECD 'Co-operative Technical Programme to Measure the Long-Range Transport of Air Pollutants' in Europe, demonstrated conclusively that transboundary pollution and acid deposition was a Europe-wide problem.

1978-   The ECE's Co-operative Programme for Monitoring and Evaluation of Long-Range Transmission of Air Pollutants in Europe (EMEP) began, involving countries in Eastern and Western Europe.

1979   The first broad international agreement covering air pollution, the framework ECE 'Convention on Long-Range Transboundary Air Pollution' was signed by 34 industrialized nations of Europe and North America to address specifically the problem of transboundary air pollution.

1982   The 1982 Stockholm Conference on Acidification of the Environment provided renewed momentum to reduce emissions of acidifying pollutants.

1985   The 'Thirty Percent Protocol' was signed by 21 signatories to the 1979 Convention, with the commitment to reduce sulphur emissions at source by at least 30% of 1980 levels by 1993 at the latest.

1988   The 'NO_x Protocol' to the Convention, signed by 27 countries, stipulated that by the end of 1994 nitrogen oxides emissions were not to exceed 1987 levels.

Research and evaluation of a number of other air pollution issues has intensified. Three particularly important examples are reviewed briefly below:

Toxics

This issue has achieved great prominence in the last two decades, particularly in industrialized countries. Of course, some countries, e.g. Sweden, Canada and the USA, have routine biomonitoring programmes, some dating back to the 1960s and early 1970s. Nevertheless, there is a particular need for global surveys of ecosystem bioaccumulation, inventories on production and use of toxics (i.e. 'cradle-to-grave' information), the establishment of routine monitoring systems, stricter pre-use testing, and regulation of production, transportation, use and disposal (Chapter 10).

Background air pollution

Background air pollution is an overall indicator of the health of the planet's environment, and so is of both scientific and practical interest. Co-ordinated by the WMO, the Background Air Pollution Monitoring Network (BAPMoN), first established in 1969, measures precipitation chemistry on a global scale. BAPMoN, currently a joint WMO/UNEP activity, now comprises almost 200 stations, three-quarters of which collect samples for measuring parameters such as precipitation chemistry, carbon dioxide and turbidity. Results indicate that between 1972 and 1984 there has been a general reduction of sulphur in precipitation. The Global Ozone Observing System (GO$_3$OS) - now incorporated into the Global Atmosphere Watch (GAW), and a number of national monitoring programmes, started in the late 1960s, are also involved in these measurements and in monitoring stratospheric ozone. However, emphasis has recently been given to greenhouse gases and tropospheric ozone. The US National Oceanic and Atmospheric Administration (NOAA) network for monitoring of greenhouse and other trace gases has also been established. At the regional level, a European monitoring network (EMEP) was set up in 1977 as a joint ECE/WMO/UNEP venture under the Convention on Long-Range Transboundary Air Pollution (see Chapter 23). The network has 102 sites throughout Europe. Recent EMEP results show that much of central and eastern Europe has high sulphate concentrations: however, the areas affected have been decreasing.

Despite some improvements in coverage during the 1980s, reliable data on air quality in remote and rural areas are still lacking for some parts of the world,

Figure 16: *The latitudinal distribution of carbon monoxide.*

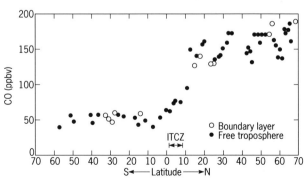

Source: Heidt et al. (1980).

Figure 17: *Annual values of sulphate and nitrate concentrations obtained from an ice core in Greenland.*

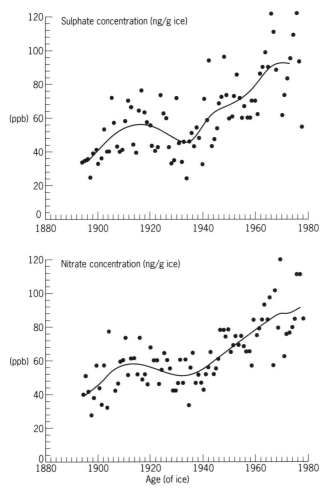

Source: Stauffer and Neftel (1988).

notably over the tropics and ocean regions. However, information from satellites, and survey data collected on a variety of expeditions, are adding greatly to the information available from fixed monitoring sites, demonstrating, for example, large interhemispheric differences for some pollutants (Heidt *et al.*, 1980) (Figure 16). It is also possible to obtain historical information from the analysis of air bubbles trapped in ice cores; see Figure 17 (Stauffer and Neftel, 1988). There is no doubt that the many exploratory surveys and background monitoring programmes of the last 20 years will provide a basis for managing the global atmosphere in the twenty-first century.

Economic assessments of air pollution damage/control costs

Twenty years ago, several methodologies were available for estimating air pollution damage costs. Some of the factors considered were: reductions in re-sale values of houses; the added costs of painting, laundry and dry-cleaning; and the increased number of hospital admittances during pollution episodes. Estimated total national costs were sometimes astronomical: however, the results could only be used subjectively to influence public policy on air pollution control.

In recent years there has been significant work relating to the cost of control, to the damages incurred, and to public policies that minimize costs. The IIASA acidic deposition studies (Alcamo *et al.*, 1987) have used models to obtain estimates of minimum costs of reducing sulphur emissions from the major European sources to meet designated target loadings of sulphur in sensitive watersheds, taking into account the varying cost of sulphur removal in different industries and different countries and at different stages of sulphur removal (see Chapter 21).

On the damage and policy side, a useful review of recent work has been given by Lave and Gruenspecht (1991). One of the new ideas is the United States EPA proposal to issue 'rights' or 'permits' for SO_2 emissions under the 1990 Clean Air Act. On the assumption that the national total of SO_2 emissions is to be controlled rather than the amounts from individual sources, 'allowances' will be issued, where 1 allowance gives an industry the right to emit 1 ton of SO_2 per year. Then, for example, if plant A could cut its emissions by one ton/yr at a cost of $US1,000 while plant B could only do this by spending $US2,000, plant A could sell its allowance for any sum of money between $US1,000 and $US2,000, and both plants would save money (Pool, 1991). There remains, however, the problem of estimating the cost of *delaying* action in reducing emissions of pollutants for various control strategies.

Concluding remarks

Society's perception of air pollution has broadened tremendously in the past two decades. In 1972, urban air pollution was the major concern; long-range transport, acidification and background pollution were just emerging as issues while indoor air pollution, radon, passive pollution from cigarette smoking and 'toxics' had not yet become front-page stories.

Today, urban pollution is worsening in many developing countries, while photochemical pollution is an example of an urban problem that has become a regional and even a global issue. Acidification damage continues even in areas where controls have been implemented, and the problem is likely to spread to parts of the developing world. 'Toxics' are widely dispersed throughout the world, even in pristine areas, and are accumulating in food chains. Background levels of most other pollutants are also increasing, although the reduction in lead concentrations in developed countries is a notable exception. These are warnings that the assimilative capacity of the global atmosphere is being overburdened. Fortunately,in some countries at least, society is heeding these signals, and increasingly strict emission controls are being introduced. Nevertheless, many developed countries still do not have comprehensive energy production policies, which must form the basis for air quality management. Unfortunately, the number of motor vehicles is increasing world-wide, and by the early twenty-first century the number is expected to rise from the current 540 million to nearly 1 billion (MacKenzie and Walsh, 1991). Air pollution control agencies in large metropolitan areas will be hard pressed to maintain present levels of air quality if this increase in vehicle population occurs with no improvement in emission standards (Swinbanks, 1991). This would place a particularly heavy burden on many developing countries, which already have serious air pollution problems. In such cases, meaningful cost-benefit analyses are urgently needed so that priorities can be assigned within national environmental and economic agendas.

Another field requiring additional investigation is that of acid rain. Some of the topics that should be given high priority are:

- Improvement of the basis for setting control targets in the context of the linkages between sulphur and nitrogen emission control strategies at the policy level.

- Improvement of the linkages between sulphur/nitrogen emission control strategies and those for oxidants, noting that nitrogen oxides also contribute to photochemical pollution.

REFERENCES

Alcamo, J., Amann, M., Hettelingh, J-P., Holmberg, M., Hordijk, L., Kamari, J., Kauppi, L., Kauppi, P., Kornai, G. and Mokkala, A. (1987) Acidification in Europe: a Simulation Model for Evaluating Control Strategies, *Ambio*, **26**, pp 232-45.

Ashby, E. and Anderson, M. (1981) *The Politics of Clean Air*, Oxford University Press, Oxford, UK.

Atlas, E. and Giam, C.S. (1986) Sea-air exchange of high molecular weight synthetic organic compounds, *In: The Role of Air-Sea Exchange in Geochemical Cycling* Buat-Menard P. (ed), D. Reidel Pub. Co., Dordrecht, the Netherlands, pp 295-329.

Boutron, C.F., Görlach, U., Candelone, J.-P., Bolshov, M.A. and Delmas, R.J. (1991) Decrease in anthropogenic lead, cadmium and zinc in Greenland snows since the late 1960s, *Nature*, **353**, 153-56.

Brimblecombe, P. (1986) *Air Composition and Chemistry*, Cambridge University Press, Cambridge, 224 pp.

Chen, B.H., Hong, C.J., Pandey, M.R. and Smith, K.R. (1990) Indoor air pollution in developing countries, *World Health Studies Quart.* **43**, 127-36.

Cleary, G.T. and Blackburn, C. R. (1968) Air Pollution in Native Huts in the Highlands of New Guinea, *Archives of Environmental Health*, **17**, 785-94.

Danilov-Danilyan V.I. and Arski, J.M. (1991) The environment in the USSR: economics and ecology, ECE ENVWA/ WG.2/R.9, Geneva.

Dignon, J. and Hameed, S. (1989) Global emissions of Nitrogen and Sulphur oxides, *J. Air Waste Management Association*, **39**, No. 2.

Duce, R.A., Liss, P.S., Merrill, J.T., Atlas, E.L., Buat-Menard, P., Hicks, B.B., Miller, J.M., Prospero, J.M., Arimoto, R., Church, T.M., Ellis, W., Galloway, J.N., Hansen, L., Jickells, T.D., Knap, A.H., Reinhart, K.H., Schneider, B., Soudine, A., Tokos, J.J., Tsunogai, S., Wollast, R. and Zhou, M. (1991) The Atmospheric Input of Trace Species to the World Ocean, *Global Biogeochemical Cycles*, **5**, 193-259.

EPA (1991) *National Air Quality and Emissions Trends Report*, EPA 450/4-91-003, US EPA.

Galloway, J.N. (1989) Atmospheric Acidification: Projections for the Future, *Ambio*, **18**, 161-66.

GESAMP (1987) *Land/sea boundary flux of contaminants: contributions from rivers*, GESAMP No. 32 (IMO/FAO/UNESCO/WMO/ WHO/IAEA/UN/UNEP).

Gunn, J.M. and Keller, W. (1990) Biological recovery of an acid lake after reductions in industrial emissions of sulphur, *Nature*, 431-33.

Hameed, S. and Dignon, J. (1992) Global Emissions of Nitrogen and Sulphur Oxides in Fossil Fuel Combustion 1970-1986, *J. Air Waste Management Assoc*, **42**, 159-63.

Heidt, L.E., Krasnec, J.P., Lueb, R.A., Pollock, W.H., Henry, B.E. and Crutzen, P.J. (1980) Latitudinal Distributions of CO and CH_4 Over the Pacific, *Journal of Geophysical Research*, **85**, 7329-36.

Hilborn, J. and Still, M. (1990) *Canadian Perspectives on Air Pollution*, SOE Report No. 90-1, Environment Canada, Ottawa, Canada, 81 pp.

Holdgate, M., Kassas, M. and White, W. (1982) *The World Environment, 1972-1982*, Tycooly International, Dublin.

Hutchinson, T.C. and Meema, K.N. (1987) *Lead, Mercury, Cadmium and Arsenic in the Environment*, SCOPE 31, John Wiley, Chichester, UK, 384 pp.

Katsoulis, B.D. and Whelpdale, D. M. (1990) Atmospheric Sulphur and Nitrogen Budgets for Southeast Europe, *Atmospheric Environment*, (in press).

Kapchella, C.E. and Hyland, M. C. (1986) *Environmental Science*, Allyn and Bacon, Inc., Boston, 589 pp.

Lave, L. and Gruenspecht, H. (1991) Increasing the efficiency and effectiveness of environmental decisions: benefit-cost analysis and effluent fees: a critical review, *J. Air and Waste Man. Assoc.* **41**, 680-93.

Leck, C. and Rodhe, H. (1989) On the Relation Between Anthropogenic SO_2 Emissions and Concentrations of Sulphate in Air and Precipitation, *Atmospheric Environment*, **23**, 959-66.

Master, K.M. (1974) Air Pollution in New Guinea, *Journal of the American Medical Association*, **228**, 1653-55.

MacKenzie, J. and Walsh, M. (1991) *Driving Forces: Motor Vehicle Trends and Their Implications for Global Warming Strategies and Transportation Planning*, World Res. Inst., Washington, D.C.

McInnes, G. (1988) Airborne road concentrations in the United Kingdom 1984-1989, Warren Springs Laboratory Report No. LR 676 (AP), Stevenage, UK.

Murozumi, M., Chow, T.J., and Patterson, C.C. (1969) Chemical concentration of pollutant lead aerosols, terrestrial dusts and sea salts in Greenland and Antarctic snow strata. *Geochim. Cosmochim. Acta*, **33**, 1247-94.

Mylona, S.N. (1989) *Detection of Sulphur Emission Reductions in Europe During the Period 1979-1986*, EMEP MSC-W Report 1/89, Norwegian Meteorological Institute, Oslo, 149 pp.

Norway (1991) *Critical Loads/Levels*, Information Booklet, Norwegian Ministry of the Environment, State Pollution Control Authority, Oslo, Norway, 8 pp.

Odén, S. (1968) The Acidification of Air and Precipitation and its Consequences in the Natural Environment, *Ecology Committee Bull. No. 1*, Swedish National Research Council, Stockholm, Sweden.

OECD (1985) *The State of the Environment, 1985*, OECD, Paris.

OECD (1991) *The State of the Environment, 1991*, OECD, Paris.

Pool, R. (1991) Polluters pay at auction, *Science*, **351**, 33.

RMCC (1990) *The 1990 Canadian Long-Range Transport of Air Pollutants and Acid Deposition Assessment Report: Part 3 Atmospheric Sciences*, Federal/Provincial Research and Monitoring Coordinating Committee, Downsview, Canada.

Rodhe, H. and Herrera, R, (eds.), (1988) *Acidification in Tropical Countries*, John Wiley & Sons, Chichester.

Semb, A. (1986) Personal communication.

Stauffer, B.R. and Neftel, A. (1988) What Have We Learned from the Ice Cores about the Atmospheric Changes in the Concentrations of Nitrous Oxide, Hydrogen Peroxide, and Other Trace Species?, In: Rowland, F. S. and Isaksen, I. S. A. (eds.), *The Changing Atmosphere*, John Wiley & Sons, Chichester, 63-77.

Sweden, 1972 *Air Pollution Across National Boundaries*. Impact on the environment of sulphur in air and precipitation. Sweden's case study to the United Nations Conference on the Human Environment, Stockholm, Royal Ministry of Foreign Affairs and Royal Ministry of Agriculture.

Swinbanks, D. (1991) Pollution on the upswing, *Nature*, **351**, 5.

UNEP (1991) *United Nations Environment Programme Environmental Data Report*, 3rd ed., Basil Blackwell, Oxford.

Volz, A. and Kley, D. (1988) Evaluation of the Montsouris Series of Ozone Measurements Made in the Nineteenth Century, *Nature*, **332**, 240-42.

WHO (1987) *Air Quality Guidelines for Europe*, Regional Pub., European Series No. 23, WHO Reg. Office for Europe, Copenhagen.

WHO/UNEP (1990) Air Quality Environment, In: *Indoor Environment: Health Aspects of Air Quality, Thermal Environment, Light and Noise*, WHO/UNEP, 43-68.

CHAPTER 2

Ozone depletion

The issue

Stratospheric ozone depletion has been a major environmental issue of the last two decades - first as an interesting hypothesis following publication of the seminal paper by Molina and Rowland (1974), and then as a matter of urgency and intergovernmental action following the discovery of the ozone 'hole' in the Antarctic stratosphere in 1984 (Farman *et al.*, 1985). The primary concern regarding ozone depletion is that a decrease in the total column content of ozone leads to an increase in the amount of UV-B radiation reaching the Earth's surface, with adverse effects on human health and ecosystems (Box 1) (UNEP, 1989). Ozone depletion may also contribute to changes in the Earth's climate (see Chapter 3).

While the largest ozone depletion is occurring at high latitudes in both hemispheres, it is happening everywhere except the tropics, and enhanced levels of UV-B radiation will have adverse effects on people of all nations, independent of their geographical position or economic status. Peoples with lightly pigmented skins are most susceptible to melanoma and non-melanoma skin cancer, but all peoples are at risk of contracting eye disorders and suppression of the immune response system. Societies in developing countries with inadequate health services are at greatest risk. Because of the impact of UV-B radiation on some plants, and hence on ecosystem functioning, ozone depletion will also reduce agricultural and fisheries productivity in the long term, and again the people most likely to be affected are those living where shortages of food already exist.

The considerable advancement in scientific knowledge since 1970 (Box 2) has been accompanied by similar advances in national and international regulatory action (Box 3). Within these past 20 years, three separate time intervals can be identified: (i) 1974 to mid-1987, including the periods during which the Vienna Convention for the Protection of the Ozone Layer (1985) and the Montreal Protocol on Substances that Deplete the Ozone Layer (1987) were negotiated; (ii) mid-1987 to mid-1990, when the London amendments to the Montreal Protocol were negotiated; and (iii) mid-1990 to the present. (Current research and monitoring studies are establishing the scientific basis for further amendments to the Montreal Protocol to be considered at the fourth meeting of the Contracting Parties in November 1992.)

Chemical processes controlling stratospheric ozone

About 90 per cent of the Earth's protective ozone layer resides in the stratosphere between 15km and 50km altitude (Figure 1). Molecular oxygen

BOX 1

Key Findings of the Impact Assessment of Stratospheric Ozone

Human health:

Non-melanoma skin cancer will increase with any long-term increase of the UV-B radiation, without a threshold value. The percentage increases will not be one-to-one: a sustained ten per cent reduction in ozone would result in a 26 per cent increase in non-melanoma skin cancer. All other things remaining constant, this would mean an increase in excess of 300,000 cases a year, world-wide. There is concern that an increase of the more dangerous cutaneous melanoma could also occur. Skin cancer would affect mainly people with little protective pigment in their skin. Exposure to increased UV-B radiation can cause suppression of the body's immune system, which might lead to an increase in the occurrence or severity of infectious diseases and a possible decrease in the effectiveness of vaccination programmes. Enhanced levels of UV-B radiation can lead to increased damage to the eyes, especially cataracts, which are estimated to increase by 0.6% per 1% of total-column ozone depletion.

Terrestrial plants:

Of the plant species investigated, about half were found to be sensitive to enhanced UV-B radiation, the impact being that plants typically exhibit reduced growth and smaller leaves. In some cases, these plants also show changes in their chemical composition, which can affect food quality and the availability of mineral nutrients. Within species, varieties have different UV-sensitivities, as demonstrated in soybeans. In certain economically important varieties, increased UV-B reduces food yield by up to 25% for exposures simulating 25% total-column ozone depletion.

Aquatic ecosystems:

Increased UV-B irradiance has been shown to have a negative influence on aquatic organisms, especially small ones such as phytoplankton, zooplankton, larval crabs and shrimp, and juvenile fish. Because many of these small organisms are at the base of the marine food web, increased UV-B exposure may have a negative influence on the productivity of fisheries. Increased exposure to UV-B radiation could lead to decreased nitrogen assimilation by prokaryotic micro-organisms and, thereby, to a possible nitrogen deficiency in rice paddies. The potential loss in yield has not yet been quantified.

Tropospheric air quality:

Enhanced levels of surface UV radiation could cause increased atmospheric abundances of several chemically reactive compounds, notably acids, hydrogen peroxide and, in polluted areas, ozone. In unpolluted regions where the concentration of NO_x is low, tropospheric ozone should decrease. It is also possible that the atmospheric abundance of particulates could be enhanced.

Materials damage:

Exposure to UV radiation is a significant cause of degradation of many materials, such as wood, plastic coatings, plastics and rubber. The impact is mainly economic. The in-creased damage will be most severe in tropical locations, where the degradation may be increased by high ambient temperatures and sunshine levels.

Source: UNEP (1989).

BOX 2

Scientific milestones: 1970 to present

1970-74 A number of papers published hypothesizing that nitrogen oxides from supersonic aircraft, and chlorine from solid rocket fuels, could catalytically reduce the level of stratospheric ozone.

1974 First postulate that CFCs would result in the catalytic destruction of stratospheric ozone.

1975 Observations of the vertical distribution of CFCs demonstrate that CFCs are removed in the stratosphere.

1977 Observations of ClO in the stratosphere demonstrate the presence of the key chlorine species responsible for the catalytic removal of ozone.

1983 Multi-year observations of CFCs 11 and 12 demonstrate that they are long-lived compounds with no significant tropospheric removal processes.

1984 First observations of the vertical extent of depletion of ozone above Antarctica based on ozone-sonde data from Syowa indicating that ozone is being depleted in the lower stratosphere between 13km and 20km altitude.

1985 Two-dimensional models are developed and used for assessment purposes to predict that CFC-induced depletion of ozone will be greatest at high altitudes (consistent with the predictions from one-dimensional models) and high latitudes.

1985 First report of the Antarctic ozone hole based on observations of total-column ozone using a ground-based Dobson spectrophotometer at Halley Bay.

1986 Spatial and temporal extent of Antarctic ozone hole established using satellite observations.

1986 Ground-based Antarctic ozone expedition (NOZE I) demonstrates that the chemical composition of the Antarctic stratosphere is highly perturbed, lending credence to the hypothesis that the Antarctic ozone hole is caused by CFCs.

1987 Aircraft campaign establishes that the Antarctic ozone hole is caused by CFCs.

1988 The International Ozone Trends Panel re-analysis of ground-based total-column ozone data shows a statistically significant decrease in ozone between 30°N and 60°N from 1969 to 1986 that is greatest in winter and that cannot be accounted for by any known natural phenomena.

1989 Aircraft campaign demonstrates that the chemical composition of the Arctic stratosphere is highly perturbed during late winter and is primed for ozone depletion.

1991 Satellite data demonstrate that ozone is being depleted at all latitudes throughout the year between 30°S and the South Pole, and from November to May between 30°N and the North Pole.

1991 Satellite (SAGE I and II), Umkehr, and balloonsonde observations of the vertical distribution of ozone have shown that ozone has decreased between 15km and 20km altitude during the last decade or so.

1992 Satellite and aircraft observations show elevated levels of stratospheric concentrations of ClO, particularly north of 50°N. At the same time, ozone depletions of 50% are found at a height of 21km in the tropics from 10°S to 20°N (due to volcanic aerosols from Mount Pinatubo).

BOX 3

National and international policy milestones: 1978 to present

1978-87   A variety of national regulatory measures introduced on CFCs 11   and 12.

1978-   UNEP Coordinating Committee for the Ozone Layer (CCOL) established.

1985   Vienna Convention for the Protection of the Ozone Layer.

1987   Entry into force of the Vienna Convention (22/9/1988): 19 nations ratified.

1987   Montreal Protocol on Substances that Deplete the Ozone Layer. This required signatory governments to regulate consumption and production:
- CFCs (11, 12, 113, 114, 115) frozen at 1986 levels in 1990, reduced by 20% in 1993, and 50% in 2000.
- Halons (1211, 1301, 2402) frozen at 1986 levels by 2005.
Developing countries that are Parties to the Montreal Protocol have a ten-year exclusionary period.

1989   Entry into force of the Montreal Protocol (1/1/1989): 23 nations ratified.

1990   Montreal Protocol on Substances that Deplete the Ozone Layer, as amended in London. This required signatory governments to regulate consumption and production:
- CFCs (11, 12, 113, 114, 115) to be reduced by 20% in 1993, 50% in 1995, 85% in 1997 and phased out in 2000. A study to be completed in 1992 will assess the feasibility of accelerating the reduction schedule.
- Halons (1211, 1301, 2402) are to be reduced by 50% by 1995 with a phase-out (except for essential uses) to occur by 2000. By decision, the parties voted to establish an ad-hoc technical group to report back to them by 1992 concerning the identification of essential uses.
- Carbon tetrachloride emissions to be reduced by 85% by 1995 and phased out by 2000.
- Methyl chloroform emissions to be frozen in 1993, reduced by 30% in 1995, 70% in 2000, and phased out in 2005.
- Other fully halogenated CFCs to be reduced by 20% in 1993, 85% in 1997, and phased out in 2000.

Jan 1991   Establishment of the Interim Multilateral Fund to provide financial assistance to developing countries in order to facilitate compliance with the control measures of the Montreal Protocol.

April 1992   82 nations have ratified the Vienna Convention.
76 nations have ratified the Montreal Protocol.
19 nations have ratified the amended Montreal Protocol.

May 1992   20 states have ratified the Amendment to the Montreal Protocol, thereby triggering the mechanism bringing it into force.

Aug 1992   The Amendment entered into force.

is broken down in the stratosphere by solar radiation to yield atomic oxygen, which then combines with molecular oxygen to produce ozone. Ozone is destroyed naturally through a series of catalytic cycles involving oxygen, nitrogen, hydrogen and to a lesser extent chlorine and bromine species. The abundance of stratospheric ozone is therefore chemically controlled by the stratospheric abundances of compounds containing hydrogen, nitrogen, chlorine and bromine. Increases in the abundances of methane and nitrous oxide (sources of hydrogen and nitrogen oxides respectively) thus affect the abundance and distribution of stratospheric ozone. Stratospheric ozone is also affected by the abundance of carbon dioxide (CO_2), because the rates of the chemical reactions that control the abundance of ozone are temperature-dependent, and the abundance of CO_2 plays a key role in determining the temperature structure of the stratosphere.

During the 1970s and early 1980s, theoretical model calculations focused on predicting the response of stratospheric ozone to changes in chlorine, assuming that the atmospheric abundances of other trace gases remained constant. However, since the early 1980s, with advances in understanding of trace gas trends and model formulation, model calculations have been used to predict the response of stratospheric ozone to simultaneous increases in chlorine (from chlorofluorocarbons, hydrochlorofluorocarbons, carbon tetrachloride, methylchloroform and methyl chloride) and bromine (from halons and methyl bromide), as well as methane, nitrous oxide and carbon dioxide.

Trends in ozone depletion and scientific understanding

Trends in chlorofluorocarbons (CFCs)

Table 1 shows the current atmospheric concentrations and trends in a number of important halocarbons, including CFCs 11, 12 and 113.

Figure 2 shows that the estimated global production of chlorofluorocarbons (primarily CFCs 11 and 12) increased during the 1960s and early 1970s, stabilized during the mid-1970s and early 1980s due to national regulations on the use of these gases as aerosol propellants and a world-wide economic recession, and increased during the mid to late 1980s as economies improved. Figure 2 also shows how the uses of CFCs have changed considerably from 1974, when aerosol propellants accounted for almost 70 per cent of the market, to 1988 when refrigerants and foam-blowing agents accounted for about 60 per cent of the market. Figure 3 shows how the atmospheric abundance of CFC11 (CCl_3F) has continued to increase over the last twenty years at about 4-5 per cent per year, even though emission rates have not continued to

Figure 1: *Temperature profile and vertical distribution of ozone in the atmosphere.*

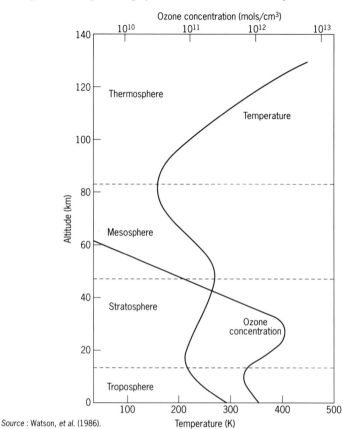

Source : Watson, et al. (1986).

Table 1: *Halocarbon abundances and trends.*

Halocarbon		Abundance (pptv)	Annual rate of increase (%)	Lifetime (years)
CCl_3F	(CFC 11)	280	4	65
CCl_2F_2	(CFC 12)	484	4	130
$CClF_3$	(CFC 13)	5	-	400
$C_2Cl_3F_3$	(CFC 113)	60	10	90
$C_2Cl_2F_4$	(CFC 114)	15	-	200
C_2ClF_5	(CFC 115)	5	-	400
CCl_4		146	1.5	50
$CHClF_2$	(HCFC 22)	122	7	15
CH_3Cl		600	-	1.5
CH_3CCl_3		158	4	7
$CBrClF_2$	(Halon 1211)	1.7	12	25
$CBrF_3$	(Halon 1301)	2.0	15	110
CH_3Br		10–15	15	1.5

increase. Figure 4 shows the estimated consumption of CFCs in 1986 by geographic region.

Model calculations 1970-87

Photochemical models are used to predict the extent of ozone depletion for a variety of assumptions concerning future emission rates of halocarbons, methane, nitrous oxide and in some cases carbon dioxide. All models predict

Figure 2: *Trends in chlorofluorocarbon emissions 1960-90 (graph) and consumption by application in 1974 and 1988 (pie diagrams).*

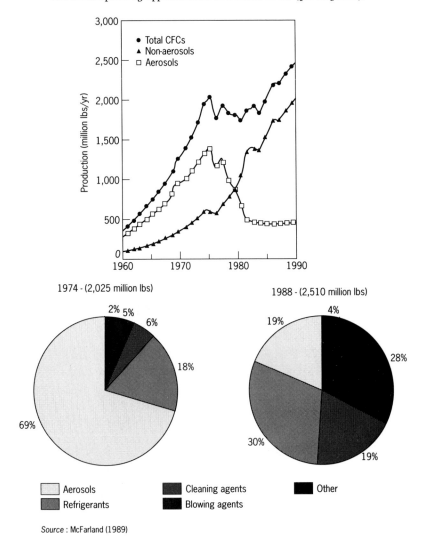

Source : McFarland (1989)

Figure 3: *Measured and predicted cumulative increases of chlorinated source gases in the troposphere from 1974 to 1990. The concentrations are given as chlorine atom equivalents. Dashed lines indicate the estimated rate of increase in total chlorine of 3.3 per cent per year based on constant emission rates at 1986 levels.*

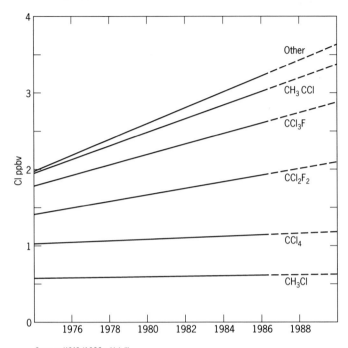

Source : WMO (1990a, Vol. II).

Figure 4: *Estimated consumption of CFCs by region in 1986.*

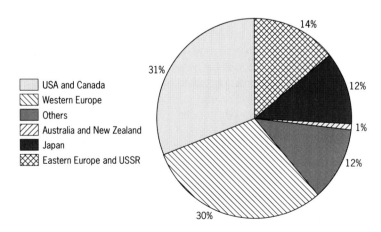

Source : M. McFarland (private communication, 1991)

Figure 5: *Model predictions for ozone profile changes (in percentages) for an extended time-dependent scenario 1985-2040 for the month of March, from the Cambridge two-dimensional model.*

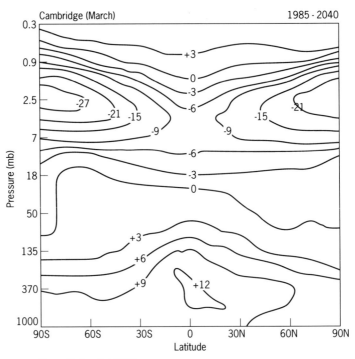

Source : Prather et al. (1988).

that total-column ozone depletion would be greatest at high latitudes, and that these changes would be greatest in the upper stratosphere at altitudes of around 40km. Models predict that the abundance of ozone: (i) decreases with increasing concentrations of chlorine, bromine and nitrous oxide, and (ii) increases with increasing atmospheric abundances of methane and carbon dioxide. Figure 5 shows a typical model calculation of the predicted change in the vertical distribution of ozone between 1985 and 2040 assuming constant emissions of CFCs (1985 emission rates), but growth in the emission rates of methane (1 per cent per year) and nitrous oxide (0.2 per cent per year). Scenarios that assume growth in the emissions of CFCs predict greater depletions of ozone.

Ozone observations 1970-87

Statistical analyses of ground-based total-column ozone measurements show that between 1970 and 1980 there was no statistically significant decrease in the annual global average content of ozone (i.e. the decrease was less than 1 per cent). Model calculations, taking into account increases in carbon dioxide, methane, nitrous oxide and halocarbons, predicted that total-column ozone should have decreased by less than 0.5 per cent at all latitudes in the summer, and between 0.5 and 1 per cent in the winter, with the largest depletions predicted at high latitudes. Thus the model results are not inconsistent with the observations.

Estimates of trends in the vertical distribution of ozone, using ground-based data, suggest that ozone decreased in the middle and upper stratosphere by 2-3 per cent in the period 1970 to 1980. While this data base is limited, both in number of stations and data quality, the trends are consistent with model predictions.

In 1985 the Antarctic ozone hole was first reported (Farman *et al.*, 1985). Observations of total-column ozone, using a ground-based spectrophotometer at Halley Bay, showed that the October level of ozone decreased from a value of typically 300 Dobson Units (DU) during the 1960s and early 1970s to about 200 DU in the early 1980s (Figure 6). This decrease of about 30-40 per cent in a decade was totally unexpected. Satellite data (Plates I and II, between pages 52 and 53) showed the spatial extent of the depletion, and balloonsonde data (Figure 7) showed that it occurred in the lower stratosphere. By 1986 three mechanisms (two involving changes in natural processes (1 and 2) and one involving human activities (3)) had been proposed to explain the observations of ozone loss:

1 solar cycle mechanism: periodic increases in the abundance of nitrogen oxides in the lower Antarctic stratosphere due to variations in solar output;

2 dynamical mechanism: a change in the circulation pattern from downwelling of ozone-rich air from the upper stratosphere to upwelling of ozone-poor air from the troposphere; and

3 halogen mechanisms: several different theories invoking the catalytic destruction of ozone due to CFCs and halons.

A ground-based study based at McMurdo, Antarctica,in 1986 showed that:

1 the abundances of nitrogen oxides were exceptionally low, which largely disproved the solar cycle mechanism;

2 the levels of long-lived tracers were such that a dynamical mechanism was unlikely; and

3 the elevated levels of active chlorine species suggested that chlorine catalysis was the most probable mechanism.

At the time that the Montreal Protocol was signed in September 1987, the cause of the Antarctic ozone hole was not well established. However, the scientific synthesis upon which the Protocol was based provided convincing evidence that the stratosphere had become perturbed:

Figure 6: *Monthly means of total ozone at Halley Bay Station, Antarctica, for October for the years 1957 to1984.*

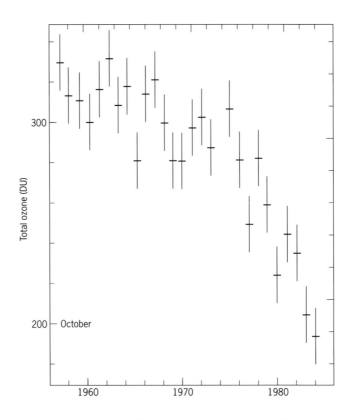

Source : Farman, *et al.* (1985). Reprinted by permission from *Nature*, **315,** pp 207-10. Copyright Macmillan Magazines Ltd.

- the levels of atmospheric chlorine were increasing at about 4-5 per cent per year because of the emissions of chlorofluorocarbons;

- theoretical models predicted a chlorine-induced ozone loss, primarily in the upper stratosphere between 30km and 45km;

- there was limited observational evidence of ozone loss near 40km; and

- a significant loss of ozone was occurring every October above Antarctica, even though its cause had not been conclusively established.

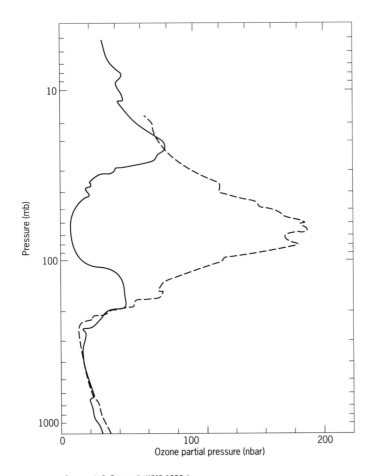

Figure 7: *Vertical distribution of ozone partial pressure (nbar) observed at Halley Bay Station on 15 August 1987 (high values), and 15 October 1987 (low values), respectively.*

Source : J. C. Farman (in WMO 1990a).

Advances in scientific understanding 1987-92

There have been tremendous advances in scientific understanding of the ozone layer in the last five years. These advances, primarily based on laboratory studies, special field campaigns, theoretical studies and a re-assessment of trends in stratospheric ozone, were reviewed during the 1989 and 1991 international scientific assessments, which were required by the Montreal Protocol. The key conclusions of these two international assessments are summarized in Boxes 4 and 5. Scientific understanding of the environmental impacts of ozone layer depletion were also reviewed at that time (Box 1).

BOX 4

Key Findings of the Scientific Assessment of Stratospheric Ozone: 1989

Remarkable progress has been made in stratospheric ozone science in the past few years. There have been highly significant advances in the understanding of the impact of human activities on the Earth's protective ozone layer. Since the Montreal Protocol was signed, there have been four major findings, each of which has heightened concern that chlorine- and bromine-containing chemicals can lead to a significant depletion of stratospheric ozone.

- **Antarctic ozone hole:**
  The weight of scientific evidence strongly indicates that chlorinated (largely man-made) and brominated chemicals are primarily responsible for the recently discovered substantial decreases in stratospheric ozone over Antarctica in the Southern Hemisphere springtime.

- **Perturbed Arctic chemistry:**
  While at present there is no measurable ozone loss over the Arctic comparable to that over the Antarctic, the same potentially ozone-destroying processes have been identified in the Arctic stratosphere. The degree of any future ozone depletion will probably depend on the particular meteorology of each Arctic winter and future atmospheric levels of chlorine and bromine.

- **Long-term ozone decreases:**
  The analysis of the total-column ozone data from ground-based Dobson instruments shows measurable downward trends from 1969 to 1988 of 3-5% (i.e. 1.8-2.7% per decade) in the Northern Hemisphere (30°N to 64°N) in the winter months which cannot be attributed to known natural processes.

- **Model limitations:**
  These findings have led to the recognition of major gaps in theoretical models used for assessment studies. Assessment models do not simulate adequately polar stratospheric cloud (PSC) chemistry or polar meteorology. The impact of these shortcomings for the prediction of ozone layer depletion at lower latitudes is uncertain.

BOX 5

Advances in the understanding of stratospheric ozone since mid-1990

Antarctic ozone losses continue:

Strong Antarctic ozone holes have continued to occur, and in four of the past five years have been deep and extensive in area. This contrasts to the situation in the mid-1980s, when the depth and area of the ozone hole exhibited a quasi-biennial modulation. Large increases in surface ultraviolet radiation have been observed in Antarctica during periods of low ozone. While no extensive ozone losses have occurred in the Arctic comparable to those observed in the Antarctic, localized Arctic ozone losses have been observed in winter concurrent with observations of elevated levels of reactive chlorine.

Larger global ozone decreases observed:

Ground-based and satellite observations continue to show decreases of total-column ozone in winter in the Northern Hemisphere. For the first time, there is evidence of significant decreases in spring and summer in both the northern and southern hemispheres at middle and high latitudes (about 3.5% at 45°N in summer for the period 1979-1991) as well as in the southern winter. No trends in ozone have been observed in the tropics. The downward trends were larger during the 1980s than in the 1970s by about 2% loss per decade. The observed ozone decreases have occurred predominantly in the lower stratosphere.

Observed ozone losses due to industrial halocarbons:

Recent laboratory research and an extended interpretation of field measurements have strengthened the evidence that the Antarctic ozone hole is primarily due to chlorine- and bromine-containing chemicals. In addition, the weight of evidence suggests that the observed middle- and high-latitude ozone losses are largely due to chlorine and bromine. Therefore, as the atmospheric abundances of chlorine and bromine increase in the future, significant additional losses of ozone are expected at middle latitudes and in the Arctic.

Future levels of ozone depletion:

Even if the control measures of the amended Montreal Protocol (London, 1990) were to be implemented by all nations, the current abundance of stratospheric chlorine (3.3-3.5 parts per billion volume (ppbv)) is estimated to increase during the next several years, reaching a peak of about 4.1 ppbv around the turn of the century. Since the middle latitude ozone losses are apparently due in large part to chlorine and bromine, the increased levels of chlorine and bromine that are estimated by the year 2000 are expected to result in additional ozone losses during the 1990s comparable to those already observed for the 1980s. There is also the possibility of incurring wide spread losses in the Arctic. Reducing these expected and possible ozone losses requires further limitations on the emissions of chlorine- and bromine-containing compounds.

Recent special field campaigns

A scientific campaign mounted from Punta Arenas, Chile, during August and September of 1987 utilizing aircraft flights over Antarctica provided good scientific evidence that anthropogenic chlorine, and to a lesser extent bromine, were the primary cause of the observed Antarctic ozone hole. Figure 8 shows that the abundance of the ClO radical, an active chlorine species, increases significantly from about 50 pptv outside the seasonal meteorological vortex, which is located between about 65 °S and 68°S, to between approximately 0.75 and 1.0 ppbv within the vortex. Figure 8 also shows how a strong anticorrelation developed between ClO and ozone from 23 August to 16 September 1987. These data, combined with the observations of: (i) the BrO radical; (ii) low abundances of water vapour, nitrogen oxides and long-lived tracers such as CFCs and nitrous oxide; and (iii) polar stratospheric clouds (PSCs), demonstrated that the ozone loss over Antarctica is initiated by chemical reactions that occur on PSCs and that convert the long-lived chlorine into chemically more-reactive forms, which in the presence of sunlight, leads to a catalytic destruction of ozone of up to 1-2 per cent per day in total column content.

A second Airborne Arctic Stratospheric Expedition (AASE-II) was launched in early October 1991. In the course of this six-month field experiment, the highest levels of chlorine monoxide (ClO) ever found in the polar stratosphere were recorded over eastern Canada and northern New England in mid to late January 1992 (1.5 ppbv) (NASA, 1992). During this period, the NASA Upper

Figure 8: *Evolution of the relationship between ozone and chlorine monoxide (ClO) from the first full penetration into the Antarctic vortex on 23 August 1987 (below left), to three weeks later on September 16, 1987 (below right).*

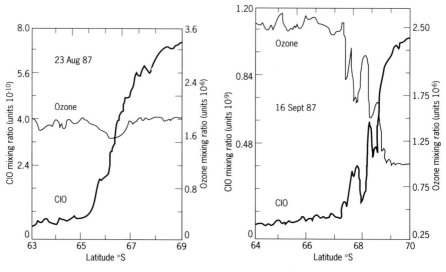

Source : Anderson *et al.* (1989).

Atmosphere Research Satellite (UARS) also detected elevated levels of ClO over large parts of Europe and Asia north of 50°N. The observations of high concentrations of ClO and also of bromine monoxide imply human-induced ozone destruction rates of 1-2 per cent per day.

At the same time, satellite measurements of ozone levels at a height of about 21km in the tropics (from 10°S to 20°N) were about 50 per cent less than usual due to the eruption of Mount Pinatubo. The presence of enhanced stratospheric concentrations of volcanic aerosols is likely to cause ozone depletion in the tropics for several years (NASA, 1992).

A similar campaign using aircraft based at Stavanger, Norway, between January and March 1989, provided compelling scientific evidence that the chemical composition of the Arctic stratosphere was highly perturbed. The chemical perturbations were similar to those found in Antarctica, namely an increase in the abundance of the ozone-depleting forms of chlorine in association with PSCs. Reactive chlorine abundances were enhanced by a factor of 50 to 100.

Recent assessments of trends in stratospheric ozone

In 1986 the International Ozone Trends Panel (IOTP) was established to analyse ground-based and satellite ozone data in terms of both total column content and vertical distribution. The most important aspect of the studies was that the ozone record was analysed for seasonal and latitudinal trends rather than for just global annual averaged trends. Some of the conclusions of the IOTP, and the more recent findings of Bishop and Bojkov (1992) and Bojkov et al. (1990) are summarized in Boxes 4 and 5. The key points included in the synthesis report of the assessment panels (November 1991) are:

- Observational records: The ground-based Dobson record has been augmented by a complete re-analysis of the data from 29 ground-based M-83/124 stations in the former USSR. This is complemented by a major new advance in the observational record of an internally calibrated Total Ozone Mapping Spectrometer (TOMS) data set that is now independent of ground-based observations. Trend analyses of the Stratospheric Aerosol and Gas Experiment (SAGE) data have been extended to the lower stratosphere and the data obtained from the global network of surface spectrophotometers have been re-analysed.

- Trends in total ozone: Ground-based (Dobson and M-83/124) and satellite (TOMS) observations of total-column ozone through March 1991 were analyzed allowing for the influence of solar cycle and quasi-biennial oscillation (Table 2). They show that:

a) the northern mid-latitude winter and summer decreases during the 1980s were larger than the average trend since 1970 by about 2 per cent per decade. A significant longitudinal variance of the trend since 1979 is observed;

b) for the first time there were statistically significant decreases in all seasons in both the northern and southern hemispheres at middle and high latitudes during the 1980s; the northern mid-latitude long-term trends (1970-91), while smaller, are also statistically significant in all seasons;

c) there has been no statistically significant decrease in tropical latitudes from 25°N to25°S.

Table 2: *Total ozone trends in per cent/decade with 95 per cent confidence limits.*

	TOMS: 1979-1991			Ground-based: 26°N-64°N	
	45°S	Equator	45°N	1979–1991	1970–1991
Dec – Mar	−5.2 ± 1.5	+0.3 ± 4.5	−5.6 ± 3.5	−4.7 ± 0.9	−2.7 ± 0.7
May – Aug	−6.2 ± 3.0	+0.1 ± 5.2	−2.9 ± 2.1	−3.3 ± 1.2	−1.3 ± 0.4
Sep – Nov	−4.4 ± 3.2	+0.3 ± 5.0	−1.7 ± 1.9	−1.2 ± 1.6	−1.2 ± 0.6

• Trends in the vertical distribution of ozone: Balloonsonde, ground-based Umkehr, and satellite SAGE observations show that:

a) ozone is decreasing in the lower stratosphere, i.e. below 25km, at about 10 per cent/decade, consistent with the observed decrease in column ozone;

b) changes in the observed vertical distribution of ozone in the upper stratosphere near 40km are qualitatively consistent with theoretical predictions, but are smaller in magnitude;

c) measurements indicate that ozone levels in the troposphere, over the few existing ozone sounding stations at northern mid-latitudes, have increased about 10 per cent per decade over the past two decades.

• Impacts of supersonic aircraft: Recent evidence has shown that reactions on sulphate aerosols can change the partitioning of nitrogen oxides. Two model studies incorporating this heterogeneous chemistry have recently re-examined the case of 500 aircraft flying at Mach 2.4 at 17-20km and found substantially less ozone change (-0.5 per cent to +0.5 per cent) than the earlier prediction (2-6 per cent). The implications of this need to be examined in detail.

• Impacts of shuttles and rockets: The increase in the abundance of stratospheric chlorine due to US annual launches of nine Space Shuttles and six Titan rockets was calculated to be less than 0.25 per cent of the annual stratospheric chlorine source from halocarbons in the present-day atmosphere (with maximum increases of 0.01 ppbv in the middle and upper stratosphere in the northern middle and high latitudes). The TOMS ozone record shows no detectable changes in column ozone immediately following launches of the US Space Shuttle.

- Ultraviolet radiation: Significant increases in ultraviolet radiation have been observed over Antarctica in conjunction with periods of intense ozone depletion. Under clear-sky conditions, these increases are consistent with theoretical predictions. Therefore, for the first time, the response of ground-level ultraviolet radiation to changes in column ozone has been observed and quantified.

- Ozone depletion and global warming: A new semi-empirical, observation-based method of calculating ODPs has better quantified the role of polar processes in this index. In addition, the direct global warming potentials (GWPs) for tropospheric, well-mixed, radiatively active species have been recalculated. However, because of the incomplete understanding of tropospheric chemical processes, the indirect GWPs of methane and other short-lived gases have not, at present, been quantified reliably. In fact, the concept of a GWP may prove inapplicable for the very short-lived gases in homogeneously mixed gases such as the nitrogen oxides. Hence, many of the *indirect* GWPs reported in 1990 by the Intergovernmental Panel on Climate Change (IPCC) are likely to be incorrect. Furthermore, the radiative cooling introduced by the lower stratospheric ozone loss may offset the radiative warming of the ozone-depleting chemicals; therefore, their GWPs may be significantly less than earlier considered because the effect cannot at present be quantified reliably (see also Chapter 3).

Responses

Policy responses

Box 3 summarizes the main international policy responses in the 1970s and 1980s to the ozone layer issue. Established by UNEP in the 1970s, the Coordinating Committee on the Ozone Layer (CCOL) has played a major role in establishing scientific research priorities, and in describing the implications of the latest scientific findings in terms understandable to policy-makers. In 1980, for example, the CCOL reported that continued releases of CFCs would eventually deplete the ozone layer, and that this could have serious impacts on the health of people and the biosphere (Holdgate *et al.*, 1982). Subsequently, the CCOL synthesized the knowledge base to provide the scientific basis for the Montreal Protocol. The policy implications of the scientific findings of the last decade have been profound (see Box 6), leading, for example, to a significant revision of the control measures of the Montreal Protocol.

The most recent scientific findings also have substantial policy implications (Box 7) and could lead to further significant revisions in the control measures of the Montreal Protocol at the Fourth meeting of the Contracting Parties, which will be held in November 1992.

BOX 6

Implications of the Key Findings of the Scientific Assessment of Stratospheric Ozone: 1989

The findings and conclusions of the intensive and extensive ozone research over the past few years have several major implications as input to public policy regarding restrictions on man-made substances that lead to stratospheric ozone depletion:

Scientific:

The scientific basis for the 1987 Montreal Protocol on Substances that Deplete the Ozone Layer was the theoretical prediction that, should CFC and halon abundances continue to grow for the next few decades, there would eventually be substantial ozone layer depletion.

The research of the last few years has demonstrated that actual ozone loss due to anthropogenic chlorine emissions (i.e. CFCs) and bromine has already occurred, i.e. the Antarctic ozone hole. Assuming that the atmospheric abundance of chlorine reaches about 9 ppbv by about 2050, ozone depletions of 0-4% in the tropics and 4-12% at high latitudes would be predicted, even without including the effects of heterogeneous chemical processes known to occur in polar regions.

The surface-induced, PSC-induced chemical reactions that cause the ozone depletion in Antarctica and that also occur in the Arctic represent additional ozone-depleting processes that were not included in the stratospheric ozone assessment models that were used to guide the Montreal Protocol. Recent laboratory studies suggest that similar reactions involving chlorine compounds may occur on sulphate particles present at lower latitudes, which could be particularly important immediately after a volcanic eruption.

Hence, future global ozone layer depletions could well be larger than originally predicted.

Policy:

Large-scale ozone depletions in Antarctica appeared to have started in the late 1970s and were initiated by atmospheric chlorine abundance of about 1.5-2.0 ppbv, compared to today's level of about 3 ppbv.

To return the Antarctic ozone layer to levels approaching its natural state, and hence to avoid the possible ozone dilution effect that the Antarctic ozone hole could have at other latitudes, one of a limited number of approaches to reduce the atmospheric abundance of chlorine and bromine is a complete phase-out of all fully halogenated CFCs, halons, carbon tetrachloride and methyl chloroform, as well as careful considerations of the HCFC substitutes. Otherwise, the Antarctic ozone hole is expected to recur seasonally, provided the present meteorological conditions continue.

Colour plates

Plate I: *Total Ozone Mapping Spectrometer (TOMS) satellite data images of October monthly means of total column ozone in the Southern Hemisphere showing progression of ozone depletion from 1982 to 1991. (NASA Goddard Space Flight Center).* ▶

Plate II: *Total Ozone Mapping Spectrometer TOMS) satellite data image of total-column ozone in the Southern Hemisphere, October 1991. (NASA Goddard Space Flight Center.)* ▶▶

Plate I

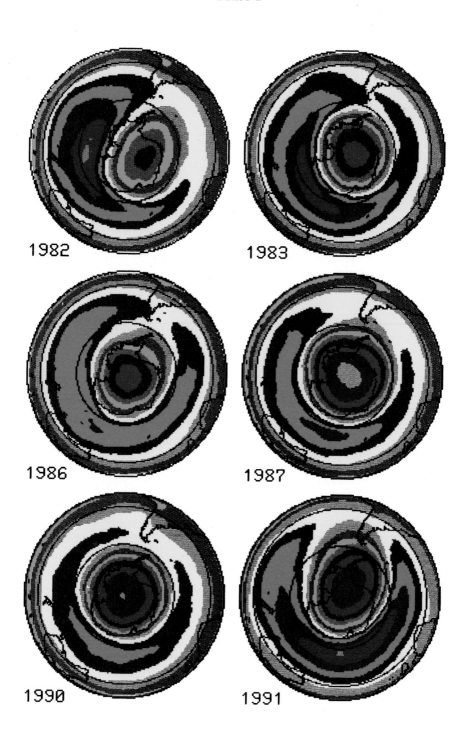

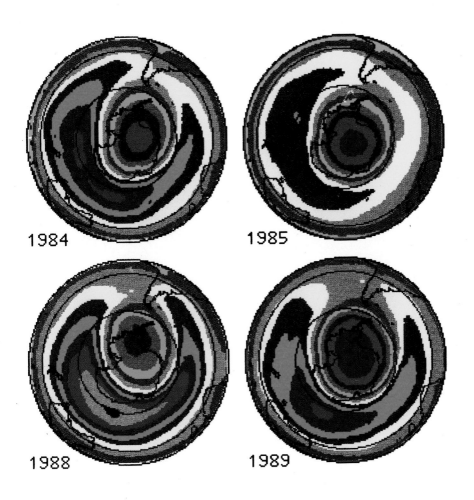

TOMS: OCTOBER MONTHLY MEANS

Plate II

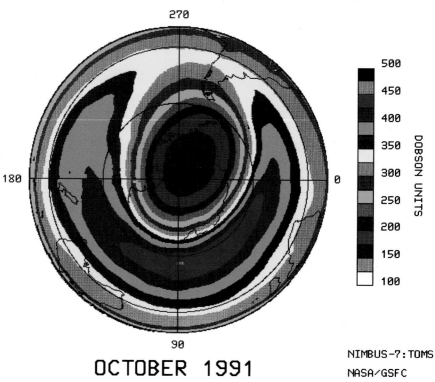

TOTAL OZONE MONTHLY MEAN

OCTOBER 1991

NIMBUS-7:TOMS
NASA/GSFC

Using simulation models to explore policy options

Nowadays, simulation models are used more and more to explore policy options for resolving the ozone depletion issue. A particular example, performed in 1989, was a sensitivity study of atmospheric chlorine-loading scenarios for a range of global emissions of compounds currently regulated under the Montreal Protocol, and for possible new HCFC substitutes. The results are summarized in Figure 9, which shows that:

- peak chlorine loading is controlled by the date of the global phase-out (assumed to be the year 2000) for long-lived CFCs, and by emission rates between 1990 and 2000;

- even with a complete phase-out of all chlorine emissions in 2000, the atmospheric level of chlorine will not decrease to 2 ppbv (the level of chlorine prior to the formation of the Antarctic ozone hole) until about 2075;

- a wide range of HCFC substitution policies can be implemented without delaying recovery of the Antarctic ozone hole, assuming that there is a phase-out of HCFC substitutes before the middle of the next century (the required phase-out date for the HCFCs is dependent upon their lifetime); and

- the additional atmospheric loading of chlorine due to HCFC substitution depends upon the HCFC emission rates and the lifetimes of the HCFCs, but is typically limited to less than 1 ppbv even with significant initial substitution rates and subsequent growth in these emission rates.

Figure 9: *Impact of chlorofluorocarbon (CFC) and hydrochlorofluorocarbon (HCFC) substitution policies on atmospheric abundance of chlorine (Cl).*

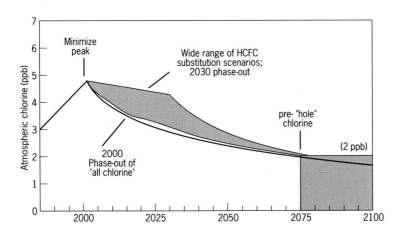

Source : Modified from Figures 3a and 3b by Prather and Watson, (1990).

BOX 7

Implications of scientific findings since mid-1990

Elimination of the Antarctic ozone hole:

The phase-out schedule of the amended Montreal Protocol, *if fully complied with by all nations and if there are no continued uses of HCFCs,* affords the opportunity to return to stratospheric chlorine abundances of 2 ppbv sometime between the middle and the end of the next century. This is the level at which the Antarctic ozone hole appeared in the late 1970s and hence is about the level that is thought to be necessary (other conditions assumed constant, including bromine loading) to eliminate the ozone hole.

Such levels could never have been reached under the provisions of the original 1987 Protocol.

Future levels of ozone:

Even if the control measures of the amended Montreal Protocol (London, 1990) were to be implemented by all nations, the current abundance of stratospheric chlorine (3.3-3.5 ppbv) is estimated to increase during the next several years, reaching a peak of about 4.1 ppbv around the turn of the century. With these increases, the additional middle-latitude ozone losses during the 1990s are expected to be comparable to those observed during the 1980s, and there is the possibility of incurring widespread losses in the Arctic.

Reducing these expected and possible ozone losses requires further limitations on the emissions of chlorine- and bromine-containing compounds.

Approaches to limiting future levels of ozone depletion:

Lowering the peak and hastening the subsequent decline of chlorine and bromine levels can be accomplished in a variety of ways, including an

---

The study concluded that in order to minimize future ozone layer depletion and to facilitate the most rapid elimination of the Antarctic ozone hole (i.e. by about 2075):

- production and emission of long-lived halocarbons (CFCs and CCl_4) should be phased out as soon as possible, with the greatest possible global compliance;

- halocarbon substitutes should be phased out sometime in the middle of the next century;

- atmospheric concentrations of bromine should be stabilized at current levels or below, and

- peak chlorine loadings should be reduced, most practicably by significant cutbacks in the emissions of abundant, short-lived halocarbons, particularly methyl chloroform.

accelerated phase-out of controlled substances and limitations on currently uncontrolled halocarbons.

Chlorine

A significant reduction in peak chlorine loading (a few tenths of a ppbv) can be achieved with accelerated phase-out schedules of CFCs, carbon tetrachloride and methyl chloroform. Even stringent controls on HCFC-22 would not significantly reduce peak chlorine loading (at most 0.03 ppbv, especially when ODP weighted), but do hasten the decline of chlorine. Specifically actions should include: a global phase-out in the emissions of long-lived chlorofluorocarbons, methylchloroform and carbon tetrachloride as soon as possible (100% compliance is essential); not-in-kind substitution of CFCs wherever practical; the substitution of long-lived CFCs with HCFCs having the shortest possible lifetimes, hence low values of ozone depletion potentials (remember that all HCFCs are not equal: those with short atmospheric lifetimes (1-5 years) pose a significantly lower threat to the ozone layer than those with moderately long lifetimes (greater than 15 years); a phase-out of halocarbon substitutes (HCFCs) sometime early in the next century (phase-out date should depend upon the atmospheric lifetime of the substitute), and possible emission rate limitations; and the recycling of HCFCs to the maximum extent possible.

Bromine

A three-year acceleration of the phase-out schedule for the halons would reduce peak bromine loading by about 1 pptv. If the anthropogenic sources of methyl bromide are significant and their emissions can be reduced, then each 10% reduction in methyl bromide would rapidly result in a decrease in stratospheric bromine of 1.5 pptv, which is equivalent to a reduction in stratospheric chlorine of 0.045 to 0.18 ppbv. This gain is comparable to that of a three-year acceleration of the scheduled phase-out of the CFCs.

Concluding remarks

Depletion of the stratospheric ozone layer by halogenated chemicals is a global problem. While the issue was primarily a matter of scientific curiosity during the 1970s and early 1980s, it has now become an urgent policy question for governments in both developed and developing countries. The weight of scientific evidence strongly indicates that chlorinated and brominated chemicals (largely man-made) are primarily responsible for the substantial decreases in stratospheric ozone (greater than 50 per cent in the total column) that occur over Antarctica every Southern Hemisphere springtime. In addition, satellite data supported by ground-based observations have demonstrated that ozone has been decreasing since the late 1960s at middle and high latitudes in both the northern and southern hemispheres, and that this decrease cannot be explained by known natural processes. Indeed, the conclusion of the 1991 International Scientific

Assessment was that 'the weight of evidence suggests that the observed middle- and high- latitude ozone losses are largely due to chlorine and bromine'.

The current and historic use of CFCs and other such chemicals in developed countries is the primary cause of the problem. However, adequate protection of the ozone layer will require a full partnership between the developed countries, whose industries and consumers have caused the problem, and the developing countries, whose people wish to use these chemicals in refrigeration and other ways.

As in the case of atmospheric concentrations of lead, which are now declining,

---

BOX 8

Main findings of the Report of the Technology and Economic Assessment Panel, (December, 1991).

Estimated phase-out progress:
- Total CFC production was cut by 40% between 1986 and 1991 with maximum reduction (70%) in the production of CFC114.
- Production of Halon 1211 and Halon 1301 peaked in 1988 and is now declining. Production of Halon 2402 has virtually ceased in OECD countries.
- Several manufacturers are moving faster than the most stringent regulations. By January 1992, halon and CFC recycling will be accepted world-wide. The first HFC-134a automobile airconditioners and domestic refrigerators will also be commercialized. Discharge testing of halon has virtually been eliminated in training and servicing equipment.

Changes in CFC global market:
- Total market has declined between 1986 and 1992 by 40%. Major declines were in propellants (58%), cleaning agents (41%), phenolic blowing agents (65%) and extruded polystyrene sheets (90%).

Technical feasibility of early phase-out:
- In *developed countries* it is technically feasible to phase-out all consumption of CFCs and halons by 1995–97; 1,1,1-trichloroethane by 1995 or 2000 at the latest and carbon tetrachloride by 1997. These dates are based on completion of toxicity tests on the transitional substances.
- The Halon Technical Options Committee did not consider a phase-out earlier than 1997. Estimates were made of halon banks to meet requirements into the next century.
- In *developing countries*, the same phase-out schedules are technically and economically feasible in many applications. More time may be needed for some applications (5–8 years). Financial assistance and training will be needed.

even in Greenland, as a result of tighter emission controls (Chapter 1), the stratosphere would gradually recover if emissions of CFCs and similar chemicals were to cease. Thus, the most urgent task is to strengthen the Montreal Protocol and to develop an international method of monitoring national programmes to control CFCs and related gases. However, scientific research and monitoring should continue into the next century to provide a basis for better policies. The findings of the December 1991 Technology and Economic Assessment Panel (Box 8) indicate that effective action for an early phase-out is both technically and economically feasible.

Implications of a 1997 phase-out:
- The costs will be higher than a year 2000 phase-out. Half the extra cost will be in retrofitting vehicle air conditioners in the USA and Japan.
- The availability of some substitutes and alternatives for small but important uses (inhalant drugs, precision cleaning, drying hi-tech products) is uncertain.

Transitional substances:
- Overall ODP impact of transitional substances will be minimized if HCFCs with lowest ODP are selected. Technologies for replacing the controlled substances should be selected so as to minimize energy consumption as well as ozone depletion.

Technical and environmental uncertainty:
- The possibility that exemptions may be needed for a 1997 phase-out cannot be ruled out. There is no perfect substitute. Each one has difficult trade-offs in performance, ODP, GWP, energy efficiency and toxicity.

Developing countries concerns:
- Successful phase-out assumes the availability of technologies, technical support and training through channels and mechanisms supported by the Interim Multilateral Fund.
- There is concern that producers in developing countries might export excess controlled substances or obsolete equipment using them to other countries.
- Recovery, recycling and management of banks of controlled substances could face difficulties due to inadequate infrastructure and training.
- Some developing countries are already entering into technology co-operation projects for phasing out controlled substances (e.g. Mexico, China, Thailand, Brazil).

References

Anderson, J.G., Brune,W.H. and Proffitt, M.H. (1989) Ozone destruction by chlorine radicals within the Antarctic Vortex: The Spatial and Temporal Evolution of ClO-O$_3$ anticorrelation based on *in situ* ER-2 data. *J. Geophys. Res.* Vol. **94**, No. D9 Special Issue on the Antarctic Airborne Ozone Experiment (AAOE), pp. 11,465-79.

AGU (1989) The airborne Antarctic ozone experiment (AAOE), *J. Geophys. Res.*, **94**, Special issues Nos. D9 (8/30/89) and D14 .

Assessment Chairs for the Parties to the Montreal Protocol (1991) Synthesis of the reports of the Scientific, Environmental Effects, and Technology and Economic Assessment Panels.

Bishop, L. and Bojkov, R.D. (1992) Total ozone change based on re-evaluated data for 1956-1991, *J. Geophys. Res.*, (in press).

Bojkov, R., Bishop, L., Hill, W.J., Reinsel, G.C. and Tiao, G.C. (1990) A statistical trend analysis of revised Dobson total ozone data over the Northern Hemisphere, *J. Geophys. Res.*, **95**, 9785-807.

Brune, W.H., Anderson, J.G., Toohey, D.W., Fahey, D.W., Kawa, S.R., Jones, R.L., McKenna, D.S. and Poole, L.R. (1991) The potential for ozone depletion in the Arctic polar stratosphere, *Science*, **252**, 1260-66.

Farman, J.C. (personal communication) in WMO (1990a) Report of the International Ozone Trends Panel: 1988, Report No. 18, 2 volumes, Vol. II, WMO, Geneva, p.688.

Farman, J.C., Gardiner, B.G. and Shanklin, J.D. (1985) Large losses of total ozone in Antarctica reveal seasonal CLOx/NOx interaction, *Nature*, **315**, 207-10.

GRL (1990) *Geophys. Res. Letters*, Special supplement on the Airborne Arctic Strato-spheric Expedition (AASE), March, 1990.

Holdgate, M., Kassas, M. and White G. (1982) *The State of the Environment 1972-1982*, Tycooly Press, Dublin.

McFarland, M. (1989) Chlorofluorocarbons and ozone, *Environ. Sci. & Technol.*, **Vol. 23**, No. 10, pp. 1203-08.

McFarland, M. (1991) Dupont Chemicals/ Fluorochemicals (private communication).

Molina, M.J. and Rowland, F.S. (1974) Stratospheric sink for chlorofluoro-methanes: chlorine atom catalyzed destruction of ozone, *Nature*, **249**, 810-14.

NASA (1992): Interim findings: Second Airborne Arctic Stratospheric Expedition, Press briefing, NASA Headquarters, 3 February 1992.

NASA Goddard Space Flight Center, Green-belt, Maryland, USA, TOMS (Total Ozone Mapping Spectrometer) satellite ozone data.

Prather, M.J. and Ad Hoc Theory Panel (1988) Model Predictions of Future Ozone Change. Present State of Knowledge of the Upper Atmosphere 1988: An Assessment Report, NASA Reference Publication 1208, p. 146.

Prather, M. J. and Watson, R.T. (1990) Stratospheric ozone depletion and future levels of atmospheric chlorine and bromine. *Nature*, **344**, No. 6268, pp 729-34.

Stolarski, R.F., Bloomfield, P., McPeters, R.D. and Herman, J.R. (1991) Total ozone trends deduced from Nimbus-7 TOMS data, *Geophys. Res. Letters*, **18**, 1015-18.

UNEP (1987) Montreal Protocol on Substances that Deplete the Ozone Layer, UNEP, Nairobi.

UNEP (1989) Environmental Effects Panel Report, UNEP, Nairobi, 64 pp.

Watson, R.T., Geller, M. A. , Stolarski, R. S. and Hampson, R. F. (1986) Present State of Knowledge of the Upper Atmosphere: An Assessment Report, NASA Reference Publication 1162, p.20.

WMO (1986) Atmospheric ozone: 1985, assessment of our understanding of the processes controlling the present distribution and change, Report No. 16, 3 volumes, WMO, Geneva.

WMO (1990a) Report of the International Ozone Trends Panel: 1988, Report No. 18, 2 volumes, WMO, Geneva.

WMO (1990b) Scientific Assessment of Stratospheric ozone: 1989, Report No. 20, in 2 volumes, WMO, Geneva.

WMO (1992) Scientific Assessment of Ozone Depletion: 1991, Report No. 25, WMO, Geneva (in press)

CHAPTER 3

Climate change

The issue

Temperature varies greatly from place to place and from time to time, yet when surface temperatures are averaged globally they differ from year to year by only fractions of a degree (Figure 1). The near-constancy of the Earth's temperature suggests the presence of homeostatic controls; and an analogy is often drawn between the Earth and the human body, which has a steady temperature of 37°C. But the scientific community, and indeed many political leaders, have become increasingly concerned since the Stockholm Conference (1972) that the Earth's temperature is likely to rise to 'fever' levels within 50 to 100 years unless remedial actions are taken quickly. The rising concern about these issues led directly to a series of national and international conferences and reports. World Climate Conferences were held in 1979 (organized by WMO and ICSU) and 1990 (organized by WMO, ICSU and UNEP in co-operation with UNESCO and FAO) (WMO, 1979, 1991). The Inter-governmental Panel on Climate Change (IPCC), with three Working Groups dealing respectively with likely climate change and its causes, environmental impacts, and possible responses, was established in 1989 by UNEP and WMO, and reported to the Second World Climate Conference (IPCC 1990, a, b, c, d). Box 1 provides a chronology of 'milestones' in the climate warming debate.

There are four reasons for the current widespread concern and the consequent elevation of the climate change and global warming issue to near the top of most scientific and political agendas:

- Atmospheric concentrations of greenhouse gases have increased significantly in the last three decades. As has been known since the last century, this could lead to global warming (see below).

- Using computer models that simulate current climate, it is calculated that for an equivalent doubling of greenhouse gas concentrations, the Earth's climate should warm significantly.

- Analogues obtained from other planetary atmospheres, particularly that of Venus, support the results of the simulations for Earth.

- There has been a global warming trend over the last 15 years. Six of the seven warmest years in this century occurred in 1981, 1983, 1987, 1988, 1989 and 1990.

None of these arguments, taken individually, is convincing evidence of climate warming: taken together, however, they are a *yellow alert*. Of course, changes in other factors might counteract at least some of the global warming. Desertification and deforestation, for example, increase the reflection of solar radiation back into space, as do sulphate aerosols from volcanic eruptions and fossil-fuel combustion. At the time of the Stockholm Conference, the relative importance of these various factors was uncertain, and some scientists believed

Figure 1: *Hemispheric and global surface temperature anomalies (in degrees C) for 1856-1990, with respect to 1951-1980. The smoothed curves show running averages over ten-year periods.*

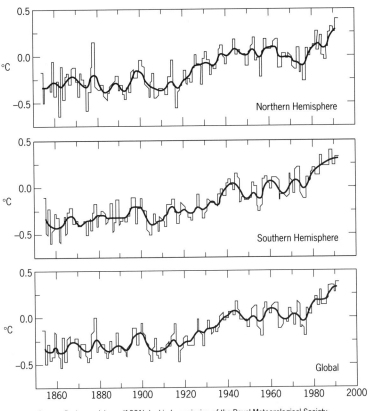

Source: Parker and Jones (1991), by kind permission of the Royal Meteorological Society.

that net cooling was likely. However, by the late 1970s most climatologists thought that tropospheric aerosols would not have a major impact on the Earth's temperature, and that in some situations, for example over snow-covered surfaces, aerosols might even result in warming, because part of the incoming solar radiation would be absorbed in the atmosphere, thus decreasing the amount lost by reflection from the surface (see, for example, Munn and Machta, 1979). At the Villach meeting of 1985 there was a general consensus that the greenhouse warming threat was indeed real, although even today a few scientists challenge this view. In this connection, as has been pointed out by Wigley (1991), even if tropospheric aerosols were to reduce climate warming significantly, this would be largely a Northern Hemisphere effect because most fossil-fuel emissions take place in that hemisphere. This would increase interhemispheric differences in the atmospheric general circulation, which could be just as disruptive to the climate system.

BOX 1

Milestones in the climate warming debate

1970 SMIC (Study of Man's Impact on Climate), Wijk, Sweden.

1974 International Study Conference on the Physical Basis of Climate and Climate Modelling, Stockholm, Sweden.

1976 Launching of GARP (Global Atmospheric Research Programme) by WMO and ICSU.

1979 (First) World Climate Conference, Geneva.

1980 Launching of World Climate Programme by WMO, UNEP and ICSU.

1985 International Conference on the Assessment of the Role of Carbon Dioxide and of the Other Greenhouse Gases, Villach, Austria.

1988 Conference on the Changing Atmosphere, Toronto.

1989 Ministerial Conference on Atmospheric Pollution and Climate Change, Nordweik, Netherlands.

1989- IPCC (Intergovernmental Panel on Climate Change).

1990 The Nairobi Declaration on Climatic Change.

1990 The Cairo Compact.

1990 Second World Climate Conference, Geneva.

1990- Intergovernmental negotiations on a climate convention.

Climate is already a limiting factor for development in many parts of the world, and in some areas water consumption (for agricultural, industrial and domestic purposes) is already beyond the capacity for natural replenishment. For example, the Aral Sea has dropped 14 metres in the last 30 years (Precoda, 1991), and many sub-Saharan cities, California and other areas are experiencing increasingly severe shortages of drinking water. If climate change leads to further desiccation of those parts of the world, the consequences would be serious.

The greenhouse gases

The principal greenhouse gases are water vapour, ozone, carbon dioxide, methane, the chloroflurocarbons (CFCs) and nitrous oxide. The natural greenhouse gases have exerted a powerful influence on the Earth's climate since

Figure 2: *A simplified diagram illustrating the greenhouse effect.*

Sun

Some solar radiation is reflected by the Earth and the atmosphere.

Atmosphere

Solar radiation passes through the clear atmosphere.

Some of the infra-red radiation is absorbed and re-emitted by the greenhouse gases. The effect of this is to warm the surface and the lower atmosphere.

Most radiation is absorbed by the Earth's surface and warms it.

Earth

Infra-red radiation is emitted from the Earth's surface.

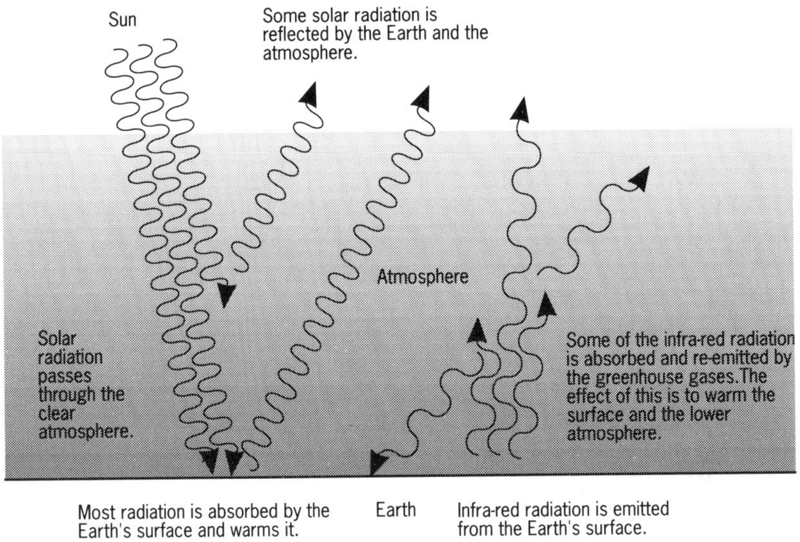

Source: IPCC (1990a).

the formation of the planetary atmosphere. In pre-industrial times, this natural effect maintained a global mean surface temperature 33 degrees C warmer than would be the case without it (Figure 2).

Because of the short turn-round time of water in the atmosphere, water vapour concentrations are in global equilibrium for current climatic conditions. Of course, global warming would cause atmospheric water vapour concentrations to rise, thus increasing the greenhouse effect (Cess, 1991). However, both the amounts and the distribution of clouds would then change, thus modifying the global radiation budget in as yet unknown ways. Lack of precise knowledge of such atmospheric readjustments is a main source of uncertainty in the predicted response of climate to increased concentrations of greenhouse gases (IPCC, 1990a).

The other greenhouse gases are relatively long-lived, and their upward trends are a matter of record (Box 2). Recent trends in carbon dioxide and methane concentrations are illustrated in Figures 3 and 4 respectively, while the main sources of these gases are shown in Box 3. Figure 5 demonstrates the steady rise in carbon dioxide emissions. On a country-to-country basis, rather large differences exist in both total and per capita emissions (Box 4).

Recent findings (IPCC, 1992) indicate that the increase in atmospheric concentration of methane has slowed down. There are also indications that emissions of methane from rice paddies are less than previously estimated.

BOX 2

Trends in key greenhouse gases affected by human activities

	Carbon dioxide	Methane	CFC-11	CFC-12	Nitrous oxide
	Atmospheric concentration				
	ppmv	ppmv	pptv	pptv	ppbv
Pre-industrial (1750–1800)	280	0.8	0	0	288
Present day (1990)	353	1.72	280	484	310
Current rate of change per year	1.8 (0.5%)	0.015 (0.9%)	9.5 (4%)	17.0 (4%)	0.8 (0.25%)
Atmospheric lifetime (years)	(50–200)*	10	65	130	150

ppmv = parts per million by volume.
ppbv = parts per billion (thousand million) by volume.
pptv = parts per trillion (million million) by volume.
* The way in which CO_2 is absorbed by the oceans and biosphere is not simple, and a single value cannot be given.

Source: IPCC (1990a).

Figure 3: *Trends in carbon dioxide concentrations at Mauna Loa, Hawaii. (See TRENDS 90 for information on sampling techniques and principal investigators.)*

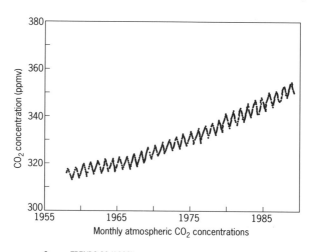

Monthly atmospheric CO_2 concentrations

Source: TRENDS 90 (1990).

Figure 4: *Trends in methane concentrations (global averages from seven representative baseline stations). (See TRENDS 90 for information on sampling techniques and principal investigators.)*

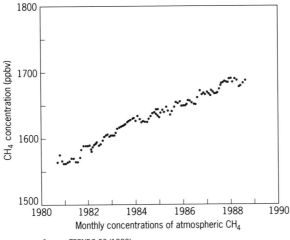

Monthly concentrations of atmospheric CH_4

Source: TRENDS 90 (1990).

Figure 5: *Trends in global carbon dioxide emissions from fossil fuel combustion, cement manufacturing and gas flaring (upper diagram) and trends in global per capita carbon dioxide emission (lower diagram). (See Marland et al. (1989) and TRENDS 90 for further details.)*

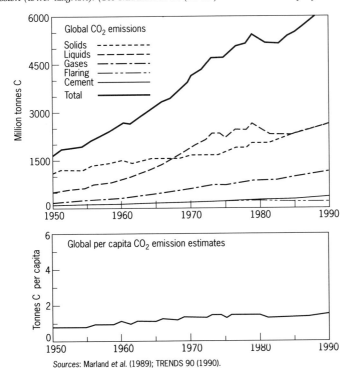

Sources: Marland *et al.* (1989); TRENDS 90 (1990).

BOX 3

The main sources of greenhouse gases

CO_2 — Fossil fuel consumption, deforestation.

CFCs — Aerosol propellants, solvents, refrigerants, foam-blowing agents, etc. (see Chapter 2).

Nitrous oxide — (1) Microbial processes in soil and water.
(2) Biomass burning and fossil fuel combustion.

Methane: — Anaerobic bacteria from natural wetlands, rice paddies, rumen of cattle and other mammals, and guts of termites and other wood-consuming insects.

Source: UNEP Environmental Data Report 1991/1992

BOX 4

1988 carbon emissions for selected countries

Country	Amount (million tonnes)	Per capita amount (tonnes)
Brazil	55	0.4
Canada	119	4.6
China	610	0.6
France	87	1.6
India	164	0.2
Japan	270	2.2
Mexico	84	1.0
Poland	125	3.3
UK	152	2.7
USA	1310	5.3
USSR	1086	3.8

Source: Marland et al. (1989) and Trends 90 (1990)

Estimating global warming

Because the greenhouse gases differ in their effectiveness in absorbing the Earth's radiation, and in their atmospheric lifetimes, they differ in their *global warming potentials* (GWP). GWP is defined as the time-integrated warming effect due to the release of one kilogram of a given greenhouse gas into today's atmosphere, relative to that of carbon dioxide. The IPCC Reports (1990a, b) provide detailed comparisons which show that over the long term, carbon dioxide, methane and the CFCs are the major concerns amongst the gases originating from human activities. However, the IPCC Scientific Assessment Working Group concluded recently (IPCC, 1992) that 'although the GWP remains a useful concept, there is increasing uncertainty in its calculation, especially regarding the indirect effect. Whilst this is likely to be significant for some gases, previous estimates of indirect GWPs are likely to be in substantial error'. While the elimination of CFCs under the Vienna Convention on the Protection of the Ozone Layer and the Montreal Protocol (Chapter 2) will contribute directly to mitigating climate change, the resulting estimated reduction in the amount of global warming may not be as great as was believed several years ago. The latest findings of the IPCC Scientific Assessment Working Group (IPCC, 1992) and Ozone Assessment Panels (see Chapter 2) indicate that the depletion of ozone in the lower stratosphere in middle and high latitudes is believed to have approximately offset the radiative forcing contribution of CFCs over the last decade or so. Furthermore, the cooling effect of sulphur emissions, recognized in the IPCC 1990 report, may have offset a significant part of the greenhouse warming in the Northern Hemisphere during the past several decades (IPCC, 1992).

It is to be noted that because of the large heat capacity of water, the oceans currently absorb a substantial proportion of the extra energy input due to the already existing increase in concentrations of greenhouse gases. As a consequence, the current 'realized' rise in surface temperature is only a fraction of the 'warming commitment' which will eventually occur. The present commitment implies a rise of about 1°C.

In order to evaluate the effectiveness of various strategies for 'managing' climate warming, it is necessary first to have quantitative estimates of the source and sink strengths of the various gases involved. Much effort has been spent over the last 20 years on this task, but there are still gaps, particularly in the case of methane whose concentrations are increasing more rapidly than models predict. (It should be mentioned that if climate warming were sufficient to cause the Arctic permafrost to melt, additional large quantities of methane would be released, thus accentuating the greenhouse effect.)

Figure 6: *Global annual net fluxes (Gt/yr) of carbon from and to the atmosphere. The net increase in the atmospheric reservoir is 3 Gt/yr. (1 gigaton (Gt) is 10^9 metric tons.)*

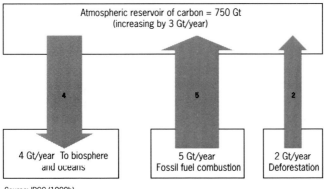

Source: IPCC (1990b).

Figure 6 provides a schematic representation of the global atmospheric carbon budget. There is currently a large measure of uncertainty in the relative amounts of carbon taken up by the land biosphere and the oceans (see for example, Tans *et al.*, 1990 and Harvey, 1991), but there is no doubt that the strength of the two source terms on the right-hand side of the figure have increased in recent decades because of increased fossil fuel emissions and deforestation (see Chapters 7 and 13).

Model simulations of climate change

Simulations obtained from climate models of the Earth-atmosphere system indicate that for a doubling of carbon dioxide from its pre-industrial level, or of any combination of greenhouse gases with an equivalent warming potential, the atmosphere would warm on average by between 1.5 and 4.5 degrees C. Furthermore, the warming would be greatest in the polar regions and least in the tropics. One example of a climate simulation is given in Figure 7 (Hansen *et al.*, 1984; IPCC, 1990b).

Figure 7: *Month-of-the-year/latitude diagram of the zonally averaged increase in surface temperature due to a doubling of carbon dioxide in a climate model. Warming greater than 4 degrees C is stippled.*

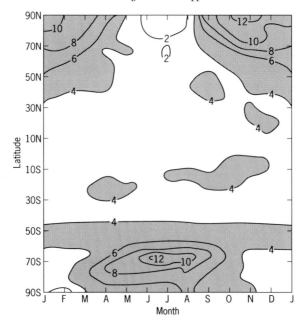

Sources: Hansen *et al.* (1984), by kind permission of the author and the American Geophysical Union.

70

The resulting decrease in the equator-to-pole temperature gradient would slow the atmospheric heat engine (i.e. the general circulation) which transfers heat from tropical to polar regions. At the same time the continents would warm more than the oceans, so that the summer monsoons would intensify as well as change their seasonal characteristics. Thus the world's future weather patterns could be quite different from those of recent years. Unfortunately, the simulations give very uncertain estimates of even globally averaged precipitation, wind and other weather elements, although it can generally be argued that warmer temperatures are likely to cause an increase in evaporation as well as rainfall. Also, the *timing* of climate change is very uncertain, even for given greenhouse gas scenarios. The Earth system contains lags (oceans and glaciers) and many feedback processes, making the rate of climate change difficult to estimate. Intermediate 'transient' climates cannot yet be predicted with any confidence, especially on the regional scale. For example, Watts and Morantine (1991) have recently suggested that the atmospheric cooling between 1940 and 1970 could be explained by changes in upwelling patterns in the world ocean circulation during that period, and may in fact have been a first response to greenhouse warming. Understanding of some climate feedbacks and their incorporation into the models has improved, particularly as regards the role of upper tropospheric water vapour. Coupled ocean-atmosphere models are capable of reproducing many features of atmospheric variability on intra-decadal time scales (IPCC, 1992). Finally, volcanic eruptions such as the recent one at Mount Pinatubo in the Philippines might spread aerosols through the stratosphere, partially masking a long-term warming trend for a few years.

Climate impacts

The assessment of impacts

Individuals and governments worry more about the climate changes to be expected in their own regions than they do about global averages. Also, they are less concerned about future temperatures, which can probably be estimated with least uncertainty of all the weather elements, than they are with other meteorological phenomena such as the frequencies of droughts and tropical storms. Unfortunately, global climate models are still too coarse to provide this kind of local information, particularly in areas where there are topographic features (for example, major lake systems and mountain ranges) not represented in the models. At the present time, much attention is being given to this problem using:

• small-scale models that cover only one grid square of a global model,

- historical analogues (the examination of climate and climate impacts in seasons or years that were abnormally warm, either regionally or globally), and

- palaeo-analogues (the examination of biospheric regimes during warm geological periods).

However, these approaches currently provide a very uncertain basis for assessing future climate conditions, and it is even difficult to quantify the degree of uncertainty. An alternative approach therefore is to begin with some assumed future climate in a region, and to examine what it would mean for the biosphere and society. This was the approach taken by the IPCC Working Group 2.

Potential impacts

Some of the potential impacts of climate change that are causing concern are listed in Box 5 (IPCC, 1990c; WMO 1991). In this regard, the idea of tallying

---

BOX 5

A sampling of some of the potential climate change impacts of current concern.

- Impacts on especially vulnerable populations (sea level rise, desertification, thawing of the permafrost) resulting in millions of environmental refugees.

- Impacts on human health (increasing heat stress and frequencies of air pollution episodes; shifts in ranges of vector-borne diseases).

- Impacts on forestry (shifts in ranges of species; increased frequencies of forest fires; economic disruptions).

- Impacts on agriculture (shifts in ranges of crops, agricultural pests and livestock diseases; shifts in demands for irrigation).

- Impacts on hydro-electric generation through declining water levels in reservoirs and other inland water systems in some regions, and possible increases elsewhere.

- Impacts on structures, roads and runways built on permafrost in the Arctic and sub-Arctic.

Sources: IPCC (1990c); WMO (1991).

---

up the 'winners' and 'losers' from climate change has to be rejected. As was pointed out at the Second World Climate Conference, and by Parry *et al.* (1988) in the agricultural context, the biosphere and society are adapted to current climatic conditions to such an extent that climate change would bring few immediate benefits - even to regions with, for example, improved rainfall or longer frost-free periods (the soil might be unsuitable or the land might already be committed to other uses). The other argument sometimes heard (e.g. NAS, 1991) is that because agriculture accounts for only a small part of the economy of an industrialized country such as the United States, a decline in crop yields would have little impact. However, Daly (1991) emphasizes that the demand

Figure 8: *Upper diagram: sea level rise predicted to result from business-as-usual greenhouse gas emissions, showing the best estimate and range. Lower diagram: model estimates of sea level rise from 1990 to 2100 due to four emission scenarios. (A: Business-as-usual. B: The energy supply mix shifts towards lower carbon fuels and large efficiency increases are achieved; CO controls are stringent, deforestation is reversed and the Montreal CFC protocol is fully implemented. C: A shift towards renewables and nuclear energy occurs in the second half of the twenty-first century and agricultural emissions are limited. D: The shift in C takes place in the first half of the twenty-first century).*

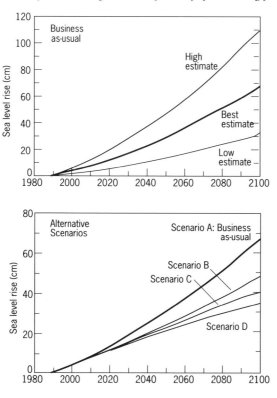

Source: IPCC (1990a).

for food is inelastic, and that a permanent decline in food supplies would be very serious, both nationally and globally. Unfortunately, developing countries would be least able to cope.

Of the various areas of concern, there is no doubt that some of the most serious impacts would occur if climate warming were sufficient to cause a significant rise in sea level. The extent of the rise can be estimated from models, given the emissions of greenhouse gases to be expected over the next century. The IPCC estimates of sea level rise are shown in Figure 8 for four different energy scenarios (IPCC, 1990a) but it should be emphasized that local rises could be considerably modified by geomorphology. The sea level rises shown in the lower graph of Figure 8 could lead to serious flooding in low-lying areas, particularly islands (e.g. the Maldives), coastal floodplains (e.g. Bangladesh and Guyana) and coastal cities, that are subsiding, (e.g. Bangkok). (See Chapter 5 for more details.) In inland waters such as the North American Great Lakes,

---

BOX 6

Estimates for changes by 2030
(IPCC Business-as-Usual scenario; changes from pre-industrial)

The numbers given below are based on high-resolution models, scaled to be consistent with our best estimate of global mean warming of 1.8°C by 2030. For values consistent with other estimates of global temperature rise, the numbers below should be reduced by 30% for the low estimate or increased by 50% for the high estimate. Precipitation estimates are also scaled in a similar way.

Confidence in these regional estimates is low.

Central North America (35° - 50°N 85° - 105°W)
The warming varies from 2 to 4°C in winter and 2 to 3°C in summer. Precipitation increases range from 0 to 15% in winter whereas there are decreases of 5 to 10% in summer. Soil moisture decreases in summer by 15 to 20%.

Southern Asia (5° - 30°N 70° - 105°E)
The warming varies from 1 to 2°C throughout the year. Precipitation changes little in winter and generally increases throughout the region by 5 to 15% in summer. Summer soil moisture increases by 5 to 10%.

Sahel (10° - 20°N 20°W - 40°E)
The warming ranges from 1 to 3°C. Area mean precipitation increases and area mean soil moisture decreases marginally in summer. However, throughout the region, there are areas of both increase and decrease in both parameters throughout the region.

on the other hand, the concern is about a possible *fall* in lake levels, of up to 1 metre due to increased summer evaporation throughout the watershed (Smith, 1991).

The IPCC estimates of regional climate change by the year 2030 for the business-as-usual scenario (i.e. the one in which energy policies are not modified) are given in Box 6 (IPCC, 1990a). As indicated, the levels of confidence in these estimates are low. Another consideration of great importance is that in addition to climate many other physical and socio-economic conditions are also expected to change significantly in the next 40 years. Climate change impacts have generally been studied one by one, assuming that all other factors remain constant. Within the last decade, however, there has been a growing recognition of the linkages between impacts and of the need to undertake integrated regional assessments, involving both the physical and the socio-economic sciences. In view of the great uncertainty involved, new

Southern Europe (35° - 50°N 10°W - 45°E)
The warming is about 2°C in winter and varies from 2 to 3°C in summer. There is some indication of increased precipitation in winter, but summer precipitation decreases by 5 to 15%, and summer soil moisture by 15 to 25%.

Australia (12° - 45°S 110° - 155°E)
The warming ranges from 1 to 2°C in summer and is about 2°C in winter. Summer precipitation increases by around 10%, but the models do not produce consistent estimates of the changes in soil moisture. The area averages hide large variations at the sub-continental level.

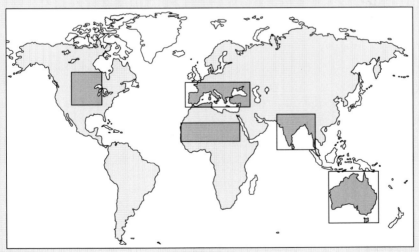

Source: IPCC (1990a).

methods of policy analysis are being developed and tested; (see Chapter 20 and Stigliani *et al.* (1989)). Steps are now being taken to construct scenarios, particularly within IPCC, for the development of more comprehensive analyses of the dependence of future greenhouse gas emissions on socio-economic assumptions and projections (IPCC, 1992).

Responses to projections of climate change

Climate change is a risk management issue, but unlike the situation with floods and earthquakes, for which records date back hundreds of years, there is no historical experience to guide the assessor. Nevertheless, many political leaders now accept the need for strong national and international responses to meet the threat: greenhouse gases and their consequences have become a major public policy issue - particularly in developed countries, island states and low-lying regions of the developing world. In the not too dissimilar case of CFCs, the detection of an ozone hole in the Antarctic stratosphere provided direct evidence of an impact, and led to rapid public policy responses. In the case of climate change, on the other hand, where global warming cannot be confirmed with certainty, most political leaders have been persuaded to act as a result of technical arguments put forward by their senior scientific advisors, largely on the basis of computer simulations. As mentioned earlier, there has been a sustained large-scale scientific effort aimed at improved monitoring of climate change, better understanding of the complex processes involved, and the development of appropriate strategies for coping with possible changes.

Principal actions to be taken

As a result of various assessments of the likely impacts of climate changes, especially those of IPCC Working Group 3 on Response Strategies (IPCC, 1990d), it is clear that the principal actions that should be taken to address the problem are:

- reduction of emissions of greenhouse gases;

- analyses of the vulnerability to climate change of current and proposed patterns of land use and development;

- assessment of the capacity of both the biosphere and society to adapt to climate change;

- research and engineering studies to reduce uncertainties and to design more efficient energy systems;

- introduction of improved early warning monitoring systems.

These actions need to be taken at both global and national levels. With respect to emission reductions, a wide range of industrial and other activities contribute to greenhouse warming, complicating the control problem. However, many nations have accepted that at least a start should be made, and a policy of 'no regrets' is being promoted, i.e. a policy under which the reduction of greenhouse gas emissions is also justified on other grounds (Box 7). Grubb *et al.* (1991) further recommend that a comprehensive approach be adopted in which, in principle at least, all the major greenhouse gases, and their risks and sources would be considered in the development of control strategies.

BOX 7

Examples of 'no-regrets' climate-warming policies

Policy	Effect on greenhouse gases	Other beneficial effects
Tree planting	Increased biosphere sink strength	Improved microclimate Reduced soil erosion Reduced seasonal peak river flows
Energy conservation	Reduced CO_2 emissions	Conservation of non-renewable resources for current and future generations
Energy efficiency	Reduced CO_2 emissions	Conservation of non-renewable resources for current and future generations
CFC emission control	Reduced CFC emissions	Reduced stratospheric ozone-layer depletion Reduced surface UV-B and skin cancer and blindness

---

Remaining questions to be addressed

There remain, however, the questions of:

- What will be the costs (of taking action, of delaying action, and of not taking action at all)?

- How should these costs be shared (within countries and amongst countries)?

- What special arrangements should be made in the case of developing countries who need to develop, and consequently to increase their greenhouse gas emissions?

The Second World Climate Conference emphasized that despite the remaining scientific uncertainties, 'nations should now take steps towards reducing sources and increasing sinks of greenhouse gases through national and regional actions' (WMO, 1991). The Conference emphasized that 'The long-term goal should be to halt the build-up of greenhouse gases at a level that minimizes risks to society and natural ecosystems', and that 'technically feasible and cost-effective opportunities exist to reduce carbon dioxide emissions in all countries', these being the major culprits in global warming. Actions needed include steps to improve the efficiency of energy use, and measures to develop renewable sources of energy (Chapter 13). Since the Conference, a number of countries and the European Community as a whole have announced actions aimed at stabilizing their emissions of carbon dioxide, generally at 1990 levels, by or close to the year 2000 (See Chapter 13).

The new findings concerning the cooling effects of CFCs and sulphur emissions are examples of the complexities that have to be dealt with in understanding the environment, and the contradictions and trade-offs that have to be considered in formulating appropriate strategies. The latest thinking is that the major culprit in global warming is CO_2, and that urgent action is essential in this area. At its 44th Session, the United Nations General Assembly agreed on the need to prepare, as a matter of urgency, a framework Convention on Climate Change. This recommendation was supported by the Ministerial Declaration of the Second World Climate Conference. Negotiating sessions of the Intergovernmental Negotiating Committee established for the purpose were held in 1991, and will continue in 1992, with the target of having a framework Convention ready for signature in June 1992 at the UN Conference on Environment and Development in Rio de Janeiro, Brazil (Box 8).

Concluding remarks

The climate change issue raises important questions of international equity. At present, the major proportion of greenhouse gases comes from the industrialized countries which contain only about 25 per cent of the world population. Agarwal and Narain (1991) have suggested that each person on Earth is entitled to an equal share of the capacity of the world's oceans and biosphere to absorb carbon dioxide. Others, including spokesmen from the developing world, insist that the developed countries must reduce their emissions so that more of the planet's capacity for assimilation of greenhouse gases will become available for the developing and newly industrializing countries - and that the latter must be assisted through the transfer of adequate new and additional

---

BOX 8

An International Convention on climate change

Under the auspices of the United Nations, an Intergovernmental Negotiating Committee was established to negotiate a global convention on climate change.

The objective of the convention is to control and counter the adverse consequences of climate change.

The key elements in the negotiations are the commitments of the parties to:

- implement suitable measures to stabilize the emission of greenhouse gases; starting with the industrialized countries;
- adopt policies and measures for reducing greenhouse gas emissions;
- update and report on national inventories of all sources and sinks of all greenhouse gases not controlled by the Montreal Protocol;
- protect existing and develop new sinks and reservoirs for greenhouse gases;
- co-operate in the development and dissemination of scientific, technical, socio-economic and legal knowledge relevant to climate change and potential responses thereto;
- co-operate in the transfer to developing countries of new technologies and techniques that consume less energy or produce less greenhouse gases;
- establish a fund for financing measures to counter the adverse consequences of climate change;
- establish a separate International Insurance Pool to provide financial insurance against the consequences of sea level rise.

The convention will also cover the institutional arrangements and procedures for implementation, review and further development of the convention, as well as managing the fund and insurance pool.

---

financial resources, through various forms of technical assistance, and through transferring to them environmentally-sound technologies in the broad sense of the term in order to pursue sustainable development pathways. This will be a major challenge in the coming decades.

To support actions and policy analyses, a strong programme of studies, applications, research and monitoring is required. High priority items in addition to the above should include:

- increased understanding of the global cycles of the greenhouse gases;
- improved capability to produce climate change scenarios at the regional scale;

- strengthening of the Global Climate Observing System and the Global Oceanic Observing System;
- analysis of the underlying ethical questions of inter-regional and inter-generational equities;
- establishing the real costs of action and inaction in the area of climate change;
- establishing and/or application of economic incentives and disincentives for achieving stabilization and reduction of CO_2 and other greenhouse gases not controlled by the Montreal Protocol;
- widespread and firm application of the 'Polluter Pays Principle' and the 'Precautionary Principle'.

References

Agarwal, A. and Narain, S. (1991) *Global warming in an unequal world*, Centre for Science and Env., New Delhi, India.

Cess, R.D. (1991) Positive about water feedback, *Nature*, **349**, 462-3.

Daly, H.E. (1991) Ecological economics, *Science*, **358**.

Grubb, M.J., Victor, D.G. and Hope, C.W. (1991) Pragmatics in the Greenhouse, *Nature*, **354**, 348-50.

Hansen, J., Lacis, A., Rind, D., Russell, L., Stone, P., Fung, I., Ruedy, R. and Karl, T. (1984) Climate Sensitivity of feedback mechanisms. In *Climate Processes and Climate Sensitivity* J. Hansen and T. Takahashi (eds.) *Geophys. Monog.* **29**, Amer. Geophysical Union, 130-63.

Harvey, L.D.D. (1991) A commentary on tropical deforestation and atmospheric carbon dioxide, *Climatic Change*, **19**, 119-21.

IPCC (1990a) Scientific Assessment of Climate Change, The Policymakers' Summary of the Report of Working Group 1, Intergovernmental Panel on Climate Change, WMO/UNEP, Geneva, 26 pp.

IPCC (1990b) *Climate Change: the IPCC Scientific Assessment*, Cambridge University Press.

IPCC (1990c) Potential Impacts of Climate Change, Report of Working Group 2, Inter-governmental Panel on Climate Change, WMO/UNEP, Geneva.

IPCC (1990d) Formulation of Response Strategies, Report of Working Group 3, Inter-governmental Panel on Climate Change, WMO/UNEP, Geneva.

IPCC (1992) Working Group I, 1992 IPCC Supplement, WMO/UNEP

Marland, G., Boden, T.A., Griffin, R.C., Huang, S.F., Kanciruk, P. and Nelson, T.R. (1989) Estimates of carbon dioxide emissions from fossil fuel burning and cement manufacturing, based on the United Nations energy statistics and the U.S. Bureau of Mines cement manufacturing data, ORNL/CDIAC-25, NDP-030, Oak Ridge Nat. Lab., Oak Ridge, TE, USA.

Munn, R.E. and Machta, L. (1979) Human activities that affect climate, *Proc. World Climate Conference*, WMO, Geneva, Switzerland.

NAS (1991) Policy Implications of Greenhouse Warming, Adaptation Panel, National Academy of Sciences and Engineering, Washington, D.C.

Ozone Assessment Panels, 1991: UNEP

Parker, D.E. and Jones, P.D. (1991) Global warmth in 1990, *Weather*, **46**, 302-11.

Parry, M.L., Carter, T.R. and Konijn, N.T. (eds.) (1988) The *Impact of Climatic Variations on Agriculture*, 2 volumes, Kluwer Academic Pub., Dordrecht, The Netherlands.

Precoda, N. (1991) Requiem for the Aral Sea, *Ambio*, **20**, 109-14.

Smith J.B. (1991) The potential impacts of climate change on the Great Lakes, *Bull. Amer. Meteorolog. Soc.*, **72**, 21-28.

Stigliani, W.M., Brouwer, F.M., Munn, R.E., Shaw, R.W. and Antonofsky, M. (1989) Future environments for Europe: some implications of alternative development paths, *Science of the Total Env.*, **80**, 1-102.

Tans, P.P., Fung, I.Y. and Takahashi, T. (1990) Observational constraints on the global atmospheric CO_2 budget, *Science*, 1431-38.

TRENDS 90 (1990) Carbon Dioxide Information Analysis Center, Oak Ridge National Laboratory, Oak Ridge, TE, USA, 257 pp.

Watts, R.G. and Morantine, M.C.(1991) Is the greenhouse gas climate signal hiding in the deep ocean? *Climatic Change*, **18**, iii-vi.

Wigley, T.M.L. (1991) Could reducing fossil-fuel emissions cause global warming? *Nature*, **349**, 503-6.

WMO (1979) *Proc. of the World Climate Conference*, WMO, Geneva.

WMO (1991) *Proc. Second World Climate Conference*, WMO, Geneva.

CHAPTER 4

Availability of fresh water

The issue

Fresh water is essential for human survival and for the maintenance of ecosystems on land. It is a key factor in development, particularly in arid and semi-arid countries where water security, like food security, is an important issue. However, concern is also growing in more humid countries where competition between the different sectoral users of water is intense. In developed countries, pollution from industry, urban waste water and agriculture reduces the fitness of rivers as sources of supply, and damages their fisheries and ecosystems. Finally, there is the hazard of too much water at certain times in many regions of the world. Floods are one of the worst killers and they cause enormous damage to property and to agricultural land (Chapter 9).

Clean water is required for drinking and for personal hygiene; irrigation is needed for agricultural production; rivers are required for hydropower generation and transportation of goods and people through inland waterways; and freshwater areas are important for recreation and as habitats (Chapters 7 and 8). Rising water consumption and rapidly growing population is causing increasing pressure on water resources in some regions, and this situation will worsen over the next 50 years unless integrated long-term water management is practised and increasing per capita water consumption is curbed. These points were made clearly in the Mar del Plata Action Plan, which stemmed from the United Nations Water Conference in 1977.

Trends in water quantity and use since 1970

Declining water availability and increasing water use

Water appears on Earth in liquid, solid and gaseous phases which are linked together in a closed cycle, with average annual precipitation equal to annual evaporation from oceans and land.

Only 2.59 per cent of the total volume of the world's water is fresh. Of this, more than 99 per cent is in the form of ice or snow in the polar regions, or as underground water. Almost half the remaining fresh water on Earth is locked in biota, soil moisture and atmospheric water vapour. Open freshwater bodies - rivers and lakes - which are the main source of water for mankind, contain the rest (WRI, 1988) with an estimated volume of 93,000km^3 (Shiklomanov, 1990).

The availability of fresh water is very unevenly distributed world-wide. Relative water availability ranges from an extremely low level of 1,000m^3/year per capita to a very high level of over 50,000 (Shiklomanov,1990).

Against this, total water use increased nearly fourfold during the last 50 years (from 1,060km³/year to 4,130km³/year), as shown in Figure 1. A good deal of this water is irretrievably lost (Table 1). Water use is still growing in the developing world, but less so in industrialized countries (Figure 2). However, in developing countries, as population increased, lifestyles changed and development projects were implemented, the per capita availability of fresh water declined (Figure 3). Thus, as new sources of water have become more scarce and more expensive to develop, competition amongst the various users has increased and will continue to do so.

There are, of course, considerable variations in water use patterns in different countries, depending upon their levels of development, relevant physical factors and other considerations. In Europe, use in industry (including energy production) is dominant, and only 37 per cent of withdrawals are for human consumption and use in agriculture. In Africa and Asia, water withdrawals for human consumption and agriculture in the 1980s represented more than 90 per cent of the total. Projections show that by the turn of the century this share will drop to around 83 per cent, although the volume will increase by a quarter as withdrawals for industrial uses will almost treble in volume during the same period (WRI, 1990).

As shown in Figure 1 and Table 1, about 70 per cent of total water use, and 90 per cent of irretrievable water losses at present are due to agriculture. World-wide, some 2,000-2,500km³ of water were used for irrigation in 1990 and the area of irrigated land has increased about 36 per cent in the last two decades. By 1990, it was clear that this rate of expansion could not be sustained. The

Figure 1*: Global water use, 1900–2000.*

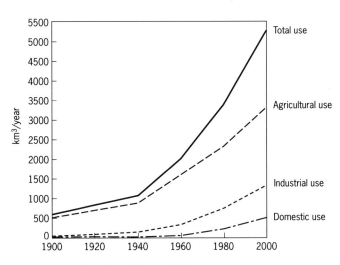

Source: Adapted from Shiklomanov (1988).

Table 1: *Irretrievable water losses (km³/year).*

Users	1900	1940	1950	1960	1970	1980	1990	2000
Agriculture	409	679	859	1180	1400	1730	2050	2500
Industry	3.5	9.7	14.5	24.9	38.0	61.9	88.5	117
Municipal supply	4.0	9.0	14	20.3	29.2	41.1	52.4	64.5
Reservoirs	0.3	3.7	6.5	23.0	66.0	120	170	220
Total	417	701	894	1250	1540	1950	2360	2900

Source: Shiklomanov (1990).

Figure 2: *Evolution of water withdrawals through the twentieth century.*

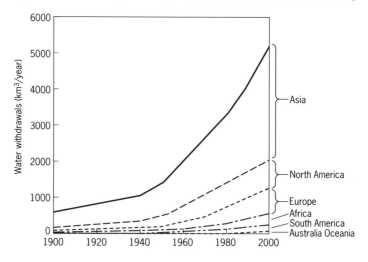

Source: da Cunha (1989).

current rate of expansion is less than one per cent per year compared with a peak rate of 2.3 per cent during 1972-75. The main constraint is the availability of water suitable for irrigation rather than the availability of land (FAO, 1990; Chapter 11).

The severity of the water shortage will be further aggravated by the steady deterioration of water quality. While the total amount of waste water produced yearly world-wide was about 1,870km³ in 1980, it is expected to reach 2,300km³ per annum by the end of the century (Shiklomanov, 1990). World freshwater resources are not enough to dilute polluted water resources.

***Figure 3**: Decreasing trends in annual per capita availability of fresh water in developing countries, 1950–2025.*

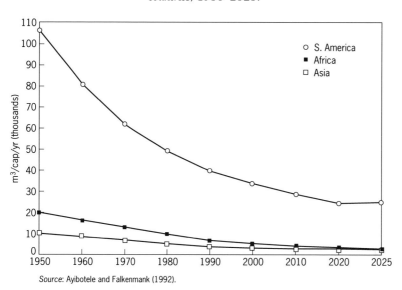

Source: Ayibotele and Falkenmank (1992).

Consequently, water pollution control is a major element in any water conservation strategy aimed at meeting future demand for fresh water.

Fresh water storage

The increasing demand for water has led to a significant increase in the number of dams and reservoirs constructed during the twentieth century in most parts of the world. Table 2 shows the number of large dams (more than 15 metres in height) constructed during the 1950-86 period (ICOLD, 1984, 1989). One country alone - China - accounted for nearly half of these dams. Collectively, the 36,327 dams existing in 1986 stored 5,500km³ of water, of which nearly two-thirds could be used, the rest being dead storage. During 1989, 45 very large dams (more than 150 metres high) were under construction, of which 20 were in Latin America and 15 in Asia (Veltrop, 1991). In 1990, reservoirs provided 3.6 per cent of total water consumption, but were responsible for 6.1 per cent of irretrievable water losses (Shiklomanov, 1990).

Associated with the increase in the number of dams, global hydroelectric power generation rose to 2,000TWh in 1986, which was nearly 20 per cent of the world's total electricity production. In 1987 an additional generating capacity of around 130GW was under construction in some 60 countries, representing an increase of about one-quarter in the hydroelectric generating capacity (UNEP, 1991). Small-scale hydroelectric generation (less than 15MW) amounted to 10GW in 1953, and was estimated to reach 29GW by 1991 (Veltrop, 1991). It is likely that such small-scale electricity generation will receive increasing attention in developing countries in coming decades.

Table 2: Number of large dams constructed or under construction, 1950–86.

Continent	1950	1982	1986	Under construction 31.12.86
Africa	133	665	763	58
Asia	1,562	22,789	22,389	613
Australasia/Oceania	151	448	492	25
Europe	1,323	3,961	4,114	222
North and Central America	2,099	7,303	6,595	39
South America			884	69
Total	5,268	35,166	36,327	1,026
of which in China	8	18,595	18,820	183

Source: ICOLD (1984 and 1989).

Trends in water quality since 1970

Monitoring the state of the world's freshwater systems

The GEMS/WATER Programme (UNEP/WHO/WMO/UNESCO) was the first of its kind to address water quality issues on a global scale. By 1989, the GEMS/WATER global water quality monitoring project comprised 450 monitoring stations for rivers, lakes, reservoirs and groundwater in 59 countries on all continents (UNEP, 1989). These collect data on about 50 physical, chemical and biological parameters of water quality.

An analysis of the implementation of the Mar del Plata Action Plan after one decade (WMO/UNESCO, 1991) indicated the unsatisfactory nature of progress in the establishment of precipitation stations in developing countries. In Africa, the number actually declined significantly (Table 3). The Latin American and Caribbean regions in contrast showed remarkable improvements. Significantly better progress can be noted in the number of discharge measurement stations between 1977 and 1989. During this period, the number of such stations nearly doubled. Even then, however, the situation continued to remain unfavourable in Africa, a continent that undoubtedly requires accelerated water resources development.

Table 3: *Stations for monitoring precipitation, discharge and water quality indicators, 1977 and 1989.*

	Number of stations for measuring					
Regions	**Precipitation**		**Discharge**		**Water quality**	
	1977	**1989**	**1977**	**1989**	**1977**	**1989**
Africa	4,047	3,596	918	1,695	123	361
Europe	49,240	48,507	9,549	23,946	15,509	42,327
Latin America and the Caribbean	12,409	19,531	3,086	5,762	218	1,439
Asia and the Pacific	20,980	20,422	5,923	7,023	3,533	2,889
West Africa	4,018	4,420	1,222	1,383	801	821
Total	90,694	96,296	20,698	39,809	20,184	47,837

Source: WMO, UNESCO (1991).

Freshwater pollution

Data collected under the GEMS/WATER project indicate that about 10 per cent of all rivers monitored in this project may be described as polluted as they have a biological oxygen demand (BOD) of more than 6.5mg/l (UNEP/WHO, 1988). The two most important nutrients, nitrogen and phosphorus, are well above natural levels in the waters measured by the network. The median nitrate level in unpolluted rivers is 100µg/l. The European rivers monitored by GEMS show a median value of 4,500µg/l. In contrast, rivers monitored by GEMS outside Europe show a much lower median value of 250µg/l. The median phosphate level in GEMS/WATER is 2.5 times the average for unpolluted rivers. The high content of nutrients in rivers has led to eutrophication in stretches of many rivers in central Europe (OECD, 1991) and elsewhere, and is a cause of algal blooms in some coastal areas (Chapter 5).

Other pollutants of fresh water which cause concern are potentially toxic substances, especially metals, and deposited acids. During the past 20 years particular efforts have been made to reduce the contamination of rivers draining industrial areas, such as the Rhine. Between 1975 and 1985 there was a steady decline in levels of arsenic, cadmium, chromium, copper, mercury, lead, nickel and zinc in the Rhine system as a whole, and the concentrations of lead , cadmium and mercury were reduced to below 20 per cent of their 1975 levels (Table 4). In some developing countries (e.g. Colombia, Malaysia and Tanzania) the levels of organochlorine pesticides measured in some rivers are higher than those recorded in European rivers.

Acid deposition results from the solution in rain and mist of sulphur and nitrogen oxides formed in fossil fuel combustion (Chapter 1). Because these gases have an average residence time in the atmosphere of several days, and hence can be transported for distances of up to thousands of kilometres, this acid deposition can occur at considerable distances from the sites of emission. This is a phenomenon of industrialized regions and is of especial concern for the major ecological changes it causes in fresh waters.

During the past two decades, freshwater acidification has become an increasingly important concern in Europe and North America (see Chapter 1). Acidity in southern Norwegian rivers increased from pH 5.0-6.5 (average 5.8) in 1940 to 4.5-5.0 in 1976-78, and in lakes from an average of 5.5 to 4.7 between 1940 and 1980. Similar trends were recorded in southern Sweden and parts of Scotland, Wales, Germany and Denmark. The process of acidification was accompanied by changes in water chemistry, with increasing concentrations of aluminium, copper, cadmium, zinc and manganese. Zinc, copper and cadmium concentrations also increased in acidified groundwater in Scandinavia (Sweden, 1982). On the basis of observations in 6,908 Swedish lakes, carried out in 1985, calculations show that some 4,000 lakes no longer contained any fish, and another 17,000 had reduced populations of acid-sensitive species (Statens Naturvardsverk, 1986). (Most of the acidified lakes were small, having an area of less than one square kilometre.) A similar situation exists in the north-eastern United States and southern Canada.

The process of acidification and associated ecological change is now reasonably well understood. The impacts are not simply the consequence of acidification, but also of the associated changes in water chemistry, with

Table 4: *Trends in metal concentrations in the Rhine at Lobith (micrograms per litre).*

	1975	1980	1983	1984	1985
Arsenic	4.5	3.0	3.6	0.3	1.8
Cadmium	2.3	1.6	0.4	0.2	0.1
Chromium	35.0	20.0	11.0	5.0	8.0
Copper	20.0	14.0	10.0	5.0	6.0
Mercury	0.4	0.2	0.1	0.07	0.07
Lead	22.0	15.0	7.0	6.0	4.0
Nickel	10.0	9.0	5.0	5.0	5.0
Zinc	135.0	102.0	57.0	37.0	50.0

Source: UNEP (1990).

increased concentrations of toxic metals. Diatoms, plant and animal plankton, fish and invertebrates such as snails, mussels and stoneflies, are highly sensitive to acidification. Fish species vary in sensitivity, the eggs and fry of salmonids being harmed when pH falls below 4.5-5.0, especially in water with low calcium, sodium and chloride concentrations while the fish themselves tend to disappear from waters acidified to below 4.0. Few if any fish survive waters more acid than pH 3.5 (Howells, 1984).

Between 1970 and 1990 considerable action has been taken to reduce emissions of the pollutants causing acidification, especially in Europe (see Chapters 1 and 13). In the meantime, direct application of lime has been used as a temporary expedient in parts of Scandinavia, pending a longer-term cure through pollution control at source. There is some evidence of recovery in some water bodies (Gunn and Keller, 1990), but the build-up of acidity and associated chemical changes is such that more time, and even tighter controls, are likely to be needed.

Contamination by agricultural chemicals

During the 1980s, water contamination by fertilizers, pesticides and agricultural wastes (especially silage liquors and slurry from intensive livestock units) was an important environmental concern, especially in Europe and North America, because these were major causes of environmental change in rivers and coastal seas and had a serious impact on groundwater quality. Substantial amounts of groundwater are currently used for domestic purposes: 73 per cent in the former Federal Republic of Germany, 70 per cent in the Netherlands and 30 per cent in Great Britain (Biswas, 1990a). In the United States, groundwater is the primary source of water for over 90 per cent of the rural population and 50 per cent of the total population.

The main concern is with nitrates, derived from fertilizers, and pesticides (Figures 4 and 5). Nitrate contamination has been studied more intensively than pesticides because standards have been set for the maximum permissible concentration in drinking water (10mg/l is the WHO recommended limit). The concentration of nitrates in surface and ground waters varies with locality. Current estimates indicate that precipitation adds 8-20kg of nitrates per hectare in the industrialized countries through the solution of nitrogen oxides released in fossil fuel combustion (Chapter 1). Some 15Mt of nitrates are annually disposed of in the USA in wastes, 40 per cent of which are from animals, 20 per cent from crop residues and 20-25 per cent from municipal wastes (Biswas, 1990b). Studies in the north-central US show annual losses of 10-20kg of NO_3N per hectare of agricultural land (Hallberg, 1989). Uncontaminated groundwater generally contains less than 3mg/l of nitrates. If this level is exceeded, human inputs can be suspected, unless there are other possible explanations (Biswas, 1990a).

The pattern of increasing nitrate concentrations in the Petite Traconne Spring, France, from about 1930 is shown in Figure 4. (Roberts and Marsh,

Figure 4: *Nitrate concentrations in the Petite Traconne Spring, France, 1930–75.*

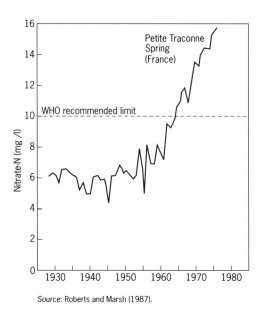

Source: Roberts and Marsh (1987).

Figure 5: *Nitrate concentration trends in four United Kingdom rivers. The three upper rivers are in agricultural regions: the lowest one is not.*

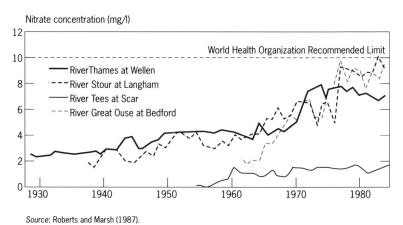

Source: Roberts and Marsh (1987).

1987). Similar trends have been noted during the past two decades in many west European countries (Figure 5), and a few irrigated areas in developing countries. Eutrophication of water courses is enhanced by the run-off of fertilizers from agricultural land.

Existing data are not sufficient to provide a clear picture of pesticide contamination even in North America or Western Europe: information is even more sparse in other parts of the world. Some 39 pesticide compounds have already been detected in groundwater in 34 states in the USA or Canada (Hallberg, 1989). GEMS data suggest that water bodies in Indonesia and Malaysia have extremely high levels of PCBs and some pesticides (Holdgate *et al.* 1982). It is highly likely that contamination of groundwater with pesticides and their metabolites will continue to increase.

Environmental impacts of fresh water uses

Impacts of uses in agriculture

Water losses in irrigation

Irrigation is essential for agricultural productivity in many areas, and significantly improves yields even where there is no overall rainfall deficit. However, poorly managed irrigation leads to salinization and degradation of soils (see Chapter 6). Where groundwater is used as the source of irrigation water, a steady decline in groundwater levels has often followed, especially in the western United States, India, China and the Middle East.

As mentioned earlier, irrigation is responsible for the majority of water losses (Table 1). Consequently, efficiency of irrigation systems has become a major issue since 40-60 per cent of water abstracted from rivers often does not reach the agricultural fields: instead it is lost, mainly to seepage. Thus, improving the present irrigation efficiency has to be a priority concern of the 1990s.

Sedimentation

Following the introduction of irrigation, cropping intensities generally increase, cropping patterns change, and fallow periods are reduced or even eliminated. These changes directly affect the soil erosion and run-off regimes of such areas. Similarly, overgrazing by livestock significantly increases soil erosion by reducing grass cover and breaking up the soil surface, as well as contributing to top-soil deterioration by compaction through trampling. Forest clearance for cultivation or timber extraction also affects river sediment loads. When the vegetation cover changes, the rates of soil losses also change (Chapters 6, 7). Much of the soil eroded contributes to higher sediment loads in watercourses draining the area, and this in turn can result in higher than expected sedimentation rates for reservoirs, with concomitant serious economic losses.

While the erosion of soil from land and its transportation by water and subsequent deposition elsewhere occurs naturally, poor agricultural practices significantly accelerate this process. Even though it is often not possible to quantify erosion reliably and to differentiate between naturally occurring erosion and soil loss due to human interventions, sedimentation is already a serious problem in many major rivers.

The sediments carried by a river are deposited upstream of a dam, a process which continually reduces reservoir capacity. For example, the construction of the Hoover Dam reduced the sediment discharge of the Colorado River at Yuma, Arizona, where it enters Mexico, from 130 million tonnes to only 0.1 million tonnes per year. Similarly, the River Nile formerly carried 100-150 million tonnes of suspended matter annually at Aswan before the construction of the Aswan High Dam. Nearly all of this sediment load is now deposited in the High Dam Lake or in the river upstream of the dam (Biswas, 1991).

Salinity and waterlogging

One of the important reasons for the decline of agricultural productivity in irrigated areas is the development of salinity and waterlogging due to inappropriate irrigation practices. As far as irrigation is concerned, salinization is usually caused by high salinity of the irrigation water, inadequate drainage, or evaporation of soil moisture due to intensive agriculture of saline soil. Waterlogging is also usually caused by bad irrigation management schemes resulting in loss of soil air content. These are discussed in Chapters 6 and 11.

Table 5: *Increase in level of water table due to irrigation.*

Irrigation project	Country	Water table (m)	
		Original depth	Rise/year
Nubariya	Egypt	15–20	2.0–3.0
Beni Amir	Morocco	15–30	1.5–3.0
Murray-Darling	Australia	30–40	0.5–1.5
Amibara	Ethiopia	10–15	1.0
Xinjang Farm 29	China	5–10	0.3–0.5
Bhatinda	India	15	0.6
SCARP 1	Pakistan	40–50	0.4
SCARP 6	Pakistan	10–15	0.2–0.4

Source: Biswas (1990b); FAO, (1990).

Impacts of domestic uses and waste water treatment

There is a whole range of diseases associated with water. These can be waterborne (bacterial, viral, parasitic), water-washed (e.g. enteric, skin), or water related (e.g. malaria, schistosomiasis). The provision of clean, safe drinking water is therefore a basic requirement for the health of any population (Chapter 18).

On the other hand, domestic waste water may contain pathogenic bacteria and viruses. It is often the carrier of viruses and diseases and has been known to result in high mortality rates in developing countries. High levels of coliforms have been known to exist in treated sewage water when sewage treatment plants are overloaded. Domestic waste water may also contain high levels of nitrogen and phosphorus which can cause eutrophication when discharged into streams.

The decade 1981-91 was declared by the UN General Assembly the Drinking Water and Sanitation Decade. Although information on treatment of domestic waste water in developing countries remains sparse, surveys were carried out by WHO on the status of community water supply and sanitation in both urban and rural areas in 1970, 1980, 1983 and 1988. While earlier data are not comparable for many reasons (Biswas, 1981) these surveys nevertheless indicate some trends.

Overall, some 1,348 million additional people (980 million in rural areas and 368 million in urban areas) had access to clean water at the end of the Decade. Progress on sanitary services was somewhat limited, since only 748 million more people (434 million in rural and 314 million in urban areas) were covered. Accordingly, the total number of people without clean water declined from 1,825 million to 1,325 million. The total number of people without sanitation facilities remained virtually unchanged. Globally, the number of urban residents without access to clean water and sanitation facilities increased by 31 million and 85 million respectively during the Decade. Progress in rural areas was much better. The number of rural people not having access to clean water and sanitation facilities declined by 624 million and 79 million respectively. The percentage of population served during 1980 and 1990, and the projections to the year 2000 by various regions are shown in Figure 6.

The lowest rate of progress was in sub-Saharan Africa where, due to high population growth, the percentage of people in urban areas not having access to clean water increased by about 29 per cent, in spite of the fact that the actual number of people receiving services doubled. Similarly, the percentage of people in urban areas not having sanitation facilities increased by 31 per cent, even though the number of people having such facilities more than doubled. Conditions in Latin America are also critical. One WHO-UNEP study (1988) points out that little, if any, urban domestic waste is treated. This means that rivers in this region generally have very high coliform counts, sometimes of more than 100,000 per 100ml (ECLAC, 1989).

By the year 2000, if present trends continue, the total number of people not

Figure 6: *Percentages of population served by clean water supplies and sanitation, by region.*

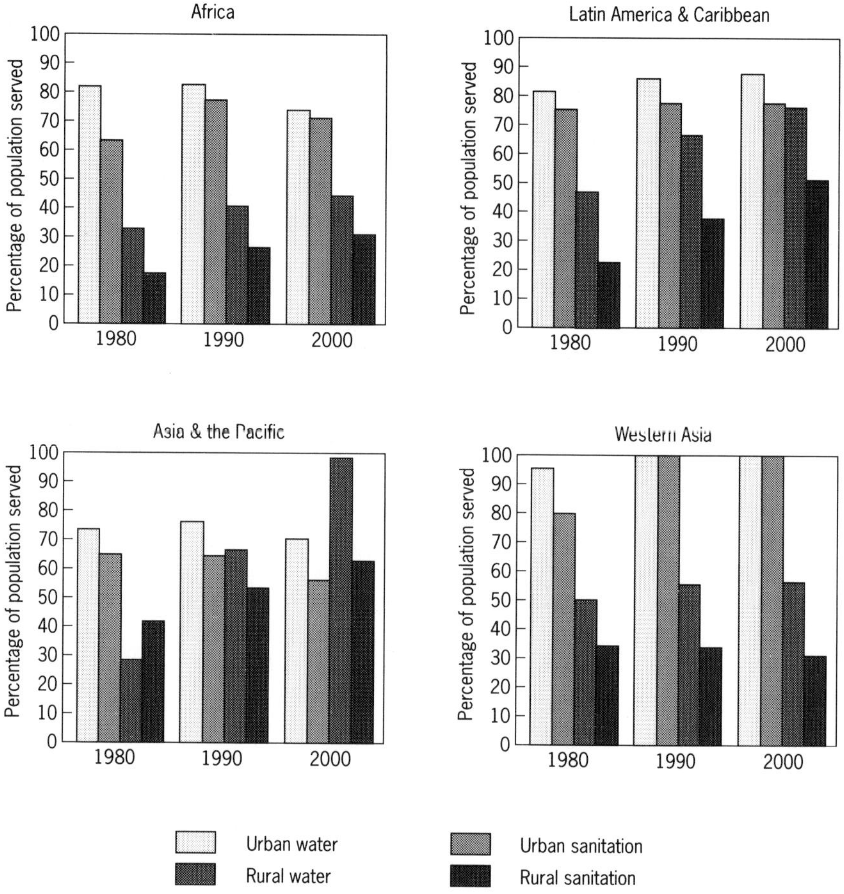

Source: Report of the Secretary-General of the UN (1990).

having access to clean water world-wide would decrease to 767 million due to continuing significant increases in coverage in rural areas. However, the number of people not having sanitation facilities would increase to 1,880 million, which would create attendant environmental health problems.

While considerable progress has been made in treating domestic discharges during the past two decades, much work remains to be done. Even in the OECD countries, where the percentage of population served by domestic waste water treatment increased from 33 per cent in 1970 to 60 per cent in 1989, some 330 million people in those countries were still not served by waste water treatment plants by 1989 (OECD, 1991). Furthermore, there were important disparities amongst OECD countries. Total population served with waste water treatment increased during 1970 to 1989 as follows: Denmark - 54 per cent to 98 per cent; France - 19 per cent to 52 per cent; Germany (FRG only) - 62 per cent to 90 per cent; Italy - 14 per cent to 60 per cent; Japan - 16 per cent to 39 per cent; and Canada-USA - 42 per cent to 73 per cent.

Impacts of freshwater storage

The environmental impacts of dams and reservoirs have been a cause of increasing concern in recent years. Their construction often inundates good-quality agricultural land and disrupts the lives of rural peoples, including indigenous people. It also destroys wetland habitats and reduces the transport of sediment essential to the maintenance of river deltas (Chapter 5). Some reservoir schemes have also led directly to an increase in diseases carried by aquatic vectors (Chapter 18). Large dams also present environmental hazards. Accidents involving dams have been known to cause considerable loss of life and property (Chapter 9). In many countries, fortunately, the issues are now well recognized and public policy is being increasingly guided by thorough environmental impact assessment, and by public consultation (Chapter 20).

Another related issue causing widespread concern during the past two decades was the possible transfer of large volumes of water between river basins. This first gained public attention when plans were announced in the former USSR to divert water from northward-flowing Siberian rivers to areas of need in the Caspian and Azov basins, and in west Siberia and Khazakstan (including areas currently irrigated from the rivers draining to the Aral Sea) (Voropaev and Velikanov, 1985). Inter-basin transfers are also being made or planned in Canada, the USA, China, Mexico and the Sudan (Golubev and Biswas, 1985). Experience shows that while such redistributions can provide overall benefits, the climatic, ecological, hydrological, social and economic impacts all need thorough analysis before they are implemented. Such analysis led to the deferment of some of the more ambitious schemes in the former USSR (Shiklomanov, 1990).

Impacts of industrial uses

This subject is discussed at length in Chapter 12. Suffice it to mention two points here. First, that the modest amounts of water used in industry, compared to uses in agriculture (see Figure 1) are due to the fact that many industries recycle water several times before finally discharging it as waste water. In the USA at present, water is re-used an average of nine times in industry before discharge. This average is expected to rise to 17 times by the end of the century (Postel, 1986). In Japan, the percentage of recycled water in the total amount of water used in industry rose from 52 per cent in 1970 to 74 per cent in 1981 (Hiraishi, 1992). The second point is that if industrial waste water, which is very heterogeneous and uneven in composition, is discharged into surface water, or mixed with municipal waste water or irrigation drainage, it creates very serious environmental problems. This is the case in many developing countries, and the problem is aggravated by the proliferation of small-scale industries whose discharges of waste water are difficult to monitor or control.

Responses

The General Situation

The three main international milestones in the field of water management during the last 15 years have been the Mar del Plata Action Plan of 1977, and the International Drinking Water and Sanitation Decade that began in 1981, and the International Conference on Water and Environment held in Dublin in January 1992. Neither of the first two was an unqualified success, and the general situation in 1991 is rather mixed, with some successes and some failures. The highlights of each are given below.

The Mar del Plata Conference recommendations, forming the Mar del Plata Action Plan (MPAP), covered eight major area (Biswas, 1978):

• Assessment of water resources.

• Water use and efficiency.

• Environment.

• Health and pollution control.

• Policy, planning and management.

• Natural hazards.

• Public information, education, training and research.

• Regional and international co-operation.

The aim of the International Drinking Water and Sanitation Decade was to provide safe drinking water and sanitation for all. Reviews of the Decade in New Delhi in 1990 showed that achievements fell far short of the target (WHO, 1990; Secretary-General, 1990). It is now evident that there is a need to maintain the impetus of the Decade until its target is achieved.

The actions taken during the International Water Supply and Sanitation Decade have had an appreciable impact on river water quality. Centralized sewer systems and waste water treatment plants have been built in a number of urban centres, so that their untreated waste water is no longer being discharged, and river water quality in some areas has already started to improve. A good example is the Ganges Action Plan in India.

In another significant advance, many arid and semi-arid countries, where water scarcity is already a serious constraint on further economic development, have realized that treated waste water is a 'new' source of water which could be used for productive purposes. Most arid countries of West Asia and North Africa have therefore embarked upon ambitious programmes for the treatment and reuse of waste water.

The Dublin Conference recommendations, forming the Dublin Statement, covered six major areas:

• Integrated water resources development and management.

• Water resources assessment and impacts of climate change on water resources, an emerging issue over the last few years.

• Protection of water resources, water quality and aquatic ecosystems.

• Water and sustainable urban development; and drinking water supply and sanitation in the urban context. This issue is stressed because of the proliferation of megacities, especially in developing countries.

• Water for sustainable food production and rural development; and drinking water supply and sanitation in the rural context - a recognition of the fact that food production is a critical factor in developing countries and that water - rather than land - is the principal constraint.

Shared water resources

Before the current changes in Eastern Europe and the former USSR, the number of rivers and lake basins shared by two or more countries was 214. The development and management of these water resources pose a special challenge, which sometimes becomes a major political issue as illustrated by the debate over dam construction on the Hungary-Czechoslovakia section of the Danube river in the 1980s and the continuing debate over the waters of the Nile, the Euphrates, the Yarmouk and Jordan rivers. As the demand for water and hydroelectric power increases, and exclusively national sources of water are fully developed, the only major new sources are likely to be international. The extent and magnitude of this global problem has not yet been generally recognized. Nearly 47 per cent of the land area of the world (excluding Antarctica) falls within international water basins that are shared by two or more countries. There are 44 countries with at least 80 per cent of their total areas within international basins (Biswas, 1991).

Until recently, no active interest was shown by the international community in the development of shared water bodies, primarily because these issues are complex and politically sensitive. However, in 1987 UNEP pioneered the promotion of an 'Environmentally-Sound Management of Inland Waters Programme (EMINWA)' (UNEP, 1989; Thanh and Biswas, 1990). Based on diagnostic studies, action plans are formulated which have the concurrence of all countries involved. Such studies have been successfully completed for the River Zambezi, (Zambezi Action Plan in 1987), and the Lake Chad Basin (the Master Plan for the development of environmentally-sound management of the natural resources of the Lake Chad conventional basin, 1990). Plans to prepare a diagnostic study of the River Nile and of the Aral Sea are now at an advanced stage. UNEP, in collaboration with the Organization of American States is providing support to the negotiations for the preparation of an integrated management plan for transboundary watersheds between Columbia and Venezuela, including the Orinoco river.

Successful models of co-operation in dealing with shared water resources are provided by the 1978 Great Lakes Water Quality Agreement of the US-Canada Joint Commission (IJC, 1987) and the establishment of the Rhine Commission in Europe.

There is no doubt that environmental quality has become firmly established as a major objective of water resource management in all parts of the world during the past two decades. This was also the period when, for the first time, a major water project in a developing country (the Silent Valley Project, in India) was rejected primarily on environmental grounds.

Water pricing and cost recovery

During the 1980s, water pricing and cost recovery became a major discussion topic in many national and international fora. The view was put forward that if economically realistic water prices could be charged, farmers and other consumers would become rational optimizers, which would contribute substantially to efficient water use. Furthermore, if government departments could receive the extra revenue generated by water pricing, they could operate and maintain their water systems much more efficiently.

By the early 1990s, it was increasingly realized that two fundamental issues have to be considered before water pricing becomes an attractive policy instrument. First, water pricing has thus far been viewed primarily as an economic instrument: its socio-political implications in developing countries have generally not been understood. Second, water has traditionally been subsidized to achieve very specific socio-political objectives of food security, provision of clean drinking water, and increasing the health and income of the rural poor, especially women. If economic water pricing is to be introduced, other policy instruments must be developed to achieve the same objectives.

The World Bank is investigating pricing issues, particularly in developing countries, within the framework of 'Integrated Water Resource Planning' (IWRP) as well as carrying out applied case studies (e.g Munasinghe, 1990).

Institutional aspects of water management

A major constraint on efficient water management during the past two decades has been the weaknesses of the institutions concerned. As a general rule, it can be said that most water institutions in developing countries - and they are often amongst the first to be established in these countries - need significant strengthening in order cope with emerging water management problems brought about by new agricultural crops, large-scale water management projects and changed social and cultural environments. In addition, in order that water can be managed in its totality in a rational fashion, inter-institutional collaboration has to be substantially improved. Currently water-related policies have been developed in a fragmented fashion by a host of institutions in nearly all these countries. For example, irrigation is considered by Ministries of Irrigation, water supply by municipalities, hydroelectric power by Ministries of Energy, navigation by Ministries of Transport, environment by Ministries of Environment and health by Ministries of Health. Lack of co-ordination, and often intense rivalries, have meant that water policies have generally been sub-optimal. The development financing institutions also suffer from the same shortcoming. Without institutional rationalization and strengthening, water management simply cannot become optimal in the future.

Emergence of citizens' groups

During the past decade there has been a major increase in the number of citizens' groups, national and international, that have attempted to stop or significantly modify large-scale water development projects. Three recent examples are the Great Whale Project in Canada, the Three Gorges Dam in China and the Narmada Valley Project in India. The two major public objections common to these three projects are resettlement of people due to reservoir inundation, and potential serious adverse environmental impacts.

Concluding remarks

During the past two decades, progress has been made on water resources development and management in different parts of the world. However, much remains to be done, especially in developing countries. Furthermore, with recent major political changes in Eastern Europe, it is becoming apparent that water quality problems in many East European countries are much more serious than many people believed. There and elsewhere, per capita water consumption is increasing. Furthermore, the share of industrial water use will probably continue to increase during the next few decades, which means that agricultural withdrawals, as a percentage of total water used, will continue to decline steadily. For these and other reasons discussed earlier it is highly likely that water, like energy in the 1970s, will become the most critical resource issue in most parts of the world by the late 1990s and the early part of the twenty-first century.

REFERENCES

Ayibotele and Falkenmark, M. (1992) Fresh Water Resources, in Proc. Int. Conf. on an Agenda for Science for the Environment and Development into the 21st Century, ICSU, Paris (in press).

Biswas, A.K. (1978) *United Nations Water Conference: Summary and Main Documents*, Pergamon Press, Oxford.

Biswas, A.K. (1981) Clean Water for the Third World, *Foreign Affairs*, **Vol. 60**, No.1, pp. 148-66.

Biswas, A.K. (1990a) Impacts of Agriculture on Water Quality: State-of-the-Art, Report to Land and Water Development Division, FAO, Rome.

Biswas, A.K. (1990b) Groundwater Quality Management: A Holistic View, IWRA Distinguished Lecture, International Seminar on Groundwater Management, Asian Institute of Technology, Bangkok.

Biswas, A.K. (1991) Land and Water Management for Sustainable Agricultural Development of Egypt: Opportunities and Constraints, Report to Economic and Social Policy Division, FAO, Rome.

Cunha, L.V. da (1989) Sustainable Development of Water Resources, International Symposium on Integrated Approaches to Water Pollution Problems, Lisbon.

ECLAC (Economic Commission for Latin America and the Caribbean) (1989) *The Water Resources of Latin America and the Caribbean* United Nations, New York.

FAO (1990) *FAO Production Yearbook*, **vol.43**, FAO, Rome.

Golubev, G. and Biswas, A.K. (eds) (1985) Large Scale Water Transfers: Emerging Environmental and Social Issues. *UNEP Water Resource Series*, **Vol. 7**. Tycooly International., Dublin.

Gunn, J.M. and Keller, W. (1990) Biological Recovery of an Acid Lake after Reductions in Industrial Emissions of Sulphur, *Nature*, **345**, pp 431-3.

Hallberg, G.R. (1989) Pesticides Pollution of Groundwater in the Humid United States, *Agriculture, Ecosystems and Environment*, **Vol. 26**, Nos. 3-4, pp. 369-89.

Hiraishi, T. (1992) Personal communication.

Holdgate, M.W., Kassas, M. and White, G. (1982) *The World Environment 1972-1982*, Tycooly, Dublin.

Howells, G. D. (1984) Fishery decline: mechanisms and predictions. *Phil Trans. R. Soc. Lond.* , **B, 305**, 529-47.

ICOLD (1984) *World Register of Dams - 1984*, Central Office, International Commission on Large Dams, Paris.

ICOLD (1989) *World Register of Dams - 1988 Updating*, Central Office, International Commission on Large Dams, Paris.

IJC (1987) Revised Great Lakes Water Quality Agreement of 1978, Joint Commission, Washington and Ottawa.

L'vovich, M.I. and White, G.F, (1991) Use and Transformation of Terrestrial Water Systems, in *The Earth As Transformed by Human Action*, Cambridge University Press, Cambridge, pp. 235-52.

Munasinghe (1990) Managing Water Resources to Avoid Environmental Degradation: Policy Analysis and Application, Environmental Working Paper No. 41, World Bank, December, 1990.

OECD (1991) *The State of the Environment*, OECD, Paris, pp. 292.

Postel, S. (1986) Increasing Water Efficiency, in L. Brown *et al.*, (eds.) *State of the World*, Norton & Co., New York.

Roberts, D. and Marsh, T. (1987) The Effects of Agricultural Practices on the Nitrate Concentrations in the Surface Water Domestic Supply Sources of Western Europe, *Water for the Future: Hydrology in Perspective*, Publn. No. 164, IAHS, pp. 365-80.

Secretary-General, United Nations (1990) Achievements of the International Drinking Water Supply and Sanitation Decade, 1981-1990, Report of the Economic and Social Council, Document, A/45/327, United Nations, New York, pp. 30.

Shiklomanov, L.A. (1990) Global Water Resources, in *Nature and Resources*, **vol.26**, pp 34-43, UNESCO, Paris.

Statens Naturvardsverk (SNV) (1986) Monitor - 1986 - Sura och Forsurade Vatten, SNV, Solna

Sweden (1982) Acidification Today and Tomorrow. Report of the Ministry of Agriculture and Environment 1982 Committee. Stockholm: Ministries of Agriculture and Environment.

Thanh, N.C. & Biswas, Asit K. (1990) *Environmentally-Sound Water Management*, Oxford University Press, New Delhi and Oxford.

UNEP (1987) Agreement on the action plan for the environmentally sound management of the common Zambezi river system, UNEP, Nairobi.

UNEP (1989,) Sustainable Water Development: A Synthesis, *International Journal of Water Resources Development*, **Vol. 5, No.4**, pp. 225-51.

UNEP (1991) *United Nations Environment Programme Environmental Data Report*, Third Edition. Basil Blackwell, Oxford.

UNEP/UNSO (1991:) Master plan for the development and environmentally sound management of the natural resources of the Lake Chad conventional basin, UNEP, Nairobi.

Voropaev, G.V. and Velikanov, A.L. (1985) Partial southward diversion of northern and Siberian rivers. In Golubev, G. and Biswas, A.K. Large Scale Water Transfers. *UNEP Water Resource Series*, **Vol.7**. Tycooly International., Dublin.

Veltrop, Jan. A. (1991) Water, Dams and Hydropower in the Coming Decades, *Water Power and Dam Construction*, **Vol.43, No.6**, pp. 37-44.

WHO and UNEP (1988) Global Pollution and Health, in *Global Environmental Monitoring System*, London, pp. 9-11.

WMO/UNESCO (1991) Report on Water Resources Assessment, UNESCO, Paris, pp. 64.

WRI (1990) *World Resources, 1990-1991*. Oxford University Press, New York and Oxford.

CHAPTER 5

Coastal and marine degradation

The issues

The oceans and seas cover 70 per cent of the Earth's surface and are active components of the global biosphere. One of the major developments of the last 20 years has been the realization that this vast sector of the environment is dynamic and interactive: thus, long-term environmental management of even a small portion of the marine environment requires an integrated approach which must include consideration of the coastal zones and also their drainage basins and the atmosphere. The main sectoral issues are given in Box 1.

The coastal zone, here defined as the region between the seaward margin of the continental shelf and the inland limit of the coastal plain, is among the regions of highest biological productivity on Earth. It is also the zone with the

---

BOX 1

The coastal and marine environment: main issues of the last 20 years.

- Coastal development (destruction of wetlands, destruction of natural habitats, upland water impoundment, urbanization, degradation of coral reefs, development of recreational beaches).

- Discharges of municipal sewage, industrial wastes, plastic litter and radionuclides into the marine environment; run-off of agricultural wastes.

- Dredging of sediments.

- Operational and accidental releases of oil from ships and oil fields.

- Transportation of hazardous wastes.

- Phytoplankton blooms and toxin outbreaks in some coastal zones and semi-enclosed seas.

- Over-exploitation of living resources in the sea, e.g. with very long nylon drift nets.

- Interpretation of trend data. (This is difficult because of the great geographical and temporal variability in the emissions of various kinds of pollutants, their environmental concentrations and bioaccumulation rates; and in the marine environment itself, e.g. El Niño.)

- The development of improved long-term strategies for managing the coastal and marine environment, recognizing the multiple uses and multiple users of these important resources.

Source: GESAMP (1990).

---

greatest human population. Although census information is imprecise, because census district boundaries rarely coincide with the zone's natural limits, more than 50 per cent of the world's population lived in coastal regions in 1970, and about 40 per cent lived in coastal urban centres with more than 10,000 inhabitants. These numbers have increased in the past two decades as a consequence of immigration and the growth in prosperity of many coastal areas as a result of industrialization and tourism. Today, about 60 per cent of humanity (or nearly three billion people) live in the coastal zone, and two-thirds of the world's cities with populations of 2.5 million or more are near estuaries. Within the next 20-30 years the population of this zone is expected to almost double (IUCN/UNEP/WWF, 1991).

This increase is inevitably altering land-use patterns in coastal zones. Other impacts there - and in the coastal regions generally - come from pollution, flooding, land subsidence and compaction, and the effects of upland water diversion. Natural habitats are being lost through reclamation for urban and industrial development, agriculture and mariculture. Nearshore regions are being degraded by eutrophication and industrial waste; public health is threatened by sewage contamination of beaches and seafood; and the marine environment is being fouled by the progressive build-up of chlorinated hydrocarbons, plastic litter and the accumulation of tar on coastlines. Some of the waste products of coastal development, augmented by discharges through coastal outfalls and rivers, spread outwards to the world oceans, carried by the atmosphere, currents and ships. The visible fingerprints of humanity (oil slicks, plastic litter and other debris) can be found everywhere. Moreover, the oceans have also been affected by many 'invisible' changes in the current century. Most commercial stocks of fish are now over-exploited and the balance of whole ecosystems is at risk. Contaminants are measurable in the open oceans, even if their concentrations do not appear high enough to damage marine life (GESAMP, 1990).

There was considerable concern over the state of the world's oceans at the time of the Stockholm Conference. Since then, action has been taken to stop the dumping of polluting wastes at sea, to eliminate damaging pollution from ships, and to limit discharges from land-based sources (see Chapters 22 and 23). The state of scientific knowledge has improved a great deal. An increase of some 50 per cent in the scientific literature on marine pollution (GESAMP, 1990), and a number of large-scale research programmes now under way, should further expand our understanding of the oceans and permit more satisfactory management. Over the same twenty years there have been some improvements in environmental quality: for example the body burden of organochlorine pesticides carried by various marine species has fallen in several areas, and action to clean up many coastal areas has been pressed forward (e.g. Oslo/Paris, 1984; DOE, 1987; ENS, 1991; Chapter 23). Nevertheless, the pressures of coastal zone development are leading to continuing environmental degradation in many parts of the world, and these conditions are likely to be exacerbated by climate change and sea-level rise within the next 50 years or so.

Changes in the coastal zone, 1970-90

Physical changes

There are two causes of change in sea-level relative to the land: crustal movement and changes in the overall volume of marine waters. Sea-level, averaged globally, has been rising 2.4 ± 0.9 mm per year in recent years (Peltier and Tushingham, 1989), and this trend is expected to accelerate. However, the levels of enclosed waters, such as the Caspian or Aral Sea (Figure 1), obviously follow separate and distinctive patterns (Precoda, 1991).

Figure 1: *Changes of level in the Caspian Sea. Between 1930 and 1977, sea-level fell by about three metres, but since 1977 it has been rising, and after a phase of emergence the bordering coasts are now again submerging. Between 1970 and 1990 there was a net rise of sea-level of about one metre. The fluctuations of level are considered to be due to climatic variations influencing fluvial discharge into the Caspian Sea, but the completion of a dam across the mouth of the Kora-Bogaz-Gol embayment, an area of high evaporation, may also have contributed.*

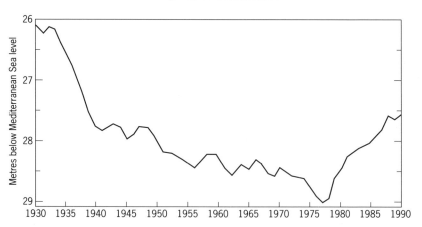

Source: Data supplied by the late Professor O.K. Leontyev and by Dr S.A. Lukyanova, Moscow State University.

On some coasts, a sea-level rise relative to the land has occurred as the result of gradual or intermittent land subsidence (Figure 2). An example of such subsidence in the Bangkok region is shown in Figure 3. The human response to coastal submergence has been either to move inland, or to try to maintain the coastline by building sea walls. The Netherlands constitutes an area of subsiding land and rising sea, with a long history of sea wall construction and maintenance. Subsidence on the south-east coast of England has resulted in high tides and storm surges attaining successively higher levels during the past two decades, and the risk of extensive flooding in the London area as a result of storm surges in the Thames estuary led to the construction of the Thames Barrier (Figure 4). In the Venice area there have been several 'aqua alta' (storm surge marine flooding events) each year as a result of continuing land subsidence, the dredging of navigation channels, and land reclamation around the Venice Lagoon.

Other coasts are rising. In some regions, such as the Baltic coast of Sweden and the Hudson Bay lowlands of Canada, this is due to a continuing elastic recovery following the removal of the loading of the massive ice sheets of the last glaciation. In other areas, such as parts of the South Island of New Zealand and the western coasts of the Americas, coastal elevation is occurring at the junction of tectonic plates.

Coastal erosion is a natural feature, but it is clearly aggravated by subsidence, and critically affected by the extent to which the sediments it removes on 'soft' coasts are replenished by new materials brought down by rivers, or supplied by drift along a coast. In a number of areas the construction of dams on major rivers has intercepted the downstream flow of sediment, and so reduced the 'nourishment' of sand banks, mud flats and other soft coast systems. Deltaic coastlines that have been maintained or advanced seaward by the supply of fluvial sediments are liable to erode as the sediment supply diminishes. The Nile delta provides a good example (Figure 5).

The building of harbours, breakwaters and barriers designed to protect a stretch of coast often interferes with the longshore drift of beach sediment and so upsets the local balance of erosion and accretion. For this reason, engineering works designed to protect vulnerable coasts have sometimes merely transferred the problem elsewhere, and started a process ending in the development of a completely artificial coastline. During the past two decades such problems have become more widespread following the construction of many new marinas for recreational boating.

A few sectors of the world's coastlines have advanced seaward during the past two decades. This has happened where rivers have continued to supply large quantities of sediment to their mouths, either to be added to growing parts of deltas as on the north coast of Java, or spread alongshore to build up the seaward fringes of coastal plains, as in southern Iceland. Yields of fluvial sediment continue to be augmented in some areas as a consequence of soil erosion in river catchments. This may be due to devegetation and agricultural development (as in the Citanduy Basin of Java: Figure 6), or to mining activities, as in Bougainville and Caledonia.

Coastal accretion has also continued where sediment, generally sand, is being washed in from the sea floor, as in Streaky Bay, Australia, and Sebastian Vizcaino Bay in Mexico. There has also been local accretion where sand dunes are spilling from the hinterland on to the shore, as in Bahia da Paracas in Peru; where material derived from cliff erosion is being deposited alongshore, as on parts of the shoreline of Puget Sound; and where drifting beach sediment accumulates alongside a headland or breakwater, as at Lagos in Nigeria (Bird, 1985).

Source: E. Bird (personal communication, 1991).

110

Fig 2: *Subsiding coasts, where the relative rise of sea-level indicated by tide gauge records has averaged more than 2.0 mm/year over the past century, and where there is evidence of continuing subsidence 1970-90. Some sectors are subsiding tectonically; others have subsided as the result of groundwater or oil extraction. These coasts show features produced by submergence and erosion which will become much more widespread around the world's coastline if there is a global sea-level rise over the coming century.*

KEY:

1   Long Beach area, Southern California
2   Columbia River delta, head of Gulf of California
3   Gulf of La Plata, Argentina
4   Amazon delta
5   Orinoco delta
6   Gulf/Atlantic coasts, Mexico and United States
7   Southern and Eastern England
8   The southern Baltic from Estonia to Poland
9   North Germany, the Netherlands, Belgium and northern France
10   Loire estuary, western France
11   Vendée, western France
12   Lisbon region, Portugal
13   Guadalquivir delta, Spain
14   Ebro delta, Spain
15   Rhône delta, France
16   Northern Adriatic from Rimini to Venice and Grado
17   Danube delta, Rumania
18   Eastern Sea of Azov
19   Poti Swamp, Soviet Black Sea coast
20   South-east Turkey
21   Nile delta to Libya
22   Tunisia
23   Nigerian coast
24   Zambezi delta
25   Tigris-Euphrates delta
26   Rann of Kutch
27   Eastern India
28   Ganges-Brahmaputra delta
29   Irrawaddy delta
30   Bangkok coastal region
31   Mekong delta
32   Eastern Sumatra
33   Northern Java deltaic coast
34   Sepik delta
35   Port Adelaide region
36   Corner Inlet region
37   Hwang-ho delta
38   Tokyo Bay
39   Niigata, Japan
40   Maizuru, Japan
41   Northern Taiwan
42   Red River delta, North Vietnam.

Figure 3: *Contours showing the extent of land subsidence (in centimetres) measured from detailed surveys in the Bangkok region, Thailand, between 1934 and 1987, due largely to groundwater extraction and compression of depleted subsurface aquifer horizons. The shaded areas of the coast near the mouth of the Chao Phraya River subsided by up to 40 centimetres during this period.*

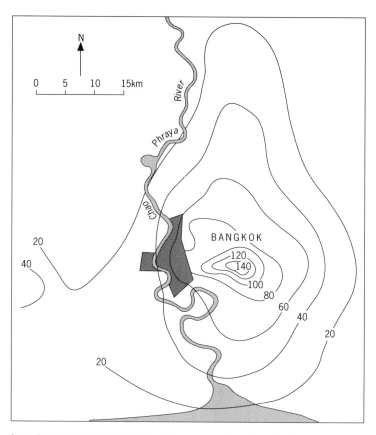

Source: Based on maps supplied by Dr Pranya Nutalaya, Asian Institute of Technology, Bangkok.

There is a long history of land reclamation from the sea, especially on accreting low-lying coasts and estuaries. Between 1970 and 1990 there has been extensive reclamation in Japan, Hong Kong, China, the Philippines, Malaysia, Singapore (Figure 7), Bahrain and Saudi Arabia. In the Netherlands, the long process of reclamation has continued with the Delta Scheme, which has shortened the coastline by 700km.

The reclamation of land for agriculture, mariculture, ports, industries and urban growth has led to a reduction in the extent of the world's coastal wetlands (mangroves, salt marshes and reed swamps) over the past century (Bird, 1985; Walker, 1988). In temperate regions, salt marshes are particularly at risk, and have been reduced considerably in both the United States and

Figure 4: *The Thames Barrier, completed in 1983, has gates that can be rotated to a vertical position to prevent storm surge flooding of the London area. When there is no such threat, the gates are rotated down to a horizontal position on the river bed, and ships can pass through. The gates had to be raised once in 1985, once in 1987, once in 1988, and six times in 1990, and on all these occasions were successful in preventing storm surge flooding of London. It is anticipated that its use will be required on an average of three or four occasions per year until the year 2000, and then more frequently as continuing land subsidence results in sea-level rise. If there is an accelerated sea-level rise due to the greenhouse effect there will eventually have to be some restructuring of the Thames Barrier to operate at a higher level.*
In this view across the Barrier, the gates are seen in the raised (i.e. flood-control) position.

Photo: National Rivers Authority, Thames Region (Thames Barrier Centre).

western Europe. (See Chapter 7 for a discussion of the impacts of the exploitation of mangroves).

The final type of physical damage to the marine waters of the coastal zone comes from dredging and dumping. Dredging of the sea bed in approaches to ports and harbours, including those in estuaries, has proceeded extensively during the past 20 years, with consequent impacts on nearby coasts and damage to marine ecosystems both directly and through the effects of increased turbidity and sediment deposition. The same is true where sea-floor sand and gravel have been dredged as a source of construction materials, or for minerals (as on the coasts of Thailand and Malaysia, where pollution by heavy metals has resulted). Damage has also occurred where materials have been dumped

Figure 5: *Changes on the coastline of the Nile delta during the past twenty years, shown by land gains and losses in metres at selected points. Erosion became more rapid and more extensive along parts of this coastline after the completion of the Aswan High Dam in 1971, and has been severe to the west of the Rosetta and Damietta mouths, but there has also been longshore spit accretion to the east of these two river mouths, and progradation west of Port Said. The shaded area indicates land more than a metre above present sea level.*

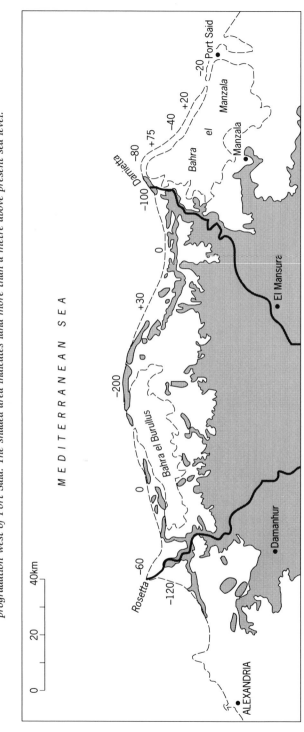

Source: Eric Bird, based on data from Dr G. Sestini and the Delft Hydraulics Institute, Marknesse, Netherlands.

Figure 6: *The Segara Anakan is an estuarine lagoon in southern Java that has been silting rapidly as a result of accelerated fluvial sedimentation. In consequence, the mangrove fringe has advanced at least 100 metres during the past two decades.*

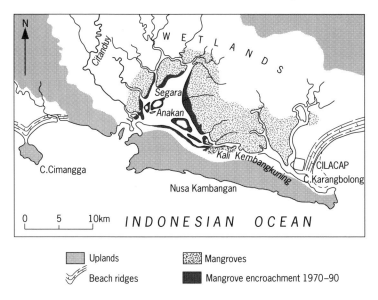

Uplands   Mangroves
Beach ridges   Mangrove encroachment 1970–90

Source: Based on data supplied by Dr O.S.R. Ongkosongo, Indonesian Institute of Sciences, Jakarta.

in order to make artificial coastlines. The quarrying of coral can also cause problems: along the south-west coast of Sri Lanka, mining for construction purposes and as a source of lime and cement has accelerated erosion along a 10km stretch of coastline and the beach has been moved 300 metres inland by erosion (WRI, 1986). In many tropical areas, a combination of sediment blanketing from dredging operations, quarrying, and fishing (using explosives) has caused major damage to coral reefs.

Chemical changes

Where the inflow of fresh waters to estuaries and lagoons is reduced, or its annual cycle suppressed, by the damming of incoming rivers or the abstraction of water for irrigation schemes upstream, salinity is likely to increase. This trend is accentuated if the navigable approaches to the estuary are at the same time dredged, facilitating the influx of marine water. Higher salinity results in the die-back of freshwater vegetation and an increase in salt-tolerant (halophytic) species, with associated changes to fauna, including fish, as freshwater species are replaced by marine ones. An example is provided by the Gippsland Lakes, a group of coastal lagoons in south-eastern Australia, which have become more brackish during the past 20 years partly because of a succession of dry periods and partly because of increased irrigation in their catchment area.

Figure 7: *Reclaimed land on the island of Singapore, much of it reclaimed in the past three decades.*

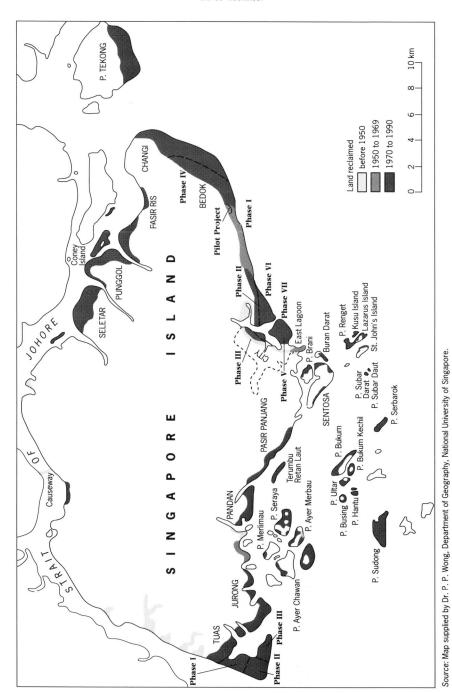

Source: Map supplied by Dr. P. P. Wong, Department of Geography, National University of Singapore.

Salt-water intrusion into underground water can be a serious problem in coastal areas where freshwater discharge has been reduced, or aquifers tapped for supplies. In the Netherlands, water management schemes have been introduced to overcome the problem, but salt-water intrusion is one of the major concerns in low-lying coral atoll states, which have only small and intensely used sources of fresh water. Salt-water intrusion may be further aggravated by sea-level rises associated with climate change.

Conversely, there are areas where the salinity of estuaries and lagoons has diminished because of the building of barrages to reduce or exclude sea penetration. Kalemitya Lagoon, Sri Lanka, is one example of a formerly brackish lagoon that has changed into a freshwater lake and swamp in recent decades following the building of a weir at its mouth. Similar changes have followed completion of the Netherlands Delta scheme.

Ecological changes

The coastal zones are areas of high biological productivity. About 95 per cent of world fishery catches come from near-shore and reef areas. Moreover, the nursery grounds for many fish species whose adult stages move into deeper waters and are fished offshore lie in the shallow, productive and biologically diverse estuaries, mangrove systems, seagrass beds and coral reefs.

Where coasts are accreting, mangroves, sea grasses and salt-marsh species are important in binding sediment and consolidating the accretion. Figure 8 illustrates one area in Indonesia where mangroves have advanced on to tidal mudflats that have been accreting rapidly as a result of increased sediment yield from the deforested catchments of inflowing rivers. Conversely, mangroves are sensitive to salinity, and where this changes their distribution will be altered.

Figure 8: *Brackish-water fish-ponds on the shores of the Citarum delta, Indonesia, have become useless as a consequence of shoreline erosion following the diversion of the river which formerly maintained a supply of muddy sediment to a protective mangrove fringe.*

Photo: Eric Bird.

There are enormous economic benefits in retaining natural vegetation as a form of coastal protection. In the United States, a hectare of intertidal marshland is estimated to be worth $US72,000 per annum as a coastal protection and as a fish nursery system. The US Army Corps of Engineers estimated that the natural protection afforded by salt marshes in Boston Harbour was worth $US17 million per year in terms of engineering costs averted (McNeely, 1988). Over much of the world there is no practicable alternative to natural protective barriers as sea defences, and the economic costs of mangrove, marsh or coral reef destruction are clearly immense. The recognition of this fact has been an important feature of the past two decades.

Marine pollution trends

GESAMP (1990) concluded that most of the world's coastal areas are polluted. The sea is the ultimate sink for most of the liquid wastes and a considerable fraction of the solid wastes resulting from human activities on land. More than three-quarters of all marine pollution comes from land-based sources, via drainage and discharges into rivers, through outfalls flowing directly to estuaries, bays and the open coast, and from the atmosphere. The rest comes from shipping, dumping and offshore mining and oil production. The greater part of this pollution passes into coastal waters, and more than 90 per cent of all chemicals, refuse and other materials entering these waters remains there in sediments, wetlands, fringing reefs and other coastal ecosystems (IUCN/UNEP/WWF, 1991).

GESAMP's review concluded that the most serious problems arise from sedimentation due to land clearance and erosion, and from nutrients discharged into coastal waters with sewage and agricultural drainage water. Human inputs of nutrients into coastal waters already equal natural sources and within 20-30 years are likely to exceed the natural background by several times. The result is likely to be a considerable extension of the kind of impact now found only in enclosed waters such as the Baltic or Japan's Inland Sea (IUCN/UNEP/WWF, 1991).

Such excessive nutrient loads bring marked ecological changes. The structure of plankton communities is altered, with preferential growth of small flagellates rather than the larger diatoms, and unusual plankton 'blooms', uncontrolled by the normal processes of grazing. The subsequent decomposition of the mass of organic matter deoxygenates the water, killing fish and invertebrates, while some species of algae produce foam and scum which interfere with fishing and reduce the amenity of beaches when washed ashore. In some cases the sea is discoloured, giving rise to the term 'red tide'. Some of the plankton species are toxic, and consumers of seafood exposed to such blooms are at risk from paralytic, diarrhoeic and amnesic shellfish poisons. The problem has become more severe in recent years and has affected the Baltic Sea, the Gulf of Mexico, the inner Adriatic and warm-water bodies such as the Gulf. In 1987 an outbreak of paralytic shellfish poisoning killed 26 people in Guatemala.

Another major problem arises from the discharge, mainly in sewage, of pathogenic organisms. Epidemiological studies have provided unequivocal evidence that swimmers in sewage-polluted seawater experience an above-average incidence of gastric disorders, and that the increase is correlated with *Enterococcus* counts in the water rather than with those of the usual 'marker' micro-organism, *Escherichia coli* (GESAMP, 1990). The consumption of contaminated seafood is firmly linked with serious illness, including viral hepatitis and cholera (Box 2).

BOX 2

Algal blooms and their effects.

- In 1976, a bloom of *Ceratium tripos* occurred in the Middle Atlantic Bight of America leading to widespread deoxygenation in the water column and the destruction of around a quarter of the total benthic bivalve stock in the area.

- In Hong Kong, the number of red tides resulting in oxygen depletion and fish kills has increased from one or two per year in the 1970s to over twenty per year in the late 1980s.

- The toxic dinoflagellate *Ptychodiscus brevis* was responsible for an extensive red tide from November 1987 to February 1988 along the coast of North Carolina, causing neurotoxic shellfish poisoning and closure of more than 400km of coastline during the peak shellfish season.

- In May-June 1988, an exceptional bloom of the flagellate *Chrysochromulina polylepsis* occurred in the waters between Denmark, Norway and Sweden, covering an area of approximately 75,000km^2. The bloom produced a toxin which killed large numbers of macroalgae, invertebrates and fish, including caged salmon in fish farms, costing the Norwegian fishing industry over $US10 million. There were no reports of illnesses in local people.

- In the North Sea, some areas in the German Bight, and off the Danish west coast, major fish and macrobenthos kills were caused in 1981, 1982 and 1983 by oxygen deficiency following large blooms of *Ceratium tripos*.

- In the northern Adriatic Sea, which has poor circulation with the open Mediterranean, extensive blooms are frequently recorded in Spring and Autumn. In 1989 a slime layer derived from the phytoplankton, up to 12 metres thick in places and weighing millions of tonnes, spread along the coast causing losses to fisheries and tourism estimated at $US1.4 million.

Sources: Ambio, 17 (1988); GESAMP Review of Potentially Harmful Substances, Nutrients (1990).

Some 6.5 million tonnes of litter finds its way into the sea each year. In the past, much of it disintegrated quickly, but resistant synthetic substances have in recent years replaced many natural, more easily degradable materials. Plastics, for example, can persist for up to 50 years, and because they are usually buoyant they are widely distributed by ocean currents and winds. Many beaches are littered with plastic waste of various kinds, from land and ships. Along the beaches of the Mediterranean, about 70 per cent of the debris examined in one investigation was plastic: in the Pacific the figure exceeded 80 per cent (Arnaudo, 1990). A major source of plastic debris is the fishing industry: it is estimated that more than 150,000 tonnes of plastic fishing gear is lost (or discarded) in the oceans each year (GESAMP, 1990). Such debris is a nuisance to the tourist industry and can be a serious hazard to marine animals such as seals. A particularly serious new problem is posed by modern plastic drift nets, which are many kilometres in length and which, if they break free from a vessel, continue to float around the oceans entrapping and killing all manner of species.

The fate of pollutants in the sea is of great importance, and has been the subject of considerable research. Some wastes are easily degraded to harmless substances, although if these have a biological impact, for example as nutrients, they can still exert a considerable ecological effect. Others, such as metals and persistent organochlorine compounds, are either not degraded or degrade only slowly, and tend to be absorbed in bottom sediment near the sources of discharge. Much of the concern over these materials in the past has arisen because some (e.g. mercury) can be converted by biological processes into highly toxic forms: methyl mercury, produced by bacterial transformation of the inorganic metal, caused the outbreak of Minamata disease in Japan (see Chapter 10). Others, notably organochlorine pesticides and polychlorinated biphenyls (PCBs) tend to accumulate in living organisms. Halogenated hydrocarbons accumulate in fatty tissue, and reach their highest concentration levels in the bodies of fairly long-lived predators high up in the food chain. Where the concentration has built up over decades, as in enclosed areas such as the Baltic and the Waddenzee in Northwest Europe, the reproductive capacity of marine mammals and birds can be affected (GESAMP, 1990).

Ten years ago, these metals and persistent halogenated hydrocarbons caused greatest concern, and figured largely in monitoring schemes (Holdgate *et al.*, 1982). In recent years, however, major efforts have been made to reduce their inputs to the sea. The recent GESAMP review (1990) concluded that pollution by heavy metals such as cadmium, lead and mercury, was chiefly of concern near sources of contamination, including mining areas and industrial centres. Similarly, in those countries where controls have been in place for some years, synthetic chlorinated hydrocarbons, though still in high concentrations in sediments and animal tissues, are now beginning to decline. Polychlorinated biphenyls, although still prominent in marine ecosystems are also declining in some developed regions, where they are banned (Figure 9), but there is evidence of an increase in their use and release to the sea in tropical regions. Among new causes of concern, tributyl tin (TBT), used as an anti-fouling agent in ship's paint, was incriminated quite early in its period of use as a cause of damage to oysters and other invertebrates, and has been the subject of stringent legal controls in France, the United Kingdom and several other developed states. Discharges of radioactive substances to the sea - a major cause of public disquiet - are also being reduced (GESAMP, 1990).

Oil pollution continues to be a highly visible form of marine pollution, and is liable to foul beaches, taint shellfish and other food species and make them unfit to eat, and kill seabirds and marine mammals. There has been a welcome decline in the number of oil spillages at sea recorded by the International Tanker

Figure 9: *Concentrations of PCBs in guillemots' eggs fell during the 1970s but there has been no further decline since 1984. PCB levels in the eggs are ten times higher than those found in herring, the guillemot's main food. This is a clear example of biomagnification, or concentration along a food chain.*

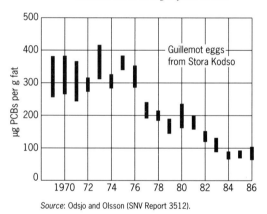

Source: Odsjo and Olsson (SNV Report 3512).

Figure 10: *The incidence of oil spills resulting from tanker accidents.*

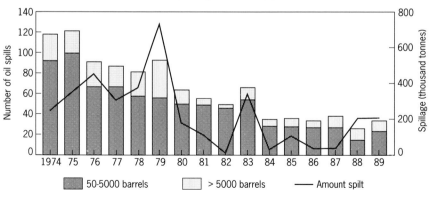

Source: GESAMP (1990), after El-Hinnawi (1991) based on data by Walder (1990) and UNEP (1991).

BOX 3

Oil inputs and impacts.

- Some 3.2 million tonnes of oil reach the sea annually. Most of this is derived from marine transportation (34.3%) and from various land-based discharges (34.4%). Atmospheric fallout accounts for 9.4% and offshore oil products for 1.6%. The remainder is from natural sources.

- An emerging problem is the decommissioning and disposal of obsolete offshore installations. Total removal can be extremely expensive but there is a potential hazard to shipping and fishing operations from partial removal.

- Accidents release oil to the seawater, affecting living resources, especially bird-life, and amenities. Large spills are dramatic and highly visible but the effects are usually local and relatively short-term.

- Offshore installations discharge oil and related contaminants which can be detected in nearby water and sediments. Adverse effects are restricted to a few kilometres round the source.

- The main global impact is due to operational discharges from shipping. These generate slicks and tar balls which, though generally harmless to marine organisms, may foul beaches and interfere with recreational activities, sometimes with major economic consequences in tourist areas.

- Interference with fishing operations by rigs, pipelines and debris can be of more concern than oil itself.

- Hydrocarbons from crude and refined oils are found in fish and shellfish but the impact of these substances on people eating unsmoked seafood is negligible, and the great sensitivity of the human palate prevents the consumption of badly tainted produce.

Sources: GESAMP (1990); A.D. McIntyre in *Environmental Protection of the North Sea* (ed. Newman and Agg), Heinemann (1988).

Owners Pollution Federation in the 1980s, from an annual average of 670 events in the first 5 years to 173 in the last 5 years (Figure 10 and Box 3).

World fisheries are discussed in Chapter 11. The environmental impacts of fishing and mariculture are given in Box 4.

BOX 4

Impacts of fishing and mariculture on the environment.

- The exploitation of living resources has physical and biological consequences for the marine environment.

- The seabed is damaged by the heavy gear used in trawling, and attached plants and animals including corals and sponges are destroyed.

- Removal of large numbers of commercially important species alters the natural ecological balance in the sea. In the Antarctic, the overfishing of whales made their main food, the krill, available to other predators so that seabirds and various species of seals increased. Similarly, deaths of penguins round the Falkland (Malvinas) Islands are attributed to overfishing of squid, and mortalities of seabirds in the north-eastern Atlantic to increased fishing of small species for oil and meal.

- Pond construction for the culture of fish and shrimps is leading to the destruction of coastal habitats around the world, for example by the cutting down of mangroves, and is seriously affecting nursery grounds for inshore species.

- Adverse effects from the culture of fish in netting and cage enclosures includes eutrophication and oxygen reduction from faeces and unused food, and pollution from food additives, therapeutic agents and antifoulants used on structures.

Source: GESAMP (1990).

Responses

The seas and oceans are classic examples of 'open access' resources, and international responses to their exploitation and conservation have gone through three stages:

- Initial international and regional agreement to regulate exploitation, through conventions and other agreements.

- Subsequent extension of coastal state authority through the creation of 200-mile (320km) exclusive economic zones (EEZs).

- Subsequent comprehensive management of the resource through an ecosystem approach which seeks to optimize sustainable resource use on the basis of sound scientific information.

International action to regulate marine pollution began at the time of the Stockholm Conference with legal instruments designed to control the dumping of wastes at sea. An Intergovernmental Intersessional Working Group on Marine Pollution established by the Preparatory Committee for the Conference elaborated the draft text of a Convention later adopted as the London Dumping Convention of 1972. Regional Conventions for the control of dumping in the north-east Atlantic (the Oslo Convention) and of discharges from land-based sources in the same region (the Paris Convention) followed (see Chapter 22).

Various measures have been taken to control marine pollution (Box 5). They range from isolated national actions to control pollution in specific sites from easily identifiable sources, to measures to curb pollution at regional levels and global approaches to controlling pollution through the general provisions of international agreements. Historically, international marine agreements dealt with the regulation of navigation and fishing: only recently has it been recognized that the world oceans should be regulated and protected as a natural resource. This important change from a 'user-oriented' to a 'resource oriented' approach became most marked in the last two decades. Most legal regimes adopted since 1970 encompass the protection, conservation and management of the marine and coastal environments and their resources (UNEP, 1991a, 1991b).

Although the importance of reducing maritime sources of ocean pollution had led to action in the 1960s, it was not until the early 1970s that land-based activities were recognized as the most significant source of marine pollution.

BOX 5

On the right track.

- Incineration of chemical wastes at sea has been practised since the late 1960s. Between 1981 and 1984, European countries burned about 624,000 tonnes of wastes at sea. In 1987, eight North Sea countries agreed to reduce waste incineration and to phase it out altogether by 1994 (WRI, 1988).

- A number of countries have dumped low-level radioactive waste in the sea. Between 1967 and 1982, about 94,000 tonnes of nuclear waste were dumped in the Atlantic Ocean. However, dumping of radioactive waste has been halted since 1982 (OECD, 1985; GESAMP, 1990).

- The 13th Consultative Meeting of the London Dumping Convention (1972) agreed that the dumping of all industrial wastes in the sea, except those of an inert nature, should be terminated by 31 December 1995.

It was also recognized that effective protection of the marine environment can be achieved only through international co-operation. This led to the adoption by UNEP of the concept of 'regional seas' in the mid-1970s.

Under the catalytic and co-ordinating role of UNEP, the Mediterranean states agreed in 1975 on an Action Plan for the Protection of the Mediterranean Sea against Pollution (MAP). In the following year, the Barcelona Convention for the Protection of the Mediterranean Sea Against Pollution, and two protocols, were signed. In the same year, a regional oil-combating centre was established in Malta as part of the MAP. In 1979 a 'Blue Plan' for the long-term management of the Mediterranean Sea was launched as part of the socio-economic component of the MAP. It was intended to integrate development plans with environmental protection measures in the Mediterranean Basin. In 1980, the Mediterranean states moved a step forward by adopting the Protocol for the Protection of the Mediterranean Sea Against Pollution from Land-Based Sources. This agreement identified measures to control coastal pollution from municipal sewage, industrial wastes and agricultural chemicals. Two years later, the Mediterranean governments also approved a protocol providing special protection for endangered species of fauna and flora as well as critical habitats. In 1985, the Mediterranean countries established ten priority targets for the decade 1985-95.

In addition to the MAP, action plans for eight other regions have also been adopted: these plans cover Kuwait (1978); the Wider Caribbean (1981); West and Central Africa (1981); Eastern Africa (1985); the South-East Pacific (1981); the Red Sea and the Gulf of Aden (1982); the South Pacific (1982) and the East Asia Seas (1981). An action plan was recently drafted for the South Asia region, and is under consideration for approval by the governments concerned. Action plans for the Black Sea and the North-West Pacific regions are also being developed. All in all, the regional seas programme involves some 130 countries, 16 UN agencies and more than 40 other international and regional organizations, all working with UNEP to protect and improve the marine environment and make better use of its resources (see Hinrichsen (1990) for a description of the regional and international seas programmes; Batisse (1990) for a description of the Mediterranean Action Plan).

In spite of the various efforts to protect the marine environment, progress has been rather slow, especially in the developing regions. The capabilities of most developing countries are still generally insufficient to cope adequately with the assessment of problems facing their marine and coastal environments and the rational management of their resources. Weak institutional structures hamper the effective participation of many countries in international efforts to protect and develop the marine and coastal environment, while the limited resources available to them place further constraints on their ability to respond to emergency situations such as accidents involving ships, and to combat the ensuing environmental threats. For countries that lack the necessary material resources and trained human resources, international agreements are consequently of limited use. Initiatives to improve this situation include the recently adopted International Convention on Oil Pollution Preparedness Response and Co-operation (OPRC Convention, 1990) which contains a mandatory requirement for oil pollution contingency plans.

The 1990 GESAMP report emphasizes that strong co-ordinated national and international action should be taken now to prevent rapid deterioration of the marine environment. At the national level in particular, the concerted application of measures to reduce discharges into the sea and to manage coastal areas in a rational and environmentally sound way is essential.

Concluding remarks

As the growth of world population continues, the movement of people and industry to the coastal zones will increase, with consequent intensification of competition for resources and space. But the coastal zones also receive inputs from the hinterlands, and in turn impact on seas and the open ocean. The need for integrated ecosystem management of the coastal zones is therefore essential.

In the coming decade, preparation must be made for the impact of the increasing population and in particular for the associated increase of effluents. The monitoring initiated by UNEP in its Regional Seas programmes, and by other agencies, will need to be extended to ensure global coverage, and the possibilities of open ocean monitoring, even if on a limited scale, should be re-examined. Chemical measurements are now being supplemented by the monitoring of biological effects on marine organisms, especially bivalves (mussels) which have been found useful as indicators and integrators of certain marine contaminants. This approach should be intensified since it offers the most relevant assessment of impacts, and the search should continue for indices of possible long-term effects arising from widespread but low level and subtle deterioration of the environment.

Oil spills, whether from wellhead blowouts or shipping accidents, represent the most dramatic, if localized, marine pollution incidents arising from human activities. Much can be done to prevent accidents and to reduce damage when they do occur, but the *Exxon Valdez* grounding in Alaska in 1989, and the massive oil spills in the Gulf in 1991, demonstrated that lessons have yet to be learned. Among other things, dangerous tanker routes should be identified and sensitive coasts protected. The arrangements made at Sullom Voe in the Shetlands, the largest oil terminal in Europe, could be usefully studied (Box 6). In the final analysis it is only through development and enforcement of national legislation and environmental laws in accordance with the regional and international agreements that effective protection of marine and coastal environments will be achieved.

Much of the damage to the marine environment caused by human activities, including the reduction of stocks by overfishing and the alteration of ecosystems by excess nutrients and oil spills, is reversible (Jernelöv, 1990). Studies should

BOX 6

Environmental protection at the Sullom Voe oil terminal, Shetland.

- Europe's biggest oil and liquified gas terminal at Sullom Voe in the Shetland Islands handles products from more than a dozen large North Sea oil fields.

- Shipping activities are meticulously controlled. Trained marine pilots bring ships in and out of the terminal, and a detailed port control system is in operation. Existing international law governing oil tankers is rigorously enforced.

- Coastal areas likely to be at risk from oil spills have been identified, and arrangements have been made for the rapid deployment of protection booms and for a wide range of appropriate clean-up methods.

- A permanent body, SOTEAG (Shetland Oil Terminal Environmental Advisory Group) was set up with authority to investigate any environmental problem associated with oil operations. It is chaired by an independent academic and composed of representatives of oil companies, local authorities, central government, scientists, NGOs and wildlife groups.

- A major aspect of SOTEAG's work is continuous surveillance of environmental conditions by a comprehensive monitoring programme which includes chemical analysis, biological sampling and wildlife surveys.

Source: Proc. Roy Soc Ed. 80B (1981).

be promoted on the recovery of damaged communities and on how this can be accelerated.

Further, relevant research should be initiated and supported on new treatments for sewage sludge and on biodegradable plastics. The sea should not be regarded as a bottomless sink or as an easy option for waste disposal. It should be utilized only if it can be shown that better environmental alternatives are not available.

The following recommendations are also made:

- *Coastal zones*: A major effort is needed to control land-based sources of marine pollution and to establish mechanisms for integrated coastal zone management if the rapid deterioration of coastal areas is to be halted and eventually reversed. There is also a need for more effective monitoring programmes. In particular, more attention should be given to the mapping and monitoring of coastal and nearshore marine species and communities, identification and explanation of changes occurring, and assessment of biodiversity as one of the elements for ecological management of coastal areas. There is also a need for systematic documentation of coastal flooding, storm surges and their effects.

- *Projected responses to climate change*: Both warmer temperatures and higher sea levels will have major impacts in the coastal zones. Except where major engineering works can be afforded, coastal realignment is bound to take place and this could be on a massive scale. In Florida, for example, a 50-cm sea-level rise could lead to a 15-km coastal retreat in some areas, while mangrove woodlands would replace the present forests in much of the Everglades (Snedaker and Parkinson, 1985). In Guyana, 90 per cent of the population live on the coastal plain behind slender sea defences which would have to be strengthened if any significant sea-level rise occurred (Commonwealth, 1989).

On many coasts the response of natural ecological systems will be crucial. Corals appear to have a potential for maximum upward growth of around 8mm per year (less in cooler waters and near their lower depth limit) (Hopley and Kinsey, 1985). There is an expectation that coral growth will revive and become extensive on flat reefs in response to a global sea-level rise (Neumann and McIntyre, 1988), but although changes on coral reefs were documented locally during 1970-90, there is no present evidence of any overall trend. On the other hand, reports of coral die-back ('bleaching'), which is said to occur when sea temperatures rise to around 30°C, have been widespread, especially in 1979-80, 1982-83, and 1986-87. This has been interpreted as an early response to global warming, with maxima possibly related to the El Niño Southern Oscillation (Williams *et al.*, 1991; Glynn and de Weerdt, 1991) although this is disputed by Ogden (1991). There have also been reports of the spread of corals poleward beyond their previous limits (for example, south to Rottnest Island off Perth, on the west coast of Western Australia) which could also be a response to rising temperatures.

Mangroves are confined to frost-free areas, and a warming climate would extend their range. Their capacity for upward growth depends on sediment accretion but they should be able to keep pace with sea-level rises of at least 8cm and possibly 10-25cm per century (Snedaker and Parkinson, 1985; Ellison, 1989). There are no signs of changes in distribution at present, but the direct human impacts on mangrove distribution are likely in any event to be overwhelming.

Thinning of the stratospheric ozone layer is a related climate change that could cause a reduction in biomass production, as well as changing species composition and biodiversity in the world oceans (UNEP, 1989) (see also Chapter 2).

REFERENCES

Arnaudo, R. (1990) The problem of persistent plastics and marine debris in the oceans. *In* GESAMP: Technical annexes to the report on the *State of the Marine Environment*, UNEP Regional Seas Reports and Studies No. 114/1:1-20, UNEP, Nairobi.

Batisse, M. (1990) Probing the Future of the Mediterranean, *Environment*, **32**, 4-9, 28-34.

Bird, E. (1985) *Coastline Changes*, Wiley, Chichester.

Commonwealth Secretariat (1989) Climate Change, Meeting the Challenge. Report by a Commonwealth Group of Experts. Commonwealth Secretariat, London.

DOE (1987) Quality Status of the North Sea. A Report by a Scientific and Technical Working Group for the Second Ministerial Conference on the North Sea. London: HMSO for Department of the Environment.

Ellison, J. (1989) Report provided to Commonwealth Secretariat and cited in Commonwealth, 1989.

ENS (1991) *Environment, Northern Seas*. Oslo: Norwegian University Press.

GESAMP (1990) *The State of the Marine Environment*, UNEP Regional Seas Report No. 115, also published by Blackwell, Oxford.

Glynn, P.W. and de Weerdt, W.H.(1991) Elimination of two reef-building hydrocorals following the 1982-83 El-Niño warming event, *Science*, **253**, 69-71.

Hinrichsen, D. (1990) *Our Common Seas: Coasts in Crisis*, Earthscan Publications Ltd. London, in association with UNEP, Nairobi.

Holdgate, M.W., Kassas, M. and White, G.F. (eds) (1992) *The World Environment, 1972-1982*. A Report by UNEP. Tycooly International, Dublin.

Hopley, D. and Kinsey, D.(1988) The effects of a rapid short-term rise on the Great Barrier Reef. *In* Pearman, G.I.(ed) *Greenhouse - Planning for Climate Change*. CSIRO, Australia.

IUCN/UNEP/WWF (1991) *Caring for the Earth: A Strategy for Sustainable Living*. IUCN, Gland, Switzerland.

Jernelöv, A. (1990) Recovery of damaged ecosystems. *In* GESAMP 385 - State of the Marine Environment, Technical annex, UNEP Regional Seas Report and Studies No. 114/2, pp.385-402.

McIntyre, A.D. (1988) Pollution in the North Sea from oil-related industry - an overview. *In Environmental Protection of the North Sea* (ed. Newman and Agg), Heinemann, London.

McNeely, J. (1988) *Economics and Biological Diversity*. IUCN, Gland, Switzerland.

Neumann, A.C. and McIntyre, A. D. (1988) Reef response to a sea-level rise: catch up, keep up or give up, *Proc. 5th International Coral Reef Congress*, 105-110.

Ogden, J.C. (1991) Premature death, *Nature* **353, 509**. Oslo/Paris 1984. The Oslo and Paris Commissions: The First Decade. London: Oslo and Paris Commissions.

Peltier, W. R. and Tushingham, A, M. (1989) Global sea-level rise and the Greenhouse Effect: might they be connected? *Science*, **244** (4906): 806-10.

Precoda, N. (1991) Requiem for the Aral Sea, *Ambio*, **20**, 109-14.

Snedaker S. and Parkinson, R.W. (1985) Potential effects of climate change on mangroves, with special reference to global warming and sea-level rise. (mimeo). University of Miami, School of Marine and Atmospheric Sciences.

UNEP (1989) Environmental Effects Panel Report pursuent to Article 6 of the Montreal Protocol, UNEP, Nairobi, 64 pp.

UNEP (1991a) Status of regional agreements negotiated in the framework of the Regional Seas Programme, Rev.3, UNEP, Nairobi.

UNEP (1991b) Register of International Treaties and Other Agreements in the Field of the Environment, UNEP, Nairobi.

UNEP (1991c) *Environmental Data Report*. Third Edition, 1991-92. Basil Blackwell, Oxford.

Walker, H.J. (ed) (1988) *Artificial Structures and Shorelines*, Kluwer Academic Publications, The Netherlands.

Williams, L. B. and Williams, E. H. (1988) Coral reef bleeding: current crises, future warming, *Sea Frontiers* **34**: 80-87.

WRI and IIED (1986) *World Resources 1986*, Basic Books, New York.

CHAPTER 6

Land degradation

The issue

Degradation of drylands (desertification) and of more humid areas is a long-standing and increasingly severe problem in many parts of the world. The global store of arable and grazing land continues to decline through urbanization, unsustainable agricultural practices and deforestation, while a significant portion of the remaining arable and grazing land is under considerable pressure from compaction by livestock and farm implements; over-use of fertilizers and pesticides; salinization, alkalinization or acidification; depletion of nutrients; water and wind erosion, and deterioration of drainage.

Three main issues are involved:

(a) The *management issue* is the need to employ integrated strategies for sustainable land management. Rural village societies have traditionally taken this integrated approach, whereas more conventional land-use strategies tend to focus on single problems (c.g. soil erosion) or single objectives (e.g. food production).

(b) The *political issue* is the failure of governments and other bodies to implement sufficiently the Plan of Action to Combat Desertification, adopted by the United Nations Conference on Desertification in 1977, as well as the World Soil Charter adopted by FAO in 1981 and the World Soils Policy adopted by UNEP in 1982.

(c) The *scientific issue* is the need for better information on the state of the world's land surfaces, using an integrated approach based on satellite and ground measurements.

This chapter describes the present status and recent trends of land degradation, and the policy responses that have been undertaken to combat the problem. The chapter is divided into two parts: *desertification* and *other forms of land degradation*.

Causes of land degradation

Land degradation is the result of complex interactions between physical, chemical, biological and socio-economic and political issues of local, national and global nature. Although often overlooked, threats to productivity and thus to the physical, chemical and biological stability of the land are closely linked to national and international economic policies. Socio-economic conditions, and thus the political framework of land tenure, taxation and trade barriers, have been particularly disadvantageous for developing countries during the last two decades. The burden placed on individual land users in poor countries can be traced in part to international policies and markets, but the problems are also rooted in transitions in local privileges and rights and in domestic priorities, often favouring the urban consumer over the rural producer. Such problems are often compounded by political and economic mismanagement in the developing countries themselves. Development policies lack poverty orientation so that marginalized peoples often get little support in breaking the vicious circle that forces them to mismanage land. Women often fail to obtain bank loans and access to advisory services that could improve their land-use practices.

Most developing countries face high population growth rates and also high rates of urbanization. In some countries in Africa, more than half the population is urbanized (see Chapter 17). The growing numbers of urban dwellers require food. There is therefore a steady stream of soil nutrients (in the form of food, fuelwood and charcoal) moving from the countryside to the towns, to end up as useless, often polluting, waste. This rapid transition from rural to urban societies has not been matched by equally rapid replenishment of soil nutrients.

Demands on production have increased the pressure on existing productive land and moved the limits of production onto increasingly marginal lands. Increased use of the world's drylands for cropping and grazing means increased dependence on rain-fed agriculture and rangelands in regions where rainfall is not only low but also highly variable. A run of dry years, as experienced in the Sahel in the seventies and early eighties, was followed by a long period of favourable rainfall when cropping and high stocking rates became common in areas previously little used. As the drought persisted, productivity fell but food demands continued to grow with growing populations. Famine resulted. As the vegetation wilted or was eaten by livestock and game, the land became denuded and open to the erosive forces of wind and water. Although the African drylands have shown remarkable resilience, returning more rapidly than was expected to productive states with subsequent wetter years, they remain vulnerable and will doubtless be subject to more droughts and famines.

The most productive forest lands in most continents have already been brought into agricultural production. Further expansion of agriculture and grazing now takes place on marginal land, often on steep slopes or on soil of poor physical structure or low inherent fertility. Wise land-use practices have yet to be developed for such conditions. Agricultural expansion onto these lands therefore often results in rapid land degradation, with a subsequent decline in production. The clearing of marginal forests and hills for food production or for timber also challenges the maintenance of biological diversity. There is an urgent need to develop land-use practices that will not lead to the type of land degradation now experienced in the wet tropics, and that will encourage restraint in clearing forest land. But as with marginal drylands, it is often hunger that causes agricultural encroachment by marginalized farmers. Unless adequate livelihoods can be created, e.g. through further intensification in fertile areas or by off-farm employment, there is little political realism in trying to stop agricultural encroachment on marginal forested land. Shortening of rest periods in traditional shifting cultivation systems has had a similar impact.

An important contribution to world food production comes from irrigated lands in the dry, semi-humid and humid tropics. Traditional irrigation practices, e.g. for paddy rice, have been maintained successfully over millennia in Asia, but recent large-scale expansion of irrigated agriculture in Asia and Africa has met with increasing problems, as waterlogging and salinity or alkalinity have built up.

Desertification

The term desertification was coined by the United Nations General Assembly when it decided to convene a conference on the subject in the wake of several years of harsh drought and famine in Africa, particularly in the Sahel region. The UN Conference on Desertification (UNCOD), convened in 1977 in Nairobi, adopted the UN Plan of Action to Combat Desertification (UN/PACD) and identified the financial resources needed to implement the plan (UN, 1977). At that time, the estimated rate of desertification was six million hectares loss per year from drylands. In 1984, the Executive Director of UNEP reported to its Governing Council on the assessment of the status of desertification and the implementation of the PACD (UNEP, 1984). The assessment showed very little progress in implementation and that the estimated rate of desertification remains the same (six million hectares/year). In 1991, UNEP again reviewed the current status of desertification and implementation of the PACD and its financing. The section on desertification in this chapter is largely based on the findings of this last assessment (UNEP, 1991).

Desertification has been defined as land degradation in arid, semi-arid and dry sub-humid areas resulting mainly from adverse human impacts (UNEP, 1991). Drylands, excluding hyper-arid areas, are defined as those having precipitation over evapotranspiration ratios between 0.05 and 0.65. Desertification is the main environmental problem of arid lands, which occupy more than 40 per cent of the total global land area (Figure 1). At present, desertification threatens about 3.6 billion hectares - 70 per cent of potentially productive drylands, or nearly one-quarter of the total land area of the world. These figures exclude natural hyper-arid deserts. About one-sixth of the world's population is affected.

Despite the severity of the problem, global statistics are poor on recent trends in the extent of the world's deserts and desertifying areas. There is, however, good information on particular regions. The most obvious symptoms of desertification relate to a reduction in the biological and economic productivity/ value of a piece of land. They include:

- reduction of crop yields (or complete failures) in irrigated or rain-fed farmland;

- reduction of biomass produced by rangeland and consequent depletion of feed material available to livestock;

- reduction of available wood biomass, and consequent extension of the distance to be journeyed to obtain fuelwood;

• reduction of available water due to decreases in river flow or groundwater resources;

• encroachment of sand bodies that may overwhelm productive land, settlements or infrastructures, and

• societal disruption due to deterioration of life-support systems, and the associated need for outside help (relief aid) or for seeking haven elsewhere (environmental refugees).

Figure 1: *Arid areas of the world.*

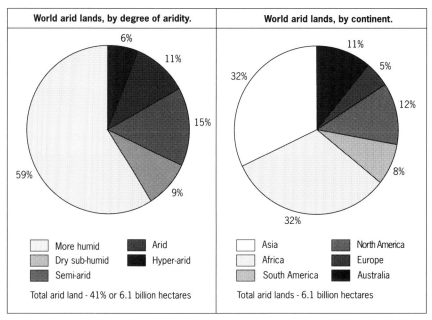

Source: UNEP (1991).

Causes of desertification

Desertification is caused by a combination of human exploitation that oversteps the natural ecological potential of the land, and the inherent ecological fragility of the resource system. Desertification is the result of excessive land-use, sometimes triggered or intensified by a naturally occurring drought. There are three principal land uses in drylands: irrigated farming, rain-fed agriculture and pastoralism. Other vegetation-related uses of the land include wood cutting, and collecting special plants for their medicinal and other values. To these uses should be added the hunting of game animals. Excessive human pressures on the resource base relate to:

• increases in human numbers and escalation of human needs;

- socio-political processes that impose pressures on rural communities to orient their production towards national and international markets;

- socio-economic processes that reduce the market value of rural products and escalate the prices of imports needed by rural people;

- processes of national development (especially programmes for the expansion of farmlands for production of cash crops and meat for export) that exacerbate conflicts of land and water use and often reduce areas available to marginalized communities; and

- political pressures that prevent seasonal migrations of pastoralists across provincial or national boundaries.

Present status of desertification

The latest assessment of the world status of desertification in drylands (Table 1) is provided by UNEP (1991).

Table 1: *The extent of desertification in the world's drylands*

		Million hectares	Percentage of total drylands
1.	Degraded irrigated lands	43	0.8
2.	Degraded rain-fed croplands	216	4.1
3.	Degraded rangelands[1] (soil *and* vegetation degradation)	757	14.6
4.	Drylands with human-induced soil degradation (1+2+3)	1016	19.5
5.	Degraded rangelands (vegetation degradation without recorded soil degradation)	2576	50.0
6.	Total degraded drylands (4+5)	3592	69.5
7.	Non-degraded drylands	1580	30.5
8.	Total area of drylands excluding hyper-arid deserts (978 million ha) (6+7)	5172	100.0

Note: 1 The term 'rangelands' in the context of desertificaion assessment includes both permanent grazing lands and other, mostly unused, drylands.
Source: UNEP (1991).

Irrigated drylands

The largest areas of such degraded lands are situated in the drylands of Asia, followed by North America, Europe, Africa, South America and Australia in descending order (Figure 2). About 43 million hectares of irrigated lands or 30 per cent of their total area in the world's drylands (145 million ha) show evidence of soil deterioration, mainly waterlogging and resulting salinization and/or alkalinization.

Figure 2: *The extent and degree of desertification in irrigated areas of the world's drylands*

Extent of desertification/land degradation in irrigated areas within the drylands of the world (thousand ha).

Continent	Slight to none	Moderate	Severe	Very severe	Total moderate, severe and very severe	Per cent desertified
Africa	8,522	1,779	122	1	1,902	18
Asia	60,208	24,335	5,788	1,690	31,813	35
Australia	1,620	100	130	20	250	13
Europe	9,993	1,340	460	105	1,905	16
N. America	15,007	4,930	730	200	5,860	28
S. America	6,998	1,047	310	60	1,517	17

Degree of degradation of irrigated lands within the world's drylands.

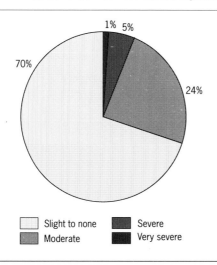

Source: UNEP (1991).

Irrigated lands in drylands constitute nearly 62 per cent of the total irrigated area of the world (240 million ha). Soil scientists estimate (IUCN/UNEP/WWF, 1991) that the world is now losing annually about 1.5 million hectares of irrigated lands due mostly to salinization, mainly in drylands.

Rain-fed croplands

Nearly 216 million hectares of rain-fed croplands or about 47 per cent of their total area within the world's drylands (457 million ha) are affected by various degradation processes, mainly water and wind erosion, depletion of nutrients, and physical deterioration (Figure 3). Rain-fed cropland in drylands constitutes nearly 36 per cent of its total area in the world (1260 million ha). An estimate has been given (IUCN/UNEP/WWF, 1991) that the world is losing annually about 7.8 million hectares of croplands due to various processes of soil degradation, mainly erosion and urbanization, more than half of it in the drylands. Therefore, about 3.5-4.0 million hectares of rain-fed croplands are currently lost every year throughout the world's drylands, being compensated by cultivating high-quality rangelands, the area of which decreases accordingly.

Figure 3: *The extent and degree of desertification in rain-fed croplands of the world's drylands*

Extent of desertification/land degradation in rain-fed croplands within the drylands of the world (thousand ha).

Continent	Slight to none	Moderate	Severe	Very severe	Total moderate, severe and very severe	Per cent desertified
Africa	30,959	43,187	5,153	523	48,863	61
Asia	95,890	100,638	18,578	3,068	122,284	56
Australia	27,800	13,900	400	20	14,320	34
Europe	10,252	8,538	3,227	89	11,854	54
N. America	62,558	10,770	721	120	11,611	16
S. America	14,711	5,950	561	124	6,635	31

Degree of degradation of rain-fed croplands within the world's drylands.

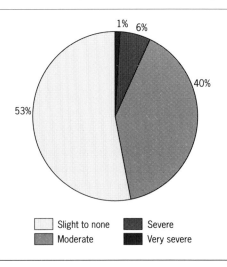

Source: UNEP (1991).

Rangelands

The situation in the rangelands of the world drylands is dramatic (Figure 4). Seventy-three per cent of their total area is affected by desertification, mostly by degradation of vegetation, although on almost a quarter of the affected area, vegetation degradation is accompanied by soil degradation, mostly erosion by wind and water. The largest area of rangelands affected by desertification is in Asia, followed by Africa.

There are no reliable data on actual losses of rangelands and their conversion into agricultural land, wasteland/badland/desert or urban lands. However, if

Figure 4: *The extent and degree of desertification in the rangelands of the world's drylands*

Extent of desertification in rangelands within the drylands of the world (thousand ha).

Continent	Slight to none	Moderate	Severe	Very severe	Total moderate, severe and very severe	Per cent desertified
Africa	347,265	273,615	716,210	5,255	995,080	74
Asia	383,630	485,221	691,602	10,787	1,187,610	75
Australia	295,873	277,040	55,310	29,000	361,350	55
Europe	31,053	27,372	51,937	1,208	80,517	72
N. America	71,987	116,102	284,858	10,194	411,154	75
S. America	93,147	88,007	184,431	15,316	297,754	76

Degree of degradation of rangelands within the world's drylands.

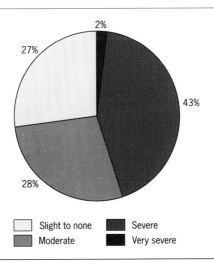

☐ Slight to none	■ Severe	
■ Moderate	■ Very severe	

the above estimates are correct, then annual losses of rangelands within the drylands could be estimated to be of the order of some 4.5-5.8 million hectares, and could be much more if unaccounted sand encroachment is considered.

Trends in desertification

Global statistics on trends in desertification are scanty. In the recent UNEP assessment (UNEP, 1991), for example, it is stated that 'accurate measurement of changes in areas of lands affected by desertification during 1977-91 at global or continental scales is not attainable as the observed changes will fall within the range of standard error'. However, estimates of trends are possible for areas where detailed assessments have been made at national or local levels (Box 1, Box 2 and Figure 5).

UNEP has recently published a *World Atlas of Desertification*, which shows the interactions amongst the variables that determine desertification (UNEP, 1992).

BOX 1

What is happening in the Sahara?

A lively scientific discussion is going on about what is happening at the southern fringes of the Sahara in a belt that is 200 kilometres wide. The dispute is not new, since the question of expansion and contraction of world deserts due to variation in rainfall has been recognized for almost a century now. Even with recent satellite data interpretation techniques, scientists are unable to reach a common perception. Some references to the most important recent publications will illustrate the point.

Scientists in the USA (Tucker and Justice, 1986; Dregne and Tucker, 1988; Tucker et al., 1991) used the analysis of vegetation productivity as revealed by interpretation of satellite imagery to show that the southern border of the Sahara was oscillating during the 1980s within a range of 130 kilometres. Their latest main conclusion is that a decade-long study is needed to determine whether long-term expansion or contraction is occurring.

British scientists, at the University of East Anglia and the Chiltworth Research Centre in Southampton, concluded that changes in the climate of Africa are causing the continent's deserts and drylands to expand faster than expected. They estimated that as much as 63 per cent of the African continent has become drier between 1931 and 1990, as rainfall in the African Sahel, the semi-arid area south of the Sahara, has declined by up to 30 per cent over this period. M. Hulme (1991) estimates that the areas classified as arid and hyper-arid have increased

---

An interesting indicator related to desertification is per capita food production in arid and semi-arid regions. Figure 6 gives the results for selected countries in Africa for the years 1980, 1985 and 1990 (UNEP, 1991), showing that per capita agricultural production has been stagnating or even declining during the last decade. Similarly, the average annual growth of GNP per capita, which in sub-Saharan Africa increased at 3 per cent between 1965 and 1973, fell by 2.8 per cent between 1980 and 1986, by 4.4 per cent in 1987 and by 0.5 per cent in 1989.

Responses

Combating desertification

In 1977 the United Nations Conference on Desertification adopted the Plan of Action to Combat Desertification (PACD) which was endorsed by the UN General Assembly in the same year. The world-wide programme was directed to stop the process of desertification and to rehabilitate affected lands. However,

by almost 54 million hectares since 1931 and that the humid zone has lost a total of 26 million hectares.

Opposing these perceptions, Swedish scientists (e.g. Hellden, 1991) working for several years in the eastern part of the southern fringes of the Sahara, using a combination of satellite data, ground observations and national food production statistics, found no evidence of long-lasting desert conditions over the period 1962-84. Changes in vegetation cover and crop productivity were identified, but could be explained simply by variations in rainfall.

However, a team of German and African scientists working in the western part of the southern fringes of the Sahara concluded that in the southern parts of Mauritania, Mali and Niger, the desert advance to the south over the period 1961-87 was of the order of 10km/year (CILSS/PAC, 1989).

The only conclusion is that within a sparsely populated belt of some 200km at the southern fringe of the Sahara, biological productivity changes from year to year according to annual rainfall fluctuations. In areas where the soil was destroyed, the decline of biological productivity would be permanent.

As the scientific debate goes on, human-induced land degradation (desertification) of the world drylands continues to threaten the resource base in arid, semi-arid and dry sub-humid areas world-wide.

BOX 2

Some examples of recent desertification trends.

Kenya   At Lake Baringo, an area of 360,000 hectares, the annual rate of land degradation/desertification between 1950 and 1981 was 0.4%. At Marsabit, an area of 1.4 million hectares, it was 1.3% for the period 1956 to 1972.

Mali   In the three localities of Nara, Mordiah and Yonfolia, with a total area of some 195,000 hectares, the average annual rate of land loss during the last 30 to 35 years was of the order of 0.1%.

Tunisia   The annual rate of desertification during the last century was of the order of 10% and about 1 million hectares were lost to the desert between 1880 and the present.

China   The present average annual rate of desertification/land degradation for the country is of the order of 0.6%, while in such places as Boakong County, north of Beijing in Hebei Province, it rises to 1.3%, and to 1.6% in Fengning County.

USSR   The annual desertification/sand encroachment rate in certain districts of Kalmykia, north-west of the Caspian Sea, was recently estimated as high as 10%, while in other localities it was 1.5% to 5.4%. The desert growth around the drying-out Aral Sea was estimated at about 100,000 hectares per year during the last 25 years, which gives an average annual desertification rate of 4%.

Syria   An annual rate of land degradation of 0.25% was found in the 500,000 hectare area of the Anti-Lebanon Range north of Damascus for the period 1958 to 1982.

Yemen   The average for the country's annual rate of cultivated land abandonment due to soil degradation has increased from 0.6% in 1970-80, to about 7% in 1980-84.

Sahara   A recent analysis, using a satellite-derived vegetation index, has shown steady expansion of the Sahara between 1980 and 1984 (an increase of approximately 1,350,000 square kilometres) followed by a partial recovery up to 1990 (Tucker *et al.*, 1991) See Figure 5.

Source: UNEP (1991).

Figure 5: *The expansion of the Sahara Desert, 1980-90, based on a recent analysis using a satellite-derived vegetation index.*

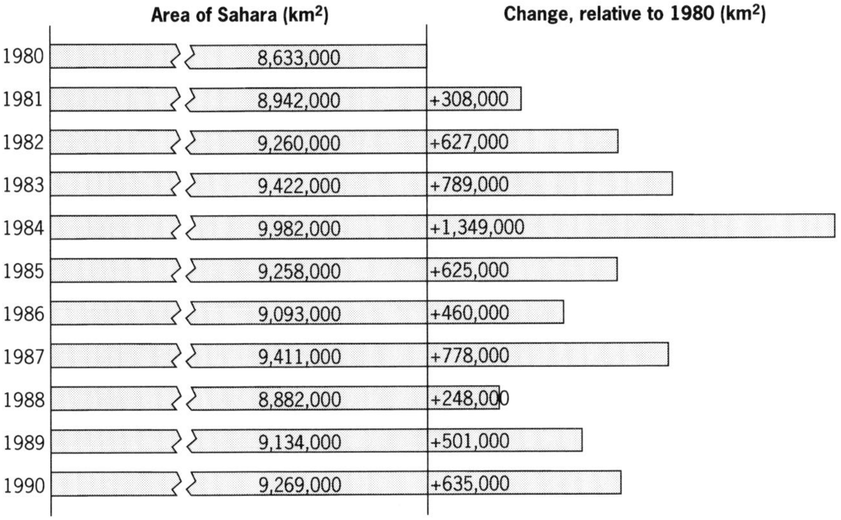

Year	Area of Sahara (km²)	Change, relative to 1980 (km²)
1980	8,633,000	
1981	8,942,000	+308,000
1982	9,260,000	+627,000
1983	9,422,000	+789,000
1984	9,982,000	+1,349,000
1985	9,258,000	+625,000
1986	9,093,000	+460,000
1987	9,411,000	+778,000
1988	8,882,000	+248,000
1989	9,134,000	+501,000
1990	9,269,000	+635,000

Source: Tucker *et al.* , Expansion and Contraction of the Sahara Desert from 1980 to 1990, *Science*, **253**, 299–301 (07.19.91. Copyright 1991 by the AAAS.

the efforts undertaken so far have not been adequate to cope with the magnitude of the problem, and desertification continues. Unfortunately, very little has been done at the local level. Civil strife over large parts of the regions affected has reduced the effectiveness of anti-desertification measures while contributing to environmental degradation, e.g. in Afghanistan, Angola, Chad, Ethiopia, Kuwait, Mozambique, Somalia and Sudan.

The principal causes of failure to implement the PACD in full were considered at several global and regional international fora, with the conclusion (UNEP, 1991) that:

- The funding and implementing agencies, both national and international, have not given priority to programmes for combating desertification, both nationally and internationally.

- Developing countries affected by desertification have been unable to cope with the problem without major external financial and technical assistance, but the necessary assistance has not been forthcoming.

- Desertification control programmes have not been fully integrated into programmes of socio-economic development and have been considered as rehabilitation measures only.

- Affected populations have not been fully involved in the planning and implementation of programmes for combating desertification.

Figure 6: *Results for some representative countries in Africa for the years 1980, 1985 and 1990 show that per capita production of food has stagnated or declined during the past decade, clearly demonstrating the deteriorating food situation in African countries affected by desertification. The upper part of the diagram shows the index of per capita food production taking 1980 as 100; the lower part shows the average annual rate of growth of per capita food production.*

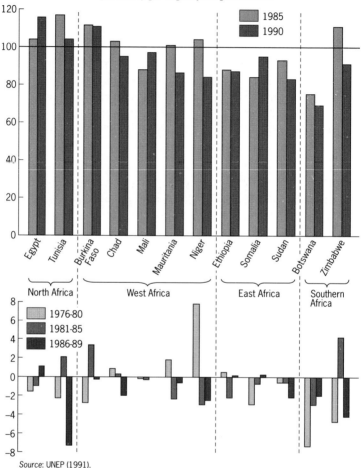

Source: UNEP (1991).

- Technical means have rarely been sought to solve these problems, leaving the solutions dependent on socio-political and socio-economic mechanisms.

A recent evaluation of the progress achieved in the implementation of the Plan of Action, and of the Plan itself, has been undertaken by UNEP (UNEP, 1991). This has resulted in the establishment of new goals, as follows:

- Recognizing that a substantial intensification of carrying capacities in rangeland can not be achieved, it is proposed to prevent deterioration of non-affected or slightly affected rangelands and to reclaim only 50 per cent of degraded rangeland as a first priority in combating desertification.

- There is an urgent need to stop soil degradation and loss of soil fertility in areas of rain-fed croplands, to reclaim them and to sustain their productivity. It is proposed to prevent degradation of non-affected or slightly affected rain-fed croplands and to reclaim 70 per cent of degraded rain-fed cropland as a first priority in combating desertification.

- There is evidence that both socio-economic and political returns from investments in rehabilitation of degraded irrigated agricultural lands, and their protection against further degradation, are the highest priority and would provide for the growth of food production to keep ahead of population growth; thus, it is proposed to prevent degradation of non-affected or slightly affected irrigated lands and to reclaim 100 per cent of degraded irrigated cropland as a first priority in combating desertification.

Despite the limited success of PACD, many actions to implement the plan have been undertaken by the United Nations system, by governments and inter-governmental regional organizations, and by peoples themselves. Several countries have adapted their national plans within the scope of the PACD implementation. Particularly significant measures have been undertaken in the countries of the Sudano-Sahelian belt of Africa, and in India, Iran, China and the former USSR. Much attention has been paid to institutional arrangements, dissemination of information, creating awareness, development of assessment methodology, and adaptive research. Many trees have been planted, partly as protective belts and partly for fuelwood; the agroforestry concept has been gradually introduced into drylands under suitable natural conditions; and shifting sands have been stabilized over large areas, particularly in China.

Costs of combating desertification

Combating desertification is costly and tends to become more expensive each year if not undertaken immediately. In 1980 it was estimated that the direct economic loss due to desertification (income foregone) was of the order of $US26 billion per year: 1990 estimates gave a corresponding figure of $US42 billion.

Similarly, according to 1980 estimates, it was necessary to spend some $US90 billion for a 20-year programme of direct corrective and rehabilitation measures in drylands that were desertified at least moderately. The corresponding figure for the 1990-91 assessment is $US270 billion, i.e., three times higher. This is due to the following factors: (a) the area affected is growing, (b) the degree of damage is increasing, and (c) world prices are rising, including the costs of land reclamation. If the cost of preventive measures is included, i.e. for the protection of 30 per cent of the drylands not affected by desertification but prone to it or only slightly affected (loss of up to 10 per cent of productivity in croplands and up to 25 per cent in rangelands), then the total cost of world-wide direct (preventive-corrective-rehabilitation) measures to combat desertification in world drylands will be of an order of $US200-448 billion for a 20-year programme

The following data (Table 2) provide a general idea of the costs involved, (in billion $US) calculated for a 20-year programme on a yearly basis and taking into account corresponding areas in world drylands (Rozanov, 1991; UNEP, 1991).

Table 2: *The costs of combating desertification (billion $US).*

	Preventive	Corrective	Rehabili- tation	Total
Total global cost	1.4 – 4.2	2.4 – 7.2	6.2–11.0	10.0 – 22.4
Cost to 18 countries not requiring assistance	0.6 – 1.8	1.0 – 3.0	2.4 – 3.0	4.0 – 7.8
Cost to 81 countries requiring external assistance	0.8 – 2.4	1.4 – 4.2	3.8 – 8.0	6.0 – 14.6

The above costs can be compared with annual economic losses (Table 3) in different categories of land use throughout world drylands, in billion $US (UNEP, 1991):

Table 3: *The economic losses caused by desertification (billion $US).*

	Annual income foregone due to desertification	Annual cost of preventive measures	Annual cost of corrective measures	Annual cost of rehabilitation measures	Total cost of direct anti-desertification measures
Irrigated croplands	10.8	0.5 – 1.6	0.9 – 2.5	1.0 – 2.0	2.4 – 6.1
Rain-fed croplands	8.2	0.6 – 1.8	0.9 – 2.8	1.1 – 3.0	2.7 – 7.5
Rangelands	23.3	0.3 – 0.9	0.7 – 1.9	2.0 – 6.0	5.0 – 8.8
Total drylands	42.3	1.4 – 4.2	2.4 – 7.2	6.2 –11.0	10.0 – 22.4

The above figures are just indicative, showing only an order of magnitude for the world as a whole. Naturally they vary between individual countries, groups of countries and continents depending on specific physio-geographic and socio-economic conditions. At the same time, they show clearly the magnitude of the problem of desertification and the difficulty of solving it, bearing in mind other world problems, needs and financial constraints.

Concluding remarks

The urgency of the desertification problem is accentuated by the following factors.

• The time for action is running short as desertification expands, threatening new areas and new societies, while anti-desertification measures tend to be long-term and time-consuming.

• The cost of anti-desertification measures escalates from year to year because the area affected is growing, the degree of damage is growing, and world prices for rehabilitative measures are rising.

• Off-site (and social) costs of desertification continue to increase.

• Other environmental and economic problems are likely to become serious, tending to distract the attention of international funding agencies to other issues (e.g. sea level rise).

• If the process of desertification is not arrested soon, the world shortage of food will increase dramatically.

It follows, therefore, that in revising the Plan of Action and outlining a new implementation strategy, aspects have to be emphasized that will provide more realistic achievement goals.

First, it is recognized that although the problem of desertification, as in the case of soil degradation under excessive human pressure, is always local and site-specific in its physical manifestations, it can only be solved by concerted international action within a framework of international co-operation because the people affected, being the poorest members of society, have no physical means of coping with the problem without external assistance.

Second, it is recognized that the problem of desertification can be solved only if addressed within the general programmes of socio-economic development of the countries concerned, with appropriate allocation of the resources available from both national and external sources.

Third, in order to solve the problem of desertification it will be necessary from the very beginning to involve the people concerned, particularly the land users at the local level - farmers, pastoralists, village communities and local governments. However, to involve these people it is imperative to provide them with incentives or alternative means of livelihood.

The problem of desertification is manageable in principle as can be seen by the examples given for some countries. Improvement of irrigation systems and water management practices; introduction of new advanced technologies, both local and imported, into rain-fed agriculture; provision of new alternative employment to local people; improvement of rangeland management and animal husbandry technologies; fixation of shifting sands and reafforestation of denuded lands - all these actions are technically feasible and practically implementable, but they need special political and economic actions with due consideration given to social aspects of the development process in order to be effective and sustainable. They also cost money. The financial support for a world-wide programme may include *inter alia* the following:

- national budgets;

- national private and co-operative state and local financial institutions;

- major international funding agencies such as the World Bank, IFAD, WFP and regional development banks;

- multilateral and bilateral aid programmes;

- loans from governments and world capital markets on a concessionary basis;

- reduction of external debts, or debt-for-nature swaps;

- the Global Environmental Facility of the World Bank/UNDP/UNEP;

- savings from disarmament; and

- specifically mobilized funds from the world community (e.g. user's fees on products extracted from drylands).

In financing the revised PACD, particular attention should be given to providing assistance to the 81 developing countries that cannot cope with the desertification problem by themselves. About 53 per cent of the global cost of combating desertification is related to these countries, and they would be able to cover from their own resources not more than half the cost: probably much less. International action on a true partnership basis is required to combat desertification.

Other forms of land degradation

Soil degradation

Soil degradation is defined as the lowering of current and/or future soil capability to produce goods or services (ISRIC/UNEP, 1988). Desertification is one cause of soil degradation but there are many others. Recognizing the seriousness of the problem, many steps have been taken to obtain a better global picture of soil degradation (Box 3) and to devise policies to reverse current trends. The data in this chapter are based on the Global Assessment of Soil Degradation (GLASOD) carried out by ISRIC and UNEP in co-operation with Winand Staring Centre - ISSS-FAO-ITC (ISRIC/UNEP, 1990).

BOX 3

A chronology of recent efforts to assess soil degradation.

1974   World Food Conference (Rome). A decision was taken that FAO, UNESCO and UNEP would prepare an assessment of lands that could still be brought under cultivation, taking into account the hazards of irreversible soil degradation.

1979   An interagency group published a provisional methodology for soil degradation assessment and prepared a first approximate identification of areas of potential degradation hazards for soil erosion by wind and water, and for salinization. Maps at a scale of 1:5,000,000 were prepared for Africa north of the Equator and for the Middle East (FAO, 1979).

1981   World Soils Charter (FAO).

1982   The International Society of Soil Science established a Subcommission on Soil Conservation and Environmental Quality.

1982   The UNEP Governing Council adopted a World Soils Policy, an element of which was the development of methodologies to monitor global soil and land resources.

1987   UNEP convened an expert meeting (ISSS, 1987) to consider the possibility of preparing, even on the basis of incomplete knowledge, a scientifically credible global assessment of soil degradation.

1987   GLASOD (Global Assessment of Soil Degradation) began when UNEP signed an agreement with the International Soil Reference and Information Centre at Wageningen, The Netherlands. GLASOD information has been digitized and all figures quoted in this chapter are derived from this data base unless otherwise stated.

1990   The GLASOD *World Map of Human Induced Soil Degradation* was published by ISRIC/UNEP at a scale of 1:10,000,000.

Displacement of soil material by water

The removal of topsoil (surface erosion, sheet erosion) accounts for about 920 million hectares of degraded land, while terrain deformation (rills and gullies - much more localized) accounts for about 173 million hectares.

The most important human interventions causing water erosion (see Chapters 4, 7 and 11) are: (1) deforestation and removal of natural vegetation (43%); (2) over-grazing of rangelands (29%); (3) poor agricultural practices (ploughing of slopes, use of heavy machinery, etc.) (24%), and (4) over-exploitation of vegetative cover for domestic use (4%).

Chemical soil degradation

Chemical soil degradation is caused by four different processes (see also Chapters 4, 5, 10 and 11).

(a) Loss of nutrients and/or loss of organic matter. This occurs usually under low-input agricultural systems on poor or moderately fertile soils, often as a result of insufficient application of manure or chemical fertilizers. Rapid loss of organic matter following the clearing of natural vegetation is also included.

(b) Salinization. Human-induced salinization is often the result of poor management of irrigation schemes: common causes are an excessively high salt content of the irrigation water used, or insufficient attention to the drainage of irrigated fields. This occurs mainly under semi-arid and arid conditions. Salinization can also occur if sea water or fossil saline ground-water intrudes into fresh ground-water reserves. A third type of salinization occurs where intensive agriculture leads to evaporation of soil moisture in soils on salt-containing parent material or with saline ground-water.

(c) Pollution. Many types of pollution are recognized, such as industrial or urban waste accumulation, excessive use of pesticides, acidification by airborne pollutants or excessive fertilizers, excessive manuring, oil spills, etc. This is generally most serious in industrialized countries with high population densities.

(d) Acidification. This can result from two completely different causes. Firstly it may occur in coastal regions, due to the drainage of pyrite-containing soils. Oxidation of pyrite will lead to the formation of, among other substances, sulphuric acid and thus to soils with very low pH values. Acidification may also be caused by over-application of acidifying fertilizers or by airborne pollutants.

The total area affected by the various causes of chemical soil degradation is about 240 million hectares.

Physical soil degradation

Three physical processes of very different natures cause soil degradation. These are:

(a) Compaction, sealing and crusting (82%). This is a deterioration of the structure of the soil. Sealing and crusting of topsoil occurs if the soil cover does

not provide sufficient protection against the impact of rain splashes, particularly in soils low in organic matter and those containing appreciable amounts of silt.

(b) Waterlogging (13%). Human intervention in natural drainage systems may result in flooding by rivers and submergence by rain. It usually results from poorly managed irrigation systems and leads to severe loss of soil air content and accumulation of toxic substances, with consequent effects on plant growth. Excluded are the submerged rice fields where human intervention deliberately creates waterlogging.

(c) Subsidence of organic soils (5%). The subsidence and subsequent oxidation of organic soils due to excessive drainage is included in this category only when agricultural potential is adversely affected. The loss of organic matter by oxidation can seriously affect soil fertility. This agricultural practice is mainly a problem in swampy areas in Southeast Asia, but it also reduces productivity in parts of the European sector of the former USSR.

It is estimated that soil degradation due to physical processes affects 83 million hectares.

Severity of soil degradation

In the approach followed by GLASOD (Global Assessment of Soil Degradation) (Oldeman *et al.*, 1990), the degree to which soil is presently degraded is related in a qualitative manner to: (1) agricultural suitability of the soil; (2) declining productivity; and (3) soil biotic functions. To obtain the best possible estimates, numerous soil scientists and environmental experts with good knowledge in their specific fields judged the degree of soil degradation in one of four categories. (Because these estimates are based on expert opinion from a world-wide group of soil scientists and agriculturists, the differences between the various degrees of soil degradation are relative and cannot be directly translated into production losses or costs for rehabilitation.) The results are given in Figure 7.

Globally, as much as 295 million hectares of land are strongly degraded, i.e. the original biotic functions are largely destroyed. This area, which is slightly less than the size of India, can be restored only through major investments and engineering works.

Of this strongly degraded area, about 113 million hectares have been damaged by deforestation, and 75 million hectares by overgrazing. Improper management of agricultural land has resulted in a strongly degraded area of 83 million hectares, over which the terrain has lost its productive capacity and is no longer reclaimable at farm level.

A much larger portion of the Earth's surface - 910 million hectares - is moderately degraded. This territory is still suitable for use in local farming systems but it suffers from a serious decline in productivity. This category needs the greatest attention of policy-makers; otherwise the land may be irretrievably lost at the farm level. Water erosion is by far the most important cause of soil degradation in this category, but wind erosion has moderately degraded about

Figure 7: *The degree of soil degradation, by continent.*

Degree of soil degradation (million ha).

	Slight	Moderate	Strong	Extreme
Africa	173.6	191.8	123.6	5.2
Asia	294.5	344.3	107.7	0.5
South America	104.8	113.5	25.0	-
North America	18.9	112.5	26.7	-
Europe	60.6	144.4	10.7	3.1
Australasia	96.6	3.9	1.9	0.47

Degree of soil degradation world-wide.

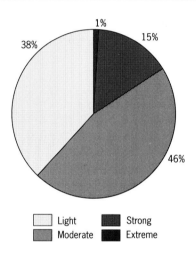

Source: Oldeman *et al.* (1990).

250 million hectares world-wide. Deforestation, mismanagement of the land and overgrazing are the main causative factors.

Recent changes in soil degradation

No systematic evaluation of the status of human-induced soil degradation has been made, although estimates of annual losses of land as a result of soil degradation have been given frequently. Current estimates differ widely and are often supported only by general statistics. As Blaikie (1985) observed: 'Statistics (on soil erosion and deforestation) are seldom in the right form, are hard to come by and even harder to believe, let alone interpret'.

The FAO production yearbooks (FAO, 1986, 1990) are the only sources from which the world coverage of land-use changes can be derived. In the period 1975-1988, the total area of arable land of the world increased by almost 38 million hectares, the area under permanent crops increased by almost 8 million hectares and the grazing land increased by almost 21 million hectares. On the other hand, forest and woodland decreased in the same period by almost 121 million hectares. The balance, called 'other land', including urban land, wasteland, parks and other forms of non-agricultural land, increased by 54 million hectares. Out of this total, in the developing countries, about 34 million hectares of arable land, 8 million hectares of land for permanent foodcrops, and 15 million hectares of grazing land were added, but no less than 147.5 million hectares of forest and woodlands disappeared and 91 million hectares were transformed into 'other land'.

The total area of agricultural land - arable land, permanent cropland and grazing land - was around 4700 million hectares in 1988 (Table 4). The GLASOD figures indicate that about 1230 million hectares of this land is degraded as a result of mismanagement of agricultural land and overgrazing of grazing land. Thus about 25 per cent of the agricultural land is affected by human-induced soil degradation.

Table 4: *Land use changes between 1975 and 1988 (million ha).*

	Arable	Permanent cropland	Permanent pasture land	Forest and woodland	Other land
Africa	9.3	1.5	0.9	-43.3	31.6
Asia (excluding China and former USSR)	5.8	2.3	-18.8	-26.3	37.0
South America	16.9	3.8	31.8	-66.0	13.5
Central America	1.6	0.3	1.0	-11.9	9.0
TOTAL DEVELOPING COUNTRIES	33.6	7.9	14.9	-147.5	91.1
Europe (excluding former USSR)	-0.5	-0.8	-4.1	4.2	1.2
USA	1.4	0.3	-0.4	-25.5	24.2
Canada	2.6	0.0	9.1	29.9	-41.6
Australasia	4.7	0.1	-31.9	-31.7	58.8
USSR	0.4	-0.2	-0.2	48.0	-48.0
China	-4.5	0.5	33.4	2.0	-31.4
WORLD	37.7	7.8	20.8	-120.6	54.3
Total area world-wide (1988)	1373	102	3212	4049	4333

Source: FAO (1986, 1990).

Other agents of land degradation

Several other kinds of land and soil degradation, many of which are irreversible, are mentioned elsewhere in this book. The causative agents are:

- rising sea level, due to either subsidence or climate warming (see Chapter 5);

- flooding of valleys for hydroelectric purposes (see Chapter 4);

- deforestation in some wet tropical and mountainous areas (see Chapter 7);

- dumping of municipal, toxic and radioactive wastes (see Chapter 10);

- tourism development along beaches and in mountains (see Chapter 15); and

- urbanization (see Chapter 17).

Responses

The responses to soil degradation must be at community and national levels, through public education, greater NGO input into local government, and better land-use management planning, including enforcement mechanisms and compensation to affected individuals, particularly the landless poor.

Concluding remarks

Buringh and Dudal (1987) estimated that about 200 million hectares of agricultural land will be transformed into non-agricultural land during the period 1975-2000, that about 50 million hectares will become seriously degraded, and that another 50 million hectares will become desertified. This loss of some 300 million hectares has to be compensated for by increasing yields and by bringing virgin land into cultivation. In order to meet the demands for food by an ever-increasing human population, these authors estimate that another 300 million hectares will be needed by the year 2000. Thus the land under forest and woodland is expected to decline from 4100 million hectares in 1975 to 3500 million hectares in 2000.

A better understanding of the consequences of human transformations of land, as well as of natural perturbations, is needed for policy formulation. For example, over the short term the conversion of good agricultural land to urban development makes good economic sense. Over the long term, such actions make no sense at all. The Global Assessment of Soil Degradation (GLASOD), the first of its kind, continues to be an important decision-support tool for policy makers.

REFERENCES

Blaikie, P.M. (1985) *The political economy of soil erosion in developing countries*. Longman, London, UK.

Buringh, P. and Dudal, R. (1987) Agricultural land use in space and time. *In: Land Transformations in Agriculture*, ed. by Wolman, M. G. and Fournier, F. SCOPE 32, John Wiley and Sons, Chichester, UK.

CILSS/PAC (1989) (Rochette, R.M., ed.) *Le Sahel en Lutte Contre La Désertification: Leçons d'Experiences*, Verlag Joseph Margraf, Weikerstein, Germany.

Dregne, H.E. and Tucker, C.J. (1988) Desert Encroachment, *Desertification Control Bulletin*, **No.16**, 16-19, UNEP, Nairobi.

FAO (1979) *A provisional methodology for soil degradation assessment*. FAO, UNEP, UNESCO, Rome.

FAO (1981) *World Soil Charter* (Resolution 8/81 of the 21st Session of the Conference of FAO, Rome, November, 1981).

FAO (1986, 1990) *FAO Production Yearbook*, **Vols. 40** and **43** (Stat. series No. 76, 94), Rome.

Hellden U. (1991) Desertification - Time for an Assessment? *Ambio*, **Vol.20**, 372-383.

Hulme, M., cited in Wright, B. (1992) Colder Winters for Northern Africa as Deserts Expand, *New Scientist*, January 18.

ISRIC/UNEP (1988) *Guidelines for general assessment for the status of human-induced soil degradation*. Ed. by Oldeman, L. R. Working Paper 88/4, ISRIC/UNEP, Wageningen, The Netherlands.

ISRIC/UNEP (1990) *World Map of the Status of Human-Induced Soil Degradation*, UNEP, Nairobi.

ISSS (1987) *Proceedings of the Second International Workshop on a Global Soil and Terrain Digital Database*, Nairobi, 18-22 May 1987, Ed. by Van de Weg, R. F. Soter Rep. No. 2, ISRIC/UNEP, Wageningen, The Netherlands.

IUCN/UNEP/WWF (1991) *Caring for the Earth - A Strategy for Environment and Development*, IUCN, Gland, Switzerland.

Oldeman, L.R., Hakkeling, R.T.A. and Sombroek, W. G. (1990) *World Map of the Status of Human-Induced Soil Degradation, An Explanatory Note*, ISRIC/UNEP, Wageningen, The Netherlands.

Rozanov, B. (1991) Personal communication.

Tucker, C.J. and Justice, C.O. (1986) Satellite Remote Sensing of Desert Spatial Extent, *Desertification Control Bulletin*, **No.13**, 2-5, UNEP, Nairobi.

Tucker, C.J., Dregne, H.E. and Newcomb, W.W. (1991) Expansion and Contraction of the Sahara Desert from 1980 to 1990, *Science*, **253**, 299-301.

UN (1977) Round-up, Plan of Action and Resolutions of UNCOD, United Nations, New York.

UNEP (1982) *World Soils Policy*, UNEP, Nairobi.

UNEP (1984) General Assessment of Progress in the Implementation of the Plan of Action to Combat Desertification, 1978-1984, GC.12/9.

UNEP (1991) *Status of Desertification and Implementation of the United Nations Plan of Action to Combat Desertification*, UNEP, Nairobi.

UNEP (1992) *World Atlas of Desertification*, Edward Arnold, London.

CHAPTER 7

Deforestation and habitat loss

The issues

The Earth's vegetation patterns are highly dynamic, and within the past million years they have changed considerably in response to changing climates (especially the recurrent Quaternary glaciations, which were associated with relatively cool and dry conditions in the tropical zones). The present extensive tropical forest cover in such regions as the Amazon and Zaire river basins is believed to have resulted from recolonization from relatively restricted refugia over the past 10,000-15,000 years (Figure 1). Within the past 3,000 years, and especially in the present century, human actions have become the predominant cause of change in the world's vegetation cover, and forests, drylands and wetlands have all been severely affected (dryland degradation has been considered in Chapter 6).

Figure 1: *Tropical forest refugia in South America during the last cold dry phase of the Pleistocene glaciation.*

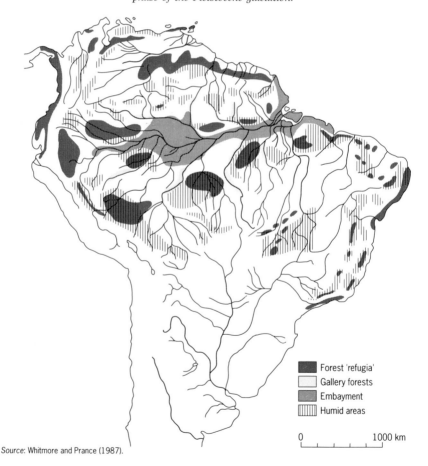

Forest 'refugia'
Gallery forests
Embayment
Humid areas

0   1000 km

Source: Whitmore and Prance (1987).

Deforestation

The rapid loss of tropical forests due to competing land uses and forms of exploitation that often prove to be unsustainable, is a major contemporary environmental issue. Severe human pressures on forests in many tropical developing countries, especially those resulting from a need to provide for the welfare of numerous poor rural dwellers, will continue to threaten the existence of these resources. In parallel, forests continue to be lost in many developed countries due to over-harvesting, inadequate regeneration, clearance for agriculture and urbanization, and air pollution.

There is a continuing debate about the optimal use of forest lands, and in particular over the areas that should be retained as natural reservoirs of biodiversity (Chapter 8), those that should be sustainably managed for timber and other products, and those that may be converted for agriculture and other uses. Management strategies are hampered by lack of information on the forest resource, by economic under-valuation of forest products harvested locally, and by the fact that most forest conversion is done by local cultivators in an essentially opportunistic way.

Wetland drainage

Wetlands (rivers, lakes, marshes, swamps, swamp woodlands, mangroves and shallow-water estuarine and marine areas) are among the most productive habitats in the world. They also occupy sites that are prime targets for reservoirs and hydroelectric dams, or for conversion for agriculture, aquaculture and urbanization.

Loss of 'ecological services'

Over the past 20 years it has become evident that the world's ecosystems play a vital role in maintaining the habitability of the planet. The economic value of the 'services' provided by natural and semi-natural vegetation types has also become apparent. The ecological principles that should govern the sustainable use of these resources for optimal human benefit are fairly well established, but practice has not kept pace with knowledge.

Ecosystems in their natural state consist of integrated and largely self-regulating communities of organisms in balance with their chemical and physical environments. Ecosystem processes dominate the global cycles of oxygen, carbon, nitrogen, phosphorus and sulphur - elements that are essential to life. Under natural conditions, carbon and nutrients are fixed into living biomass at the same rate as they are released from dead biomass. When ecosystems are placed under stress (for example by drainage, cultivation, fire or timber harvesting) they tend to simplify, releasing carbon and nutrients, and losing species. Ecologists have described the trends expected in stressed ecosystems in some detail, and these are summarized in Table 1.

Table 1: *Trends expected in stressed ecosystems.*

PROCESS	RESULT
Photosynthesis/respiration becomes unbalanced (usually <1.0) and community respiration increases.	Carbon dioxide output grows and contributes to global warming. Less biomass is maintained; options for utilization are reduced.
Nutrient turnover increases, but is less efficient and nutrients are lost to the system. Ecosystem becomes more open as internal cycling is reduced.	Soils lose their fertility, nutrients are lost into river systems, siltation of rivers occurs.
Proportion of species with prolific, dispersive reproduction strategies increases (r-strategists). Parasites increase, mutualists decrease.	Pest, weed and vermin outbreaks occur.
Food chains shorten, species diversity and size of organisms decreases; dominance increases.	Species extinctions occur, especially in larger animals and plants.
Succession reverts to earlier stages.	Fire takes a hold. Forests become grasslands.

Source: Simplified and adapted from Odum (1985).

By altering the global carbon cycle, deforestation and other vegetation changes may augment greenhouse warming (Chapter 3). The amount of carbon held in ecosystems varies with vegetation type (Olson *et al.*, 1983; Prentice and Fung, 1990; Groot, 1990). Models based on FAO deforestation statistics indicate that carbon releases from tropical land-use changes amounted to 0.58 +/- 0.06 Gt per year in 1980 (Hall and Uhlig, 1991). Judging from the latest FAO statistics, current rates could be twice as high. Studies on tropical grasslands suggest that the burning of savanna could contribute three times as much CO_2 to the atmosphere as comes from rain forest clearance.

Deforestation by human action

The initial extent of forests

The distribution of forests in pre-agricultural times is difficult to evaluate from potential vegetation maps because they vary so much in their classifications and boundaries. An alternative is to use climate models to predict 'original' vegetation cover, as Matthews (1983) has done. She constructed global data bases of vegetation and land-use patterns in 1° x 1° latitude/longitude grid cells, and calculated the decrease in major vegetation types from their pre-agricultural extent (Table 2). This study led to an estimate that 6,150 million hectares of forest and woodland had by 1970 been reduced by 14.8 per cent to 5,200 million hectares. Of this, 700 million hectares had been taken from closed canopy forests, mostly (650 million ha) in the temperate regions (reduced by 15.5 per cent). Woodlands and shrublands, such as dry African forest (miombo), Mediterranean maquis and garrigue woodland, and the eucalyptus and mallee woodland of Australia, had declined by 3 million square kilometres.

Table 2: *Estimates of pre-agricultural and present areas of major ecosystems (million square kilometres).*

Ecosystem	Pre-agricultural	Present	Reduction
Tropical closed forest	12.77	12.29	0.48
Other forest	33.51	26.98	6.53
Total forest	46.28	39.27	7.01
Other woodland	15.23	13.10	2.13
Shrubland	12.99	12.12	0.87
Grassland	33.90	27.43	6.47
Tundra	7.34	7.34	—
Desert	15.82	15.57	0.25
Cultivation	0.93	17.56	+16.63

Source: Matthews (1983), by permission of the American Meteorological Society.

The history of deforestation

The scale and extent of deforestation in prehistoric and early historic times clearly varied from continent to continent and zone to zone. In North America, humans began clearing lowlands at least 12,000 years ago. Even in rain forest areas, the record of impact goes back for millennia, possibly as far back as 23,000 years in the Peruvian Amazon. Middle European forests were extensively cleared between 7,000 and 3,000 years ago (Williams, 1989, 1991), and in the Mediterranean basin were further thinned by Greek and Roman civilizations which used timber for metal smelting, ship-building and other purposes, adding to the substantial agricultural and pastoral pressures. In contrast, the destruction of perhaps 25 per cent of the lowland forests of New Zealand was delayed until the centuries after AD950, when the Maori settlers arrived (Glasby, 1991). Fire has been an important agent of deforestation from early times.

The spread of European settlement around the globe from AD1500 onwards led to much destruction of forest. In the USA, 60,000 square kilometres were cleared by 1850, and 660,000 square kilometres by 1910. In Canada, New

Zealand, South Africa and Australia a total of perhaps 400,000 square kilometres of forest and woodland were cleared by the early twentieth century. In the tropics, cash crop plantations and subsistence agriculture took over huge land areas. It has been estimated that 2.4 million square kilometres of forest were cleared there between 1860 and 1978, together with 1.5 million square kilometres of open woodland.

In the temperate zones, the total area of forest has changed little in recent decades. This overall statistic, however, conceals substantial declines in natural forest in the USA in the late 1970s, and between 1950 and 1980 in Australia, which were offset only by afforestation in the temperate zones of the former USSR, China and New Zealand.

Between 1923 and 1985 there have been at least 23 estimates of the area of remaining closed forest land, ranging from 60.5 million square kilometres to 23.9 million square kilometres. The most recent (and most accurate) data on forest extent are the FAO and ECE calculations to 1980 and 1990 baselines. The FAO data, in Table 3, are continually being refined.

Table 3: *Areas of forests and woodlands at the end of 1980 (million square kilometres).*

	Boreal[1] countries	Temperate countries	Tropical countries	All countries
Africa	–	0.081	7.012	7.093
America	2.03	3.108	8.898	14.036
Asia[2] (excluding former USSR)	–	1.884	3.034	4.918
Pacific	–	0.487	0.426	0.913
Europe (including former USSR)	7.17	2.115	–	9.285
World Total	9.20	7.675	19.37[3]	36.245

Notes: 1. Including Canada, Finland, Norway, Sweden, USA and former USSR.

2. Asia includes the Middle and Near East.

3. Estimates suggest that by 1990 this figure was reduced by 2 million km[2].

Source: Adapted from FAO (1991).

Contemporary deforestation and its causes

Human population growth, agriculture and resettlement

In the developing world, the forest edges are being pushed back to create arable land, while in the forest itself the shifting cultivators, whose cyclic movements are sustainable when spread widely, find their populations rising and their land areas dwindling. They are being forced to shorten rotations, which can lead to permanent nutrient loss and degradation to non-forest ecosystems. The forest edge succumbs first, the change often easily detectable by remote sensing. Thinning of the forest interior is more difficult to detect, and impossible to map at a small scale. By the time it can be detected, the forest may already have lost much of its regenerative capacity and be well on its way to becoming 'wasteland'.

Heavy deforestation may stop abruptly at the edges of forest reserves, as it does in Ghana (Whitmore, 1990). In many countries, however, deforestation rolls over the boundaries of reserves and National Parks with impunity. Parks are being over-run in the Philippines, and the forest reserve systems of India have declined in the face of relentless human pressure (Collins et al., 1991).

Forest degradation and loss from spontaneous expansion of people into forest lands is notoriously difficult to quantify, and is the primary cause of confusion about rates of loss (Melillo et al., 1985; Molofsky et al., 1986). Shifting agriculture is the primary cause of deforestation, accounting for about 45 per cent of the 7.5 million hectares loss of tropical forest in 1976-80. In 1980 it accounted for 35 per cent of deforestation in Latin America (notably in Mexico, Central America and the Andean countries), for 70 per cent in Africa (notably in West Africa and Madagascar) and for 49 per cent in Southeast Asia (notably in Sri Lanka, Thailand, Northeast India, Laos, Malaysia and the Philippines) (Lanly, 1982).

People do not move into forests from choice, but from lack of it. Economic patterns in the developing world, with grossly distorted distribution of wealth and inequitable land tenure systems, mean that the forests are the only hope of subsistence for many people. In a number of tropical countries, governments have taken advantage of the peasant farmer's willingness to settle new lands. Schemes to promote resettlement have attracted massive numbers of people, but have often failed because the soils and other conditions were inappropriate. In Brazil, economic incentives for such settlement and for forest clearance for pasture were reinforced by road building which facilitated access: in Rondonia in the late 1970s some 5,000 new settlers arrived every month. Satellite imagery (Figure 2) reveals the massive extent of clearance. Unfortunately the soils in this area are poor, and harvests have declined causing settlers to move on, selling their plots for unsustainable cattle ranching (Fearnside, 1986; Whitmore, 1990). The economic policies that promoted this development pattern were ended in 1990 (Pearce, 1991).

The problem is not unique to developing countries, or to recent decades. In the early part of the twentieth century, promises of land and seed were given to prospective British immigrants to settle in the 'clay belt' of Northern Ontario: unfortunately, the frost-free 'summer' period was less than 60 days in length and few crops could be grown.

Grazing and ranching

Domestic animals in tropical woodlands and forests reduce regeneration through grazing, browsing and trampling. The problem is serious in the African

Figure 2: *Satellite imagery shows progressive clearance of forest farms (black areas) in southwestern Amazonia in the period 1973-80, along roads spaced five kilometres apart. The main national highway,* **BR 364**, *runs across the southwest corner of the area shown.*

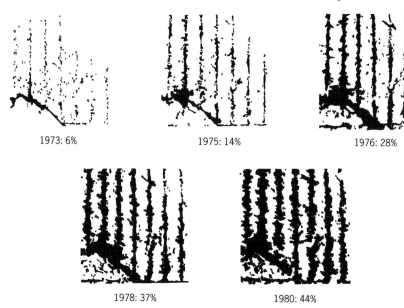

1973: 6%   1975: 14%   1976: 28%

1978: 37%   1980: 44%

Source: Fearnside (1986) and Whitmore (1990).

Figure 3: *Loss of primary forest in Costa Rica, 1940-83. Percentages show forest land as a proportion of total land area.*

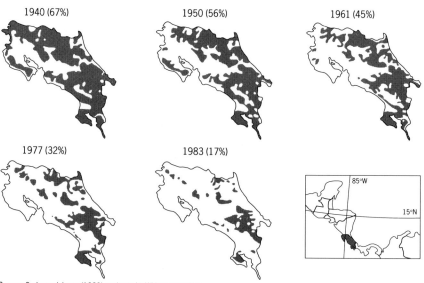

1940 (67%)   1950 (56%)   1961 (45%)

1977 (32%)   1983 (17%)

Source: Sader and Joyce (1988), redrawn by Whitmore (1990).

savanna woodlands and also in India, which has about 15 per cent of the world's cattle, 46 per cent of its buffaloes and 17 per cent of its goats. The spread of irrigated and cultivated land in India has forced livestock owners into forest areas, where 90 million of the estimated 400 million cattle now reside. Carrying capacity is estimated at only 31 million (Forest Survey of India, 1987).

In South America, resettlement of farmers onto small holdings for agricultural development tends by default to result in eventual conversion to pasture. Pastures are easier to maintain (grazing prevents reversion to secondary forest), and it is easier to claim title of ownership (land registration is often chaotic). Until recently, such clearance was encouraged by tax laws (Pearce, 1991).

In Central America, however, pasture development is a primary cause of deforestation, carried out by a minority of ranchers who own disproportionately large areas. In the late 1970s some two million hectares per year were converted to pasture in Latin America. In Costa Rica, 67 per cent of the country was forested in 1940, but by 1983 only about 17 per cent remained under forest (Sader and Joyce, 1988; Whitmore, 1990) (Figure 3). Much of the converted land was used for beef production by large landholders.

Fuelwood and charcoal

Exploitation for fuelwood and charcoal is mainly a problem of tropical and sub-tropical woodlands, although there are examples of closed forests being severely affected (notably in Sri Lanka, India and Thailand). Removals of fuelwood exceed the replacement capacity of woodlands around towns, cities and roads, ultimately leading to deforestation. Production of charcoal causes degradation over larger areas, away from roads. Examples occur in Brazil's cerrado, which produces charcoal for the steel industry, and in the north of Thailand, which provides charcoal to Bangladesh.

In 1980 fuelwood accounted for about 58 per cent of the energy consumption in Africa, 17 per cent in Asia and 8 per cent in Latin America. Some African countries are highly dependent on it, for example Kenya (68%), Ethiopia (96%) and Mozambique (98%). Between 30 and 40 per cent of the world's population relies on fuelwood for warmth and cooking. By 1980, 1.3 billion people were short of fuelwood: by 2000 the figure is expected to reach 2.7 billion (Lanly, 1982).

Total roundwood production figures for the developing and developed world do not differ much (Table 4). However, in the developed world 82 per cent is used for industrial purposes and 18 per cent for fuelwood, while in the developing world the figures are reversed, at 20 per cent and 80 per cent respectively (1989 data). Between 1977 and 1988 fuelwood production in the developing world rose by almost 30 per cent, and it will continue to do so until deforestation is so widespread that output can no longer be maintained, and the industry declines. Paradoxically, the world's two billion fuelwood burners would consume only 3.5 per cent of world petroleum production if such a switch were feasible, but the cash is not usually available. Whilst difficult to evaluate accurately, about 20,000–25,000 square kilometres of woodland and forest are cleared annually for fuelwood (Williams, 1989).

Table 4: *Production of wood, rounded up to the nearest million cubic metres.*

	Developed countries		Developing countries		World total	
	1978	1989	1978	1989	1978	1989
Fuelwood	187	268	1183	1538	1370	1786
Industrial wood	1118	1274	307	403	1425	1677
Total	1305	1542	1490	1941	2795	3463

Source: after FAO (1990).

Timber exploitation

On a global scale, industrial roundwood accounts for marginally less exploitation than fuelwood, although it remains high at around 1.7 billion cubic metres in 1989 (Table 4). In industrialized countries extraction and regeneration are roughly in equilibrium, regrowth exceeding extraction in Canada, New Zealand, the former USSR and Scandinavia, although there are unsustainable wood harvests in some sub-regions (UNECE, 1985). However, in much of the tropical world timber is being 'mined' unsustainably for the export market. As Table 5 demonstrates, roundwood exports from some of the major tropical timber-producing countries have declined significantly in recent years, in some cases due to past over-exploitation and in others as a result of policies to direct logs away from export and towards domestic processing. By contrast, exports from the major producers of the developed world have reached ever-higher peaks, except in Czechoslovakia. In several areas, however, the most valuable species and old-growth forests have been over-cut, and the forest industry is harvesting lower quality wood.

In Sarawak the rate of extraction is claimed to be two or three times the rate of natural regrowth, resulting in denudation of up to 40 per cent of the logged area. Furthermore, studies indicate that more trees are broken or damaged than are cut for extraction (Mattson *et al.*, 1980). Roads built to extract logs allow slash and burn cultivators to move into new forest lands. The impact of this is illustrated by the fact that in 1980 logged-over forests were being cleared eight times more quickly than undisturbed forests (2.06% versus 0.27% per year) (Lanly, 1982). Fire is the main tool of shifting agriculturalists, but there are cases where the fire has gone out of control and destroyed vast areas of forest unintentionally. The most dramatic of these, in 1982-83, burned some 38,000 square kilometres of East Kalimantan, Indonesia, an area about the size of The Netherlands (Malingreau *et al.*, 1985).

Logging need not cause deforestation. A combination of careful economic

Table 5: *Export figures for nations that have exported over three million cubic metres of industrial roundwood in any year, 1977-88 (in thousand cubic metres).*

	Production in 1977	Production in 1988	Peak production (Year)
	Developed world		
Australia	4788	8497	8497 (88)
Canada	3478	6499	6499 (88)
Czechoslovakia	2699	1176	3445 (80)
France	2932	5423	5819 (86)
Germany (former F.R.)	3171	5002	5002 (88)
USA	21155	26980	26980 (88)
USSR	19081	20505	20505 (88)
	Developing world		
Côte d'Ivoire	3229	550	3229 (77)
Indonesia	20127	932	20694 (78)
Malaysia	16873	20776	23057 (87)
Philippines	2047	334	2200 (78)

Source: after FAO (1990).

management, maintenance of forest security and application of known environmental technologies can result in rapid regeneration to near-climax forest. The reality, however, is that less than one per cent of tropical rain forest is under sustainable management (Poore, 1989). In 1980 FAO estimated that 4.7 per cent of productive closed tropical forest was under intensive management, of which three-quarters was moist deciduous and dry forest, in India and, to a lesser extent, Myanmar (Burma) (Lanly, 1982; Gadgil, 1991). The widespread processes of degradation and deforestation in the tropics created the need to introduce and apply practices of sustainable use, and to give preferential market support for their products.

Plantations

A smaller but none the less significant reason for the removal of natural forest is the planting of tropical tree crops such as rubber, oil palm, eucalyptus, *Gmelina* and *Acacia mangium*. Plantations can, of course, also be a valuable element in environmental restoration. However, the rate of plantation establishment is small compared to deforestation rates (about 10% in 1976-90), but it is increasing. On hills and marginal soils, plantations maintain a forest cover and offer high yields (at least in theory). There are few data on afforestation, and monitoring is needed.

Atmospheric pollution

Sulphur dioxide, oxides of nitrogen, and ozone are believed to be the main air pollutants that damage forests. Their impact is exerted directly through deposition on foliage, and indirectly via acidification of the soil. The interactions are complex: for example, soil acidification may mobilize aluminium, which is toxic to tree roots, while the overall stress of pollution may bring about nutrient deficiencies and increase the vulnerability of the trees to drought, insects and disease. The reverse process - increased sensitivity to pollutants as a consequence of stress caused by drought - is also likely. Forest damage was first recognized on a significant scale in the 1970s in Central Europe, particularly at high elevations. Monitoring has been undertaken under the EMEP programme of the ECE Convention on Long Range Trans-Boundary Air Pollution in Europe. Europe remains the worst-affected continent, but forests in North America, China, Mexico, Australia and Japan have been affected to varying extents, and the problem is increasing around urban and industrial centres in other parts of the developing world.

A five-year study by the International Institute for Applied Systems Analysis (IIASA) indicates that damaging levels of sulphur deposition are being inflicted on 75 per cent of European forests, while 60 per cent are affected by excessive nitrogen deposition. The conclusion is that annual losses of timber will be 118 million cubic metres every year for the next century; 48 million cubic metres in Western Europe, 35 million cubic metres in Eastern Europe and a further 35 million cubic metres in the European regions of the former USSR (Nilsson *et al.*, 1992 a, b). The loss has been estimated at $US30 billion per year, made up of $US6.3 billion worth of timber, $US7.2 billion of processing and $US16.9 billion worth of non-timber and social benefits (including tourism, recreation, wildlife habitat and protection of soil and water). The damage is particularly severe at high elevations (Nilsson and Pitt, 1991).

Forest trends

In 1980, a joint FAO/UNEP assessment of forest resources (FAO/UNEP, 1981) gave the total forest area world-wide as 1,935 million hectares, and the annual rate of deforestation as 11.3 million hectares or 0.6 per cent over the period 1976-80. Table 6 provides the latest provisional estimates of forest cover and deforestation for Africa, Latin America and Asia for the years 1980 and 1990. In all three regions, the forest area is decreasing. Comparison of the two estimates for 76 countries present in both assessments shows that the annual rate of deforestation was higher in the latter assessment (16.9 million hectares against 11.3 million hectares, or 0.9 per cent against 0.6 per cent) (Dembner, 1991).

In the temperate and boreal zones, deforestation rates have not been systematically assessed. The process there is much more restricted, and is

Table 6: *Latest provisional FAO estimates of forest cover and deforestation for 87 countries in the tropical regions, revised 15 October 1991.*

Continent	Number of countries studied	Forest area 1980	Forest area 1990	Annual deforestation 1981–90	Rate of change 1981–90 (%/year)
		(millions of hectares)			
Africa	40	650	600	5.0	-0.8
Latin America and Caribbean	32	923	840	8.3	-0.9
Asia	15	321	275	3.6	-1.2
Total	87	1,894	1,715	16.9	

Notes: Countries include almost all of the moist forest zone, along with some dry areas. Figures are indicative, and should not be taken as regional averages.

Figures may not tally due to rounding.

Souce: Dembner (1991).

generally compensated by afforestation and reforestation. In parts of Europe and the eastern United States, forests have expanded over land that was previously cleared and farmed. Much of the developed world has apparently extended its forests, particularly Europe and the former USSR.

A trend towards deforestation in areas with rapidly increasing human populations has been apparent (Allen and Barnes, 1985; Mather, 1990). Conversely, forests (usually plantations) are expanding where population is relatively stable. The implication is that agricultural expansion related to population growth is the principal cause of deforestation. However, Westoby (1989) suggested that it is not so much the number of human beings that has the crucial impact on forests as the way in which society is organized. Different social groups and classes use the forest for different purposes; sometimes they coexist, but at other times they are in conflict.

Drainage of wetlands

Freshwater wetlands

The most recent global analysis, based on Gore (1983) and a wide variety of other map sources, is that somewhere between 530 million and 860 million hectares of freshwater wetlands remain in the world (Aselmann and Crutzen, 1989; Matthews and Fung, 1987; WRI, 1987). Of these, alkaline mineral swamps and marshes account for some 200-400 million hectares while acid bogs, fens and peatlands cover about 150-360 million hectares. There are no data on their global historical extent, or the rate at which they are being destroyed globally. However, it is commonly stated that the United States has lost some 54 per cent (87 million ha) of its original wetlands, while in various regions of Europe losses are estimated at between 90 per cent and 60 per cent. In developing countries, rice paddies are estimated to cover 1.3 million square kilometres, much of which would originally have been natural wetland. Coastal wetlands have not been altered as extensively as inland ones, but some 300,000 hectares (67%) of the mangrove resources in the Philippines were lost between 1920 and 1980 (Zamora, 1984), and estuarine and coastal marshes in regions as far apart as western France, Brazil and Nigeria have been degraded by pollution, reclaimed or otherwise modified (Dugan, 1990).

Table 7 and Figure 4 display summaries of wetland distribution by region and by 10° bands of latitude respectively. The largest areas occur between 50°N and 70°N, where the peatlands of the former USSR, Canada and Alaska cover over three million square kilometres. The largest swamps and flood plains occur in the Amazon region, in southern Brazil and Argentina, and on the African peneplain.

Table 7: *Global freshwater wetland areas (thousand square kilometres).*

REGION	Bogs	Fens	Swamps	Marshes	Flood plains	Lakes	Total
Former USSR	917	531	25	39	–	–	1,512
Europe	54	93	1	4	1	1	154
Near East	–	–	–	8	–	–	8
Far East	–	–	11	–	–	–	11
China	11	–	3	18	–	–	32
Southeast Asia	197	–	44	–	–	–	241
Australia/ New Zealand	2	3	1	–	9	–	15
Africa	–	–	85	57	174	39	355
Alaska	?	250-400	?	?	?	?	(325)
Canada	673	531	14	44	–	6	1,268
USA (excluding Alaska)	13	–	80	40	95	–	228
Central America	–	–	15	2	1	–	18
South America	–	–	851	62	543	68	1,524
Total:	1,867	1,483	1,130	274	823	114	5,691

Source: Aselmann and Crutzen (1989), by permission of Kluwer Academic Publishers.

Figure 4: *Distribution of natural wetlands along 10° latitude bands.*

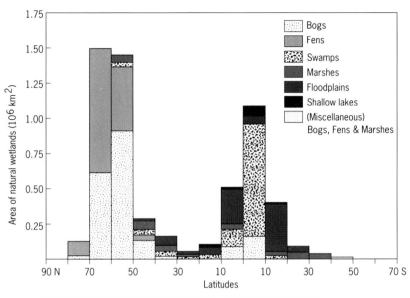

Source: Aselmann and Crutzen (1989), by permission of Kluwer Academic Publishers.

The rate at which wetlands are now being converted to other uses is not known but it is undoubtedly high. The only well documented loss of wetlands for an entire region is that of the USA, where-one third of the 500,000 - 1 million square kilometre area outside Alaska had been converted by 1950, and one-half by 1985 (OECD, 1991) (Table 8). Losses are probably even higher in Europe but the data are not available except in particular areas (Box 1). There is little information on rates of wetland loss in the developing world, and the situation can only be described by a series of examples (Box 2).

Mangroves

A global status report published in 1983 concluded that vast areas of mangrove were being destroyed either intentionally or as a consequence of other activities, but no global statistics were presented (Saenger *et al.*, 1983). It is surprising that such an easily identifiable ecosystem is not monitored more effectively, particularly in view of its economic and ecological importance.

In Southeast Asia, conversion to brackish-water fish and prawn ponds has been a major cause of loss of mangroves. In the Philippines, 80 per cent of the substantial destruction of mangroves recorded between 1952 and 1977 was for mariculture; in Thailand the mangrove area has diminished by more than 40 per cent, and between 1966 and 1982 an estimated 16 per cent of the mangroves in the Gulf of Guayaquil, Ecuador, were reclaimed for shrimp production (WRI, 1986). Elsewhere, cutting of mangrove poles and reclamation

Table 8: *Wetland losses in selected US states and regions.*

State or region	Original wetlands (ha)	Remaining wetlands (ha)	Percentage loss	Date
Iowa (natural marshes)	930,000	10,700	99	1981
California	2,000,000	182,000	91	1977
Nebraska (rainwater basin)	38,000	3,400	91	1982
Mississippi (alluvial plain)	9,700,000	2,100,000	78	1979
Michigan	4,500,000	1,300,000	71	1982
North Dakota	2,000,000	810,000	60	1983
Minnesota	7,450,000	3,500,000	53	1981
Louisiana (forested wetlands)	4,570,000	2,300,000	50	1980
Connecticut (coastal marshes)	12,000	6,000	50	1982
North Carolina (pocosins)	1,000,000	610,000	40	1981
South Dakota	810,000	525,000	35	1983
Wisconsin	4,000,000	2,700,000	32	1976

Source: Tiner (1984).

for urban or industrial uses have been significant. In the Gambia, a proposed dam on the Gambia River is estimated to be likely to reduce the mangrove area by about 70 per cent. There are similar examples in many other parts of the developing world, and the consequences are serious because mangrove habitats not only protect coasts from storm surges but are also major fish nurseries: taking the Gambia example, the loss of mangrove could well lead to a 32 per cent reduction in the fish catches in the estuary (WRI, 1986). Virtually all the mangrove habitat in Singapore was reclaimed between 1950 and 1980.

Mangroves are also exploited to meet fuelwood and industrial wood needs. At a local scale, mangrove can sustain fuelwood offtake, but charcoal production is more damaging. Mangrove timber is rarely used directly, being more often turned into woodchips or pulp. Clear-felling is the usual approach, and attempts to encourage natural regeneration (e.g. in Sarawak and Sabah, Malaysia) have not generally succeeded. Artificial replanting has met with greater success, for example, in Vietnam, where war damage to mangroves was extensive.

Clear-felling is now acknowledged to be happening on a very large scale, and is the major threat to mangroves throughout the tropics. It brings consequent risks to coastlines, which will be aggravated if sea level rises.

Comparable impacts arise from the diversion of fresh waters upstream of mangrove swamps. Mangroves grow best where freshwater run-off is significant;

BOX 1

Loss of wetlands in selected parts of Europe and New Zealand.

- In Finland, 104,000km² (30% of land area) were once covered by mires, but 55,000km² have been reclaimed for forestry (Gore, 1983).
- In Portugal, some 70% of the wetlands of the Western Algarve, including 60% of estuarine habitats, have been converted for agricultural and industrial development.
- In Brittany, 40% of the coastal wetlands have been lost since 1960 and two-thirds of the remainder are seriously affected by drainage and similar activities.
- Over 20% of Sweden's 85,000 medium to large lakes are now acidified by acid rain, 4,000 of them seriously. In Norway, fish had disappeared from 13,000km² of lake area by 1980, and been depleted in a further 20,000km² (McCormick, 1985).
- Canalization of Europe's River Rhine in the nineteenth century reduced its length by over 100km. Stream velocities rose by up to 30%, the water table fell by 3-4 metres up to 3km from the river, and damage to agriculture, forestry and fisheries amounted to $US171 million (Braakhekke and Marchand, 1987).
- In New Zealand, it is thought that over 90% of the natural wetlands have been destroyed since European settlement, and drainage continues today.

BOX 2

Examples of losses of wetlands in developing countries.

- In the Philippines, some 3,000km² (67%) of the country's mangroves were lost between 1920 and 1980 (Zamora, 1984), over half of these being converted to culture ponds for shrimp and milkfish (Hamilton and Snedaker, 1984).
- In Nigeria, the flood plain of the Hadejia River alone has been reduced by over 300km² as a result of dam construction (Adams and Hollis, 1988). Dam construction on the Sokoto River caused losses of crops worth $US7 million at 1974 prices (Adams, 1986). Such projects have already harmed wetlands and more damage is likely. Productive flood plains have dried out, forcing farmers onto marginal lands and causing nomads to graze their cattle on smaller areas of flood plain pasture, leading in turn to overgrazing, rising livestock mortality and emigration of many herding communities into surrounding arid rangelands where the degradation continues.
- In Malaysia, the Wetlands Working Party calculated that nearly half the total area of what were originally wetlands in the peninsula had already been converted to agriculture by 1986 (Scott and Poole, 1989) and areas continue to be cleared for development.

and settlements, irrigation and dams can cause major reductions in supply. The Indus River in Pakistan is so heavily utilized that for nine months of the year there is no seaward discharge of fresh water, and as a result the surviving mangroves in the delta are sparse and stunted. Similarly, the Sundarbans Forest in India and Bangladesh is in decline because of irrigation projects, barrages and other uses of the fresh waters of the River Ganges.

The third main cause of mangrove destruction is conversion for non-forest purposes, including agriculture, aquaculture, salt ponds, urban development, harbours and mining.

Significance of wetland loss

Wetlands serve a wide variety of functions: they assist flood control, purify water supplies, protect shorelines from erosion, and trap sediments that can pollute waterways. Many people depend on wetlands for their livelihood. For example, in Mali, the inner delta of the River Niger covers 30,000 square kilometres in the Sahelian region and supports some 550,000 people (7-8% of the country's population). The post-flood pastures of the delta provide dry season grazing for about a million cattle and as many sheep and goats. Fishermen also depend on the flood for their catch, while more than half the rice-growing area of Mali is in this region. Wetlands are also important for providing wildlife habitats as, for instance in West Africa where the flood plains of the Senegal, Niger and Chad basins support over a million waterfowl, many of them migratory species. In many countries the products of wetlands are essential to the local people, for instance the Rwandans harvest papyrus and compress it into briquettes that have a calorific content equal to that of fuelwood, and in the Northern Hemisphere, peat has been used as a fuel for at least 2,000 years. Wetlands may also be important to the country for generating income - as in Malaysia where the 40,000 - hectare Matang Forest Reserve annually yields mangrove timber worth $US9 million, of which some $US424,000 accrued directly to the government in 1976. Coastal wetlands are also of great value as fish nursery areas and as storm protection (Chapter 9): one estimate by the US Army Corps of Engineers was that retaining a wetland complex near Boston, Mass, gave an annual saving of $US17 million through flood protection alone (Hair, 1988).

Despite the importance of the goods and services which wetlands provide (Table 9), they have tended to be taken for granted. As a result, maintenance of natural wetlands has received low priority in most countries. Indeed, wetlands are often regarded as a hindrance to development. Wetlands everywhere have been lost or altered because of the disruption of natural processes by agricultural intensification, urbanization, pollution, dam construction, regional water transfers and other forms of intervention in the ecological and hydrological system (Table 10). In many instances where wetlands have been destroyed by dams and other river basin schemes, the benefits have fallen far short of those predicted (Goldsmith and Hildyard, 1984).

Table 9: *Wetland values.*

	Estuaries (without mangroves)	Mangroves	Open coasts	Flood plains	Freshwater marshes	Lakes	Peatlands	Swamp forest
Functions								
1. Groundwater recharge	O	O	O	■	■	■	●	●
2. Groundwater discharge	●	●	●	●	■	●	●	■
3. Flood control	●	■	O	■	■	■	●	■
4. Shoreline stabilization/ erosion control	●	■	●	●	■	O	O	O
5. Sediment/toxicant retention	●	■	●	■	■	■	■	■
6. Nutrient retention	●	■	●	■	■	●	■	■
7. Biomass export	●	■	●	■	●	●	O	●
8. Storm protection/ windbreak	●	■	●	O	O	O	O	●
9. Micro-climate stabilization	O	●	O	●	●	●	O	●
10. Water transport	●	●	O	●	O	●	O	O
11. Recreation/tourism	●	●	■	●	●	●	●	●
Products								
1. Forest resources	O	■	O	●	O	O	O	■
2. Wildlife resources	■	●	●	■	■	●	●	●
3. Fisheries	■	■	●	■	■	■	O	●
4. Forage resources	●	●	O	■	■	O	O	O
5. Agricultural resources	O	O	O	■	●	●	●	O
6. Water supply	O	O	O	●	●	■	●	●
Attributes								
1. Biological diversity	■	●	●	■	●	■	●	●
2. Uniqueness to culture/ heritage	●	●	●	●	●	●	●	●

Key: O –Absent or exceptional   ● –Present   ■ –Common and important value of that wetland type.

Eradication of disease has been a principal argument for draining wetlands. It is true that malaria, yellow fever, dengue, filariasis and encephalitis are among the tropical diseases associated with wetlands but they are not specific to them (Giglioli, 1980) and neither drainage nor conventional forms of intensive development necessarily provide healthy alternatives.

Changes in other habitats

The state of the world's major biomes was reviewed in the UNEP report *The World Environment, 1972-82* (Holdgate, Kassas and White, 1982), and that account remains broadly applicable a decade later. With the greatest extension of human populations occurring in tropical zones, there is a tendency to concentrate on issues there (Chapter 8).

Table 10: *The causes of wetland loss*

Human Actions	Estuaries	Open coasts	Flood plains	Freshwater marshes	Lakes	Peatlands	Swamp forest
Direct causes							
Drainage for agriculture, forestry and mosquito control.	■	■	■	■	●	■	■
Dredging and stream channelization for navigation and flood protection.	■	○	○	●	○	○	○
Filling for solid waste disposal, roads, and commercial, residential and industrial development.	■	■	■	■	●	○	○
Conversion for aquaculture/mariculture	■	●	●	●	●	○	○
Construction of dykes, dams, levees, and seawalls for flood control, water supply, irrigation and storm protection.	■	■	■	■	●	○	○
Discharges of pesticides, herbicides, nutrients from domestic sewage and agricultural runoff, and sediment.	■	■	■	■	■	○	○
Mining of wetland soils for peat, coal, gravel, phosphate and other materials.	●	●	●	○	■	■	■
Groundwater abstraction.	○	○	●	■	○	○	○
Indirect							
Sediment diversion by dams, deep channels and other structures.	■	■	■	■	○	○	○
Hydrological alterations by canals, roads and other structures.	■	■	■	■	■	○	○
Subsidence due to extraction of ground water, oil, gas and other minerals.	■	●	■	■	○	○	○
Natural causes							
Subsidence	●	●	○	○	●	●	●
Sea-level rise	■	■	○	○	○	○	■
Drought	■	■	■	■	●	●	●
Hurricanes and other storms	■	■	○	○	○	●	●
Erosion	■	■	●	○	○	●	○
Biotic effects	○	○	■	■	■	○	○

Key: ○ –Absent or exceptional: ● –Present, but not a major cause of loss: ■ –Common and important cause of wetland degradation and loss.

Responses

During the past two decades, the economic value of forests, wetlands and other natural or semi-natural systems has been increasingly appreciated. One detailed calculation indicated that a hectare of forest in Brazil, harvested sustainably for fruit, latex and a small amount of timber, could yield $US8,890 per annum whereas its value if harvested for pulpwood was $US3,184 and as cattle pasture, $US2,960 (Peters *et al.*, 1989). It is also accepted that destruction of natural vegetation may accelerate global climate change and turn potentially useful ecosystems into wastelands. These issues have now been recognized by economists, development aid agencies and governments, as well as by scientists.

The problem of deforestation has attracted particular emphasis, and stimulated both the protection of areas of forest and the implementation of reafforestation schemes. Brazil has set aside some 15 million hectares in a series of forest parks and conservation areas, while Costa Rica has protected about 80 per cent of its remaining wildlands. Virtually all the remaining rain forests in Queensland, Australia, have been protected. Many countries have taken steps to improve forest management: some have restricted the harvesting of timber while others have improved harvesting techniques. However, considerable tracts of land in many tropical countries continue to be converted for agriculture and pasture regardless of the fact that nutrient levels in the soils of the cleared areas are low, and the resulting land uses unsustainable in consequence. Even in Europe and North America, ancient ('old growth') forests continue to be destroyed, despite their importance as reservoirs of biodiversity and the fact that what remains is just a residual fragment of their original extent. The South African fynbos and areas of ancient forest types in Southern Chile and Tasmania continue to be eroded.

Reforestation and afforestation activities are in progress in many countries. FAO estimates the annual rate of successful tree planting at 1.1 million hectares (FAO, 1991). The total area of man-made forests in tropical countries alone is estimated to have reached 25 million hectares in 1990. China, Zambia and Cyprus are among the countries with substantial reforestation programmes. In Argentina, Brazil, India and the Philippines, 'energy' plantations have been established to fuel power stations: the projection in the Philippines was that in the year 2000, over 700,000 hectares of plantation would produce some 2,000 megawatts of electricity (Harlow and Adriano, 1980). Yet despite these efforts the reforestation programme remains an order of magnitude less than the annual rate of deforestation.

There has also been considerable action at regional and global levels. The ECE Convention on Long Range Trans-Boundary Air Pollution, signed in 1979, has been the focus for much of the effort to cut emissions of sulphur and nitrogen oxides in Europe and so reduce damage to forests and lakes. In 1985 a Tropical Forestry Action Plan (TFAP) was launched by FAO, the World Bank, UNDP and the World Resources Institute, and this has now been adopted by 81 countries (FAO, 1991). The International Tropical Timber Agreement, which came into force in 1985 under the auspices of UNCTAD, is now being implemented through the International Tropical Timber Organization (ITTO) based at Yokohama in Japan. Through ITTO, producing and consuming countries are working together to develop sustainable management of tropical forests, and to improve techniques for reforestation.

The World Conservation Strategy (IUCN/UNEP/WWF, 1980) called on states to prepare national conservation strategies, essentially plans for the optimal and sustainable use of their natural environmental resources, and over sixty have now done so. *Caring for the Earth*, the new Strategy for Sustainable Living (IUCN/UNEP/WWF, 1991), calls for a wide range of actions for sustainable use of the world's forests, for protection of areas of natural forest, for increased reafforestation, for management systems that give a larger role to local communities, and to marketing systems that promote sustainability. A main component of national strategies is the rational use of forests, wetlands and mangroves. The increasing recognition of the economic value of forests, wetlands and other ecosystems, as well as of their significance in maintaining essential life-support services and as reservoirs of biological diversity (Chapter 8) has added impetus to their conservation. Many states have extended their national laws to regulate land use and to protect important areas of natural habitat.

Concluding remarks

The efforts exerted in response to the challenges of deforestation and habitat loss are coming together in a new global initiative to promote the development of an international legal instrument on forests. At the UN Conference on Environment and Development (Rio de Janiero, 1992) lengthy negotiations led to agreement on a set of non-binding principles on forests (Box 3) which may be considered a first step in this direction. The Convention on Biological Diversity (Chapter 8) and the Convention on Climate Change (Chapter 3) will

---

BOX 3

Forestry principles.

1. Forests are essential to economic development and the maintenance of all forms of life.
2. All aspects of environmental protection, social and economic development as they relate to forests and forest lands should be integrated and comprehensive.
3. National forest policies should recognize and duly support the identity, culture and rights of indigenous people, their communities and other communities and forest dwellers.
4. Efforts should be made towards the greening of the world.
5. Access to biological resources, including genetic material, shall be with due regard to the sovereign rights of the countries where the forests are located, and trade in forest products should be based on non-discriminatory and multilaterally agreed rules and procedures.

Source: UNCED (1992).

---

also be important vehicles for promoting the conservation of forests and other habitats. Wetlands are already the subject of an international Convention, the Ramsar Convention on Wetlands of International Importance, Especially as Waterfowl Habitat (Box 4) adopted in 1974 and having 55 Contracting Parties at the beginning of 1990 (Ramsar, 1990). Under this convention, projects for the wise use of wetlands are supported and a special fund has been established to support surveys and the development of management plans for designated wetland areas.

BOX 4

Principal Obligations of the Contracting Parties to the Ramsar Convention

1. To designate wetlands for the List of Wetlands of International importance (Article 2.1), to formulate and implement planning so as to promote conservation of listed sites (Article 3.1) and to advise the Bureau of any change in their ecological character (Article 3.2), to compensate for any loss of wetland resources if a listed wetland is deleted or restricted (Article 4.2), to use criteria for identifying wetlands of international importance and to establish 'shadow' lists.

2. To formulate and implement planning so as to promote the wise use of wetlands (Article 3.1), to make environmental impact assessments before transformations of wetlands, and to make national wetland inventories.

3. To establish nature reserves on wetlands and provide adequately for their wardening (Article 4.1), and through management to increase waterfowl populations on appropriate wetlands.

4. To train personnel competent in wetland research, management and wardening (Article 4.5).

5. To promote conservation of wetlands by combining far-sighted national policies with co-ordinated international action, to consult with other Contracting Parties about implementing obligations arising from the Convention, especially about shared wetlands and water systems (Article 5).

6. To promote wetland conservation concerns with development aid agencies.

7. To encourage research and exchange of data (Article 4.3).

Source: Ramsar, 1989.

REFERENCES

Adams, W. M. (1986) Traditonal agriculture and water use, Sokoto Valley, Nigeria. *Geographical Journal*, **152**, 30–44.

Adams, W.M. and Hollis, G.E. (1988) *The Hadejia-Nguru Wetland Project*. Report to IUCN, ICBP and RSPB.

Allen, J.C. and Barnes, D.F. (1985) The causes of deforestation in developing countries. *Annals of the Association of American Geographers*, **75**, 163-84.

Aselmann, I. and Crutzen, P.J. (1989) Global distribution of natural freshwater wetlands and rice paddies, their net primary productivity, scasonality and possible methane emissions. *Journal of Atmospheric Chemistry*, **8**, 307-58.

Braakhekke, W.G. and Marchand, M. (1987) *Wetlands: The Community's Wealth*. European Environment Bureau, Brussels. 24 pp.

Carrier, J.-G. and Krippl, E. (1990) Comprehensive study of European forests assesses damage and economic losses from air pollution. *Environmental Conservation*, **17**, 365-66.

Collins, N.M., Sayer, J.A. and Whitmore, T.C. (1991) *The Conservation Atlas of Tropical Forests: Asia and the Pacific*, Macmillan, London.

Dembner, S. (1991) Provisional data from the Forest Resources Assessment 1990 Project. *Unasylva*, **164(42)** 40-44, updated December 1991, personal communication.

Dugan, P.J. (ed.) (1990) *Wetland Conservation: a Review of Current Issues and Required Action*. IUCN, Gland, Switzerland.

FAO (1990) *FAO Yearbook. Forest Products 1977-1988*. FAO Forestry Series No. 23, FAO Statistics Series No. 90. FAO, Rome.

FAO (1991) Protection of Land Resources: Deforestation, Prepcomm. UNCED, 2nd Session, Doc. A/CONF.151/PC/27.

FAO/UNEP (1981) *Tropical Forest Resources Assessment Project* , 4 volumes. FAO, Rome.

Fearnside, P.M. (1986) Spatial concentration of deforestation in the Brazilian Amazon. *Ambio*, **15**, 74-81.

Forest Survey of India (1987) *The State of the Forest Report*. Government of India, Ministry of Environment and Forests.

Gadgil, M. (1991) Restoring India's forest wealth, *Nature and Resources*, **27**, 12-20.

Giglioli, M.E.C. (1980) Population demography and health. *In: Estudio Científico e Impacto Humano en el Ecosistema de Manglares*. Proceedings of a seminar, Cali, Colombia, 27 November -1 December 1978.

Glasby, G.F. (1991) A review of the concept of sustainable management as applied to New Zealand. *J. Roy. Soc. New Zealand*, **Vol. 21**, No. 2, June 1991, 61-81.

Goldsmith, E. and Hildyard, N. (1984) *The Social and Environmental Effects of Large Dams, Vol. 1: Overview*. Wadebridge Ecological Centre, Camelford, UK.

Gore, A.J.P. (1983) (ed). *Ecosystems of the World* (4A). Mires: Swamp, Bog, Fen and Moor. Vol. 1. Elsevier, Amsterdam.

Groot, P. de (1990) Are we missing the grass for the trees? *New Scientist* 6th January.

Hall, C.A.S. and Uhlig, J. (1991) Refining estimates of carbon released from tropical land-use changes. *Canadian Journal of Forest Research*, **21(1)** 118-31.

Hair, J. (1988) The economics of conserving wetlands: a widening circle. Paper presented at workshop on economics, IUCN General Assembly, Costa Rica, 1988. Gland, Switzerland, IUCN.

Hamilton, L.S. and Snedaker, S.C. (eds) (1984) *Handbook for Mangrove Area Management*. IUCN, Gland, Switzerland; UNESCO, Paris, France; East-West-Center, Hawaii, USA.

Harlow, C.S. and A.S. Adriano (1980) The Philippines dendrothermal power program. Proceedings of Bioenergy '80 Congress. Bioenergy Council, Washington D.C.

Holdgate, M.W., Kassas, M. and White, G.F. (1982) *The World Environment, 1972-1982*. A report by UNEP published by Tycooly International, Ireland.

IUCN/UNEP/WWF (1980) *The World Conservation Strategy*. IUCN, Gland, Switzerland.

IUCN/UNEP/WWF (1991) *Caring for the Earth: A strategy for Sustainable Living*. IUCN, Gland, Switzerland.

Lanly, J.-P. (1982) *Tropical Forest Resources* (Technical Report 4). FAO, Rome. 106pp.

Malingreau, J.P., Stephens, G. and Fellows, L. (1985) Remote sensing of forest fires: Kalimantan and North Borneo in 1982-3. *Ambio*, **14**, 314-21.

Matthews, E. (1983) Global vegetation and land use: new high-resolution databases for climatic studies. *J. Climate and Meteorology*, **22**, 474-87.

Matthews, E. and Fung, I. (1987) Methane emissions from natural wetlands: Global distribution, area and environmental characteristics of sources. *Global Biogeographical Cycles*, **1**, 61-86.

Mather, A.S. (1990) *Global Forest Resources*. Belhaven Press, London. 341 pp.

Mattson Marn, H. and Jonkers, W. (1980) Logging damage in tropical high forest. International Forestry Seminar, Kuala Lumpur, 11-15 November 1980.

McCormick, J. (1985) *Acid Earth: the Global Threat of Acid Pollution*. Earthscan, London. 191 pp.

Melillo, J.M, Palm, C.A., Houghton, R.A., Woodwell, G.M. and Myers, N. (1985) A comparison of recent estimates of disturbance in tropical forests. *Environmental Conservation*, **12**, 37-40.

Molofsky, J., Hall, C.A.S. and Myers, N. (1986) *A Comparison of Tropical Forest Surveys*. United States Department of Energy, Washington D.C. 66pp.

Nilsson, S. and Pitt, D. (1991) *Mountain World in Danger*. Earthscan Pubs. UK, London, 196 pp.

Nilsson, S., Salluäs, O. and Duinker, P. (1992a) *Future Forest Resources of Western and Eastern Europe*. Parthenon Publ. Group.

Nilsson, S., Salluäs, O., Hugosson, M. and Svidenko, A. (1992b) *Future Forest Resources of the European USSR*. Parthenon Publ. Group. (in press).

Odum E.P.(1985) Trends expected in stressed ecosystems. *BioScience*, **35**, 419-22.

OECD (1991) *The State of the Environment*. Organization for Economic Co-operation and Development, Paris.

Olson, J.S., Watts, J.A. and Allinson, L.J. (1983) *Carbon in Live Vegetation in Major World Ecosystems*. Environmental Sciences Division Publication No. 1997, Oak Ridge National Laboratory, Oak Ridge, Tennessee.

Pearce, D. (1991) Deforesting the Amazon: Toward an Economic Solution. *Ecodecision*, **Vol. 1**, No. 1, 40-49.

Poore, D. (ed.)(1989) *No Timber Without Trees*. IIED/Earthscan. London.

Prentice, K.C. and Fung, I.Z. (1990) The sensitivity of terrestrial carbon storage to climate change. *Nature*, **346**, 48-51.

Ramsar Convention Bureau (1990) *Directory of Wetlands of International Importance*. Gland, Switzerland. 796 pp.

Sader and Joyce, (1988) Deforestation rates and trends in Costa Rica, 1940-1983. *Biotropica*, **20**, 11-19.

Saenger, P., Hegerl, E.J. and Davie, J.D.S. (1983) *Global Status of Mangrove Ecosystems*. IUCN Commission on Ecology Papers Number 3, IUCN, Gland, Switzerland. 88 pp.

Scott, D.A. and Poole, C.M. (1989) *A Status Overview of Asian Wetlands*. Asian Wetland Bureau, Kuala Lumpur, Malaysia. No. 53.

Tiner, R.W. (1984) *Wetlands of the United States: Current Status and Trends*. US Fish and Wildlife Service, Washington D.C. 159 pp.

UNECE (1985) *The Forest Resources of the ECE Region. (Europe, the USSR, North America)*. UN-ECE, Geneva.

UNECE/UNEP (1990) *Forest Damage and Air Pollution; Report of the 1989 Forest Damage Survey in Europe*. UN Economic Commission for Europe and UN Environment Programme, Geneva and Nairobi.

UNEP (1991) *Environmental Data Report*. Third Edition. Basil Blackwell, Oxford.

Westoby, J. (1989) *Introduction to World Forestry*. Blackwell, Oxford. 228p.

Whitmore, T.C. (1990) *An Introduction to Tropical Forests*. Clarendon Press, Oxford.

Whitmore, T.C. and Prance, G.T. (1987) *Biogeography and Quaternary History in Tropical America*. Clarendon Press, Oxford.

Williams, M. (1989) Deforestation, past and present. *In: Progress in Human Geography*. Edward Arnold. Pp 176-208.

Williams, M. (1991) Forests. *In*: B.L. Turner II et al (eds). *The Earth as Transformed by Human Action*. Cambridge University Press, New York.

WRI (1986) *World Resources 1986*, Basic Books, New York.

WRI (1987) *World Resources, 1987* Basic Books, New York.

WRI (1990) *World Resources 1990-1991*. Oxford University Press, New York and Oxford. 3883 pp.

Zamora P.M. (1984) Philippine mangrove: assessment of status, environmental problems, conservation and management strategies. *In*: Soepadmo E., Rao, A.N. and McIntosh D.J. (eds). *Proceedings of the Asian Symposium on Mangrove Environmental Research and Management*. Pp.696-707. University of Malaya and UNESCO, Kuala Lumpur.

CHAPTER 8

Loss of biological diversity

The issue

The Earth's genes, species and ecosystems are the product of over three thousand million years of evolution, and are the basis for the survival of our own species. But the available evidence indicates that human activities are now leading to the loss of the planet's biological diversity, and as a consequence are eroding biological resources essential for future development. Given the projected growth in both human population and economic activity, the rate of loss of biodiversity is far more likely to increase than stabilize. As humans use more of nature's energy, natural life-support systems could begin to deteriorate on a global scale. According to one estimate, almost 40 per cent of the Earth's net primary terrestrial photosynthetic productivity is now directly consumed, diverted or wasted as a result of human activities (Vitousek *et al.*, 1986), an excellent indicator of the power of our ecological influence. The only way to reverse this trend is through the conservation of biological diversity.

Biological diversity means the variability among living organisms from all sources, including, *inter alia*, terrestrial, marine and other aquatic ecosystems and the ecological complexes of which they are part. This includes diversity within species, between species, and of ecosystems. *Biological resources* include genetic resources, organisms or parts thereof, populations, or any other biotic component of any ecosystem with actual or potential use or value to humanity. *Conservation of biological resources* means the preservation, maintenance, sustainable use, recovery and enhancement of the components of biological diversity. *Ex situ conservation* means the conservation of the components of biological diversity outside their natural habitats. *In situ conservation* means the conservation of ecosystems and natural habitats and the maintenance and recovery of viable populations of species in their natural surroundings and, in the case of domesticated or cultivated species, in the surroundings in which they have developed their distinctive properties.

Conserving biological diversity is important for reasons of both principle and human self-interest. The World Charter for Nature, adopted by the United Nations General Assembly in 1984, states that all species warrant respect regardless of their usefulness to humanity. Human self-interest is involved because ecosystems function as the planetary life-support system, renewing atmospheric oxygen and playing a central part in the biochemical cycle (Chapter 20). They are a source of food, fibre, timber, natural drugs and other products; they conserve soil; and they shelter genetic strains to which crop breeders continually return in order to improve cultivated varieties.

Biological diversity must be conserved to help humanity adapt to a changing environment. The importance assigned to conservation will depend on the values ascribed to species and ecosystems, and hence on economic and political judgements as much as on scientific understanding. Ecologists and other scientists must be ready to provide the technologies that will be required to implement the political decisions that are taken.

The past two decades have seen a considerable growth in understanding of the evolutionary processes that have created the biological diversity on Earth today, and the contemporary factors that are leading to its reduction. Trends in that reduction have been inferred and observed in many regions, and the importance of genetic conservation has been more and more widely accepted. The economic valuation of living natural resources has advanced greatly. New international activities have been undertaken, and new international legal instruments are being prepared. Many of these key activities are, however, far from completion: indeed they have not been proceeding fast enough to keep pace with the losses. Intensified international and national action is therefore essential in the coming decades. These actions should be seen as part of the wider programmes in public health, nutrition, sanitation, education and security that are essential to long-term sustainable development of global natural resources.

An overview of *biological diversity* is included in the UNEP 1991 *State of the World Environment* report (UNEP, 1991).

Historical perspective

Biological diversity is the variety and variability of life. *Genetic diversity* refers to the variation of genes within species, as expressed, for example, in the thousands of traditional rice varieties in Asia. *Species diversity* refers to the variety of species within a region, measured either as the total number of species present (sometimes called 'species richness') or as a combination of species numbers and distinctiveness ('taxonomic diversity'). *Ecosystems* are dynamic complexes of plant, animal and micro-organism communities and their non-living environment, interacting as an ecological unit. Ecosystems vary from place to place, and their diversity can be evaluated provided that they are described using a consistent set of criteria.

Over the past 3,000 million years, biological diversity has been steadily increasing but this expansion did not proceed smoothly. The fossil record over the past 600 million years has shown several 'mass extinctions', during which the general trend of growing diversity was rapidly reversed. Figure 1 shows both the trend and the mass extinctions for families of marine organisms.

Throughout the Earth's history, major taxonomic groups have emerged and diversified, only to decline and be replaced by other groups (Figure 2). The causes of this growth in diversity and the periodic extinctions and replacements that occurred are still only partly understood, but took place as continents drifted apart and rejoined, climates changed, new taxa evolved, massive volcanoes erupted, and asteroids hit the Earth (Quinn and Signor, 1989; Moses, 1989). The species that now dominate the Earth belong to genera that appeared during the Cenozoic era (which began about 65 million years ago and has lasted until the present time). For most of that period, the appearance and extinction of genera have been roughly in balance.

Extinctions have always been a fact of life: indeed, over 99 per cent of the species that have ever existed are now gone (Ehrlich and Ehrlich, 1982). But

Figure 1: *Increase in species diversity over geological time. The diversity of species, as shown in the fossil record of marine vertebrates and invertebrates, has increased remarkably over geological time, interrupted only by the five major extinction episodes shown on this graph.*

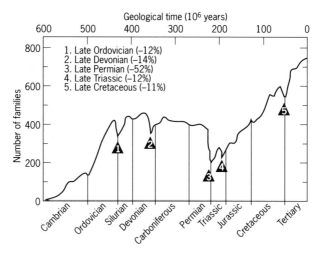

Source: Raup and Sepkoski (1982), by permission of the American Association for the Advancement of Science.

Figure 2: *Expanding numbers of families of terrestrial four-legged animals. Three assemblages of families, marked by roman numerals, succeeded each other through geological time: I. primitive reptiles; II. dinosaurs; III. modern forms. Six mass extinctions are indicated by the numbers 1 - 6.*

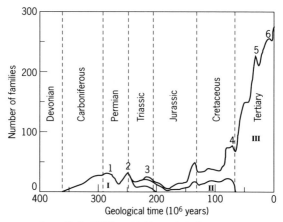

Source: Benton (1985), reprinted by permission from *Nature*, copyright Macmillan Magazines Ltd.

it seems clear that as people became more numerous and moved into new areas, they caused significant numbers of extinctions. Australia, the Americas, Madagascar and New Zealand appear to have suffered much greater species losses in the past 100,000 years, as humans settled in those continents and islands, than has Africa, where our species evolved (Figure 3).

The number of species now living on Earth can only be estimated. About 1.4 million have been described, and of these some 750,000 are insects, 250,000 are plants and only 41,000 are vertebrates. The best calculations suggest that the actual total is likely to be around ten million, but estimates range from two to one hundred million.

Species extinctions have increased steadily since 1600, mostly due directly or indirectly to human activity (Figure 4). About 75 per cent of the mammals and birds that have become extinct in the past 400 years were island-dwelling species, and these are known to be especially vulnerable to introduced species and other 'new' evolutionary pressures. Similarly, island floras tend to be far

Figure 3: *Losses of species of large animals on three continents and two large islands. These graphs, based on a logarithmic scale, show that the intensity and sequence of extinctions follow the development of human cultures. Extinction was less severe in Africa, where humans evolved, than in North America, Australia, New Zealand, and Madagascar, each of which was settled relatively late by humans and suffered heavy losses.*

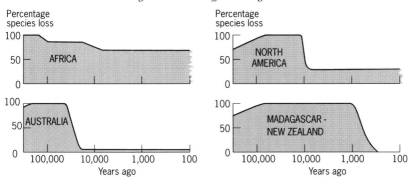

Source: Martin (1984), by permission of the University of Arizona Press.

Figure 4: *Number of species of birds and mammals known to have become extinct between 1600 and 1950. These extinctions appear to have increased remarkably since 1850, coinciding with growing human populations and industrial development.*

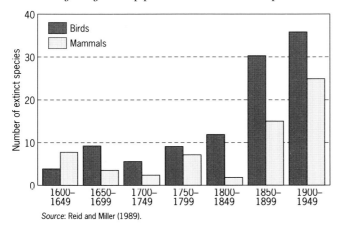

Source: Reid and Miller (1989).

187

more endangered than continental ones, and on several islands more than 90 per cent of the endemic plant species are rare, threatened or extinct.

Recent developments: 1972-92

Trends in biological diversity need to be considered at all three levels - ecosystems, species and genes - because different factors operate in each case, and the implications of diversity losses are different.

Ecosystems

Many people would say that major habitat changes are the inevitable price we pay for progress as humans become an ever more dominant species. But society has cause for concern when habitats are degraded to lower productivity, especially when those habitat losses can have world-wide ramifications.

Not all ecosystems and habitats are equally diverse. Deserts tend to have relatively low diversity while tropical forests have by far the highest numbers of species. Coral reefs may have far higher taxonomic diversity, because they support a wider range of major animal groups (phyla). Nor are all habitats of the same broad type equally important. Myers (1988, 1990), for example, has identified 18 'hot-spots' which together support about 50,000 endemic plant species (about 20% of the planet's total) in just 0.5 per cent of the Earth's land surface (Figure 5). Most of these habitats in tropical forest and Mediterranean climate ecosystems are under very considerable threat.

The loss of forests and wetlands has been described in Chapter 7, while Chapter 6 dealt with the desertification and land degradation that particularly affect the world's rangelands. Tropical forests support well over half the planet's species on only about six per cent of its land area. After Amazonia, the second largest block of rain forest in the world is the Indo-Malayan realm, which supports an estimated 25,000 species of flowering plants representing almost ten per cent of the world's flora. The rapid conversion of these forests has particularly serious implications for biological diversity. Table 1 gives estimates of the loss of original habitat in the Indo-Malayan realm by country (MacKinnon and MacKinnon, 1986). While wetlands are not noted for high species diversity or endemism, they are very complex and highly productive ecosystems and the effects of their loss are felt widely through disruptions of the hydrological cycle, destruction of habitats for migratory birds, and reduction of the productivity of fisheries. The degradation of coral reefs in many parts of the tropics has implications for biological diversity because, as noted above, they are probably the most diverse of all marine habitats.

Mediterranean-climate ecosystems - which often have high degrees of endemism (notably Southwest Australia with about 2,500 endemic plants, the Cape Region of South Africa with almost 6,000 endemic plants, the Mediterranean basin with about 12,500 plants endemic to the basin and over 2,500 endemic to individual countries, and California, with some 1,500 endemic plants) - are

Figure 5: *Hotspots in diverse habitats. This map shows 'hotspot' areas in tropical forest and Mediterranean habitats. These areas show both extremely high species diversity and high percentages of endemic species. Relatively small amounts of the original habitat remain in most of these areas.*

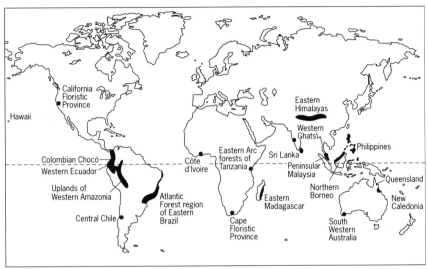

Source: Myers (1988 and 1990).

Table 1: *Loss of original habitat in the Indo-Malayan Realm.*

Country	Loss of original habitat (%)	Country	Loss of original habitat (%)
Bangladesh	94	Myanmar	71
Sri Lanka	83	Laos, PDR	71
India	80	Nepal	54
Vietnam	80	Indonesia	49
Philippines	79	Malaysia	41
Cambodia	76	Bhutan	34
Thailand	74		

Source: MacKinnon and MacKinnon (1986).

under threat from land developers and from introduced species. In fact, virtually all of the world's major habitat types are losing diversity.

Species

Calculations based on the rates of habitat transformation and the numbers of organisms restricted to particular areas, especially of tropical forest, lead to much higher extinction figures than those derived from direct observation.

189

Such estimates imply that species extinctions are now at an all-time high (see, for example, papers in Wilson, 1988). These calculations predict that if present trends continue, up to 25 per cent of the world's species will become extinct in the next several decades, and that there will be an equally alarming degradation of habitats and ecosystems.

A very considerable body of work in the field of conservation biology over the past several decades has shown that reducing the area of habitat reduces not only the population of each species, but also the number of species the habitat can hold. As a broad general rule, reducing the size of a habitat by 90 per cent will reduce the number of species that can be supported in the long run by about 50 per cent (Figure 6).

The size of an island is closely related to the rate of extinction of the species on the island, with small islands having a much greater chance of losing species (Figure 7). This general rule applies for most vertebrates, though bird

Figure 6: *Species-area curve. Since the number of species found in a region increases with the area of habitat in a predictable manner, it is possible to predict the effect of habitat loss on the number of species found in a region through the use of a 'Species-Area Curve'. The two curves show the difference in the relationship found in continental (upper curve) and island (lower curve) habitats.*

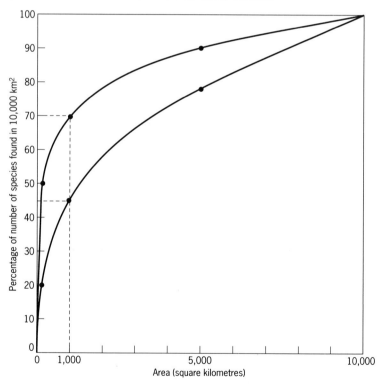

Source: Reid and Miller, (1989), copyright World Resources Institute.

Figure 7: *Extinction and island size. Based on data for breeding land birds of Northern European islands, on a time scale of one year, the risk of extinction is plotted as a function of island area. Risk of extinction decreases with increasing island area, a relationship that many ecologists believe also holds true with 'islands' of habitat on continents.*

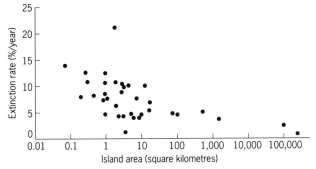

Source: Diamond (1984), by permission of the University of Arizona Press.

populations become extinct more than ten times faster than lizard populations on an island of the same area (Diamond, 1984).

Many ecologists believe that this relationship also holds true when formerly large areas of continuous habitat are fragmented into smaller habitat 'islands', separated by land that supports other kinds of ecosystems. Even relatively large National Parks are expected to suffer over a period of years from this kind of isolation, with the degree of species loss depending on the initial number of species as well as on the size of the protected area (Figure 8).

Figure 8: *Loss of species in isolated habitats. Projected losses of species have been estimated for large mammals in three East African National Parks, all of which are becoming increasingly isolated from the surrounding lands. The shaded area covers the probable range within which the trajectory will fall for each protected area.*

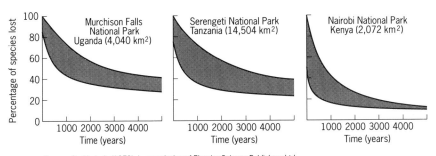

Source: Soulé et al., (1979), by permission of Elsevier Science Publishers Ltd.

As with other natural resources, the distribution of living species in the world is not uniform. Species richness tends to increase from the poles to the Equator. The species richness of freshwater insects, for example, is three to six times

191

higher in tropical areas than in temperate zones. Tropical regions also have the highest richness of mammal species per unit area, and vascular plant species diversity is much richer at lower latitudes (though this general pattern varies locally and regionally) (Reid and Miller, 1989; Box 1). Tropical forests are the most species-rich land ecosystems, while coral reefs, because of their complex ecological systems, support tremendous species diversity per unit area.

BOX 1

Species richness in different regions

Tropical regions have both more species and a higher diversity in a given area than temperate regions.

The total vascular plant flora of Denmark is only a little over 1,000 species, while the British Isles support about 1,700. Canada has about 3,270 native vascular plant species and 140 native tree species. In contrast, Peninsular Malaya alone contains nearly 8,000 species of flowering plants, from 1,500 genera, and the Malaysian region which includes that Peninsula and the Indonesian and Philippine archipelagos supports around 25,000 flowering plants - about 10% of the world's total. Colombia, which occupies less than 1% of the world's land surface, contains 10% of all species of animals and plants.

Forty species of trees may occur on one hectare of temperate forest in eastern North America, whereas a similar area in lowland Malaysia may support over 550 species with a stem diameter greater than 2cm. Some 700 species of tree have been identified in one 15-hectare area of rain forest in Borneo (Gentry, 1988).

A similar pattern can be seen in the marine environment. The number of tunicate (sea squirt) species increases from 103 in the Arctic to some 629 in the tropics. Planktonic foraminifera (a group of marine micro-organisms whose shells form chalk rocks) increase from only two species near the poles to some 16 in tropical waters. Canada has 1,100 species of fishes, while the Philippines has 2,200 marine and freshwater species.

Source: Reid and Miller (1989).

Because habitats are being reduced in all parts of the world, the populations of many species are being dramatically reduced, and this increases the rate of extinction (Soulé; 1989; Wilson, 1988). Figure 9 provides a real-life example, based on the breeding land birds of the California Channel Islands. Drawing on data collected over the past eight decades, breeding populations of each species on each island were grouped into abundance classes and compared with the percentage of the populations in that class that have become extinct during the twentieth century. Not surprisingly, the risk of extinction is highest for the rarest species.

Figure 9: *The risk of extinction. This curve shows the risk of extinction as a function of abundance, using data from breeding land birds of the California Channel Islands. The percentage of birds that have become extinct is directly related to their initial population size.*

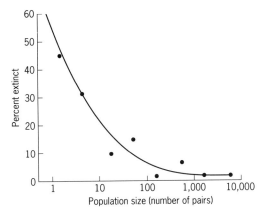

Source: Jones and Diamond (1976).

Many species that are not in immediate danger of extinction are suffering from declining populations and declining genetic variability. While some wild animal species - sparrows, pigeons, crows, starlings, opossums, rats, hedgehogs, raccoons, foxes, coyotes, several deer, and other opportunists - are expanding their ranges and populations, far more are suffering population declines. Low populations make species far more vulnerable to inbreeding, disease, habitat alteration and environmental stress.

Estimating precise rates of loss, or even the current status of species, is challenging because no systematic monitoring system is in place and much of the baseline information is lacking. For example, in order to say how much is being lost, we first need to know how much we have, yet, as noted above, estimates for the total number of species existing in the world vary from two million to a hundred million - most of them beetles living in tropical forests. Wilson (1985) has estimated the loss of species attributable to the loss of rain forests at somewhere between 0.2 and 0.3 per cent per year: if 1 million species are present, this amounts to 2,000 to 3,000 species a year; but if 10 million species are present, we could be losing as many as 30,000 per year.

While such figures are alarming, the best current knowledge indicates that we have not yet begun to experience the full impacts of species extinction. According to the latest IUCN figures, we have lost only about 0.15 per cent of plants since 1600, 1.2 per cent of birds, and 2 per cent of mammals (Table 2). These figures include only those species that can reasonably be assumed to have become extinct; many others may have been lost without our knowledge.

Many other species have been reduced to very low population levels and may be on an all-but-inevitable slide toward extinction. Some 10 per cent of mammals and birds are under a significant degree of threat, along with well over 7 per cent of the plants. All of these 'Red Data Book species' could disappear within the next few decades unless effective conservation measures are instituted.

Table 2: *Current status of species of plants and animals.*

	Total species	Number extinct	Percentage extinct	Number threatened	Percentage threatened
Plants	250,000	384	0.15	18,694	7.4
Fish	19,056	23	0.12	320	1.6
Amphibians	4,184	2	0.05	48	1.1
Reptiles	6,300	21	0.33	1,355	21.5
Birds	9,198	113	1.23	924	10.0
Mammals	4,170	83	1.99	414	10.0

Notes: 'Extinct' means species known to have become extinct since AD1600, subject to the limitations in knowledge quoted in the text.

Source: McNeely *et al.* (1990).

The emergence of biotechnology

For thousands of years, people have been manipulating the genetic wealth of biodiversity by selecting and breeding crops and livestock to meet their needs. The exploitation of plant and animal resources has been the mainstay of agriculture, forestry and fisheries activities, from which a vast variety of domesticated animals and plants has emerged. Fermentation has similarly been used for centuries around the world. But today, new biotechnologies are emerging that enable great increases to be achieved in the efficiency of traditional breeding programmes and that allow the modification of organisms in ways that were impossible using traditional techniques. Some of these new technologies, such as tissue culture, already have a record of application, but the most novel techniques, such as genetic engineering, are only today yielding their first commercial products.

The emergence of biotechnology and recent developments in recombinant DNA technology present potential for a link between conservation and sustainable utilization of genetic diversity. This relationship is in fact one of mutual dependence. On the one hand, biotechnology has much to offer to conservation of biological diversity. It could lead to new and improved methods of preservation of plant and animal genetic resources and speed the evaluation of germplasm collections for specific traits. On the other hand, maintenance of a wide array of biological diversity, and hence a wide genetic base, is important for the future of biotechnology and sustainable development. The genetic material contained in domesticated varieties of crop plants, trees and animals, and their wild relatives, is essential for breeding programmes by which genes are incorporated into commercial lines for the improvement of yields, nutritional quality, flavour, pest and disease resistance, and responsiveness to different soils and climates.

Accurate measurements of the genetic diversity within a population or species can be made, but limitations of data generally restrict such measurements to domesticated species and populations held in zoos or botanic gardens. Genetic diversity provides the variability with which a species can adapt to changing conditions. While this is important to all species, biodiversity in the form of genetic variability in cultivated and domesticated species has become a significant socio-economic resource. To provide just one example, new varieties of sugar cane in Hawaii have been required about every ten years, to adapt to pests and to maintain productivity (Figure 10). Without the genetic variability which enables plant breeders to develop new varieties, global food production would be far less than it is at present.

Figure 10: *Genetic diversity in sugar cane. The pattern of replacement of sugar cane varieties in Hawaii demonstrates the importance of frequent infusions of new genetic diversity in maintaining productivity, pest resistance and other desirable traits.*

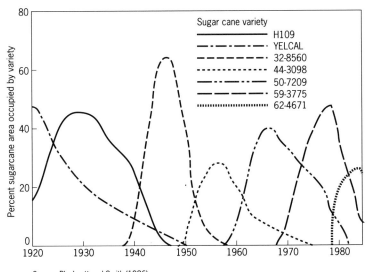

Source: Plucknett and Smith (1986).

The genetic variety inherent in a single species can be seen in the many races of dogs, cats or sheep, or the many specialized types of rice, potatoes, apples or maize developed by breeders. But whole races of cultivars are losing their genetic variability and thus their ability to withstand environmental change. The remaining gene pools in crops such as maize and rice amount to only a fraction of the genetic diversity they harboured only a few decades ago, even though the species themselves are not threatened and the various seed banks still retain many of the previously-cultivated forms. Little evolution and adaptation can take place in a seed bank. Thus in terms of biological diversity, both loss of species and loss of gene reservoirs are significant, and many agriculturalists argue that the loss of genetic diversity among domestic plants and animals looms as an even greater threat to human welfare than does the loss of wild species, because that diversity is what will enable crops to adapt to future changes.

The importance of genetic diversity in wild species should not, however, be ignored, for it is likely to prove crucial in their adaptation to changing climates and habitats. In fish, and probably other organisms, extinction of local races

Figure 11: *Vavilov centres of crop genetic diversity. The shaded areas indicate regions of high current diversity of crop varieties.*

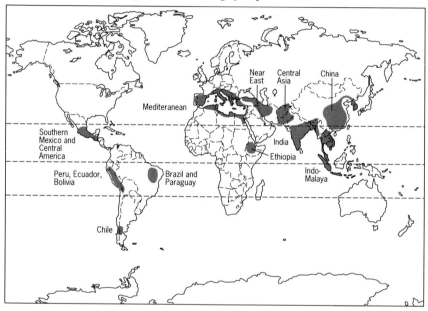

Examples of crops with high diversity in each area include:

China: Naked oat, soybean, adzuki bean, leaf mustard, apricot, peach, orange, sesame, China tea.
India: Rice, African millet, cucumber, tree cotton, pepper, jute, indigo.
Indo-Malaya: Yam, pomelo, banana, coconut.
Central Asia: Wheat (bread, club, shot), rye, pea, lentil, chickpea, sesame, flax, safflower, carrot, radish, pear, apple, walnut.
Near East: Wheat (einkorn, durum, bread), barley, rye, red oat, chickpea, pea, lentil, blue alfalfa, sesame, flax, melon, almond, fig, pomegranate, grape, apricot, pistachio.
Mediterranean: Durum wheat, hulled oats, broad bean, cabbage, olive, lettuce.
Ethiopia: Wheat (durum, Poulard, Emmer), barley, chickpea, lentil, pea, teff, African millet, flax, sesame, castor bean, coffee.
Southern Mexico and Central America: Corn, common bean, pepper, upland cotton, sisal hemp, squash, pumpkin, gourd.
Peru, Ecuador, Bolivia: Sweet potato, potato, lima bean, tomato, sea island cotton, papaya, tobacco.
Chile: Potato.
Brazil and Paraguay: Cassava (manioc), peanut, cacao, rubber tree, pineapple, purple granadilla.

Source: Reid and Miller (1989), copyright World Resources Institute.

through pollution, environmental change and over-exploitation is common, and the forms lost are not necessarily replaced by other forms of the same species. Cropping strategies - and especially destructive methods - alter community structures and may reduce the resilience of the exploited populations.

Not all parts of the planet are equally important for maintaining the genetic diversity of crops. Figure 11 shows 'Vavilov Centres' which have an exceptional level of crop genetic diversity, and where exceptional measures for conserving this diversity are justified.

Responses

The past 20 years have seen a resurgent concern over the loss of biodiversity with several kinds of response, outlined below.

The precautionary principle

From an ethical standpoint, many people would argue that our own species has no right to destroy other species, ecosystems, or even genes. But what about smallpox; and in ecological or more utilitarian terms, does humanity really need so much biological diversity? Nature certainly has some built-in redundancy, and some diversity could be lost without anybody noticing. But few data are available on which genes or species are particularly important in the functioning of ecosystems, so it is difficult to estimate the extent to which society is suffering from the loss of biological diversity. The impact of the loss of top predators may be obvious enough, but the ecological roles played by many species or populations are still only partly known. Because of this uncertainty about the roles of different components in determining the functioning of ecosystems, the wisest course is to apply the 'precautionary principle' and avoid actions that needlessly reduce biological diversity.

Improved information and data

Decisions for planning conservation, identifying priorities, and formulating management policies must be based on a rational analysis of the most complete and up-to-date factual information. Since the publication of the World Conservation Strategy in 1980 by IUCN, UNEP and WWF, some 50 national-level conservation analyses and strategies have been produced. What these planning documents show is the general paucity of reliable detailed information suitable for planning and management purposes. Moreover, baseline information on the status and distribution of species and ecosystems that can serve as a benchmark for monitoring is virtually non-existent in developing countries. Efforts such as country Tropical Forestry Action Plans (TFAPs), Environmental

Action Plans (World Bank), Environmental Profiles (US-AID) and National State of the Environment Reports (UNEP) have been partially successful in remedying the situation. Recognizing this information deficiency, the three partners in the World Conservation Strategy in 1988 established the World Conservation Monitoring Centre (WCMC) to gather, analyse and disseminate data on global biodiversity, and WCMC has recently produced a major report on the status, management and utilization of the Earth's living resources (WCMC, 1992).

Both this information and the series of in-depth country studies of biodiversity that UNEP is initiating will contribute to the effective implementation of the Convention on Biodiversity that is currently being negotiated. The information-gathering and monitoring capabilities of national institutions are being expanded, and are being linked into a global biodiversity information network. IUCN's major programmes on species conservation (through the Species Survival Commission), habitat management, and protected areas (through the Commission on National Parks and Protected Areas) are all augmenting the information base essential for biodiversity conservation. UNESCO, IUBS and SCOPE have recently launched a new programme on the relationship between diversity and ecosystem function (Box 2).

BOX 2

IUBS/SCOPE/UNESCO programme on biodiversity.

In 1990-91 the International Union of Biological Sciences (IUBS), the Scientific Committee on Problems of the Environment (SCOPE) and UNESCO launched a collaborative programme on biodiversity. The main aim is to improve knowledge of the functional role of biodiversity at the species, community, ecosystem and landscape levels and, in consequence, to strengthen the scientific basis for their management.

The programme has four themes, involving syntheses of existing knowledge, research design, and long-term studies of:

- the ecosystem function of biodiversity;

- origins and maintenance of biodiversity;

- inventory and monitoring of biodiversity;

- biodiversity of wild relatives of domesticated species.

Marine and microbial aspects are considered in all themes. The programme will have an initial life of ten years, although the monitoring activities will last for a longer period.

Source: Solbrig (1991).

Improved economic evaluations

In addition to the well-known values of agriculture, fisheries and forestry, wild species and the genetic variation within them make contributions to agriculture, medicine and industry worth many billions of dollars per year. Nevertheless, biodiversity tends to be perceived largely in scientific and conservationist terms rather than in economic and resource terms. Thus the issue lacks 'political clout' (WCED, 1987). This situation is partly the result of the fact that biodiversity researchers and managers have often been unable to provide sufficient socio-economic insight to quantify the socio-economic benefits of nature conservation. Economics can help demonstrate the importance of biological diversity to human society by assigning values to the full range of goods and services it provides. Careful analysis of the full costs and benefits of conservation action will often justify much greater investments in conservation.

'Consumptive use values' - from resources that are consumed directly, without passing through a market - are often the foundation of community welfare in rural areas. For example, firewood and dung provide over 90 per cent of the total primary energy needs in Nepal, Tanzania and Malawi, and exceed 80 per cent in many other countries. One study of four indigenous Amazonian Indian groups found that they used from half to two-thirds of all forest trees as food, construction material, raw material for other technology, medicinals and trade goods (virtually all species were used as firewood or as food for harvested animals). Conventional measures of economic performance, such as GNP, have tended to ignore this very extensive use when calculating the annual income of such groups, even though the value of replacing such goods from other sources would be considerable.

In Africa, harvested species help feed rural people, especially the poorest villagers living in the most remote areas. In Botswana, over 50 species of wild animals provide animal protein exceeding 90 kilograms per person per year in some areas: over 3 million kilograms of meat are obtained yearly from springhare (*Pedetes capensis*) alone. In Ghana, about 75 per cent of the population depends largely on traditional sources of protein supply, mainly wildlife, including fish, caterpillars and snails. In Nigeria, game constitutes about 20 per cent of the mean annual consumption of animal protein by people in rural areas (including 100,000 tonnes of the two giant rat species (genus *Thrynomys*) known as 'grasscutters'), while 75 per cent of the animal protein consumed in Zaire comes from wild sources (McNeely, 1988).

'Productive use value' is assigned to products that are harvested commercially for exchange in formal markets, and is therefore often the only value of biological resources reflected in national income accounts. Productive use of such biological products as fuelwood, timber, fish, animal skins, musk, ivory, medicinal plants, honey, beeswax, fibres, gums, resins, rattans, construction materials, ornamentals, animals sold as game meat, fodder, mushrooms, fruits, dyes, and so forth can have a major impact on national economies.

Such values can be remarkably high. Some 4.5 per cent of GDP in the USA is attributable to the harvest of wild species, amounting to some $US87 billion per year from 1976 to 1980 (Prescott-Allen and Prescott-Allen, 1986). The percentage contribution of wild species and ecosystems to the economies of developing agrarian countries is usually far greater than it is for an industrialized country. Timber from wild forests, for example, is the second leading foreign exchange earner for Indonesia (after petroleum), and throughout the humid tropics, governments have based their economies on the harvest of wild trees: total exports of wood products from Asia, Africa and South America averaged $US8.1 billion per year between 1981 and 1983.

While market prices represented by productive use value can be an important indicator of value, the market price is not always an accurate representation of the true economic value of the resource, and does not deal effectively with questions of distribution and equity. It is also apparent that consumers may value resources differently: for example, tropical forests are valued by consumers of scenic beauty very differently from consumers of lumber products. The methodology for defining and relating these different valuations is still being developed (Filion, 1990; McNeely, 1988).

Species without consumptive or productive use may nevertheless play important roles in an ecosystem, supporting species that do have such uses. In Sabah, Malaysia, for example, recent studies suggest that high densities of wild birds in commercial *Albizia* plantations limit the abundance of caterpillars that would otherwise defoliate the trees: the birds require natural forest for nesting.

All species are parts of ecosystems which provide services of very considerable - but seldom calculated - value to humans (Box 3). These services are often 'public goods' which benefit the entire community or the whole world, without a cost being assessed.

Moreover, as has been demonstrated in Nepal (Wells, 1991), while these benefits may be enjoyed within the country itself, many benefits from conservation are realized outside the country's borders, in forms as diverse as reduced flooding because of the protection of upland forests, the supply of medicinal plants and genetic material, or the pleasure given to international tourists. For these reasons, the costs of conserving biological diversity need to be shared internationally. Current evidence of the impacts of human activities on natural ecosystems suggests that far greater investments are required in maintaining the continued productivity of these ecosystem services.

Information is gradually accumulating on the economic benefits derived from using genetic diversity to improve crop production by conventional breeding, or the use of plant-derived drugs. There are signs of great promise and profits.

The examples in Box 4 indicate the relevance of economic evaluation of the added value of biological resources. In order to facilitate the integration of economic considerations into conservation concerns and firmly anchor nature conservation programmes into national development plans, managers need to adopt a proactive approach in their tasks. Figure 12 contrasts the traditional defensive approach, essentially typified by a stress approach in which the main

---

BOX 3

Indirect benefits of natural ecosystems.

Many natural ecosystems provide benefits that are indirect, having economic value through services rather than products. Most such benefits will fall into one or another of the following categories:

- Photosynthetic fixation of solar energy, transferring this energy through green plants into natural food chains, and thereby providing the support system for species that are harvested.
- Ecosystem functions involving reproduction, including pollination, gene flow, cross-fertilization; maintenance of environmental forces and species that influence the acquisition of useful genetic traits in economic species; and maintenance of evolutionary processes, leading to constant dynamic tension among competitors in ecosystems.
- Maintaining water cycles, including recharging groundwater, protecting watersheds, and buffering extreme water conditions (such as flood and drought).
- Regulation of climate conditions, both macro-climatic and micro-climatic (including influences on temperature, precipitation and air turbulence).
- Production of soil and protection of soil from erosion, including protection of coastlines from erosion by the sea.
- Storage and cycling of essential nutrients, e.g. carbon, nitrogen and oxygen; and maintenance of the oxygen–carbon dioxide balance.
- Absorption and breakdown of pollutants, including the decomposition of organic wastes, pesticides and air and water pollutants.
- Provision of recreational-aesthetic, socio-cultural, scientific, educational, spiritual and historical values of natural environments.

Source: McNeely (1988).

---

role of managers is to mitigate adverse impacts, with the alternative proactive one in which managers strategically employ the benefits of flora, fauna and their habitats as incentives to promote conservation initiatives in accordance with the concept of sustainable development. As the diagram shows, both approaches should be considered essential and complementary.

The increasing role of biotechnology

Recent advances in biotechnology research and development permit a better understanding of how genes are expressed in the plant or animal concerned. This knowledge is now being used to speed the development and use of germplasm in modern crop varieties and to develop new varieties. Using this knowledge and germplasm, biotechnology offers new possibilities for increasing

BOX 4

Economic benefits of biodiversity.

In agriculture:

• In Asia, by the mid-1970s, improvements using genetics had increased wheat production by $US2 billion and rice production by $US1.5 billion per year by incorporating dwarfism into both crops.

• A 'useless' wild wheat plant from Turkey was used to give disease resistance to commercial wheat varieties worth $US50 million annually to the United States alone.

• One gene from a single Ethiopian barley plant now protects California's $US160 million annual barley crop from yellow dwarf virus. Major cultivars of crops improved by wild genes have a combined farm sales import value of $US6 billion a year in the United States.

• An ancient wild relative of corn from Mexico - a perennial, resistant to seven major corn diseases and which can grow at high elevations in marginal soils - can be crossed with modern annual corn varieties with potential savings to farmers estimated at $US4.4 billion annually world-wide (Witt, 1985).

In medicine:

• Of all useful plant-derived drugs, only ten are synthesized in the laboratory: the rest are still extracted from plants.

• In the industrialized world, the retail value of plant-derived drugs was estimated at $US43 billion in 1985 and it is estimated that the Western market for herbal drugs could reach $US47 billion by the year 2000 (McNeely, 1988).

• In 1960, a child suffering from leukaemia had only one chance in five of survival. Now, the child has four chances in five, due to treatment with drugs containing active substances discovered in the rosy periwinkle (*Catharanthus roseus*), a tropical forest plant originating from Madagascar (Akerle, Heywood and Synge, 1991). Commercial sales of drugs from this plant now total around $US100 million a year world-wide.

• Today, with advances in plant biotechnology and the availability of new and precise screening tools, interest in plants as a source of raw materials for developing new medicinal products is expanding.

Figure 12: *Complementary approaches for conserving flora, fauna and their habitats for sustainable development.*

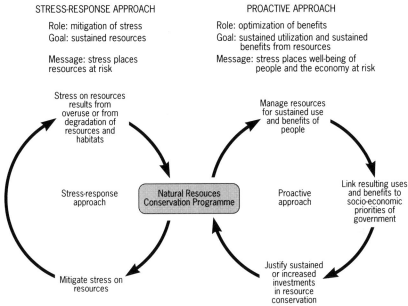

STRESS-RESPONSE APPROACH

    Role: mitigation of stress
    Goal: sustained resources

    Message: stress places
    resources at risk

PROACTIVE APPROACH

    Role: optimization of benefits
    Goal: sustained utilization and sustained
        benefits from resources

    Message: stress places well-being of
        people and the economy at risk

Stress on resources results from overuse or from degradation of resources and habitats

Manage resources for sustained use and benefits of people

Stress-response approach

Natural Resouces Conservation Programme

Proactive approach

Link resulting uses and benefits to socio-economic priorities of government

Mitigate stress on resources

Justify sustained or increased investments in resource conservation

Source: Adapted from Runka and Filion (1989).

the production of food, medicines, energy, speciality chemicals and other raw materials and for improving environmental management. This reinforces rather than diminishes the need to maintain the richest possible pool of genes. As the field of biotechnology develops, the future needs for germplasm will be far greater than is currently the case. The high stake which the biotechnology industry holds in the conservation of biodiversity should not be underestimated - the projected loss of diversity could cripple the genetic base required for the continued improvement and maintenance of currently utilized species and deprive us of the potential to develop new ones.

The issues relating to access to biological diversity and biotechnology are important and complex. The potential of genetic diversity can best be exploited when genes remain freely accessible to all users and when the information and technology on how to use them is equally freely transferable to all. Having a particularly valuable genetic resource or the technical capabilities to develop new varieties must not imply exclusive rights of ownership. In this regard, free access does not in any sense mean free of charge. Although two-thirds of all species are found in developing countries, particularly in the tropics, it is the developed nations that have most of the biotechnological tools needed to exploit them. While possession and custody of a potential genetic resource might be limited to one nation, benefits can accrue to all nations. Accordingly, a fair balance of benefits between owner and consumer needs to be found, involving co-operation with reciprocal benefits between developing countries and industrialized countries. In recent years, biotechnology has been seen as the direct channel by which developing countries can, in a practical way, tap their enormous biodiversity for economic development.

The new biotechnologies increase the value of the world's biodiversity because they allow increased use of the genetic diversity of wild and domesticated species. But biotechnologies also pose significant ecological and economic risks that could ultimately undermine their potential contribution to biodiversity conservation. The introduction of any 'novel' organism poses a risk to the environment. The majority of the world's known extinctions have probably been caused primarily by the introduction of exotic species. The release of genetically engineered organisms into the environment thus deserves the most careful oversight and monitoring (Mooney and Bernardi, 1990).

Measures to conserve biological diversity

Protected areas

Protected areas provide the most effective mechanism for conserving wild biodiversity, and most countries today have established at least some such areas. Many countries today have reasonably adequate protected area systems, with several having over 15 per cent of their territory under protected status. Over 6.5 million square kilometres have been established under national legislation, amounting to some 4.9 per cent of the world's land area (oceanic protected areas lag somewhat behind this figure).

Figure 13 shows the remarkable growth that has occurred since 1972, with the total number of protected areas nearly doubling and the area protected increasing by over 60 per cent. The increases are evenly distributed over the different categories of protected areas, but it appears that the rate of increase has begun to slow, and the options for additional protected areas - especially in the more strictly protected categories - seem to be closing quickly in the face of competition from other forms of land use. However, the percentage of protected land can be a misleading statistic, not necessarily reflecting accurately the degree or efficiency of the protection provided, as many of these protected areas have been established and managed for some other reason. For example, in Sri Lanka, only 10 per cent of the land in protected areas is located in the region of highest concentration of biological diversity - the wet lowlands and mountain forests in the southwest. Another difficulty arises because in many countries protected areas do not receive the active protection and management they require. Encroachment by settlers, poachers and other exploiters is widespread. In other areas, armed conflict has severely damaged protected areas.

Figure 13: *Growth of the world's protected area network. The graph line shows the number of sites (over 1,000 ha, in IUCN Categories I to V); the bars show the area in thousand square kilometres.*

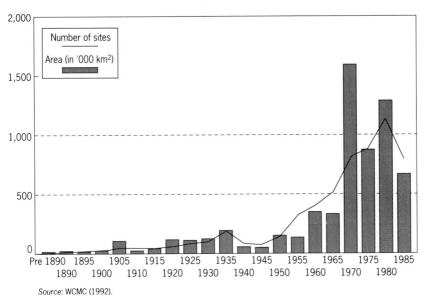

Source: WCMC (1992).

A variation on the traditional park approach to conservation is the UNESCO Biosphere Reserve Programme, introduced in the early 1970s under its Man and Biosphere Programme (MAB). The goal was to establish a series of protected areas, linked together by a co-ordinated international network, with the objective of demonstrating the benefits of conservation to society and the development process. This goal was to be achieved by emphasizing the combination of different biosphere reserve management components, such as scientific research, environmental education, training, environmental monitoring and local participation. Multiple use and zoning of the reserve into different areas are central to the concept (Figure 14). The first biosphere reserves were designated in 1976. The current emphasis is on three major thrusts: improving and expanding the network; using the network locations as permanent sites for research and monitoring; and making conservation in biosphere reserves socially acceptable by combining conservation with development and emphasizing the participation of local people. By late 1991, the international network included 300 biosphere reserves in 75 countries, with a total area of somewhat over 1.5 million square kilometres, and the concept was gaining increasingly wide support (Batisse, 1990).

Several existing international Conventions contribute to the maintenance of protected areas. Under the Ramsar Convention, (see Chapter 7) Contracting Parties list sites of international importance as wetlands and waterfowl habitat, undertake an obligation to safeguard them, and may benefit from an international fund established to assist developing countries to survey sites and

Figure 14: *A model biosphere reserve. A totally protected core zone is surrounded by a buffer zone within which scientific research, carefully controlled tourism and limited traditional land uses are permitted.*

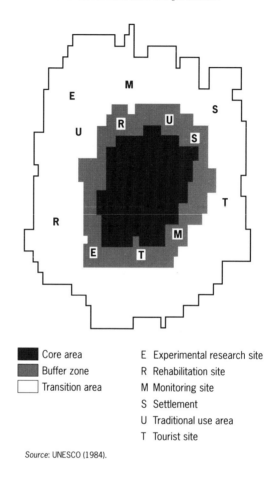

■ Core area	E Experimental research site
▨ Buffer zone	R Rehabilitation site
□ Transition area	M Monitoring site
	S Settlement
	U Traditional use area
	T Tourist site

Source: UNESCO (1984).

work out appropriate management plans. Under the Convention on the Protection of the World Cultural and Natural Heritage, adopted by UNESCO in 1972 and now with 119 nations as Parties, outstanding natural and cultural sites are evaluated and, if approved, added to a World Heritage list. IUCN provides technical advice on proposed natural sites, and under the Convention assistance is available for survey and management, and the status of sites is monitored.

IUCN, especially through its Commission on National Parks and Protected Areas (CNPPA), has a central role in defining various categories of protected area and for developing methods for their management. It is also devoting increasing attention to the question of how biological diversity can best be safeguarded outside specific protected areas.

Species and genetic conservation

The protection of species - especially those that people find particularly attractive, or that are in evident danger - has been a central theme of conservation, especially in developed countries, for many years. The work of bodies such as the Species Survival Commission of IUCN, and their publications, have had a world-wide impact. National legislation to protect species of mammals, birds, reptiles, amphibia, plants and some other taxonomic groups are widespread, and extended by international Conventions such as those on International Trade in Endangered Species (CITES) and on Migratory Species (Bonn) (see Chapter 23).

In recent decades the scientific community has begun to respond vigorously to the scientific challenges posed by issues of biodiversity. For example, whether the true basis for specific diversity is the species itself or its race or variety, the management practices to be instituted constitute an issue for scientific clarification. Another practical point on which intensive research is under way is understanding the extent to which cropping and diversification strategies can destroy or conserve biodiversity, through, for example, increased heterozygosity (Soulé, 1989; Woodruff, 1989).

In response to the threat of the loss of genetic diversity, two important international institutional initiatives have been launched since 1972. The International Board for Plant Genetic Resources (IBPGR) was established in 1974 under the umbrella of the Consultative Group on International Agricultural Research (CGIAR). IBPGR has played a catalytic role in developing effective national and international crop genetic resource conservation efforts. Focusing largely on major international crops such as wheat, rice and corn, IBPGR provided technical assistance and funding to establish national and international seedbanks and to collect a large fraction of the varieties of these crops. More recently, IBPGR has shifted its priorities to crops of regional or national importance and has emphasized training and human capacity building.

In the early 1980s, the Commission on Plant Genetic Resources was established in FAO. This Commission and the associated International Undertaking on Plant Genetic Resources have been instrumental in elevating concern over genetic resource ownership and access to diplomatic levels. Guided by the work of the Commission, the debates now taking place over intellectual property rights and genetic resources reflect a positive new thrust in international conservation. The debates reflect the growing awareness of the potential values of genetic resources and an active interest in the steps needed both to conserve those resources and to benefit from their conservation. After a decade of discussion, the principle of 'farmers' rights' is now widely accepted as a useful counterpoint to the concept of 'breeders' rights'.

It is often forgotten that human welfare is heavily dependent upon the microbial biodiversity pool. The Stockholm Conference recognized this and recommended concerted international action to tap the enormous potential of microorganisms. One response was the establishment jointly by UNESCO and

UNEP of a series of interlinked regional microbiological resource centres (MIRCENs), each consisting of a regional network of collaborating institutions. By 1985 such MIRCENs had been established in Bangkok (Thailand), Cairo (Egypt), Dakar (Senegal), Guatemala City (Guatemala), Nairobi (Kenya) and Porto Alegre (Brazil). These developing country MIRCENs are linked directly for supportive services to well-established and reputable microbiological research centres and institutions in developed countries. In addition, a World Data Centre for Microorganisms (WDC) exists at Riken, Japan, as a global register of computerized storage, retrieval and distribution of microbial genetic resources information. By 1985 the WDC had registered 380 culture collections from 53 countries. The International Microbial Strain Data Network (MSDN) established in the United Kingdom also serves as a global referral system for information on microbial strains and collections. (World Data Centre, 1985; Zedan and Olembo, 1988). Further progress was made in international efforts to harness microbiology and biotechnology for development when, at the initiative of UNIDO, the International Centre for Genetic Engineering and Biotechnology (ICGEB) was established in 1984 with components in New Delhi (India) and Trieste (Italy).

The challenge that these and other institutions now face is to develop new instruments, intellectual property regimes, funding mechanisms and technologies that will provide effective incentives for encouraging the conservation of genetic resources, encouraging local and national innovation related to biological resources, and justly compensating the custodians of the world's genetic resources. While genetic resources are of common interest to all people, nations have a right to treat them as sovereign resources; in addition, local and private innovators of biological diversity have a right to treat them as local resources. Reaching a concensus by which these interests are balanced for the benefit of people everywhere is a major challenge to the negotiators for a Global Convention on Biodiversity.

Increasing financial resources for conservation

Over the past two decades, very considerable increases in funding have become available. For example, so-called 'debt-for-nature swaps' involve the purchase of a country's debt notes which are discounted on the secondary market. These notes are presented to the debtor country in exchange for local currency in the amount of the face value of the debt, with the local currency being invested in conservation. Dogse and von Droste (1990) have provided guidelines to assist developing countries seize this opportunity to finance their conservation programmes (Box 5). As of mid-1990, debt-for-nature programmes have been established in Costa Rica, Ecuador, Bolivia, Dominican Republic, Argentina, Peru, Madagascar, Zambia, the Philippines, Sudan and Poland. However, added together, the exchange sums barely exceed $US300 million face value of debt, of which $US96.01 million has been exchanged for a total of $US58.16 million in conservation funds.

BOX 5

Summary of debt-for-nature guidelines for the debtor country, the conservation investor and the creditor bank.

Debtor government

- Establish debt-for-nature exchange programmes.

- Try to keep track of who owns the country's debts.

- Support local management debt-for-nature concept.

- Include representatives of residents and interest groups in the planning process.
- Inform the public about the functioning of debt-for-nature exchanges, including the sovereignty issue.
- Minimize inflationary risks.

Conservation investor

- Build good working relations with banks and financial institutions.
- Initiate debt-for-nature exchange discussions as part of regular management consultations with governments.
- Calculate and compare net present values for different financing arrangements.
- Try to safeguard the real value of swap proceeds.

- Consider potential resource use conflicts carefully, and safeguard the rights and needs of local people.
- Develop working relationships with local partners.
- Co-ordinate programmes with those of other debt-for-nature investors.
- Do not forget the purpose of the exchange.
- Work with a long-term perspective.

Creditor bank

- Initiate discussions with interested debt exchange investors.
- Inform shareholders about benefits from debt-for-nature swaps.
- Investigate the possibility of combining debt-equity negotiations with debt-for-nature exchanges.
- Try to maximize the debt portfolio's debt-for-nature exchange potential.

- Co-operate with other banks so as to minimize the risks of competing banks acting as free-riders, and of increased moral hazard reducing the debt service discipline.

- Consider adopting a green policy, including making an assessment of the bank's credit's environmental impacts.

Source: Dogse and von Droste (1990).

Biological diversity conservation components are also being included as part of larger projects funded by international or bilateral development agencies, with the World Bank being a leader. The conservation of biological diversity is one of four international issues eligible for support through the Global Environmental Facility, established on a three-year pilot basis by a number of nations in 1989, and administered by the World Bank, UNEP and UNDP. The GEF has a total pool of resources of around $US1.3 billion to allocate, and biodiversity conservation projects are receiving a significant fraction of this sum. The programmes of other organizations working in this field, including IUCN and a number of national and international conservation NGOs, have also become expanded in response to increased governmental and public concern and support. Expenditure by NGOs in biodiversity activities over the past 20 years has amounted to some $US2 billion, and the magnitude of NGO investment in biodiversity continues to grow.

Development of strategies and legal instruments

The major problems of conserving biological diversity lie not in the biology of the species concerned but rather in the social, economic and political arenas within which people operate. Biological diversity has global, national and local dimensions, but actions at these various levels are not well co-ordinated. Concerted action by governments, international organizations, NGOs and private citizens is therefore required.

As stated earlier, several existing regional and international Conventions play an important part in the conservation of biological diversity. The Ramsar and World Heritage Conventions are important instruments for safeguarding key sites. The Convention on International Trade in Endangered Species (CITES), adopted in Washington in 1973, the Convention on the Conservation of Migratory Species of Wild Animals (Bonn, 1979), the Convention on the Conservation of Antarctic Marine Living Resources, and the International Convention for the Regulation of Whaling are examples of measures that contribute by safeguarding particular categories of species.

Concluding remarks

Current measures to slow down diversity losses are valuable but insufficient. A more co-ordinated campaign is required, backed by a more comprehensive legal instrument. Working together, UNEP, IUCN and WWF have prepared a new version of the World Conservation Strategy, designed to meet the needs of the 1990s (*Caring for the Earth: A Strategy for Sustainable Living*: IUCN/ UNEP/ WWF, 1991). In order to carry out the broad prescriptions of this new strategy,

UNEP, WRI and IUCN, in consultation with FAO, UNESCO and a large number of government agencies, NGOs and individual experts, have prepared a Global Biodiversity Strategy (WRI/UNEP/IUCN, 1992) which includes guidelines for action to save, study and use the Earth's biotic wealth sustainably and equitably. This document includes 85 specific actions in 5 main fields:

- To develop national and international policy frameworks that promote the sustainable use of biological resources and the conservation of biodiversity.

- To create conditions and incentives for effective conservation by local communities.

- To increase the number and effectiveness of protected areas, gene banks, zoos and botanic gardens.

- To develop environmental awareness and strengthen the human skills and training needed to conserve biodiversity, particularly in developing countries.

- To catalyse conservation through international agreements and national planning.

BOX 6

An International Convention on Biological Diversity.

- Under the auspices of UNEP, an Intergovernmental Negotiating Committee (INC) was established to negotiate a treaty for the conservation and rational use of biological diversity.

- The objective of the convention is to conserve biological diversity, to the maximum extent possible, for the benefit of present and future generations and for its intrinsic value.

- Key elements of the convention are:
  1. *In situ* and *ex situ* conservation of wild and domestic species.
  2. Promotion and enhancement of research, education, training, public awareness and scientific and technical co-operation.
  3. Carrying out surveys on and inventories of biological diversity.
  4. Ensuring access to genetic resources and to relevant technologies.
  5. Handling biotechnology and the distribution of its benefits.
  6. Benefiting from and rewarding traditional indigenous and local knowledge.
  7. Securing new and additional financial resources for the implementation of the convention.

The approach adopted in the Global Biodiversity Strategy recognizes that in the coming years, improvements in the conservation of biological diversity will increasingly come from improved management of land and water outside protected areas, and that the areas themselves will need to be managed as part of wider land use programmes. These wider programmes will include diversification of agriculture and cropping strategies, encouragement to the spread of new varieties and non-traditional crops, rationalization of cropping techniques so as to minimize ecological damage, and strategies for the rehabilitation and diversification of damaged habitats.

It is also clear that the conservation of global biodiversity will require new financial mechanisms. Some of the large profits made from biodiversity must be ploughed back into the conservation of that diversity. Taxes, royalties, donations, sponsorships, bilateral agreements and one-off cash payments are promising mechanisms to be explored. Despite the substantial growth in resources allocated to this field in recent years, funding levels are still far too low in proportion to the magnitude of the task. In developing action further, it is important that local communities are involved, especially in developing countries, as active participants in the new land use and protected area strategies that are required.

An International Convention on Biological Diversity was first proposed as long ago as 1974, and has been in the process of active development since 1983 (Box 6). It has been negotiated by governments under the leadership of UNEP throughout 1990 and 1991, and was presented at the United Nations Conference on Environment and Development in June 1992.

REFERENCES

Akerle, O., Heywood, V. and Synge, H. (1991) *The Conservation of Medicinal Plants.* Cambridge University Press, Cambridge.

Batisse, M. (1990) Development and Implantation of the Biosphere Reserve Concept and Applicability to Coastal Regions, *Env. Conservation*, **17**, 111-16.

Benton, Michael J. (1985) Mass Extinction Among Non-Marine Tetrapods. *Nature* , **316**, 811-13.

Diamond, Gerard M. (1984) Historic Extinctions: A Rosetta Stone for Understanding Prehistoric Extinctions. pp824-862 in: Martin, P.S. and Klein, R.G. (eds.) *Quaternary Extinctions: A prehistoric revolution.* The University of Arizona Press, Tucson. 892pp.

Dogse, P. and von Droste, B. (1990) *Debt-for-nature exchanges and biosphere reserves: Experiences and Potential. MAB Digest 6.* UNESCO, Paris, FRANCE.

Ehrlich, Paul and Ehrlich, A. (1982) *Extinction: The causes and consequence of the disappearance of species.* Victor Gollancz, Ltd. London.

Filion, F.L. *et. al.*, (1990) *The importance of wildlife to Canadians in 1987: The economic significance of wildlife-related recreational activities.* Environment Canada.

Gentry, A.M. (1988) Tree species richness of upper Amazonian forests. *Proc. US National Academy of Science*, **vol. 85**, p.156.

IUCN/UNEP/WWF (1991) *Caring for the Earth: A Strategy for Sustainable Living.* IUCN, Gland, Switzerland.

Jones, H.L. and Diamond, J. M. (1976) Short-time Base Studies of Turnover in Breeding Bird Populations on the California Channel Islands. *Condor*, **78**, 526-49.

Martin, Paul S. (1984) Prehistoric Overkill: The global model. pp 354-403 in Martin, P.S. and Klein, R. G. (eds.) *Quaternary Extinctions: A prehistoric revolution.* The University of Arizona Press, Tucson.

MacKinnon, J. and MacKinnon, K. (1986) *Review of the Protected Areas System in the Indo-Malaysian Realm*, IUCN, Gland, Switzerland.

McNeely, J.A., 1988. The Economics of Biological Diversity: Developing and using Economic Incentives to conserve Biological Resources. IUCN: Gland, Switzerland.

McNeely, J.A. Miller, K. R., Reid, W. Mittermeier, R. and Werner, T. (1990) *Conserving the World's Biological Diversity.* IUCN, World Resources Institute, World Bank, WWF-US, and Conservation International, Washington, D.C.

Mooney, H.A. and Bernardi, G. (1990) *Introduction of Genetically Modified Organisms into the Environment*, SCOPE 44, John Wiley, Chichester, UK.

Moses, Karl O. (1989) A Geochemical Perspective on the Causes and Periodicity of Mass Extinctions. *Ecology*, **70(4)**, 812-32.

Myers, N. (1988) Threatened Biotas: 'Hotspots' in Tropical Forests. *The Environmentalist*, **8**, 187-208.

Myers, N. (1990) The Biodiversity Challenge: Expanded Hot-spot Analysis. *The Environmentalist*, **10(4)**, 243-256.

Plucknett, D.L. and Smith, N.J.H. (1986) Sustaining Agricultural Yields. *BioScience*, **36**, 40-45.

Prescott-Allen, Robert, and Prescott-Allen, Christine, (1986) *The First Resource: Wild Species in the North American Economy.* Yale University Press, New Haven.

Quinn, James F. and Signor, P.W. (1989) Death Stars, Ecology, and Mass Extinctions. *Ecology*, **70(4)**, 824-34.

Raup, D.M. and Sepkoski, J. J. Jr. (1982) Mass Extinctions in the Marine Fossil Record. *Science.*, **215**, 1501-03.

Reid, Walter V. and Miller, K. R. (1989) *Keeping Options Alive: The scientific basis for conserving biodiversity.* World Resources Institute, Washington D.C.

Runka, G. and Filion, F. L. (1988) *Natural Resource Socio-economized Project Design.* CIDA, Ottawa, Canada.

Solbrig, O.T. (1991) Biodiversity: Scientific Issues and Collaborative Research Proposals. MAB Digest No.9, UNESCO, Paris.

Soulé, M.E.,Wilcock, B.A. and Holtby, C. (1979) Benign Neglect: A Model of Faunal Collapse in the Game Reserves of East Africa, *Biological Conservation*, **15**, 259-72.

Soulé, M.E. (1989) Conservation Biology in the Twenty-first Century: Summary and Outlook. In: Western, D. and Pearl, M.C. (eds.) *Conservation for the Twenty-first Century.* Oxford University Press, New York, and Oxford.

UNEP (1991) Biological diversity. *In: The State of the World Environment*, UNEP, Nairobi, Kenya, pp 19-26.

UNESCO (1984) African Plan for Biosphere Reserves, *Nature and Resources*, **20**, 11-22.

Vitousek, P.M., Ehrlich, P.R., Ehrlich, A.H. and Matson, P.A. (1986) Human appropriation of the products of photosynthesis. *BioScience*, **36**, 368-73.

Wells, M.P. (1991) *Conserving Biological Diversity in Nepal: The Social, Economic and Institutional Issues.* Policy and Research Division, Environment Department, The World Bank.

Wilson, E.O. (1985) The biological diversity crisis: A challenge to science. *Issues in Science and Technology* (Fall): 20-29.

Wilson, E.O. (ed.) (1988) *Biodiversity.* National Academy Press, Washington D.C.

Witt, S.C. (1985) Biotechnology and Genetic Diversity, California Agricultural Lands Project, San Francisco, California.

Woodruff, D.S. (1989) The problems of conserving genes and species. In: Western, D. and Pearl, M.C. (eds.) *Conservation for the Twenty-first Century.* Oxford University Press, New York, and Oxford.

WCED (1987) *Our Common Future.* Report of the World Commission on Environment and Development. Oxford University Press, Oxford.

WCMC (1992) Biodiversity Status Report. World Conservation Monitoring Centre, Cambridge.

World Data Centre (1985) *World Directory of Collections of Cultures of Microorganisms.* Brisbane and Riken.

WRI/UNEP/IUCN (1992) *Global Biodiversity Strategy,* WRI/UNEP/IUCN.

Zedan, H. and Olembo, R. (1988) A network of microbiological resources centres (MIRCENs) for environmental management and increased bioproductivity in developing countries. *In:* Rao, D.N. *et. al.,* (eds.) *Perspectives in Environmental Botany 2.* Today and Tomorrow's Printers and Publishers, New Delhi.

CHAPTER 9

Environmental hazards

The issue

Mankind has always faced hazards, either by necessity (e.g. earthquakes) or by choice (e.g. mountain climbing). Societies seek - with varying degrees of success - to reduce the likelihood of harmful consequences of these hazards by taking actions such as building flood control structures and prohibiting harmful substances. Despite such measures, however, the number of accidental deaths and the scale of property losses have increased in recent years, particularly in developing countries. Figure 1 suggests that frequencies of major natural disasters have increased in the last two decades (UNEP, 1991). At the same time, there has been a threefold increase in disaster losses. For this 'worst-case' category of accident, three million lives have been lost in the last 20 years, and nearly two billion people have been adversely affected.

There are several reasons for this. At one end of the social scale, poverty and population pressures have caused more and more people to live in hazardous regions: on mountain slopes, on hurricane-prone shores, on flood plains, or near industrial plants or stores of dangerous wastes. In some cases these people intensify the risks to which they are exposed, for example by cutting trees on the slopes of valleys, increasing the possibility of mudslides and avalanches.

At the other end of the scale, hazardous situations are also created or intensified by increasing affluence. Automobile numbers and associated road accidents are rising while homes and recreational facilities are increasingly being constructed in vulnerable areas such as tropical shorelines, mountains and earthquake zones. Furthermore, the increasing use of toxic chemicals, radionuclides and unproven technology, as well as extreme operating conditions in industry, are introducing new environmental hazards.

Generally speaking, there are relatively few fatalities but very high economic losses from accidents and disasters in developed countries, while the reverse is usually the case in developing countries. Developed countries have invested in insurance, physical defences, and various forms of 'hazard monitoring and mapping'. This protects potentially fragile systems against most contingencies. When these defences fail, financial, technical and other resources are readily available to minimize or repair the damage. Not only do many developing countries lack these resources, but they also import technologies that come without the complete package of surrounding safety measures. The Bhopal incident is perhaps the best-known example: there, the failure of internal safety systems was exacerbated by a lack of enforcement of local zoning regulations.

Adopting the definitions of the International Programme on Chemical Safety (WHO, 1989), a *hazard* is a source of danger: the word is a qualitative term expressing the potential for harm in a given situation. A *risk* is the probability that a potentially harmful event will occur. During the last two decades, a broad field of study called *risk assessment* has grown up, the objectives of which are to identify hazards, quantify the risks involved, reduce the uncertainties in these estimates, understand public perception of these risks, design methods to manage risk, and develop accident/disaster relief procedures (see Chapter 20).

Figure 1: *Frequency of major natural disasters, (1960-89).*

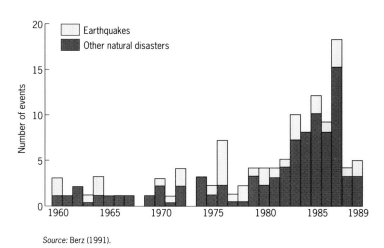

Source: Berz (1991).

Trends

Natural disasters

In the last three decades both the frequency and magnitude of natural disasters have increased dramatically. Records of major occurrences (OFDA, 1987; Berz, 1991) indicate that the number of such events was 16 in the 1960s, increased to 29 in the 1970s, and reached 70 in the 1980s (Figure 2). Although the number of these events in developed countries in the three decades 1960-90 was higher than those in the developing countries in the same period (63 to 50), absolute mortality rates were far lower (about 35,000 were killed in the developed countries compared to about 800,000 in the developing countries). Furthermore, between 1974 and 1984, droughts have caused the death of some 500,000 people, nearly all of them in the developing countries (OFDA, 1987). This illustrates the vulnerability of these latter countries to the effects of natural hazards. The economic losses due to natural hazards have also increased world-wide. In the 1960s, the losses were about $US10 billion, in the 1970s $US30 billion and in the 1980s, $US93 billion. Adjusted for inflation, the losses averaged $US3.7 billion per year in the 1960s and $US11.4 billion per year in the 1980s (Berz, 1991).

Figure 2: *Number of events, death toll and overall economic losses from major natural disasters, (1960s - 1980s).*

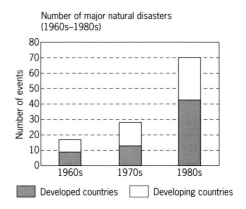

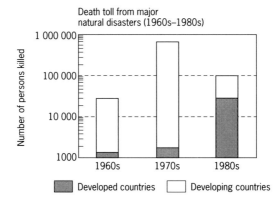

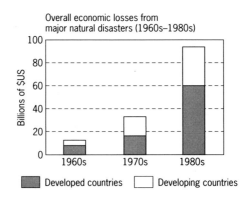

Source: Data from OFDA (1987); Berz (1991).

This upward trend is primarily due to the continuing steady growth of world population and the increasing concentration of people and investments in marginal, disaster-prone areas. For example, coastal areas which are generally more exposed to natural disasters are also areas where population is growing rapidly in many parts of the world. Examples range from the hotels concentrated along the hurricane-exposed coasts of the USA and the Caribbean to the offshore oil and gas industry operating in the North Sea. Similar examples are also found in developing countries, e.g. in the Philippines and Thailand. Island developing countries are particularly vulnerable to natural disasters.

Earthquakes

Earthquakes are among the deadliest and most destructive of natural disasters: between 1970 and 1990 they killed more than 400,000 people world-wide and caused overall economic losses estimated at $US65 billion (OFDA, 1987; Berz, 1991). Two-thirds of the world's large earthquakes occur in the circum-Pacific belt. The next most important earthquake zone stretches from Indonesia, through the Himalayas and along the axis of the Mediterranean. Some 75 per cent of the world's earthquake deaths occur in this zone, which is more densely populated than the circum-Pacific belt. The primary effects of earthquakes are violent ground motion accompanied by fracturing which may shear or collapse large buildings, bridges, dams, tunnels and other rigid structures. Secondary effects include short-range events such as fires, landslides, tsunami and floods, and long-range effects, such as regional subsidence, uplift of landmasses, and regional changes in groundwater hydrology.

Human activity has increased the occurrence of earthquakes in three main ways. First, the Earth's crust has been loaded with increasing numbers of large water reservoirs and this has created local minor earthquakes (El-Hinnawi, 1981; El-Hinnawi and Biswas, 1981). Second, the disposal of liquid waste in deep wells has caused an increase in fluid pressures in rocks in certain regions, facilitating movements along fractures. Third, the underground testing of nuclear devices creates pressure which could affect the stability of parts of the Earth's crust.

Earthquakes affect mostly poor people. Of the ten deadliest earthquakes that occurred between 1970 and 1990 (Box 1), nine occurred in developing countries. Most of the people who died or who were injured lived in rural areas or slums. Many lived in ravines or gorges which are highly susceptible to landslides whenever earth movements occur. A number of recent earthquakes such as the 1985 Mexico City earthquake, the 1986 El Salvador earthquake and the 1988 earthquake in Armenia demonstrated that the collapse of reinforced concrete buildings is a significant problem that hinders rescue operations and may increase the death toll in urban areas.

BOX 1

The ten deadliest earthquakes

1970–90

Date	Location	Number killed
27 July 1976	China	242,000
31 May 1970	Peru	67,000
21 June 1990	Iran	40,000
7 Dec. 1988	USSR	25,000
4 Feb.1976	Guatemala	22,778
16 Sept. 1978	Iran	20,000
19 Sept. 1985	Mexico	10,000
10 April 1972	Iran	5,400
23 Dec. 1972	Nicaragua	5,000
24 Nov. 1976	Turkey	3,626

Source: OFDA (1987) Berz (1991).

Volcanic activity

On a world-wide basis, the frequency of disastrous volcanic activity is low. Most of the Earth's 540 historically active volcanoes lie along the famous Ring of Fire, a crescent of volcanic activity that runs around the rim of the Pacific Ocean through the edges of Asia, North America and South America. Many have been dormant for decades, though some occasionally release limited amounts of gases and/or ash, and a few explode violently releasing huge amounts of ash and/or molten rocks (lava). Historically, the number and nature of volcanic eruptions have not changed. Most disastrous effects occur because more and more people have inhabited villages on the slopes of the volcanoes so that when a strong eruption occurs it leads to widespread local damage (as is the case, for example, on Mt Etna in Sicily, Italy). Although the death toll from volcanic eruptions is usually limited, a few recent volcanic incidents have caused substantial numbers of deaths. For example, the eruption of the volcano Nevado del Ruiz in Colombia on 13 November 1985 caused the death of about 23,000 people. Not all the death or damage is caused, however, as a result of the eruptive phase itself: some results from secondary effects. Lava flows, for example, may move as fast as 100km/h down the sides of a volcano and can be catastrophic if a populated area lies in the path of the flow. When the loose ash becomes saturated with water from the rain, it produces mudflows which are unstable and move suddenly downslope (Box 2).

BOX 2

Mount St Helens

On 18 May 1980 Mount St Helens, Washington, erupted with a lateral blast from the side of the mountain and a large vertical blast from the top. A cloud of volcanic ash rose to an altitude of approximately 19 kilometres. The eruption continued for more than nine hours, and large volumes of ash fell on wide areas of Washington, northern Idaho, and western and central Montana. The first of several mudflows occurred minutes after the start of the eruption. The flows and accompanying floods raced down the valleys of the north and south forks of the Toutle River at estimated speeds of 29-55 km/h. Water levels in the river reached about 4 m above flood stage and nearly all bridges along the river were destroyed. As a result, about 40 million cubic metres of material were dumped into the Columbia River, reducing the depth of the shipping channel from a normal 12 m to 4.3 m for a distance of 6 kilometres.

The eruption of Mount St Helens caused damage over an area of about 400 square kilometres and killed 60 people. The total damage has been estimated at about $US1 billion.

Source: Keller (1985).

Volcanic activity releases into the atmosphere ash of various sizes and several gaseous products. It has been estimated that volcanic activity contributes about 20 million tonnes of sulphur into the atmosphere each year in the form of sulphur dioxide, hydrogen sulphide and sulphates. This is equivalent to about 5-7 per cent of the total global sulphur emissions into the atmosphere (Berresheim & Jaeschke, 1983; Brimblecombe & Lein, 1989). The injection of large quantities of fine dust into the upper atmosphere from explosive volcanic eruptions could contribute to climate change (Robock, 1983; Toon & Pollac, 1982). The National Research Council (1985) indicated that model studies predict that if one million tonnes of ash were emitted into the stratosphere by a volcanic eruption, this would produce a world-wide drop in average temperature of about 10°C for several months. However, most of the ash emitted from the eruptions of Mount St Helens, Washington, in 1980 (about 100 million tonnes) or El Chinchón in Mexico in 1982 (about 200 million tonnes) contained predominantly large ash particles; the submicron mass (which could stay in the atmosphere for longer periods of time) was relatively small. The recent eruption of Mount Pinatubo in the Philippines, however, could lead to a drop in the average global temperature.

A peculiar event that took place in 1986 was the release of a cloud of gas from Lake Nyos in Cameroon. The gas, which was later found to be mainly carbon dioxide, asphyxiated 1734 people who lived near the lake. It is claimed that the carbon dioxide came from a chamber of molten rocks deep under Lake Nyos and accumulated in sediments on the bottom of the lake. When the pressure of the gas became too high it erupted to the surface in the form of a lethal cloud. This is the first known event of its kind.

Tropical storms

Tropical storms are rivalled only by earthquakes as the most devastating of natural hazards. Between 1970 and 1990 they killed about 350,000 people world-wide and caused economic losses estimated at about $US34 billion (Berz, 1991). Again, most of the devastation occurred in developing countries. In Bangladesh alone two major cyclones - one in 1970 and the other in 1985 - killed about 310,000 people, i.e. 90 per cent of those killed by cyclones in the world between 1970 and 1990. Another cyclone hit Bangladesh on 29 April 1991, killing 132,000 people (Box 3). The Philippines suffered 43 tropical storms (typhoons) between 1970 and 1990, with about 8000 people killed; and in November 1991, over 5000 people were killed in tropical storm Thelma. These are much lower figures than those for Bangladesh, although the Bangladesh coastline is much shorter than that of the Philippines. This is due to the fact that the population density in Bangladesh is much higher. It should be noted that the economic losses due to cyclones in developing countries - and indeed those of other natural hazards - are grossly underestimated. The damage figures seldom reflect the full extent of human suffering, such as that resulting from the loss of the means of livelihood, while the setbacks to social and economic development caused by a single cyclone are often measured in years.

Flooding due to tropical cyclones frequently constitutes a much greater threat than the wind. Abnormally high tides may cause large areas to be inundated, and these effects have accounted for some of the greatest disasters associated with tropical cyclones. In Bangladesh, a storm surge in 1970 variously estimated at between three and nine metres in height, caused the death of 300,000 people. Most of the devastation caused by the cyclone that hit Bangladesh in April 1991 was also due to storm surges. Altering the environment can make people and property more vulnerable to the effects of tropical storms. The destruction of coral reefs, mangroves and other seafront forests, and the levelling of coastal sand dunes, remove the shoreline's natural protection and allow storm surges to reach people and their property more quickly and forcefully. About 20 million people in Bangladesh are exposed to the effects of cyclones because they inhabit low flood plains vulnerable to storm surges. The coastal zones in the Philippines are more protected by reefs, mangroves and trees than those of Bangladesh.

BOX 3

The deadly cyclone

On the night of 29 April 1991, winds reaching 225 km/h swept through the villages of eight coastal districts in the Bay of Bengal, Bangladesh. The winds created waves up to 7 m high which razed more than 860,000 houses, drowned 440,000 head of cattle and destroyed hundreds of thousands of trees. The cyclone killed 132,000 people and injured 458,000, and has left millions of destitute survivors, many of them homeless. In Chittagong district alone, farmers lost about 58,000 hectares of land to sea water, and the salt level in the remaining soils increased tenfold. More than 16,000 ha of fish farms in the southeastern coastal belt were lost, and Chittagong port, which deals with 80 per cent of Bangladesh's foreign trade, was brought to a standstill. The overall economic losses are not known, but have been estimated at billions of US dollars.

Source: Sattaur (1991).

Floods

Floods occur in many developing and developed countries (OECD, 1991; UNEP, 1991), and although many of them do not result in fatalities, major floods can kill thousands of people. This is especially the case in developing countries. Between 1960 and 1990, severe floods caused the death of about 65,000 people world-wide. Some 51,000 of these deaths, or 78 per cent of the world total occurred, however, in five developing countries: Bangladesh, China, Colombia, India and Pakistan. One flood in Bangladesh in 1974 accounted for the death of about 29,000 people (OFDA, 1987). Estimated economic losses from floods vary widely from one country to another, but have been estimated at a conservative figure of $US50 billion between 1970 and 1990.

Despite the risk, hundreds of millions of people who have been adversely affected by floods continue to inhabit flood plains; and are occupying such areas in increasing numbers. Many of Asia's squatters live on flood plains. Much of the expansion of Delhi has been onto the flood plain of the Yamuna River. Many of the city's 600,000 squatters, plus 700,000 people living in unauthorized areas, and a further 150,000 to 200,000 in campsites, are vulnerable to flooding. In Bangkok, at least 1.2 million people live in slums on swampy ground prone to flooding.

As with other natural disasters, the protective measures employed or available depend not only on the physical phenomena concerned but also on the prevailing social circumstances and the degree and nature of the human response anticipated. People have always lived near rivers and estuaries so as to benefit from them as a means of transport and a source of water and fish, and to grow crops on their rich flood plains. In spite of the vast investment throughout the world in flood control dams, diversion channels, levee banks and other installations to reduce the risk of floods, increasing population pressure has forced people to live on the flood plains themselves, especially in developing countries, to cultivate crops on potentially flood-prone land.

Increases in population combined with poor resource management have also resulted in the conversion of forests to pasture and arable land. Less water is stored in the upper reaches of catchment areas, and it flows more rapidly to the plains. Floods thus become more frequent, more severe and arrive with less warning. For example, in the Indian subcontinent the Himalayas send water southward through the three great rivers: the Indus, the Ganges and the Brahmaputra. The mountains water a vast stretch of northern India and much of Pakistan, Burma and Bangladesh. Growing populations are stripping the forests from the habitable areas on the southern slopes of these mountains. The slopes can no longer hold the water, and floods are increasing throughout the Himalayan watershed. Annual flood losses in India today are 14 times those of the 1950s. In 1988, flood waters covered 60 per cent of Bangladesh, and abnormal levels of flooding are now an annual event. During the floods, vast amounts of sediment brought down from the Himalayas cause extensive silting in river channels and reservoirs, shortening their useful life.

Drought

Drought is a sustained and regionally extensive deficiency in precipitation (rain and/or snow), often leading to desertification. The magnitude of the impact of drought on people depends to a large extent on the economic status of the affected population (Wilhite et al., 1987). Pastoralists are often the first to feel the impact of a drought because they usually live along the desert's edge. In the Sahel in the 1980s, hundreds of thousands of nomadic pastoralists were driven southward after they had consumed the last shreds of dried-up vegetation. Many Sahelians moved to coastal West African countries where they took menial jobs and swelled shanty towns and slums. Thousands of Mauritanian nomads arrived in Mali, searching for water and pastures and swamping the country with their cattle. Others travelled beyond Mali, towards Burkina Faso and Niger. Ivory Coast was the principal destination for more than one million of these environmental refugees (El-Hinnawi, 1985). Because of recurrent and persistent drought, in early 1984 more than 150 million people in 24 western, eastern and southern African countries were on the brink of starvation (Figure 3). Ethiopia and Somalia were the most seriously affected countries in eastern Africa: by the end of 1984, more than 6 million people in 14 regions in Ethiopia were affected by drought. In southern Africa, persistent drought caused an equally critical situation with about 8 million people affected and at least 1.5 million in need of food aid. There is a close parallel here with floods, in that the growth in the number of people and their livestock has contributed significantly to the severity of these disasters.

Figure 3: *Countries affected by drought in Africa.*

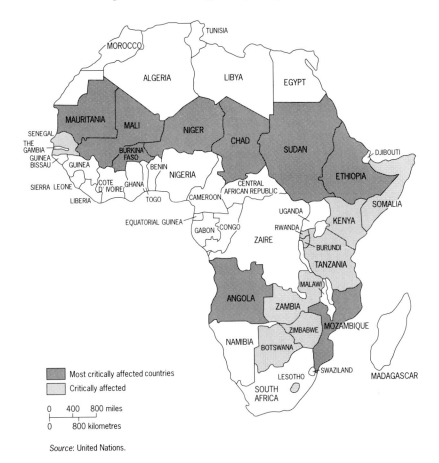

Source: United Nations.

Droughts force people out of their traditional habitats. Some migrate to areas within their national borders but others cross international boundaries. Governments, trying to cope with starving populations, establish transit and refugee camps which rely heavily on assistance from the international community. However, many environmental refugees - women, children and the elderly - do not survive. In the case of the Sahelian disaster of the 1980s, starvation, dehydration and infectious diseases combined to accelerate the death of hundreds of thousands. Conservative estimates indicate that the death toll in Africa directly linked to droughts was about 500,000 between 1974 and 1984 (OFDA, 1987). However, the figure for 1970 to 1990 may be twice or three times as high.

Other natural hazards

Landslides, mudslides, storms, avalanches and tsunami are common in certain parts of the world (UNEP, 1989; 1991), but their effects are more limited. It has been claimed that the damage caused by landslides is, on aggregate, as great as that caused by other more widely publicized natural disasters. However, most landslides are associated with volcanic activity, earthquakes or heavy rainfall, and the damage caused by landslides is normally included in assessments of damage caused by these other events. For example, many of the deaths in the November 1991 storm in the Philippines were from mudslides down deforested slopes. Similar events have been reported from Tanzania, Brazil and Colombia; and mudslides are almost endemic in parts of China (Rapp *et al*, 1991). In Switzerland and Austria there has been an increase in mudslides and avalanches due to the construction of more and more ski slopes and ski villages and due to forest decline at high elevations.

Human-induced hazards

Accidents occur primarily due to human error and to technology failures. Many accidents have demonstrated that there is no such thing as a fool-proof technology, and that there is no absolute safeguard against human error. Thousands of incidents occur daily as 'routine' accidents which are quickly contained. However, some events lead on to major accidents, and even disasters. The main human and technology-induced hazards are described in the following sections.

Forest fires

Forest fires - or wildland fires - are caused by lightning (a natural cause), negligence, accidents, reckless burning as an agricultural or grazing practice, poor forest management techniques (leading to accumulation of litter, often, paradoxically, as a consequence of successful efforts to prevent fires) and arson.

Fire is the main cause of forest destruction in the Mediterranean basin (Figure 4). About 50,000 fires sweep through 700,000 to 1,000,000 hectares of Mediterranean forest each year, causing enormous economic damage as well as loss of human life (Velez, 1990). However, fires also release essential nutrients and can thus be an important agent for forest regeneration. The majority of these fires were caused by people, although lightning was the cause of a fire that burnt more than 30,000 hectares in Ayora-Enguera, Spain, in 1979. In 1985, the economic losses due to forest fires ranged from $US17 million in Portugal to $US111 million in Spain.

Fire damage is by no means confined to southern Europe. For example, wildfires burn uncounted millions of hectares of African savannah each year, and in Asia, a single fire in Kalimantan, Indonesia, damaged more than 3.6 million hectares in 1982 (Malingreau *et al.*, 1985). In North America, notwithstanding extensive control efforts, more than 2.3 million hectares of forest land still burn each year. During 1988, nearly 75,000 fires burned more than 2 million hectares of wildland in the USA (McCleese *et al.*, 1991).

Figure 4: *Forest fires in the Mediterranean area.*

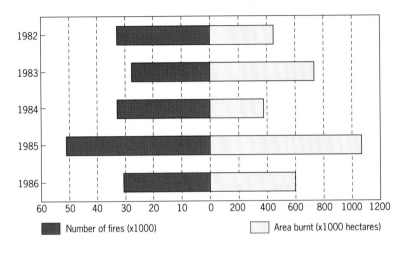

Source: Based on data by Velez (1990).

Oil spills

Accidental releases of oil (and oil products) occur on land and in the sea. The latter are the ones that normally make headline news (Box 4), although data show that spills on land can release significant amounts of oil. Table 1 shows that the number of oil spills exceeding 6,800 tonnes dropped 74 per cent between 1974 and 1986. This is partly because world exports of crude and finished oil products declined 25 per cent during that period. (See also Chapter 5).

Accidental oil spills can be very costly. The principal costs involved include containment, clean-up and environmental restoration costs, damages to fisheries, and losses suffered in tourism profits and labour earnings.

Slow and cumulative hazards

Chapter 10 discusses the cumulative and long-term effects of chemical substances on human beings and the environment. To define the potential hazards of a great many chemical substances with any degree of confidence takes a long time and considerable resources. Risk assessment, in its turn, requires knowledge of the concentration of a chemical of known hazards, and the duration and extent of exposure to which human beings or ecosystems are subjected.

BOX 4

The ten largest oil spills from tanker accidents.
1970-90

Date	Tanker	Country affected	Oil spilled
July 1979	Atlantic Express	Tobago	276,000 tonnes
Aug. 1983	Castello Belver	South Africa	256,000 tonnes
Mar. 1978	Amoco Cadiz	France	228,000 tonnes
Dec. 1972	Sea Star	Gulf of Oman	120,000 tonnes
Feb. 1980	Irenes Serenade	Greece	102,000 tonnes
May 1976	Urquiola	Spain	101,000 tonnes
Feb. 1977	Hawaiian Patriot	Hawaii	99,000 tonnes
Nov. 1979	Independenta	Turkey	95,000 tonnes
Jan. 1975	Jacob Maesk	Portugal	84,000 tonnes
Dec. 1985	Nova	Iran	71,000 tonnes

Source: OECD (1991); Oil Spills Intelligence Report (1991).

Table 1: *The number of oil spills world-wide (1974-86).*

Year	Number of spills		
	All sizes	6,800—68,000 tonnes	>68,000 tonnes
1974	1,450	91	26
1975	1,350	98	23
1976	1,099	66	25
1977	956	66	20
1978	746	57	24
1979	695	54	37
1980	554	48	13
1981	401	48	5
1982	247	44	3
1983	216	52	11
1984	146	25	7
1985	137	25	8
1986	118	24	6
Total	8,115	698	208

Source: WRI (1990).

BOX 5

Chemical time bombs

Ecological system	Chemical stored	Threshold mechanism	Delayed effect
Forest soils	Acid (from deposition)	Depletion of buffering capacities	Acidification of soils and lakes; leaching of heavy metals
Agricultural soils	Phosphate fertilizer	Saturation of phosphate sorption capacities	Leaching of phosphate to aquatic systems (eutrophication)
Agricultural soils (abandoned)	Heavy metals (e.g. Cd)	Lowered sorption capacities on cessation of liming	Leaching of metals to water bodies; plant uptake
Coastal waters	a)	Depletion of oxygen; generation of H_2S; mixing of deep water during storm events	Leaching of metals to water bodies; plant uptake
Estuary sediments	Heavy metals	Changes in redox potential; resuspension of sediments (sea level rise)	Release of metals; fish poisoning
Wetlands	Sulphur; heavy metals	Drying from climate change (causing exposure to air)	Release of sulphuric acid; heavy metals

Note: a) No chemicals stored *per se*, but 'over-fertilization' of coastal waters from, for example, runoff of agricultural fertilizers can lead to sudden episodes of anoxia and H_2S generation.

Source: Stigliani (1991).

229

Recently, the concept of 'chemical time bombs' (CTBs) (Box 5) was proposed. This concept goes beyond the usual view of chemical hazards to emphasize the time element in hazardous situations. A CTB refers to a situation where a chain of events results in the delayed sudden occurrence of harmful effects due to the mobilization of chemicals stored in soils and sediments, in response to alterations in the environment (Stigliani, 1991). Box 6 gives some examples, cited by Stigliani, of the hazards inherent in the storage and/or disposal of chemicals.

Chemical accidents

Between 1970 and 1990, about 180 accidents occurred world-wide, releasing various chemical compounds into the environment. These accidents, caused

---

BOX 6

Chemical accidents that made headline news (1970–90).

SEVESO

On 10 July 1976 an explosion at the ICMESA chemical factory in the North Italian town of Seveso released a cloud of chemicals into the atmosphere, contaminating the surrounding area.

The chemicals contained 2kg of dioxin, a potentially toxic compound. The cause of the accident is believed to have been a 'runaway reaction' in the reactor producing sodium trichlorophenate, a main product.

There were no deaths but 200 people suffered slight injuries. The main victims were domestic animals. Contamination of the land affected some 37,000 people. Restrictions were imposed for 6 years on an area of 1800 ha. The worst affected area covered 110 ha. The estimated direct costs of the accident were about $US250 million.

Source: Hay (1977, 1978); Smets (1988).

BHOPAL

On the night of 2-3 December 1984, a sudden release of about 30 tonnes of methyl isocyanate (HIC) occurred at the Union Carbide pesticide plant at Bhopal, India. The accident was a result of poor safety management practices, poor early warning systems, and the lack of community preparedness.

The accident led to the death of over 2,800 people living in the vicinity and caused respiratory problems and eye damage to over 20,000 others. At least 200,000 people fled Bhopal during the week after the accident.

Estimates of the damage vary widely from $US350 million to as high as $US3 billion.

Source: Bowander (1985, 1987); Weir (1987); Smets (1988).

---

mainly by fires, explosions or transport accidents, killed about 8,000 people, injured more than 20,000, and led to hundreds of evacuations involving hundreds of thousands of people (OECD, 1991). The most disastrous chemical accident of all occurred in Bhopal in India in 1984 (Box 6), while the Basel accident (see also Box 6) made it clear that industrial accidents can have harmful transboundary impacts. The massive explosion at the liquefied petroleum gas storage facility in the crowded San Juanico neighbourhood of Mexico City in November 1984 killed 452 people, injured 4,248 and displaced 31,000. The blast illustrated the precarious nature of a city where many of the 17 million inhabitants live cheek by jowl with a variety of potentially dangerous installations.

Although it is rather difficult to establish accurate trends, there has been a

BASEL

On 1 November 1986 a fire broke out at a Sandoz storehouse near Basel, Switzerland. The storehouse contained about 1,300 tonnes of at least 90 different chemicals. The majority of these chemicals were destroyed in the fire, but large quantities were introduced into the atmosphere, into the Rhine River through runoff of fire-fighting water (about 10,000 to 15,000 cubic metres), and into the soil and groundwater at the site. The mass of chemicals that entered the Rhine has been estimated at 13 to 30 tonnes.

Following the accident the biota in the Rhine was heavily damaged for several hundred kilometres. Most strongly affected were benthic organisms and eels, which were completely eradicated for a distance of about 400 km (an estimated 220 tonnes of eels were killed). Several compounds were detected in the sediments of the Rhine after the accident.

Within a few months the Rhine River had purged itself of all the chemicals released from the accident (with the possible exception of mercury and endosulfan). One year after the accident most aquatic life had returned to the situation that existed before the accident. However, the groundwaters in the extensive Rhine alluvial aquifer are still polluted.

The damage caused by the Basel accident has been estimated at $US50 million.

Source: Capel et al. (1988); Smets (1988).

general increase in industrial accidents, especially in developing countries. Between 1974 and 1978, five major accidents occurred world-wide (a major accident being defined as one with at least 100 deaths, or 400 injured, or 35,000 people evacuated). Between 1984 and 1988, the figure was 16 (Smets, 1988).

The risk of major accidents has increased hand in hand with technological advances, since such developments are generally less tolerant of human failure. Risks have also risen with the increase in the number, size and age of installations in use, as well as with the greater toxicity of substances involved in production, reprocessing and storage. Furthermore, more and more factories have been built in flood-prone areas, or in earthquake zones, and more importantly on the borders of cities which have later expanded to enclose them. As long as strict safeguards and standards are not implemented, and as long as industrial installations are not located far from dense population centres, major accidents are likely to increase, particularly in developing countries.

Nuclear accidents

At the end of December 1990, there were 423 nuclear reactors operating in 24 countries, 112 of them in the USA (IAEA, 1991). 'Routine' accidents - referred to as 'unusual events' - frequently occur during the operation of these reactors. These may include spontaneous failures of equipment, and deviations discovered through surveillance and maintenance activities. These unusual events are classified by IAEA into: events unrelated to safety (with an average frequency of 0.5 to 1 event/week/reactor); safety-related events (0.5 to 1 event/month/reactor); and events of safety significance (0.5 to 1 event/year/reactor) (Franzen, 1987). Although the IAEA established an Incident Reporting System (IAEA-IRS) in the early 1980s, reporting of such events to the IRS has been rather uneven and incomplete. This situation should be improved by the adoption in 1986 of the Convention on Early Notification of A Nuclear Accident (the convention entered into force on 27 October 1986, and by the end of December 1990 had been signed by 49 countries).

Several studies have tried to establish the probabilities of reactor accidents of varying degrees of severity. The much publicized Reactor Safety Study (also known as the Rasmussen or WASH-1400 report) published in 1975, estimated the probability of a meltdown in a pressurized water reactor at 1 in 20,000 per reactor per year, and that most meltdowns would not breach the main containment above the reactor. The worst accident, which WASH-1400 estimated might happen once per 10 million years of reactor operation, might cause 3300 early fatalities, about ten times that number of early illnesses, additional genetic effects and long-term cancers, and perhaps some $US14 billion in property damage (WASH-1400, 1975). The WASH-1400 estimations have been widely criticized (Lewis *et al.*, 1978; NAS, 1979), and it is generally conceded that there is a large range of uncertainty in the numerical results quantifying the risks of an accident with major consequences (El-Hinnawi, 1980).

The occurrence in 1975 of a fire at Browns Ferry nuclear power plant (USA) as a result of a human error, fuelled the debate on the safety of nuclear installations and the validity of studies such as WASH-1400. Four years later, the Three Mile Island accident occurred (Box 7) after just 1500 years of reactor operation, indicating that the probabilities outlined in WASH-1400 were indeed inaccurate. Although not a light-water reactor, the Chernobyl disaster in 1986 (Box 8) followed after another 1900 reactor years. If this 'historical' accident rate continues, three additional accidents would occur by the year 2000, at which point - with over 500 reactors in operation world-wide - core-damaging accidents would happen every four years (Flavin, 1987). But no one knows for certain how often nuclear disasters will happen, and no one knows the extent of the damage that might occur to people and to the environment. This uncertainty, together with increasing doubts about the safest means of disposal of nuclear wastes, has been a factor slowing down investment in new nuclear plants since 1985.

BOX 7

The Three Mile Island accident.

Early on the morning of 28 March 1979, the 880 MWe Three Mile Island Unit 2 (TMI-2) pressurized water reactor (PWR), which was operating at nearly full power, experienced a loss of normal feedwater supply that led to a turbine trip and later to a reactor trip. Subsequently, a series of events took place that resulted in serious reactor core damage. Core temperatures locally reached fuel melt. The accident occurred because of a combination of design, training, regulatory and mechanical failures, and human error.

From 28 March through 7 April 1979, radioactive fission products were released into the environment. The release consisted mainly of noble gases (Xenon-133, Xenon-135) and traces of iodine-131. Approximately 8 per cent of the core inventory of Xe—133 was released. Another major release of noble gas radionuclides was during the controlled purge of the reactor building about 15 months after the accident. Approximately 46 per cent of the Krypton-85 inventory was discharged into the atmosphere.

No one was killed as a result of the TMI accident and there were no noticeable public health effects from the radiation released. The accident led to the evacuation of some 220,000 people from around the site, for varying periods of time. Some stress and psychological disorders were reported among the population.

The total cost of the TMI accident has been at least $US2 billion.

Source: El-Hinnawi (1980); Toth *et al.* (1986).

BOX 8

The Chernobyl disaster.

The accident at the Chernobyl nuclear power station in the Ukraine, USSR, on 26 April 1986, was the most serious accident ever to have occurred at a nuclear power reactor. The cause was human error: safety systems had been shut off during a low-power operation test. When control of the reactor was lost, two rapid steam explosions destroyed the reactor and allowed radioactive debris to be released. A fire burned for ten days until the reactor core could be completely smothered and cooled. During this period a substantial portion of the more volatile radionuclides in the core, including iodine-131 and caesium-137, escaped and were widely dispersed throughout Central and Western Europe and then in trace amounts to the entire Northern Hemisphere.

As a result of the accident, 30 people died. They were members of the reactor operating staff and of the fire fighting crew. One of these died in the explosion and the others of severe thermal and radiation burns. A further 209 emergency workers received relatively high radiation exposures giving some signs of acute radiation sickness, but they have recovered. These workers will be closely monitored and immediately treated if longer-term radiation-induced health effects should eventually occur.

A striking feature of the accident was the widespread contamination produced, especially throughout European countries. High radionuclide deposition occurred in areas in which rainfall occurred during passage of the contaminated cloud. A highly inhomogeneous deposition pattern resulted. Areas of Austria, Bulgaria, Finland, Germany, Romania, Sweden, Switzerland and Yugoslavia were among the regions receiving the highest contamination. Large numbers of measurements were made of the resultant radiation levels. From these data, estimates of radiation doses were made. Although authorities were not prepared to give consistent guidance to the public, general precautions to control radiation levels in foods were effective, and radiation levels declined rapidly during the course of May 1986.

The United Nations Scientific Committee on the Effects of Atomic Radiation (UNSCEAR) made a detailed assessment of global radiation exposures from the accident[1]. The highest average country-wide radiation doses to people were of the order of 0.6 to 0.8 mSv during 1986 in Austria, Bulgaria, Greece and Romania. Average doses of about 2 mSv were estimated to have occurred in the first year following the accident in subregions of Romania and Switzerland. A dose of 2 mSv is comparable to the annual dose received by individuals each year from natural background radiation.

The doses to residents of settlements in the Soviet Union near the reactor have been evaluated by Soviet authorities and have been reviewed by scientists of the International Review Team in a project organized by the International Atomic Energy Agency and completed in 1991[2]. The disruption of agricultural activities, relocation of people and the consequent psychological stress have been shown to be the main consequences of the accident. During the first few weeks after the accident 116,000 people were evacuated from the region. The primary concern for these individuals is the dose to the thyroid gland from iodine-131 in air and milk. This radionuclide disappears quickly: its radioactive half-life is 8 days. Longer-term but low-level exposures were received by residents in villages not evacuated. Caesium-137 has a radioactive half-life of 30 years. Exposure occurs from external irradiation from caesium on the ground and internal exposure from caesium in foods. Measurements of body contents of caesium in affected settlements in Byelorussia, the Russian Federation and the Ukraine have shown that food monitoring and food control have reduced internal exposures. The International Review Team estimated that life-time doses to these residents would be of the order of 80 to 160 mSv which is comparable to the average life-time doses from natural background radiation (2 mSv per year over 70 years). Doses of the magnitude received by these individuals would be expected to lead to statistically imperceptible increases in the normal cancer incidence rates in populations. The International Review Team could find no differences in the state of health of residents of contaminated and uncontaminated villages in examinations performed in 1990: 1356 individuals of all ages were included in the study.

The Chernobyl accident was unique in its seriousness and the large areas that were affected by the released radioactive materials. It has been a costly lesson that has heightened the safety consciousness of reactor operation. With complacency apart and basic safety systems in place, safe operation of nuclear reactors can and must be assured so that even human errors cannot be allowed to cause serious accidents.

Sources: 1. UN (1988); 2. IAEA (1991).

Although accidents at nuclear facilities have been responsible for the majority of deaths and radiation overexposures (Figure 5), accidents related to the use of radio-isotopes in industry, research and medical facilities account for a significant number of casualties from radiation accidents. The number of such accidents has recently increased. For example, there were 8 fatal accidents between 1970 and 1987 as compared to 9 such accidents between 1945 and 1970. Most people derive the greatest part of their non-natural exposure to radionuclides from medical sources, with less than 0.05 per cent from nuclear power stations.

The radiation accident at Goiania, Brazil, in 1987 (Box 9) has demonstrated that public awareness of the potential danger of radiation sources is an important factor in reducing the likelihood of radiological accidents, and in reducing the severity of the consequences of such accidents if they occur. Unfortunately, the curtain of secrecy that is drawn in many countries, especially the developing ones, on information about radiation, nuclear safety and related topics, has complicated and confused both the public's perception of such issues and the regulatory system itself, including the questions of preparedness for and mitigation of possible accidents.

Figure 5: *Serious radiation accidents reported, 1945–87.*

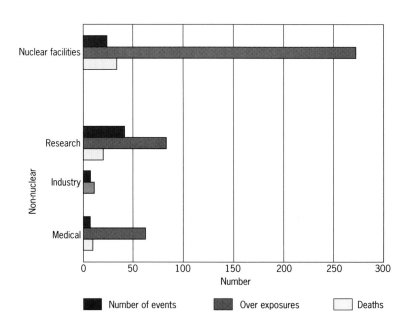

Source: Based on data from IAEA (1988).

BOX 9

The radiological accident in Goiania, Brazil.

On 13 September 1987, a shielded, strongly radioactive caesium-137 medical source was stolen from its housing in a teletherapy machine at an abandoned clinic in Goiania, State of Goias, Brazil. The two people who took the assembly tried to dismantle it and in their attempt the source capsule was ruptured. The remnants of the assembly were sold for scrap to a junkyard owner. He noticed that the source material glowed blue in the dark. Several persons were fascinated by this and, over a period of days, friends and relatives came and saw the phenomenon. Fragments of the source the size of rice grains were distributed to several families. A few days later a number of people were showing gastro-intestinal symptoms arising from their exposure to radiation from the source.

On 28 September, a physician in Goiania recognized the characteristic symptoms of radiation over-exposure. Soon an emergency response centre was set up and more than 112,000 people were screened for possible contamination. Eventually, 249 people were found to be contaminated. Twenty persons were hospitalized, four of them died.

A survey of 67 square kilometres of the area of Goiania showed that 8 locations were contaminated. In total, 85 houses were found to have significant contamination, and 200 individuals were evacuated from 41 of them. Decontamination of the affected sites continued until the end of December, 1987. As a result, some 3,500 cubic metres of radioactive waste had to be stored in a temporary site 20 km from Goiania.

Source: IAEA (1988).

Responses

Although scientific ability to predict most environmental hazards has improved substantially in recent years, many parts of the world are becoming more vulnerable to losses of life and property. Many of the benefits of reliable prediction - identification of hazards, speedy and accurate transmission of warning to appropriate targets, and availability of alternative refuges - are available only to developed countries; and even then are not always fully used.

Policies are needed to improve the ability of societies to understand and detect hazards before accidents occur; as well as to improve society's ability to cope. The risk profile would have to take into account the difference between what the risk assessment literature refers to as 'adjustment' and 'adaptation' (Burton *et al*, 1978). Adjustment to risk is a short-term response, often to a risk with very low frequency of occurrence, or to a risk for which adjustment is relatively painless. Adaptation, on the other hand, indicates a longer-term, more permanent response. This may be a cumulative set of adjustments (Box 10), or may be the result of a senior policy decision which in turn shapes more local responses.

BOX 10

National risk profiles

One initial step in the formulation of a risk management policy would be a 'risk profile' for each country. This would include:

- Identification and estimates of the major risks or hazards facing a given society, systematically treating those arising from nature, environmental degradation, infectious diseases, technology, and urbanization/industrialization.;
- Actuarial evidence on the numbers and incidence of risk events and the types and magnitudes of associated consequences.
- Trend lines for risks, risk events and consequences.
- Distribution of risks among geographical regions, economic sectors and social groups.
- Identification and estimates of the population at risk from particular hazards, including subgroups that are demonstrably or potentially particularly vulnerable.
- An account of the coverage by existing data bases of major risk areas and indicators.

Source: Kasperson and Kasperson (1989).

Governments

One of the major contributions that governments can make to reducing risks is to facilitate adaptive responses to long-term environmental risks. Governments are the only agencies capable of setting down rules and regulations that can minimize exposures to potential hazards. In 1988, UNEP in co-operation with industry launched the Awareness and Preparedness for Emergencies at Local Level (APELL) programme (UNEP, 1988). It aims to alert communities to industrial hazards and to help them to develop emergency response plans, involving industry, the local authorities and the public. The ILO has recently issued a code of practice to provide guidance in the setting up of an administrative, legal and technical system for the control of major hazard installations. It seeks to protect workers, the public and the environment by: (a) preventing major accidents from occurring; and (b) minimizing the consequences of major accidents on and off site (ILO, 1991).

Governments also have an important role once an accident has occurred. In the Mississauga, Canada, chlorine car derailment in 1979, for example, the subsequent evacuation (Box 11) showed that it is possible for governments to respond to large-scale urban chemical accidents, but that it requires careful emergency planning and appropriate resources.

A number of administrative and technical steps have recently been taken to improve the prevention of industrial accidents and mitigate their consequences. One example is the European Economic Community's directive on the major hazards of certain industrial activities (the 'Seveso' directive). The directive obliges manufacturers within the Community to identify potential danger areas in the manufacturing process and to take all necessary measures to prevent major accidents as well as to limit their consequences should they occur. The OECD has embarked on the development of common principles, procedures and policy guidelines related to accidents, as well as the establishment of mechanisms for effective exchange and provision of information.

The Basel accident prompted the UN Economic Commission for Europe to initiate work towards the formulation of a regional convention on the transboundary impacts of industrial accidents. Such an instrument would establish agreed policies, measures and procedures for reducing the risk of such hazards and for co-ordinating emergency action in response to their unintended occurrence. It would cover issues such as early warning and alarm systems; joint or co-ordinated contingency planning; preventive and remedial measures; risk assessment, damage evaluation and monitoring; mitigation and containment of damage; resolution of questions of mutual assistance, compensation and rehabilitation; and post-accident surveillance.

The International Decade for Natural Disaster Reduction (IDNDR), was launched by the General Assembly of the United Nations on 1 January 1990. It is promoting an integrated approach to disasters by improving warning systems and disaster preparedness and by changing the sometimes fatalistic attitudes of local people towards disasters. Increased community participation and better training will be very important components of the Decade. Achieving the aims of the IDNDR would cause a change in the basic approach to disasters, from the present concentration on post-disaster relief to a future emphasis on preparedness.

National and international committees for IDNDR have been established in many jurisdictions. ICSU has established a Special Committee on IDNDR, one of whose objectives is to improve predictions of potential disasters. One recent success is the prediction by a team of vulcanologists of the eruption of Mount Pinatubo in the Philippines: their persuasiveness led to the evacuation of 90,000 people just in time to escape the devastating event of 15 June 1991 (ICSU, 1991).

The issue of improving the international response to disasters through UNDRO and/or other agencies is complex, since it involves potential infringements of national sovereignty as well as raising questions about the effectiveness of external aid during emergencies. Selective acceptance of post-disaster relief is gradually coming to be seen as the optimum strategy (PAHO, 1991).

BOX 11

The train derailment in Mississauga, Canada.

The largest peacetime evacuation in North American history occurred following a freight train derailment late on 10 November 1979 in the suburban city of Mississauga, Ontario, Canada. The derailment littered twisted and punctured freight cars around a crossing area, among them propane cars which immediately exploded into intense fires called BLEVE's (Boiling Liquid Expanding Vapour Explosions), as well as a mixture of chemical cars, one of which contained chlorine gas. At densities of 3 parts per million or more, chlorine becomes a health threat, irritating eyes and causing respiratory problems. At 900 ppm or above it causes rapid death. Because chlorine gas hugs the ground as it spreads, it was feared that the leaking chlorine gas might suddenly release a dense cloud of gas, trapping emergency workers or evacuees. Initial evacuation of neighbouring areas began almost immediately, and by the end of 24 hours a quarter of a million people had departed. This exodus was made easier by the fact that the derailment occurred on a weekend, and the population of Mississauga has a high percentage of automobile owners: 95 thousand cars left the area within those 24 hours.

A cordon was placed around most of the city, approximately 180 square kilometres in extent. Evacuees fled to a ring of evacuation centres, volunteer residences, and to extended families in other parts of southern Ontario. Residents of hospitals and old people's homes, and the disabled, were of special concern early on, as it began to be realized that

Industry

Industries bear substantial responsibilities for the prevention of human-induced technological hazards. Industry - mainly in industrialized countries - has gradually moved from a fragmented approach in treating environmental hazards towards a holistic approach, encompassing the refinement of risk analysis techniques, sophisticated safety measures, detailed planning of preparedness and effective response to accidents, in order to achieve prevention of accidents and damage control should they happen.

As it became evident that the environmental impacts of serious technological accidents extend beyond the boundaries of the industrial plant, industrial associations in the USA, Canada and Europe took the initiative of exhorting their members to take the lead in seeking to co-ordinate their prevention and preparedness measures and plans with the local community, and with regional and national authorities. Detailed guidelines, codes of practice and charters were developed and applied. Industry now professes an ethical responsibility to the evacuation could potentially cause more casualties than the disaster itself.

Authorities and chemical experts then struggled to put out the propane fires and to drain the chlorine car. As time passed they came under intense pressure from the media, the large number of evacuees, and the loss of business revenues amounting to 13 million dollars a day. Finally, a week after the derailment, the last evacuees were able to return home. A major inquiry into the accident was held, and Canadian Pacific Railway compensated 50,000 claimants approximately 10 million dollars in total.

The extraordinary success of this evacuation - there was no loss of life and only one recorded injury - has made the Mississauga derailment a model of emergency response in urban areas ever since. It has been of particular interest to nuclear emergency planners because the derailment was a 'point-source' emergency, and it took place in a highly urbanized environment. Among the important lessons learned were: the importance of the local emergency plans of the immediate response agencies - fire, police and ambulance - as the foundation for a larger response; the need to create a small Control Group of authorities and experts to oversee resolution of the emergency; and the need to foster the often under-recognized potential for unofficial support of an emergency from volunteer agencies, extended families and others.

Source:: Burton *et al.*, (1981).

care for the environment and to protect it from technological accidents (Chapters 12 and 21).

Public responses

The public response to environmental hazards is usually mediated through what is often called 'environmental perception', an area that has been the subject of extensive research over the past twenty years. Efforts at environmental protection are likely to be unsuccessful unless these ideas are taken into account. For example, in many cases populations devastated by a natural disaster refuse to leave their home-sites, and when forced to be relocated, return as soon as conditions permit (Oliver-Smith, 1991). Although this has been described as 'non-rational human behaviour', this kind of response complicates the whole spectrum of disaster mitigation. Nor is this behaviour always irrational. Inadequate relocation sites (post-disaster settlements) and inadequate new constructions are behind the failure of several resettlement schemes in Turkey, Iran and Peru (Oliver-Smith, 1991). From the perspective of displaced peoples, forced resettlement is always a disaster.

The complexity of public perception and attitudes is well illustrated by the case of Bangladesh. It is easier to agree on an aid package to build embankments that may hold back the flood waters, than to understand and solve the question of why it is that so many millions of Bangladeshis continue to live on the islands of that country's coastal delta, on the permanent brink of disaster (Pearce, 1991). Closer examination reveals a society dependent on 'normal' flooding for its survival, and for which the hazard of abnormal flooding 'is less of a deterrent to settlement than it might be in an economically advanced country' (Johnson, 1982).

Concluding remarks

Gilbert White, in reviewing developments in the field of coping with environmental hazards (White, 1992) notes that freedom from disastrous events has become an important and essential aspect of sustainable benefits from development projects. Gradually, continuing or new developments are not embarked upon without consideration of the sustainability of society in a particular environment. Consideration of environmental hazards has now expanded to include slow and cumulative events, biological events as well as geophysical events and technological hazards, while benefiting from carefully designed retrospective studies. In the meantime, the complexity of hazardous situations has increased, while decision-making processes have shifted from focusing on individuals to social decision-making as it relates to response to hazards and the conditions of social vulnerability. Consequently, theory has broadened to encompass political and economic aspects of the problem, even though its impact on public theory is still not clear.

The historical distinction between natural and human-induced hazards, generally acceptable twenty years ago, has given way to emphasis on the interdependence of human activities and natural events in shaping the frequencies and intensities of accidents and disasters.

It is likely that substantial improvements in responses to environmental hazards will come only from a mixture of expert-designed, top-down strategies for those hazards that require expert understanding and handling, and other, bottom-up strategies that will depend upon improvements in the understanding of social systems, in environmental risk interpretation, and in broader support for appropriate indigenous adaptive responses to hazards (Rasid and Paul, 1987).

Top-down strategies

Expert systems for the assessment of environmental risk, for modelling environmental emergencies and for the improvement of disaster warning, mitigation and reconstruction, are already available, and will surely improve. They will be more widely distributed with the gradual spread of computerization into the management of complex systems. Management of environmental risks in developed countries is already quite advanced, both in terms of what is regularly monitored and in how society responds to exceptional emergencies.

Bottom-up strategies

Traditional myth-making and other ways of coping with life's expected and unexpected disasters are important factors to consider. For example, a study of the traditional Beja pastoralists of Sudan during the drought of the 1980s showed that they placed blame for the drought on God's anger or on the moral qualities of the central regime, rather than on such factors as increased competition for grazing land or water (Dahl, 1991).

Similarly, more attention should be paid to indigenous strategies of adjustment and adaptation to environmental calamities. In emergency planning and disaster preparedness, the most effective responses are those which better orchestrate what the public is going to do in any case. That is, one should build upon natural reactions to hazard: first, by trying to change those reactions that would be correct in other situations, but are incorrect in one type of emergency (for example, flight from home in a nuclear accident may be more hazardous than remaining indoors); and, second, by streamlining more efficiently other natural reactions (for example, designating familiar roads as escape routes).

Rapidly developing countries will have to cope with both the increasing pressures of natural hazards that have been amplified by human pressures, and new technological hazards for which the people have little understanding or experience. To reduce either form of hazard will require major new initiatives both locally and globally.

REFERENCES

Anspauch, L.R. *et al.* (1988) The global impact of the Chernobyl reactor accident. *Science*, **242**, 1513.

Bennett, B and Bouville, G. (1988) Radiation doses in countries of the northern hemisphere from the Chernobyl nuclear reactor accident. *Environment International*, **14**, 75.

Berresheim, H. and Jaeschke, W. (1983) The Contribution of Volcanoes to the Global Atmospheric Sulphur Budget, *J. Geophys. Res.*, **vol.88**, C6, 3732.

Berz, G.A. (1991) Global Warming and the Insurance Industry, *Nature and Resources*, UNESCO, **Vol.27**, p.19.

Bojcun, M. (1991) The legacy of Chernobyl, *New Scientist*, 20 April.

Briblecombe, P. and Jaeschke, W. (1983) The contribution of volcanoes to the global atmospheric sulphur budget, *J. Geophys. Res.*, **vol.88**, C6.

Briblecombe, P. and Lein, A.W. (1989) *Evolution of the Global Biogeochemical Sulphur Cycle*, SCOPE 39, John Wiley & Sons, Chichester.

Burton, I., Victor, P. and Whyte, A. (eds.) (1981) The Mississauga Evacuation, Final Report to the Ontario Ministry of the Solicitor General, Institute of Environmental Studies, University of Toronto.

Burton, I. Kates, R. W. and White, G.F. (1978) *The Environment as Hazard*, Oxford University Press, New York.

Bowander, B. (985) The Bhopal Accident: Implications for Developing Countries, *The Environmentalist*, **Vol.5**, p.89.

Bowander, B. (1987) An Analysis of the Bhopal Accident, Project Appraisal, **Vol.2**, p.157.

Capel, P.D. *et al.* (1988) Accidental Input of Insecticides into the Rhine River, *Environmental Science and Technology*, **Vol.22**, p. 992.

Dahl, G. (1991) The Beja of Sudan and the Famine of 1984-1986, *Ambio*, **Vol.20**, pp 189-191.

El-Hinnawi, E. (1980) *Nuclear Energy and Environment*, Pergamon Press, Oxford.

El-Hinnawi, E. (1981) *The Environmental Impacts of Production and Use of Energy*, Tycooly International, Dublin.

El-Hinnawi, E. (1985) *Environmental Refugees*, UNEP, Nairobi.

El-Hinnawi, E. and Biswas, A. (1981) *Renewable Sources of Energy and the Environment*, Tycooly International, Dublin.

Flavin, C. (1987) Reassessing Nuclear Power, *In*: L.R. Brown *et al.*, (editors), State of the World, Norton and Company, New York.

Franzen, F. (1987) Reviewing the Operational Safety of Nuclear Power Plants, *IAEA Bulletin*, **4**, 13.

Hay, A.W.M. (1977) Tetrachlorodibenzo-p-dioxin Release at Seveso, *Disasters*, **Vol.1**, p.289.

Hay, A. (1978) Seveso: No Answers Yet, *Disasters*, **Vol.2**, p.163.

Hohenemser, C. and Renn, O. (1988) Chernobyl's other legacy, *Environment*, **30**, 5.

IAEA (1988) Radiation Sources: Lessons from Goiania, *IAEA Bulletin*, **Vol.30(4)**, p.10.

IAEA (1991) Nuclear Power States Around the World, *IAEA Bulletin*, **Vol.33**, p.43.

IAEA (1991a). *The International Chernobyl Project. Assessment of the radiological consequences and evaluation of protective measures.* IAEA, Vienna.

ICSU (1991) Third report of ICSU Special Committee for IDNDR, ICSU, Paris.

ILO (1991) *Prevention of Major Industrial Accidents*, ILO, Geneva.

Ilyin, L.A. and Pavlovskij, O.A. (1987) Radiological consequences of the Chernobyl accident, *IAEA Bull*, **29(4)**, 17.

Johnson, B.L.C. (1982) *Bangladesh* (2nd edition), Heineman Educational Books, London.

Kasperson, J. and Kasperson, R.E. (1989) *Priorities in Profile: Managing Risks in Developing Countries*, CENTED Reprint No.63, Center for Technology, Environment and Development, Worcester, Mass.

Keller, E.A. (1985) *Environmental Geology*, C.E. Merril & Co., Columbus, Ohio.

Lewis, H.W. *et al.* 1(978) Risk Assessment Review Group. Report NURES/CR-400, Washington, D.C.

Maki, A.W. (1991) The Exxon Oil Spill: Initial Environmental Impact Assessment, *Environmental Science and Technology*, **Vol.25**, p.24.

Malingreau, J.P., Stephens, G. and Fellows, L. (1985) Remote Sensing of Forest Fires: Kalimantan and North Borneo 1982-3, *Ambio*, **14**, 314-21.

McCleese, W.L. *et al.* (1991) Real-time Detection, Mapping and Analysis of Wildland Fire Information, *Environment International*, **Vol.17**, p.111.

NAS (1979) *Risks Associated with Nuclear Power*, National Academy of Science, Washington, D.C.

OECD (1991) *The State of the Environment - 1991*, OECD, Paris.

OFDA (1987) *Disaster History: Major Disasters World-wide*, Office of Disaster Assistance, USAID, Washington, D.C.

Oil Spills Intelligence Report (1991) **Vol. XIV**, No.12.

Oliver-Smith (1991) Successes and Failures in Post Disaster Resettlement, *Disasters*, **Vol.15**, p.12.

PAHO (Pan American Health Organisation), (1991) The International System of Humanitarian Assistance, *Disaster Preparedness in the Americas*, **No.48**, October.

Pearce, F. (1991) Acts of God, Acts of Man? *New Scientist*, 18 May, 20-21.

Rapp, A., Li, J. and Nyberg, N. (1991) Mudflow Disasters in Mountainous Areas, *Ambio*, **Vol. XX**, No.6.

Rasid, H. and Paul, B.K. (1987) Flood Problems in Bangladesh: Is There an Indigenous Solution?, *Environmental Management*, **Vol.11**, No.2.

Rich. V. (1991) An ill wind from Chernobyl, *New Scientist*, **20** June, 26.

Robock, A. (1983) Internally and Externally Caused Climate Change, *J. Atmospheric Sci.*, **35**, 1111.

Sattaur, O. (1991) Counting the Cost of Catastrophe, *New Scientist*, 29 June, p.21.

Savchenko, V. K. (1991) The Chernobyl catastrophe and the biosphere, *Nature and Resources*, UNESCO, 27, 37.

Smets, H. (1988) *The Cost of Accidental Pollution, Industry and Environment*, **Vol.11**, p.28, UNEP, Paris.

Stigliani, W.M. (ed.) (1991) *Chemical Time Bombs: Definitions, Concepts and Examples.* IIASA, Laxenburg, Austria.

Toon, O.A. and Pollack, J. B. (1982) Stratospheric Aerosols and Climate, In: R.C. Witten (ed.) *The Stratospheric Aerosol Layer.*

Toth, L.M. *et al.* (1986) The Three Mile Island Accident, American Chemical Society Symposium, Series 293, Washington, D.C.

UNEP (1988) *APELL Handbook*, UNEP, Paris.

UNEP (1989) *Environmental Data Report, 1989-1990*, Blackwell, Oxford.

UNEP (1991) *Environmental Data Report, 1991-1992*, Blackwell, Oxford.

United Nations (1988) *Sources, Effects and Risks of Ionizing Radiation.* United Nations Scientic Committee on the Effects of Atomic Radition, 1988 Report to the General Assembly, with annexes. United Nations sales publication E.88IX.7. United Nations, New York.

Velez, R. (1990) Mediterranean Forest Fires: a Regional Perspective, *Unasylva*, **Vol.41**, p.3.

WASH-1400 (1975) *Reactors Safety Study*, US Atomic Energy Commission, Washington, D.C.

Webb, J. (1991) Chernobyl findings, *New Scientist*, 1 June, 17.

Weir, D. (1987) *The Bhopal Syndrome*, Earthscan, London.

White, G.F. (1992) Natural Hazards Research, In: *Natural Hazards Observer*, **Vol. XVI**, No.3, pp. 1-2, January, 1992.

WHO (1989) *Glossary of Terms on Chemical Safety for Use in IPCS Publications*, WHO, Geneva.

Wilhite, D.A. (1990) The Enigma of Drought Management and Policy Issues for the 1990s, *International Journal of Environmental Studies*, **Vol. 36**, p.41.

Wilhite, D.A. *et al.* (1987) *Planning for Drought*, Westview Press, Boulder, Co.

WRI (1990) *World Resources*, World Resources Institute, Washington, D.C., Oxford University Press.

CHAPTER 10

Toxic chemicals and hazardous wastes

Introduction

Although chemicals in the form of natural products have been used by mankind from prehistoric times, man-made chemicals, particularly man-made organic chemicals, came into common use only relatively recently. The first breakthrough occurred during the second half of the nineteenth century when synthetic dyes began to be manufactured on a large scale and synthetic pharmaceuticals came into use.

New discoveries immediately before or during World War II, in conjunction with new techniques, formed the point of departure for an unprecedented development in the chemical field. A great variety of new products became available. Synthetic fibres such as nylon and terylene for use in various kinds of textiles; plastics such as PVC and polythene for use as packaging materials, in furniture, and in cars; insecticides, herbicides and other pesticides; an array of new pharmaceuticals, and many other chemically based new products came into widespread use. In addition, a great many new process chemicals were synthesized and used in industry. In the homes, new detergents, types of paints and other household chemicals marked the beginning of a new era.

Almost all human activities produce wastes which have to be disposed of, either as a matter of convenience or in conformity with tradition or legislation. Wastes occur in the primary sector (mining, agriculture), the secondary sector (manufacturing industry) and the service sector (which has recently grown considerably in the developed regions), as well as in the consumer sector (the end users in some cases). Such wastes, and their environmental and human health impacts, are dealt with in different parts of this book (Chapters 1-5, 17 and 18). Although the production and consumption of goods is currently responsible for the major share of wastes, today even agricultural wastes, which were traditionally recycled and used as raw materials for other productive activities, have become a problem in many areas. Their volume has increased considerably with the spread of intensive agriculture and a decline in their age-old uses as traditional life-styles have changed.

The main concern in this chapter is with 'hazardous' wastes. The definition of what constitutes a hazardous waste is still not universally agreed upon: however, it is generally recognized that such wastes contain toxic chemicals, micro-organisms or radioactive material. The annexes to the 'Basel Convention on the Control of Transboundary Movements of Hazardous Wastes and their Disposal', drawn up in March 1989, list 45 categories of non-radioactive wastes that are considered hazardous. Radioactive wastes are generally regarded as separate from other forms of hazardous wastes, and are specifically regulated by the International Atomic Energy Agency (IAEA). These are dealt with in Chapters 9 and 13. Most important among the wastes dealt with in the Basel Convention are compounds of heavy metals, organic cyanides, phenols and phosphorous compounds.

Thus it is clear that the proliferation of toxic chemicals in many walks of life is one of the major sources of hazard from wastes generated today world-wide.

The issue (1): toxic chemicals

World-wide, about ten million chemical compounds have been synthesized in laboratories since the beginning of this century. The European Inventory of Existing Commercial Chemical Substances (EINECS) lists 110,000 chemicals. In 1982 it was estimated that there were 60,000 chemical substances on the market and that the production of synthetic materials had increased some 350 times since 1940 (Holdgate *et al.*, 1982). This trend continued until the number of commercially-available chemicals reached the 100,000 mark of today, with 1,000 new substances becoming available every year. Existing testing facilities world-wide only can test 500 substances each year - and only then at great cost (OECD, 1991).

In contrast to its small beginnings before World War II, the chemical industry in the late 1970s produced about 400 million tonnes of products a year, and employed about four million people (OECD, 1979). However, many more people come in contact with chemicals at the work-place. In agriculture alone, around one billion workers are employed world-wide, accounting for 50 per cent of the total world work-force (ILO, 1988). The new products of the rapidly expanding chemical industry were for a long time looked upon as tokens of a prosperous new development. However, several events - particularly in the 1960s - indicated that chemicals could be a threat to the environment and to human health, and to a much greater extent than had been imagined previously.

Many questions arose concerning knowledge of the properties of the new chemicals, and the risks they posed to human health and the environment. Mainly during the last twenty years, answers have been found for some of these questions. To a certain extent the risks have been assessed, and ways have been designed to manage them, but much remains to be done to meet the challenges of the 'chemical age.'

The distribution by volume of different chemicals is extremely uneven. Figures illustrating this were available for the first time in the early 1980s. A study based on the US inventory of existing chemicals showed that less than 10 per cent of them accounted for more than 99 per cent of the total volume produced (Blair, 1991). As the volume produced of a chemical is a good surrogate for exposure, these figures underscore the urgent need for good safety information on high production-volume chemicals. A study by OECD indicates that such information is missing in several cases. For example, out of 1,338 high production-volume chemicals there is little or no available safety data for 147 (OECD, 1990).

The impacts of toxic chemicals on human health and the environment

During the 1960s a number of events occurred that indicated that public health and the environment were subject to new threats caused by chemicals (Box 1).

BOX 1

Landmarks in the revelation of the threats posed by toxic chemicals.

1958   Revelation in Sweden of mercury poisoning of birds feeding on seed treated with seed-dressing Panogen (a chemical based on an alkyl-mercury compound) introduced during the 1940s (Borg, 1958).

1962   Rachel Carson's book *Silent Spring* revealed the threats that excessive use of pesticides and other agrochemicals pose to flora and fauna, and consequently to human health.

1966   Jensen demonstrated that PCBs appear as pollutants. Later convincing evidence showed that they are wide-spread in the seas and accumulate in the biosphere (OECD, 1973). PCBs were available in the late 1920s and were widely used in electrical and other equipment.

1966   Evidence of the presence of high levels of mercury in fish from Swedish lakes. Sources traced to chlorine alkali plants and the pulp and paper industry using mercury compounds as preservatives (Westö, 1966).

1968   Minamata report published. Methyl-mercury generated from an inorganic mercury catalyst used in the manufacture of acetaldehyde was discharged into the sea and ingested by fish and shellfish. People consuming the seafood were poisoned: 1,500 became sick and more than 200 died.

The accumulating evidence on chemicals as a threat to the environment underscored the need for a reorientation of the control of chemicals to take into account also the environmental risks. However, it also became increasingly obvious that chemicals as a potential risk to human health had dimensions other than acute damage. Two types of long-term effects came into focus, chemical carcinogenicity and the ability of chemicals to cause congenital defects.

A large number of scientific reports during the 1950s and 1960s increased the understanding of chemical carcinogenicity and promoted interest in testing chemicals for carcinogenicity as a means of reducing the health risks involved. The establishment of the International Agency for Research in Cancer (IARC) and a consensus report on Carcinogenicity Testing published in 1969 by the International Union Against Cancer (UICC) (Borenblum, 1969) contributed considerably to widening the scope of chemicals control.

While awareness of chemical carcinogenicity was developed over a very long period of time, the signal that chemicals can cause congenital abnormalities came suddenly and quite dramatically. In 1961 the German physician W. Lenz published observations indicating that pregnant women taking a new sedative,

Thalidomide, which had been considered as a mild and relatively harmless drug, risked giving birth to children with severe congenital abnormalities (Lenz, 1961). This was the first indication of what came to be regarded as the greatest catastrophe in the history of pharmaceuticals. Testing for teratogenicity is now performed not only for pharmaceuticals but also for many other chemicals, particularly those to which pregnant women may be exposed. Examples of other long-term risks include damage to the reproductive functions, disturbance of the immune system and neurological injury.

Full understanding of the potential hazards of a chemical requires a large volume of data, including identification (structural formula, degree of purity), estimated production and intended uses, analytical methods, physical-chemical properties (e.g. melting and boiling point, vapour pressure, solubility), toxicity (acute oral, dermal and inhalation toxicity, repeated dose toxicity), mutagenicity, carcinogenicity, ecotoxicity, ability for bio-degradation and bio-accumulation. The generation of a full set of data is an expensive and time-consuming procedure: it has been estimated (IPCS-IRPTC, 1990) that testing a chemical for multiple toxicological endpoints takes up to 64 months and that in 1985 a pre-chronic study cost $US575,000 and the cost of a chronic toxicity and carcinogenicity study amounted to $US1,300,000.

The impacts in developing regions, though less well documented, are considered to have caused more damage to human health and the environment. This is particularly the case in agricultural communities that started using pesticides and herbicides at rapidly increasing rates without much knowledge of their health hazards, and consequently without even the minimum necessary precautions. Cases of poisoning of human beings and livestock, resulting in serious illness and sometimes death, were reported from many countries in Asia, Africa and the Middle East (Chapter 11). The WHO estimated that the number of unintentional acute poisonings due to exposure to pesticides was half a million in 1972. With the increasing use of pesticides this figure doubled by 1985 (WHO, 1990).

Another characteristic of the situation in developing countries is the increase in traditional occupational diseases at a time when they are declining in the developed regions. This is particularly the case in small-scale industries that suffer from the combined effects of ignorance of the hazards of toxic chemicals and absence of protective measures. As mentioned in Chapter 12, small-scale industry is still responsible for a considerable part of total industrial output in many developing countries.

Responses

Scientific responses

Sustained, in-depth scientific effort is necessary to identify clearly the dangers involved in the use and handling of toxic chemicals and to devise feasible approaches to the mitigation of these dangers. Without this foundation of solid scientific knowledge, monitoring and regulation will be ineffective or even counter-productive.

Risk assessment

The distinction between hazards and risks has been addressed in Chapter 9. In assessing the risks brought about by toxic chemicals, we need to distinguish between the absolute risk, which is the excess risk due to exposure, and the relative risk, which is the ratio between the risk in the exposed population and the risk in the unexposed population (WHO, 1978). To evaluate the risk of a chemical requires knowledge of the intrinsic properties of the chemical (hazard), and the exposure in the situation under consideration. Consequently the risk can vary widely depending on the local situation, while the hazard is constant and characteristic for each chemical.

It is seldom possible to determine exactly the total level of exposure to a chemical. Usually it occurs at the work-place, in the home or via the environment. Often it is an indirect exposure originating from food, air and water. The route of exposure is also of importance (inhalation, dermal contact, oral ingestion), as is the frequency and the duration of exposure. It is also important to take into account the exposure of particular groups of the population such as pregnant women, children and elderly people.

Testing and generation of data

Increased efforts were made in the 1980s to develop test methods to replace those using vertebrate animals (US OTA, 1986). In order to reduce costs and avoid technical barriers to trade it is important that the methods used for the testing and generation of data are internationally recognized. A first set of the OECD *Guidelines for Testing of Chemicals* was published in 1981. These have since been elaborated and updated. In December 1991 a total of 78 guidelines had been published, covering physico-chemical properties, ecotoxicology, degradation/ accumulation and health effects. Comprehensive principles of good laboratory practice (GLP) were also published. The OECD Test Guidelines and Principles of GLP are used in many non-OECD countries. In a formal agreement on co-operation between the OECD Chemicals Programme and IPCS, chemicals control tools such as the Test Guidelines and GLP-principles are specifically mentioned.

Within the framework of IPCS, work continues on the development, improvement, validation and use of laboratory tests, ecological and epidemiological studies, and other methods suitable for the evaluation of health and environmental risks and hazards from chemicals. So far this work has resulted in 13 volumes in the *Environmental Health Criteria* series. They cover evaluation of chemical hazards (4 volumes), general methodology applicable in chemical safety studies (3 volumes), disease-oriented approaches to specific chemical safety problems (1 volume), chemicals in food (2 volumes) and short-term tests (3 volumes).

Jointly with the Scientific Committee on Problems of the Environment (SCOPE) of the International Council of Scientific Unions (ICSU), IPCS sponsors scientific work on the development and validation of methods for evaluation of chemical risks. This is carried out by the SGOMSEC (Scientific Group on Methodologies for the Safety Evaluation of Chemicals). SGOMSEC has published a number of comprehensive reports, based on symposia. They include volumes on methods for assessing the effects of chemicals on reproductive functions, methods for assessing the effects of mixtures of chemicals and methods of reducing injury from chemical accidents.

Twenty years ago, very few data were available other than those indicating acute toxicity. This led to a dilemma in the 1970s when new legislation in many countries required more information. In 1980 the US National Toxicology Program commissioned the National Research Council and the National Academy of Sciences to characterize the toxicity-testing needs for substances to which there is known or anticipated human exposure, as well as to develop and validate uniformly applicable and wide-ranging criteria by which to set priorities for research on substances with potentially adverse public-health impact. The result of the US study (US National Research Council, 1984) was that for about 80 per cent of the commercially available chemicals no toxicity information whatsoever was available. For none of them was information available for a complete health hazard assessment. The situation was better for pesticides, food additives, cosmetic ingredients, and in particular for drugs. However, even for these groups the available information was often insufficient for a complete evaluation (Figure 1).

Figure 1: *Percentage of chemicals for which toxicity data are available.*

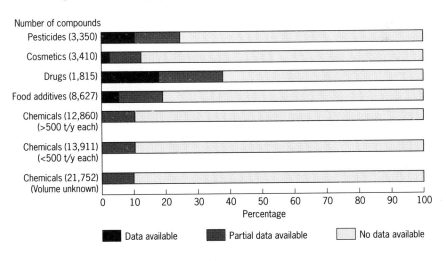

Source: Based on data from US NRC (1984).

253

The situation is slowly improving. First, a number of chemicals have been assessed, completely or partly, by IPCS, JECFA, JMPR and IARC during the last decade. Moreover, within the European Community countries, Japan, the United States and a few other countries all new chemicals have been subject to an initial, yet meaningful evaluation. In several countries systematic work is also in progress to review selected existing chemicals (ECETOC, 1991).

New approaches in national and regional legislation

Until the end of the 1960s, national legislation on chemicals was usually limited to substances known to be hazardous, but from 1969 this attitude began to change. Switzerland was the first country to promulgate new legislation encompassing all chemicals. Among other things the new laws stipulated that chemicals could not be put on the market unless the class of hazard they represent had been determined. Also, in Japan, a law promulgated in 1973 stipulated the classification of new chemicals before they could be put on the market. The new law introduced a number of new concepts. An inventory of all 'existing chemicals' was established. For 'new chemicals' information had to be given on structural formula, physical-chemical properties, intended use, and the foreseen amount to be manufactured or imported during the first three years. Testing for persistency and bio-accumulation became compulsory. Test methods for an array of long-term effects were described. Canada, France, Norway, Sweden and the United Kingdom are other countries in which new legislation was enacted in the early 1970s with the aim of broadening the scope of the control of chemicals, including the incorporation of environmental issues in the quality requirements. In the United States the new Toxic Substances Control Act (TSCA) entered into force in 1977. It included provisions for the testing and assessment of existing chemicals, disclosure of all health and safety data on chemicals, regulation of exports and imports, etc.

Similar approaches had been adopted in the EC Council 1979 Directive on chemicals, known as the 'sixth amendment'. An innovation in the Directive was that the notification of new chemicals by manufacturers or importers should be accompanied by a comprehensive set of information ('Base Set') necessary for evaluating the foreseeable risks, whether immediate or delayed, which the substance may entail for man and the environment. According to the Directive, the Member States of the European Communities should modify national legislation in order to be compatible with the Directive.

The international response

International responses have included programmes for chemical safety, information exchange and concerted action for risk reduction and emergency response.

Chemical safety

The attention of the international community was first focused on insecticides, and at the request of the First World Health Assembly in 1948, WHO set up an expert group on insecticides. One of the subsequent highlights of the many WHO contributions in the pesticides field was the Recommended Classification of Pesticides by Hazard, approved by the World Health Assembly in 1975. This classification gained wide acceptance.

In the 1970s, FAO broadened its work on chemical safety, and convened governmental consultations aiming at the harmonization of national requirements for registration of pesticides. These activities culminated in the elaboration of the International Code of Conduct on the Distribution and Use of Pesticides, adopted by the FAO Conference in 1985. The Code was amended in 1989 and is the basis of a great number of guidelines issued by FAO (see later on information exchange).

As regards food safety, WHO in co-operation with FAO organized the first FAO/WHO Joint Meeting on Pesticide Residues (JMPR) in 1963. The Meeting is convened annually and has evaluated more than 200 pesticides. Concurrently, FAO/WHO also hold annual meetings of the Joint Expert Committee on Food Additives (JECFA) which was established in 1956. More than 700 food additives and numerous food contaminants have been evaluated by JECFA. During the last few years the Committee has also evaluated certain veterinary drug residues in food.

In 1971, OECD set up a Chemicals Group. The OECD work on chemicals was substantially expanded in 1978 and has played an important role in the promotion of chemical safety.

In 1976, UNEP established the International Register of Potentially Toxic Chemicals (IRPTC) in response to the Stockholm Recommendation No 74 (Box 2). The Register now contains data profiles of about 800 chemicals, including detailed scientific data collected mainly from secondary literature. A legal file includes more than 42,000 records from 12 countries and 6 international organizations on about 8,000 chemicals.

The Stockholm Conference also gave impetus to the launching of the WHO Environment Health Criteria programme. This activity is now part of the International Programme on Chemical Safety (IPCS), established in 1980 by UNEP, ILO and WHO (Box 3). Within the framework of IPCS, risk evaluation of chemicals is carried out, methodology for testing and evaluation is developed and validated, technical co-operation with Member States is promoted and training of required manpower is supported.

In addition to the establishing of IRPTC, UNEP has initiated two other far-reaching activities in the chemical safety area, namely the efforts to phase out or diminish the use of chemicals threatening the ozone layer (see Chapter 2) and the London Guidelines for the Exchange of Information on Chemicals in International Trade (see next section).

ILO has considerably strengthened its activities in this field from the mid 1970s when it launched the International Programme for the Improvement of Working Conditions and Environment (PIACT) (Chapter 18). This has resulted in the Convention on Safety in the Use of Chemicals at Work, adopted by the International Labour Conference in 1990.

BOX 2

IRPTC

Adequate information to assess the potential hazards posed to human health and to the environment is a prerequisite to their safe use and disposal. In 1976 UNEP established the International Register of Potentially Toxic Chemicals (IRPTC) which collects and disseminates information on hazardous chemicals, including national laws and regulations controlling their use.

IRPTC operates through a network of national and international organizations, industries and external contractors, and national correspondents for information exchange which have now been appointed in 112 countries. IRPTC's computerized central data files contain data profiles for over 800 chemicals. In addition, special files are available on waste management and disposal, on chemicals currently being tested for toxic effects, and on national regulations covering over 8,000 substances.

BOX 3

IPCS

In 1980, WHO, UNEP and ILO set up the International Programme on Chemical Safety (IPCS) to assess the risks that specific chemicals pose to human health and the environment.

IPCS publishes its evaluations in four forms: as detailed '*Environmental Health Criteria*' for scientific experts; as short, non-technical '*Health and Safety Guides*' for administrators, managers and decision-makers; as '*Chemical Safety Cards*' for ready reference in the work-place; and as '*Poisons Information Monographs*' for medical use.

In the 1950s, the UN issued recommendations on the transport of dangerous goods including chemicals. Based on these recommendations, compatible safety rules now govern the transport of hazardous chemicals by sea, air, road and rail. The International Maritime Organization (IMO) has also carried out work to prevent chemical pollution of the sea (see second part of this chapter on **Hazardous wastes**). This has been supported by the Group of Experts for the Scientific Aspects of Marine Pollution (GESAMP), established jointly by UNEP, FAO, UNESCO, WHO, WMO, IMO and IAEA.

In 1982, a ministerial meeting of GATT Contracting Parties requested the Secretariat to examine measures to bring under control the export of products the sale of which is prohibited in the domestic markets of exporting countries because they are harmful to humans, animals, plants or the environment. A draft decision prepared by a working group covers trade in all products (including hazardous wastes) that are determined by a Contracting Party to present such serious and direct dangers that they are banned or severely restricted in the exporting country. For the sake of transparency, the draft recommends the creation of a notification system for such products and that the Secretariat would immediately use the information to respond to enquiries. In 1991, UNEP convened an Intergovernmental Group of Experts to develop procedures for establishing an intergovernmental mechanism for risk assessment of chemicals.

Information exchange

Information exchange between producers and users, both in industrialized countries and in developing countries, is crucial for sound chemical risk management. Adequate labelling of chemicals is the simplest way to transmit basic facts. For more detailed information, chemical safety data sheets and other guides are essential. Internationally harmonized labelling systems now exist for application during transport of chemicals. For other purposes no similar harmonization has been made, but the labelling system for industrial chemicals, originally developed by the Council of Europe and adopted by the European Communities, is gaining recognition. The development of Chemical Safety Cards contributes to harmonization of data sheets.

The most far-reaching activities for exchange of information on chemicals have been developed by UNEP. The Query Response Service operated by UNEP/IRPTC since its inception in 1976 now processes over 500 queries a year concerning 1,500 chemicals. A broad basis for information exchange is provided by the London Guidelines for the Exchange of Information in International Trade, adopted by the UNEP Governing Council in 1987. Two years later the Guidelines were amended by introducing the 'Prior Informed Consent' (PIC) procedure. This is an addition to the original system for information exchange and for notification of chemicals that are banned or severely restricted in the country of export. It is based on the principle that international shipment of a chemical that is banned or severely restricted to protect human health or the environment should not proceed without the agreement, or contrary to the wishes, of the designated national authority in the importing country. The PIC procedure has also been incorporated in the FAO International Code of Conduct on the Distribution and Use of Pesticides. To facilitate the implementation of PIC (both in the Guidelines and in the Code) UNEP and FAO co-operate in the elaboration of PIC Decision Guidance Documents including summary information on the chemicals concerned. At the request of the UN General Assembly the *Consolidated List of Products Whose Consumption, and/or Sale Have Been Banned, Withdrawn, Severely Restricted or Not Approved by Governments* was first issued in 1984. The fourth edition of the list appeared in 1991. Attempts are continuing for the development of a convention that would make PIC obligatory for contracting parties.

Within the OECD a formal procedure for notification and consultation on measures for control of substances affecting man and his environment was adopted in 1971. In addition, a Complementary Information Exchange Procedure was introduced in 1977. A voluntary scheme concerning information exchange related to export of banned or severely restricted chemicals was launched in 1984.

Concerted actions for risk reduction

As environmental damage by chemical pollutants occurs without respect for national borders it is important that actions against severe pollutants be taken on a broad international basis. Integrated pest management (IPM), practised in successful FAO projects in rice fields in South and Southeast Asia, has increased harvests without, or with substantially decreased amounts of, pesticides (Chapter 11). The ILO has been active since its creation in promoting concerted actions, starting with the Lead Convention of 1921. More recent ones are the Benzene Convention of 1971 and the Asbestos Convention of 1986 which were international attempts at limiting the exposure of workers to benzene and asbestos. Another example of concerted action against a harmful chemical is the 1973 OECD Decision entitled 'Protection of the Environment by Control of Polychlorinated Biphenyls'. This Decision substantially limited the use of PCBs for industrial and commercial purposes in OECD countries. The most far-reaching concerted action on chemicals is the Montreal Protocol to protect the ozone layer by phasing out the use of CFCs and some other halogenated chemicals (see Chapter 2). Such actions are based on the principle that whenever harmless or less harmful substitute products are available they should be used.

Accident prevention and emergency response

In a recent listing of major industrial disasters in this century, 22 events out of 31 accounted for during the last 20 years were chemical accidents (Friedrich Naumann Foundation, 1987). The environmental impacts of these accidents are discussed in Chapter 9. As a reaction to these and other accidents, intensive international work has been carried out during the last decade in order to improve accident prevention and preparedness for emergency response. One outcome of such work is the so-called 'Seveso Directive' issued by the Council of the European Communities in 1982. In 1988, UNEP launched the APELL Programme (Chapters 9, 12 and 21). ILO, after having published a manual on major hazard control (ILO, 1988), has issued a Code of Practice on the prevention of major industrial accidents, and is preparing a convention on the prevention of industrial disasters for possible adoption in 1993. In 1985, the World Bank published *Guidelines for Identifying, Analysing and Controlling Major Hazard Installations* and a *Manual of Industrial Hazard Techniques*. On an initiative of the French Government, the OECD in 1988 organized a Conference on Accidents Involving Hazardous Substances, following which a comprehensive programme on accident prevention and response was launched (OECD, 1989). One of the outcomes of this programme is an OECD/UNEP International Directory of Regional Response Centres.

Concluding remarks

While the last two decades have witnessed concerted efforts at the national and international levels to safeguard life and the environment against the dangers of a relentless proliferation of toxic chemicals, there is still an urgent need for action on several fronts. On the scientific front, a large number of chemicals handled by many people in different walks of life are still without a complete assessment of the hazards and risks they present in a variety of situations. We are still without internationally agreed reliable methods of testing, and harmonized methods of reporting on the toxicity of chemicals. On the regulatory front, the long-awaited global convention for the exchange of information on chemicals in international trade is yet to be concluded and an intergovernmental mechanism for chemical risk assessment and management set up.

The issue (2): hazardous wastes

Hazardous wastes represent considerable potential risks to human health and the environment, the level of risk depending on the type of substance and the volume. This complicates the problem of managing hazardous wastes, which are usually very heterogeneous mixtures (Box 4).

Modern industry is the main source of hazardous wastes, particularly in developed regions. The generation of hazardous wastes is not, however, confined to large-scale industrial plants: small-scale industry, small workshops, garages and very small production units collectively produce large quantities of hazardous wastes. Their volume is usually difficult to monitor and quantify. Furthermore, transport services, hospitals, research laboratories and even households are sources of sometimes quite dangerous wastes. Within the industrial sector itself, the chemical industry is by far the main source of hazardous wastes in the developed regions (e.g. Figure 2 for the USA).

BOX 4

Some illustrative examples of hazardous wastes.

Sector	Source	Hazardous waste
Commerce, services and agriculture	Vehicle servicing	Waste oils
	Airports	Oils, hydraulic fluids, etc.
	Dry cleaning	Halogenated solvents
	Electrical transformers	PCBs
	Hospitals	Pathogenic/infectious wastes
	Farms/municipal parks etc.	Unused pesticides, 'empty' containers
Small-scale industry	Metal treating (electro-plating etching, anodizing, galvanizing)	Acids, heavy metals
	Photofinishing	Solvents, acids, silver
	Textile processing	Cadmium, mineral acids
	Printing	Solvents, inks and dyes
	Leather tanning	Solvents, chromium
Large-scale industry	Bauxite processing	Red muds
	Oil refining (Petrochemical manufacture)	Spent catalysts Oily wastes
	Chemical/pharmaceutical manufacture	Tarry residues, solvents
	Chlorine production	Mercury

Source: World Bank (1989).

Figure 2: *Hazardous wastes generated in the USA, by type (in percentage).*

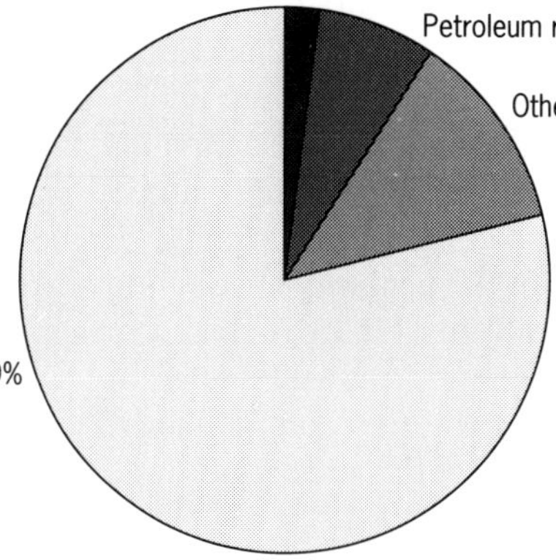

Metal industries 2%
Petroleum refining 7%
Others 12%
Chemical industries 79%

Source: Based on data from EPA (1988).

The management of wastes

Waste management in general could be broadly classified into two main categories: the preventive (addressing the causes of waste generation), and the curative (controlling the harmful effects of wastes). Both could be considered within a hierarchy of approaches:

- dematerialization of production (reducing material and energy inputs);

- recycling wastes back into the production process;

- recovery of some ingredients, and/or treatment of wastes, and

- dispersal, dumping or storage.

Dematerialization of production, recycling, and the concepts of cleaner production, industrial metabolism and industrial ecosystems, addressing both the preventive and the curative approaches, are dealt with in Chapter 12. However, no matter what efforts are made to address the causes of waste generation within the current state of technology there will always be some waste discharge into the environment: the ideal of zero waste generation is a long way from realization. When such wastes are of the hazardous variety, they pose serious problems.

The treatment of hazardous wastes

Treatment of hazardous wastes has so far involved physical, thermal, chemical or biological processes (Box 5). More recently, biotechnology has provided new options for dealing with some hazardous wastes, but some of these methods, using micro-organisms, have not yet been applied on a scale that would allow proper assessment of their effectiveness. Serious pollution can be caused by badly designed or badly operated waste treatment plants and incinerators - a situation that is fairly common in developing regions.

Finally, there remains the necessity of dumping or storing hazardous wastes, which again has its problems. Figure 3 gives a typical example of the volumes of hazardous chemical wastes treated in different ways. It shows that disposal (on land or at sea) is by far the most common approach for managing hazardous wastes. The consequences of this are serious indeed. Landfills contaminate the

---

BOX 5

Waste treatment and disposal technologies.

General division	Subdivision
Recycling	Gravity separation
	Filtration
	Distillation
	Solvent extraction
	Chemical regeneration
Physical/chemical	Neutralization
	Precipitation/separation
	Detoxification (chemical)
Biological	Aerobic reactor
	Anaerobic reactor
	Soil culture
Incineration	High temperature
	Medium temperature
	Co-incineration
Immobilization	Chemical fixation
	Encapsulation
	Stabilization
	Solidification

Source: UNEP (IE/PAC), Paris.

Figure 3: *Management of chemical wastes in the United Kingdom (1985).*

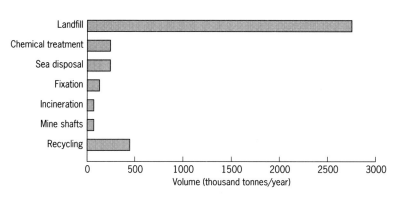

Source: Based on data from Burns (1988).

soil causing changes in its micro-ecology. Leachates from landfills contaminate surface and ground water with their harmful compounds, making such water unusable and a threat to biota and human health. The classic example of the dangers of landfill disposal of hazardous wastes is the case of Love Canal, a small town in the USA, close to Niagara Falls. Buildings were erected on a site previously used as a dump for waste chemicals. Heavy rains, years later, created ponds of contaminated water, forcing the relocation of residents and site remedial measures costing about $US100 million - a good example of a 'chemical time bomb' (see Chapter 9).

Disposal in the marine environment has diverse harmful effects such as the reduction, alteration or contamination of the fauna, or oxygen depletion. The classic example of the dangers of dumping at sea is the 'Minamata' disease, caused by mercury poisoning. A chemical factory had been using inorganic mercury and discharging effluent containing methyl mercuric chloride, a highly toxic substance, into the sea at Minamata Bay, Japan. People eating fish from the region suffered from neurological disorders. More than 1,500 people were poisoned and more than 200 died. The plant had been operating since 1932, but it was not until 24 years later that the chemical was identified as the cause of the disease. The effects were still being felt more than ten years after the discharges were halted

Hazardous wastes tend to 'run downhill' to the least regulated and least expensive disposal option unless market forces are such that they direct hazardous waste management to more appropriate options. The main attraction of landfills is their low cost. For centuries now, highly corrosive and toxic residues of mining and metal refining processes have been accumulating in the neighbourhood of these operations in many parts of the world - without due consideration of their devastating impacts on the local environment, or the dangers to human health, even though the devastation was clearly there to see. Increases in the GDP of a country result in increases of all wastes, including hazardous wastes (Chapter 12, Table 6). With the continuing increase in the production of goods, there is not much space left in most developed countries for new landfills. As urban settlement also expands into previously uninhabited areas, the 'not in my backyard' (NIMBY) syndrome has intensified the pressure to find new approaches to hazardous waste management.

As hazardous wastes accumulated in the industrialized regions, export of such wastes to other countries began, and has continued until it has now reached disturbing levels. Some wastes are exported to other developed countries that have special treatment facilities, but considerable amounts are spirited away to less developed regions, mainly in Africa and Asia, where they are dumped under highly dangerous conditions. Thus the problem of transboundary movement of hazardous wastes has emerged in the last decade as a serious environmental issue.

Magnitude and trends of the hazardous wastes problem

Estimates of waste quantities depend on national definitions of waste. Countries with the most all-embracing and stringent definitions register large quantities of wastes that are ignored by other countries. Consequently, inter-country comparisons are often misleading. World-wide, it has been estimated that no less than 338 million tonnes of hazardous wastes are produced annually, representing about 16 per cent of industrial waste (OECD, 1991). Of this, about 80 per cent is produced in the USA alone (Figure 4). The figures for South Korea are 12.172 million t/y in 1985, rising to 21.041 million tonnes per year in 1989. For Japan, the figure given by Japanese participants in a recent workshop was as high as 312 million t/y for 1985 (UNEP, 1990). The discrepancies in these figures quoted in different contexts highlight the absence of universally accepted definitions of hazardous wastes, as well as the scarcity of reliable methodologies for estimating their volume. At present, IMO is co-operating with a number of international organizations in an attempt to carry out a reliable assessment of hazardous wastes on a global scale.

Figure 4: *Hazardous wastes generated, in million tonnes/year (late 1980s).*

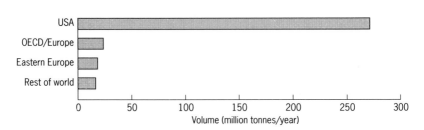

Source: Based on data from OECD (1991).

Comparison of the hazardous waste figures for some OECD countries in the early 1980s and the late 1980s indicates that hazardous wastes are increasing (Table 1). The production of hazardous wastes in the USA, for example, increased from 9 million tonnes per year in 1970 to about 264 million tonnes at present, more than in any other country (WRI, 1987).

Figures for the developing regions are scarce. It is estimated that hazardous waste generated in Malaysia amounted to 417,000 t/y, for Thailand 22,000 t/y, and for Singapore 28,000 t/y (Uriarte, 1989). These are certainly conservative estimates. Apart from the absence of reliable systems of monitoring wastes, and the ambiguity surrounding the definition of hazardous wastes, the proliferation of small-scale industries in developing countries makes it difficult to compile accurate data on hazardous wastes in these countries.

Table 1: *Hazardous waste generation in some OECD countries in the early 1980s and late 1980s.*

Country	Hazardous and special wastes (in 10^3 tonnes)	
	Early 1980s[1]	Late 1980s[2]
Canada	3,290	3,300[3]
USA[4]	264,000	275,000
Japan	768	n.a.
France	2,000	3,000
Germany[5]	4,892	6,000
Italy	n.a.	3,800
United Kingdom	1,500	4,500

Notes: 1. OECD Environmental Data Compendium (1985).
2. OECD The State of the Environment (1991).
3. 1980.
4. USA totals include liquid waste.
5. For Western Germany only.

There are considerable variations in national definitions of hazardous and special wastes, and comparisons between different countries can therefore be very misleading.

A breakdown of the generation of hazardous wastes in terms of amount generated per capita and per one thousand dollars of GDP (Table 2) shows wide variations between countries. The Netherlands generates the largest amount of hazardous waste per capita (102.3kg); Ireland and Greece the lowest. Greece produces the largest amount per unit GDP, and Luxembourg the lowest. While these figures reflect differences in production and consumption patterns, they are also influenced by the discrepancies in definitions of hazardous wastes.

Landfills containing hazardous wastes have proliferated under the impact of intensive industrialization. For many years no records were kept of those sites,

Table 2: *National generation of hazardous wastes per capita and per unit gross domestic product in OECD countries.*

Country	Hazardous waste (tonnes x 10³)	Generation in terms of population (kg/capita)	Generation in terms of Gross Domestic Product (kg/thousand $US)
Austria	200	26.4	1.71
Belgium	915	93.0	6.59
Denmark	112	21.8	1.11
Finland	270	54.7	3.02
France	3,000	53.9	3.41
Germany	6,000	98.0	5.37
Greece	423	42.3	8.96
Iceland	5	20.3	0.93
Ireland	20	5.6	0.68
Italy	3,800	66.3	5.01
Luxembourg	4	10.7	0.67
Netherlands	1,500	102.3	7.04
Norway	200	47.7	2.42
Portugal	165	16.1	4.50
Spain	1,708	44.0	5.91
Sweden	500	59.5	3.15
Switzerland	400	60.4	2.34
Turkey	300	5.7	4.45
United Kingdom	4,500	79.0	6.72
OECD Europe	24,022	-	-
Canada	3,290[1]	128.3	6.86
United States	275,000[2]	1,127.4	57.23
Australia	300	18.4	1.24
Japan	666	5.5	0.23
New Zealand	60	18.1	1.42

Notes: Generation figures are taken from the OECD *State of Environment* reports (various reference years).
For Population and GDP figures *OECD in Figures* was used (reference year -1987).
1. Wetweight
2. The difference between US and Europe figures for waste generation arises in large measure because the United States manages large quantities of dilute waste waters as hazardous wastes while in Europe these materials are managed under water protection regulations.

Source: H. Yakowitz in Chemical Industry Conference, Zurich (1991).

and it is only recently that a start has been made on determining their location and extent. In the USA, no fewer than 32,000 sites have been categorized as potentially hazardous. Of these, some 1,200 need immediate remedial action (Schweitzer, 1991). In the Netherlands, 4,000 sites have been identified as potentially hazardous; in Denmark the number is 3,200; and in western Germany it is 50,000. The cost of remedial action is estimated at $US20-100 billion for the USA; $US6 billion in the Netherlands and $US30 billion for western Germany.

Dumping at sea is controlled by the London Dumping Convention of 1972 and its Annexes (Box 6), as well as a number of regional legal instruments (see section on **International responses**). These stipulate licensing of waste dumping at sea by the appropriate enforcement authority. The conventions also ban the dumping of certain hazardous wastes. It is estimated that in Europe, one million tonnes of hazardous wastes are dumped at sea. This is 2.5 - 4 times the amount recycled and 1.5 times that disposed of in a country other than the country of origin (OECD, 1991).

Thermal processes have been developed considerably as a means of reducing the volume of wastes to be disposed of, particularly in Europe. However, incineration has been criticized for its adverse environmental effects, particularly the discharge of dioxins and furans when burning is carried out at the wrong temperatures. A more recent concern is the discharge of acidifying effluents that contribute to the leaching of heavy metals and aluminium in soils. This could be remedied by advanced techniques of emission treatment, but at appreciable cost increases. On the positive side, incineration reduces the volume of wastes and consequently the need for landfills that are becoming difficult to accommodate in developed regions. The public outcry against incineration in the North Sea led to the decision in a Ministerial meeting in 1987 to bring to an end incineration in that area. The tendency now is to build large-scale carefully controlled incinerators on land (DOE, 1987; Bradshaw *et al.*, 1992).

BOX 6

The London Dumping Convention (1972).

- 'Dumping' means deliberate disposal at sea of wastes and other matter on board vessels, aircraft, platforms or other man-made structures at sea.
- Annex I of the Convention (the 'Black List') lists the substances that may not be dumped at sea, unless it can be demonstrated that they are present in wastes only as trace contaminants or that they become rapidly harmless at sea. Exemption needs a suite of biological tests to ascertain that they would not harm the marine environment. Annex I does not apply to their disposal by incineration at sea.
- Annex II lists substances requiring special care when dumped at sea, and consequently, special authorization by national administrations.
- The lists in the Annexes were intended to be reviewed and updated. However, there is reluctance to amend the Annexes. In view of this difficulty, allocation guidelines for assigning substances to the Annexes were adopted in 1980 and 1982.
- Annex III includes all factors to be considered in issuing permits for disposal at sea. It incorporates many of the key elements of a waste management plan.

It is thus crucial to investigate the full implication of adopting a particular disposal method. The final choice has to take into consideration a number of political, economic and geomorphological considerations, as well as public attitudes and perceptions. Some countries have guidelines in the choice of technology for a particular type of waste (World Bank/WHO/UNEP, 1989). The last two decades have witnessed the emergence of a number of economic instruments that favour more environmentally-sound waste management options. These are dealt with in Chapter 12.

In Europe, some 700,000 tonnes of hazardous wastes are exported to a country other than the country of origin every year (OECD, 1991), sometimes passing through other countries in transit. More than 2,000,000 tonnes of hazardous wastes are moved across national frontiers in OECD-Europe every year, representing more than 100,000 frontier crossings. Europe also exports legally about 120,000 tonnes of hazardous wastes to the developing countries every year (Crump, 1991). For North America, the figures are 200,000 tonnes of hazardous wastes exported and 9,000 crossings in 1988 (OECD, 1991). A general idea of such transboundary movements of hazardous wastes is shown in Figure 5. A report of the UN Secretary General to the General Assembly's forty-fourth session (A/44/362) reviews the world situation in the illegal traffic of hazardous wastes. The report gives examples of illegal waste exports from the USA to Latin America, Asia and Africa; from Western to Eastern Europe; from Europe to Africa, Latin America, the Middle East and Asia; and from Australia to New Zealand and between the Southeast Asian countries. It concludes that 'there is clear growth in the number of proposals from the industrialized world to construct in the developing world so-called waste-to-energy plants or to provide supposedly non-hazardous waste landfills or incineration facilities'. It concludes that the escalation in disposal costs and the great difficulty in obtaining government approval for the construction and operation of incineration facilities will increase the pressure on the developing world to accept these proposals.

Responses

Technological responses

Technological responses have been in two main directions: minimizing the quantities of hazardous wastes, and minimizing their adverse impacts on human health and the environment. The approaches to minimizing waste generation through cleaner production in its widest sense (the so-called 'cradle-to-grave' approach) are discussed in Chapter 12, together with some achievements in waste minimization and recycling in general.

In the field of minimizing the adverse impacts of hazardous wastes themselves, work has proceeded along two main lines: remedial action on landfills, and improvements in incineration technologies. One example is the

Figure 5: *Some major transfers of hazardous wastes.*

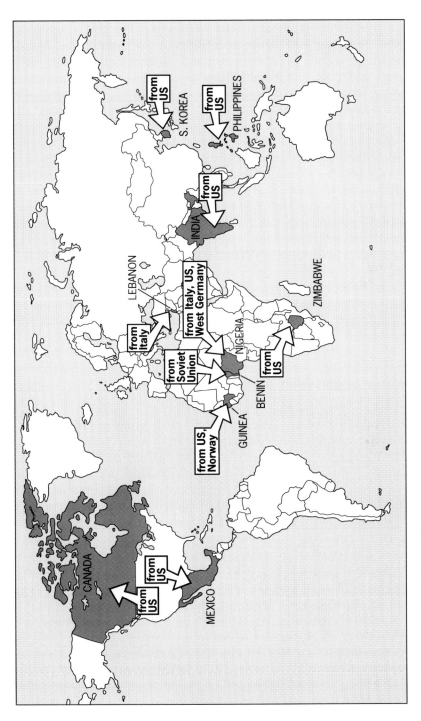

Source: Development and Co-operation **No. 6** (1988).

Super Fund Innovative Technology Evaluation Program (SITE) funded by the US EPA. Currently, there are over 50 field-scale tests undertaken by 49 technology developers. Some of the technologies are now commercially available. There has been particular emphasis in most programmes on *in situ* treatment of landfills by different methods. A clear understanding of the processes taking place in a landfill, and the manner in which leachates are generated, has also been developed. Biotechnological remediation, using naturally occurring micro-organisms, as well as genetically-engineered ones, is a promising new field of research and development for *in situ* remediation.

As the availability of landfills has decreased there has been a significant increase in the number of incinerators and more stringent conditions governing their operation. The present tendency is to diversify incinerator designs to suit particular types of waste, so as to eliminate dangerous emissions on the one hand, and reduce the toxicity of the residual ashes on the other. Modern incinerators provide more than 99.999 per cent destruction efficiency of PCBs, chlorobenzenes and chlorophenols without increasing emissions of dioxin and furan. Development is currently going on to reduce levels of these emissions by several orders of magnitude below the levels of existing clean plants (Kempa, 1991). As for the presence of dioxins and furans in incinerator ash, it is worth noting that ash from modern industrial hazardous waste incinerators contains less dioxins and furans than that from common municipal solid waste incinerators.

Regulatory responses at the national level

At the national level, responses have to include provisions of a technical, legislative, monitoring and enforcement nature. Several countries, mainly in the developed regions, have already formulated their approach to the regulation of hazardous wastes. At the technical level, national policy may stipulate specific techniques for the treatment and disposal of specific types of wastes. The problem with the formulation of national legislation is that it has to be clear, fairly straightforward in application and practical to enforce. Given the complexity of hazardous wastes this is not always easy to achieve. In particular, monitoring and enforcement requires the establishment of competent inspection and enforcement organizations that can achieve the aim of environmental protection without hampering national economic development. This has proved to be difficult in many developing countries, particularly when standards formulated in other countries have been adopted without due consideration of the variety and complexity of the issues involved (Table 3). A workshop organized recently by the International Solid Wastes and Public Cleaning Association (ISWA) reviewed the experiences of a number of countries in establishing hazardous wastes control measures (Waste Management and Research, 1990).

Table 3: *Aspects of hazardous waste legislation.*

Purpose and scope

Public health	Raise funds
Pollution prevention	Empower a corporation
Workplace safety	Co-ordinate agencies
	Control operations
	Establish liabilities

Type

Framework law (Enabling legislation)	Subordinate regulations
Subordinate law	Schedules
Primary laws on hazardous waste	
Complementary laws on	
hazardous waste	
Waste provisions in other laws	

Application

Waste handling	Waste information
Chemical use	Definitions
Waste generation	Assessments/studies
Storage	Research
Transport	Measurements
Treatment/recycling	Training
Disposal	
Dumping	
Clean-up	

Instruments

Notification	Standards
Certification	Guidelines/codes, policies
Labelling	Monitoring
Orders to act/not to act	Release of information
Bans	Assistance measures
Licensing/permitting	Fiscal measures

Fiscal measures

Fees and charges	Compensations
Fines	Tax concessions
Subsidies	Levies

Powers

Obligations	Giving directions/orders
Responsibilities	Right of refusal
Offences	Right to know
Authorizations/delegations	Right to secrecy

Enforcement

Which agencies	Possible conflict of interest
Extent of proof	(self licensing of agencies)

Source: UNEP, IE/PAC, Paris.

International responses

As early as 1982, UNEP's Governing Council recognized hazardous wastes as one of the major areas where global legal instruments had to be developed. A working group was convened, and in 1985 produced the Cairo Guidelines and Principles for Environmentally Sound Management of Hazardous Wastes adopted by UNEP Governing Council in 1987. These Guidelines outlined certain principles guiding the environmentally-sound management of the transboundary movement of hazardous wastes. This paved the way for the drafting and adoption of the Basel Convention on the Control of Transboundary Movement of Hazardous Wastes and Their Disposal in March 1989 (Box 7). The convention was scheduled to enter into force ninety days after the deposition of the twentieth instrument of ratification. This has now been achieved. Draft national legislation addressing the problems of management, disposal and transboundary movement of hazardous wastes has been prepared by UNEP as called for by the Basel Convention.

BOX 7

The Basel Convention (adopted by 116 countries and the European Community on 22 March 1989).

The main points in the Basel Convention are:

- A signatory state cannot send hazardous waste to another signatory state that bans its import or to any other country that has not signed the treaty.
- Every country has the sovereign right to refuse to accept a shipment of hazardous waste.
- The exporting country must first provide detailed information on the intended export to the importing country.
- Shipments of hazardous waste must be packaged, labelled and transported in accordance with international rules and standards.
- The consent of the importing country must be obtained before shipment.
- Should the importing country be unable to dispose of the imported waste in an environmentally sound manner, the exporting country has a duty either to take it back or find another way for the safe disposal of the shipment.
- Illegal traffic in hazardous waste is criminal.
- A secretariat is set up to supervise and facilitate implementation of the treaty.

The treaty entered into effect on 5 May 1992.

At the regional level, the OECD established the Waste Management Policy Group in 1974. The work of the group, particularly on transboundary movements, contributed to the preparation and adoption of the Basel Convention. The Organization of African Unity (OAU) drew up an African Convention on the Ban on Imports of All Forms of Hazardous Wastes Into Africa and the Control of Transboundary Movements of Such Wastes Generated in Africa (the Bamako Convention), which was adopted in January 1991. Unlike the Basel Convention this also covers radioactive waste. The European Community has also been implementing hazardous wastes management policies, and has developed a body of legislation on waste management as part of its Fourth Environmental Action Programme. Of particular importance to hazardous waste disposal is the Directive of the Council 78/319/EEC, later replaced by another directive giving a new definition of hazardous wastes, and a proposal for a regulation on the supervision and control of waste movement within and without the Community to replace an earlier Directive of the Council (84/631/EEC).

New legal instruments of a regional nature on the dumping of hazardous wastes at sea were also formulated: these included the Oslo Convention of 1972; the Baltic Convention of 1974; the South Pacific Protocol, 1986; and others within some of UNEP's Regional Seas Programmes (e.g. the Mediterranean, 1976, and the Caribbean, 1983).

Concluding remarks

R.K. Turner reminded us in 1991 that from an ethical point of view we have a duty to manage our wastes so as to leave future generations what he calls 'interesting archaeological sites' (i.e. formerly properly controlled landfill sites) rather than a huge clean-up cost burden. There are still several areas requiring future action in minimizing and managing hazardous wastes, particularly in developing countries where small-scale industry is a main source of ill-defined, heterogeneous quantities of such wastes, and where capabilities for standard-setting, formulation of legislation, monitoring and enforcement are quite weak. Another major problem in developing countries is the scarcity of resources that could be allocated to good hazardous waste management practices. These costs will only increase as new, more sophisticated and more effective treatment techniques are developed. The problem could become worse as chemical industries proliferate in developing regions. It remains to be seen also how the entry into force of the Basel Convention will affect illegal traffic in hazardous wastes. Furthermore, the pursuit of cleaner production and waste minimization is still in its early stages. Further determined efforts in this direction promise to yield considerable achievements both in waste minimization, and in reducing the dangers that these wastes present to the environment. Finally, it should be remembered that hazardous wastes are but one component of the total waste problem - a problem that calls for an integrated approach dealing with all types of wastes.

REFERENCES

Blair, E. (1991) *In: Proceedings of the Workshop on the Control of Existing Chemicals*, pp.252-60, Berlin.

Borenblum, J. (ed.) (1969) *Carcinogenicity Testing*. UICC Technical Report Series, Volume 2. Geneva.

Borg, K. (1958) *In: Proceedings of the Eighth Nordiska Veterinärsmötet*, Helsinki.

Bradshaw, A.D., Southwood, R. and Warner, F. (eds.) (1992) *The Treatment and Handling of Wastes*. Chapman and Hall, London.

Burns, P. (1988) Hazardous Wastes Management - The Way Forward. *In: Journal of the Institute of Water and Environmental Management*.

Crump, A. (1991) *Dictionary of Environmental Development*. Earthscan Publications Ltd., London.

DOE (1987) Second International Conference on the North Sea, London, 24-25 November, 1987. Ministerial Declaration. Department of the Environment, London.

ECETOC (1991) *Existing Chemicals: Literature Reviews and Evaluations*. Technical Report No. 30.

EPA (1988) *Environmental Progress and Challenges: EPA's update*, EPA-230-07-88-033, USEPA, Washington, D.C.

Friedrich Naumann Foundation (1987) *Industrial Hazards in a Transnational World*. New York.

IPCS-IRPTC (1990) *Computerized Listing of Chemicals Being Tested for Toxic Effects*. Geneva.

ILO (1988) *Major Hazards Control*, ILO, Geneva.

Johnels, K., Westermark, T., Berg, W., Persson, P.J. and Sjöstrand, B. (1967) Pike and some other aquatic organisms in Sweden as indicators of mercury contamination in the environment, *Oikos*.

Kempa, E.S. (ed.) (1991): Environmental Impact of Hazardous Wastes, Polish Association of Sanitary Engineers and Technicians (PZITS). Warsaw.

Kutsuna, M. (Ed.) (1968) *Minamata Disease*. (Minamata Report). Kumamoto University, Japan.

Lenz, W. (1961) Kindliche Missbildungen nach Medikamenteinnahme während der Gravidität? *Deutsche Med. Wochenschrift*, **86**.

OECD (1973) *Polychlorinated Biphenyls, Their Use and Control*, OECD, Paris.

OECD (1979) *Regulation and Innovation in the Chemical Industry*, OECD, Paris.

OECD (1985) *OECD Environmental Data, Compendium 1985*, OECD, Paris.

OECD (1989) Accidents Involving Hazardous Substances, *Environment Monograph Series* **No.24**, OECD, Paris.

OECD (1990) Press Release A (90) 21, April 1990, OECD, Paris.

OECD (1991) *The State of the Environment*, OECD, Paris.

Schweitzer, G.E. (1991) *Borrowed Earth, Borrowed Time: Healing America's Chemical Wounds*. Plenum Press, New York.

Turner, R.K. (1991) *Towards an integrated waste management strategy*, British Petroleum, London.

UNEP (1990) Country Papers in UNEP Regional Workshop for Asia and the Pacific on Hazardous Waste Management. Kyoto (unpublished).

Uriarte, F.A. (1989) Hazardous Waste Management in ASEAN. *In*: Maltezou S.P.*et al.* (eds.) *Hazardous Waste Management*, Tycooly. London.

US NRC (1984) *Toxicity Testing*. National Research Council, Washington, D.C.

US OTA (1986) *Alternatives to Animal Use in Research, Testing and Education*. Office of Technology Assessment, Washington, D.C.

Waste Management and Research (1990) Adapting Hazardous Wastes Management to the Needs of Developing Countries, A special edition of Waste Management and Research. **Vol.8**, March 1990.

Westö, G. (1966) Determination of Mercury and Methylmercury in Fish and Crayfish, *Var Föda* **3**, 19.

WHO (1978) *Principles and Methods for Evaluating the Toxicity of Chemicals.* WHO, Geneva.

WHO (1990) *Public Health Impact of Pesticides Used in Agriculture,* WHO, Geneva.

World Bank/WHO/UNEP (1989) *Safe Disposal of Hazardous Wastes: The Special Needs and Problems of Developing Countries.*

WRI (1987) *World Resources,* Basic Book, New York.

PART TWO

CAUSES AND CONSEQUENCES

An introductory overview

The first five chapters of this section consider the main sectors of economic activity, their development and their environmental impacts during the period 1972 to 1992. They also outline how society has responded to these impacts and what developments are likely in the 1990s.

It is clear that the remarkable economic growth achieved over the last two decades was not only uneven (comparing the developed and developing countries) but that it also caused considerable environmental damage in addition to contributing to improved standards of living. The technological innovations central to much of this economic growth, for example synthetic pesticides, CFC refrigerants and efficient low-cost motor transport, often brought with them unforeseen environmental hazards, some of which have now assumed global proportions.

Concern about the condition of the environment stems newly from the desire of people to live without discomfort and with the greatest attainable security. This means a life with sufficient resources to ensure not merely the absence of hunger but access to a diet that is adequate in quantity and quality; shelter that is comfortable and from which they cannot be arbitrarily displaced; access to safe and sufficient supplies of water, and proper sanitation, and freedom from the threat of injury, disease, war and oppression.

People generally also hope that their children and descendants will live in circumstances as good as, if not better than, their own. In short, environmental concern at the individual level is concern for personal well-being or quality of life.

Chapters 16 to 19 consider several related aspects of the human condition that contribute to well-being and the overall quality of life. These are: health and life-expectancy, a place to live, and peace and security.

As each of these is related to population and development, this section also deals with the question of population growth and access to resources, expressed in the comparative states of affluence and poverty. These chapters do not address the implications of the prevailing world economic order and such issues as barriers to trade, commodity pricing and the operations of financial markets. These are considered in Chapter 24. However, all these issues have important influences on human well-being and on the environment, sometimes for the better, sometimes for the worse. The interrelationship between economic policies and the state of the environment warrants more intensive study and continued monitoring as it is at the very heart of the quest for sustainability.

Population growth is perhaps the most pervasive problem confronting the world. How can we provide an adequate quality of life for a human population which is about to pass the six billion mark and will certainly exceed ten billion within a few decades?

Of course, absolute numbers of people do not, alone, create a problem. Many people who live at high population densities have achieved high levels of development and great wealth despite having few natural resources of their own. However, they have done this only by drawing resources from other parts of the world through their control of economic forces, their level of education and technical know-how, and their political power.

It is clearly possible for countries with large and increasing numbers of people to attain a high level of development, as some of the newly industrialized economies of Asia have demonstrated. However, the rate at which the number of human beings is increasing has for more than a decade exceeded the ability of many countries to develop their economies. Indeed, rather than enjoying an improving quality of life, people in some countries are facing declines that are even beginning to spill over into social instability and civil strife.

While it may be that the world could, in theory, adequately sustain its present population, it cannot do so given the present imbalances in population density, access to resources, and life-style expectations. It certainly cannot do so if adequate sustenance is taken to mean the extravagant life-styles and consumption levels common in the developed countries; and each additional billion people will make it even more difficult to redress the present inequities.

It would be easy to throw up one's hands in despair; to say that there is nothing that can be done about it; or even that this is the natural order of things - that it is God's will. Such an attitude, common a century or more ago, would be unconscionable in the last decade of the twentieth century. A more equitable distribution of the world's resources is a *sine qua non* for any possible future that makes environmental sense - not just for the poorest countries of the world but for the wealthy countries as well.

CHAPTER 11

Agriculture and fisheries

Introduction

Humanity has relied solely on hunting, fishing and gathering food for most of its two million years of existence. Agriculture - the domestication of plants and animals - appeared only about 10,000 years ago.

As long as agriculture used renewable resources at a rate compatible with their natural regeneration, and sought increased efficiency in the mechanisms of their renewal, agriculture remained a sustainable process as regards its natural resource base. But with the mounting demands of growing populations, agriculture has been undergoing accelerated change over the last century. More and more fragile land has been cultivated, ever more intensively, resulting in environmental damage, especially erosion. Resources from outside the agricultural ecosystem have been brought in (water through irrigation, chemical pesticides and fertilizers, fossil fuel energy) and this has allowed more plants and animals to be produced per unit area, although it has also undermined the resource base and caused environmental problems.

At present, agricultural production systems can be viewed as consisting of three interrelated components: resources, technology and environment. The quantity, quality and terms of availability of resources condition the kinds of technology available to farmers and their choices among them. The technologies employed may, in turn, damage the environment, generating demands for policies to reduce the damage, and they may also affect the future terms of availability of resources. The realization that resource availability may place significant constraints on the future of croplands sharpens the need for care in selecting agricultural technologies that might lead to yield improvements.

This chapter examines the main environmental impacts associated with agricultural practices, and the ways and means of reducing these impacts to achieve sustainable agriculture and food production.

Food production

Superficially at least, world agriculture appears to be doing well. At the beginning of the 1990s, a world-wide average of 2,670 calories of food products per capita were consumed - a level considered nutritionally adequate (FAO, 1991a). However, since food production, buying power and consumption are not distributed evenly over the world's population, large surpluses and deficits exist at the regional, national and local levels. A nutritional gap of 965 calories per capita exists between the developed and developing countries, and wider gaps exist between and within the developing countries themselves. In fact, the rate of increase in per capita food availability in the developing countries as a whole slowed in the 1980s compared with the 1970s and 1960s. The situation for some countries, for example the sub-Saharan African countries, worsened to the extent that per capita food availability in 1989 was less than it was in 1970 (WFC, 1991).

This world-wide disparity has been created and aggravated by a combination of social, economic, environmental and political factors. Western Europe worries about surpluses, particularly of dairy products, and the United States still idles cropland to control production. Grain-exporting countries use subsidies to compete for markets that never seem large enough (Brown, 1991). Furthermore, agricultural trade barriers, inequitable access to resources and products, and the often primitive conditions of production and processing of agricultural output, affect food availability and distribution especially in developing countries. As a result, the number of chronically hungry people in the world increased from about 460 million in 1970 to about 550 million in 1990 and is expected to reach 600-650 million by the year 2000 (WFC, 1991). Close to 60 per cent of the hungry people in the developing world live in Asia, about 25 per cent in sub-Saharan Africa and some 10 per cent in Latin America and the Caribbean. These figures indicate that the goal enunciated at the World Food Conference convened in Rome in 1974 - namely that 'no child, woman or man should go to bed hungry and no human being's physical or mental potential should be stunted by malnutrition' - remains unfulfilled.

Agricultural output and food production increased in both developed and developing countries in the period 1970-90. The annual rate of increase was higher in the developing countries (about 3.0 per cent) than in the developed countries (about 2.0 per cent). In the latter countries there was near stagnation in per capita food production in the 1980s, with marked drops in 1983 and 1988 due to unfavourable weather conditions, particularly in North America. In the developing countries there were major increases in Asia, near stagnation in Latin America and a marked drop in Africa (Figure 1). The rate of increase in cereals production (Figure 2) was higher in the developing countries than in developed countries (about 32 per cent and 15 per cent respectively, between 1970 and 1990). In the developed countries, the annual rate of cereals production was higher than population growth (about twice as much), but in the developing countries it was much lower (about one-fifth as much). A wide gap, currently 529 kilograms per capita, continues to exist between annual cereals output of the developed and developing countries as a whole (777kg per capita and 248kg per capita respectively in 1990) (FAO, 1991b).

In the last two decades, two important changes in the pattern of crop production have occurred in several countries. The first is the growing shift towards production of cash crops (whether for export or for urban consumers). This has been at the expense of subsistence food crops, as part of the structural adjustment policy reforms. The argument has been that cash crops will generate more income for the farmers and, in the case of export crops, more hard currency for the country. It has been argued that part of that hard currency could be used to import cheaper cereals and other staple food crops from those countries with surpluses. This policy has led to an imbalance in agricultural systems in many countries. Large producers have benefited more, while small and poor farmers have been detrimentally affected. Some countries (e.g. Morocco) even turned into net importers of basic crops, though they were

Figure 1: *Indices of per capita food production, 1978-89, for developed countries, developing countries and developing regions of the world.*

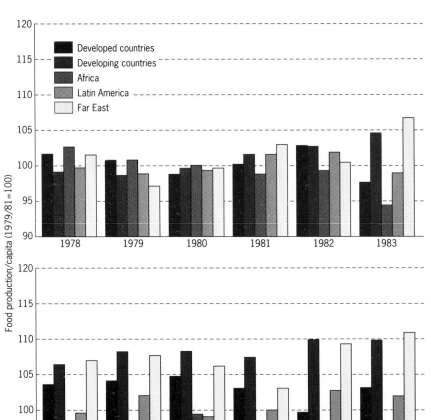

Source: FAO (1990).

formerly net exporters (George, 1988). A number of African countries have been encouraged to develop export crops in which they enjoy a comparative advantage, on the assumption that their agricultural labour productivity is generally substantially higher in export than in food crop production (Conway and Barbier, 1990). In some instances, the expansion of cash cropping for export - such as in the southern Volta region of Ghana and the Cauca Valley of Colombia - was in the most fertile land, pushing food production and subsistence farming on to marginal lands. In other countries, government policies have deliberately encouraged the production of food crops in marginal areas, often without simultaneously encouraging proper management techniques

Figure 2: *Crop production in developed and developing countries, 1970-90.*

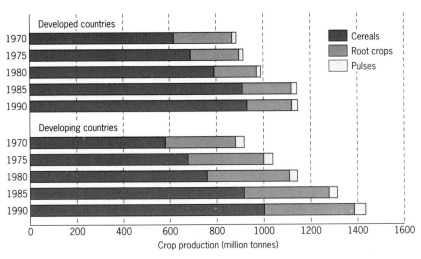

Source: Based on data from FAO (1988, 1991a).

and agricultural practices which can reduce environmental and soil erosion problems. In Haiti, pricing policies have encouraged the growing of maize and sorghum in hilly areas at the expense of coffee and other tree crops, thus increasing soil runoff and erosion. Similarly, the planned extension of maize, sorghum and millet into dryland areas has tended to exacerbate problems of soil degradation in the developing countries (Conway and Barbier, 1990).

The second change in the pattern of crop production has been the increase in areas growing grain for feed, fodder and forage. Roughly 38 per cent of the world's grain - especially corn, barley, sorghum and oats - is fed to livestock. In the USA, animals account for 70 per cent of domestic grain consumption, while India and sub-Saharan Africa offer just 2 per cent of their cereal harvest to livestock (Durning and Brough, 1992). In Egypt, over the past quarter-century, corn for animal feed has taken over cropland from wheat, rice, sorghum and millet - all staple grains. The share of grain fed to livestock rose from 10 to 36 per cent. Likewise, the area in Mexico planted to corn, rice, wheat and beans has declined steadily since 1965, while that planted to sorghum has grown phenomenally. The expansion in cultivation of fodder crops has been encouraged by the fact that their market is free. Accordingly, fodder crops are more profitable than food and other crops, which have controlled market prices (El-Hinnawi, 1991).

Livestock production

Domesticated animals have played a prominent and largely beneficial role in the human economy for thousands of years, providing food, fuel, fertilizer, transport and clothing. About twelve per cent of the world's population is entirely dependent on livestock production. On average, one-quarter of the gross value of agricultural production is attributed to livestock production, but when the non-monetized contribution of livestock is taken into account (through the provision of draught power and manure), this proportion amounts to 44 per cent (FAO, 1991c).

Animal proteins are not indispensable in the diet, but they complement staple diets of many carbohydrates and provide a more balanced food diet for vulnerable groups, such as children, pregnant and nursing women, and sick and elderly people. World-wide, about 24 per cent of the supply of protein per capita originates from animal products (FAO, 1991c), with strong differences between regions and countries. On average, the daily per capita animal protein supply in developed countries is 60 grams per day, while that in developing countries is only 13 grams per day (FAO, 1990).

The largest share of the world's livestock population is found in the developing countries - 99.5 per cent of buffaloes, 98.5 per cent of camels, 94.0 per cent of goats, 68.5 per cent of cattle, 57.8 per cent of pigs and 52.5 per cent of sheep (1989 figures). Global meat production has nearly quadrupled since 1950. In 1970, world meat production was 108 million tonnes, and in 1990 it was about 171 million tonnes. The production of meat in developing countries is much lower than in developed ones (68.7 million tonnes and 103.2 million tonnes respectively in 1990) (FAO, 1991b). This is attributed mainly to the fact that most of the livestock in the developing countries is in traditional, small-scale farming systems where it is a source of subsistence and additional

Figure 3: *Distribution of human labour, animal power and tractors in agriculture.*

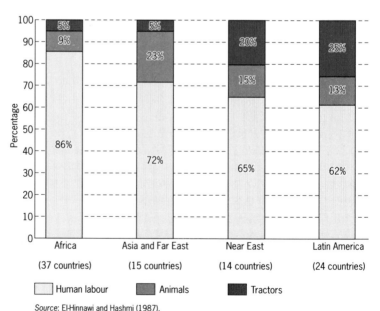

Source: El-Hinnawi and Hashmi (1987).

income generated by selling animal products. Also, animals supply draught power. It is estimated that 20 per cent of the world's population depends on animals for transport (FAO, 1987), while animal power for agricultural production varies from 9 to 23 per cent (Figure 3).

Fisheries production

Fisheries produce 16 per cent of the total animal protein available in the world - a contribution to protein supply that is approximately the same as that from beef or pork (FAO, 1991d). Most of the world's fish production comes from marine areas (Figure 4), which accounted for about 86 per cent of the estimated fish production in 1990. Of this amount, some 90 per cent is estimated to be from coastal areas. About 14 per cent of the world catch comes from inland (fresh) waters. Approximately 7 million tonnes of the total world fish catch are from freshwater aquaculture, compared to about 5 million tonnes from mariculture. In all, about 11 per cent of global fish production comes from aquaculture, and at present growth rates, aquaculture production by the end of the century should be almost doubled (FAO, 1991d), and may reach 25 million tonnes per year in 2010 (Figure 5). Most of the aquaculture is in Asia, where production from freshwater operations amounts to about 4 million tonnes per year (Bailey and Skladany, 1991). Coastal aquaculture of shrimp in Asia accounted for 82 per cent of world cultured shrimp in 1990 (about 400,000 tonnes from a world-wide production of about 471,000 tonnes). It should be noted that most freshwater aquaculture in Asia is for local consumption in rural areas. Small-scale systems such as rice-fish culture and integration of aquaculture with livestock are common in many Asian countries.

Figure 4: *Fish production in developed and developing countries, 1970-90.*

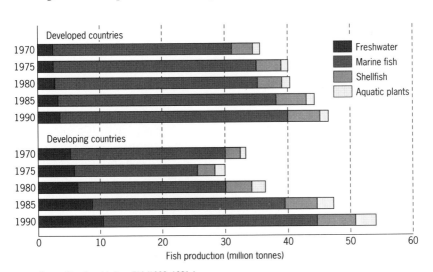

Fish production (million tonnes)

Source: Based on data from FAO (1988, 1991a).

Figure 5: *Recorded and projected growth in world aquaculture production.*

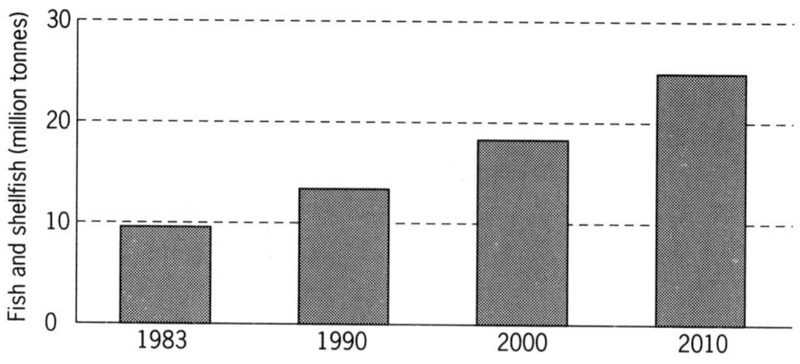

Source: El-Hinnawi and Hashmi (1987).

The environmental impacts of crop production

Impacts on land

Of the total land area in the world (about 13,382 million hectares, 13,069 million of which are ice-free), only 11 per cent (about 1,475 million ha) is currently under cultivation; another 24 per cent is permanent pasture. The world's potentially cultivable land has been estimated at about 3,200 million hectares, more than twice the area currently used as cropland. On a regional basis, crop areas range from as low as 6 per cent of Africa and 8 per cent of South America to 17 per cent of Asia and 30 per cent of Europe. About 15 per cent of all cropland is irrigated, ranging from 6 per cent in Africa and South America to 31 per cent in Asia (FAO, 1990).

World-wide, the area of arable land increased by only 4.8 per cent over the period 1970-90; the increase in developed countries was 0.3 per cent and that in the developing countries was 9 per cent (Figure 6). However, per capita arable land decreased from a world-wide average of 0.38 hectares in 1970 to 0.28 hectares in 1990, largely due to population growth (Chapter 6). The decrease was most noticeable in the developing countries, from 0.28 hectares per capita to 0.20 hectares per capita: in the developed countries the decrease was from 0.64 hectares per capita to 0.56 hectares per capita (FAO, 1988, 1990). It has been estimated that if the arable land area is maintained at the present level (1,475 million ha world-wide) and assuming that no new land is brought under cultivation and no existing land goes out of production due to degradation, the per capita arable land in the world will progressively decline to 0.23 hectares in 2000, 0.15 hectares in 2050 and 0.14 hectares in the year 2100 (FAO, 1991e).

Figure 6: *Changes in area of arable and permanent cropland, and in arable land per capita, 1970-90*

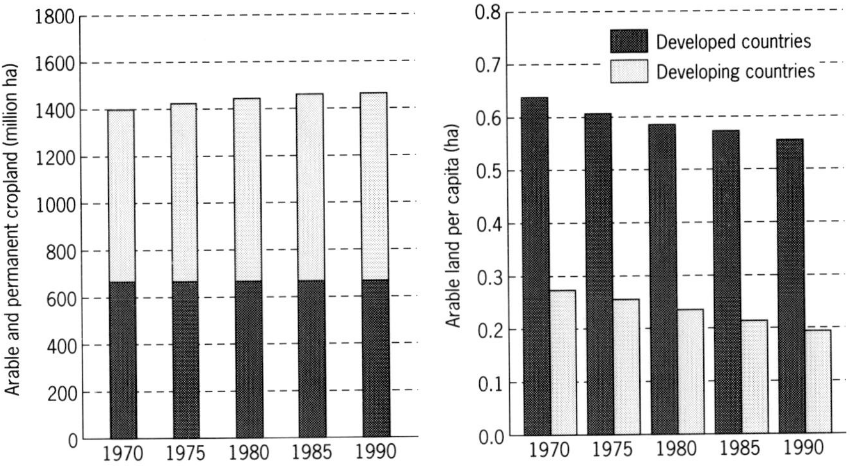

Source: Based on data from FAO (1988, 1990).

It has been argued that large areas of new land could be brought under cultivation (Gale Johnson, 1984; Revelle, 1984). However, reserves of potentially arable land are hard to find and are unevenly distributed. For example, densely populated Asia, home to 75 per cent of the world's population, has little additional land to convert into arable use. Similarly, comparatively little is available in Europe. Furthermore, in many parts of the world, further expansion of agricultural land is constrained by various environmental factors. Most of the available land in Africa and South America is located within fragile and ecologically sensitive regions, e.g. tropical rain forests, acid savannas and the drought-prone Sahel. Bringing new land under production through the clearing of tropical rain forest has severe ecological, environmental and socio-political implications. In addition to loss of biological diversity and potentially valuable genetic resources (Chapter 8), conversion of tropical rain forest contributes to total global emissions of greenhouse gases (Chapter 3).

The productivity of farmland depends principally on the capacity of the soil to respond to management. Soil is not an inert mass but a very delicately balanced assemblage of mineral particles, organic matter and living organisms in dynamic equilibrium. Soils are formed over very long periods of time, generally from a few thousand to millions of years; yet excessive human pressure or misguided human activity can destroy them in a few years or decades, and the destruction is often irreversible.

Agricultural practices have had the greatest impact on soil degradation. Traditionally, farming practices had been well balanced with soil sustainability. In recent decades, however, human management of agro-ecosystems has been steadily intensified, through irrigation and drainage, heavy inputs of energy and chemicals, and improved crop varieties increasingly grown as monocultures. Although bringing some general growth in agricultural production, this process has made agro-ecosystems more and more artificial and often unstable and prone to rapid degradation.

Soil erosion is a serious problem in several ecologically sensitive areas, e.g. the Himalayan-Tibetan ecosystem, the Andean region and large parts of Africa. Steeply sloping land, comprising a large percentage of the total land area in these regions, is over-exploited and misused, and this has led to accelerated soil erosion (Chapter 7). Many soils cultivated by shifting cultivators and subsistence farmers in the tropics and sub-tropics are subject to fertility depletion through decline in soil organic matter, reduction in nutrient reserves by crop removal, leaching and acidification. Waterlogging and soil salinization are major causes of declining yields on irrigated land. Salt-affected soils are widely distributed throughout the arid and semi-arid regions. The problems are particularly severe in China (where 7 million hectares are affected), India (20 million ha), Pakistan (3.2 million ha), and the Near East (El-Hinnawi, 1991). However, they are not limited to developing countries: they also occur in the USA (5.2 million ha) and southern European countries (FAO, 1991e).

Land degradation leads to a drop in agricultural production. It has been estimated that land degradation causes a world-wide loss of about 12 million tonnes of grain each year, estimated at a cost of $US3000 million (FAO, 1991f). This is just one example; the actual cost of land degradation to society and the environment has yet to be estimated. Soil misuse and over-exploitation, causing rapid depletion of soil organic matter, can lead to large emissions of greenhouse gases into the atmosphere. Lal (1990) estimated that if one per cent of the organic carbon content of the top 15-centimetre layer of soils of the tropics is lost, this can lead to an annual emission of about 128 billion tons of carbon into the atmosphere (compared to about 5.7 billion tons per year due to fossil fuel burning; see Chapter 3).

Impact on water

Agriculture is a major consumer of water, particularly in developing countries (Chapter 4). The world's irrigated land increased from 168 million hectares in 1970 to 228 million hectares in 1990, an increase of about 36 per cent in two decades (Figure 7). Although irrigated land at present accounts for one-sixth of cultivated land, it produces one-third of the world's food (over twice the productivity of average rain-fed land). Although there is still potential for expansion of irrigation in some regions, there are several economic and environmental constraints. Irrigation schemes are becoming more and more expensive, and the damming or diversion of rivers can create a number of environmental impacts.

Many dams have been built since the 1950s for water management or as multipurpose dams for both water management and electricity generation

Figure 7: *Changes in area of irrigated croplands in developed and developing countries, 1970-90.*

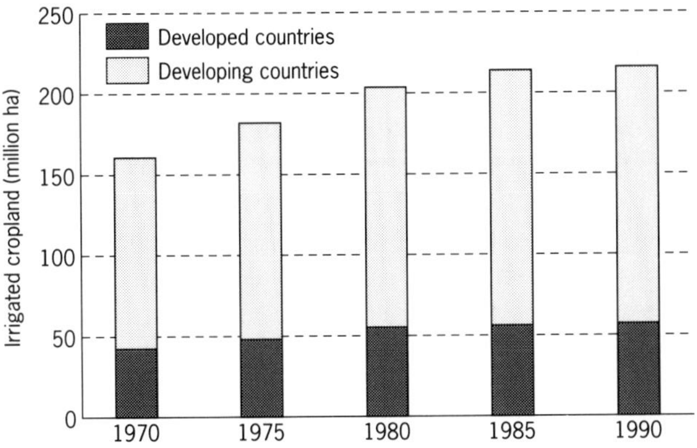

Source: Based on data from FAO (1988, 1990).

(Chapter 4). The environmental impacts of dams (especially of large multipurpose dams) have been a matter of discussion over the last two decades. Upstream, the impoundments created often cover vast areas of land, and while they may create additional wetlands habitat, they eliminate existing ecosystems and in some cases displace large numbers of people. The reservoirs themselves are often affected by excessive siltation from upstream land-use practices such as overgrazing, deforestation and intensive cropping. In India, the expected siltation rate of the Nizamsagar reservoir in Andra Pradesh was 5.2 million cubic metres per annum, but the actual rate has reached 13.8 million cubic metres per annum, mainly due to such practices (ESCAP, 1990). Downstream effects of dams vary according to the geology and hydrology of the areas (for detailed reviews of the environmental impacts of dams, see El-Hinnawi, 1981; El-Hinnawi and Biswas, 1981; White, 1988).

Other water management projects include the diversion of rivers and the construction of by-passes to reduce water losses in specific terrains. An example of this is the Jonglei Canal project which aims at the construction of a 200 kilometre long canal in southern Sudan to by-pass the swamps of the Sudd Region, thus reducing the amount of water lost in the swamps. The project would increase the water available to both Sudan and Egypt, but some environmentalists think that it will affect the ecosystems in the Sudd region in a detrimental way (see Abdel Mageed, 1985, for a review of the Jonglei Canal Project).

The improper construction and management of irrigation schemes has in many cases led to an increase of malaria, schistosomiasis and other vector-borne diseases, particularly on the African continent (FAO, 1991e). However, recent studies in Egypt and in South and Southeast Asia show that much of present-day schistosomiasis infection results not from irrigation water, but from

291

inadequate domestic water supply and sanitation, and that provision of these services combined with education is reducing the incidence of the disease (Biswas, 1991).

Water use efficiency is generally low, and the overall performance of many irrigation projects often does not measure up to expectations. Inefficient operation and maintenance, combined with poor management, contribute to agricultural and environmental problems. The rise in groundwater levels, leading to waterlogging, depressed yields and soil salinity, is a major concern. It is not unusual to find that 60 per cent of the water diverted or pumped for irrigation does not actually reach the plants (FAO, 1991f). This excess input from seepage and deep percolation on farm fields is the major cause of waterlogging (Chapter 4).

Over-exploitation of ground water for irrigation has led to the depletion of groundwater resources in some arid areas, e.g. in the Middle East, and in coastal zones it has resulted in excessive intrusion of salt water from the sea into groundwater aquifers. There are fears that the extensive use of ground water for agriculture in large schemes such as those in Saudi Arabia and Libya may lead to the rapid depletion of such non-renewable resources (El-Hinnawi, 1991).

Unless carefully managed and monitored, the growing tendency in some countries to use water of marginal quality for irrigation could lead to a considerable increase in the salinization of land, and to a deterioration of the quality of ground water in the aquifers near irrigated lands. Although the use of municipal waste water for irrigation has been practised for centuries, a conservative approach in utilizing this source is essential. Pathogenic bacteria, parasites, and viruses are all found in sewage water and may survive treatment processes. Once released in the environment, many of them can exist for prolonged periods of time and outbreaks of cholera, typhoid, etc. associated with waste water irrigation have been documented (Rose, 1986).

Impacts of agrochemicals

Fertilizers

The increased application of chemical fertilizers to supply plant nutrients (nitrogen, phosphorus and potassium) is an essential component of the green revolution. World consumption of chemical fertilizers more than doubled over the past two decades, rising from about 69 million tonnes in 1970 to about 146 million tonnes in 1990 (Figure 8). The rate of increase in consumption was much higher in the developing countries (360 per cent) than in the developed countries (61 per cent). Most fertilizers used are nitrogenous fertilizers, followed by phosphates and potash. The use of fertilizers per hectare in the developed countries has been much higher than in the developing countries, although the rate of application in the latter countries has been rising fast (327 per cent increase from 1970 to 1989) as a result of the increasing introduction of green revolution packages. About 50 per cent of the fertilizer used benefits the plants; the remainder is lost from the soil system by leaching, runoff and volatilization (Engelstad, 1984).

Figure 8: *Total fertilizer use, and fertilizer use per hectare, in developed and developing countries, 1970-90.*

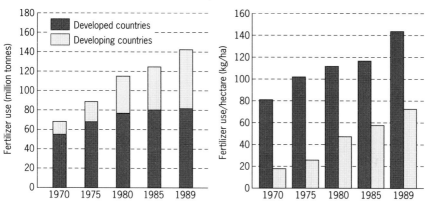

Source: Based on data from FAO (1990).

Fertilizers are indispensable for increasing food production. They compensate for regular soil nutrient losses that occur through annual cropping, leaching and soil erosion. A study of soil nutrient depletion (net removal) in sub-Saharan Africa concluded that depletion amounts to approximately 20 kilograms of N, 10 kilograms of P_2O_5 and 20 kilograms of K_2O per hectare per year on average. Nutrient depletion was found to be twice as high in East Africa (FAO, 1991f). To compensate for this nutrient loss, imported fertilizers have to be used, and this accounts for up to 40 per cent of farmers' net income in some countries. In several developing countries, fertilizers have been subsidized to different degrees to enable small farmers to use them, but these subsidies have often led to excessive and inefficient application, with consequent economic losses and increased environmental damage both on and off the farms.

Fertilizers can easily be leached into drainage water, and when such water is discharged into rivers or the sea the leached nutrients (nitrogen and phosphorus) create widespread eutrophication. Nitrate and phosphate have been responsible for generating dense algal growths which have harmed fish and other aquatic life (Chapter 4).

The contamination of ground water with nitrates is a major problem in several European countries and in North America, and is becoming a problem in some developing countries. Ground water is extensively used in many countries for domestic purposes and for drinking. Nitrate itself is not very toxic, but bacteria in the mouth and elsewhere can convert nitrate into nitrite, which can induce methaemoglobinaemia (a reduction in the oxygen-carrying capacity of the blood), especially in infants. According to WHO (1977), water becomes unpotable when nitrate concentration exceeds 45 ppm. The EC has issued a directive requiring any area where the nitrate concentration in surface or ground water exceeded 50 ppm to be declared a 'vulnerable zone', in which compulsory restrictions on farming are automatic.

Pesticides

Crops are affected by many different pests and by competition from weeds. Changes in cropping practices, such as shortened fallow periods, narrow rotation and replacement of mixed cropping by large-scale monocultures of genetically uniform varieties, have for a number of crops resulted in the escalation of pest problems. In North America, Europe and Japan, crop losses caused by pests are estimated to be in the range of 10-30 per cent. In the developing countries such losses are of the order of 40 per cent, but losses as high as 75 per cent have been reported, for example for maize in Africa. Furthermore, pests do not only affect the quantitative yields of crops: both pre-harvest and post-harvest infestations also seriously affect food and feed quality.

Pesticides have long been used to control pests, and about 90 per cent of pesticides sold are used in agriculture; the remainder are used in public health programmes (WHO, 1990). The growth of world pesticide use is normally measured in terms of world sales rather than in tonnage, because information on production in terms of weight or volume of active ingredients is scarce. It has been estimated that the total sales of pesticides increased from $US7,700 million in 1972 to $US15,900 million in 1985, and reached about $US25,000 million in 1990 (1985 $). The major groups of pesticides used are herbicides (46 per cent), insecticides (31 per cent), fungicides (18 per cent) and others (5 per cent). About 80 per cent of the pesticides used in the world are used in the developed countries. However, the rate of increase in their use in developing countries (7-8 per cent per year) is faster than that in the developed countries (2-4 per cent per year). The amounts and types of pesticides used differ greatly from one country to another. Of the estimated 434 million kilograms of pesticides used annually in the USA, 69 per cent are herbicides, 19 per cent insecticides and 12 per cent fungicides. Of the herbicides, 75 per cent is applied to cotton and corn. Fungicides are used mainly on fruit and vegetables, and about 95 per cent of grapes and potatoes are treated in this way (Pimentel et al., 1991).

It has been estimated that more than 90 per cent of pesticides do not reach the target pests (Pimentel and Levitan, 1986) and that these chemicals contaminate land, water and air. The increased use and misuse of pesticides has led to the build-up of resistance in insect pests, a problem that has grown enormously over the last two decades (Figure 9). Resistance is also increasingly found in plant pathogens and weeds. The consequences of resistance development are a matter of great concern, from both the economic and environmental points of view. Consequently, increasing amounts of pesticides are used to achieve control, or alternative products are required. The steady growth of multiple resistance in particular is bound to become a major problem, as this results in pests that can no longer be controlled by any pesticide.

Figure 9: *Number of pest species resistant to pesticides, 1908-88.*

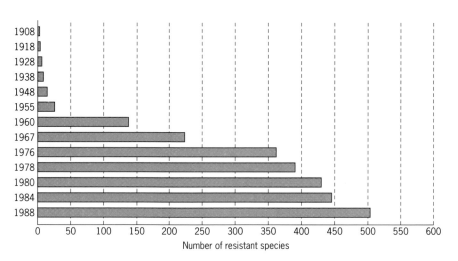

Number of resistant species

Source: Based on data from Trappe (1985) and Georghiu (1989).

The injudicious use of pesticides results in the elimination of natural enemies that keep insect pests partly under control, causing pest resurgence and secondary pest outbreaks. The role of natural enemies is of great importance, particularly in the humid tropics. This explains why the (mis)use of pesticides in such climates can have a more dramatic effect than in other areas (FAO, 1991g). Metcalf (1986) gives examples of thirteen insect pests in cotton, rice, deciduous fruits and citrus that have become a serious problem as a consequence of pesticide use, and states that 'the widespread development of these (man-made) pest outbreaks is one of the most serious indictments of our present day pest technology'.

In addition to ecological disturbances, the increased use of pesticides poses environmental and toxicological problems. It has been shown that even relatively non-volatile pesticides such as DDT evaporate into the atmosphere quite rapidly, particularly in hot climates, and can be transported over long distances, contributing to what has been known as global chemical pollution. Despite their generally low solubility, pesticides can be leached into drainage water causing pollution of surface and coastal waters into which drainage water is discharged. Groundwater contamination is also common in areas where pesticides are heavily used. For example, Aldicarb, a nematocide, has become a common contaminant of aquifers below potato fields and citrus orchards in several countries (Conway and Pretty, 1991). The pollution of surface and ground water with pesticides can affect aquatic life and human health. Pesticides, especially persistent ones, can build up through the food chain, with consequent risks to humans. For example, in populations exposed to high levels of pesticides (or repeated applications of pesticides), high levels have been found in human milk. Studies (e.g. Slorach *et al.*, 1983; Jensen, 1983 and Karakaya *et al.*, 1987) have revealed that the concentration of DDT and DDE in human milk in some countries is higher than the acceptable daily intake criteria and maximum residue limits established by WHO/FAO.

The WHO has estimated that the number of unintentional acute poisonings due to exposure to pesticides, numbering about half a million in 1972, increased to one million in 1985 due to increased pesticide use. About 60-70 per cent of these cases are due to occupational exposure. Some 20,000 deaths per year occur as a result of pesticide poisoning (WHO, 1990). The problem is particularly acute in cotton-producing areas. It should be noted that these figures are only a rough estimate and that the problem is probably much more severe as most toxicity cases go unreported or are confused with other health problems.

Impacts of the use of high-yield varieties of seeds

Agricultural development thinking in the 1960s and 1970s was preoccupied with the problem of feeding a rapidly increasing world population. The obvious solution then was to increase per capita food production. The resulting Green Revolution has had a dramatic impact on the Third World, particularly in terms of increasing the yields of the staple cereals - wheat, rice and maize. The greater part of Asia and Latin America has managed to avoid declines in per capita food availability due mainly to the introduction of improved wheat and rice varieties (high-yield varieties - HYVs). India, for example, doubled its wheat production over a six-year period. Many other countries, including Mexico, Pakistan, Turkey, Indonesia and the Philippines, also increased cereal production dramatically (FAO, 1991g). It was estimated that in 1983 the developing countries were planting half of their wheat areas and 60 per cent of their rice areas with HYVs. At present, approximately 80 per cent of the wheat areas in Latin America and India and 95 per cent of the rice areas in China are planted with HYVs (FAO, 1991g).

None the less, these efforts in high-production agriculture have not been without adverse side-effects. The nature of the HYVs is that in areas with conditions favourable to them, they tend to be adopted quickly and spread widely. They have frequently supplanted native varieties which are often more resistant to pests and diseases and in some cases the native varieties are now in danger of extinction. For example, the spread of HYVs of wheat and rice since the mid-1960s has inadvertently caused a loss of the gene pools in centres of crop diversity such as Afghanistan, Iran, Iraq, Pakistan and Turkey (El-Hinnawi, 1991). In 1980 there were as many as 30,000 varieties of rice in

India, yet it is estimated that by the turn of the century as few as 12 varieties will dominate 75 per cent of that country. The use of HYVs is generally accompanied by a trend away from crop rotation and towards raising a single crop of the HYV, which can mean that virtually the entire crop comes from a narrow genetic base and is therefore very vulnerable to changed circumstances. For example, in 1970 a virulent fungus plague swept through the United States corn belt, spreading at up to 150 kilometres a day. United States corn production was reduced by 15 per cent as a result of the fungus, which was well matched to the 'T-cytoplasm' that had been incorporated into most of the hybrid seed corn planted that year. By the following year, seed producers had re-established a more variable genetic basis for their corn (Crosson and Rosenberg, 1989).

The widespread cultivation of the HYVs has also favoured an increase in pests and diseases and this in turn has led to increased pesticide use. The HYVs also generally require more fertilizers, irrigation and machinery in order to realize their full potential. All this has contributed to increases in production costs, environmental pollution and human health hazards.

The use of HYVs of seeds has depended on the presence of a whole package of inputs - water, fertilizer, pesticides and machinery. The technology is, therefore, less suitable to resource-poor environments. Farmers with small or marginal holdings have, on the whole, benefited less than farmers with large holdings, and now there is growing evidence of diminishing returns from intensive production with HYVs. These are not simply second or third generation problems capable of being solved by further technological adjustment. They require an approach that is equally revolutionary, yet very different in its conceptual and operational style (Conway and Barbier, 1990).

The environmental impacts of livestock production

Livestock production systems have been changing rapidly. Nelson (1987) pointed out that the classification of livestock production systems can be based on a number of factors, for example the relative importance of livestock (animal-based, mixed crop-animal, crop-based), scale of operation (large vs small), utilization of output (subsistence vs commercial), levels of development (traditional vs modern), source of feed (uncultivated vs cultivated lands), etc. Each of these systems has its own environmental impacts. Broadly speaking, however, livestock production systems are classified into migratory and sedentary. The main features of migratory (pastoral) systems are extensive grazing on communal land, mobility and maximization of herd size. Sedentary livestock production systems vary from subsistence level to large intensive commercial farms.

In pastoralist systems, livestock owners exploit natural pastures. Nomads move with their herds from place to place, their movements determined by the availability of food and water and the presence/absence of disease. In contrast, those practising transhumance have a permanent residence and migrate seasonally with their herds to suitable grazing grounds. Traditional pastoral societies have developed skills to manage large herds in spite of short grazing seasons, limited supplies of drinking water, irregular rainfall and periodic drought conditions. However, traditional pastoral systems have now become destabilized and unsustainable due to increasing pressure on land resources and unfavourable environmental conditions. Rangeland has been degraded in many parts of the world as a result of mismanagement and overgrazing (Chapter 6). One of the causes of overgrazing in semi-arid regions is population pressure, which is forcing farmers to cultivate marginal lands, often traditional grazing lands.

For hundreds of years, and even today in some areas, livestock rearing served as a critical counterpart to crop production, thus keeping farms ecologically balanced. Growing hay, legumes and other fodder for farm animals calls for environmentally sound crop rotation. Pastures and fodder fields suffer less soil erosion and absorb more water than raw-crop fields, and nitrogen-fixing fodder plants such as alfalfa improve soil fertility (Durning and Brough, 1992). Conversely, ecological burdens arise from modern intensive livestock production systems such as ranching; intensively-managed poultry, pig and dairy units; and beef feedlots.

A major concern in some countries is the rapid disappearance of indigenous genes due to the use of imported stock through breed substitution and crossbreeding. Many indigenous breeds have special adaptive traits, for example disease resistance, climate tolerance, ability to use poor-quality feed and to survive with reduced and/or irregular supplies of feed and water. The conservation of these indigenous breeds is important if the sustainability of livestock development is to be maintained.

Livestock contributes to climate change by emitting methane, the second most important greenhouse gas (Chapter 3). Ruminant animals release perhaps 80 million tonnes of this gas each year in belches and flatulence, while animal wastes at feedlots and factory-style farms emit another 35 million tons. This occurs when the wastes are stored in the oxygen-short environments of sewage lagoons and manure piles, where methane forms during decomposition. Livestock account for 15-20 per cent of global methane emissions - about 3 per cent of global warming from all gases (Durning and Brough, 1992).

Livestock world-wide produced about 1,500 million tonnes of dung in 1970 and about 2,200 million tonnes of dung (air-dry) in 1990 (El-Hinnawi, 1991). This waste constitutes a major source of pollution, especially in developed countries in the neighbourhood of animal farms. Nitrates from these feedlot wastes are becoming a major source of surface and groundwater pollution in several countries. For example, in England and Wales, 20 per cent of the annual number of pollution incidents recorded by the water authorities in 1988 were from feedlot wastes (Conway and Pretty, 1991). In the Netherlands, Belgium and parts of France, manure has created several ecological problems. Nitrogen from manure escapes into the air as gaseous ammonia, and ammonia from the livestock industry is the largest source of acidic deposition on Dutch soils (Durning and Brough, 1992).

The environmental impacts of fish production

About 60 per cent of the population of the developing world derive 40 per cent or more of their total annual protein supplies from fish. Fish and fish products are not only highly nutritious, with protein content varying between 15 per cent and 20 per cent, but their biochemistry and amino acid characteristics also make them particularly efficient in supplementing the cereal and tuber diets widely consumed in Asia and Africa. The total world fisheries catch has increased considerably over the past two decades (Figure 4), and FAO estimates that the world catch from marine resources ought not to exceed 100 million tonnes per year if the risk of a substantial depletion of fish stocks is to be avoided. However, pressures on stocks in certain areas is already causing overfishing. In regions close to the industrial areas of the Northern Hemisphere, for example, this has resulted in a decline in the size and quality of some species of fish and in increasing scarcity of others. Overfishing has led to a sharp drop in catches of cod and herring in particular.

Fishing techniques have affected the aquatic environment in a number of ways. In some countries fishermen use inappropriate nets, or even explosives, to increase their catch. This has resulted in the killing of small traits and non-target organisms. The huge nets used by very large trawlers scrape the ocean floor over vast areas, causing the death of large numbers of non-target species through habitat destruction and through being accidentally engulfed by the net. The very large nylon drift-nets (some as long as 100 kilometres and with a depth of 30 metres or more) used by a number of countries cause severe collateral damage to non-target species, especially marine mammals, turtles and sea-birds, and their use has been banned by a number of countries. A major source of plastic debris in the marine environment is the fishing industry. It has been estimated that more than 150,000 tonnes of fishing gear made of plastic is lost (or discarded) in the oceans each year (GESAMP, 1990).

Traditional extensive culture of milkfish, mullets, molluscs and shrimp in Asian and Mediterranean countries accounts for the conversion of moderately large areas of coastal lowlands into ponds and shallow water areas for use as shellfish farms. Countries with coastal areas under extensive pond culture are the Philippines, Indonesia, Thailand and India. Recently, some of the traditional ponds have been upgraded for more intensive culture, and large areas of coastal mangroves and marshlands are being converted to ponds. The environmental costs and benefits of such conversion have to be thoroughly evaluated. On the one hand, most of the local vegetation and wildlife is destroyed in the process of pond construction, while water courses and dykes change or restrict the natural movement of water in the areas. On the other hand, the establishment of ponds leads to socio-economic development in their neighbourhood. In addition, some parasites that breed in swampy areas could be eliminated or reduced, (e.g. by conversion of swamps into ponds, mosquito populations could be reduced through the consumption of larvae by fish or shrimp).

Freshwater aquaculture also has a number of environmental impacts. Ponds occupy land that could be used for other purposes, and aquaculture consumes water by way of evaporation and seepage from ponds. Aquaculture could, therefore, compete with agriculture for such resources. In some countries, because of the low prices offered for agricultural products (especially those controlled by governments), farmers have converted increasing areas of agricultural land into fish ponds to gain more income. This has resulted in the loss of agricultural land in addition to creating hydrogcological problems in adjacent fields due to the increase in the level of ground-water as a result of seepage from the ponds. Some countries, for example Egypt, have prohibited the conversion of agricultural land into fish ponds.

The quality of the water available has a substantial bearing on aquaculture. Pollution of pond water can wipe out some fish species, while toxic substances and pathogens can build up in fish tissues and constitute health hazards to consumers. Pathogens are of particular concern where human excreta are used in aquaculture. The latter has been common practice in China, Malaysia, Thailand, Indonesia and Bangladesh (Edwards, 1985). The country with the largest number and largest area of sewage-fed fish ponds is India. There are also sewage fish systems in Germany (Munich), Hungary and Israel. Although there is little danger of disease from eating well-cooked fish from such ponds (as heat destroys pathogens), there is the possibility of such fish being contaminated with toxic chemicals present in the sewage. Eating poorly fermented or raw fish (widely consumed in parts of Asia) from such sewage-fed ponds could cause serious health hazards.

Agricultural and agro-industrial residues

Residues from agriculture and agro-industries are the non-product outputs from the growing and processing of raw agricultural products such as fruit, vegetables, meat, poultry, fish, milk and grain. These residues - in solid, liquid or slurry form - are usually organic and biodegradable, and hence are amenable to conversion by biological, chemical and physical processes into energy, animal feed, food, organic fertilizer and other beneficial products. However, although many agricultural residues have been used for centuries, especially in the developing countries, as fuel, animal feed or fertilizer, they are still widely considered as 'wastes'.

Throughout the world, farm crops leave substantial residues, the extent and scale of which are rarely realized. Residues from many crops are often more massive than the products themselves. For example, most cereal crops give between one and three tonnes of straw per tonne of grain, while cotton gives 3.5 to 5.0 tonnes of stalks per tonne of cotton (Barnard and Kristoferson, 1985). The amount of agricultural residues generated in the world has been estimated at about 930 million tonnes in 1970 and about 1,500 million tonnes in 1990 (El-Hinnawi, 1991), about 75 per cent of which was cereal straw and residues from maize and barley crops. Agro-industries also produce vast quantities of residues. The sugar cane industry, for example, creates 50 million tonnes of sugar cane tops and 67 million tonnes of bagasse each year, as well as molasses and press mud. In pineapple production less than 20 per cent of the fruit is used in the canning process, the remainder is discharged as waste (UNEP, 1982; El-Hinnawi and Hashmi, 1987).

Discharged into the environment, these residues can poison the soil, kill fish, cause eutrophication of lakes, pollute rivers and streams, create unpleasant smells and cause air pollution harmful to human health. As mentioned earlier, symptoms of this pollution are especially evident near high-density animal stocking areas. The straw left after removal of the grain must be lifted from the ground in order to control pests and diseases and to prevent fouling of the soil for the next crop. In some countries, mainly developed ones, most of the straw produced is burnt in the field, causing smoke and fire hazards and other ecological problems. Agro-industrial residues are often extremely polluting. For instance, the waste waters created when starch is extracted from tapioca are up to 20 times more polluting than municipal sewage, those from palm oil production up to 200 times more polluting, and those from whey up to 300 times more polluting (UNEP/FAO, 1977).

Agricultural residues have been used as fuel since ancient times, especially in rural areas of the developing countries where they still constitute a considerable share of the fuel used. Several technologies have been developed to use these residues more efficiently as fuel. Such technologies include gasification, pyrolysis into liquid fuels, charcoal production and production of biogas, etc. (El-Hinnawi and Biswas, 1981). There are several advantages to the conversion of residues into biogas. In addition to the gas produced, the slurry formed in the biogas plants is rich in nutrients and can be used as organic fertilizer and/or for feeding fish in ponds. Several opportunities exist for the use of agro-industrial residues. In meat production, the edible and non-edible residues can often be converted into useful products. Rice bran contains about 15-20 per cent oil, vitamin B and amino acids, which are extracted and used. Rice straw is used as raw material for paper, board and animal feed. Opportunities to recycle agricultural and agro-industrial residues are numerous and limited only by lack of incentives and of appropriate research and development.

Agriculture, international trade and economic relations

The last two decades have highlighted the manner in which the prevailing international trade climate has affected agricultural development, both in the developed and the developing countries. This can be seen in four main areas: enforcement of environmental standards on agricultural imports; protectionism; agricultural subsidies, and the provision of financial and technical assistance from the developed countries to the agricultural sectors of the least developed countries.

The report of the preparatory meeting held in Founex, Switzerland, one year before the Stockholm Conference, expressed a fear that the insistence of the developed countries on rigorous environmental standards of products exchanged in international trade would give rise to a 'neo-protectionism' (UNEP, 1981). As a recent example, the residual pesticides in agricultural products exported to developed countries prompted them to impose standards on the levels of such contaminants in the imported products. The World Bank has pointed out that such restrictions end up by hindering agricultural diversification in many developing countries and affecting poorer farmers in particular (World Bank, 1992). While larger farmers receive assistance to ensure that their products do not violate the limits on pesticide residues, small independent producers, who sometimes apply three times as much pesticides as do larger growers, do not receive such assistance. It would be ironic, the World Bank report adds, if concern about health in industrial countries impoverished the poorest farmers in developing countries (World Bank, 1992). Quite recently, the US banned imports of tuna from Mexico and several other countries, because dolphins were being snared in the drift nets. New Zealand is also imposing restrictions on timber imports from some countries on the basis of environmental considerations.

Developing countries rely heavily on the export of primary, and mainly agricultural, commodities (Chapter 16). This has exposed them to the effects of the steep drop in commodity prices experienced over the last decade. The prices of tea and coffee, for example, were lower in 1991 in real terms than at any time since 1950. Coffee, the largest primary export of developing countries after oil, suffered an estimated loss of between $US4 and $US7 billion a year. This general trend was aggravated by fluctuations in prices in response to the smallest changes in supply and demand. Together with the high sensitivity of agricultural commodities to weather conditions, market speculation and world recession, this has meant a highly unstable market for agricultural products that was at least difficult for the producers to adjust to, if not ruinous for any long-term development plans (UNDP, 1992).

If trade restrictions, quotas and internal taxes on tropical products, and other tariff and non-tariff barriers were removed, the revenues of agricultural producers would benefit considerably. It is estimated that the initial benefit to net exporters of agricultural products to industrial countries would be $US984 million for Latin America and the Caribbean (in 1985-87 prices) and $US428 million for Asia and the Pacific. By the end of 1990, GATT members had instituted hundreds of restraint arrangements covering a variety of imports. As far as agricultural products were concerned, GATT members had instituted no fewer than 59 actual or potential export restraint arrangements on agricultural products (UNDP, 1992).

At the same time, the industrial countries, particularly the USA and Europe, have been subsidizing their farmers to the tune of several hundred billion dollars to produce mountains of surplus food (Figure 10) and even subsidizing their export at well below the cost of production. In 1987, Thailand, who had managed to become a major producer of rice, was complaining of what it called 'unfair competition' from the USA. Such high subsidies have distorted international price levels and created a major barrier for agricultural exports, forcing down prices and constraining production in developing countries where the same commodities are grown using far less energy and other inputs per unit product.

Figure 10: *Producer subsidy equivalents for all agricultural products, 1982-86.*

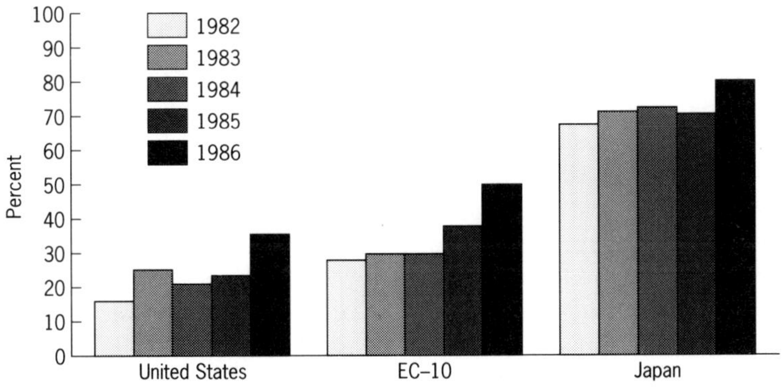

Source: US Department of Agriculture (1989).

Financial and technical assistance to the least developed countries has been biased in favour of the production of cash crops for export while these countries were unable to feed themselves and were receiving large quantities of staple foods. The drought period in the Sahel provides some peculiar anomalies. Although the weather has been blamed for the Sahel's poor food crop production, the weather does not seem to have affected export cash crops in the same way. During the period 1967-72, a period of most severe drought in Mali, peanut production increased by 70 per cent. In 1971, Senegal produced some 22,000 tonnes of cotton, although it grew no cotton at all when it became independent in 1960. These anomalies have been attributed by Oxfam to causes such as the fact that 40 per cent of rural development funds under the first Lomé Convention of the EEC (1975) went to development of crops for export. Irrigation programmes have been almost exclusively used to produce rice for urban consumption and export (Oxfam, undated). Governments became dependent on these export crops in paying for their imports, while the prices of these crops were deteriorating steadily (Chapter 16).

This pattern does not seem to have changed over the years. In 1944, when one and a half million hungry people died in Bengal, the British Government allowed the export of 200,000 tonnes of rice. After the floods of 1974 in Bangladesh, when many people were again dying of hunger, the granaries were brimming with some four million tonnes of rice that was too expensive and thus beyond the reach of hungry destitute people (Moorelappe and Collins, 1980).

Responses

Over the last two decades, the environmental impacts of food production have received considerable attention, not only in clarifying the issues involved but also in finding solutions. The realization that the arable lands of the world are finite prompted development of the science of land capability analysis to the point where it is now possible to predict the likely consequences of agricultural development and the management requirements to safeguard the ecological base. Some of the factors in crop land-use planning are illustrated in Figure 11.

There is now widespread recognition that the world's forests and woodlands are important, both as carbon sinks and as storehouses of biological and genetic diversity. The clearing of forest and woodland is no longer considered desirable in many countries, and a range of measures is being adopted to prevent unnecessary land clearing and to encourage afforestation.

Unsustainable pastoralism is still an intractable problem in dryland countries, in spite of efforts to introduce more productive breeds, to reduce stock densities, and to improve water availability and rangeland management. In many cases it has not been possible to overcome traditional attitudes and practices. There is strong resistance to reducing herds to sizes that the range can support, even though the social, economic and environmental circumstances that made traditional pastoralism sustainable no longer exist (IUCN/UNEP/WWF, 1991).

The availability of more water resources is a *sine qua non* for increased food production to meet the needs of a growing world population. Two main trends have been followed: efficiency in the use of water, and the development of new sources. As mentioned earlier, gigantic water impoundment schemes are no longer considered the blessing they once were. Brackish and drainage water, as well as municipal waste water, are used more extensively nowadays. There is better understanding of the hazards involved in such practices, and while improved water treatment techniques have been developed, the problem of heavy metal accumulation in the food chain remains.

Figure 11: *Crop land-use planning in an agro-ecological zone.*

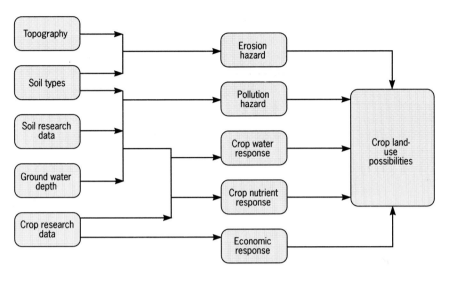

Source: Swaminathan (1991).

More attention is now being paid to the efficient use of water in agriculture through both pricing mechanisms (Chapter 4) and systems of delivery and use. Pricing of water for irrigation is a delicate issue in developing countries. It would be of little benefit if policies aimed at saving water were to provoke social turmoil. The challenge policy-makers face is to design policies that carry the correct signals without provoking unacceptable levels of conflict (Crosson and Rosenberg, 1989).

The efficiency of the gravity-flow irrigation system, which is probably the most widely-used system, particularly in developing countries, can be improved by levelling the fields, sometimes using modern laser-based instruments, so that they can be flooded quickly and uniformly. Trickle (or 'drip') systems that deliver water in the proper amounts continuously and directly to plant roots are also becoming widely adopted (Crosson and Rosenberg, 1989). These are areas where international co-operation in technology transfer and the provision of modest additional resources can produce good results.

The problems arising from the use of pesticides were recognized and addressed as far back as 1973, when WHO, in co-operation with UNEP, established an environmental health criteria programme to evaluate the effects of chemicals - including pesticides - on human health and the environment. In 1975, the Recommended Classification of Pesticides by Hazard was approved by the World Health Assembly. In 1976, UNEP established the International Register for Potentially Toxic Chemicals (IRPTC) (see Chapter 10). In 1980, the environmental health criteria were incorporated in the International Programme for Chemical Safety (IPCS) established by UNEP, ILO and WHO. One particularly significant development was the introduction of the principle of 'Prior Informed Consent' (PIC) by UNEP in 1989. Later in the same year, the PIC procedure was incorporated in the FAO International Code of Conduct on the Distribution and Use of Pesticides (Chapter 10). FAO and UNEP agreed to share the operational responsibility for the implementation of PIC procedures, setting up a joint expert group to provide advice on implementation procedures. These have included the appointment of 'designated national authorities' (DNAs); the preparation of a *Guidance for Governments* document on the operation of PIC procedures; and the preparation of an initial list of chemicals to be covered by PIC procedures. Activities are also underway to provide technical assistance for implementation, including training and institutional arrangements.

There are five alternative approaches to chemical pest control that address the problems arising from the use of pesticides: (i) environmental control; (ii) genetic and sterile male techniques; (iii) biological control; (iv) behavioural control; (v) resistance breeding. Each approach has its advantages and disadvantages. Consequently, the concept of 'integrated pest management' (IPM) is currently the focus of attention. IPM strategies may include the selective use of pesticides and rely on the use of biological methods, genetic resistance and appropriate management practices (Box 1). Although the application of IPM has been slow, especially in relation to food crops, many success stories have demonstrated its viability. In the USA, IPM is now used on about 15 per cent of the total area of cultivated land (Conway and Pretty, 1991), and its use is growing, for example in Central America and some Asian countries. If IPM strategies are implemented in combination with the application of the International Code of Conduct on the Distribution and Use of Pesticides, and the training of farmers, a great deal will be achieved in reducing the environmental impacts of pesticides. Although IPM is now well developed for agricultural pests (Box 2, Box 3), there are still difficulties in its application to public health vector control. Furthermore, subsidizing pesticides in developing countries has undermined the promotion of more cost-effective IPM techniques (Repetto, 1985). Although various IPM approaches hold great potential, other feasible measures such as crop rotation; varying sowing, harvesting and irrigation times; changing the location of crops to different fields; destroying crop remains, and choice of more resistant plant varieties have proved successful in some cases. Pesticide fees or taxes encourage farmers to be more selective in their choices of chemicals and to switch to other alternatives as they become relatively cost-effective (Zilberman *et al*, 1991). It should be noted, in conclusion, that the sophistication of IPM calls for training and upgrading of existing extension services. Some Asian countries seem to have achieved various degrees of success in its introduction (Kenmore *et al.*, 1987).

Many simple ways and technologies have been developed to increase the efficient use of different inputs in agricultural systems. Adjusting the timing of fertilizer application and the amounts used has led to a considerable saving of fertilizer, with both economic and environmental benefits. The use of sulphur-coated urea (SCU) on rice has led to control of nitrogen release and hence a

BOX 1

Biological control of pests

	Pest management using chemicals	Pest management using biological control
Number of times insecticide used in rice season	4.5 applications	0.5 applications
Cost to farmers per hectare	7.5 rupiah	2.5 rupiah
Cost to government per hectare	27.5 rupiah	2.5 rupiah
Rice yield per hectare	6 tonnes	7.5 tonnes

The advantages of integrated pest management (IPM) to Indonesia. Farmers switched from chemicals to encouraging natural predators of the pest destroying their rice crops.

In the 1970s, the development of high-yield strains of rice and the increasing use of fertilizers and pesticides allowed farmers in Indonesia to grow two rice crops each year instead of one. Unfortunately, this led to an enormous growth in the population of brown planthoppers. Farmers were spraying up to eight times in the rice-growing season to try to reduce the damage done by this pest, and the government was providing huge subsidies to help the farmers pay for the expensive pesticides.

Then scientists showed that spraying had caused the problem in the first place. The sprays had wiped out all the natural predators of the brown planthoppers, particularly spiders, and yet had only a limited effect on the pest itself.

In response, the Indonesian Government introduced an integrated pest management (IPM) system. First, it reduced the subsidies on chemical sprays, and banned farmers from using 57 insecticides on rice. It then set up a nationwide training programme to show farmers how to conserve natural predators such as spiders. Spraying was to be considered only as a last resort.

Within three years, farmers were using 90 per cent less pesticide, with large savings in cost both for them and the government. Yields of rice were increasing, and less harm was done to the environment.

Similar IPM programmes for rice are being introduced in Bangladesh and India.

Source: Fullick and Fullick (1991).

BOX 2

Biological pest control

A lethal pest - the New World screwworm fly - has recently killed more than 12,000 animals in the Libyan Arab Jamahiriya. Unchecked, the larvae of the fly could have eaten their way through 70 million head of livestock in five North African countries. The outbreak began in the Libyan Arab Jamahiriya in 1988 and the flies had infected about 40,000 square kilometres there. Female screwworm flies lay their eggs in open animal wounds. The maggots that develop eat living tissue and can eventually kill the host animal. FAO and IFAD started an eradication programme that relied on swamping the female screwworm flies with male flies that had been irradiated to make them sterile. The males mate with females whose eggs then fail to hatch, and the population eventually dies out. More than a billion sterilized Mexican flies were flown into the area and used in the control operation. No infected animals have been found since April 1991.

Source: New Scientist (12 October 1991).

lowering of the concentration of nitrogen in both the soil and water at any given time. Another technique that is gaining acceptance in greenhouses is the use of biodegradable mulches that release nutrients - and sometimes pesticides - at controlled rates. Recourse to biological processes for fertilization (nitrogen-fixing plants, crop rotations, use of trees as a 'nutrient pump', recycling of wastes) is growing in several countries, especially in industrialized countries where it is sometimes referred to as 'ecological' agriculture. There is also a growing tendency to apply the concept of integrated plant nutrition systems (IPNS) which involves the use of carefully derived combinations of mineral and organic fertilizers which are applied in combination with complimentary crop practices, such as tillage, rotation and moisture conservation. As a result, soil quality is conserved and pollution is reduced to a minimum.

The environmental problems that arise from more intensive farming and harvesting methods are not easily overcome. It seems likely that farming will become more intensive rather than less so in the developed countries. However, the OECD concluded that despite overall statistics indicating a continuing trend to intensification 'they conceal the changing nature of some technologies towards a better integration of environmental concerns. Farm machinery or irrigation system designs, fertilizer application rates and methods, pesticide spectra, composition or dosage changes associated with a better ability to measure environmental stress have tended to shift the damage potential of agrochemical input use, specialization or intensification in the direction of improvement. To the extent that the new technologies embody environmental concerns of society, albeit with a lag, aggregate indicators might underestimate such achievements' (OECD, 1991).

Several biotechnologies have recently been developed to solve specific agricultural problems. For example, the herbicide atrazine is used to kill weeds in maize fields. Maize can tolerate atrazine (UNEP, 1985). However, where maize is planted in rotation with soybeans, the latter are susceptible to residues of atrazine and their yield is affected. In response to this, an atrazine-resistant soybean has now been developed for growing in rotation with maize. Marked progress has been made in transferring the genes for nitrogen fixation present in certain bacteria to some crops, which would lead to dramatic improvements in biological nitrogen fixation and decrease the dependence on chemical fertilizers (El-Hinnawi and Hashmi, 1987). However, biotechnology, stimulated by the search for maximum profits for both biotechnology companies and commercial farmers, has not so far been geared to solving the problems of poor Third World farmers. Moreover, few Third World countries are equipped to apply high-technology solutions to their food and agriculture problems and none is even remotely able to evolve its own biotechnology research and technology capacity (FAO, 1991h).

The introduction of genetically engineered plants, like the introduction of cultivars through normal plant breeding, should have beneficial environmental effects - for example, the reduction of chemical fertilizer and pesticide use, increased tolerance to salt and drought, and so on. However, if drought tolerance leads to the expansion of dry-land cropping into ever-drier regions where rainfall variations from year to year are extreme, increased wind and water erosion could result in severe soil degradation during dry years. The introduction of salt-tolerant cultivars could prompt increased use of saline water for irrigation, with the subsequent contamination of shallow ground water and increased salinization of soils. This in turn would narrow the choice of crops for cultivation in rotation. Increased tolerance to herbicides could make it difficult to eradicate crop plants that have become weeds. Although our knowledge of the environmental consequences of plant genetic engineering is in its infancy, researchers and the public are expressing some concern about the safety of the new technology. One cause for such concern is the risk posed by the release of novel organisms into the environment. The introduction of any species into an ecosystem it does not normally inhabit can have unexpected negative results. Guiding or perhaps accelerating the course of evolution can lead to changes that disrupt an ecosystem and, hence, may undermine man's reverence for life (UNEP, 1985).

Tissue culture - the multiplication of plants through *in vitro* micropropagation - holds special promise. It permits much faster multiplication rates than seeding or other traditional propagation techniques and gives uniformity of yield, quality and rate of ripening. The last decade has witnessed considerable development and application of tissue cultures for more and more plants and in more countries, including developing countries.

As mentioned earlier, yields from natural fish stocks are not likely to rise much further without substantial depletion of stocks due to over-fishing.

BOX 3

Examples of Integrated Pest Management Programmes

Rice in Orissa, India (Brader, 1979)

Pests: gall midge, brown planthopper, stemborers.

1 Establish thresholds below which spraying is uneconomic.
2 Develop forecast system based on rainfall patterns.
3 Monitor pest densities in fields.
4 Plant early-maturing, short-duration varieties; avoid times of high pest density.
5 Select varieties resistant to the pests.
6 Plough in crop residues after harvest, thus preventing carry-over of pests.
7 No application of pesticides when predators are abundant.

Result: insecticide applications cut in half; increase in use of modern varieties.

Rice in Kwangtung, China (Brader, 1979)

Pests: weeds, leafhoppers, planthoppers, leafrollers, stemborers.

1 Insect forecasting and monitoring in national network.
2 Herd ducks through fields to eat pests.
3 Flood rice fields to drown stemborer larvae.
4 Catch flying insects at night in light-traps.
5 Plant resistant varieties.
6 Rear and release parasites of leafroller eggs.
7 Conserve frogs to prey on pests.
8 Apply *Bacillus thuringiensis*.

Result: pesticide costs reduced by 30 per cent; but more labour required for rearing predators, maintaining traps and monitoring.

Cotton in Texas, USA (Dover, 1985)

Pests: boll weevil, pink bollworm, tobacco budworm.

1 Establish uniform planting period and short duration varieties, so that emergence of bollworm moths occurs when there is no cotton fruit on which to lay eggs.
2 Fine uncooperative farmers.
3 Irrigate desert areas prior to planting.

4 Apply insecticides only in areas where high boll weevil populations are expected.

5 Selectively apply organophosphates during harvest to kill adult boll weevils before they emigrate to overwintering sites in wood and field margins.

6 Defoliate mature crop so all cotton bolls open together.

7 Use mechanical strippers for harvest, which kill larvae.

8 Shred stalks and plough under remnants immediately after harvest, thus denying pests food during winter.

Result: dramatic increase in area of cotton harvested; decrease in production costs; increase in yields.

Apple in Nova Scotia, Canada (Corbet, 1974; Flint and van den Bosch 1981)

Pests: codling moth, European red mite, winter moth, apple maggot.

1 Conserve natural enemies by using selective pesticides.

2 Carefully time applications.

3 Monitor and spray only when absolutely necessary.

4 Introduce the parasite *Cyzenis* to control winter moth, although only successful where orchards are near forests.

5 Remove wild hosts and dispose of infested drop-fruits to control maggots.

6 Use sticky traps.

7 Selective use of pheromones and attractants.

Results: ninety per cent of apple growers have been using this programme continuously since the 1950s

Alfalfa in California, USA (Flint and van den Bosch, 1981)

Pests: alfalfa caterpillar, spotted alfalfa aphid.

1 Monitoring of numbers of healthy caterpillars by skilled entomologists: if unhealthy, probably parasitized by wasp or dying from viral disease.

2 Timely spraying if caterpillar numbers above economic threshold.

3 Three species of wasp parasites imported from the Near East and Europe to control the aphid.

4 Introduce strip harvesting to provide refuges for lady beetle natural enemies.

BOX 3 CONTINUED

5 Timely irrigation to enhance activity of virulent fungal diseases of the aphid.

6 Establish monitoring of predators, to prevent spraying when abundant.

7 Use selective aphicide demeton.

8 Introduce reistant varieties of alfalfa.

Result: rapid reduction in losses; pests now rarely seriously injurious.

Tobacco in Canada (Corbet, 1974)

Pests: hornworm, cutworm.

1 Control hornworm by using *B.t.*

2 Use viruses to control cutworms.

3 Plant trap plants of tobacco among the rye cover crop, on which larvae develop and overwinter.

4 Spray trap plants with virus.

Result: scientifically successful, but uneconomic.

Cotton in Central America (ICAITI, 1977)

Pests: boll weevil, leaf-eaters, sap-suckers, shoot-pruners, bollworm, whitefly.

1 Leave small cotton islands of old stock standing after May to concentrate boll weevil remaining from previous season.

2 Plant trap crops at field margins early where weevils known to invade.

3 Selectively spray islands and trap crops when required.

4 Establish uniform planting date.

5 Add nitrogen fertilizer not before 60 days old, unless signs of deficiency, to reduce attraction to leaf-eaters and sap-suckers.

6 *Trichogramma* predators released when leaf-damage greater than 50 per cent.

7 Selective use of pesticides, including methyl parathion, monocroptofos, chlordimeform, metamidophos.

8 Defoliate when 50 per cent of bolls open.

9 Harvest promptly to avoid whitefly problems.

Results: yields still same, but relative profits up threefold; number of pesticide treatments fell; control costs and total variable costs also fell.

Source: Conway and Pretty (1991).

Additional catches will need to come from species not currently regarded with favour and from unconventional species, as well as from expansion of mariculture to new areas or species. If the present yield is to be maintained, better management practices at the international level are needed, and regular monitoring of stocks is required on a much wider scale. This is currently undertaken adequately in only a few areas such as the northern parts of the Atlantic and the Pacific. Generally speaking, new policy approaches may be necessary in the fishing industry.

The seas and oceans are classic examples of an 'open access' resource, and the destruction of seal, whale and fish stocks through competitive over-fishing is an example of the unsustainable nature of the traditional use of such resources. As the extent of this over-exploitation became generally recognized, the responses have gone through three stages:

- Initial international agreement to regulate exploitation, through fisheries conventions and other agreements, e.g. the North-east Atlantic quota system in the 1970s, and the subsequent banning altogether of fishing of certain stocks to allow them to recuperate, or the banning of fishing of certain species for particular periods of time every year.

- Subsequent extension of coastal state authority through the creation of 200-mile exclusive economic zones.

- The recent recognition that regulation must change from a focus on the users of the resource to a 'resource-oriented' approach in which the regulation is consciously designed to optimize sustainable resource use on the basis of sound scientific information about the whole ecosystem.

A number of international legal measures adopt the latter approach. One example is the Convention on the Conservation of Antarctic Marine Living Resources (CCAMLR) which demands that all catches shall be regulated on the basis of the overriding need to protect the ecosystem as a whole. Another example is the banning of drift netting by 1 June 1992 (UN General Assembly resolution, December, 1989).

It should be recognized that small-scale, community-based fisheries account for almost half the world fish catch for human consumption, employ more than 95 per cent of the people in fisheries and use only 10 per cent of the energy of large-scale corporate fisheries (Figure 12). Governments should review those policies that encourage large-scale and sport fisheries at the expense of small-scale fisheries (because the former two have more effective lobbies and generate foreign exchange). They should allocate marine resource user rights more equitably, giving more weight to the interests of local communities and organizations (IUCN et al., 1991). Every effort should be made to bring an end to the use of all forms of destructive fishing gear.

A great deal of research activity is under way in many research centres around the world to study ways and means of increasing agricultural

Figure 12: *The world's two marine fishing industries - how they compare.*

	Large scale	Small scale
Number of fishermen employed	Around 500,000	Around 12,000,000
Annual catch of marine fish for human consumption	Around 29 million tonnes	Around 24 million tonnes
Capital cost of each job on fishing vessels	\$US 30,000–\$US 300,000	\$US 250–\$US 2500
Annual catch of marine fish for industrial reduction to meal and oil, etc.	Around 22 million tonnes	Almost none
Annual fuel oil consumption	14–19 million tonnes	1.5–2.0 million tonnes
Fish caught per tonne of fuel consumed	2–5 tonnes	10–20 tonnes
Fishermen employed for each \$US 1 million invested in fishing vessels	5–30	500–4000
Fish destroyed at sea each year as by-catch in shrimp fisheries	6–16 million tonnes	None

Source: IUCN *et al.* (1991).

productivity. A number of international and regional organizations are also supporting various research and development activities to achieve the same goal. The activities of FAO, the International Fund for Agricultural Development, UNEP and bodies such as the International Board for Plant Genetic Resources (IBPGR), the Consultative Group on International Agricultural Research (CGIAR), the International Rice Research Institute (IRRI), the International Centre for Maize and Wheat Improvement, the International Centre for Insect Physiology and Ecology (ICIPE) and several others are well documented. However, there is still a great need for accelerated research and development efforts, at the national and international levels, to develop appropriate and environmentally sound agricultural practices and technologies. Special emphasis should be given to the modernization of indigenous technologies.

Concluding remarks

In his 1985 *State of the Environment Report*, the Executive Director of UNEP took up the concept of a 'Third Agricultural Revolution' put forward by the Director General of FAO in 1980. Developing this concept, the Executive Director of UNEP wrote:

'Agricultural practices which lead to environmental degradation will trigger or exacerbate the neglect of land and of rural development, prompting an increase in rural-urban migration. This in turn will aggravate the already dire problems of urban areas, and - most importantly - will negatively affect indigenous food production, thereby increasing national dependence on imported food. Ultimately, this will create or aggravate national instability. It is therefore in the interest of national stability and security that countries should pursue the development and implementation of environmentally sound agricultural development plans' (UNEP, 1985).

There are two schools of thought regarding future possibilities of meeting agriculture needs sustainably. The first (the optimistic) considers that current yield and productivity gaps are large enough to allow sufficient food and livestock products to be produced from the Earth's limited resources, with sufficient injections of inputs and other technology, as well as incentives and knowledge for farmers. The FAO/IIASA/UNFPA work on agro-ecological zoning and population-carrying capacities arrives at similar conclusions: potential agricultural production from existing technologies is largely sufficient to feed growing populations; the real problem is one of transferring the technology through the right mix of institutions and incentives. The second school (the pessimistic) considers that, in the long run, current development patterns are not sustainable. Neo-Malthusian studies, such as that of the first Club of Rome report, belong to this category, as does the World Bank's *1990 Development Report* which includes a 'nightmare scenario' for Africa, extrapolating current trends in population and food production, resulting in an ever widening gap.

Future sustainability of agricultural production will depend on global, regional and local environmental conditions. Little is known about the combined impact of climate changes, higher levels of atmospheric carbon dioxide, increased ultraviolet radiation, and increasing acid deposition and air pollution on crops, animals, and plant and animal pests and diseases. Not much is known also about the potential impacts of changing atmospheric and oceanic conditions on marine ecosystems and fisheries. Even more poorly understood are the many complex links and feedbacks that are likely to exist between (a) food system activities that contribute significantly to environmental change, (b) food system activities that would be directly or indirectly affected by environmental change, (c) impacts in related activities such as energy production and transportation, and (d) actions taken to reduce or modify the effects of any of these activities (Chen, 1990). Thus, it is clear that the problem of providing more food to more people during the next several decades is greatly dependent on the interactions between food-producing systems and the environment.

Sustainable agriculture is not simply a matter of adopting environmentally sound practices. It is also a matter of changing the attitudes of society. Success in achieving sustainable agriculture needs to reconcile individual and societal interests. Environmentally-benign technologies will only be introduced if they benefit the individual farmer on the one hand, and ensure the rational use of resources such as water and genetic diversity on the other. The challenge is to achieve this in a market system in which property rights for such resources are difficult to establish. This is essentially an institutional rather than technological problem. Devising institutional mechanisms that give the correct market signals of the emergence of scarcities of national resources of land, water and genetic diversity is the challenge facing agricultural development (Crosson and Rosenberg, 1989).

This challenge was addressed at the FAO/Netherlands Conference on Agriculture and the Environment in April 1991. An important outcome of the Conference was the introduction of an International Cooperative Programme Framework for Sustainable Agriculture and Rural Development (ICPF/SARD). ICPF/SARD is a flexible process being launched at the international, regional and national levels, which aims to start a transition toward sustainable agriculture and rural development. It will result in more emphasis on redirecting or reinforcing agricultural policy and planning procedures, greater participation of local people in decision-making, and the promotion of integrated production systems (FAO and Netherlands Ministry for Agriculture, Nature Management and Fisheries, 1991).

SARD strategies, developed for each country (and regionally where appropriate) must meet the essential requirements of ensuring food security, eradicating poverty and conserving natural resources. In intensifying food production, undesirable environmental effects should be avoided, especially in the use of external inputs. Mixed farming systems, traditional and local know-how, and waste recycling will also be promoted as key components of ICPF/SARD.

Action will be required on many fronts, including the international level, to improve international economic relations, eliminate unfair terms of trade, protectionist barriers and price distorting subsidies, and to alleviate debt. At the national level the focus will be on the pressing need to alleviate hunger and malnutrition, on land reform, people's participation, employing market processes, and in particular, enhancing the role of women. While additional human and financial resources will be required to implement the ICPF/SARD, many adjustments, particularly in the policy and planning area, can be undertaken immediately. The FAO in collaboration with other international organizations (including NGOs) is assisting countries in initiating this process.

'Agenda 21' presented to UNCED, confirms that 'the priority must be on maintaining and improving the capacity of the higher-potential agricultural lands to support an expanding population. However, conserving and rehabilitating the natural resources on lower potential lands in order to maintain sustainable man/land ratios is also necessary. The main tools of SARD are policy and agrarian reform, participation, income diversification, land conservation and improved management of inputs. The success of SARD will depend largely on the support and participation of rural people, national governments and the private sector, and on international co-operation, including technical and scientific co-operation' (UNCED, 1992).

For many developing nations, agricultural commodities are the main export products and the primary source of the foreign exchange they require to import both agricultural and other inputs (especially oil and food). High agricultural subsidies have been the primary cause of the lack of agreement in the Uruguay Round of Negotiations under the General Agreement on Tariffs and Trade (GATT). It is important that the resumed negotiations reach a successful conclusion.

REFERENCES

Abdel Mageed, Y. (1985) The Jonglei Canal: a conservation project of the Nile. *In*: Golubev, G. and Biswas, G. (eds) *Large Scale Water Transfers*. Tycooly International, Oxford.

Bailey, C. and Skladany, M. (1991) *Aquaculture development in tropical Asia*. Natural Resources Forum, Feb. 1991, p.66.

Barnard, G. and Kristoferson, L. (1985) *Agricultural residues as fuel in the third world*. Technical Report No. 4, Earthscan, IIED, London.

Biswas, A.K. (1991) *Land and water development for sustainable agricultural development of Egypt: opportunities and constraints*. FAO Economic & Social Policy Division, Rome.

Brown, L.R. (1991) The New World Order. *In*: L.R. Brown *et al.*, (eds) *State of the World*. Earthscan Publications, London.

Chen, R.S. (1990) Global agriculture, environment and hunger. *Environ. Impact Assessment*. Rev.10, p. 335.

Chen, R.S. and Fiering, M. B. (1988) Feeding ten billion - can it be sustained? The Future of Hunger - Occasional Paper OP-88-1. Providence RI: Alan Shawn Feinstein World Hunger Program, Brown University.

Conway, G.R. and Barbier, R. (1990) *After the Green Revolution*. Earthscan Publications, London.

Conway, G.R. and Pretty, J. N. (1991) *Unwelcome Harvest*. Earthscan Publications, London.

Crosson, P.R. and Rosenberg, N J. (1989) Strategies for Agriculture. *Scientific American*, **261 (3)**.

Durning, A.T. and Brough, H. B. (1992) Reforming the livestock economy. *In*: L.R. Brown *et al.*, (eds) *State of the World*. Earthscan Publications, London.

Edwards, P. (1985) Aquaculture: a component of low cost sanitation technology. *World Bank Technical Paper No. 36*. World Bank, Washington, D.C.

El-Hinnawi, E. (1981) *The Environmental Impacts of Production and Use of Energy*. Tycooly International, Dublin.

El-Hinnawi, E. (1991) *Sustainable agricultural and rural development in the Near East*. Regional Document No. 4. FAO/Netherlands Conference on Agriculture and Environment. FAO, Rome.

El-Hinnawi, E. and Biswas, A. (1981) *Renewable Sources of Energy and Environment*. Tycooly International, Dublin.

El-Hinnawi, E. and Hashmi, M. (1987) *The State of the Environment*. Butterworth, London.

Engelstad, O.P. (1984) Crop nutrition technology. *In*: B.C. English *et al.*, (eds) *Future Agricultural Technology and Resource Conservation*. Iowa State Univ. Press.

ESCAP (1990) *The State of the Environment in the ESCAP Region*. ESCAP, Bangkok.

FAO (1987) *Agriculture: Toward 2000*. FAO, Rome.

FAO (1988) *Country Tables*. FAO, Rome.

FAO (1990) *FAO Production Yearbook* **vol. 43**. FAO, Rome.

FAO (1991a) *The State of Food and Agriculture - 1990*. FAO, Rome.

FAO (1991b) *Current World Food Situation*. Doc. CL/99/2. FAO, Rome.

FAO (1991c) Livestock production and health for sustainable agriculture and rural development. Background Document No.3 FAO/Netherlands Conference on Agriculture and Environment. FAO, Rome.

FAO (1991d) *Environment and sustainability in fisheries*. Document COFI/91/3. FAO, Rome.

FAO (1991e) *Sustainable development and management of land and water resources*. Background Document No. 1. FAO/Netherlands Conference on Agriculture and Environment. FAO, Rome.

FAO (1991f) *Issues and perspectives in sustainable agriculture and rural development.* Main Document No.1. FAO/Netherlands Conference on Agriculture and Environment. FAO, Rome.

FAO (1991g) *Sustainable crop production and protection.* Background Document No. 2. FAO/ Netherlands Conference on Agriculture and Environment. FAO, Rome.

FAO (1991h) *Technological options and requirements for sustainable agriculture and rural development.* Main Document No.2. FAO/Netherlands Conference on Agriculture and Environment. FAO, Rome.

Fullick, A. and Fullick, P. (1991) Biological pest control. *New Scientist,* 9 March, *Inside Science* No. 43.

Gale Johnson, D. (1984) World Food and Agriculture. *In*: J.L. Simon and H. Khan (eds) *The Resourceful Earth.* Blackwell, Oxford.

George, S. (1988) *A Fate Worse than Debt.* Penguin Books, London.

Georghiu, G.P. (1989) *Pest Resistance to Pesticides.* Plenum Press, New York.

IUCN/UNEP/WWF (1991) *Caring for the Earth,* IUCN, Gland, Switzerland.

GESAMP (1990) *The State of the Marine Environment.* UNEP, Nairobi.

Jensen, A.A. (1983) Chemical contaminants in human milk. *Residue Review,* **vol. 89**, p.1.

Karakaya, A.E. *et al.* (1987) Organochlorine pesticide contaminants in human milk from different regions of Turkey. *Bull. Environmental Contamination and Toxicology,* **vol.39**, p. 506.

Kenmore, P., Litsinger, J.A., Bandong, J.P., Santiago, A.C. and Salac, M. M. (1987) Philippine rice farmers and insecticides: thirty years of growing dependency and new options for change, *In*: (E.J. Tait and B. Napompeth, eds) *Management of Pests and Pesticides: Farmers' Perceptions and Practices,* West View Press, London.

Lal, R. (1990) Managing soil carbon in tropical agro-ecosystems. EPA Workshop on Sequestering Carbon in soils. Corvallis, Oregon, Feb. 1990.

Metcalf, R.L. (1986) The ecology of insecticides and the chemical control of insects. *In*: Kogan, M. (eds) *Ecological Theory and Integrated Pest Management Practice.* John Wiley & Sons, New York.

Moorelappe, F. and Collins, J. (1980) *Food First: The Myth of Scarcity.* Candor Book, Souvenir Press.

Myers, N. (1989) Loss of biological diversity and its potential impact on agriculture and food production. *In*: Pimentel, D. and Hall, C.W. (eds) *Food and Natural Resources.* Academic Press, New York.

Nelson, R. (1987) *Strategies for the development of the livestock sector.* Economic and Policy Division, Agriculture and Rural Development Department. World Bank, Washington, D.C.

OECD, 1991 *The State of the Environment,* OECD, Paris.

Oxfam (undated) *Behind the weather: Why the Poor Suffer Most.* Oxfam Public Affairs Unit, Oxford.

Pimentel, D. and Levitan, L. (1986) Pesticides: amounts applied and amounts reaching pests. *BioScience,* **vol.36**, p.86.

Pimentel, D. *et al.* (1989) Ecological resource management for a productive sustainable agriculture. *In*: Pimentel, D. and Hall, C.W. (eds) *Food and Natural Resources.* Academic Press, New York.

Pimentel, D. *et al.* (1991) Environmental and economic effects of reducing pesticide use. *BioScience,* **vol.41**, p. 402.

Repetto, R. (1985) *Paying the price : pesticide subsidies in developing countries,* Res. Rept No.2, World Resources Institute.

Revelle, R. (1984) The world supply of agricultural land. *In*: Simon, J. L. and Khan, H. (eds) *The Resourceful Earth.* Blackwell, Oxford.

Rose, J.B. (1986) Microbial aspects of wastewater reuse for agriculture. *CRC Critical Reviews in Environmental Control,* **vol. 16**, p.231.

Slorach, S.A. *et al.* (1983) *Assessment of human exposure to selected organochlorine compounds through biological monitoring.* Swedish National Food Administration, Uppsala.

Swaminathan, M.S. (1991) Personal communication.

Trappe, A.Z. (1985) The impact of agrochemicals on human health and environment, *Industry and Environment,* **Vol.8**, p.10, UNEP, Paris.

UNCED 1992 *Promoting sustainable agriculture and rural development,* A/CONF.115/PC/100/Add.19, discussed in the fourth session of the Preparatory Committee for the United Nations Conference on Environment and Development, New York, 2 March - 3 April, 1992.

UNDP 1992 *Human Development Report 1992,* Oxford University Press, New York, Oxford.

UNEP/FAO (1977) *Residue utilization - Management of agricultural and agro-industrial wastes.* UNEP, Nairobi.

UNEP (1981) Founex Report *In Defence Of The Earth,* UNEP, Nairobi.

UNEP (1982) *Guidelines on Management of agricultural and agro-industrial residue utilization.* UNEP, IEO, Paris.

UNEP (1985) *State of the Environment 1985,* UNEP, Nairobi.

US Department of Agriculture (1989) *GATT and Agriculture : The Concept of PSEs and CSEs,* USDA, Washington, D.C.

WFC (1991) *Hunger and malnutrition in the world.* World Food Council. Document WFC/1991/2, WFC, Rome.

White, G.F. (1988) The environmental effects of the High Dam at Aswan. *Environment,* **vol. 30**, p.5.

WHO (1977) *Nitrates, Nitrites and N-Nitroso compounds.* Environmental Health Criteria No. 5, WHO, Geneva.

WHO (1990) *Public Health Impact of Pesticides Used in Agriculture.* WHO, Geneva.

World Bank (1992) *World Development Report 1992 : Development and the Environment,* Oxford University Press, New York, Oxford.

Zilberman, D., Schmitz, A., Casterline, G., Lichtenberg, E. and Siebert, J. B. (1991) The economics of pesticide use and regulation. *Science,* **253**: 518-22.

CHAPTER 12

Industry

Introduction

The developing countries have long considered industry to be a dynamic instrument of growth essential to their rapid economic and social development, and their share of total world industrial output has risen steadily, particularly since World War II. For developing countries, industrialization is not just an important engine of economic growth: it is viewed as an effective means of modernizing society, promoting new and more appropriate work habits and value systems, and reducing dependence on the export of unprocessed raw materials and natural resources. No wonder then that the Second General Conference of UNIDO, held in Lima, Peru, in 1975 (less than three years after the Stockholm Conference), declared that industrial production in developing countries 'should be increased to the maximum possible extent and as far as possible to at least 25 per cent of total world industrial production by the year 2000' (UNIDO, 1976). In the developed countries, on the other hand, industry has been - and still is - responsible for a substantial part (30-40 per cent) of GNP, even though the industrial scene in these countries has been changing over the last two decades in response to the requirements of environmental protection, changes in social demand, scientific and technological advances, and developments in international relations and world markets.

The attitude of developing countries towards the environmental impact of industrialization was markedly different from that of the developed countries in the early 1970s. For the developing countries, rapid development and economic growth were the top priorities. The general view held by most developing countries at the time was that industrial pollution - a major concern of the developed countries at the Stockholm Conference - was not, as yet, their problem and that it would be quite some time before it became a cause for serious concern. Poverty was the main polluter and industrial expansion would foster high rates of economic growth and improvements in standards of living.

The 1980s witnessed significant changes in these early perceptions as the two groups drew closer together in their views of the industry-environment nexus. This was brought about mainly by first-hand experiences in the developing countries themselves of the deleterious effects of certain industries - whether on the physical environment, on human health and well-being, or on social stability. A second major influence was the impact of a number of large-scale industrial accidents which had drastic environmental consequences in both developed and developing countries.

The attitude of industry in the developed countries has changed dramatically since Stockholm. Industry's attitude towards the 1972 Conference ranged from downright hostility to disregard, or - at best - to lukewarm support. This attitude has been gradually shifting, and industry has now come to assume a visible role in addressing environmental concerns. The last few years have witnessed two international conferences on environmental management (WICEM) organized by the business community (Versailles 1984, in co-operation with UNEP, and Rotterdam 1991), both of which included programmes to foster closer ties with society as well as commitments by leading industrial conglomerates to cut pollutants by certain percentages over relatively short periods of time.

The attitude of industry (both national and transnational) in developing countries as a whole has also changed over the last two decades, particularly in the wake of a number of serious industrial accidents. Yet it is true to say that industry in most developing countries has been less keen on addressing environmental problems than it has in developed countries. There are many reasons for this, notably a lack of expertise within the industrial sector, the use of obsolete polluting technologies (often imported from developed countries), the absence or lenience of environmental regulations and means of enforcement, low priority for environmental protection and lack of public pressure. Furthermore, in many developing countries the state is a major industrial entrepreneur and the public and parastatal sectors are major industrial polluters.

Main developments in the industrial scene

It is estimated (UNIDO, 1990) that the total world-wide manufacturing value added (MVA) has increased from about $US2,500 billion in 1975 to a little under $US4,000 billion in 1990, at constant 1980 prices. Manufacturing value added can be defined as the difference between the value of industrial products and the cost of the inputs that go into producing them. The share of the developing countries is estimated to have risen from 10 per cent in 1970 to around 14 per cent in 1990 (Figure 1). This growth has occurred in almost all industrial sub-sectors. The annual growth rates of MVA in both the developed and developing regions have been rather erratic (Figure 2), twice dropping to negative values in the developed regions. However, the rates of growth in the two regions have moved together. As might be expected, the rate of growth was higher in the developing regions almost throughout this period.

These aggregated trends do not reveal the variations in the distribution of industry amongst the different regions and different countries, whether developed or developing, nor do they highlight the discrepancies in the rates of growth of different industries. These have important implications when reviewing the impact of industry on the environment. There have been, and still are, considerable differences in the average growth rates of MVA in different regions (Figure 3).

In terms of industrial outputs, the OECD countries witnessed a drop in the period 1970-82 (OECD, 1991a) yet this trend has been halted in recent years. The share of industry in the GNP has indeed fallen from 40 to 30 per cent between 1960 and 1988, yet at constant prices the share of manufacturing

Figure 1: *Manufacturing share of the developing countries in world industrial output, 1970-91.*

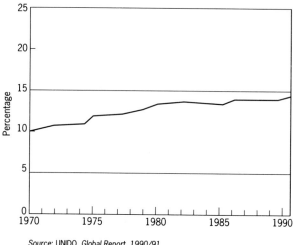

Source: UNIDO, *Global Report, 1990/91.*

Figure 2: *Growth rates of MVA (manufacturing value added) in developed and developing regions, 1970-91.*

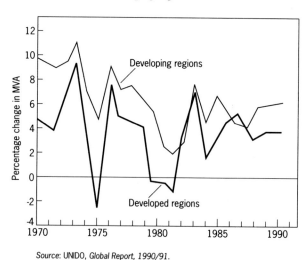

Source: UNIDO, *Global Report, 1990/91.*

industry in GDP remained relatively stable. The proportionate decline is mainly due to the rapid expansion of the service sector.

The European centrally-planned economies at first experienced rapid and sustained growth in industrial output, particularly in iron and steel, metal

Figure 3: *Average growth rates of MVA in the major regions, 1970s and 1980s.*

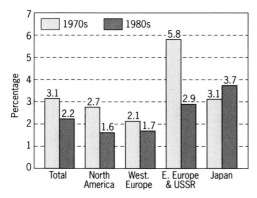

Developed Regions

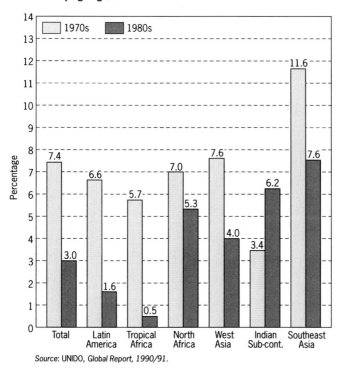

Developing Regions

Source: UNIDO, *Global Report, 1990/91.*

refining, chemicals and heavy machinery. Recent years, however, have witnessed an almost 50 per cent drop in their growth rates, from 4 per cent in 1988 to 2.1 per cent in 1989 (UNIDO, 1990). The whole industrial structure of these countries is now being redesigned.

Throughout the developing regions, rates of growth in MVA have tended to drop, with the largest decline occurring in tropical Africa (from 5.7 per cent in the 1970s, to a mere 0.5 per cent in the 1980s). But even this disaggregation does not show the sharp differences between countries in the same region. Despite the overall drop in growth rate in tropical Africa, Ghana, for example, had a 13.7 per cent growth rate in 1988, and 13.4 per cent in 1989. Although Southeast Asia has also witnessed a marked drop, the rate of growth of MVA was around 12 per cent in Indonesia, Malaysia and Thailand for 1988 and 1989. Moreover, as can be seen from Figure 4, the developing countries' share of total world production has varied considerably for different industries (UNIDO, 1990). On the basis of UNIDO forecasts for 1991, the developing countries will have:

- a 25 per cent share or more, of world production in tobacco manufacture, leather and fur products, textiles and petroleum refineries.

- a 20 per cent to 25 per cent share in beverages, non-metallic mineral products and footwear.

However, over the same period, the industries that grew fastest in the world as a whole were those in which the share of the developing countries was less than 10 per cent of world MVA (furniture and fixtures, printing and publishing, non-electrical machinery and professional and scientific goods). In fact, as Figure 4 shows, the developing countries' share in most of these industries has declined over the last fifteen years.

As expected, the developing countries have the largest shares in those industries where natural resources play the most important role (e.g. petroleum refining in the manufacturing sector, and mining in the primary sector), or where labour cost is most important (textiles, leather and fur products), (Table 1). In spite of remarkable structural changes and phenomenal growth in some countries, particularly in Southeast Asia and Latin America, the developing countries as a whole have yet to make significant encroachments into 'hi-tech', knowledge-intensive industries such as microelectronics, information technology and biotechnology.

During the period 1970-88, four of the five manufacturing sub-sectors known to be most energy-, material-, and pollution-intensive (iron and steel, non-ferrous metals, non-metallic minerals, chemicals, pulp and paper) have grown faster in developing countries (UNIDO, 1990). In fact, during the period 1980-85, they grew twice as fast in developing countries as in developed countries (UNIDO, 1989). The developing countries, as a whole, are expected to continue to increase their share of 'smoke stack' industries and consequently increase their levels of industrial pollution (UNIDO, 1990). This is the challenge that industry, governments and society at large have to face in developing countries if the benefits of industrialization are not to be jeopardized by serious environmental deterioration.

Figure 4: *Developing countries' share of world industrial production, 1975-91.*

Developing countries' percentage share of world industrial production, 1975-91

Legend:
- 1975
- 1991 (estimated)

Industry	1975	1991 (estimated)
Tobacco manufactures	32.1	36.8
Textiles	18.7	25.9
Leather and fur products	18.1	31.5
Petroleum refineries	30.5	45.4
Beverages	19.3	24.1
Footwear (excluding rubber and plastics)	18.8	24.5
Other non-metallic mineral products	13.9	21.2
Furniture and fixtures	10.2	8.4
Printing and publishing	8.8	6.2
Non-electrical machinery	5.0	4.4
Professional and scientific goods	2.6	4.3

Industries with 25% share or more | Industries with more than 20%, less than 25% share | Industries with less than 10% share

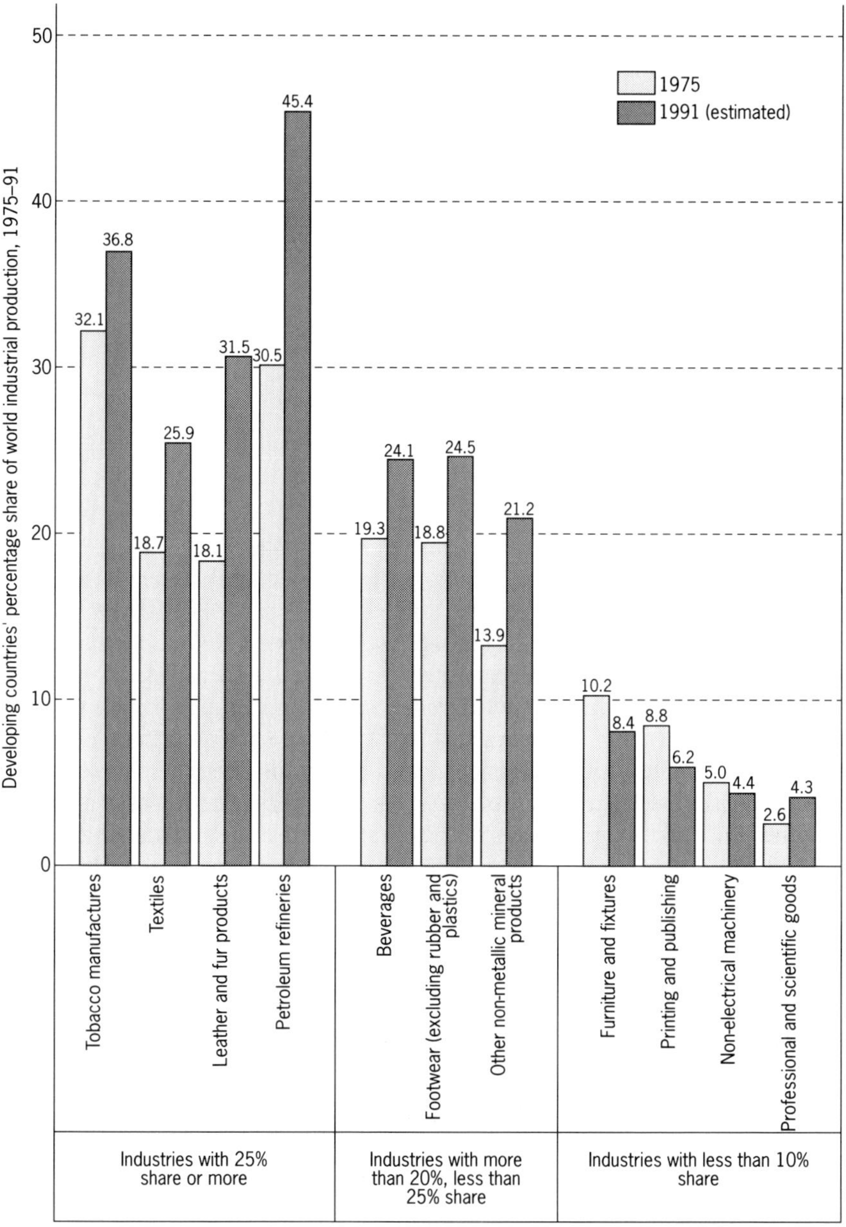

Source: UNIDO, *Global Report, 1990/91*.

Table 1: *Fast-growing industries in developing countries.*

Industry	Developing countries' share of world production (%)	
	1975	1991 (projected)
Industries where the share has doubled, or more:		
Petroleum and coal products	6.9	17.5
Iron and steel	9.9	19.5
Industries whose share has increased by 50% or more:		
Wearing apparel	11.9	18.0
Leather and fur products	18.1	31.5
Petroleum refineries	30.5	45.4
Rubber products	13.0	19.2
Other non metallic mineral products	13.9	21.2

Source: UNIDO statistics.

In the developed countries, the most significant change over the last two decades, and particularly in the 1980s, has been the shift from heavy industries to processing, 'hi-tech' and service industries. In the OECD countries, the growth of these sectors in less than ten years is more than double that of the traditional sectors. While iron and steel production, and petroleum refining, hardly increased over the last two decades, and the maximum growth in paper and pulp and chemical industries in any OECD country was to double output, electrical machinery output doubled in the USA and increased almost sevenfold in Japan (OECD, 1991a). The same tendency could be seen in Eastern Europe and the USSR. However, structural changes in these countries were hampered by difficulties associated with a deepening economic crisis, the change to market economies, and a marked drop in the rates of growth of GDP and MVA from those of just five years earlier.

The environmental impact of industry

The relative importance of the environmental impact of industry varies considerably with place and time. It depends not only on the type and scale of the industrial activity (Box 1), and the local climatic and physical conditions, but also on the type and extent of environmental impacts of other anthropogenic activities and social conditions.

Broadly speaking, we can distinguish between three different categories of environmental impacts: those brought about by the introduction of new 'hi-tech' industries in industrialized countries; those occurring in the NIEs (Newly

Industrializing Economies) and oil rich countries; and those affecting fragile ecosystems where poverty, low standards of living and overpopulation are rife. Although most of the new technologies are less material- and energy-intensive than the traditional ones, some (e.g. microelectronics) are heavy users of complex new toxic materials that produce pollutants such as heavy metals, toxic gases, water pollutants and hazardous wastes. Furthermore, the full environmental impacts of some of the new biotechnologies, and some of the new materials, are yet to be defined comprehensively. Most NIEs and oil-rich countries are suffering from environmental problems already known in the industrialized countries. In the least developed countries, appropriate industrialization could contribute to the alleviation of poverty and environmental degradation. Furthermore, the picture is likely to change rapidly as the structural changes outlined earlier gather momentum, in both the developed and developing regions.

The impact of industrial activities (Figure 5) extends over the entire chain of events, from raw material extraction or pre-processing, through the manufacturing processes themselves, right up to the disposal of wastes and discarded products. Industry is a major consumer of natural resources (mineral and non-mineral ores, agricultural produce and energy in different forms). Industrial processes involve the release of harmful gases, solid wastes and numerous other effluents, some of which are highly toxic, and this can occur either during the processes themselves or later during the use of or disposal of the products. Many industrial products themselves become inconvenient, or even hazardous, wastes when they are disposed of. Finally, some industrial processes are themselves hazardous and could result - and have been known to result - in serious accidents with harmful effects on the local environment, or even on an international scale.

The following is a concise review of the impacts of industrial processes on various constituents of the physical and human environment (Box 1).

Industrial consumption of non-renewable resources

Relentless industrial development places heavy demands on the world's non-renewable resources, particularly fossil fuels and mineral resources. Figure 6 is a graphical representation of the scale of energy consumption world-wide in 1985, and the share of each of the three main consuming sectors (Industry - Transport - Residential/Commercial/Agriculture). While industry's share of total consumption hovers around 33 per cent in the USA and EEC countries, it reaches 50 per cent or more in other industrialized countries (e.g. Japan), industrializing countries (e.g. Mexico, India), and the former USSR, Central and Southern Europe, and China. Table 2 shows that industry's share of total energy consumption has decreased from 39.3 per cent in 1970 to 33.7 per cent in 1987, with the maximum reduction achieved in Japan (from 61.5 per cent to 46.9 per cent).

As for the energy sources themselves (Figure 7), oil and natural gas are the

BOX 1

Environmental effects of selected industrial sectors

	Raw material used	Air	Water resources Quantity
Textiles	Wool, synthetic fibres, chemicals for treating	Particulates, odours, SO_2, HC	Process water
Leather	Hides, chemicals for treating and tanning	Odour	Process water
Iron and steel	Iron ore, limestone, recycled scrap	Major polluter: SO_2, particulates, NO_x, HC, CO, hydrogen sulphide, acid mists	Process water, scrubber effluent
Petrochemical refineries	Inorganic chemicals	Major polluter: SO_2, HC, NO_x, CO, particulates, odours	Cooling water, process water, scrubber effluent
Chemicals	Inorganic and organic chemicals	Major polluter: organic chemicals (benzene, toluene), odours, CFCs	
Non-ferrous metals (eg aluminium)	Bauxite	Major local polluter: fluoride, CO, SO_2, particulates	Scrubber effluents
Micro-electronics	Chemicals (eg solvents), acids	Toxic gases	
Bio-technologies			Process water

Source: Based on OECD, The State of the Environment (1991a).

Water resources Quality	Solid wastes and soil	Risks of accidents	Others: noise, workers' health and safety, consumer products
BOD, suspended solids, salts, toxic metals, sulphates	Sludges from effluent treatment		Noise from machines, inhalations of dust
BOD, suspended solids, sulphates, chromium	Chromium sludges		
BOD, suspended solids, oil, metals, acids, phenol, sulphides, sulphates, ammonia, cyanides, effluents from wetgas scrubbers	Slag, wastes from finishing operations, sludges from effluent treatment	Risk of explosions and fires	Accidents, exposure to toxic substances and dust, noise
BOD, COD, oil, phenols, chromium, effluent from gas scrubbers	Sludges from effluent treatment, spent catalysts, tars	Risk of explosions and fires	Risk of accidents, noise, visual impact
Organic chemicals, heavy metals, suspended solids, COD, cyanide	Major polluter: sludges from air and water pollution treatment, chemical process wastes	Risk of explosions, fires and spills	Exposure to toxic substances, potentially hazardous products
Gas scrubber effluents containing fluorine, solids and hydrocarbons	Sludges from effluent treatment, spent coatings from electrolysis cells (containing carbon and fluoride)		
Contaminations of soils and ground-water by toxic chemicals (e.g. chlorinated solvents). Accidental spillage of toxic material	Sludges		Risk of exposure to toxic substances due to spills, leaks
Used effluent treatment, modified organic species	Used for clean-up of contaminated land		Fears of hazards from the release of micro-organisms into the environment

Figure 5: *Environmental impacts of industrial production.*

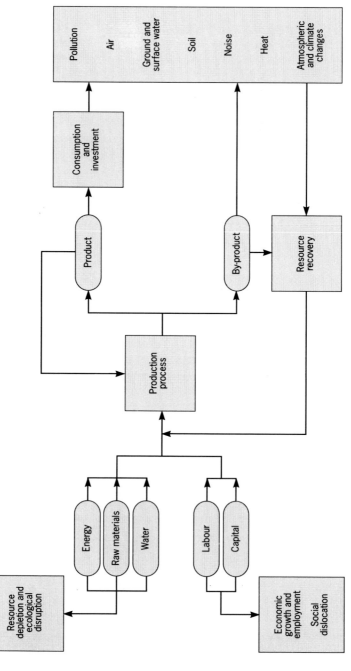

Source: Adapted from United Nations Environment Programme, *Industry and the Environment,* Environmental Brief No. 7, 1988.

Figure 6: *Energy consumption by end use, 1985.*

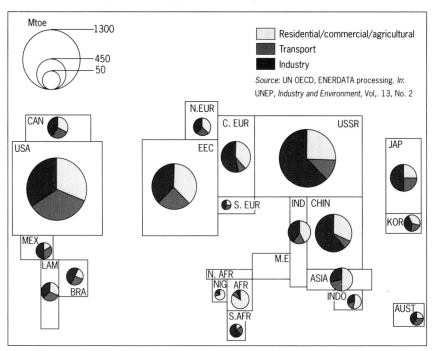

Table 2: *Energy consumption by end use, 1970 and 1987 (percentage).*

	Industry		Transport		Other	
	1970	1987	1970	1987	1970	1987
Total WEC countries	39.3	33.7	23.6	28.8	37.0	37.2
USA	36.8	31.1	29.0	35.4	34.3	33.4
UK	44.3	30.4	17.9	27.2	37.8	42.4
Japan	61.5	46.9	16.2	22.1	22.3	31.0

Source: World Energy Council, *Survey of Energy Resources* (1989)

two main sources world-wide, with some notable exceptions in Central Europe, India, China and South Africa, where coal is used extensively.

Apart from the depletion of these non-renewable energy resources, power generation plants themselves have adverse environmental impacts, as discussed in Chapter 13. In many developing countries, small-scale and cottage industries, responsible for a sizeable part of total industrial production, depend mainly on non-commercial energy sources that are notorious for their air polluting properties.

Similarly, industrialization has continuously increased the demand for non-renewable mineral resources, particularly iron ores, and consequently has also

Figure 7: *World primary energy consumption structure by source of energy, 1985.*

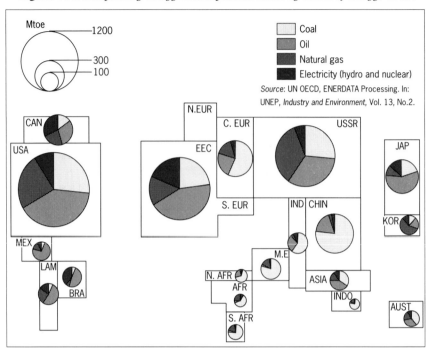

Table 3: *Replacement of natural materials by plastics in selected industries, 1985 and 2000 (projected).*

Process	Natural material	End product
Motor vehicle manufacturing	Steel	Selected vehicles Automobiles and vans Compact and light pickup trucks
Passenger aircraft manufacturing	Aluminium	Commercial aircraft
Building construction	Iron and steel	Selected iron and steel building materials Sewer lines Drain and waste vent pipes Conduits
	Aluminium	Selected aluminium building materials Sidings for buildings Window frames
Machinery and equip-ment manufacturing	Metal	Heavy machinery

Source: UNEP, Environmental Data Report 1990/91.

increased the drive to recycle wastes and recover such metals. Mineral resources are unevenly distributed world-wide, so that the highest producers are not necessarily the highest consumers (UNEP, 1991b; WRI, 1990). Production and utilization of most non-ferrous minerals have serious adverse environmental impacts, and reducing these impacts in a variety of ways has checked the rise in their consumption to a certain extent. With the exception of mercury, production of most other materials has increased, even though at decreasing rates (Figure 8). Some of the slowing down of the rates of increase in consumption of certain minerals is due to their replacement by other materials (e.g. plastics instead of aluminium, composite materials replacing metal alloys, or optical fibres in place of copper conductors). In fact, as Table 3 shows, plastics have replaced almost 10 per cent of natural materials in certain industrial sectors (OECD, 1991) and the percentage is expected to increase dramatically. This trend has had dire economic consequences in those developing countries whose economies have depended largely on exports of mineral ores to earn foreign currency (e.g. Zambia after the slump in the copper market).

The environmental problems associated with mining operations vary greatly with geographical and climatic conditions, as well as with the type of mining operation. In all cases, considerable care is needed to control waste waters from ore processing and rainwater runoff contaminated with toxic minerals. Discharge, particularly in sluggish rivers, can be disastrous. Non-ferrous metals in particular, and in spite of their relatively low-volume production, are the

1985		2000	
Quantity replaced (10³t)	% natural material replaced	Quantity replaced (10³t)	% natural material replaced
1,179.4	7 - 9	2,499.4 - 6,622.6	8 - 20
		208.7	20 - 60
0.5	3	3.6 - 10.0	10 - 13[a]
2,258.9	9	3,901.0 - 5,533.9[a]	
1,769.0		3,311.3 - 4,626.7[a]	
226.8		308.4 - 390.1[a]	
263.1		308.4 - 526.2[a]	
33.8		58.1[a]	
23.0			
10.9		21.8[a]	
			5[b]
635.0	5	752.8[b]	

[a] Forecast for 1995   [b] Forecast for 1990

Figure 8: *World production of some environmentally-important materials, 1965-86.*

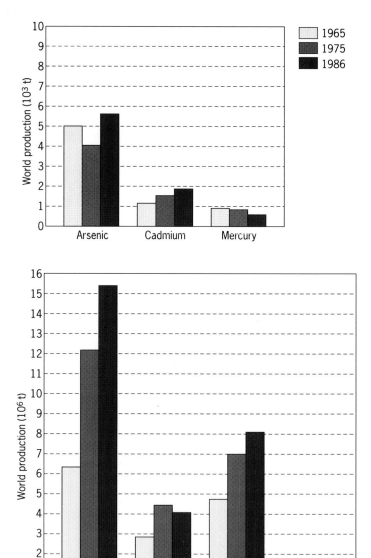

Source: UNEP, Environmental Data Report 1989/90

most dangerous during mining and refining. Apart from toxicity of the metals, especially in aquatic environments, the low grade of the ores requires that extremely large quantities of rock and overburden have to be removed and processed. Furthermore, potentially toxic chemicals (e.g. xanthates), as well as inorganic reagents (zinc and copper sulphates, sodium cyanide and sodium dichromate) are used in processing certain ores. The tailings left after metal extraction contain residues of both the minerals and the flotation agents. Some ores contain considerable quantities of sulphides which become transformed - through various oxidizing processes - into sulphuric acid. The slurries cause leaching of some mineral residues. Other problems associated with mining operations, besides physical change, are noise, dust and high radon levels. Worker health is a serious consideration in all mining operations. Consequently, the real environmental problem is not so much the rapid depletion of scarce non-renewal resources (with the exception of oil), as the degradation of air, water, and soil quality, biological diversity or human health and well-being. It is worth noting here that an appreciable amount of mining and ore processing is carried out in developing regions, where environmental monitoring and protection are at their weakest.

Air pollution

Global statistics on pollutants by source are scanty. One conservative estimate (UNIDO, 1990) is that industrial operations as a whole are responsible for about 20 per cent of the total air pollution. Furthermore, many manufacturing industries consume large amounts of electricity and, as such, should bear their share of the air pollution involved in electricity generation. This share obviously depends on industry's share of the total power generated and the source of energy. As we have already seen, industry accounts for 50 per cent or more of total energy consumption in the former USSR, Eastern Europe, China and India (Figure 6). Unfortunately, these are the countries where coal, a more polluting source of energy than petroleum or natural gas, is the main fuel used in power generation (Figure 7).

In western industrialized countries, the last two decades have seen a marked drop in the absolute quantities and the percentage share of almost all air pollutants produced by industry. Figure 9 shows the percentage share of industry in the USA of the total emission of air pollutants. While industry remains the main source of particulates and volatile organic compounds, it has managed to reduce emissions of other gases appreciably over time.

Information on industry's share of air pollution in developing countries is also scarce. As an example, a recent report on the state of the environment in Asia and the Pacific (UN ESCAP, 1990) gives the estimated share of industry in Thailand of the total emissions from combustion sources (Figure 10).

In Eastern Europe and China - and in spite of growing concern and remedial actions in these countries over the last few years - industrial air pollution has reached very high levels. For example, heavy industry is the second largest source of sulphur dioxide in Czechoslovakia (ERL, 1990). In Bulgaria, the 'Kremikovtsi' metallurgical plant outside Sofia is the biggest single polluter in the country. The Government has announced plans to close the plant, clean

Figure 9: *Percentage share of industry in nationwide air pollution in the USA, 1970-86.*

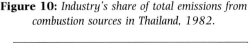

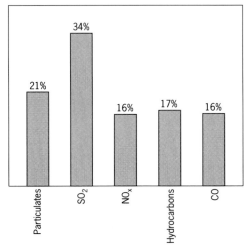

Source: Bureau of the Census, *Statistical Abstracts of the United States, 1989.*

Figure 10: *Industry's share of total emissions from combustion sources in Thailand, 1982.*

Source: ESCAP, *State of the Environment in Asia and the Pacific* (1990).

it up, and switch it to natural gas as a power source. In Poland, the 'Nova Huta' steel works outside Kracow have become notorious as a symbol of industrial pollution. The Benxi and Shenyang area in northeastern Liaoning Province has the most polluted air in China. It has long been a centre of heavy industry using coal-fired furnaces in smelters and for power generation, and in this region total suspended particulate concentrations and sulphur dioxide levels are among the highest in the world.

A recent World Bank study of market demand for commercial energy in eight developing countries (Brazil, China, India, Indonesia, Malaysia, Pakistan, Philippines and Thailand) (Imran and Barnes) forecasts that industry will remain the largest energy consumer and that by the year 2010 (Figure 11) will have increased its share to nearly 56 per cent. One study estimates that opportunities for energy conservation in the industrial sector in developing countries may be between 10 and 30 per cent. For example, energy consumption per unit product in the Indian iron and steel industry could be reduced by 12 per cent if capacity utilization were to increase by 10 per cent (Touche Ross, 1991). Subsidizing industrial energy to promote industrialization in developing regions has not encouraged energy conservation in those regions. A study in Thailand (Jasiewicz, 1990) has shown that good house-keeping alone could lead to a 12 per cent improvement in energy efficiency, while process improvement would lead to a further 16 per cent improvement. Another study in Egypt (El-Hinnawi, 1990) concludes that these measures could lead to 20 per cent reduction in industrial energy consumption. Some potential targets for reducing energy use in industry are in energy-intensive industries (steel, pulp and paper, petrochemicals) which account for more than 60 per cent of industrial energy consumption.

Industrial activities contribute to the environmental damage caused by acid deposition of sulphates and nitrates, a problem that has reached serious proportions in developed regions as well as in some developing regions (e.g. large parts of Eastern China). In addition, industry is the main source of highly toxic heavy metals (arsenic, cadmium, mercury and lead) emitted as particulates (see also Chapters 1 and10).

Figure 11: *Market demand for commercial energy in eight developing countries: Brazil, China, India, Indonesia, Malaysia, Pakistan, Philippines and Thailand, 1988 and 2010.*

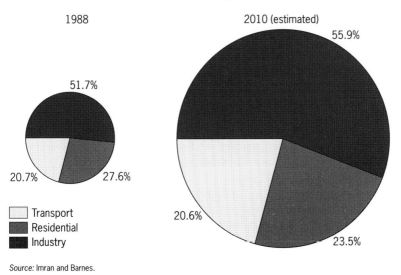

Source: Imran and Barnes.

Water pollution

Industry is both a consumer of water and a generator of waste water that is polluted in one way or another. Figures 12 and 13 show the relative share of industry in water withdrawals and discharges in different regions of the world. As expected, agriculture is responsible for the bulk of water withdrawals in all regions, except in Europe where industry's share is 60 per cent more than that of agriculture. Waste water generated by industry exceeds that from domestic sources almost everywhere, being considerably larger in highly-industrialized regions (North America, Europe and the former USSR).

Unlike domestic waste waters which contain more or less the same contaminants wherever they occur, industrial waste water varies considerably in the type of contaminants it contains. Broadly speaking, these contaminants are either floating, suspended or dissolved matter. Another source of water pollution, for which industry is responsible to some extent, is thermal pollution due to the large quantities of water used for cooling reaction vessels in a variety of industrial processes. However, electric power generation is the main source of thermal pollution world-wide.

Although it is quite difficult to estimate exactly the share of industry in overall water pollution, a surprisingly small number of industries account for the major share of pollution. As an example for developed regions, the USA EPA figures for 1987 (Table 4) show that only three industries (industrial chemicals, primary metals and pulp and paper products) account for more than 80 per cent of the total discharge to water. In developing countries - as might be expected - the contribution of individual industries is very varied, reflecting the structural differences discussed earlier. Table 5 is an example of the volume and characteristics of industrial effluents in an industrial region in Egypt. In this case, only 57 industrial plants out of more than one thousand constituted the main sources of water pollution (Hamza, 1983).

Stratospheric ozone layer depletion

Industrial operations contribute to the destruction of the stratospheric ozone layer. Such operations include refrigeration, foam production and the use of industrial solvents.

Refrigeration systems are common in industrial plants, particularly in the chemical, petrochemical and pharmaceutical industries; the oil and gas industries; and the metallurgical industries. Refrigeration, air conditioning and heat pump installations were estimated to represent approximately 25 per cent of global consumption of controlled CFCs (UNEP, 1989). The controlled refrigerants used in industrial refrigeration (some 3,500 tonnes per annum globally) account for approximately 25 per cent of global consumption in refrigeration. Their impact is considered negligible compared with domestic use.

The foam industry used approximately 267,000 tonnes of CFCs world-wide in 1986, representing 25-30 per cent of the total global use of controlled CFCs.

Figure 12: *Percentage annual water withdrawals for domestic, industrial and agricultural uses, by region, 1987.*

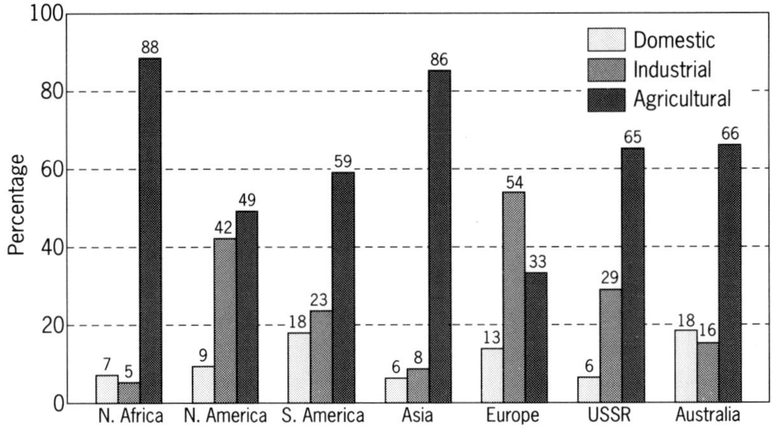

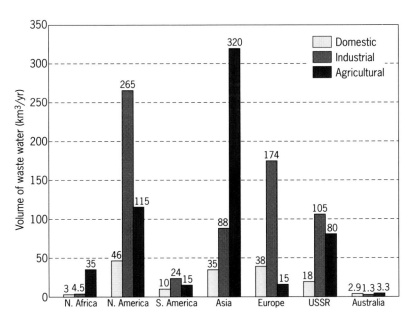

Source: Compiled by Kaltenbrunner from data in WRI, *World Resources 1990/91*.

Figure 13: *Annual volume of waste water generated by domestic, industrial and agricultural users, by region, 1987.*

Source: Compiled by Kaltenbrunner from data in WRI, *World Resources 1990/91*.

341

Table 4: *Amount of all categories of chemical releases, and percentage going to water, by type of industry for the ten industries ranked highest for releases, 1987-89.*

	Total release					
	1987		**1988**		**1989**	
	10^3 *lbs*	%	10^3 *lbs*	%	10^3 *lbs*	%
Industrial chemicals	**2,742,360**	**53.6**	**2,487,884**	**52.1**	**2,103,088**	**49.6**
Primary metals (iron, steel and non-ferrous)	**722,938**	**14.1**	**668,185**	**14.0**	**614,572**	**14.5**
Paper and paper products	**343,994**	**6.7**	**315,678**	**6.6**	**315,325**	**7.4**
Transport equipment	244,522	4.8	235,126	4.9	219,814	5.2
Rubber and plastic products	179,452	3.5	191,726	4.0	187,185	4.4
Metal products (fabricated metals)	143,787	2.8	148,108	3.1	143,136	3.4
Electrical machinery	132,767	2.6	127,466	2.7	104,887	2.5
Petroleum refining	130,795	2.6	119,714	2.5	116,561	2.7
Printing and publishing	63,538	1.2	64,601	1.4	57,254	1.3
Professional and scientific equipment	61,458	1.2	58,857	1.2	53,440	1.3
Total release (all industries)	**5,114,271**	**100**	**4,777,560**	**100**	**4,242,270**	**100**

Note: Transfers not included.

Source: Data released by the USA EPA, Office of Toxic Substances, May 1991.

The gases are mainly used as blowing agents in foam manufacturing because of their suitable boiling points, vapour pressures and low thermal conductivity.

The use of CFCs as solvents represents approximately 16 per cent of global consumption of controlled CFCs. Their use is mainly in the microelectronics industry. A large variety of non-HCFC alternatives exists, covering both product and process substitutes.

Solid wastes

The subject of solid wastes from various sources, and hazardous wastes in particular, is dealt with in Chapter 10. Data on industrial solid wastes world-wide is scanty and difficult to use as a basis for comparative studies due to variations in the definitions of wastes in different countries. Most available data are for OECD countries. Table 6 shows the contribution of municipal and

1987		1988		1989		
10^3 lbs	%	10^3 lbs	%	10^3 lbs	%	
291,735	71	231,252	75	110,432	58	**Industrial chemicals**
30,671	7.5	20,738	6.7	19,741	10.4	**Primary metals (iron, steel and non-ferrous)**
63,653	15.5	40,323	13.0	44,072	23	**Paper and paper products**
591	0.1	467	0.1	168	0.1	Transport equipment
407	0.1	636	0.2	702	0.4	Rubber and plastic products
3,107	0.8	1,759	0.6	510	0.3	Metal products (fabricated metals)
1,153	0.3	805	0.3	698	0.4	Electrical machinery
5,605	1.4	3,922	1.3	5,361	2.8	Petroleum refining
4	-	32	-	5	-	Printing and publishing
468	0.1	691	0.2	423	0.2	Professional and scientific equipment
411,414	100	309,506	100	188,994	100	**Total release (all industries)**

Discharge to water (header spanning 1987, 1988, 1989)

industrial sources of solid wastes in a number of selected countries. Generally speaking, in those countries where there are major mining operations, mining is by far the main source of solid wastes, while manufacturing industry wastes are less than mining or agricultural wastes and greater than municipal wastes.

Industrial solid waste is very heterogeneous and thus poses serious problems in disposal and management in general. With developments in the chemical industry and the proliferation of chemical products, hazardous wastes have come to represent an increasing proportion of industrial wastes. Table 7, covering more countries, highlights the quantities of industrial waste classified as 'hazardous and special'. This shows extremely wide variations which are obviously due to widely differing regulatory definitions of wastes. The percentage of wastes classified as hazardous ranges from 41 per cent in the USA and 33.5 per cent in Hungary to around 6 per cent in France and the Netherlands, 3 per cent in the Republic of Korea and the United Kingdom, and 0.3 per cent in Japan and Italy.

Table 5: *Estimated waste loads of pollution-contributing industries in Alexandria Metropolitan Area, Egypt, 1982.*

Industry	Number of plants	Discharge	Flow ML/d	BOD
			(Kg/day)	
Pulp and paper	2	S	93.0	83462
Paper conversion	3	L, S, D	5.0	3679
Textiles	13	L, S, D	37.0	19895
Dyes	1	S	4	983
Fertilizers	1	S	30	252
Steel	1	Se	13	520
Oil and soap	8	Se.C	32.5	30935
Tyres	1	Se	4.3	504
Refineries	2	S.L	230	12615
Chemicals (inorganic)	1	S	35	10850
Tanneries	6	S	1.6	2688
Power	2	C	324	7662
Matches	2	L	1.1	496
Electronics	1	D	0.5	138
Refractories	1	D	0.5	147
Plastics	1	D	2.5	788
Bottling	2	Se	1.9	484
Canning	2	D	4	3000
Dairy	1	D	0.8	1240
Yeast and starch	3	Se.L	3.2	2440
Brewery	1	Se	1.2	386
Poultry	1	D	0.5	429
Pharmaceuticals	1	Se	0.9	576
Total	57		828.0	184235

BOD = Biological Oxygen Demand	VR= Volatile Residues	S = Sea
COD = Chemical Oxygen Demand	P = Phenols	Se = Sewer
O&G = Oil and Gas	N = Aqueous Ammonia	C = Canals
SR = Suspended Residue	L = Lake	D = Drain

Source: UNEP, Industry and Environment, Special Issue No.4, 1983.

As mentioned earlier, these figures highlight the confusion caused by different definitions of wastes. The data available seem to indicate that industry is the main source of hazardous waste. In the United States, 85 per cent of hazardous waste is accounted for by manufacturing industry (Piaseck and Gravarder, 1985).

In the developing countries, the trends already noted towards expansion in chemical industries will pose an increasingly serious threat to the environment and human life unless adequate measures are taken to establish a minimum level of identification, monitoring and control of the toxicity of chemicals, products and wastes. Furthermore, as some developing countries, particularly

(Kg/day)					
COD	O&G	SR	VR	P	N
103356	1817	56069	80635	302	210
7379	1996	7543	7454	43	12.5
37877	3114	29949	41312	116	123
580	48	366	447	3.1	2.5
1392	276	558	1032		
1430	170	585	890	8.6	4.3
61943	9800	44685	51202	6.7	56
1260	286	940	1092	5.4	4.3
41875	10740	24370	44770	36.6	37.4
22035	3215	39050	35600	74.1	195
4109	405	13600	11424	43.1	24.3
12022	11248	15606	12987	135	128
862	98	1085	1452	8.2	28.6
269	59	320	356		2.1
297	171	806	716		
725	395	713	905	11.3	19.4
693	89	256	432	6.0	9.3
4264	177	1137	2258	5.4	3.1
3660	950	2982	6055	2.5	3.0
3360	106	1950	2130	8.6	5.1
184	41	160	192	1.6	0.6
583	51	681	693	2.5	3.1
936	39	108	475	4.5	0.7
311591	45330	243519	304509	886	1094

in Southeast Asia and Latin America, establish and expand electronics industries, new toxic chemicals and materials will be added to those already handled in traditional industries. The Department of Environment of Malaysia reported, as far back as 1984, that more than half the toxic and hazardous wastes generated in the country came from the nascent electronics industry - an industry that uses a huge variety of toxic chemicals and heavy metals in vapour deposition, etching and cleaning, and in the production of printed circuit boards and semiconductors.

This trend is a cause for concern about our ability to monitor and control the toxicity of these new substances and products in a systematic way. For many years, UNEP has been compiling a register of potentially-toxic chemicals (IRPTC) which now lists some 8,000 substances. Other UN organizations and regional bodies have also been active in this field (see Chapter 10 for more details). The US EPA lists some 500 substances as hazardous and the EEC a mere 30 items (OECD, 1991a). In 1987, the US EPA started a programme of reporting on 328 chemicals and chemical categories by manufacturing establishments

Table 6: *Amounts of municipal, industrial and nuclear waste generated in selected countries, late 1980s.*

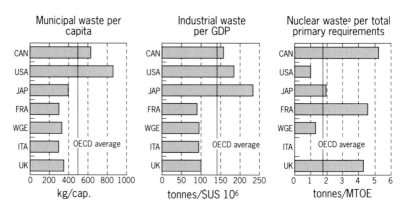

Municipal waste per capita — kg/cap.

Industrial waste per GDP — tonnes/$US 10⁶

Nuclear wasteª per total primary requirements — tonnes/MTOE

Source: OECD (1991b)

	Amounts of waste generated, late 1980s		
	Municipal waste		**Industrial waste**
	Total	per capita	Total
	(1000 tonnes)	(kg/cap.)	(1000 tonnes)
Canada	16400	632	61000
USA	208800	864	760000
Japan	48300	394	312300
Australia	10000	681	20000
New Zealand	2110	662	300
Austria	1730	228	13260
Belgium	3080	313	8000
Denmark	2400	469	2400
Finland	3000	608	12700
France	17000	304	50000
West Germany	20230	331	61400
Greece	3150	314	4300
Ireland	1100	311	1580
Italy	17300	301	43700
Netherlands	6900	467	6690
Norway	2000	475	2190
Portugal	2350	231	6620
Spain	12550	322	5110
Sweden	2650	317	4000
Switzerland	2850	427	-
UK	17700ᶜ	353ᶜ	50000
OECD	420000	513	1430000

Notes: a) Amounts of spent fuel expressed in tonnes of heavy metal.   b) Tonnes of heavy metal per unit of total primary energy
Source: OECD (1991b).

BOX 2

Chemicals

One particular aspect of modern industry that has had a profound impact on the issue of waste management is the proliferation of chemical products and inputs. In 1982, it was estimated that there were 60,000 chemical substances on the market, and that the production of synthetic materials had increased by a factor of 350 since 1940 (Holdgate, Kassas and White, 1982). This trend has continued, and in 1991 it was estimated that some 100,000 chemical substances were on the market, and that about 1,000 new substances were becoming available every year. Available testing facilities world-wide can only test 500 substances each year, and at great cost. The United States and Europe each produce over $US200 billion worth of chemical products. If present trends continue, in fifteen years' time, half the chemical products in use will have appeared *since* this report was written (OECD, 1991a).

Industrial waste		Nuclear waste	
per unit of GDP	of which Hazardous waste	Total[a]	per unit of energy[b]
(t/10⁶ $US)	(1000 tonnes)	(tonnes HM)	t/MTOE
155	3300	1300	5.2
186	275000	1900	1.0
235	-	770	1.9
146	300	-	-
15	60	-	-
211	200	-	-
104	920	122	2.7
41	90	-	-
221	270	77	2.6
89	3000	950	4.5
95	6000	360	1.3
123	423	-	-
87	20	-	-
94	3800	-	-
50	1500	15	0.2
35	200	-	-
292	170	-	-
27	1710	270	3.2
37	500	240	4.3
-	400	85	3.0
97	4500	900	4.3
146	303000	6990	1 7

requirements.   c) England and Wales only.

Table 7: *Waste generated in selected countries, 1980-87.*

		Annual industrial waste generation			
		Year of estimate	Total (10³t)	Per million $US of industrial GDP(t)	Per unit area (t per km²)
Africa	Guinea	x	x	x	x
	Nigeria	x	x	x	x
	South Africa	x	x	x	x
	Zimbabwe	x	x	x	x
Americas	Brazil	x	x	x	x
	Canada	1980	61,000	730	6.6
	Costa Rica	x	x	x	x
	Haiti	x	x	x	x
	Mexico	1986	192	4	0.1
	United States	1985	628,000	513	68.5
	Venezuela	x	x	x	x
Asia	Cyprus	1985	56	x	6.0
	Hong Kong	1987	6	1	6.1
	India	x	x	x	x
	Israel	x	x	x	x
	Japan	1985	312,000	573	828.6
	Korea, Rep	1981	7,030	274	71.2
	Lebanon	x	x	x	x
	Malaysia	x	x	x	x
	Singapore	x	x	x	x
Europe	Austria	1983	13,258	510	160.3
	Belgium	1980	8,000	186	243.8
	Bulgaria	x	x	x	x
	Czechoslovakia	1982	80,910	x	645.8
	Denmark	1985	1,317	95	31.1
	Finland	1985	15,000	841	49.2
	France	1984	50,000	301	90.9
	German Dem Rep	x	x	x	x
	German Fed Rep	1984	55,932	198	229.0
	Greece	1980	3,904	378	29.8
	Hungary	1985	21,146	2,509	229.0
	Iceland	1985	105	x	1.0
	Ireland	1984	1,580	346	22.9
	Italy	1980	35,000	207	119.0
	Luxembourg	1985	135	x	52.2
	Netherlands	1986	3,942	66	116.2
	Norway	1980	2,186	93	7.1
	Poland	1985	274,885	x	902.8
	Portugal	1980	11,200	1,110	121.8
	Romania	x	x	x	x
	Spain	1986	5,108	60	10.2
	Sweden	1980	4,000	102	9.7
	Switzerland	x	x	x	x
	United Kingdom	1984	50,000	327	207.0
	Yugoslavia	x	x	x	x
USSR		1985	306,311	x	13.8
Oceania	Australia	1980	20,000	386	2.6
	New Zealand	1982	300	38	1.1

Notes: **a** Refers to 1983   **c** Thousand cubic metres per year   **d** Cubic metres per year   **h** Refers to 1985   **j** Refers to 1984   **m** Refers to 1980   **n** Refers to 1986   **x** Not available   **t** Metric tons   **km²** Square kilometres

Annual hazardous and special waste generation

Year of estimate	Total (10^3t per year)	Per unit area (t per km^2)	Imports (10^3t)		Exports (10^3t)	
x	x	x	15.0		x	
x	x	x	4.0		x	
x	x	x	0.0		x	
x	x	x	6.9	a	x	
x	x	x	40.0		x	
1980	3,290	0.4	130.0		65.0	
x	x	x	x		x	
x	x	x	4.5		x	
x	x	x	7.0		x	
1985	265,000	28.9	45.3		203.4	
x	x	x	2.0		x	
x	x	x	x		x	
x	x	x	x		x	
1980	35,722	12.0	x		x	
x	30	1.5	x		x	
1986	666	1.8	3.0	a	x	
1981	180	1.8	x		x	
x	x	x	2.4		x	
1985	419 c	1.3 d	x		x	
x	x	x	x		x	
1983	200	2.4	0.3		3.4	
1980	915	27.9	914.1	h	13.2	
x	x	x	x		x	
x	x	x	x		x	
1985	125	3.0	x		20.0	a
1985	124	0.4	x		2.8	a
1984	2,000	3.6	95.9		25.0	h
x	x	x	814.3		x	
1985	5,000	20.5	75.0	h	1,695.6	h
x	x	x	x		x	
1984	7,081	76.7	1.5		x	
x	x	x	28.6		x	
1984	20	0.3	x		20.0	j
1980	2,000	6.8	x		22.8	
1985	4	1.5	x		4.0	a
1986	1,500	44.2	320.0	a	250.0	a
1980	120	0.4	x		0.3	
x	x	x	x		1.4	
1986	1,049	11.4	x		x	
x	x	x	4.0	a	x	
1987	1,708	3.4	x		2.6	
1980	500	1.2	x		15.0	m
1987	120	3.0	7.1		68.0	n
1986	3,900	16.1	82.5		x	
x	x	x	2.4		x	
x	x	x	x		x	
1980	300	0.0	x		0.7	a
1982	45	0.2	x		0.1	

Source: WRI, *World Resources 1990/91*.

BOX 3

The Toxics Release Inventory (TRI) of the United States

The Emergency Planning and Community Right-to-Know Act of 1986, usually referred to as Title III of the Superfund Amendments and Reauthorization Act (SARA) provides for the collection and public release of information on the presence and release of hazardous or toxic chemicals nationwide. It calls upon the EPA to establish a Toxics Release Inventory (TRI). Starting in 1987, data were compiled on the release of more than 200 chemicals and 20 chemical categories into the air, water or soils by manufacturing facilities with ten or more employees, manufacturing or processing more than 75,000 pounds of any of the reportable chemicals, or using more than 10,000 pounds of any of them. The threshold for chemicals manufactured or processed was later reduced twice, first to 50,000 pounds and later on to 25,000 pounds. Several chemicals have also been deleted from the list because they do not meet EPA's toxicity criteria to warrant further reporting. Nine new chemicals were added for 1990 reporting and seven CFCs and halons for 1991 reports. It should be noted that:

- The data do not cover other sources of releases.
- Releases are not exposures of the public to those releases.

- The Toxics Release Inventory Programme (TRI) (Box 3). The toxicity of these substances and the threats they pose vary within very wide limits. Some are lethal, some are highly toxic and some insignificantly toxic. Consequently, minute releases of the first are far more harmful than the discharge of fairly large quantities of the last. On the other hand, large releases of relatively non-toxic chemicals could be more serious than low volume releases of highly toxic chemicals under certain circumstances. Mention should be made here of the considerable advances achieved during the past two decades in defining the environmental toxicity of chemicals and in developing effective detection techniques, both of which are necessary for any system of pre-marketing testing and risk assessment (Chapter 10).

As might be expected, the chemical industry accounts for more than half the total releases. Together with the primary metals production and paper and paper products sectors, they account for almost three-quarters of all releases (Table 4). Figure 14 shows the percentage releases and transfers to various destinations in the USA.

The environmental impacts of new technologies

Within the scope of manufacturing industry, the more prominent new technologies are microelectronics, composite materials and biotechnology. The widespread application of computer-based technologies has produced substantial environmental benefits in industrial practices (e.g. computer-aided design and

Figure 14: *Total release and transfer of TRI (Toxics Release Inventory) chemicals in the USA, 1987-89.*

	1987	1988	1989
Totals	7.015 billion lbs	6.436 billion lbs	5.740 billion lbs
Air	37.7%	40.4%	42.4%
Water	5.9%	4.8%	3.3%
Land	10.5%	8.3%	7.8%
Underground injection	18.8%	20.7%	20.6%
POTW	8.9%	8.9%	9.7%
Off-site	18.3%	16.9%	16.3%

POTW = Public-Owned Treatment Works

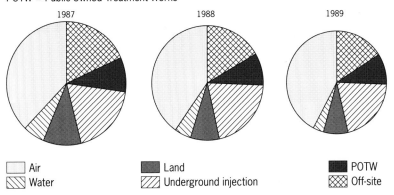

| Air | Land | POTW |
| Water | Underground injection | Off-site |

Source: Data released by US EPA office of Toxic Substances, May 1991

manufacturing (CAD/CAM), while improved maintenance practices and process design have resulted in savings in material and energy inputs and reduced wastes. On the debit side, the manufacture of microelectronics components involves the use of highly toxic substances. To this may be added the fact that - unlike earlier predictions of a 'paperless society' - the consumption of paper has increased markedly with the introduction of computers in the work-place. The composite materials, although offering considerable advantages in weight-saving and strength, are inconvenient to handle as waste. They are heterogeneous in composition and not degradable. Perhaps the greatest threats come from modern biotechnology based on recombinant DNA techniques which are finding applications in many sectors other than industrial production. As far as industry is concerned, the applications are basically in processes using micro-organisms to produce a product (such as feedstock, food, pharmaceuticals or chemicals) in a reactor. As such, the situation is basically different from other less contained applications or uses of recombinant DNA techniques (e.g. in agriculture). The technology poses no direct threat to the environment other than that to the work force involved in the production and handling of products, the risks of accidents (minor or major) or those inherent in the product itself. With proper technology assessment, and the in-depth consideration of environmental impacts, the adverse effects of introducing new technologies would be greatly reduced or avoided altogether (UNEP, 1991c).

Industrial accidents

The subject of the environmental impacts of some industrial accidents is dealt with in Chapter 9.

Social impacts of industrialization

It is generally recognized that industrialization is a major factor in economic development and in improving living conditions and standards. Industry offers employment opportunities, and produces a wide range of intermediate goods and products necessary for other economic sectors and for consumers at large. Furthermore, it is generally associated in developing countries with improvements in health standards, mortality rates and health conditions, at least at the early stages of industrialization.

On the other hand, industrialization affects the environment in three ways: by the release of emissions, effluents and wastes that pollute; in the proliferation of products that replace environmentally-sound traditional products; and by the destruction and transformation of the natural resource base to obtain raw materials. This last aspect of interaction creates serious social tensions in many countries (OECD, 1985b). The manner in which the natural resource base has been exploited in the least developed countries could - in some cases - intensify poverty and unemployment, by increasing the rates of deforestation, desertification, and soil erosion and thus reducing productivity. The reason is that the majority of people in these countries live in a biomass-based subsistence economy. Their fundamental needs are for food, fuel and shelter; and most of the raw materials used in their crafts are derived from one biomass source or another. Industry depletes the stock of biomass through its increasing demand for the biomass resources themselves. Nearly half the industrial output in developing countries is biomass-based (e.g. sugar, rubber, jute, textiles and processed food).

Furthermore, in developing countries industrialization has largely gone hand in hand with urbanization, since the concentration of population and economic activity in metropolitan areas usually stimulates faster economic growth. This trend is self-sustaining for a few decades because the concentration of economic activity and population is a powerful attraction for new investments and more people. However, in the long run this results in massive internal migrations from rural areas to the fast-expanding urban centres - a movement that such centres are ill-prepared to cope with in terms of providing the minimum infrastructural facilities needed, or in generating employment opportunities. The final result is a slowing down of economic development, increased social tensions, health problems and degradation of the urban environment. Apart from the pollution caused by the slum areas that mushroom around the metropolitan centres, the concentration of industry results in high concentrations of harmful emissions and toxic wastes. This is further aggravated by the spread of highly polluting cottage and small-scale industries within the densely-populated areas.

Government policies play a crucial role in the speed with which industrialization fuels urbanization. The availability of energy, transport and communication facilities, and public services, as well as the scale of bureaucratic intervention in the conduct of industrial operations, are all important factors in the siting of industry. It is only in the case of resource-based industries such as mining, quarrying or agro-industrial complexes that industrialization is not closely linked with urbanization. Another major influence has been that of the transnational corporations (TNCs) and their affiliates in the developing countries. Without belittling the political and economic factors, both on the side of the government and the TNC, industrial plants set up by TNCs have been, more often than not, enclaves rather isolated from their surroundings.

Although the last two decades have witnessed several examples of attempts at luring industry away from urban centres in developing countries, the main problem has been the high cost involved in providing the necessary infrastructural facilities needed for industrial operations. In Southeast Asia, where industry has been growing rapidly for the last two decades, Greater Bangkok is home for one-quarter of all the factories registered with the Industrial Works Department. In Indonesia, East Java, with an area of 48,000 square kilometres and a population of more than 30 million, had 425,036 registered industrial firms in 1986, of which 418,078 were small-scale industries (UN ESCAP, 1990). However, as industrialization takes root and industrial production becomes more specialized, industrial facilities become more widespread geographically. This will be an important factor in all future attempts at reducing industrial pollution in developing countries. While large-scale and some medium-scale industries could undertake pollution reduction measures, small-scale industries in developing countries, and even in some developed countries, generally have neither the knowledge base nor the funds to undertake such measures.

Working environment and human health

The impacts of industrial development on human health and well-being are dealt with in Chapters 10 and 18. Suffice it to state here that with the decline of the older industrial diseases of the last century, the development of new technologies and new materials, and advances in medical science and practice, new diseases that were unknown a century ago have now emerged.

During this century, the focus has widened from the identification of industrial diseases to the broader concept of occupational health, defined by ILO and WHO in 1950 as 'the promotion and maintenance of the highest degree of physical, mental and social well-being of workers in all occupations'. Adverse health effects result from exposure to heat, pressure, electromagnetic radiation, noise, vibration and particularly chemicals. While diseases can be related to several causes, distinction needs to be made between occupational diseases and those relating to the work environment itself. The latter can be an important factor in causing or aggravating some common diseases such as hypertension, locomotor diseases and stress-related health problems.

In 1976, the 'International Programme for the Improvement of Working Conditions and Environment' (PIACT) was launched. In 1982 a review of the results of PIACT in its first five years was carried out and the results were discussed at the ILO Conference of 1984. The last decade has also witnessed closer co-operation between UNEP, ILO and WHO in the 'International Programme on Chemical Safety' (IPCS) (Chapter 10). The first Thematic Joint Programming Meeting on the Working Environment convened by UNEP was held in 1979.

Responses to environmental impacts of industrial development

Broadly speaking, responses to the concerns raised by the harmful impacts of industrial development have been at a number of different levels, technological, regulatory, economic and social, and these have interacted vigorously over the last two decades.

Public pressure has prompted governments to intervene in different ways to curb the adverse impacts of industrialization on the environment. Government regulations, and sometimes public pressure, have moved industrial enterprises to pay increasing attention to environmental concerns. Scientific research, in its turn, has clarified or revealed the causes and extent of environmental degradation caused by industry. Technological developments have pointed the way to means of eliminating or reducing undesirable impacts. Funding for such research and technological developments has been provided by both governments and enterprises.

These complex interactions have had their successes and failures. The result has generally been a reduction, particularly in developed countries, of the discharge of pollutants, savings in energy and material consumption, and the identification of new concerns hitherto unsuspected or, at best, ambiguous. All this stems from a growing understanding of the multi-faceted manner in which industrial development interacts with the environment. Furthermore, there is now a general appreciation that most technologies go through a life cycle during which unsuspected adverse impacts on the environment gradually become clearer, and the need for more environmentally-sound technological alternatives becomes a necessity. However, by the time the dangers are scientifically proven beyond reasonable doubt, and alternatives are developed, the social and economic cost of changing to new processes or new products is substantial, if not downright prohibitive.

Chapter 20 deals with the development of tools in the natural and social sciences for understanding, monitoring and managing the complex processes shaping the environment, and the impacts on it of human activities. Chapters 22 and 23 trace the evolution of environmental policies and institutions at the national, regional and international levels, as well as the instruments of social control introduced to protect the environment and harmonize development activities with environmental considerations. Here, we restrict ourselves to the specific actions in the industrial sector and the extent of their success, or otherwise, in maintaining a healthy environment.

Technological developments

As the full impacts of industrial activities on the environment have gradually revealed themselves, technological responses have been moving steadily from a rather fragmented approach tackling particular issues (emissions, discharges, resource conservation) to a more holistic approach that encompasses the whole sequence of industrial operations, as well as the total life cycle of industrial products (the 'cradle to grave' concept).

Dealing with discharges

Dilution has long been the usual method of dealing with polluting discharges: high chimneys to disperse smoke and emissions; long outfalls to secure dispersal of effluent discharges in large water bodies (e.g. lakes, the open sea), and preferential discharge to fast-flowing rivers and running streams. During the seventies and early eighties - as the extent of harm caused by such practices began to show up - emphasis on minimizing the discharges of harmful emissions, effluents and wastes has mainly been by the so-called 'end-of-pipe' approach. Here, the by-products are treated so as to extract or neutralize their harmful ingredients before discharge into the environment, rather than modifying the production processes or products themselves to reduce the production of pollutants. Such solutions are essentially 'add-ons' to the production process. This has gone on side by side with the much older practice of collecting discarded materials and products and feeding them back, in one form or another, into the production process.

Apart from its economic benefits, recycling also has environmental benefits: savings in virgin raw materials, reduced energy and water consumption, and a reduction in air and water pollution and mining wastes. Table 8 gives estimates of the substantial benefits resulting from recycling (UNEP, 1991b).

Table 8: *Potential savings in recycling.*

	Potential saving (%)			
	Aluminium	Steel	Paper	Glass
Energy used	90-97	47-74	23-74	4-32
Water used	-	40	58	50
Air pollution emissions	95	85	74	20
Water pollution discharges	97	76	35	-
Mining wastes	-	97	-	80

Source: UNEP, *Environmental Data Report*, 3rd Edition, 1991.

During the past two decades, many countries have shown a significant increase in their waste paper recovery rates. The amount of aluminium recycled, as a percentage of consumption, has also increased world-wide (Figure 15). Steel recycling occurs partially 'in-house' as process scrap is recycled. A good deal of scrapped steel products (cars, consumer durables etc.) are reclaimed, sorted out and recycled. Figure 16 shows the trend in steel recycling over the period 1976-88 and indicates that the percentage of recycled steel has not changed much in most countries.

In 1985, the European Community issued Directive 85/339, promoting the recycling of beverage containers and the development of new containers that would reduce raw material and energy consumption. A second directive, currently under discussion, deals with all kinds of packages. In April 1991, Germany ratified a new packaging law based on the principle that those who use packaging (producers and distributors) should be responsible for taking it back and recycling certain percentages of it. Implementation began in December 1991, with full implementation scheduled for January 1993. The Netherlands has adopted an alternative approach to regulation based on co-operation between business and government, and a packaging covenant was signed on 6 June 1991 with the goal that no packaging is to go to landfills by the year 2000. Other EC, North American and OECD countries, and even some developing countries are following suit. Such actions have prompted significant changes in the sizes, designs and materials of packaging used for a variety of goods and consumer products (Lund University/UNEP, 1991; OECD Waste Management Policy Group, 1991; Fishbein, 1991).

The percentage of plastics in solid wastes has been increasing steadily, and solid wastes containing non-biodegradable plastics often cause damage to, and difficulties in operating, municipal solid waste incinerators. The plastics content in solid waste is expected to reach 10 per cent in the USA by the year 2000, thus making it a candidate for recycling. However, plastic wastes are highly heterogeneous and difficult to identify for sorting purposes. Some have been successfully recycled and attempts continue to recycle others. More recently, some car manufacturers using large quantities of plastics in their products have adopted the practice of labelling the plastic parts, thus facilitating the process

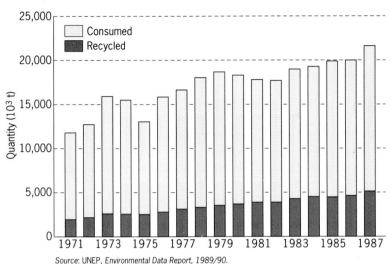

Figure 15: *Trends in world consumption and recycling of aluminium.*

Source: UNEP, *Environmental Data Report, 1989/90.*

Figure 16: *Trends in steel recycling in relation to consumption in selected countries.*

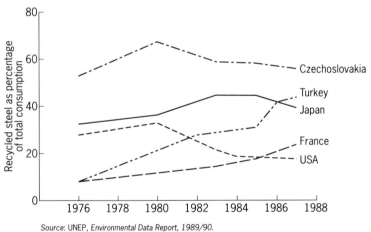

Source: UNEP, *Environmental Data Report, 1989/90.*

of sorting them for recycling. BMW claims that 80 per cent of the 20,000 parts in its cars are recyclable, and that the company aims to produce a car that is 100 per cent recyclable. Volkswagen claim that they can strip down a car in 20 minutes, and that they will take back and recycle their latest Golf model free of charge.

From treatment of discharges to cleaner production

Attempts at pollution control and prevention have continued to intensify in scope and depth throughout the last two decades. The major industrialized countries, east and west, were reported in the seventies to be spending billions of dollars on the control of industrial pollution (Holdgate *et al.*, 1982). By the early eighties there were several noteworthy examples of considerable reductions in polluting discharges, combined with savings in the consumption of raw materials and energy, particularly in the chemical, petroleum, pulp and paper, iron and steel and other metal industries.

During the latter half of the seventies, the concept of 'low and non-waste technologies' (LNWT) emerged, and one of the first meetings on the subject was organized by the EEC in 1976. The emphasis then was on the development and promotion of technologies that produce less harmful discharges, or none at all. Compendia of such technologies were being published as early as 1978.

Data on the cost of pollution control in industry are limited and cover mainly the industrialized countries. In Europe, for example, it is reported (UNIDO, 1990) that the share of investment in industrial pollution control in the Federal Republic of Germany between 1971 and 1977 was more than 5 per cent of total industrial investment. According to the Ministry of International Trade and Industry, in Japan the ratio was as high as 10.6 per cent in 1973, rising to 17.1 per cent in 1975, under the influence of newly promulgated and stringent legislation, before dropping to 5 per cent in the early eighties (OECD, 1985a). In the USA, the ratio was 5.8 per cent in 1975. Figure 17 gives some information on private (industrial) pollution control expenditure in some industrialized countries over the period 1972-86. This expenditure has varied between 0.8 and 1.7 per cent of GNP (OECD, 1990). On average, countries with the most stringent environmental programmes spend about 1.5 per cent of their GNP. Data from the USA and Germany indicate that the manufacturing sector accounts for about 25 per cent of the total expenditure, or about 0.4 per cent of GNP. UNIDO expects that the new approach, based on changing processes and products rather than abatement, and emphasizing ambient rather than discharge standards, would result in reasonable spending on pollutant reductions in developing countries (UNIDO, 1991). A small number of industrial sectors account for a large proportion of expenditure. These are chemicals, metal products and machinery. In Sweden and Finland, paper and printing stand out, as do wood products and non-ferrous metals in Norway, and petrochemicals in Japan (OECD, 1990). In general, the operating costs of pollution abatement are much higher than the capital costs (almost five times as much in the USA) (Figure 18).

Information on the split of this expenditure between the 'end-of-pipe' approach and recourse to less polluting technologies is difficult to come by. Unfortunately, available data do not cover many parts of the world. The most comprehensive data available are from the USA. It was reported (OECD, 1985a) that over the period 1973-80 the USA invested some 20 per cent of the total expenditure on air and water pollution abatement on changes in production processes. A detailed breakdown of expenditure (Bureau of the Census, 1987) shows that the fabricated metal and transport equipment industries top the list

Figure 17: *Private investment expenditure on pollution control, at 1980 prices[1] and exchange rates.*

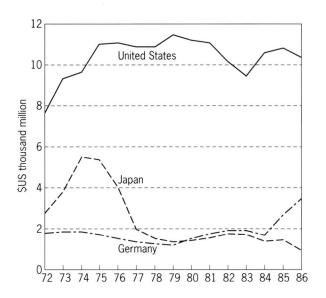

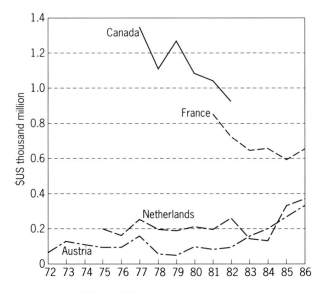

Notes: 1. Deflated by GDP price index.
Private household expenditures excluded.
Source: OECD

Figure 18: *Capital expenditure and operating costs of pollution abatement measures, by source of pollution in the USA, 1985 and 1988.*

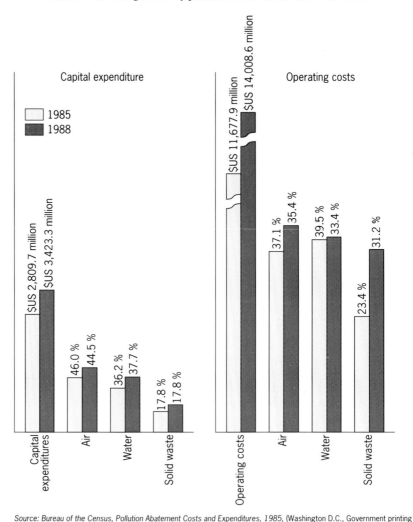

Source: Bureau of the Census, Pollution Abatement Costs and Expenditures, 1985, (Washington D.C., Government printing office) 1987 (for 1985 figures) & 1990 (for 1988 figures).

of percentage expenditure on new processes. In France, the penetration of 'clean' technologies was considered to be small, while in Denmark about one-third of the firms adopted new production processes for pollution abatement between 1975 and 1980 (OECD, 1985a). Table 9 provides some interesting comparisons of the cost of pollution control and that of pollution damage for three cases in Japan.

As more and more attention was focused on the development of 'clean' technologies - and as their economic and risk reduction benefits were demonstrated - the concept of 'cleaner production' finally emerged. The

Table 9: *Comparative data on pollution control cost and damage cost.*

Pollution damage case	Annualized pollution control cost	Annualized damage cost	Note
Yokkaichi	9,347 million Yen	1,322 million Yen (49,648 million Yen if no early action)	Control measures have been taken at relatively early stage. 1,231 patients.
Minamata (Kumamoto)	94 Million Yen	1,196 Million Yen	2,248 certified patients ca, 2000 applications. 31 March 1991. Dredging work.
Itaiitai Disease	54 million Yen	251 million Yen	129 certified patients. 1,500 hectares. Designated farmland for reclamation.

Source: Economic Study Group on Global Environment, July 1991.

comparative degree is now preferred since there is no clear-cut distinction between 'clean' and 'dirty' production methods or products. This concept now encompasses the whole life cycle of a product (from 'cradle to grave'), covering product design, production process, and management practices right up to the disposal of the discarded product (Box 4). The concept could be articulated through a hierarchy of substitutions at different levels that contribute to achieving the goal (Huisingh, 1989):

- The process level: modifying the process to be more environmentally sound (e.g. producing less pollution or waste).

- The component level: adding a new component providing an advantage without changing the overall process (e.g. 'end-of-pipe' treatment plants).

- The sub-system level: substituting a better sub-system for an old one (e.g. an electric motor in place of a petrol engine in a car).

- The system level: substituting a whole system or one function for another (e.g. mass transport system for private cars).

- The value level: questioning the very premises on which products or services are provided. This is the most difficult since it impacts on culture, social organization and value systems.

This hierarchy of substitutions emphasizes the much broader way in which the problem is now viewed, and the new system approach was recently presented in the methodological framework of 'industrial metabolism'. This is based on the analogy between biological metabolism, in which biological systems take in nutrients, metabolize them to fulfil certain vital functions and discharge the by-products as waste, and the manner in which production, consumption and

---

BOX 4

Cleaner production

Cleaner production means the continuous application of an integrated *preventive* environmental strategy to processes and products to reduce risks to humans and the environment.

For *production processes* cleaner production includes conserving raw materials and energy, eliminating toxic raw materials, and reducing the quantity and toxicity of all emissions and wastes before they leave a process.

For *products* the strategy focuses on reducing impacts along the entire life cycle of the product, from raw material extraction to ultimate disposal of the product.

The goal of *cleaner production* is not to generate waste in the first place. Cleaner production is achieved by applying know-how, improving technology, and/or changing attitudes.

Source: UNEP IE/PAC

---

waste occur in human societies (Ayres, 1989). The concept is useful in dealing with all material flows within an interconnected industrial system that forms an integral part of a larger environmental system. Practical application of this concept, however, requires a detailed materials balance of all stocks, flows and losses to the environment of selected environmentally relevant materials - a condition that is not satisfied completely by the currently available data on industrial products life cycles, particularly data on recycling and disposal emissions.

Industrial systems, however, are not as efficient at recycling waste as high-level biological systems in the biosphere. An alternative analogy (Frosch and Gallopoulos, 1991) is that of an 'ecosystem' in which industry is one part of a larger whole, involving producers and consumers. This is analogous to a community of living organisms and their environment. This analogy emphasizes the importance of minimizing the inputs of energy and scarce materials, and of utilizing wastes of all sorts as inputs to a whole spectrum of processes, in a manner similar to the recycling of nutrients by various organisms in the food chain in a biological system.

On the practical level it is claimed that a review of the results of applying the cleaner production approach in 500 case studies of industrial firms has shown 70 to 100 per cent reductions in some air emissions, water emissions and/or the production of hazardous and non-hazardous solid wastes, and that the payback period was quite short, ranging from three years to less than one year.

The last few years have seen an increasing number of programmes in many industrial countries, east and west, based on the cleaner production approach outlined above. One such recent project in Europe is PRISMA (Project on Industrial Successes with Waste Prevention) initiated in 1985 by the Netherlands

Organization of Technology Assessment. Preliminary results of field research started in ten companies in 1988 are now available. No less than 164 'prevention options' were identified, and these included minor changes in procedures and installations as well as drastic innovations in products or processes. In 30 per cent of the cases these changes were in the category of good house-keeping; another 30 per cent in material and raw materials; 30 per cent in changes in equipment; and the rest in process modifications (Sybren de Hoo et al.,1991). Some good house-keeping measures effected 25 to 30 per cent reductions in chemicals used. In a number of companies, technological changes resulted in reductions in waste and emissions of 30 to 80 per cent. Occasionally, it was possible to eliminate a noxious waste flow completely. The use of alternative raw materials resulted in a 100 per cent reduction in emission of substances such as cyanide (in zinc plating) and solvents (in degreasing parts).

The cost-benefit analysis showed a heterogeneous picture with a few extremely favourable peaks. In one extreme case an investment of Dfl2500 saved the company Dfl240,000 per annum! Another company recouped an investment of Dfl800,000 within a year (saving more than Dfl one million per annum). Of the 45 options actually implemented so far, 20 turned out to be cost saving, and 19 to be neutral. It was found, moreover, that these measures had further indirect benefits, such as improvement in product quality.

Within EUREKA (the co-operative research programme in Europe), the EUROENVIRON programme began promoting PREPARE (Preventive Environmental Protection Approaches in Europe) in 1990 on the basis of the Dutch PRISMA project. PREPARE seeks to catalyse industry/government co-operation and the transfer of information on pollution prevention and cleaner production.

In the United States, 'prevention teams' have been in operation for some years as part of the 'Technical Assistance Programme', helping small businesses in transferring pollution prevention technologies. These are groups of external consultants who assist industrial firms in formulating and implementing a pollution prevention programme. Evaluation of their work indicates that these programmes are not only commercially profitable and environmentally beneficial, but also useful at the macroeconomic level, as well as helpful in clarifying gaps in current knowledge and identifying areas for research.

Mexico and Brazil have achieved some noteworthy successes in organizing efforts for pollution control at various levels of government. The programme in Cubatao, home of a large industrial complex in the state of Sao Paulo in Brazil, is one important example of a unique institutional system. Enforcement powers were delegated by contract to the Environment Management Company (CETESB), and an integrated control approach was started in 1984. Some of the results achieved by the end of 1989 in air pollution abatement are summarized in Table 10. Similar success was also achieved in reducing water pollutants.

In China, Kwantung Provincial Government has achieved good progress in controlling pollution. Similar examples can be cited in Singapore, Malaysia and Thailand (UN ESCAP, 1990). Data in UNEP's International Cleaner Production

Table 10: *Reduction in atmospheric pollution in Cubatao, Brazil.*

Pollutant	Initial emissions in July 1984 (kg/day)	Emissions in December 1989 (kg/day)	Actual reduction(%)
Particulates	316,350	77,949	75
Fluorides	2,620	400	92
NH_3	8,736	205	97
NO_x	61,085	47,561	24
HC	90,000	11,970	86
SO_2	78,353	49,527	37

Source: The Action of CETESB in Cubatao, January 1990.

Information Clearing House (ICPIC) include encouraging results in several textile mills in India. Payback periods ranging between a few months and a few years have been reported from some countries (e.g. five months in a meat factory in Poland from reduced water consumption, and one year in the case of heat recovery).

A study of the obstacles that stood in the way of implementing maximum waste reduction activities in the USA in the early eighties (Gardner and Huisingh, 1987), found that 60 per cent were political, 30 per cent financial, and only 10 per cent technical.

The last decade has witnessed a proliferation of data bases on pollution abatement, clean technologies, and cleaner production in general, at the national, regional and international levels. At the national level, such data bases and information systems exist in many industrialized countries. In India, the Centre for Environmental Science and Engineering (CESE) of the Indian Institute of Technology (IIT) in Bombay decided in 1988 to build a data base of 'low and non-waste' technologies in the textile, distillery and sugar industries. Most technologies involve recycling and reuse; but out of 100 items approximately one-third are case studies of genuinely cleaner technologies. The Centre has already established effective working relations with local industry.

The European Community has set up a Network for Environmental Technology Transfer (NETT) as part of its 'Fourth Environmental Action Programme' and the 'European Year of the Environment' (1987-88). UNEP has recently established, and is currently operating, the 'International Cleaner Production Clearing House' (ICPIC), based on the US 'Pollution Prevention Clearing House'. ICPIC is a computer-based information exchange system, which now has over 400 technology and programme case studies, a directory of experts, a bibliography of hundreds of basic reference publications and descriptions of low- and non-waste technologies.

The regulatory level

Regulating industrial impacts on the environment can be traced back to the last century. The main advantage of direct regulation is that the machinery of government is used to it, and that - if properly implemented - it provides a reliable picture of its impact on the environment. On the other hand it is criticized for being inflexible and bureaucratic to a degree that can be counter-productive (Bohm and Russell, 1985). The scale and extent of enforcement of control measures has been increasing steadily under the influence of public pressure, better understanding of the manner in which industrial impacts affect human life and well-being, technological developments, and - lately - economic benefits. Furthermore, the drive to regulate and control the adverse environmental impacts of industry has been intensified by new threats brought about by the proliferation of highly dangerous synthetic chemicals in the last few decades, and the increasingly drastic damage caused by industrial accidents involving these substances. In direct contrast to the progress made in developed countries, the experience of many developing countries with direct regulation in environmental matters is far from satisfactory, mainly because of the adoption of imported standards that are almost impossible to meet, and the inability of government machinery to monitor and enforce the regulations.

The subject of policies, institutions and policy instruments is dealt with in Chapter 22, while developments in clarifying the dangers inherent in industrial operations are discussed in Chapter 20. We restrict ourselves here to tracing in general terms the development of regulatory and control mechanisms directly related to the industrial sector.

As might be expected, the industrialized countries have led the way in the scope and strictness of regulation and control instruments and measures. This has necessitated the elaboration of the standards to be applied. Air pollution was the first concern, whether in the USA, Europe or Japan. Clean air bills/acts, concerned mainly with particulates and sulphur oxides, were already in place in the early sixties and the limits set for emissions became progressively more stringent and extended in scope to cover more types of emissions. Japan provides a striking example of the environmental and economic benefits of stringent controls (originally opposed by industry) resulting in considerable reductions in industrial and transport emissions, substantial energy savings (Table 2) and increased recycling. Measures were also enacted to control other harmful emissions and discharges, particularly to water resources (rivers, lakes, seas and underground reservoirs). Starting in the mid-seventies, a new generation of chemicals management legislation has emerged in the developed regions (Chapter 10). Unlike the older type of legislation, subjecting only a number of 'gazetted' chemicals to regulatory measures, the new legislations require pre-marketing or pre-manufacturing assessments of new industrial chemicals. However, many chemicals in actual use still await satisfactory assessment. Similar measures gradually spread to the newly-industrialized and developing countries, particularly in Asia and Latin America. Many such countries, however, have faced enormous difficulties in implementing these measures due to a lack of human and technical capabilities.

Industrial accidents have also led to the tightening of standards and the development of new guidelines and directives. In 1984, and as a direct result of a major industrial accident in Seveso, Italy, in 1976, the so-called 'Seveso Directive' on major hazards in industrial activities was developed by the EEC. Since then it has been revised twice and amended to strengthen its provisions on the storage of dangerous chemicals and on public information. A 'Community Documentation Centre on Industrial Risks' was also set up to collect, classify, review and publish information on technical rules, guidelines, accident reports and all other relevant matters.

The problem of waste disposal - particularly of hazardous waste - (Chapter 10) has emerged of late and prompted new legislation at the international level (the Basel Convention in 1989). Other impacts, currently treated on a global scale, are those of CFCs and halons (the Montreal Protocol in 1987 and the London Amendments in 1990). UNEP has been regularly compiling and updating a compendium of environmental protection legal instruments at the international and regional levels (UNEP, 1991a).

The economic level

Economic instruments have come to be considered as alternative to, or complementary to, regulatory instruments. A comparative analysis of the results of direct regulation and economic incentives in the USA concludes that the former is more expensive than the latter (Luken, 1990).

The main thrust of economic instruments is to 'internalize' those external environmental costs that are not usually taken into consideration in the cost-benefit analyses on which investment decisions are based. Generally speaking, economic instruments are (OECD, 1989):

- charges, based on the so-called 'polluter pays principle' (PPP) voiced soon after Stockholm (OECD, 1975) and imposed mainly on effluents, products or users, sometimes as a tax differentiation;

- subsidies, as grants, soft loans or tax exemptions, to encourage compliance with the regulations (particularly useful with small-scale enterprises);

- deposit-refund systems that promote better waste collection and better management;

- market-creating instruments, e.g. trade in pollution permits, in recyclable waste or insurance policies.

Apart from these regulatory-cum-economic instruments which together have played an important role in directing industry to tackle its worst pollution problems even if this has entailed increased investments and higher operating costs, the last decade has witnessed clear and direct economic benefits in pollution prevention. In fact the '3Ps' came to stand for 'pollution prevention pays' by the time the world industry organized, jointly with UNEP and in co-operation with the International Chamber of Commerce (ICC), the World Industry Conference on Environmental Management (WICEM) in 1984

(Sallada and Doyle, 1985). This is in marked contrast with the thrust and conclusion of previous discussions on the impacts of instituting pollution control measures on the prices of industrial products (Holdgate, Kassas and White, 1982), when price increases were estimated to range between 1 per cent for food products and almost 5 per cent on refined petroleum products.

The economic implications of environmental protection, particularly pollution control and prevention have been discussed in various fora and publications, particularly in the last decade (OECD, 1989, 1985b, 1978). In general, adopting 'end-of-pipe' approaches results in higher costs due to internalization of the cost of environmental damage, and may result in reduced productivity and reduced productive investment. Cleaner production could result in economic advantages due to economies in energy and material consumption as well as labour. OECD studies seem to indicate that short-term impacts on output and employment are favourable, unlike long-term ones which are mixed, although mainly rather insignificant.

The cleaner production approach seems to have swung general opinion towards the view that it really can offer economic benefits. As outlined earlier, there are many cases in which pollution prevention has proved to be economically beneficial at the microeconomic level of the enterprise and regardless of macroeconomic or social benefits.

The social context

Perhaps the most important development on the social front has been the enthusiasm with which the business community has come to embrace environmental issues (Chapter 21). WICEM II, held in Rotterdam in April 1991, was an occasion for confirming the commitment of big business to sustainable development, and it has resulted in a charter. Industrialists could see no contradiction between the operation of market forces and the pursuit of sustainable development, thus reiterating the outcome of a previous meeting held in Bergen a year earlier. Several major industrial enterprises (3M, Du Pont, Chevron, AMOCO, ICI, Dow and AT&T among others) have now committed themselves publicly to achieving considerable reductions in the wastes produced.

Industry has also become much more sensitive to the need to improve its public image, particularly after the series of major accidents causing grave losses in life and property (Chapter 21). Responsible industrial enterprises operate ongoing programmes of informing the public about their operations, the hazards and risks involved and the lengths to which they go to prevent accidents and cope with them effectively should they happen. The Chemical Manufacturers Association (CMA) in the USA started its 'Community Awareness and Emergency Response' (CAER) programme which is now implemented on a large scale throughout the Americas and elsewhere. In Europe, the European Federation of Chemical Industries (CEFIC) established a similar programme (CICERO). At the international level, industrial associations supported UNEP in starting its 'Awareness and Preparedness at the Local Level' (APELL) Programme in 1987, which is gradually being adopted in developing countries in Latin America, Africa and Asia (see Chapter 10).

This growing concern of industry for closer contacts with society, and its greater understanding and appreciation of industry's moral responsibility towards the safety and well-being of people and the environment, has recently resulted in the 'Responsible Care' programme, which originally started in Canada in the mid-eighties and has now been introduced in the USA, Europe, Japan, Australia and New Zealand (see Chapter 21).

At the professional level, concern for the environment is demonstrated by the 'Code of Environmental Ethics for Engineers', drafted by the 'Committee on Engineering and the Environment' of the 'World Federation of Engineering Organizations' (WFEO) and adopted in a plenary session held in Delhi in 1985 (see Chapter 21).

Concluding remarks

The challenge facing industry now is one of how to restructure itself so as to move rapidly towards environmentally-sound and sustainable industrial production. The developed countries have placed increasing emphasis on reducing the adverse impacts of industrial activities. This could not generally be said of most developing or newly-industrialized economies. They have, no doubt, begun to pay more attention to the problem, particularly during the last decade: however, if the pace of industrial development and that of remedial action continue in developing countries at present rates, resource depletion and industrial pollution are likely to become a major issue on a global scale.

Recent examples, in both developed and developing countries, have clearly demonstrated that good house-keeping practices, such as proper maintenance, sustained optimum operating conditions of industrial plants and products, and proper training of operators, can produce significant results over short periods of time, combining environmental benefits with micro-, and sometimes macro-, economic benefits. However, in the long run, and if concern for the environment is not to become a brake on industrial expansion, new environmentally-benign technological alternatives have to be developed. Some approaches that offer such advantages are already available. Decreasing raw material and energy inputs, applying information technology to processes and products, finding biotechnological alternatives to physical and chemical processes and designing products offering less environmental damage or greater opportunities for recycling are but a few examples of such developments. However, the resulting design/production/consumption/disposal sequence will not be entirely 'clean'. There will always be scope for 'cleaner' production. Past experience strongly indicates the need for vigilance in monitoring environmental impacts over long time periods, as well as sustained efforts in developing and deploying yet cleaner and cleaner production world-wide if there is to be sustainable industrial development.

The present mix of regulation (command and control) and economic incentives (positive and negative) has not had a marked effect on fostering the adoption of existing promising technologies, let alone encouraging investment in the development of new technologies. The main reason is that the current costing and pricing mechanisms of industrial products do not internalize the costs of environmental damage and rehabilitation. This has been aggravated by the concentration of research and enforcement efforts on end-of-pipe approaches, rather than on cleaner production. There is need for a sustained research effort directed toward finding socially-acceptable mixes of regulation, incentives and technical support to industry, and toward encouraging the development and use of cleaner production technologies throughout industry. It is obvious that the optimum mix will vary considerably from one socio-economic environment to another. While at one end of the scale the transnational corporations are now spending more money on R&D to develop new technologies and turn results into environmental and economic benefits, small-scale industries, particularly in developing countries where they are responsible for a considerable proportion of industrial output, will need substantial technical and financial support if they are to adopt cleaner production techniques.

At the social level, such measures need to be coupled with sustained efforts to effect changes in the attitudes of industrial management, government machinery and the public at large. Understandable inertia and resistance to change by managers needs to be overcome. Government officials need to achieve far greater integration of environmental considerations in the formulation of industrial and technological policies and plans, so as to promote the adoption of the 'cleaner production' concept. In some developed countries, public pressure is already forcing the production of 'environment-friendly' goods and 'eco-labelling', as well as cleaner production methods. This needs to spread to the developing countries as their industries grow.

At the global level, industrial associations and the business community at large in the major developed countries have recently adopted guidelines for the transfer of technology to developing countries, to ensure minimum environmental damage. This movement needs to be strengthened and developed into a code of globally-accepted and applied rules governing technology transfer. For their part, the developing countries need to pay serious attention to the environmental impacts of their industrial technology choices, as well as taking full account of the measures that need to be taken from the outset to ensure proper siting, operation and monitoring of industrial plant and waste disposal facilities.

The new technological breakthroughs, while providing promising new approaches to environmental protection, also bring with them new environmental hazards. This calls for alertness and sustained effort - through legislative and institutional technology assessment - in clearly identifying their full impacts, over the short and long term, as well as in preventing, or at least minimizing, their adverse effects on the environment.

References

Ayres, R.V. (1989) Industrial Metabolism. In: J.H. Ausubel and H.E. Sladovich (eds.), *Technology and the Environment*, National Academy Press, Washington D.C.

Bohm, P. and Russell, C.S. (1985) Alternative Policy Instruments. In: A.V. Kneese (ed.) *Handbook of Natural Resources and Energy Economics* (Vol.1), North Holland, Amsterdam.

Bureau of the Census (1987) *Pollution Abatement Costs and Expenditures, 1985* Government Printing Office, Washington D.C.

Economic Study Group on Global Environment and Economics (1991) *Pollution in Japan - Our Tragic Experiences*, Tokyo.

El-Hinnawi, E. (1990) *Energy conservation in Industry*, National Research Centre, Cairo.

Environmental Resources Ltd. (1990) *Eastern Europe : Environmental Briefing*, London.

Fishbein, B.K. (1991) Reducing Packaging Waste: Europe Takes The Lead,In Third Annual Recycling Conference, November 12-13, 1991, Rochester, New York.

Frosch, R.A. and Gallopoulos, N. E. (1991) Towards an Industrial Ecology. In: Bradshaw, A.D. *et al.*, (eds.) *The Treatment and Handling of Wastes*, Chapman and Hall, London

Gardner, L. and Huisingh, D. (1987) Waste Reduction through Material and Process Substitutions: Progress and Problems Encountered in Industrial Implementation. In: *Hazardous Waste and Hazardous Materials*, **Vol.4**, No.1.

Hamza, A. (1983) Management of Industrial Hazardous Wastes in Egypt, *Industry and Environment*, Special Issue No.4, UNEP, Paris.

Holdgate, M., Kassas, M and G. White (eds) (1982) *The World Environment 1972-1982*, Tycooly, Dublin.

Huisingh, D. (1989) Cleaner Technologies through Process Modifications, Material Substitutions and Ecologically-based Ethical Values, *Industry and Environment*, Vol.12, No.1, 1989.

Imran, M. and Barnes, M. Energy demand in the developing countries: Prospects for the Future, IBRD Commodity working Paper 23.

Jasiewicz, J. (1990) Comparison of the Cost Effectiveness of Industrial Energy Conservation, Natural Resources Forum, Feb. 1990

Luken, R.A. (1990) *Efficiency in Environmental Regulation: A Cost-Benefit Analysis of Alternative Approaches*, Kluwer Academic Publishers, Boston, USA.

Lund University/UNEP Report of the Invitational Expert Seminar on Packaging And The Environment - Policies, Strategies and Instruments, Trolleholm Castle, Sweden, February 7-8, 1991, Department of Industrial Environmental Economics, Lund University, Lund, Sweden.

OECD (1975) *The Polluter-Pays Principle*, OECD, Paris.

OECD (1978 and 1985) *Macro-economic Impact of Environmental Expenditures*, OECD, Paris.

OECD (1985a) *Environmental Policy and Technical Change*, OECD, Paris.

OECD (1985b) *Environment and Economics*, OECD, Paris.

OECD (1989) *Economic Instruments for Environmental Protection*, OECD, Paris.

OECD (1990) Pollution Control and Abatement Expenditure in OECD countries, *Environment Monograph No.38*.

OECD (1991a) *The State of the Environment*, OECD, Paris.

OECD (1991b) *Environmental Indicators*, OECD, Paris.

OECD (1991c) Waste Management Policy Group Reduction and Recycling of Packaging Waste, draft report, OECD, Paris.

Piaseck, B. and Gravarder, J. (1985) The Missing Links : Restructuring Hazardous Waste Controls in America, *Technical Review*, October, 1985.

Sallada, L.H. and Doyle, Brendan G. (eds.) (1985) *The Spirit of Versailles, The Business of Environmental Management*, ICC, Geneva.

Sybren de Hoo, Brezet, H. Crul, M. and Dieleman, H. (1991) *Prisma: Industrial Success With Pollution Prevention*, Nederlandse Organisatie voor Technologisch Aspectenonderzoek (NOTA), Den Haag.

Touche Ross Management Consultants (1991) *Global climate change, the Role of Technology Transfer*, a report financed by the UK Department of Trade and Industry and Overseas Development Administration.

UNEP (1989)Report of the Technology Review Panel, Technical Progress on Protecting the Ozone Layer, UNEP, Nairobi.

UNEP (1991a) *Register of International Treaties and Other Agreements in the Field of the Environment*, UNEP, Nairobi, 1991.

UNEP (1991b) *United Nations Environment Programme, Environmental Data Report*, Basil Blackwell, Oxford.

UNEP (1991c) *State of the Environment*, UNEP, Nairobi.

UN ESCAP (1990) *State of the Environment in Asia and the Pacific*, UNEP, Bangkok.

UNIDO (1976) *Lima Declaration*, UNIDO, Vienna.

UNIDO (1989) *Industry and Development, Global Report 1989/90*, UNIDO, Vienna.

UNIDO (1990) *Industry and Development, Global Report 1990/91*, UNIDO, Vienna.

UNIDO (1991) Conference on Ecologically Sustainable Industrial Development, Copenhagen, 14-18 October 1991, Working Paper No. 1.

World Energy Council (1989) *Survey of Energy Resources.*

World Resources Institute (1990) *World Resources, 1990-91*, Basic Books, New York.

CHAPTER 13
Energy

Introduction

All forms of human activity and development require energy. A century ago, non-commercial sources (fuelwood, agricultural residues, dung, etc.) constituted about 52 per cent of total energy used. This share dropped significantly as fossil fuels became the predominant source of energy. In 1930 the share dropped to 25 per cent, in 1950 to 21 per cent and in 1970 to 12 per cent, where it has remained almost unchanged ever since, although more than two billion people in the developing countries depend on non-commercial fuels (Chapter 7).

This chapter is concerned mainly with the impacts on the environment of energy transformation, supply and use, and the manner in which the interactions between energy-related activities and the environment have developed over the last two decades - a period during which the energy sector has witnessed, on the one hand, some momentous changes, while on the other our knowledge of the environmental impacts of energy-related activities has deepened. Modern economies have become very dependent on reliable and diverse supplies of energy to satisfy different and changing needs. While some countries are endowed with a relatively abundant resource base, others, both industrialized and developing, are heavily dependent on imports to meet their needs.

Energy production and consumption

Consumption patterns

Prior to the 1950s, total commercial energy consumption in the world grew at an annual rate of 2.2 per cent. Between 1950 and 1960 the figure more than doubled to 4.9 per cent, reaching 5.6 per cent annually in the following decade. The increases in oil prices in the early and late 1970s and the general recognition of the limited nature of fossil fuel resources triggered a slow-down in energy consumption. In the 1970s, the annual growth rate of commercial energy consumption was 3.5 per cent, and in the 1980s it went down to 2.0 per cent annually.

The historical increase in commercial energy consumption was accompanied by a major change in primary energy sources. In the 1920s, coal accounted for about 80 per cent of the world's total commercial energy consumption, but its share in later years was greatly reduced due to the increase in oil discoveries and numerous technological changes. A whole set of new technologies was developed, not only to improve methods of oil supply but also to expand the range of its end uses. The result was to reduce the share of coal in the world's commercial energy consumption.

In 1950, the posted price of a barrel of oil in the Middle East was about $US2;

after which it dropped to $US1.8 until 1970. The consequences of the shift to cheap oil were far-reaching. It implied not only the creation of physical and economic structures based on the use of oil but also the spread of particular social institutions and cultural values that made life dependent on its ready availability. Obviously it is in the developed countries, where most of the technological changes and new patterns of energy consumption took place, that the shift from coal to oil and its consequences are most visible. The developing countries started industrialization and development efforts after the shift from coal to oil had been well established. Not only were oil-dependent structures built up in these countries, but the oil-based pattern of life was also taken over from the developed countries to a significant degree.

In 1970 total commercial energy consumption in the world was 5,000 million tonnes of oil equivalent (toe); in 1990 it was 8,100 million toe, i.e. an increase of about 62 per cent over the two decades (Figure 1). This rate of increase is much slower than that between 1950 and 1970. In 1950, total commercial energy consumption was 1,650 million toe and in 1970 it was 5,000 million toe, i.e. an increase of 203 per cent.

Figure 1: *Total commercial energy consumption by source (in thousand million tonnes oil equivalent).*

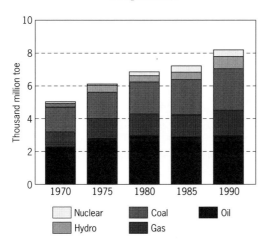

Source: El-Hinnawi (1991).

The energy mix also varied from 1970 to 1990, with the share of oil decreasing from 47 per cent in 1970 to 36 per cent in 1990. On the other hand, the share of natural gas increased from 18 to 19 per cent, that of coal from 30 to 32 per cent and that of nuclear energy from 1.6 to 4.3 per cent. The slow-down of energy consumption from 1970 to 1990 has, therefore, been primarily due to the slight decrease in oil consumption.

Consumption of the world's commercial energy resources is heavily concentrated in the developed regions - the industrial market economies and the former centrally planned economies (East Europe and the USSR). These regions, with about 22 per cent of the total world population, account for about 82 per cent of the total world consumption of commercial energy; the other 78 per cent of the world population, living in the developing countries, consume only about 18 per cent of commercial energy (Figure 2). The per capita commercial energy consumption in the OECD countries is about ten times that of the per capita energy consumption in the low- and middle-income developing countries (Figure 3).

Figure 2: *Total commercial energy consumption by region (in thousand million tonnes oil equivalent).*

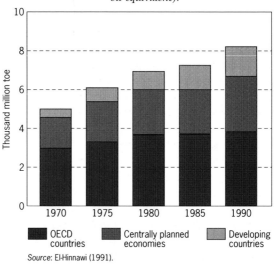

Source: El-Hinnawi (1991).

Figure 3: *Per capita commercial energy consumption (in kilograms oil equivalent).*

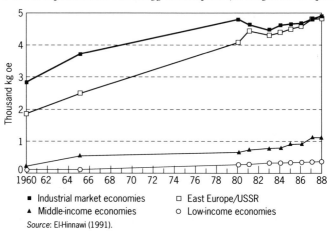

Source: El-Hinnawi (1991).

Energy consumption and economic growth

In the developing countries, commercial energy consumption remains relatively low, reflecting the low level of economic and social development. Consumption of commercial energy in these countries is typically concentrated in urban areas, where the beneficiaries are mainly industry, the commercial sector and a minority in the upper and upper-middle classes. In rural areas and in urban slums, where the vast majority of the population live, most household energy needs are still met by non-commercial energy sources such as fuelwood, agricultural residues and animal manure, as well as human and animal power, rather than by commercial sources of energy such as electricity and oil. Although no reliable national and/or global statistics exist on the current use of non-commercial sources of energy, estimates indicate that they constitute an often neglected yet very important share of the total energy budget of many developing countries (Chapter 7). This is an important consideration in assessing the full environmental impacts of energy supply at the global level.

The relationship between economic growth and energy consumption is a complex one, influenced by many factors. It was formerly assumed that there is a consistent positive relationship between gross national product (GNP) and energy consumption. However, detailed studies (for example, El-Hinnawi and Biswas, 1981 and El-Hinnawi and Hashmi, 1987) have demonstrated that the *structure* of the economy will affect the correlation between GNP and energy consumption. The rate of increase of GNP and energy use is somewhat similar for countries with energy-intensive exports. For non-energy-intensive export countries, however, GNP increases at a faster rate than energy use. Gradual and steady improvements in the efficiencies of energy conversion and utilization have shown up as historical trends indicating decreasing ratios between per capita energy consumption and per capita real GNP. The life-style preferred by a given society will also enter into the equation. The level of comfortable temperatures; illumination in residential, commercial and industrial structures; the preference for suburban living; mobility, and decreased reliance on mass transit systems are all relevant factors that can lead to increases in per capita energy consumption without a compensating increase in economic productivity. Finally, the growing energy intensiveness of agriculture and food production, and the displacement of natural products by synthetics, are two further examples of factors influencing the relation between GNP and energy consumption.

The non-dependence of GNP on energy consumption can be illustrated by the fact that while GNP per capita in Japan is higher than that in the USA ($US21,000 and $US19,800 in 1988 respectively), the per capita energy consumption in Japan is less than half that in the USA (3,306 and 7,655 kg oil equivalent in 1988 respectively). Furthermore, with the sustained emphasis on energy conservation over the last two decades, energy consumption has been increasing at slower rates than those of GNP, particularly in developed regions (see **Increasing the efficiency of energy utilization** below).

The energy alerts

The first large increase in oil prices in 1973-74 triggered world-wide disturbance. At that time the term 'energy crisis' came into common use. In fact the events of the early 1970s were more in the nature of an alert rather than a crisis. The 1970s brought into focus:

- a general realization of the finite nature of fossil fuels (especially oil and natural gas);

- a general realization that the era of cheap energy was over and that all economies would have to adapt to higher energy prices;

- the importance of increasing the efficiency of energy utilization;

- the importance of development of energy strategies to reduce dependence on energy imports and to exploit indigenous energy resources;

- the importance of establishing energy 'mixes', including renewables, to meet the demands for future development;

- a realization that environmentally-sound development must be based on environmentally-sound energy policies; and finally,

- that energy policy questions are not only technical, but are also social, environmental and political.

The energy alert of the early 1970s triggered many responses. Exploration for oil intensified in many regions, and proven crude oil reserves continued to increase. In spite of the steady increase in oil reserves, however, there was a marked instability in oil prices. A second sharp increase in oil prices occurred in 1979-80 and prevailed until 1985. The oil glut at that time resulted in a sharp drop in 1986. The Gulf crisis which started in August 1990 led to a slight increase in oil prices, but since then the oil market has remained erratic, but more or less stable.

Another response to the energy alert of the 1970s was the use of other energy sources as substitutes for oil, to reduce dependence on imported oil. Considerable fuel substitution (from oil to natural gas and coal) took place in some energy-intensive industries - notably iron and steel and cement - which were able to switch relatively easily. Oil substitution also occurred in other energy-intensive industries, e.g. paper and pulp which increased its use of biomass (wood waste and tree stumps) particularly in North America and Scandinavia (WRI, 1986). The use of heavy fuel oil in OECD countries for heating and electricity generation has fallen by 54 per cent since 1978 (OECD, 1988a). As mentioned earlier, more coal and natural gas are used instead, and the share of electricity generated from nuclear power also increased (OECD, 1989a). Substitute fuels for road transport were introduced in some countries. The best-known example is the use of ethanol fuel on a large scale in Brazil. Synthetic gasoline (from natural gas) is used in New Zealand on a commercial scale and other fuels are in the experimental stage (Chapter 14).

One could speak of a second energy 'alert' in the late 1980s due to the increasing concern over climate change and global warming and the role of fossil fuel burning in the accumulation of carbon dioxide - a major greenhouse gas. The last two decades have witnessed scientific consensus that the burning of fossil fuels has to be capped and eventually reduced (Chapter 3). One of the main issues in the intergovernmental negotiations to draft a convention on climate change concerned the changes that have to be brought about in energy sources, and technologies of transformation and transmission, as well as means of effecting savings, without imposing severe constraints on socio-economic development.

Increasing the efficiency of energy utilization

An important development that has been taking place since the mid-1970s is the institution of different measures - regulatory as well as technical - to increase the efficiency of energy utilization, especially in developed countries (IEA, 1987). Studies carried out in the 1970s (for example, ECE, 1976) demonstrated that more than one-half of the daily energy consumption is wasted due to losses induced by technology and human negligence. A more recent study (IEA, 1989) of electricity use in industrial and commercial sectors concluded that based on investment criteria, real rates of return on investment in increasing the efficiency of energy use are likely to exceed 30 per cent. The term 'energy conservation' came into common use in this period, and has been defined as the strategy of adjusting and optimizing energy use so as to reduce energy requirements per unit of output (or well-being) without affecting socio-economic development or causing disruption in life-styles. Apart from its economic benefits, energy conservation also has its environmental benefits. Not only does it stretch the time-span of availability of non-renewable high-quality fuels, but it also contributes to the curtailment of the adverse environmental impacts of energy production, distribution and utilization.

The relationship between energy demand and economic growth is measured by 'energy intensity', defined as energy consumption divided by gross domestic product (GDP). In developed market economies, conservation and greater energy efficiency have resulted from the switch to less energy-intensive capital goods and consumer durables, and from the application of energy-efficient machinery in industry, heating and air conditioning. As a result, energy intensity in the developed market economies declined by 29 per cent between 1970 and 1990. In Eastern Europe and the former USSR, energy intensity fell by 20 per cent over the same period (Figure 4). This trend, which cannot be attributed to a single government or industry initiative, lies behind the decoupling of the level of energy use and economic well-being (Goldemberg et al., 1987).

It should be noted that the above-mentioned average trends conceal marked differences between different countries (Figure 5). For example, Japan tops the list in increasing energy efficiency in industry (Chapter 12). The energy

Figure 4: *Energy intensities (in barrels oil equivalent per $US1,000 GDP).*

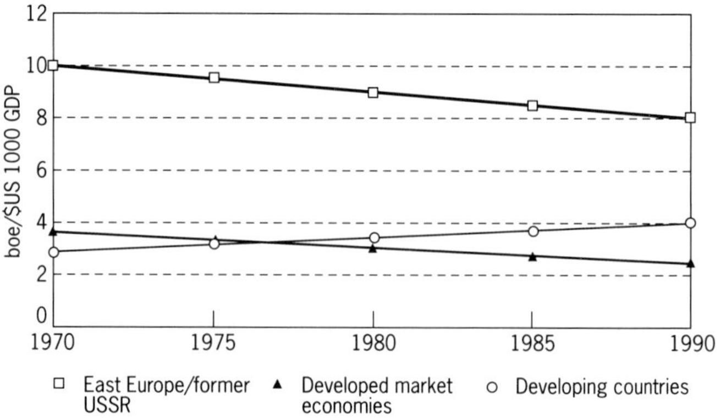

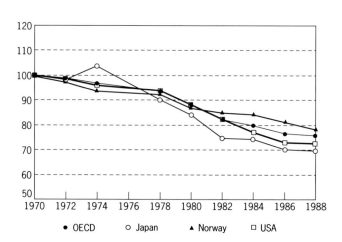

Source: El-Hinnawi (1991) based on data by UN (1990).

intensities of chemical and steel production have dropped by 38 and 16 per cent respectively since 1973 (Chandler, 1985). The French industrial sector also ranks among the most energy efficient. Energy intensities in textiles, building materials, rubber and plastics, and mechanical construction have fallen by more than 30 per cent since 1973. In comparison, the energy intensity of production of paper, aluminium, steel and cement in the USA fell by only 17 per cent. In general, the energy intensity in Japan fell by 30 per cent between 1970 and 1988, higher than the average of 25 per cent for the OECD countries.

Figure 5: *Changes in energy intensity in OECD countries, 1970-88 (based on 1970 = 100).*

Source: El-Hinnawi (1991) based on data by OECD (1990).

Measures to encourage energy conservation and to improve energy efficiency have been rather limited in developing countries, where industry often consumes two to five times as much fuel for a given process due the use of old industrial equipment (Flavin, 1986; World Bank, 1986). In some countries factories are often required to produce a fixed quota of goods at a fixed price regardless of cost. Consequently, there is little or no incentive to reduce energy consumption. Many buildings and transport systems in developing countries are old and lack proper maintenance, and hence are energy-inefficient. The situation has been aggravated by heavy subsidization of energy prices to all types of consumers, poor metering of consumption (or its total absence in some cases) and high losses in energy transmission. Consequently, the amount of energy consumed per unit of output is high. Another main factor in this increase is the progressive replacement of non-commercial energy (fuelwood, agricultural residues, etc.) by commercial sources such as oil products and electricity, as a result of the transition of rural areas into semi-urban industrial and commercial societies, without a corresponding increase in GDP (a common feature in development patterns in many developing countries). This has led to rapid growth in energy intensity, accentuated by growth in energy-intensive industries in the early phases of industrialization. As a result, energy intensity in developing countries has risen by 30 per cent between 1970 and 1990 (Figure 4).

In the mid-1970s, when the problem of fuelwood scarcity was brought into focus, several efforts were made to increase the efficiency of utilization of such non-commercial sources of energy. The commonly used cooking stoves in developing countries have efficiencies of between 6 and 10 per cent. More efficient stoves have been developed in China, Indonesia, India, Guatemala, Nepal, Kenya and other countries. Efficiencies as high as 30 per cent have been achieved by some of these stoves (El-Hinnawi and Hashmi, 1987). However, the adoption of such stoves has been slow and has encountered a number of economic, social and cultural problems. In spite of the progress made in the past two decades on many fronts there is still an enormous potential for further improvement in energy efficiency, both in developed and developing countries.

Future energy demand and supply

Various computations of future world primary energy demand have been made in the past two decades (e.g. WRI 1986, Goldemberg, *et al.*, 1987). Suffice it to note here that the projections given in these studies vary markedly. Many have been criticized for being based on uncertain assumptions, or for ignoring important factors such as non-commercial sources of energy, the substitution of these sources by other fuels, or the widespread use of more energy-efficient products and systems.

The main point of concern here is that the 1970s brought into focus the necessity of developing appropriate energy 'mixes' to meet future energy demand. Several scenarios have been proposed: expanded utilization of coal resources which are more abundant than those of oil and natural gas; development of non-conventional fossil fuels such as oil shales and tar sands; further development of nuclear power, and development of renewable sources of energy. However, the question of how to meet future energy demand by an environmentally sound and appropriate energy mix still remains largely unanswered. Uncertainties prevail about the availability of energy resources, future energy prices, degree of market penetration of new technologies, environmental aspects of new energy technologies, etc. Such uncertainties can best be illustrated by projections made for nuclear power. In 1975, the IAEA estimated that by the year 2000 nuclear power would contribute 2,600 gigawatts electric (GWe). Two years later this figure was reduced to 2,000 GWe and in 1981 it dropped to 1,075 GWe, and then to 444 GWe in an IAEA 1987 projection (Semenov *et al.* 1989).

Estimates of proven recoverable reserves of fossil fuels have undergone several revisions in the past two decades. Table 1 gives a comparison of estimates made in 1980 and in 1991. The figures show marked increases in the estimates over a period of about ten years. Even so, these resources are non-renewable and cannot, therefore, be relied upon for sustainable development. Although the proven recoverable reserves of non-conventional fossil fuels, oil-shales and tar sands, have been estimated at about 46,000 million toe and 40,000 million toe respectively (WEC, 1980), the development of such sources of energy is constrained by technical, economic and environmental factors (UNEP, 1983).

Table 1: *World resources of fossil fuels.*

	Proven recoverable reserves (in million tonnes of oil equivalent)		
As estimated by	Coal	Oil	Natural gas
WEC (1980)	452,000	89,000	62,000
British Petroleum (1991)	572,000	136,000	109,000
Increase	27%	53%	76%

The world's proven reserves of uranium have been estimated (NEA/IAEA, 1990) at four million tonnes (a resource base equivalent to about 42,000 million toe). Although the development of alternative nuclear fuel cycles, including reprocessing and the deployment of the breeder reactor, could extend the resource base, it is not evident that such developments will occur in the near future, at least not on a large scale within the next two to three decades (El-Hinnawi, 1980).

The resource base of renewable sources of energy is extremely large. With the present state of technology it is, however, difficult to estimate how much of the resource base can be technically and economically exploited, and what would be the contribution of renewable sources of energy to the overall future world energy supply.

As has been mentioned earlier, some renewable sources of energy (non-commercial fuelwood, agricultural residues, etc) have been used extensively in many developing countries for ages. Most of this energy is used at home and for small industries in a primitive and inefficient way. Such renewable sources of energy could become non-renewable if over-exploited.

Interest in the use of renewable energy sources (especially wind, solar, biomass and small hydro) gained momentum after the first major oil price increase in the early 1970s. Several research and development projects were initiated, some by UNEP, and pilot projects were conducted to establish all-electric systems for rural areas on the basis of renewable sources of energy. Some authors went even further and called for 'soft energy paths' (Lovins, 1977) based exclusively on solar and other renewable sources of energy. This enthusiasm culminated in the United Nations Conference on New and Renewable Sources of Energy, convened in Nairobi in 1982. At that time it was estimated that renewable sources of energy accounted for 15 per cent of world energy use and that this percentage would increase to 18-30 per cent by the year 2000 (El-Hinnawi et al. 1983).

However, there were many constraints facing the development of renewable sources of energy. Most research was - and still is - concentrated in the developed countries, and many of the technologies developed are in many cases not suited to the needs of developing countries because of cost or complexity. In addition, funds allocated to R&D were inadequate to match the enthusiasm for the development of renewable sources of energy. For example, in Japan in 1980 the national R&D expenditure on renewable sources of energy was about 6.7 per cent of that spent on all sources of energy. In developing countries, the situation was even more unsatisfactory. The wider utilization of renewable energy systems has also been hampered there by technical, economic and social factors. Some systems were unreliable, capital costs have been rather high, and there has been a lack of necessary infrastructure in many countries to build renewable energy systems. Decentralized renewable energy systems have not been widely accepted by the public, and the decrease in oil prices in 1986 had a negative effect on the development of renewable sources of energy. Because the costs of conventional energies now seem unlikely to rise as far, or as fast, as was anticipated in the 1970s, and because of the long lead time for development and market penetration of renewable systems, government support and industry interest in developing alternatives to oil have weakened.

The uncertainties that emerged in the 1970s about future energy supply and demand still prevail, particularly with the new concern over climate change. There is need now to rekindle the enthusiasm for the formulation of national energy policies that emerged in the 1970s and which has since faded away, leaving several large energy consumers (e.g. the USA) without a definite energy policy. Should this attitude continue, there will not only be serious implications for future energy supply and demand, but also for various environmental problems that are closely associated with energy sources, production and wasteful consumption.

Energy and environment

The interface between energy and environment is complex and constantly evolving. The environmental impacts of energy production, transformation, transport and use have been of major concern at the local, national and international levels, and this concern has been focused on problems such as:

- transboundary acidic deposition;

- the global build-upoof carbon dioxide and the likely effects on climate;

- the transboundary release of radioactivity after nuclear accidents, (Chapter 9) and

- the safe disposal of wastes.

In the early 1970s, discussion of the environmental impacts of different energy sources focused on direct and immediate impacts on occupational and public health and/or the physical environment, rather than on long-term socio-economic and environmental effects. However, in the 1980s increasing efforts were devoted to the analysis of these less tangible long-term impacts. Considerable methodological problems were apparent: the criteria for judging and comparing environmental impacts were disputed; the methods for quantifying impacts and costs were poorly developed and in some cases manifestly flawed; and the values placed upon different environmental impacts and different degrees of protection from various perceived hazards varied within societies and over time. Nevertheless, it soon became clear that all energy technologies have the potential to perturb critical environmental processes as well as threaten human health; and that no energy technology is so free of environmental risk that its adoption brings only benefits.

Experience gained in the past two decades shows that energy production and use, and the protection of the environment, can be reconciled. Energy production, conversion, transport and use need to be carried out in an environmentally acceptable manner, but for this to happen, energy and environmental policy objectives should be assessed in the context of one another. In formulating energy policies, due consideration should be given to environmental factors, while in the formulation of environmental policies due weight has to be given to energy considerations. Finally, the more efficient use of energy prompted by economic considerations is of primary importance in achieving the objectives of both energy and environmental policies.

Assessing the environmental impacts of different energy systems must include all impacts on air, land, water and biota, as well as for the entire sequence of operations from the extraction of raw material, through transportation, processing, storage, transformation, distribution and end use, to the management of the wastes produced (Box 1). The contribution of the energy sector to the major environmental issues discussed in Part One of this book, and the manner in which it has changed over the last two decades, is briefly discussed below. (The impacts of energy used for transport are discussed in Chapter 14).

Air pollution

Most energy systems emit a variety of gases, volatile organic compounds and particulate matter. These vary considerably with the amount and type of fuel used and the method of burning (Table 2).

Table 2: *Emissions from fossil-fuel-operated power stations (in tonnes/1000 MW(e)y).*

		Coal[1]	Oil[2]	Natural Gas[3]
I.	Airborne effluents:			
	Sulphur oxides (SO_x)	110,000	37,000	20
	Nitrogen oxides (NO_x)	27,000	25,000	20,000
	Carbon monoxide (CO)	2,000	710	-
	Particulates	3,000	1,200	510
	Hydrocarbons	400	470	34
	Aldehydes	-	240	-
II.	Solid Wastes:			
	Ash	360,000	9,000	

Notes: 1. Assuming power plant burns 3×10^6 tonnes coal, sulphur content 2 per cent; energy content of coal 2.74×10^7 J/tonne; thermal efficiency of power plant 38 per cent; fly ash removal efficiency 99 per cent; no flue gas desulphurization. Solid wastes: bottom ash + recovered fly ash.

2. The power plant uses 2×10^6 tonnes residual fuel (1 per cent S, 0.5 per cent ash).

3. The power plant uses $2.2 \times 10^9 m^3$ of natural gas with energy value of 37,000 kJ/m^3.

Source: UNEP (1981).

BOX 1

Selected environmental effects of the energy sector[1]

Energy sources	Air	Waters (surface, undergound, inland and marine)	Land and soils
FOSSIL FUELS EXTRACTION, TREATMENT, TRANSPORT, WASTE DISPOSAL			
COAL	SO_x, NO_x, particulates CO_2, CH_4	Acid and salted mine drainage. Mine liquid waste disposal. Water availability. Wash water treatment. Water pollution from storage heaps.	Land subsidence. Land use for mines and heaps. Land reclamation of open cast mines.
PETROLEUM PRODUCTS	H_2S production SO_4, NO_4, CO_4, CO_3, HC_2, CH_4, ammonia, particulates, trace elements CO_2	Oil spills Water availability	Land use for facilities and pipes.
GAS	HC emission (mainly methane) NO_x, CO_2, H_2S and combustion emissions	Liquid residual disposal.	Land use for facilities and pipes
ELECTRICITY GENERATION FROM FOSSIL FUELS (EXCLUDING NUCLEAR ENERGY)			
	SO_2, NO_2, CO, CO_2, HC, trace elements particulates, radionuclides. Long-range transport and disposition of pollutants. 'Greenhouse effect'.	Water availability. Thermal releases.	Land requirement
URANIUM FUEL CYCLE AND ELECTRICITY FROM NUCLEAR POWER PLANTS			
	Radioactive dust. Gaseous effluent. (radionuclides F. NO_x). Noble gas, H-3, I-131, C-14. Local climatic impact of cooling towers.	Mine drainage. Underground water contamination. Water availability. Thermal releases Liquid radionuclide emission (H-3, CO-60, Sr-90, I-131, Ru-106, Cs-136 and 137.	Land subsidence (mine). Land reclamation of open cast mines. Land use for mines.
RENEWABLE ENERGY			
HYDROPOWER	Local climatic effects of large installations.	Effect on hydrological cycles, Water quality and resources.	Land irreversibly flooded. Landslide ri
OTHERS	Biomass combustion: air pollution, particulates. Geothermal: air pollution.	Biomass conversion: water pollution; water availability. Geothermal: water pollution.	Land use for energy plantations. Land requirement of solar energy.

Note: 1. Excluding energy use in transport and agriculture.
Source: OECD (1991).

Wild life	Risks	Others: wastes, human health, noise, visual impacts
Natural habitat disturbed. Exploitation of wilderness or natural areas for surface mining.	Occupational risks	Noise of rail transport of coal. Dust emission. Visual impact of coal heaps.
Natural habitat disturbed. Pipeline impact on wild life. Wild life polluted through leaks or spills.	Blowouts, explosions and fires.	Odour. Pipeline leaks. Spills (accidental and operational). Visual impacts of pipelines.
Natural habitat disturbed Impacts of pipelines on wild life.	Blowouts. High leak potential. General safety.	Spills and explosions Visual impacts of pipelines.
Secondary effects on water, air and land.	Occupational risks	Visual impact of cooling towers and power lines. Solid wastes. Ash disposal. Noise
Secondary effects of impacts on water, land and air.	Occupational risks Plant accidents. Disposal of high level radioactive wastes.	Radioactive products. Mine water. Mill tailing water (toxic metal, liquid and solid chemical wastes, radiological wastes). Recycled fission products. Visual impact of cooling towers and power lines. Noise. Decontamination and decommissioning of nuclear power plants.
Wild life habitat of rivers. Change in ecosystems. Fish migration affected.	Risk of dam rupture	Visual impacts
Biomass ecosystem disruption by energy plantations.	Biomass risk to workers	Noise of wind generators. Visual impact of wind generators. Photovoltaic toxic pollution when decommissioning.

Furthermore, the energy sector's share of emissions varies widely from one country to another. The contribution of energy systems to global emissions, (excluding those for transport), are as follows:

1 Carbon dioxide (CO_2)

Energy-related emissions world-wide have increased over the last two decades by almost a quarter (Figure 6). Furthermore, while the increase in OECD countries was well below the global average, energy transformation's share of total emissions has increased, reaching around one-third in 1988.

Figure 6: *Trends in man-made carbon dioxide emissions (in million tonnes of carbon).*

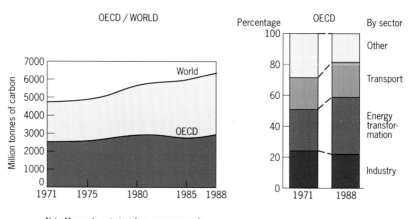

Note: Man-made emissions from energy use only.
Source: OECD (1991).

2 Sulphur oxides

Energy systems are the main source of sulphur oxide emissions (estimated as 90% of the total). As shown in Chapter 1, emissions in developed regions have decreased over the last two decades, while those in developing regions have been increasing. However, while emissions in developed regions have decreased over the last two decades, the share of energy from power stations has increased appreciably (Figure 7, for the UK, is typical of this trend).

3 Nitrogen oxides

Latest figures indicate that fossil fuel combustion is not a major source world-wide of nitrogen oxides emissions (Table 3). Nitrogen fertilizers and microbial processes in soils and water are the main sources. In countries with large numbers of cars, about half of the man-made emissions come from cars, followed by power plants (Figure 8, for the Federal Republic of Germany, is typical of this trend in industrialized countries).

Figure 7: *Trends in sulphur dioxide emissions in the UK (million tonnes per year).*

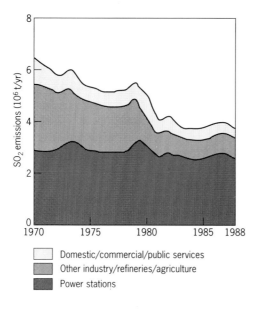

Source: UNEP *Environmental Data Report, 1991/92.*

Figure 8: *Nitrogen oxides emissions in the former Federal Republic of Germany (million tonnes per year).*

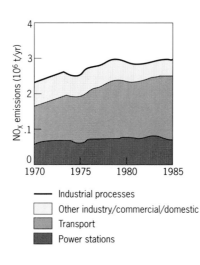

Source: UNEP *Environmental Data Report, 1991/92.*

Table 3: *Sources of nitrogen oxides.*

Source	Range of releases 10^6 tonnes/year		
Oceans	1.4	-	2.6
Soils (tropical forests)	2.2	-	3.7
Soils (temperate forests)	0.7	-	1.5
Fossil fuel combustion	0.1	-	0.3
Biomass burning	0.02	-	0.2
Fertilizers	0.01	-	2.2

Source: UNEP (1991).

4 Methane

While leakages of natural gas account for some emissions, and biomass burning being a not insignificant source of methane emissions, rice paddies, wetlands and livestock are the main sources (Figure 9).

Figure 9: *Trends in estimated global methane emissions, 1940-80 (million tonnes per year).*

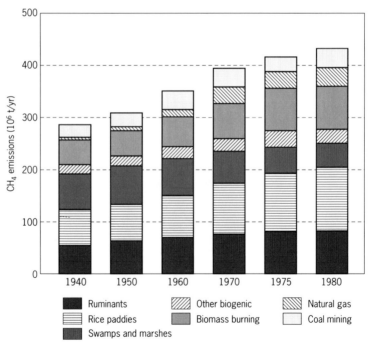

Source: UNEP *Environmental Data Report, 1989/90.*

5 Particulate matter

Globally, the release of particulates has remained more or less constant (5-6x10^9 tonnes/year). Natural sources are estimated to release two to twenty times the volume of man-made releases (UNEP, 1989). The latter are usually toxic. Fuel burning is the main source of these emissions. Consequently, with the installation of dust removal equipment in coal-fired utilities, and the decline in the use of solid fuels in domestic heating, considerable reductions of emissions were achieved in many OECD countries. However, lower emissions have been offset in some cases by growing emissions from the proliferation of diesel-powered vehicles (OECD, 1991b).

6 Carbon Monoxide

In industrialized regions, road traffic and industry rather than power stations are the main sources of CO emissions. Combustion of fossil fuels and the burning of wood as fuel contribute just under one-third of total man-made emissions (Table 4). In developing regions, inefficient combustion in primitive stoves, furnaces and boilers is a main source. In those countries where emission limits have become progressively more stringent (e.g. USA, Canada, Western Germany) CO emissions decreased by more than 25 per cent between the mid-seventies and late 1980s (OECD, 1991b).

Table 4: *Estimates of CO releases ($10^6 t/yr$ as CO).*

Origin	Source	Amount
	Fossil fuel combustion	190
	Wood burning as fuel	20
Man-made	Sub-total combustion	210
	Forest clearing, savanna burning, oxidation of methane	460
	Total man-made	670
Natural	Oxidation of hydro-carbons, plant emissions, oceans, methane oxidation, forest fires.	500
Total		1170

Source: UNEP (1989).

7 Trace elements

Lead, the predominant trace element in air pollution, comes from the use of leaded petrol in cars, which is now being phased out (Chapter 14). Coal and oil combustion in power plants, and wood burning, are not among the main sources of trace elements released to the atmosphere. The highest levels of releases from these sources are chromium, mercury and nickel.

8 A number of other hazardous air pollutants are emitted through energy-related activities. For example, hydrocarbons such as benzene are emitted during oil and gas extraction and industrial processing and trace quantities of radionuclides can be released during the combustion of coal and heavy fuel oil in power plants and industrial boilers.

Water pollution

Water pollution can result from several energy-related activities. The most evident is marine pollution due to oil, resulting from normal shipping operations and from accidental oil spills (see Chapters 5, 9 and 14). Power plants and refineries also produce effluents containing hazardous material that can contaminate surface and coastal waters. Onshore oil and geothermal energy production produce huge amounts of brines that pose several problems in receiving waters. Acid drainage from existing or abandoned coal mines and from coal preparation plants can contaminate surface and ground water. Thermal pollution from the discharges of cooling systems of power plants, or from geothermal facilities, can threaten aquatic life. A further source of groundwater pollution in many countries is oil leaks from underground storage tanks.

Solid wastes

Solid wastes resulting from energy activities can be hazardous (e.g. wastes from coal or uranium mining and milling) or non-hazardous (e.g. bottom ash from power plants). Such wastes require large tracts of land for disposal, with adequate containment practices to avoid water contamination. A major and growing source of solid waste has developed along with air pollution control in power plants. Sludge from flue gas desulphurization devices and collected fly ash from particulate control devices require special management to avoid land, air and water pollution in the vicinity of disposal sites.

Land use

The siting of power stations (especially nuclear power stations) has been a matter of debate and controversy in many countries. Concern has also been voiced about the large land areas that might be needed for the large-scale exploitation of renewable sources of energy, such as wind and solar power stations, or biomass production. This could compete with other land uses. In addition, growing siting problems are now being faced in the disposal of solid wastes, ranging from those generated by mining to high-level radioactive waste generated by nuclear power stations.

Impacts of renewable sources of energy

Renewable sources of energy vary widely in their impact on the environment (OECD, 1988b). The environmental impacts of hydropower and geothermal energy are better known than those of other renewable sources of energy (El-Hinnawi and Biswas, 1981). The burning of biomass fuels in primitive unvented stoves poses serious health problems (Chapter 1). Much attention has been given to the problem of fuelwood and its relation to deforestation (Chapter 7). In 1980, about 200 million people in the developing countries were acutely short of fuelwood and could not meet their needs, while another 1,283 million people lived in areas where they could still get enough wood but only by cutting trees in an unsustainable way. It has been estimated that by the year 2000, about 3,000 million people in the developing countries will face acute scarcity and deficit of fuelwood (Table 5).

Table 5: *Populations experiencing fuelwood deficit in 1980 and 2000 (millions).*

| Region | 1980 | | | | 2000 | |
| | Acute scarcity | | Deficit | | Acute scarcity or deficit | |
	Total population	Rural population	Total population	Rural population	Total population	Rural population
Africa	55	49	146	131	535	464
Near East & North Africa			104	69	268	158
Asia & Pacific	31	29	832	710	1,671	1,434
Latin America	26	18	201	143	512	342
Total	112	96	1,283	1,053	2,986	2,398

Note: Rural population = total population less that of towns with more than 100,000 inhabitants in zones whose fuelwood situation has been classified.

Source: Adapted from FAO (1983).

Impacts of nuclear energy

There are three major hazards involved in nuclear energy systems.

1. **Radiation** from the nuclear fuel cycle remains an issue of major concern although about 90 per cent of exposure to radiation is due to natural causes. Uranium mining and milling release radon and radon daughters which are potential occupational hazards. Normal reactor operation produces low-level radioactive emissions which are not considered to be harmful. In all, the normal operation of the nuclear fuel cycle contributes about 0.04 per cent of all radiation to which humans are exposed. Natural sources, in comparison, account for 83 per cent and man-made medical sources for about 17 per cent. People living near nuclear installations do, of course, receive higher doses than the average. Even so, typical doses around nuclear reactors constitute a fraction of one per cent of doses from natural sources. However, the potential risk of a failure and the environmental effects of an accidental leak remain major areas of concern. The Chernobyl accident (see Chapter 9) demonstrated that the impact extends beyond reactor safety and accident prevention to include problems such as emergency planning and decommissioning. It has also demonstrated the international dimension of the problem, with heightened public and political awareness of the transboundary risks of accidents involving nuclear facilities. Many uncertainties still prevail on the effects of low-level radiation (UNEP, 1981). The difficulty of proving cause and effect is also a problem in studying links between human genetic effects and irradiation. However, it has recently been suggested that the incidence of leukaemia was higher in children born near the Sellafield nuclear plant in the United Kingdom and in children of fathers employed at the plant, particularly those with high radiation dose recordings before their children's conception.

2 **Nuclear waste disposal** involves varying degrees of hazard depending on the characteristics of the wastes and whether or not they are released into the environment or isolated from the biosphere. There are large differences in the perceived potential for damage and the corresponding level of concern about low-level waste containing short-lived radionuclides and high-level waste containing significant amounts of long-lived radionuclides which need to be isolated for thousands of years. At present, wastes from reprocessing plants and spent fuel from reactors amount to about 420,000 cubic metres (Zhu and Chan, 1989) and are expected to rise to 8 million cubic metres by the end of the century. The techniques of managing nuclear wastes have ranged from storage on the surface to shallow or deep burial underground, containment in cement, bitumen or resins, or vitrification and storage in stable geological formations on land or on the seabed for high-level wastes (OECD,

1991b). All these approaches present serious environmental hazards which will probably become more serious as time goes on, particularly for high-level wastes which could reach the one million cubic metre mark by the year 2000.

3 Decommissioning of nuclear installations is technically feasible. However, the issues involved are complex, and will not be resolved until more experience is gained in the complete dismantling of large facilities. According to the IAEA, 64 commercial nuclear reactors and 256 research reactors could need decommissioning by the year 2000. There is increasing worry that the risk of exposure to radiation would be high throughout the dismantling of the components, especially the reactor vessel, let alone the high cost involved. The cost of decommissioning a 1,000 MW plant has been estimated to be $US480 million. One approach to meeting these costs is to levy a fee on energy consumed, to be put aside in a 'Nuclear Waste Fund'.

Accidents

Accidents related to energy activities, and their environmental consequences, have been brought into focus in the past two decades. Major oil tanker and off-shore drilling platform accidents and the resulting oil spills and fatalities are well documented (Chapters 5 and 9). On-shore and off-shore blow-outs, explosions of gas storage tanks and pipelines, and fires at refineries and oil rigs have occurred. But nuclear accidents, such as those at Three Mile Island and Chernobyl (Chapter 9), have been of major concern. These accidents again brought into focus the problem of public perception of hazards, which is heavily weighted by the severity of an accident and very little by its frequency, as well as by the substantial difference between voluntary and involuntary risks.

In conclusion, the environmental impacts of energy production and use present an urgent challenge highlighting the need to provide for energy needs in a way that is best for future development. Some actions depend on increased public awareness and/or minor investments (such as changing habits leading to energy conservation, and measures to increase efficiency of energy use). Others will result in future internationalization of environmental costs. This will bring about significant changes in the relative costs of energy supply and use, with corresponding shifts in fuel choice as well as changes in overall demand. Fossil fuels could see resultant major cost increases because of the problems of air quality, acid rain and global climate change. For nuclear energy too, substantial cost increases can be expected for safe decommissioning and long-term storage of high-level radioactive wastes. Concern about climate change due to the accumulation of greenhouse gases caused mainly by the extraction and burning of fossil fuels is potentially the most important emerging environmental problem related to energy.

National and international responses

As the previous discussion shows, the environmental impacts of the production and use of energy have been reduced in many developed countries over the last two decades. This has resulted from developments on the technological, regulatory and institutional levels, leading to more rational use of energy, structural adjustment in the energy mix, and emission control measures. However, progress has not been uniform across the world. Least progress has been seen in the developing countries, where the technological infrastructure is less developed, conservation measures are rather weak, and financial constraints have prevented investments in emission control equipment or in cleaner energy production techniques. In some cases (e.g. some Eastern European countries, China and India) the resource base consists largely of highly-polluting fuels.

Restructuring of the energy mix since 1973 has had an important impact on sulphur and nitrogen oxides emissions in several OECD countries. The restructuring has been most pronounced in electricity generation, which has been marked by a drop in oil use and an increase in the use of natural gas, low-sulphur solid fuels and nuclear energy (a development that has raised widespread public concern). This restructuring together with emission control measures, especially flue gas desulphurization and low NO_x combustion systems, has resulted in marked decreases in emissions in most OECD countries.

The signing of the ECE Convention on Long-range Transboundary Air Pollution in 1979 demonstrated the determination of European countries to work together to cut back sulphur and nitrogen oxides emissions to acceptable levels. In 1987, the Protocol to the Convention - which requires participating nations to reduce either national sulphur emissions or their transboundary flows by 30 per cent from 1980 levels by 1993 - entered into force. The Nitrogen Oxides Protocol, signed in November 1988, calls for a freeze on emissions at 1987 levels in 1994, as well as further discussions beginning in 1996 aimed at actual reductions. Some countries have made commitments to go beyond both protocols. At least nine countries have pledged to bring sulphur dioxide levels down to less than half their 1980 levels by 1995. Austria, Sweden, and the Federal Republic of Germany are committed to reducing them by two-thirds. On nitrogen oxides, twelve Western European nations have agreed to go beyond the freeze and reduce emissions by 30 per cent by 1998. A November 1988 directive by the European Economic Community represents a binding commitment by the members to reduce sulphur and nitrogen oxides emissions significantly. The directive will lower community-wide emissions of sulphur dioxide from existing power plants by a total of 57 per cent from 1980 levels by 2003, and of nitrogen oxides by 30 per cent by 1998.

Several countries have instituted ambient air quality standards, fuel quality standards and emission standards. Although there is a trend towards strict regulation of new large combustion facilities, regulation of existing power plants is less consistent. The high costs involved have worked against the adoption of programmes of retrofitting the existing combustion facilities. However, Germany, Japan, the Netherlands, Sweden, Denmark and the United Kingdom have all adopted programmes requiring major retrofit investments. In contrast, in the USA, retrofitting of existing power plants with sulphur dioxide and nitrogen oxides control devices is controversial (OECD, 1988a). While regulations in OECD countries are difficult to compare, the actual investments in flue gas desulphurization, selective catalytic reduction (to reduce NO_x emissions) and low-NO_x burners, show major control efforts are under way. The situation in Eastern Europe is very different. The investments involved in instituting similar measures are considerable, and are not easy to provide under present economic conditions. The problem is aggravated by the fact that the fuel resources available in some countries are highly polluting.

Advanced generating technologies, such as fluidized bed combustion, integrated gasification, combined cycle, and high-efficiency gas turbines, offer a number of advantages over conventional technologies, such as low emissions of SO_2 and NO_x, high thermal efficiencies and fuel flexibility. These technologies should continue to promote the role of coal and natural gas for electricity generation. Interest in these technologies has been shown in countries expecting continued reliance on coal in electricity generation (Germany, the United Kingdom and the USA). However, the environmental impacts of such new technologies, not yet commercialized, remain to be assessed.

Development of combined heat and power (CHP), propagation of automated processes in industrial production, use of industrial waste heat and development of CHP/district heating systems can have overall beneficial economic and environmental effects. It should be noted, however, that the net environmental effects of CHP are complex. They may include an increase in industrial activity and in some cases a change in combustion technology. The environmental effects of both should be evaluated, as such effects might require additional control measures.

Energy-related accidents have prompted the ratification of a number of international agreements for dealing with oil spills, notification of nuclear accidents with transboundary impacts, and assistance at the international level to contain such impacts (see Chapters 5 and 9).

In the late 1980s, attention turned to greenhouse gas abatement, and especially to reductions in CO_2 emissions. The European Community has now committed itself to stabilize its output of CO_2 at 1990 levels by the year 2000. In September 1991, the Community environment ministers approved the principle of imposing an energy or carbon tax. The proposed tax is to be increased in steps so as to reach $US10/barrel of oil by the year 2000 (first by $US3 in 1993, increasing by one dollar per year up to the end of the century). This would allow their economies to provide the necessary investments for energy savings without severe impacts on economic growth or sharp price rises. The tax is made up of two components, half on carbon content and half on all energy types regardless of the pollution they cause (excluding renewable energies). The tax has been critized for favouring the use of coal, a highly polluting fuel which is subsidized in some EC countries, being the most abundant source there.

Other approaches currently proposed (IUCN/UNEP/WWF, 1991) are reductions in wastage in energy distribution, and in energy use in the home, industry, business and transport; developing renewable and other non-fossil fuel energy sources; conducting public awareness campaigns to promote energy conservation, and the use of energy-efficient products. In short, to develop comprehensive and explicit national energy strategies.

The problems of the use of non-commercial sources of energy remain without tangible solutions. As mentioned earlier, more and more people are facing fuelwood deficit, and non-commercial sources of energy are still predominantly used in inefficient ways. Efforts to halt and reverse tropical deforestation are progressing very slowly (Chapter 7). Many of these efforts have been made at the grassroots level and by non-governmental organizations.

Energy for sustainable development

Patterns in world energy use are marked by striking contrasts and inequities. Differences in wealth, in economic development and in life-styles create huge disparities in the amounts of energy consumed not only from country to country but also from one sector of society to another. A knowledge of how energy is used - and how much is used - is essential for dealing with inefficiencies and for ensuring the availability of supplies fundamental to sustainable development.

Several attempts have been made to quantify basic energy needs (Reddy and Prasad, 1977; Palmedo et al., 1978; Krugmann and Goldemberg, 1980). However, because of the considerable divergence in assumptions, most values given cannot be used for long-term planning. Energy needs are not static. They are subject to considerable variations with time, especially in developing countries. The question of how much energy is needed per unit of output (i.e. what is the optimum energy intensity) is also a subject of widely divergent opinions. Energy intensity depends on the structure of the economy and the state of human development.

The World Commission on Environment and Development (WCED, 1987) concluded that, 'A generally acceptable pathway to a safe and sustainable energy future has not yet been found'. However, it should be stressed that the energy problem over the next two decades should be seen as one of transition, in which countries need to ensure the rational, non-wasteful use of non-renewable sources of energy, and to diversify their sources of energy, establishing appropriate and environmentally realistic energy mixes to meet their incremental needs. Such energy mixes will vary from one country to another. By 'appropriate' is meant all energy systems that economically match production with utilization. For example, if low heat is required for a process, it would be more appropriate to generate such heat with solar energy than with conventional boilers. By 'environmentally realistic' is meant energy systems that are constructed with the 'best available technology' to minimize their environmental impacts. Finally, such a national energy mix should be based on detailed assessments of national energy supply and demand and should maximize the use of suitable indigenous sources to achieve national energy security. This is a pre-condition for achieving independence in development strategies and decisions. However, self-sufficiency in energy sources could turn out to be counterproductive, both economically and environmentally in an interdependent world, if the true cost of using indigenous polluting energy sources is taken into account.

The emphasis on sustainable development, and the emergence of global warming as an environmental threat, have rekindled interest in non-fossil sources of energy. Geothermal, tidal and wave energy do not - as yet - represent energy sources of worthwhile magnitude. The two main sources for 'commercial energy' that are being debated and pursued with varying interest and perseverance, are nuclear and solar energy. The problems with the first are safety and the cost of energy: for solar energy, it is cost and the efficiency of conversion.

Fission inevitably generates radioactive isotopes, and the complexity of current generations of nuclear reactors makes them prone to catastrophic failures. New approaches to safety explore approaches based on the use of passive safety systems, modular designs for smaller units and the use of working fluids other than water (WRI, 1989). However, none of these novel approaches has proved its viability in actual operation. The twin problems of managing nuclear waste and decommissioning of old plant remain without satisfactory solutions. The problem is partly attributed to public perceptions of the hazards involved. If nuclear energy is to contribute to sustainable development, fusion technology is the hope for the future. It is clean - or at least less contaminating than fission technologies - and it provides unlimited resources of energy. The recent announcement (November 1991) of a significant breakthrough is a landmark. However, it will probably be several decades before the commercialization of fusion technologies arrives.

The situation on the solar front is promising, even though solar-based devices do not yet contribute to a great extent to commercial energy needs anywhere in the world. A view that is gaining support is that solar-based energy systems may well prove to be a viable basis for sustainable energy supplies. It is already widely used all over the world in water heating. The most promising development in electricity generation is the increase in the conversion efficiencies of photovoltaic (PV) systems. From as low as 2-3 per cent conversion efficiency some twenty years ago, efficiencies have reached 20 per cent, and more, in sizeable panels and over longer periods. Developments have covered cell materials, methods of manufacture, packaging of panels, and systems for specific applications. The last decade has also witnessed substantial and continuous decreases in costs, coupled with improved reliability. One forecast (Ogden and Williams, 1989) is that the cost of PVs will decline from $US4-5/Wp (peak watt) to $US0.2-0.4/Wp over 10-15 years. This is particularly suited to providing the energy needs of small, widely dispersed communities, once the storage problems are appropriately solved. Furthermore, some twenty utilities in the USA already incorporate PV systems in their operations (UNEP 1991). Sizeable experimental installations are in operation in Europe, the Middle East, Asia and Australia. The USA Solar Energy Research Institute (SERI) maintains that PVs have the potential of becoming the primary source of electricity world-wide by the end of the next century.

One interesting concept, which has been pursued since the mid-1970s, envisages a hydrogen-based economy. Hydrogen is a 'clean' fuel that would eliminate all the known undesirable environmental impacts of fossil and biomass fuels. The main source of hydrogen is water electrolysis, which needs electric energy. Hydro-electric schemes are an obvious clean source of electricity. The hydrogen 'lobby' claims that piping hydrogen from electrolysis installations close to hydro-electric power stations is more efficient and less expensive than transmitting electrical energy over high-tension grids, transformer stations and switchgear. On the consumption side, recent developments in fuel cells are claimed to produce electricity more efficiently and cleanly, providing pure water for domestic and other purposes. Producing steam directly from hydrogen with 100 per cent efficiency of conversion has been demonstrated on a pilot scale in Germany. The use of hydrogen in transport vehicles is also being actively pursued (Chapter 14). In the long term other means of producing hydrogen (e.g. thermolysis, catalytic photolysis, bio-photolysis) are also being investigated. A recent initiative launched by the German Aerospace Agency envisages the use of PVs covering vast areas of desert to produce hydrogen. The project proposes setting up a solar farm in southern Algeria and using existing pipelines across the Mediterranean for the transport of hydrogen to consumption centres in Europe. The International Association for Hydrogen Energy has been holding international conferences every two years, and also publishes the *International Journal of Hydrogen Energy*.

The fuel cell was invented more than a century and half ago, but the revival of interest over the last few decades was pioneered by R&D for manned space flights. A fuel cell consumes fuel and produces electricity without changes in its electrolyte or electrodes. In principle it can operate on a variety of gaseous fuels: in practice, however, only hydrogen and carbon monoxide give useful efficiencies. When hydrogen is used as a fuel, the cell also yields water. Fuel cells are fast becoming serious competitors to heat engines. As mentioned earlier, they are more efficient (claims have been made for a possible 40-70 per cent reduction in primary energy consumption). From an environmental point of view, their main attraction is that they do not produce air polluting, global-warming, flue gases. Current designs use reformers that are fed on hydrocarbon fuels and supply the cell with hydrogen-rich gases. Several 50 to 200 KW units are already on the market, while a Japanese utility is now using an 11 MW plant of mixed American-Japanese design. Japan's MITI is supporting the development for a target of 10 GW by the year 2010. A 'World Fuel Cell Council' was formed in 1991 to promote the development and use of fuel cells.

Wind energy is currently used for commercial power generation. Unit costs have decreased steadily with advanced design and operation techniques, and reliability has improved. Units of more than 1,000 MW are already in operation in developed countries where the climatic conditions are suitable. It was estimated, at the 1989 European Wind Energy Conference, that 1600 MW of wind turbine power are already installed world-wide and that they were reducing CO_2 emissions by three million tonnes per year. Experimentation is actively pursued with different configurations of windmills (vertical axis, single blade designs, Darius rotors).

REFERENCES

BP (1991) *BP Statistical Review of World Energy*. London.

Chandler, W.V. (1985) Increasing energy efficiency. In: Brown et al., (eds) *State of the World - 1985*. W.W. Norton, New York.

ECE (1976) *Increased Energy Economy and Efficiency in the ECE Region*. E/ECE/883/REv.1, Economic Commission of Europe, Geneva.

El-Hinnawi, E. (1991) Personal communication.

El-Hinnawi, E. (1980) *Nuclear Energy and Environment*. Pergamon Press, Oxford.

El-Hinnawi, E. and Biswas, A. (1981) *Renewable Sources of Energy and the Environment*, Tycooly International, Dublin.

El-Hinnawi, et al., (eds) (1983) *New and Renewable Sources of Energy*, Tycooly International, Dublin.

El-Hinnawi, E. and Hashmi, M. (1987) *The State of the Environment*. Butterworth, London.

FAO (1983) *Fuelwood Supplies in Developing Countries*. FAO, Rome.

Flavin, C. (1986) *Electricity for a developing world*. Worldwatch paper 70. Worldwatch Institute, Washington, D.C.

Goldemberg, J. et al., (1987) *Energy for Development*. World Resources Institute, Washington, D.C.

Holdgate, M., Kassas, M. and White, G. *The World Environment, 1972-1982*, Tycooly International, Dublin.

IEA (1987) *Energy conservation in IEA countries*. IEA, Paris.

IEA (1989) *Electricity end-use efficiency*. IEA, Paris.

IUCN/UNEP/WWF (1991) *Caring for the Earth*. IUCN, Gland, Switzerland.

Krugmann, H. and Goldemberg, J. (1980) *The energy cost of satisfying basic human needs*. Instituto de fisica, Sao Paulo University.

Lovins, A.B. (1977) *Soft Energy Paths*. Penguin, Harmondsworth.

NEA/IAEA (1990) *Uranium Resources, Production and Demand*, Paris.

OECD (1985) *The State of the environment - 1985*. OECD, Paris.

OECD (1988a) *Emission Control in Electricity Generation*. OECD, Paris.

OECD (1988b) *Environmental Impacts of Renewable Energy*. OECD, Paris.

OECD (1989a) *Energy and the Environment*. OECD, Paris.

OECD (1989b) *Energy balances of OECD countries*. OECD, Paris.

OECD (1991a) *Environmental Indicators*. OECD, Paris.

OECD (1991b) *State of the Environment*. OECD, Paris

Ogden, J.M. and Williams, R. H. (1989) *Solar Hydrogen*, WRI, Washington, D.C.

Palmedo, P.F. et al. (1978) *Energy Needs and Resources in Developing countries*. Brookhaven National Lab. BNL,50784.

Reddy, A. and Prasad, K. (1977) Technological alternatives and the Indian energy crisis. *Economic & Political Weekly*, p.1496.

Semenov, B. et al. (1989) Growth Projections and Development Trends for Nuclear Power, *IAEA Bull.*, **Vol. 31**.

UNEP (1981) *Environmental Impacts of Production and Use of Energy*, Tycooly International, Dublin.

UNEP (1989) *UNEP Environmental Data Report 1989/1990*, Blackwell.

UNEP (1990) *UNEP Environmental Data Report 1990/1991*, Blackwell.

UNEP (1991) *UNEP Environment Data Report 1991/92*, Blackwell.

WCED (1987) *Our Common Future*, World Commission on Environment and Development. Oxford University Press, Oxford.

WEC (1980) *Survey of Energy Resources*, World Energy Conference, Munich, 1980.

World Bank (1986) *World Development Report*. World Bank, Washington, D.C.

WRI (1986) *World Resources*, Basic Books, New York.

WRI (1989) *World Resources*, Basic Books, New York.

Zhu, J.L. and Chan, C. Y. (1989) Radioactive Waste Management: World Overview, *IAEA Bull.*, **Vol 31**.

CHAPTER 14

Transport

The issue - transport, environment and development

Transport services play a major role in trade and in the national economy, but most of all in the everyday lives of people. They are essential to a modern society and people expect that they will be there when required and that they will function properly, even though they know that transport imposes a substantial cost burden on the community. However, it is not only a question of cost: transport also has significant effects on the quality of the environment and has both direct and indirect health impacts, causing millions of deaths and injuries each year in car accidents alone (Table 1). These side effects result in enormous real costs to society which so far have been virtually ignored in economic and development planning. Not only is it necessary to integrate the transport sector into all aspects of planning, but it is also necessary for the transport sector to accept liability for these environmental and health costs.

Table 1: *Selected environmental effects of principal transport modes.*

Principal transport modes	Air	Water resources	Land resources
Marine and inland water transport.		Discharge of ballast water, oil spills, etc. Modification of water systems during port construction and canal cutting and dredging.	Land taken for infrastructures; dereliction of obsolete port facilities and canals.
Rail transport			Land taken for rights of way and terminals; dereliction of obsolete facilities.
Road transport	Air pollution (CO, HC, NO_x, particulates and fuel additives such as lead). Global pollution (CO_2, CFCs).	Pollution of surface water and ground-water by surface run-off; modification of water systems by road building.	Land taken for infrastructures; extraction of road building materials.
Air transport	Air pollution	Modification of water tables, river courses and field drainage in airport construction.	Land taken for infrastructures; dereliction of obsolete facilities.

Source: Based on OECD (1991).

Despite the fact that the development of an adequate transport system and its associated infrastructure has been slow in most developing countries, and that the number of cars and other vehicles is relatively small, these countries also suffer from major environmental problems, especially in urban areas, arising from the transport sector.

During the 1980s, many developing countries faced a major development dilemma in the transport sector. While the curtailment of investments in transport infrastructure, the attenuation of maintenance and the restriction of imports of vehicles and spare parts (as well as fuels) may have helped to achieve internal and external fiscal equilibrium, such measures conflicted with growth and development objectives. This dilemma was particularly acute since transport infrastructures were generally not well developed in these countries. The demand for transport is a 'derived demand' which mainly reflects the level of economic activity. The market for passenger transport depends particularly on the travel demands of households, while the market for freight transport depends on production and trade. The future trend for the transport sector (and the related environmental impacts) in the developing countries is therefore

Solid waste	Noise	Risk of accidents	Other impacts
Vessels and craft withdrawn from service.		Bulk transport of fuels and hazardous substances.	
Abandoned lines, equipment and rolling stock.	Noise and vibration around terminals and along railway lines.	Derailment or collision of freight trains carrying hazardous substances.	Partition or destruction of neighbourhoods, farmland and wildlife habitats.
Abandoned spoil tips and rubble from road works; road vehicles withdrawn from service; waste oil.	Noise and vibration from cars, motorcycles and lorries in cities and along main roads.	Deaths, injuries and property damage from road accidents; risk of transport of hazardous substances; risks of structural failure in old or worn road facilities.	Partition or destruction of neighbourhoods, farmland and wildlife habitats; congestion.
Aircraft withdrawn from service.	Noise around airports.	Deaths, injuries, property damage, from aircraft accidents; but slight compared with road transport.	

directly related to overall development and economic progress in these countries (UNCTAD, 1990).

By comparison, current transport trends for the OECD countries indicate that both freight and passenger traffic are likely to continue to grow steadily, as has been the case for the last few decades (Figure 1). This is a reflection of the expected economic growth in these countries. For Europe, the liberalization of road haulage, the introduction of cabotage (freedom for a carrier to operate in the domestic market of another country) and the removal of frontier barriers in 1992 are expected to reduce the costs of transport by truck by up to 25 per cent and to increase truck kilometres by up to 100 per cent, unless actions to limit this growth are taken. The growth rate of freight transport in terms of tonne-kilometres is not expected to be very high, because of changes in demand pattern. However, in terms of transported goods-value or wagon-kilometres, a more substantial increase is anticipated (ECMT/OECD, 1990).

Figure 1: *Traffic trends for OECD countries, 1970–87: (left) passenger traffic trends and (right) freight traffic trends.*

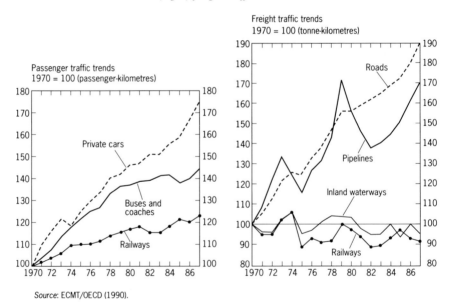

Source: ECMT/OECD (1990).

Because road transport is by far the biggest energy consumer of all transport modes, it is responsible for the major part of the environmental impacts caused by transport. Cars, buses and trucks have become the largest single source of carbon monoxide, nitrogen oxides and hydrocarbons in most areas in both developed and developing countries, and they produce other pollutants of great concern, for instance toxins such as benzene, particulates, poly-aromatic compounds, and dioxins related to chlorinated scavengers used in leaded petrol.

Emissions of these pollutants cause or contribute to adverse health effects in addition to harming natural and managed ecosystems. There are other health impacts from motor vehicles. Millions of people are injured or killed in car accidents every year and a major part of the population, especially in urban areas, is affected by traffic noise. In addition to such local and regional impacts, a significant development during the last few years has been the emergence of global warming as a matter of major international concern, to which the transport sector is one of the major contributors. As can be seen from Table 2, energy use and related emissions from transport are highly dependent on the choices that are made between various modes of transport, although for some people the choice of modes may be limited (for example, most long-distance travellers in practice have only one option - air transport).

Table 2: *Equivalent emissions (in grams per person-kilometre) and equivalent energy use (in kWh per person-kilometre) for long-distance domestic transport in Sweden.*

Type of vehicle	HC	NO$_x$	CO	Energy use	Load factor (number of passengers)	Reference[3]
Car *without* catalytic converter	0.6	1.1	4.6	0.4	2	(A)
Car *with* catalytic converter	0.04	0.05	0.3	0.4	2	(B)
Bus model 1989: engine type ECE R49–20%	0.05	1.21	0.09	0.23	11	(B)
Bus model 1989: engine type ECE R49–50%	0.05	0.76	0.09	0.23	11	(B)
Passenger train[1]	0	0	0	0.11	150	(A)
Air Fokker[2] F28-4000	0.15	0.53	0.55	0.69	57	(C,D)
Air DC9–41	0.18	0.72	0.58	0.8	82	(D,E)
Air MD–82	0.07	0.71	0.23	0.55	105	(D,E)
Air Saab 340A	0.1	0.21	0.6	0.57	23	(B)
Air Saab 2000	0.03	0.24	0.1	0.58	36	(B)

Notes: 1) Electric train and electricity produced by PFBC (pressurized fluid bed combustion: NO$_x$ = 0.04; equivalent energy 0.25- ref: ABB).

   2) For all airplanes, emissions are estimated for an air transportation length of 380 kilometres.

   3) A = Swedish Road and Traffic Research Institute.

   B = Saab-Scania.

   C = Linjeflyg (domestic air company).

   D = Swedish Air Transport Authority.

   E = Scandinavian Airlines System, SAS.

Source: T. Mansson, personal communication.

Transport trends of environmental significance

Motorized road transport

About half the world's oil production is consumed by road vehicles, whose growth has consistently outpaced that of the human population. The 'car explosion' may prove to be as severe a problem as the population explosion has ever been. According to some estimates, the fleet's annual increase has averaged 10 million cars and 5 million buses and trucks world-wide. If the trend continues, a billion vehicles will use the world's roads by 2030 (Figure 2) (Walsh, 1990a; Pemberton, 1988). Table 3 shows the wide variations in fleets between developed and developing regions and within the developing regions themselves. According to another estimate almost 80 per cent of the global car population is to be found in the industrialized world. The table also illustrates the dominance of world motor vehicle markets by the United States and Western Europe. However, developing regions will account for much of the growth in fleet numbers as vehicle ownership in the developing regions stabilizes (Faiz *et al.*, 1990)

Although the great majority of the vehicles on the road today are in the industrialized world, a significant increase during the next 50 years is likely to occur in Central and Eastern Europe and the developing countries. However, the markets in the industrialized countries are so large that they will continue to dominate the total world market. The environmental consequence of this is that to a large extent, vehicle characteristics in the industrialized countries will continue to have a significant impact on vehicle characteristics in other parts of the world (Walsh, 1990a).

While in absolute terms most of the automobile growth has occurred in highly industrialized countries, many developing countries are experiencing very high rates of growth especially in urban areas. In extreme cases, such as the city of Seoul, South Korea, vehicle populations are doubling about every four years. Rising incomes, increased demand for personal mobility, and travel are the factors behind the very rapid increase in both automobile ownership and bus transportation in much of Asia, the Middle East, Central and Eastern Europe and parts of Africa. The numbers of motorcycles, scooters and three-wheelers are also growing rapidly in these countries. In Taiwan, for example, motorcycles accounted for 76 per cent of the registered motor vehicle fleet in 1988 (Faiz *et al.*, 1990). However, there are few statistics available on these vehicles for other parts of the world.

In urban areas there are growing numbers of various kinds of buses and taxis which account for a large proportion of the passenger vehicle-kilometres travelled. In many developing countries, intermediate-size vehicle transport

Figure 2: *The growth in the global population of vehicles: (upper graph) as total numbers of cars, trucks and buses 1930-90 and (lower graph) as a comparison between population growth and passenger car growth in the world (with 1950 = 100).*

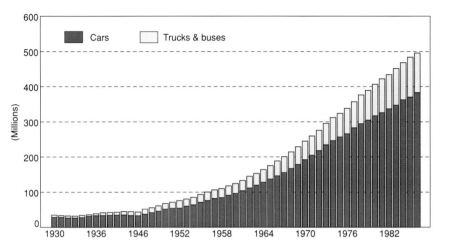

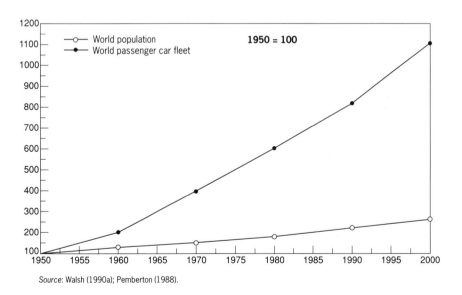

Source: Walsh (1990a); Pemberton (1988).

services (known as paratransit systems) are among the most important means of transporting people and goods. Paratransit vehicles flourish in different forms and under various names, such as Tempos, Jeepneys and Matatus; see Table 4 (UNCHS, 1984).

411

Table 3: *Car density and car fleet in 1986, by region.*

Region or country	Density (people per car)	Fleet (million vehicles)
USA	1.8	135
Western Europe	2.8	125
Oceania	2.8	8
Canada	2.2	11
Japan	4.2	29
South Africa	11	3
Eastern Europe	11	17
Latin America	15	26
Soviet Union	24	12
Asia[1]	62	12
Africa[2]	110	5
India	554	1.4
China	1,374	0.8
World	12	386[3]

Notes: 1. Excluding Japan, China and India.

2. Excluding South Africa.

3. Column does not add to total due to rounding.

Source: Brown, L. (ed.) (1989).

Motor vehicle use is inextricably linked to the degree of urbanization. The Latin America and Caribbean region, for example, is highly urbanized (70%) and is also the most motorized among developing regions, accounting for over 60 per cent of the automobiles and about 30 per cent of the buses and trucks in developing countries. In 1986, Brazil, Mexico, Argentina and Venezuela collectively had almost twice as many automobiles as all the developing countries in East and South Asia and sub-Saharan Africa. Because urbanization remains very rapid in these regions, the environmental problems related to transport are also likely to increase rapidly, and effective measures to deal with them are urgently required.

The demand for fast and reliable distribution of goods, coupled with the increasing pace of containerization and the increasing value of transported goods, all favour using transport options that provide reliable 'just-in-time' services. This is likely to increase the reliance on trucks in all parts of the world.

The global car explosion has not been fully matched by expansion of the road network, consequently traffic congestion in many parts of the world, especially in the big cities, is an ever-increasing problem. In OECD countries the road network has continued to expand since 1970, but at a much slower rate than in earlier decades. However, the motorway network increased at a high rate: over the period 1970 to 1988 it more than doubled in Europe and increased almost sixfold in Japan, whereas the North American network increased by only 60 per cent, partly due to citizens' protests against new construction projects. In total length, however, the United States remains in the lead with the construction of almost 1,600 kilometres of motorways annually. In the oil-producing countries and some of the Newly Industrialized Economies such as South Korea and Brazil, the level of investment in roads has also been increasing very rapidly during the period. By contrast, nineteen of the Least Developed Countries have a road network density of less than 60km per 1000km². This is significantly lower than the level of many other developing countries (OECD, 1991; UNCTAD, 1990).

Fuel consumption and motor vehicles

After the oil crises in the 1970s, all car manufacturers started to look for ways of increasing fuel efficiency. As a result, new passenger cars in the United States today are much more fuel-efficient than the 'gas-guzzlers' of the early seventies, though still less so than those produced in most other OECD countries. However, as illustrated in Figure 3, the results for the other OECD countries are quite discouraging. In these countries, the power and size of new cars continued to increase, offsetting any fuel-efficiency gains and resulting in only marginal effects on the fleet average (Schipper, 1991). It must be remembered that there are limits to the possibility of reducing average fuel consumption without compromising vehicle safety. However, efforts to improve efficiency continue and it is currently possible to approach 4 litres per 100 kilometres, even though actual results are strongly influenced by driver behaviour.

Figure 3: *OECD automobile energy intensities (in mJ per vehicle-kilometre) and fuel economy (in litres per 100 kilometres), as on-the-road fleet averages, 1970-88.*

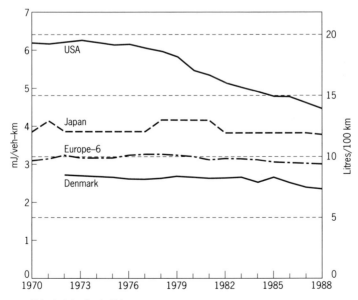

Notes: Includes diesel vehicles.
US data include personal light trucks.
Source: Stockholm Environment Institute, based on Schipper and Meyers (1992).

Despite the fact that the oil crises of the 1970s resulted in increased energy-efficiency in many households and in the manufacturing and service industries, energy demand for transport continued to increase. While energy intensities of passenger transport fell between 1970 and 1987 in a few countries, growth in the volume of travel and shifts towards more energy-intensive modes resulted in increased energy use for transport in many OECD countries. As a result the total oil consumption in the world, and transport's share of it, continued to increase (Figures 4, 5, 6 and 7). These statistics suggest that the prospects for reduced energy use by transport are not good; after all, if little or no energy was saved during the years when fuel prices were high, little can be expected in the 1990s if prices remain at close to their 1973 levels (Schipper and Meyers, 1992).

Figure 4: *Transport's share of total world oil consumption, 1974–86.*

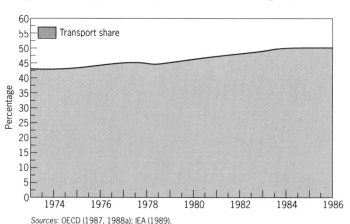

Sources: OECD (1987, 1988a); IEA (1989).

Figure 5: *World oil consumption for motor gasoline, 1950–90.*

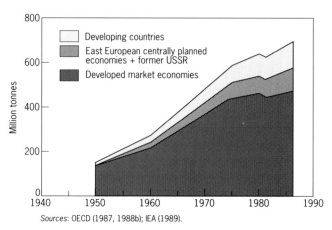

Sources: OECD (1987, 1988b); IEA (1989).

Non-motorized transport

Despite the fact that there has been a global explosion in the number of cars and a dramatic increase in overall transport activity, this is very unevenly distributed among people. Motorcycles and mopeds are increasingly popular in much of the developing world, but in addition to being heavily polluting, their initial cost is too great for many low-income earners and they require expensive (and in many cases scarce) fossil fuels to operate. Even buses are out of reach for many: it is estimated that one-fourth of households in Third World cities cannot afford public transport (Pemberton, 1988).

Table 4: *Importance of paratransit services, illustrated by selected locations.*

	Vehicles per 1,000 population	Percentage of public transport trips
Bus-type paratransit		
Tempos (Jaipur, India)	0.3	72
Jeepneys (Manila, Philippines)	3.2	64
Bemos (Surabaya, Indonesia)	0.9	39
Taxi-type paratransit		
Cycle-rickshaws (Kampur, India)	33.0	88
Auto-rickshaws (Jakarta, Indonesia)	0.9	20

Source: UNCHS (1984).

The traditional means of getting about such as walking, or using bicycles or animals, remain the major modes of transport for most people in the world. For example, in 1981 it was estimated that two-thirds of Indian rural transportation was by animal-drawn vehicles carrying perhaps 15 billion tonne-kilometres of freight per year. Walking and cycling are the most common forms of individual transport, but are usually limited to short journeys. The proportion of walking-trips is especially high for low-income groups, exceeding 80 per cent of low-income trips in Bombay, compared with about 25 per cent for the highest income group. Bicycles are a particularly important form of personal transport in many developed and developing countries: they are both environmentally friendly and, for most people, healthy, and do not require complex infrastructures for either their manufacture or servicing (Table 5).

Table 5: *Bicycles and automobiles in selected countries.*

Country	Bicycles (millions)	Autos (millions)	cycle/auto ratio
China[1]	300.0	1.2	250.0
India	45.0	1.5	30.0
South Korea	6.0	0.3	20.0
Egypt	1.5	0.5	3.0
Mexico	12.0	4.8	2.5
Netherlands	11.0	4.9	2.2
Japan[1]	60.0	30.7	2.0
West Germany	45.0	26.0	1.7
Argentina	4.5	3.4	1.3
Tanzania	0.5	0.5	1.0
Australia[1]	6.8	7.1	1.0
USA[1]	103.0	139.0	0.7

Note: 1. Data for 1988.
Source: Lowe (1990).

Railways

Internationally comparable statistics on railway traffic are very limited, especially those dealing with passenger traffic, but Table 6 gives some indication of trends. It is evident that railway passenger transport has increased substantially since 1975 in the developing countries, but only slightly in the rest of the world. On the other hand there has been a substantial increase in railway freight in all regions. The former USSR and other countries in Central and Eastern Europe have dominated railway freight transport, accounting for more than 50 per cent of world total railway freight (in tonne-kilometres) though it seems possible that this dominance will decline with the shift to market economies and the freeing-up of trade with Western Europe.

Further development of the railway systems of Europe is likely to be one outcome of the integration of the European economies in 1992. There is likely to be an increase in the average distance over which goods and passengers will be transported, and because of the limited capacity of the road infrastructure and the generally adverse environmental effects of motor transport, unrestricted growth of automobile and truck usage is likely to prove unacceptable. The best alternative seems to be the train for the longer distances, probably in combination with intermodal transport. While it is increasingly recognized that substantial amounts of money need to be directed towards expansion of rail infrastructure if this is to be fully effective, significant organizational adaptations will also be necessary since almost every nation has its own railway company, using different standards for signalling and electrical equipment. Not all the railways are connected to the European railway information system (HERMES) and consequently data communication between the different railways is

Table 6: *Railway freight and passenger transport in various regions.*

	Railway freight net ton-kilometres (1,000 millions)				Railway passenger-kilometres (millions)		
	1970	1975	1980	1985	1975	1980	1985
Developing countries	556	747	907	1,116	363,000	500,000	638,000
E. Europe + former USSR	2,784	3,596	3,828	4,096	439,000	471,000	518,000
Developed market economies	1,687	1,639	1,980	2,007	617,000	627,000	632,000
World total	5,027	5,982	6,715	7,219	1,419,000	1,598,000	1,788,000

Source: UNCTAD (1990); UN (1986).

inadequate, making it difficult at times to get information concerning the whereabouts of a train and its expected time of arrival (Nijkamp *et al.*, 1990).

As a result of the closure of unprofitable lines, the total length of the rail network diminished by 4.1 per cent in the European OECD countries and by 1.8 per cent in Japan between 1970 and 1985. However, the length of electrified lines increased and the proportion of electrified lines in the overall network has risen from 29.3 per cent to 38.6 per cent in European OECD countries and from 42.1 per cent to 52.8 per cent in Japan in the same period.

In developing countries, most of the railways were built during colonial times and were designed for colonial needs, and usually for an extractive economy. They were built for low speeds and light (20 to 30 tonne) axle weights, and in recent years tracks, ways and bridges have been undermaintained. This does not mean that these transport systems cannot be adapted to serve present-day and future national needs. However, there is a substantial need for rehabilitation and maintenance of existing infrastructure (ECMT/OECD, 1990). The rail transport problem is particularly acute in the Least Developed Countries. Available data for eighteen of them indicate that eight have a railway network density of less than 2km per 1000km². This is well below the average level for other developing countries. In view of the investments required, very few such countries have plans to build or expand their railway transport networks (UNCTAD, 1990).

Air transport

Transport by air is growing very rapidly in all parts of the world. Airlines registered in developing countries have greatly expanded their services, both domestic and international, and airport construction or expansion has occurred or is under way in many of them (Meyers, 1988). Judging from historic trends (see Figures 6 and 7) developing-country air travel could therefore have a significant effect on world oil demand in the 1990s.

Figure 6: *World oil consumption for aviation gasoline and jet fuel, 1972–87, for three major country groupings.*

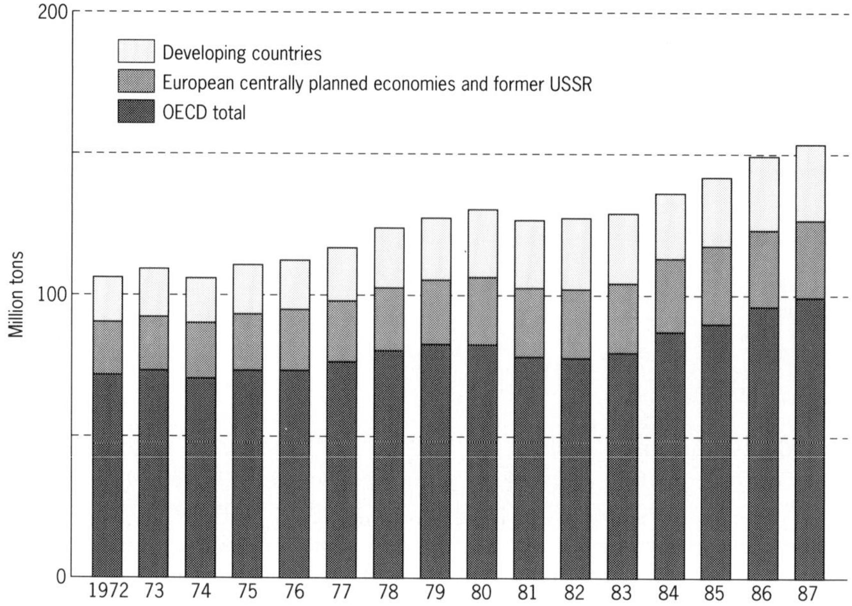

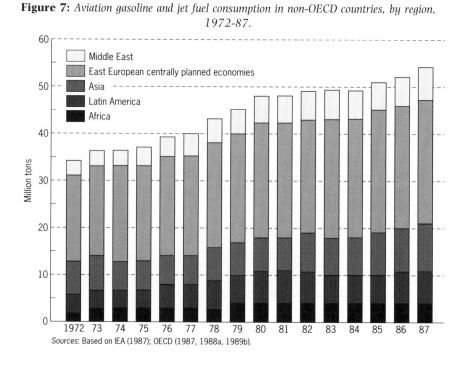

Sources: Stockholm Environment Institute, based on OECD (1987, 1988a, 1989b).

Figure 7: *Aviation gasoline and jet fuel consumption in non-OECD countries, by region, 1972-87.*

Sources: Based on IEA (1987); OECD (1987, 1988a, 1989b).

Because aircraft have been getting larger, the number of aircraft movements has not increased as rapidly as has the number of passengers and the volume of freight traffic; and this trend is expected to continue. During the 1980s the so-called wide-bodied jets increased in number at a rate greater than for other types of aircraft. Of all aircraft in the world, about 70 per cent belong to United States companies or individuals, and account for about 65 per cent of total aircraft hours flown (ICAO, 1988).

Between 1972 and 1990, the number of passenger-kilometres of air travel more than tripled, while air freight has increased from 25,940 million tonne-kilometres in 1978 to 53,450 million tonne-kilometres in 1988. During this time, the growth in railway passenger transport has been relatively slight. Therefore it seems likely that air transportation will, eventually, surpass railways as a mass passenger carrier (Figure 8). However, there is a growing congestion problem in air corridors around many of the larger cities of Europe, the United States and Japan, arising from the rapid increase in air traffic which, in combination with the introduction of high-speed trains, may change the situation for some countries. The experience of Japan and France with high-speed trains is encouraging, demonstrating that there are technologically feasible alternatives to air travel that are also acceptable to the travelling public.

Figure 8: *Comparison in growth of air and rail transport for passenger traffic (in billions of passenger-kilometres).*

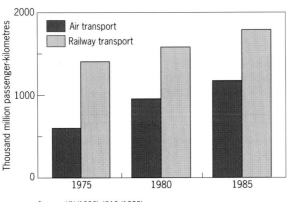

Sources UN (1986); ICAO (1988).

The transport by air of dangerous goods is relatively limited, although substantial increases have occurred in recent years. The Convention on International Civil Aviation (the Chicago Convention) contains a set of standards for the safe transport of dangerous goods by air. The updating of these standards is the responsibility of the Dangerous Goods Panel (DGP) of ICAO. The International Air Transport Association (IATA) issues precise guidelines which are updated regularly. In addition, IATA levels significantly high charges for the transport of dangerous goods.

Transport by ship

The transport of passengers over long distances by sea virtually ceased in the decades following the 1960s when intercontinental passenger jets entered service, but during this period there has been a dramatic increase in the movement of goods by sea, especially those transported in bulk carriers. In 1990 there were over 3,000 tankers and combined carriers in service worldwide, and although this represents a decline of about 7 per cent in number since

Table 7: *World tanker and combined carrier fleets, 1970-90.*

	Tankers[1]			Combined carriers[2]		
Year	Number	Total capacity[3] (10^7 dwt)	Average capacity[4] (dwt)	Number	Total capacity[3] (10^7 dwt)	Average capacity[4] (dwt)
1970	3,235	152	46,897	212	15	70,733
1971	3,331	171	51,446	246	20	82,493
1972	3,359	190	56,485	299	29	95,355
1973	3,458	216	62,346	351	37	104,852
1974	3,638	256	70,305	381	41	108,248
1975	3,674	291	79,323	394	44	110,728
1976	3,636	321	88,209	410	46	111,413
1977	3,564	332	93,287	421	48	113,259
1978	3,370	328	97,472	417	48	114,821
1979	3,320	328	98,760	409	47	115,084
1980	3,338	325	97,302	406	47	115,291
1981	3,351	320	95,564	390	45	115,205
1982	3,264	304	93,049	366	42	115,671
1983	3,115	283	90,921	344	40	114,920
1984	3,012	270	89,537	329	38	115,119
1985	2,874	247	85,840	307	35	113,244
1986	2,874	241	83,846	291	33	114,064
1987	2,886	239	82,940	293	34	114,972
1988	2,923	243	83,295	294	34	115,656
1989	2,978	250	84,097	291	34	115,197
1990	3,015	255	84,474	290	33	115,333

Notes: 1. Commercial tankers include oil company or other privately owned vessels, and those owned by governments. Vessels that are owned or are under long-term charter to governments for military use, or tankers that were originally built or have been subsequently converted or classed primarily for non-transportation service are not included.
2. Vessels retaining the ability to carry both wet and dry cargoes.
3. Total fleet carrying capacity in deadweight tons (dwt).
4. Average carrying capacity per vessel in deadweight tons (dwt).

Source: UNEP (1991).

1970, their combined carrying capacity has increased by about 80 per cent because of an increase in the size of vessels. This trend has now been reversed, however, with the average size of both tankers and combined carriers significantly less than it was in 1978 (Table 7). Furthermore, the total tonnage of goods transported by sea declined from 3,436 million tonnes in 1977 to 3,125 million tonnes in 1985, while the average length of haul declined from 5,242 to 4,034 nautical miles over the same period (UNEP, 1989a).

As Figure 9 indicates, during the late 1970s and early 1980s there was a decline in world seaborne trade related to decreased trade in oil. However, there has been a steady growth of seaborne trade in dry cargo and it is expected that this will continue to increase (OECD, 1989a).

Figure 9: *Development of international seaborne trade, 1970–90.*

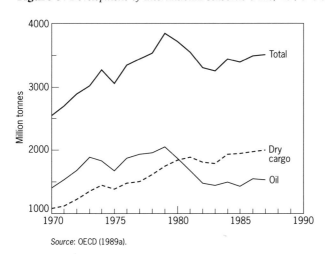

Source: OECD (1989a).

Despite the decline in oil trade in recent years, much of present-day shipping activity still involves the transport of oil. Figure 10 shows the main inter-area oil movements by sea in 1987 (in millions of barrels daily). Not surprisingly, the greatest density of oil spills occurs along the busiest sea routes, for example in the Persian Gulf and Southeast Asia. While oil spills account for only a small proportion of the total input of hydrocarbons into the marine environment, the grounding of the *Exxon Valdez* in Prince William Sound, Alaska, in March 1989 was a reminder that oil spillages on a catastrophic scale can occur virtually anywhere along the main tanker routes (Chapter 9).

A very significant amount of marine pollution is caused by marine transport, and the substances involved vary enormously in quantities transported and their potential harm to the marine environment. In tonnage terms, the most important pollutant resulting from shipping operations is oil. The best known cause of oil pollution is that arising from tanker accidents, although the most common pollution incidents occur during terminal operations when oil is being

Figure 10: *Movement of oil by sea in 1987 (in millions of barrels per day). The map shows movements in excess of 0.2 million barrels per day: regional totals therefore do not necessarily equal the sum of movements between the regions.*

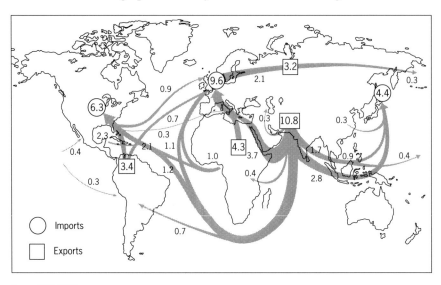

Source: UNEP (1991).

loaded or discharged (IMO, 1989). A great quantity of oil also enters the sea as a result of normal tanker operations, usually associated with the cleaning of cargo residues. Although the volume of dangerous goods carried at sea is only a fraction of the amount of oil transported every year, many of these goods - particularly chemicals and noxious substances - are far more dangerous to the marine environment than oil.

Coastal shipping is an important mode of transport in a number of developing countries, as well as in the developed world. It is a mode that is particularly suitable for bulky freight that does not require rapid delivery since it requires less capital investment and has lower operating costs than either rail or road transport.

In some countries, considerable use is made of inland waterways for the transport of goods. In a number of countries extensive canal systems exist, most of which were built in the eighteenth and early nineteenth centuries, but which generally declined in use following the establishment of the railway. Many of these canal systems have been inadequately maintained in the intervening years but they represent a substantial capital investment, and in some cases are being rehabilitated, in part for recreational use, at comparatively little cost. Table 8 indicates that for some countries, inland waterways remain very important for the surface transport of goods.

Effective development of ship transport is, of course, dependent on the availability of adequate port facilities. Most countries have ports but their

Table 8: *Inland transportation of goods: a comparison of modes in selected countries.*

Region/country	Year	By road (10^6 t)	By road (10^6 t-km)	By inland waterways (10^6 t)	By inland waterways (10^6 t-km)	By rail (10^6 t)	By rail (10^6 t-km)
AFRICA							
Cameroon	1986	11[1]				2	884
Egypt	1988	186	30,137	4	2,333	14	6,315
Madagascar	1988	1[2]	326[2]	1		1	204
South Africa	1987	317		94		174	93,158
Tunisia	1988		975[3]	15		12	2,190
NORTH AMERICA							
USA	1987		1,081,024		713,332		1,589,024
SOUTH AMERICA							
Brazil	1986		238,862[2]	61	79,303[2]	221	111,263
Chile	1987			14		14	2,105
Colombia	1986	24	10,007	4[4]	2,849	1	807
Ecuador	1988	29	264			55	8,311
ASIA							
Hong Kong	1988	7		11		2	71
Japan	1987	5,287	230,048[5]	470	204,608	283	21,106
Korea	1988	188	8,783	49	16,883	62	14,005
Turkey	1988		63,480				
EUROPE							
Austria	1988			9	1,626	56	11,392
Belgium	1985	353	19,430	96	5,144	73	8,386
Bulgaria	1988	933	14,961	1	114	71	15,484
Czechoslovakia	1988	344	13,288	15	5,489	300	76,499
Denmark	1988	209[3,6]	10,262[3,7]			3	1,016
Finland	1988	427	23,876	13	4,166	34	7,940
France	1988	1,514	139,802	62	7,417	141	51,511
Germany, Fed. Rep.	1988	2,657	155,956	237	53,746	307	60,960
Hungary	1987	565	12,969	11	10,868	118	22,079
Italy	1986		153,058		164		18,566
Luxembourg	1987	14[8]	243[8]	2		3	602
Netherlands	1988	370[3]	20,549[3]	91	7,564	5	1,075
Norway	1986	250	7,182	76	12,428	9	1,862
Poland	1988	1,440	39,417	16	1,416	435	124,159
Portugal	1987					6	1,640
Spain	1988		126,229			37	14,697
Sweden	1988	342[3]	25,400			54	18,456
UK	1987	1,567	115,113	145	55,067	143	17,577
Yugoslavia	1986	176	22,697	20	4,623	91	28,014
USSR	1987	28,034	499,826	684	256,743	4,132	3,885,895
OCEANIA							
New Zealand	1985			10[9]		11	3,241

Notes: 1. 1984.

2. 1985.

3. 1987.

4. Including coastal trade.

5. Excluding vehicle-km by light motor vehicles (less than 550 cc).

6. Domestic trucks over 6 tonnes total weight.

7. Trucks over 2 tonnes total weight.

8. 1986.

9. Coastal shipping.

Source: UNEP (1991).

adequacy in terms of depth, ease of entry, and standard of facilities varies greatly. In many developing countries more than 90 per cent of all imports and exports are moved by sea. Consequently, these countries have become increasingly involved in port development, and projects to increase the efficiency of port operations and to rehabilitate facilities have been undertaken by several of them during the last two decades. Most attention has been paid to construction of quays, dredging, cargo handling equipment and warehouses but, with the exception of manpower training, relatively little attention has been paid to the non-physical aspects of improving port efficiency (OECD, 1989a). In 1987 there were some 2,448 ports in the world, of which 1,059 were in developing countries (Lloyds, 1987).

The environmental impacts of transport systems

Transport systems can adversely affect the environment in a variety of ways: through their emission of noise and pollutants, especially gases and particulates; through their virtually exclusive use of large areas of land to the detriment of other uses; and through the direct impacts of their construction, including resource consumption and waste generation. Of these, the most pervasive impacts arise from their exhaust gas and particulate emissions. Furthermore, all forms of transport, and particularly road transport, present a significant hazard to human life and property, as well as to the environment.

Transport emissions

Of all the environmental impacts of transport systems, those resulting from emissions of gases and particulate matter, the products of combustion, have received most attention. Emissions from the transport sector represent a growing share of overall emissions arising from human activity. On average for OECD countries, emissions from mobile sources have increased by between 20 per cent and 75 per cent since 1975. Today, in the industrialized countries in general, from 70 per cent to 90 per cent of all carbon monoxide (CO) emissions originate from the transport sector, mainly from motor vehicles. Between 40 per cent and 70 per cent of nitrogen oxide (NO_x) emissions stem from the transport sector and almost 50 per cent of total hydrocarbons (HCs) are emitted by motor vehicles. Around 80 per cent of all benzene emissions originate from gasoline-powered motor vehicles while at least 50 per cent of atmospheric lead emissions are due to automobiles (ECMT/OECD, 1990).

While all emissions result in some change in air quality, their environmental consequences manifest themselves in various ways, and the results are not always directly proportional to the emission levels observed. In the case of pollutants directly emitted from mobile sources (CO, NO_x, HCs and lead) air quality changes are generally observed in the immediate vicinity of the emissions, for example along highways, in streets, in tunnels and in urban areas: these are local impacts.

With respect to secondary pollution effects arising from the conversion of precursor emissions (photochemical smog, acid deposition and eutrophication), both precursors and products can be transported over long distances and their effects are felt not only in the source vicinity but also in relatively remote areas: these are regional impacts.

In the case of carbon dioxide and other gases causing climate change, the scale of the impact is very large and the effects can be felt over the whole planet: these are global impacts.

Local environmental impacts of emissions

At the local level, the consequences of air pollution originating from mobile sources vary considerably according to the pollutant concerned and the urban area affected. Furthermore, as motor vehicles usually emit their pollutants in close proximity to people, their emissions have a greater impact on health than do many other sources of air pollution. The problem is exacerbated in some cities such as Bangkok, Mexico City, Los Angeles and Athens when gridlock (the inability of traffic to move due to occupation of virtually all available street space by vehicles) sets in during peak traffic periods. At these times, air pollution adjacent to the stationary traffic can approach crisis proportions. A further major factor in some places (e.g. Los Angeles) is a combination of local topographical and meteorological conditions that leads to temperature inversions which trap pollutants near the ground for extended periods of time.

Local environmental impacts vary as a consequence of the different fuel and vehicle characteristics that apply in various parts of the world. In the urban areas of many developing countries, the combination of densely congested traffic, poor vehicle maintenance and large numbers of diesel vehicles and two-stroke engined motorcycles has contributed to elevated air pollution and consequent health problems. Many of these problems are also related to high levels of diesel particulates, as the proportion of diesel-powered vehicles (trucks and buses) is often higher in developing countries. For example, many of the paratransit vehicles, such as Jeepneys in the Philippines, are fuelled with low-quality diesel, thus emitting excessive smoke. Several Eastern European countries also have a high proportion of cars and motor cycles with two-stroke engines which are very polluting because of their elevated emissions of hydrocarbons - for example, one-third of the Hungarian car fleet in 1986 (Hinrichsen and Enyedi, 1990).

Particulates in diesel exhausts consist of chain aggregates of carbon microspheres coated with a variety of organic compounds that comprise 15 to 65 per cent of the total particle mass. Several hundred organic compounds have been identified in diesel particulates, some of which are known to be carcinogenic. The particles are small and easily inhaled causing respiratory malfunctions, increased risk of infection and respiratory tract cancers (El Hinnawi and Hashmi, 1987). Since diesel engines are more fuel-efficient and diesel fuel costs less than gasoline, it seems likely that the use of diesel powered vehicles will increase. In Europe, about 60 per cent of particulate emissions come from heavy commercial vehicles, and European countries are moving to regulate the allowable emission levels. Recent technological advances have resulted in significant reductions in particulate levels, including a durable and efficient self-regenerating particle trap that has been installed on city buses in Athens, where particulate air pollution is extremely bad. These have been supported by advances in fuel quality (diesel with very low sulphur content) which have made a significant difference to air quality in cities such as Athens and Mexico City.

Despite having the toughest emission standards in the world, in 1988 over 120 million Americans were living in areas that failed to attain national air quality standards for ozone, carbon monoxide and particulates (US EPA, 1988). If these same standards were in force elsewhere they would routinely be exceeded in many cities. The carbon monoxide content of the air in Budapest, for example, is two-and-a-half times the permissible level in Hungary while smog in Athens is reckoned to claim as many as six lives a day. Sao Paulo, Mexico City, Cairo and New Delhi are among the cities with the world's worst air pollution problems. In Calcutta, an estimated 60 per cent of residents are believed to suffer from respiratory diseases related to air pollution. Motor vehicles contribute substantially to all these air pollution problems (Renner, 1988).

As the global vehicle population is expected to increase to about one billion over the next 40 years, it seems certain that future global CO, HC and NO_x emissions will also increase. Figure 11 shows what the trend for these emissions would be if today's state-of-the-art emission controls were introduced on all new vehicles across the entire planet (Walsh, 1990b).

Of all the automobile-generated air pollutants, it is lead that has been dealt with most successfully by the industrialized countries (Chapter 1). By coincidence it was found that catalytic converters (the equipment first introduced in the mid-seventies to reduce emissions of hydrocarbons and carbon monoxide) can only function properly on lead-free gasoline. Thus, both health and technical considerations made the removal of lead from gasoline imperative. However, in Central and Eastern Europe and in many developing countries, the reduction or removal of lead has not yet been given high priority, and atmospheric lead levels are still increasing.

The United States and Japan led the effort to reduce the use of lead, and a large fraction of their car fleets now runs on unleaded gasoline. Between 1976 and 1986, total annual lead emissions in the United States decreased by 94 per cent. The health benefits are unequivocal: over the same period, the average lead level in Americans' blood dropped by more than a third. Steps towards phasing out lead, both by prohibiting the manufacture and sale of cars designed

Figure 11: *Trends in vehicle emissions, in millions of tons per year: (top left) for hydrocarbons, (top right) for carbon monoxide and (bottom) for nitrogen oxides.*

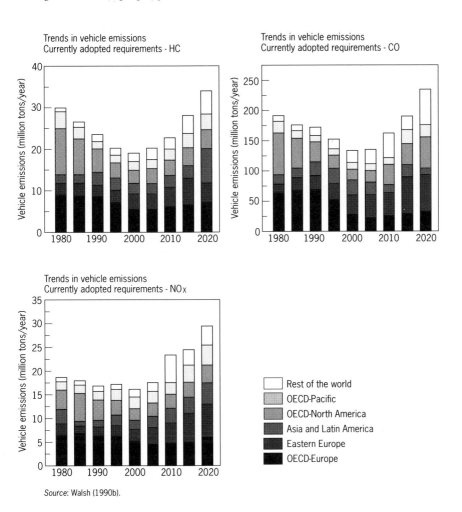

Source: Walsh (1990b).

to use leaded fuel and by making unleaded gasoline widely available and affordable, have also been taken in the other OECD countries.

The overall pattern of emissions from road transport can be changed by introducing new fuels. Brazil's alcohol programme for light-duty vehicles demonstrates this. Following its introduction, lead concentrations in urban areas decreased from 1.6 milligrams per cubic metre in 1978 to 0.3 milligrams per cubic metre in 1983. Carbon monoxide also decreased to a lesser extent, but oxidant levels have risen, probably as a consequence of aldehyde emissions which are five times higher in the exhaust emissions of alcohol-powered vehicles (Renner, 1988).

Programmes for the control of other automotive emissions began in California and were introduced at federal level in the United States in 1963. Japan followed in 1966, while Canada, Australia and most European countries established control programmes in the early 1970s. Initially the main concern was with carbon monoxide (CO) in urban areas. Later, hydrocarbons (HC) were also regulated, but it was not until 1973-74 that control of nitrogen oxides (NO_x) was required in the United States, Japan and Canada, and not until 1976 in Sweden. Collectively, the regulations introduced since about 1970 have substantially reduced the allowable level of exhaust emissions from new cars. As a result of these regulations, which have been gradually strengthened, the use of catalytic converters and of unleaded fuel is now the accepted standard for all cars in OECD countries. At present, more than 40 per cent of all new cars sold in the OECD countries use catalytic converters (ECMT/OECD, 1990).

Progress is also being made in other parts of the world. For example, Mexico decided in 1988 to adopt United States new car standards by the 1993 model year. As a result, it is hoped that air quality in Mexico City, one of the most polluted cities in the world, will gradually improve (Walsh, 1990a), but at present the situation is getting worse; the fleet of three million automobiles increased by an estimated 500,000 in the last two years and gasoline consumption increased by eight per cent a year over the same period.

There are some local environmental impacts caused by the emissions arising from sea transport, although they are generally less of a problem than for the other modes of transport described above. The combustion of marine fuel oil (used by most ships) gives rise to emissions that as yet are not regulated. Marine fuel oil is an unusual compound, which may contain admixtures of what are in reality chemical wastes. However, it should be remembered that transport by ship tends to be more energy efficient per tonne-kilometres than transport by either motor vehicle or aircraft.

Regional environmental impacts of emissions

A regional environmental problem which is in part caused by nitrogen oxide emissions from the transport sector is the phenomenon commonly known as acid rain (see Chapter 1). As has been mentioned previously the global trend is towards an increase, after 2000, of NO_x levels arising from motor vehicle emissions. To these should be added the NO_x emissions from increasing numbers of aircraft, which are estimated to total around 3 million tonnes annually (equivalent to about 15 per cent of automobile NO_x emissions). In contrast to near ground level emissions, where the nitrogen oxides are usually washed out by rain within days, they persist in the upper atmosphere for long periods. Between one and two million tonnes of the NO_x is emitted in the particularly sensitive layers of the atmosphere between 9,000 and 13,000 metres, from where it diffuses gradually to both lower and higher levels (Egli 1991a, 1991b) with both regional and global consequences. Recent modelling studies reveal that NO_x released into the troposphere between 8,000 and

12,000 metres leads to an increase in ozone concentration (Beck *et al.*, 1992) whereas at higher altitudes it contributes, albeit to a small extent, to ozone destruction.

Global environmental impacts of emissions

In Chapter 3 the phenomenon of global climatic change due to the increase in greenhouse gases has been described. At present, the global fleet of motor vehicles is responsible for between 14 per cent and 16 per cent of the world's CO_2 output. However, as illustrated in Figure 12, on a national or regional basis, the share of CO_2 produced by transport varies between 5 per cent and 40 per cent. Figure 13 shows that the United States accounts for more than one-third of the global transport CO_2 budget, emitting almost as much CO_2 from transport as does Central and Eastern Europe, Asia, China, Africa, Latin America and the Middle East combined (Walsh, 1990a).

As previously indicated, although there were significant increases in the fuel efficiency of US-manufactured motor vehicles over the last two decades this was not the case for European and Japanese manufactured vehicles (which of course were much more efficient than US cars in the beginning - see Figure 3). Some manufacturers are achieving substantial improvements in fuel economy from experimental vehicles, while the recently introduced 'lean-burn' technology which uses an air/fuel ratio of about 25:1 rather than the standard 14.6:1, reportedly offers significant fuel-economy gains, although it will take many years for the global vehicle fleet to reflect these developments. The significance of this for global CO_2 is illustrated in Figure 14. In effect, even with only a 2–3 per cent annual growth world-wide in vehicle kilometres travelled, motor vehicle CO_2 emissions will increase enormously over the next 30 to 40 years (Walsh, 1990a).

Figure 12: *The share of a region's CO_2 produced by transport, in selected regions of the world.*

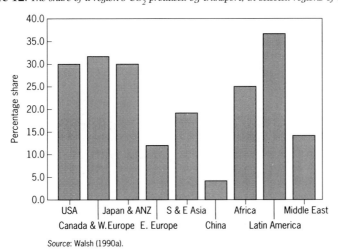

Source: Walsh (1990a).

Figure 13: *Share of the world's transport-generated CO_2 held by various regions.*

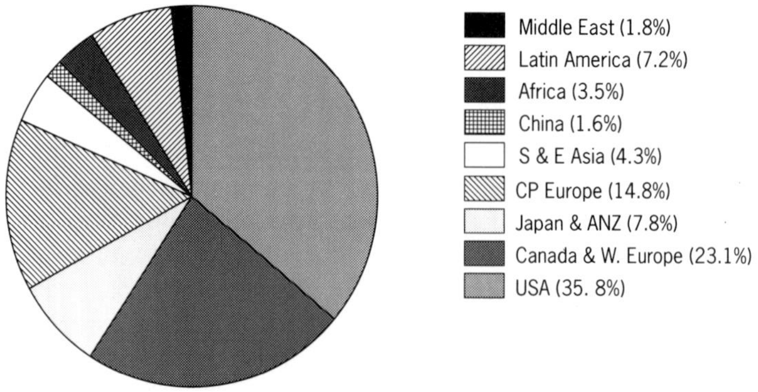

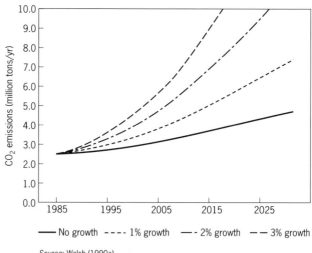

Middle East (1.8%)
Latin America (7.2%)
Africa (3.5%)
China (1.6%)
S & E Asia (4.3%)
CP Europe (14.8%)
Japan & ANZ (7.8%)
Canada & W. Europe (23.1%)
USA (35. 8%)

Source: Walsh (1990a).

Figure 14: *Global automobile CO_2 emissions for various growth rates in vehicle-kilometres travelled, 1985–2025.*

10.0
9.0
8.0
7.0
6.0
5.0
4.0
3.0
2.0
1.0
0.0
CO2 emissions (million tons/yr)
1985
1995
2005
2015
2025
No growth
1% growth
2% growth
3% growth

Source: Walsh (1990a).

Until recently it was assumed that the share of the global CO_2 budget contributed by forms of transport other than motor vehicles was quite small, and not much attention had been paid to them. However, recent studies on the emissions from aircraft indicate that they are in fact significant contributors to global CO_2 (some 13 per cent of traffic-generated CO_2 emissions world-wide on the basis of the fuel consumed, which is increasing rapidly). One study estimates that this represents 2.7 per cent of CO_2 from fossil fuel use world-wide. However, possibly more important in its greenhouse effects is the water vapour emitted by aircraft at high altitudes. Most of this water vapour forms ice-crystal cirrus clouds, which could contribute to global warming.

The other global environmental problem - the depletion of the stratospheric ozone layer - is caused mainly by the release of chlorofluorocarbons (CFCs) into the atmosphere, including those from air conditioners, refrigerators and heat exchangers. A major source of released CFCs is motor-vehicle air conditioning: in 1987 approximately 48 per cent of all new cars, trucks and buses manufactured world-wide were equipped with air conditioners. The amount of CFC-12 required for this is about 30,000 tonnes annually, while almost 90,000 tonnes is required to service existing air-conditioned vehicles (UNEP, 1989b). This is a little more than 10 per cent of total world production.

Nitrogen oxides emitted from high-flying aircraft can contribute to the ozone destruction process. It is estimated that in 1989, air traffic emitted around three million tonnes of NO_x, of which about one million tonnes was at altitudes above 9,000 metres. Above 12,000 metres, NO_x react with water vapour to form nitric acid which crystallizes at altitudes between 12,000 and 26,000 metres to form polar stratospheric clouds at temperatures below -80°C, in which the surface-catalysed chemical reactions leading to ozone destruction occur (Egli, 1991; Barret, 1991).

Measurements in the stratosphere above Sweden have revealed clouds containing aluminium oxide particles that come from rocket exhausts and burning space-craft debris which may contribute to the destruction of ozone. The ozone in the stratosphere can also be destroyed by the use of solid fuel in booster rockets which is considered to be more harmful than liquid fuel (Colucci, 1990). It is estimated that nine USA space shuttles and six Titan rockets a year would put 0.725 kilotons of chlorine into the stratosphere, the effect of which would be equivalent to 300 kilotons from industrial sources (NASA, 1990).

The impact of nitrogen oxide emissions from high-flying aircraft, as well as the emissions from rockets and space vehicles, is a subject of intensive study and analysis at present. The synthesis report of the assessment panels of the Montreal Protocol has found substantially less ozone change due to supersonic flight than earlier predictions had suggested. They also note that there were no detectable changes in total column ozone immediately following each of several launches of USA space shuttles (Chapter 2). The impact of air-breathing aero-engines is the subject of continuing study in several countries, both by national research establishments and by engine manufacturers, particularly as interest in supersonic flight at high altitudes is reviving (Aerospace, 1992).

Since 1958, about 3,000 artificial satellites and space shuttles have been launched, of which about 200 were still operative at the beginning of 1990. Some of these satellites are equipped with nuclear power generators and can cause radioactive pollution, as was demonstrated in 1964 when an American nuclear-powered satellite failed to go into orbit and broke up on re-entry, releasing significant amounts of Plutonium 238 into the atmosphere. Again in 1978, the Soviet satellite Cosmos 954 fell to Earth, contaminating the atmosphere and Canadian territory with radioactive materials (SIPRI, 1990).

Transport noise

Noise from various modes of transport is a form of pollution that causes concern everywhere. Noise disrupts sleep and can generate stress; and in its severest form it can be permanently harmful to hearing. It can be disruptive to conversation and concentration, hence to work and leisure. As the numbers of motor vehicles and aircraft increase, so does transport-generated noise, which affects increasing number of people throughout the world. Figure 15 indicates that in the early 1980s, between 40 and 80 per cent of people in selected OECD countries were subject to unsatisfactory levels of traffic noise and between 7 and 42 per cent to unacceptable levels (above 65 dBA, which is the maximum outdoor noise level considered acceptable for indoor comfort). Since then substantial restrictions on noise emissions from motor vehicles have been imposed (Figure 16) (UNCHS, 1984). Recent data indicate that 130 million people in the OECD countries are exposed to road, rail or air traffic noise levels in excess of 65 dBA (OECD, 1991).

Figure 15: *Percentage of residents disturbed by traffic noise of various intensities in selected countries in the early 1980s.*

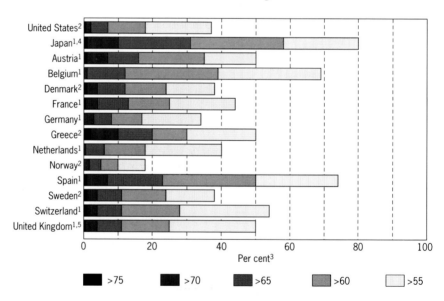

Notes:
1. Daytime Leq (6–22 hours)
2. Leq over 24 hours
3. Percentages are cumulative and not additive (e.g. percentage of persons exposed to >55 dB(A) includes percentage of persons exposed to >60 dB(A), etc.
4. OECD estimates
5. Road traffic noise: Leq averaged (06.00–24.00), 1973 survey, England only.
Source: Based on OECD (1988a).

Figure 16: *Changes in permitted drive-by noise levels for EEC countries.*

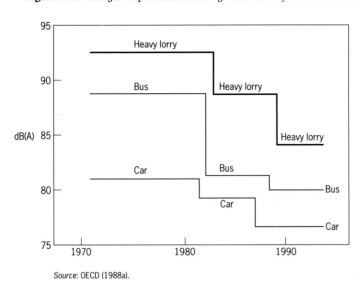

Source: OECD (1988a).

Aircraft noise is a particularly irritating form of pollution, since the noise from a single aircraft flying over a town or city can reach into all parts of it at disturbing levels. About 0.5 per cent of the population in European countries and Japan are exposed to noise levels from aircraft in excess of 65 dB(A), while in the United States about two per cent of the population (about five million people) are affected (OECD, 1991). The problem is particularly acute in the vicinity of airports, especially those located in urban or suburban areas. It has been estimated that in the United States in 1990, some 2.7 million individuals were within the 65 dB(A) contour zone around airports, though this number is expected to drop significantly as older, noisier aircraft are phased out of service.

As a result of the growth in air traffic and the increased size of aircraft, airports have become larger and often have several runways which may be operated simultaneously. This has led to hostile reactions from the public in a number of countries - both to airport operations and especially to proposals for airport capacity increase, for example through the construction of additional runways which is usually less expensive than building a new airport. A wide range of measures have been taken to reduce or limit the level of aircraft noise, including the design of quieter engines, operating restrictions at certain airports on aircraft that do not meet ICAO standards, the adoption of noise abatement measures during landing and take-off, and in many cities a prohibition on normal operations during specified night-time hours. Where the noise level cannot be effectively reduced, sound-proofing of buildings or even relocation of residents may be necessary. The system of environmental countermeasures for airports adopted in Japan is illustrated in Figure 17 (Japan Environment Agency, 1985).

Figure 17: *System of environmental countermeasures for Japanese airports.*

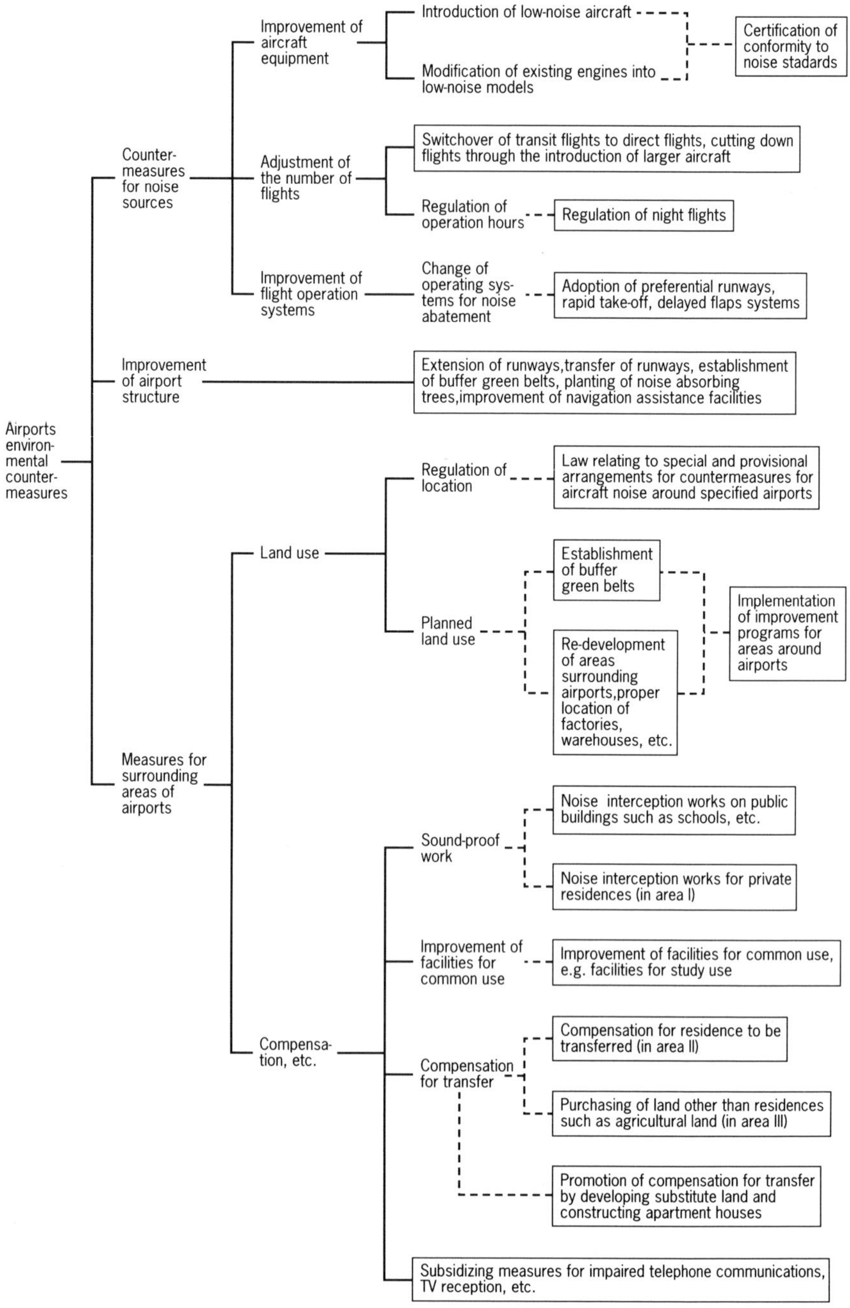

```
Airports environmental countermeasures
├── Countermeasures for noise sources
│   ├── Improvement of aircraft equipment
│   │   ├── Introduction of low-noise aircraft ─ ─ ─ ┐
│   │   │                                            ├─ ─ Certification of conformity to noise stadards
│   │   └── Modification of existing engines into low-noise models ─ ┘
│   ├── Adjustment of the number of flights
│   │   ├── Switchover of transit flights to direct flights, cutting down flights through the introduction of larger aircraft
│   │   └── Regulation of operation hours ─ ─ Regulation of night flights
│   └── Improvement of flight operation systems
│       └── Change of operating systems for noise abatement ─ ─ Adoption of preferential runways, rapid take-off, delayed flaps systems
├── Improvement of airport structure
│       └── Extension of runways, transfer of runways, establishment of buffer green belts, planting of noise absorbing trees, improvement of navigation assistance facilities
└── Measures for surrounding areas of airports
    ├── Land use
    │   ├── Regulation of location ─ ─ ─ Law relating to special and provisional arrangements for countermeasures for aircraft noise around specified airports
    │   └── Planned land use
    │       ├── Establishment of buffer green belts ─ ─ ─ ┐
    │       │                                             ├─ Implementation of improvement programs for areas around airports
    │       └── Re-development of areas surrounding airports, proper location of factories, warehouses, etc. ─ ┘
    └── Compensation, etc.
        ├── Sound-proof work
        │   ├── Noise interception works on public buildings such as schools, etc.
        │   └── Noise interception works for private residences (in area I)
        ├── Improvement of facilities for common use ─ ─ Improvement of facilities for common use, e.g. facilities for study use
        ├── Compensation for transfer
        │   ├── Compensation for residence to be transferred (in area II)
        │   ├── Purchasing of land other than residences such as agricultural land (in area III)
        │   └── Promotion of compensation for transfer by developing substitute land and constructing apartment houses
        └── Subsidizing measures for impaired telephone communications, TV reception, etc.
```

Source: Japan Environment Agency (1985).

Transport system construction and land use

The construction of transport infrastructure consumes substantial areas of land, virtually to the exclusion of any other use. Roadways and especially multi-lane highways, railways, airports, and ports (not to mention parking areas for motor vehicles) occupy vast areas of land, especially in the developed countries, and the amount is growing rapidly. In some cases, valuable agricultural land is taken over, while in other cases areas of natural habitat are devoured, or split into isolated fragments.

Evaluation of the environmental impact of proposed new or expanded transport infrastructure projects has been undertaken in various ways in most developed countries for the last two decades. Experience in the use of environmental impact assessment has helped to bring about considerable changes in attitude on the part of project planners and designers, as well as leading to much better methods of measuring and evaluating environmental characteristics. However, despite the use of environmental assessments there is a genuine and growing need for land for transport infrastructure which conflicts with other forms of land use. These problems are steadily increasing in the industrialized countries. In the European Community countries, roads occupy up to 1.3 per cent of the total area of the region whereas the rail networks use about forty times less land. In Germany, for example, the average overall land-take per kilometre of length is of the order of 9 hectares for motorways and 3.5 hectares for new railway lines. To give an idea of the land-take of airports, the Paris-Lyons TGV railway-line occupies less land area than the Roissy-Charles de Gaulle airport (ECMT/OECD 1990).

In addition to the land it consumes, transport system infrastructure uses vast quantities of natural resources and energy during construction. It gives rise to a range of impacts on the environment similar to those of other major construction projects, including the generation of noise, dust, heavy truck traffic, and sometimes the dredging of rivers or estuaries, the pollution of water courses and the disruption of the social life of nearby communities. All these effects should be addressed during the environmental impact assessment process and solutions should be devised for them (to the greatest extent possible) before construction proceeds.

Construction activities can also generate large quantities of waste materials for disposal. In the case of port construction there are often special problems arising from dredging. A port facility is likely to require dredging of bottom sediments for both construction and maintenance, and the dredging itself may lead to environmental problems from the plume of resuspended sediment trailing downstream from the site. However, the main problem is likely to be the disposal of dredging spoil, which is frequently contaminated with both inorganic chemicals (e.g. zinc or lead) and organic chemicals (e.g. pesticides, PCBs and petroleum hydrocarbons) as well as biological contaminants such as pathogenic bacteria. Much of this contamination consists of industrial effluents, residential sewage and the waste products of agricultural practices on adjacent lands, and the very nature of estuaries (on which most ports are located) is for them to be zones of sediment deposition, which means that these materials tend to accumulate there.

Traditionally, dredged materials have been disposed of at sea by dumping, but this has recently become an increasing concern because of its effects on the marine environment. The definition and degree of contamination of dredged spoil varies from nation to nation, but is often based on sediments regarded as uncontaminated. In the Netherlands for example, sediments with contamination levels lying equal to or below specified 'reference' values can, in general, be deposited without restriction. When the levels lie between 'reference' and specified 'testing' values, open water disposal is permitted under certain circumstances. If the dredged materials have concentrations higher than the 'testing' value, they must be disposed of in controlled containment facilities subject to constant monitoring (Davis *et al.*, 1990).

Responses

Vehicle characteristics

The oil crises during the 1970s revealed the energy vulnerability of both developed and developing countries that were reliant on oil imports, especially from destabilized regions. The average fraction of developing country export earnings used to pay for oil imports tripled during that decade. By 1981, Kenya, South Korea and Thailand spent close to one-third, Brazil over one-half and Bangladesh two-thirds of such earnings on imported oil (Renner, 1988). Throughout the world the search for alternative transport fuels was intensified.

Within the next 50 years the role of oil as the major fuel is likely to diminish and alternative fuels for transport will have to be found. Already about 50 million cars (about every eighth car globally) are driven by some form of alcohol fuel, either pure ethanol or methanol or various mixes of these with gasoline. At present, about four million vehicles are using ethanol from sugar cane as fuel. In 1988, about 90 per cent of all cars sold in Brazil were equipped to use ethanol but there has been a dramatic fall since then and in 1991 only 20 per cent of new cars sold were ethanol users. In the United States there is great interest in using alcohol fuels, but in this case in the form of methanol. The OECD International Energy Agency (IEA) has found that ethanol from sugar and starch can be substituted for about 10 per cent of all fuel for vehicles. The percentage would be even higher if biomass and wood were to be used as sources (IEA, 1990).

However, as long as the price of conventional fuels remains low (under $US20 per barrel for oil), and in the absence of substantial government intervention and support, the use of alternative fuels is likely to remain confined to specific circumstances (e.g. highly polluted urban environments). The alcohol programme in Brazil has been worthwhile, but not many developing nations could afford to provide a similar level of support (Trindade and de Carvalho, 1989; Trindade, 1991).

New technologies such as solar-electric vehicles, and new fuels such as hydrogen, are under development and a number of fleet-tests around the world may lead to solutions for the future (IEA, 1990). Hydrogen as a fuel has been the subject of considerable research and development, with emphasis on adapting internal combustion engines for the combustion characteristics of hydrogen, and on means of storing hydrogen fuel (Sprengel and Hoyer, 1990). Other researchers are exploring hydrogen fuel cells (which are much more efficient than internal combustion engines) for automotive applications and also the development of hydrogen-fuelled aircraft. Another particularly promising option presently under development is the concept of 'hybrid' cars which combine electric motors with a fuel-driven motor designed to cut in at higher speed and power requirements, or to recharge the storage batteries.

The State of California has once again taken the lead in dealing with automobile-generated air pollution, by requiring that two per cent of new cars sold in that state should be zero-emission vehicles (ZEVs) by 1998 and ten per cent by 2003. A number of other states have adopted similar programmes while others are considering doing so. If they do it will mean a sales level of 500,000 ZEVs per annum (mainly electric-powered cars) by 2003. Already a wide range of electric-powered vehicles has been developed, and several have performance and range that is more than adequate for city and suburban use. However, they still rely on heavy battery packs that need frequent and time-consuming recharging, and at present they are still more costly per kilometre than gasoline-fuelled cars. Clearly the future of electric cars lies in reducing battery weight and charging time, while improving performance and especially range.

New techniques in enforcing emission standards may also be an important factor in meeting higher air quality standards. One recent innovation is a remote sensing device that can measure the ratio of selected gases in the exhaust plume of a passing vehicle. The system, developed at the University of Denver and known as FEAT (Fuel Efficiency Automobile Testing) includes a video picture of the monitored vehicle, which permits correlation of the exhaust data with licence number. It is particularly suited to the identification of grossly polluting vehicles which studies indicate may be responsible for a very high percentage of total emissions (Peterson et al., 1991).

A further development in the United States is the requirement for the use, from January 1995, of reformulated gasoline to reduce ozone emissions in specified cities which have ozone pollution problems and of oxygenated gasoline in cities with carbon monoxide problems, beginning in 1992. The former programme will provide flexibility for gasoline refiners, blenders and importers by allowing them to meet the emission performance standards through a variety of reformulations designed to limit levels of benzene, reduce emissions of ozone-forming hydrocarbons, and remove heavy metals (US EPA, 1991). Already almost 10 per cent of the world automobile fleet (some 60 million vehicles) uses oxygenated fuels. The oxygenates most commonly used are methanol and ethanol but methyl-ter-butyl ether (MTBE) is the oxygenate with the most potential. Oxygenates serve three basic objectives - extending gasoline stock or serving as a fuel substitute (e.g. ethanol in Brazil), boosting octane values, and providing an effective means of reducing harmful emissions (Faiz *et al.*, 1990). In particular the extra oxygen enhances fuel combustion (which tends to proceed less efficiently in cold weather) leading to significantly lower CO emissions.

Meanwhile, manufacturers of motor vehicles are exploring other ways and means of reducing their environmental impacts. One approach being used is to ensure that as many parts of old automobiles are recycled as is possible. While metal parts have always been easy to collect and turn to scrap for re-use, fluids and plastics left in the metal lower its value. Some companies are now arranging for their dealers to collect engine oil, brake fluids and the like from traded-in cars and to strip out the many thousands of parts for reuse or reconditioning.

Measures to reduce the impact of car air-conditioning systems using ozone-depleting substances currently include reducing leakages from car systems, recycling of refrigerants and conversion to hydrofluorocarbons in new cars. These measures could lead to a reduction in the use of CFCs of around 70,000 tonnes annually during the 1990s.

Beginning with the 1972 Oslo Convention for the Prevention of Marine Pollution by Dumping from Ships and Aircraft, the European nations sought to limit the input of contaminants to the adjacent marine international waters, and it was accepted that disposal of dredged materials could occur provided the materials contained only trace quantities of contaminants. This was followed by the 1972 Convention on the Dumping of Wastes at Sea (the London Dumping Convention or LDC) which was designed primarily to regulate the dumping of chemical or industrial wastes in the marine environment. The LDC Guidelines on dredged material disposal recommend representative sampling, measuring of general characteristics, measuring of priority contaminants and biological testing if necessary (see Chapter 10).

Infrastructure development

In both Europe and the United States, the annual investment in new transport infrastructure is decreasing while expenditure on maintenance is increasing, partly reflecting competition for the limited financial resources usually allocated to transport infrastructure (OECD, 1991). Investment in transport infrastructure for the European Community as a whole declined by 25 per cent between 1975 and 1984 (ECMT/OECD 1990).

Despite this, some large-scale infrastructure projects have been completed or started in Europe during the last decade, including several high-speed train lines in France, the Channel Tunnel between the United Kingdom and France, and connections through the Alps and between Danish islands. In Japan, the double-deck bridge connecting Honshu and Shikoku islands by road and rail was completed in 1988, as was the Tsugaru Strait Tunnel connecting the islands of Honshu and Hokkaido by railway, while the Kansai International Airport is under construction on reclaimed land in Osaka Bay (OECD, 1991). The world's longest land tunnel (25.8km) is presently being constructed in Japan to link Iwate and Aomori prefectures by bullet train. In Germany, the 670-kilometre Rhine-Main-Danube Canal linking the Danube and Rhine river systems is expected to carry ten million metric tonnes of cargo when it opens for use in 1992.

In the developing countries, overall progress in the development of virtually all transport sectors has been sluggish. However, during the 1980s, some of the LDCs started to attach greater priority to transport infrastructure development (UNCTAD, 1990).

Public transport plays a central role in any efficient urban transport system. Over the past two decades, some 21 large cities including Mexico City, Beijing, Shanghai and Cairo have built metro systems. Recently Manila, Tunis and other cities have opted for the smaller and less expensive light rail technologies for their urban systems. Table 9 illustrates the extent of use of urban public transport systems in a number of cities in 1983 (Lowe, 1991).

Table 9: *Dependence on public transport facilities, in selected cities, 1983.*

City	Population (million)	Mode[1]	Trips per person per year
Moscow	8.0	bus, tram, metro	713
Tokyo	11.6	bus, tram, metro, rail	650
East Berlin	1.2	bus, tram, metro, rail	540
Seoul	8.7	bus, metro	457
West Berlin[2]	1.9	bus, metro	389
Buenos Aires[3]	9.0	bus, metro	248
Kuala Lumpur[4]	1.0	bus, minibus	224
Toronto	2.8	bus, tram, metro	200
Nairobi	1.2	bus, minibus	151
Abidjan	1.8	bus, boat	132
Beijing	8.7	bus, metro	107
Chicago[5]	6.8	bus, metro, rail	101
Melbourne[5]	2.7	bus, tram rail	95
Dallas[3]	1.4	bus	22

Notes: 1. In this table, 'rail' refers to suburban rail.

2. 1982.

3. Metropolitan area.

4. Excludes cycle rickshaws and private minibuses.

5. 1980.

Source: Lowe (1991).

A number of countries have been experimenting with magnetic levitation (maglev) transportation systems which involve lifting, guiding and propelling a vehicle along a guideway by use of magnets. Because there is no friction (as with steel-wheeled high-speed trains) maintenance costs can be substantially reduced while speeds of 290kph to 480kph are feasible. The development of several full-scale prototypes by both Japan and Germany has provided a large measure of confidence in the performance capabilities of maglev technologies. For example, the German Transrapid 06 vehicle attained a speed of 412.6kph on a test track facility at Emsland, Germany, in January 1988. While several technical problems remain to be solved, none appears substantial enough to preclude construction of an operational first-generation maglev system (Johnson, 1991).

Physical planning and transport management

Many of the local environmental problems caused by motor traffic can be dealt with by improved urban planning and a variety of transport management measures. Table 10 indicates a range of such measures that have been used by local authorities around the world, together with some possible environmental and social consequences.

A city's capacity to expand public transport and facilitate cycling and walking depends on much more than providing buses, trains and safe streets. The layout of a city helps determine whether or not these transport options are appropriate or even feasible. Many urban areas are designed for the automobile - for example Los Angeles where two-thirds of urban space is paved for cars. However, where it is possible, planners in a number of cities are finding ways to provide cycle-path networks separated to a significant extent from the road system to encourage cycling and to increase its safety.

Stockholm exemplifies the successful integration of land-use planning and development with transport planning. The city is ringed with satellite communities of 25,000 to 50,000 people, each linked closely by a rail and expressway network. Paris and other cities have followed similar development policies. Toronto's experience is that car dependence can be lessened by combining public transport expansion with planned land use to create higher densities and shorter travel distances. Today, Toronto's overall density is comparable to several major European cities (see Table 11) and despite increasing car ownership, public transport use has increased 80 per cent in a little more than two decades (Lowe, 1991).

Transport safety

All transport vehicles have to satisfy certain safety standards set up and monitored by regular inspections, at the national and international levels. Historically this started with shipping organizations such as Lloyds. Veritas and the American Bureau of Shipping set, and supervise conformity with, rules of

Table 10: *Examples of traffic management measures and the results.*

Measure taken	Town	Result
Encouragement of fleet renewal.	Los Angeles	Up to 95% of emissions reduction due to this measure.
Petrol gas recovery devices in service stations.	Los Angeles	Reduction of emissions equivalent to 19% of total HC emitted in California State in 1980.
Inspection and maintenance programmes for cars (+ fleet renewal).	New York	Reduction of 38% in NO_x and 34% in CO emissions between 1980 and 1987 (+28% and +11% respectively).
Parking permits, access permits and taxes on vehicles.	Singapore	The share of private cars in total vehicle traffic at peak hours in the town centre decreased from 50-60% to 23%. Commuting journeys towards the town centre have shifted from cars to buses from a 56%/33% share ratio to 46%/46%. Accidents in the town centre have fallen by 25%.
Road pricing in urban motorways (three projects).	Hong Kong	Bottle-necks have diminished by 14,16 and 17% representing time gains of respectively 98,000, 113,000 and 124,000 hours.
Traffic regulation in residential areas	Osaka	A 9% decrease in accidents.
Generalized improvement of public transport towards an integrated public transport system.	München	A 30% increase in public transport users. The share public transportation/cars has changed from a 37/63 distribution to 46/54 between 1970 and 1980.
Simplified and integrated fares ('travel card').	London	A 16% increase in public transport users. Cars arriving at central London in peak hours have diminished by 10-15% with similar effects on air quality. Costs of 75 million pounds against estimated savings of 171 million pounds.
Simplified and integrated fares ('travel card').	Paris	A 1/3 increase in public transport users and 2-3% reduction in car utilization during the whole day.
Traffic restricted to given partitions in central areas.	Göteborg	Victims of accidents decreased by 40-45%. Net noise reductions of about 4 dB(A) for 1/3 of inhabitants. 7% increase in the average length of journeys outside the areas compensated in part by increased average speed within the areas.
Banning of cars on alternate days according to their plate number.	Athens	A 20% reduction in overall traffic within the essay area.

Source: OECD (1988a).

Table 11: *Urban densities and commuting choices in some selected cities.*

City	Land use intensity (pop + jobs/ha)	Percentage of workers using		
		Private car	Public transport	Walking/ cycling
Phoenix	13	93	3	3
Perth	15	84	12	4
Washington	21	81	14	5
Sydney	25	65	30	5
Toronto	59	63	31	6
Hamburg	66	44	41	15
Amsterdam	74	58	14	28
Stockholm	85	34	46	20
Munich	91	38	42	20
Vienna	111	40	45	15
Tokyo	171	16	59	25
Hong Kong	403	3	62	35

Source: Lowe (1991).

design, manufacture and operation. There has recently been a review of tanker design to minimize the environmental hazards and risks of accidents through the use of double hulls or midship decks. Perhaps the most stringent are those in the design, manufacture, operation and maintenance of aircraft. Regulations for road vehicles exist in industrialized countries while all countries world-wide have requisites of varying degrees of detail for licensing road vehicles and small seaborne and inland water craft. Licensing also covers personnel operating ships and aeroplanes and drivers of road vehicles. Aircraft maintenance personnel are also licensed world-wide.

The International Maritime Organization (IMO) is the main international organization providing a forum for the development of international standards for the prevention of pollution resulting from transport by ships. The most important international instrument in the field of marine environment protection is the Convention for the Prevention of Pollution from Ships, 1973, and the protocol of 1978 relating thereto - usually abbreviated to MARPOL 73/ 78. MARPOL 73/78 contains measures designed to prevent or reduce accidental as well as operational pollution, to reduce the consequences of accidents, to provide compensation and to help governments in developing contingency plans for countering pollution.

The International Convention for the Safety of Life at Sea (SOLAS 74/78) was also elaborated within IMO. One chapter deals exclusively with the carriage of dangerous goods, which is prohibited unless the goods are carried in accordance with the provisions of the Convention. Each Contracting Party is required to issue detailed instructions on safe packing and storage of dangerous goods. The United Nations Convention on the Law of the Sea, which has not yet entered into force, also deals with the protection and preservation of the marine environment in its Part XII.

Besides these two international conventions, there are a number of regional conventions, among them the Barcelona Convention concerning co-operation in combating pollution of the Mediterranean Sea, the Convention on the protection of the marine environment of the Baltic Sea area, and the Agreement for co-operation in dealing with pollution of the North Sea, (Chapter 10).

Communication without transport

During the last two decades there has been a dramatic improvement in telecommunications which it was thought would lead to a decrease in the use of transport, but experience to date indicates that improved telecommunication systems in fact generate new needs and opportunities for travel as fast as they eliminate old ones. While it is still too soon to be certain, it is now generally thought that telecommunications will not provide a solution to urban congestion in the foreseeable future through a drastic reduction in travel (ECMT/OECD, 1990).

Concluding remarks

Because both transport volumes world-wide and the related environmental effects seem certain to increase, it is essential that the relationship between transport and the environment be analysed from a global perspective. Since the transport sector is one of the major contributors to environmental degradation, especially through its emissions to the atmosphere, it seems reasonable that it should also be required to meet the costs and to assume the responsibility for changing this situation. The aim should be to reduce emissions to levels that do not have significant harmful effects on the environment. In general, this means that the transport sector has to decrease its total emissions. However, in practice the most cost-effective solutions have to be chosen, which means that strategies may differ between countries and regions in response to local situations.

It is important that common emission requirements, based on best available technology, are imposed for all modes of transport. The introduction of catalytic conversion for motor vehicles is proceeding well, but there is still more to be done because technology can achieve much more than was initially thought possible. Given the co-operation of the many industries involved, it should eventually be possible to produce clean, energy-efficient and safe vehicles. For the industrialized countries, dialogue between governments and industry with this objective in mind has already started within ECMT/OECD. However, the process has to be broadened to take into account the growing transport needs of the developing countries and of Central and Eastern Europe.

In view of the significant impact that transport is having on the environment locally, regionally and globally, it is essential to set considerably more ambitious goals than those currently adopted. A co-ordinated global strategy for the transport sector would include the following features.

- The overall use of fossil fuels for transport must be reduced where feasible. Economic incentives, such as environmental taxes and charges, should be used in a balanced mix with other measures to stimulate more efficient forms of transport and patterns of transport use. The initiatives already taken in some countries specifying certain percentages of cars with zero emissions by specific dates need to be pursued and followed in other localities and countries world-wide.

- Global emission standards, based on best available technology, should be adopted as soon as possible for all modes of transport where they do not already exist. Through common action in this area the industrialized countries can actively promote the development of new technology which can be transferred to other countries.

- Each country should adopt and incorporate a comprehensive emissions control programme addressing emissions from all sources in all vehicle categories. Such a programme should include retrofitting of in-use vehicles as well as new models with emission control devices such as catalytic converters, and should extend to monitoring, inspection, maintenance and enforcement, as well as alternative fuels, traffic management and fuel pricing.

- Special measures should be taken in urban areas in developed as well as in developing countries where both the transport and the environmental situation are critical. In these areas, more cars and lorries are not the solution. Instead, heavy investment is needed to create clean, safe and reliable public transport systems. Physical planning and traffic management are also important instruments for promoting an acceptable urban development with an environmentally sound transport infrastructure and low transport demand.

- While there is a need for substantial investment in transport infrastructure in most countries of the world, and especially in developing countries, there is also a need for a more integrated approach, with a change of priorities towards more environmentally friendly modes of transport such as rail and water transportation. As an integral part of the planning and decision-making process for new infrastructure projects, environmental impact assessments should always be done.

Programmes such as those suggested above require public support if they are to be successful, and that means there needs to be adequate public participation, both in their development and in their implementation. Public participation is the best means of creating an awareness of what individuals can do to improve the relationship between their own transport needs and the way these impact on the environment. Such awareness would include an understanding of the importance of environmentally-sensitive driving behaviour and on the advantages of alternative modes of transport.

In order to further develop the proposals made above, there is a need for closer international co-operation. Various possibilities for strengthening the institutional framework should be explored, and the balance between institutional arrangements for different modes of transport should be examined.

REFERENCES

Aerospace (1992) Green clean power, in Vol. 19 No. 1, January 1992 Royal Aeronautical Society, London.

Barrett, M. (1991) *Aircraft Pollution - Environmental Impacts and Future Solutions.* WWF Research Paper, August 1991.

Beck, J. P., Reeves, C. E., de Leeuw, F. A. A. M. and Penkett, S, A. (1992) The effect of aircraft emissions on tropospheric ozone in the northern hemisphere. *Atmospheric Environment* **26A**, 17-29.

Brown, L., *et al.* (eds) (1989) *State of the World, 1989,* Worldwatch Institute, Norton & Co., New York and London.

Colucci, F. (1990) Launching Clean. *Space,* **6 (5),** Sept-Oct 1990.

Davis, John D., MacKnight, S., IMO Staff and Others. (1990) Environmental Considerations for Port and Harbour Developments. *World Bank Technical Paper No. 126, Transport and Environment Series.*

ECMT/OECD (1990) *Background papers for the European Conference of Ministers of Transport on Transport Policy and the Environment.* ECMT Ministerial Session. November 1989.

Egli, R. A. (1991a) Air Traffic and Changing Climate. *Environmental Conservation,* **18,** 73-74 & 44.

Egli, R.A. (1991b) Climate Air Traffic Emissions, *Environment,* **33(9),** pp 2-5.

El Hinnawi, E. and Hashmi, M. H. (1987) *The State of the Environment.* UNEP and Butterworths, UK.

Faiz, A., Sinha, K. Walsh, M., and Varma, A. (1990) *Automotive Air Pollution. Issues and Options for Developing Countries.* World Bank PRE Working Papers (Transport). WPS 492.

Hinrichsen, D. and Enyedi, G. (1990) *State of the Hungarian Environment.* Hungarian Academy of Sciences, Ministry for Environment and Water Management and the Hungarian Central Statistical Office, Budapest.

ICAO (1988) *Digest of Statistics Series (Financial Data) 1982 and 1988.* ICAO, Montreal.

IEA (1989) *World Energy Statistics and Balances, 1971-1987.* OECD, Paris.

IEA (1990) *Substitute fuels for road transport. A technical assessment.* OECD/IEA, Paris.

IMO (1989) *The Environmental threat, Focus on IMO,* International Maritime Organization, September 1989.

Japan Environment Agency (1985) *Quality of the Environment in Japan - 1985.* Environment Agency, Government of Japan, Tokyo.

Johnson, L. R. (1991) Magnetic Levitation Transport Technology: History and Status. *ATAS Bulletin,* **No. 6** (December 1991).

Liou, K. N., Ou, S. C. and Koenig, G. (1990) An investigation of the climatic effect of contrail cirrus. In: *Air Traffic and the Environment* (U. Schumann ed). Springer Verlag, Berlin and Heidelberg, Germany.

Lloyds (1987) *Maritime Atlas 1987. III. List of Ports and Shipping Places of the World.*

Lowe, Marcia D. (1990) Cycling into the Future. In: *State of the World 1990* (L. Brown ed) Worldwatch Institute and W. W. Norton, New York.

Lowe, Marcia D. (1991) Rethinking Urban Transport. In: *State of the World 1991* (L. Brown ed) Worldwatch Institute and W. W. Norton, New York.

Meyers, S. (1988) *Transportation in the LDCs: A Major Area of Growth in World Oil Demand.* Lawrence Berkeley Laboratory, University of California.

Nijkamp, P., Perrels, A., and Schippers, L. (1990) Strategic evaluation of new infrastructure in Europe. *Vrije Universiteit Amsterdam, Facuklteit der Economische Wetenschappen en Econometrie, Research Memorandum,* 1990-74, December 1990.

OECD (1987) *Energy Statistics 1970-1985, 1986/87*. OECD, Paris.

OECD (1988a) *Energy Statistics, 1986-87*. OECD, Paris.

OECD (1988b) *Transport and Environment*. OECD, Paris.

OECD (1989a) *Maritime Transport 1988*. OECD, Paris.

OECD (1989b) *Energy Statistics, 1987-88*. OECD, Paris.

OECD (1991) The State of the Environment. OECD, Paris.

Pemberton, M. (1988) *The World Car Industry to the Year 2000*. Economist Intelligence Unit, London.

Peterson, J. E., Stedman, D. H., and Bishop, G. A. (1991) Remote sensing of automotive emissions in Toronto. *Currents* - Newsletter, Ontario Section, Air and Wastewater Association, August 1991.

Renner, M. (1988) Rethinking the Role of the Automobile. *Worldwatch Paper* **84**, Worldwatch Institute.

Schipper, L. (1991) Improved energy efficiency in the industrialized countries - Past achievements, CO_2 emission prospects. *Energy Policy* March 1991.

Schipper, L. and Meyers, S. (1992) *Energy Efficiency and Human Activity: Past Trends, Future Prospects*. Stockholm Environment Institute, Stockholm.

SIPRI (1990) *World Armaments and Disarmament SIPRI Yearbook, 1990*. Oxford University Press, Oxford.

Sprengel, U., and Hoyer, U. (eds) (1990) *Solar Hydrogen - Energy Carrier for the Future*. DLR, ZSW and Ministry of Economic Affairs and Technology, State of Baden-Wurtemberg, Stuttgart, Germany.

Trindade, S. C. and de Carvalho, A. V. (1989) Transportation fuels policy issues and options. The case of ethanol fuels in Brazil. *In*: Sperling, (ed) *Alternative Transportation Fuels*, Quorum Books.

Trindade, S. C. (1991) Non-fossil transportation fuels: The Brazilian sugar cane ethanol experience. *In: Energy and the Environment in the 21st Century*. MIT Press.

UN (1986) *Statistical Yearbook 1985-86*. UN, New York.

UNCHS (1984) *Transportation Strategies for Human Settlements in Developing Countries*. United Nations Commission for Human Settlements - Habitat, Nairobi.

UNCTAD (1990) *The Least Developed Countries. 1989 Report*. UN, New York.

UNEP (1989a) *Environmental Data Report, 1989/1990*, p.439, Blackwell, Oxford.

UNEP (1989b) *Technical Progress on Protecting the Ozone Layer - Refrigerators, Air Conditioning and Heat Pumps - Technical Options Report*. UNEP, Nairobi.

UNEP (1991) *Environmental Data Report, 1991-92 (3rd Edition)*. Blackwell, Oxford.

US EPA (1988) *National air quality trend report*, March, 1988. Doc. No. EPA 450/4-90-002.

US EPA (1991) Agreement reached on clean vehicle fuels. *EPA Environmental News*, 16 August, 1991.

Walsh, M. (1990a) Global Trends in Motor Vehicle Use and Emissions. *Annual Review of Energy* **15**, 217-43

Walsh, M. (1990b) *System Solutions and the Environment - Energy and Transport, Fuel Economy and Emissions*. Paper presented at Ecology and Transport Conference (Swedish Environment Institute, November 27-29 1990, Gothenburg, Sweden).

CHAPTER 15

Tourism

Introduction

The emergence of tourism as a major industry is one of the most remarkable changes to have taken place in economic activity in the years since World War II. It is a sizeable and complex service industry, governed by the laws of supply and demand. It exhibits strong seasonality of demand in most areas and can be affected by relatively unpredictable changes in consumer preferences, in addition to being influenced by political events.

Despite this, tourism has been more stable than many other sectors. The past 30 years have seen rapid and continuous growth: both the number of tourists and tourism receipts have increased significantly throughout the world. Developments in the transport sector, especially air transport, have significantly improved access to tourist destinations, although these same developments have created their own environmental problems (Chapter 14).

The quality of the environment, or some particular feature of it, is frequently the primary attraction for tourists. This can lead to considerable pressure on the environment that attracted tourists in the first place and in particular on the local environment where tourists are staying. At the same time, tourism can have positive effects since those responsible for tourism development have an interest in working with those who are concerned with protection of the environment. Income from tourism can also assist in the development and improvement of facilities for permanent residents as well as for tourists - such as better water-supply systems, sewerage services and the like.

Because of its financial benefits to individuals as well as to national and regional economies, the desirability of tourism has seldom been questioned by governments, but it has drawn criticism in recent years from those concerned about the adverse environmental impacts of some inadequately controlled tourism developments, as well as from local communities whose way of life has been adversely affected by them. The tourism industry generally does not question the need for appropriate environmental standards and controls for the protection of those environmental assets that support tourism. Accordingly, maximizing the positive impacts and minimizing or avoiding the negative impacts must be a major goal of any tourism development strategy. In effect, the concept of sustainable development must be applied to tourism. Consequently, where tourism has already caused environmental damage, policies have been directed, whenever possible, towards corrective action.

Tourism development 1972-92

In most countries since the late 1970s, tourism has grown from being a marginal aspect of national economic life to an important socio-economic asset. At present, tourism is the second largest item in world trade, surpassed only by oil. It is, moreover, one of the fastest-growing sectors of economic activity.

In 1970 there were about 160 million international tourist arrivals. International arrivals are defined as individual international travel events involving a temporary absence from home with a duration greater than 24 hours. By 1980 the number had increased to about 285 million, and by 1990 to about 439 million. Forty years ago, there were only 25 million international travellers - about six percent of the present number (Figure 1). However, to understand the potential impact of tourism on the environment, it is necessary also to take account of domestic tourism which, in some developed countries, increases the number two- or threefold.

Figure 1: *The growing number of international tourist arrivals, 1950–90.*

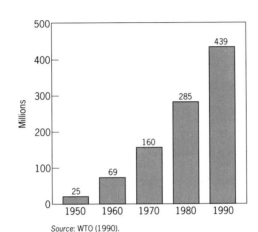

Source: WTO (1990).

The amount of money spent is a more reliable indicator of the level of international tourist activity than travel or border-crossing statistics, since the latter includes much non-tourist travel. Table 1 indicates that international tourism receipts have increased almost elevenfold since 1972, to a sum close to $US260 billion. (International tourism receipts are the receipts of countries in the form of consumption expenditure, i.e. payments for goods and services made by foreign tourists out of foreign currency resources. They exclude international air fare receipts.)

The post-war development of tourism as a mass phenomenon in the industrialized countries, and its subsequent geographical expansion, can be ascribed mainly to the growth in real income per capita and to an increase in both leisure time and paid holidays. Additionally, significantly increased discretionary spending, coupled with substantial reductions in the real cost of car ownership and air travel, have caused most families from the industrialized countries to abandon railways and coaches as the preferred means of transport for holiday purposes. These factors have been largely responsible for the spectacular growth of holiday tourism in southern European countries such as Spain, Italy and Yugoslavia.

Table 1: *International tourist arrivals (in millions) and international tourism receipts (in billions of $US), 1972-90.*

Year	Arrivals (millions)	Index (1972=100)	Receipts (billions$US)	Index (1972=100)
1972	181.9	100	24.6	100
1974	197.1	108	33.8	137
1976	220.7	121	44.4	180
1978	257.4	142	68.8	280
1980	284.8	157	102.4	416
1982	286.8	158	98.6	401
1984	312.4	172	109.8	446
1986	330.5	182	139.2	566
1988	391.9	215	196.5	799
1990	438.6	241	256.9	1,044

Source: WTO (1990).

Product innovation in the aircraft industry, manifesting itself mainly in faster planes with a longer range and a greater seating capacity, resulted in lower costs, as a result of which the price-sensitive holiday segment of the travel market was largely won by air travel (Figure 2). This development enabled tourism to 'take off' at the beginning of the 1960s, initially in a number of countries around the Mediterranean, including several in northern Africa. Since then, long-haul mass tourism has expanded geographically to many destinations further afield. In particular, countries in East Asia and the Pacific have experienced tremendous increases both in arrivals and in receipts. International tourist arrivals in the East Asia-Pacific region increased from 6.3 million in 1972 to 49.4 million in 1988 (Figure 3); in the same period receipts increased from $US1.45 billion to $US32.4 billion.

Figure 3 indicates that the bulk of international arrivals are into European countries, which can be attributed to the continuing importance of intra-regional tourism (arrivals from countries within the region). This in turn reflects the size of the region and the mix of travel-generating and travel-receiving countries it contains. Scale and level of economic development attained also explain the relative importance of intra-regional tourism in the Americas. However, between 1967 and 1983, the importance of intra-regional tourist arrivals in the world as a whole declined relative to that of inter-regional arrivals (arrivals from countries in other regions) (Table 2).

These figures reveal a trend towards more long-haul travel. Between 1970 and 1990, inter-regional arrivals surged in the Americas, the Middle East and Europe. In absolute numbers, most inter-regional arrivals are into Europe. Notwithstanding this growth, the vast majority of total travel in each of the regions is for domestic tourism (Table 3).

Figure 2: *Indices of passenger miles, seat miles, GNP and airline ticket prices for US air travel, 1965-86 (1965=100).*

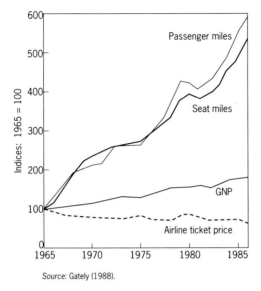

Source: Gately (1988).

Figure 3: *International tourist arrivals by region, 1972–90.*

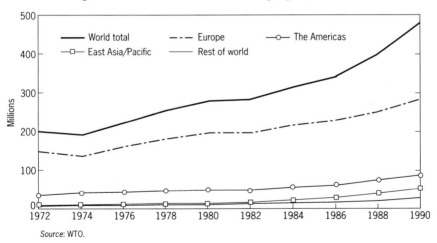

Source: WTO.

A significant recent phenomenon in the tourism industry, 'ecotourism' or 'nature tourism', has been increasing rapidly in a number of countries, although there is still no standard definition of it (Figure 4). The important distinction is between mass tourism, in which leisure activities are the main purpose, usually in a developed or resort environment, and nature tourism, in which the focus is on observing and enjoying nature and natural environments, often in relatively remote places. More and more people are choosing to spend

Table 2: *Intra-regional and inter-regional world tourist arrivals index, 1977-83 (1967=100).*

Year	Intra-regional arrivals index (1967=100)	Inter-regional arrivals index (1967=100)
1977	168	204
1979	181	270
1981	194	292
1983	196	304

Source: WTO (1990).

Table 3: *Inter-regional, intra-regional and domestic tourist arrivals by region, 1983 (millions).*

Region	Inter-regional	Intra-regional	Domestic
Europe	36.1	164.6	1,700
Americas	11.5	40.8	1,300
Africa	5.5	1.5	10
East Asia/Pacific	8.1	17.0	150
South Asia	1.5	1.0	60
Middle East	2.8	3.4	10
World	65.5	228.4	3,230

Source: WTO (1990).

their vacations in pristine natural surroundings, which are often enhanced in appeal by a distinctive local culture. In this kind of tourism, the concept of sustainability is of paramount importance.

There is a growing awareness world-wide that unspoiled nature and, from the point of view of the tourist, 'exotic' ecosystems, have a real economic value. However, most of these tourist assets are still very much underpriced. For example, it has been estimated that each lion in an African National Park has an annual visitor-attraction value of $US27,000 and that each elephant herd is worth $US610,000. Yet at present the entrance fees for visitors to these parks are far below what the visitors would be willing to pay to watch these animals. Furthermore, the absence of adequate charges could lead to over-use of popular attractions to the point where their value is reduced or even eliminated. Many studies indicate that tourists will be willing to pay more if they know that the extra money will be used to help protect the special features that they have come so far to see. In Nepal, for example, six out of ten trekkers in the Annapurna area said they would pay $US5 to $US10 more than current government fees if they knew the money would be used in programmes to conserve the area (Lindberg 1991).

Figure 4: *Growth in numbers of visitors at the Galapagos National Park (Ecuador), the Amboseli National Park (Kenya) and for Nepal, 1967–89.*

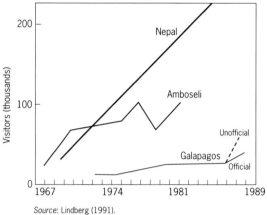

Source: Lindberg (1991).

For many countries, tourism is a source of much-needed foreign exchange as well as contributing significantly to gross domestic product. Tourism employment can be important, especially at the local level. Tourism of all sorts earned developing countries an impressive $US55 billion in 1988. Nature tourism's share of this ranged from an estimated $US2 billion to $US12 billion. The relative impact can be much larger than these figures suggest. For instance, in some parts of the Caribbean and in countries such as Kenya, Rwanda, Costa Rica, Ecuador and Nepal, nature tourism is a leading foreign exchange earner. Caribbean marine areas bring in close to $US1 billion a year from scuba divers alone. Tourism generates about 30 per cent of Kenya's foreign exchange, more than either coffee or tea. In Rwanda, 'gorilla tourism' in the Parc National des Volcans brings in roughly $US1 million a year in entrance fees and generates up to $US9 million indirectly. Nepal earned roughly $US45 million in 1983 from visitors attracted primarily by Himalayan geography and culture (Lindberg, 1991).

However, recent studies have shown that in some developing countries, a significant part of the foreign exchange earned is used to pay for the cost of the imported goods and services demanded by tourists, as well as repaying the cost of capital investment in tourism facilities. Consequently the balance of foreign exchange accruing to those countries may be relatively small (Ascher, 1985).

Industry experts expect tourist demand to remain firm for a number of reasons. General tourism, currently growing at four per cent annually, will continue to expand as population, leisure time and discretionary income levels increase while the real cost of travel decreases. Nature tourism and other forms of specialized tourism are expected to grow faster than general tourism, as people become more environment-conscious and tire of crowded beaches and urban destinations (Figure 5). Some observers estimate that specialized tourism will grow 10 to 15 per cent per year over the next five years (Lindberg, 1991).

455

Figure 5: *Peak summer tourism at Büsum on the North Sea coast. Overcrowding at traditional vacation centres is one factor in the growth of specialized forms of tourism - notably nature tourism (see text).*

Photo: Pfeiffer (Deutsche Presse-Agentur).

Figure 6: *Comparison of changes in foreign exchange earnings from tourism and other sources: the value of coffee and tourism 'exports' in Kenya, 1963–80.*

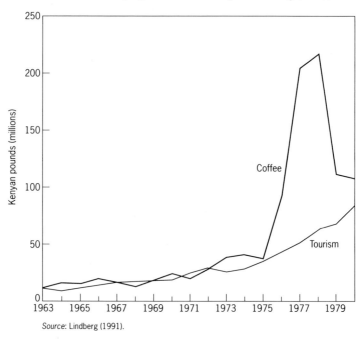

Source: Lindberg (1991).

However, when considering tourist potential for an individual country, possible threats to the continuity of demand have to be born in mind. Political instability and natural and environmental catastrophes are examples of factors affecting tourist demand. The impact on tourism world-wide of the 1991 war over Kuwait was dramatic and long-lasting. Never the less, these risks should always be compared with the uncertainties that affect other sectors. In Kenya, for example, the value of coffee exports rose rapidly between 1975 and 1978, but crashed between 1978 and 1980. Tourism 'exports' grew more slowly, but more steadily (Figure 6).

Environmental impact of tourism development

The World Tourism Organization's Manila Declaration (WTO, 1980) stated that, 'Tourism development at both the national and international level can make a positive contribution to the life of the nation provided the supply is well planned and of a high standard and protects and respects the cultural heritage, the values of tourism and the natural, social and human environment'.

Unspoiled nature (especially the outstanding scenery often associated with the coast, islands, lakes, rivers and mountainous regions) and historical sites and monuments, constitute the stock of natural and man-made resources on which the tourism industry is largely based. Until recently, the availability and permanence of such resources were taken for granted. During the 1970s, however, it became increasingly apparent that these resources can be quite fragile, with limited resilience and carrying capacity.

While the concept of 'carrying capacity' has its origins in livestock husbandry, it has been adapted by conservation biologists as a measure of the ability of ecosystems to sustain populations of particular species that live within them and it is equally applicable to the capacity of natural areas to withstand human use. The concept has been further extended in relation to tourism to cover the capacity of particular societies to 'carry' tourist impact without adverse social consequences. Box 1 illustrates the use of the concept in managing Canadian National Parks.

The environmental impact of tourism depends on how the developments and activities are managed (Figure 7). Regrettably there are numerous examples of environmentally destructive tourism development, resulting in such problems as the depletion of groundwater reserves, destabilization, erosion and salinization of soils, the despoliation of scenic vistas, and the destruction of natural areas and habitat, even to the extent that survival of some species has been threatened. Sensitive ecosystems and landscapes as well as buildings of character and distinction have in some cases been destroyed because they did not suit tourist needs. The transport systems associated with tourism, especially

BOX 1

Managing carrying capacity in a natural environment: the Canadian National Parks.

The Canadian Parks Service manages 34 National Parks and one National Marine Park, covering over 180,000km². These parks are major tourist attractions, with about 13 million visits per year with an average stay of 2.3 days.

A 'Visitor Activity Management Process' (VAMP) has been established. It defines objectives, constraints and opportunities for visitor use of the parks. This information feeds into the management plan, a key element of which is the zoning plan which expresses the carrying capacity of various parts of the park.

Based on an analysis of the resource data and visitor objectives, normally two or three overall plan concepts are produced which are then the subject of comprehensive consultation with the public.

The concepts adopted typically classify a park into five different land-use zones. These are:

Zone I   **Special preservation**: these areas contain the rare and endangered habitats and are strictly protected with access either prohibited or controlled.

Zone II   **Wilderness**: this is the best representation of the natural region of the park (about 60–90% of it). The objective is resource preservation and use is dispersed with only limited facilities.

Zone III   **Natural environment**: access is primarily non-motorized and the area acts typically as a buffer zone.

Zone IV   **Recreation**: major overnight facilities e.g. camp sites, are concentrated in this area.

Zone V   **Park services**: this area is characterized by highly modified landscapes but usually comprise less than 1% of a park area.

An array of management and design techniques is used to establish carrying capacities at acceptable levels. These include the greater use of public transport to reduce the intrusion of private vehicles, the use of elaborate boardwalks over sensitive areas where the biological carrying capacity is very low and the use of arbitrary methods where psychological carrying capacity is the priority criterion.

It is essential that objectives are set to which carrying capacity values can be related. Monitoring is equally essential to provide the feedback for potential adjustment.

Source: UNEP (based on a presentation of D. Lohnes to a Workshop jointly organized by UNEP/WTO/French Ministry of the Environment, Paris, June 1990.)

roads and airports, can give rise to noise and air pollution, while inadequate sewage and waste disposal facilities can cause surface-water, groundwater and coastal pollution.

Figure 7: *The costs and benefits of tourism. This growth industry can be very attractive for developing countries, but there are social, economic and environmental disadvantages which must be taken into account.*

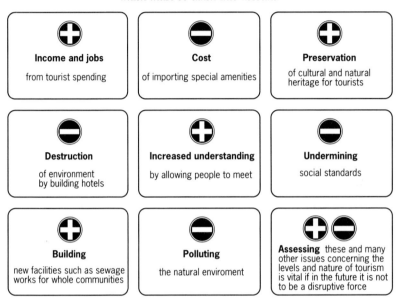

Source: El-Hinnawi and Hashmi (1982).

The environmental short- and long-term impacts of tourism or its excesses can be classified as follows (OECD, 1980):

(a)   Effects of pollution

i.   Air pollution mainly due to motor traffic and to the production and use of energy.

ii.   Water pollution (sea, lakes, rivers, springs), due to:

- discharge of untreated waste water due to the absence or malfunction of sewage treatment plants;

- discharge of solid waste from pleasure boats;

- discharge of hydrocarbons from motor-boats.

iii.   Pollution of sites by littering (picnics, etc.) and the absence or inadequacy of waste disposal facilities (mainly household waste).

iv.   Noise pollution, due mainly to motor traffic or the use of certain vehicles used for recreational purposes (snow-mobiles, cross-country motor cycles, motor-boats, private planes, etc.), but also to the crowds of tourists themselves and the entertainment provided for them (publicity stands, beach contests, etc).

(b)   Loss of natural landscape: agricultural and pastoral lands

i.   The growth of tourism brings with it the construction of housing facilities and infrastructure for tourists which inevitably encroach on previously open spaces, i.e. natural landscape or agricultural or pastoral lands.

ii.   Some valuable natural sites (beaches, forests) are often barred to public access because they become privately owned by hotels or individuals.

(c)   Destruction of flora and fauna

i.   The various kinds of pollution mentioned above, together with loss of natural landscape and agricultural and pastoral lands, are responsible for the disappearance of some of the local flora and fauna.

ii.   Excessive access to and use of natural sites also result in the disappearance of various plant and animal species, owing to tourist behaviour (trampling, excessive picking of fruit or flowers, carelessness, vandalism, or the kind of thoughtless conduct sometimes leading to forest fires, for example).

(d)   Degradation of landscape and of historic sites and monuments

i.   The installation of modern tourist-related facilities and infrastructure often leads to aesthetic degradation of the landscape or sites: the style and architecture of such new installations may not always be in harmony or on a scale with traditional buildings; moreover tourist facility development is often disorderly and scattered, giving the landscape a 'moth-eaten' look.

ii.   An excessive number of visitors to historical or exceptional natural sites may also result in degradation (graffiti, pilfering, etc).

(e)   Effects of congestion

i.   The concentration in time and space of tourists on holiday leads to congestion of beaches, ski slopes, resorts etc. and overloading of tourist amenities and infrastructure, thus causing considerable harm to the environment and detracting from the quality of life.

ii.   One major consequence is traffic congestion on roads at week-ends and at the beginning and end of peak holiday periods, leading to loss of leisure time, high fuel consumption, and heavier air and noise pollution.

(f)   Effects of conflict

During the tourist season, the resident population not only has to put up with the effects of such congestion, unknown during the rest of the year, but often has to change its way of life completely (faster work pace, an extra occupation, etc.) and to live cheek by jowl with people of a different, largely urban kind in search of leisure pursuits. This 'co-existence' is by no means always easy, and social tensions may occur, particularly in places where there are many tourists.

(g)   Effects of competition

Since the development of tourism uses up a great deal of space and siphons off a fairly large proportion of local labour, competition is bound to occur, usually to the detriment of traditional activities, (for instance, less manpower and less land under cultivation means less agriculture).

Competition of this kind generally tends to result in the exclusive practice of tourist-related activities, which may be economically undesirable to the regions concerned.

In some regions, the sheer volume of tourists is alone sufficient to generate massive environmental problems, and when these combine with the impacts arising from the resident population and their normal urban and industrial activities the environment can be seriously endangered (Box 2).

Among the environmentally-sensitive areas that have been most affected by tourism are the high mountains and especially the Austrian Alps. Austria has the highest tourist intensity in the world, with 123 million guest/nights in 1989/90 on a land area of only 84,000 square kilometres, compared with

---

BOX 2

Environmental problems in the Mediterranean Basin.

According to the study *Futures for the Mediterranean Basin* (The Blue Plan) the first effect of increasing tourism will be congestion which can be measured in terms of numbers of visitors and the space occupied by tourism facilities. At present, the Mediterranean Basin accounts for 35% of the international tourist trade and is the world's leading tourist area. By 2025, the study forecasts that there will be about 380 million tourists in all the countries around the Mediterranean if economic growth is weak and about 760 million tourists (eleven billion nights' lodging) if it is strong. Almost half the tourists will be found along the coastline.

The area occupied by tourist accommodation and associated infrastructure in 1984 was approximately 4,400km² and 90% of it was in three countries — Spain, France and Italy. This could double to 8,000km² by the year 2025. The solid waste generated by the tourists, currently 2.8 million tonnes per year, would be between 8 and 12 million tonnes by 2025, while waste water would increase from 0.4 billion m³ to as much as 1.5 billion m³. These figures are in addition to those arising from the local population; 350 million people in 1985 but between 530 and 570 million by 2025.

Source: Based on Grenon and Batisse (1989).

about 10 million guest/nights in 1950 when the industry was less demanding of resources such as land for housing, skiing facilities and roads, water and energy. It is estimated that the resource-use intensity of tourism is now about 50 times what it was in 1950 (Breiling, 1991).

High-mountain tourism directly impacts on both the traditional agricultural way of life and on the alpine ecosystems which are especially sensitive to disturbance so that erosion sets in quickly (Briand et al.,1989). A modelling study of the village of Obergurgl in the Austrian Tyrol indicated that in the absence of any controls, growth of recreation (in the face of essentially infinite potential demand) has been limited by the rate of local population growth, but that the amount of safe land for development is disappearing rapidly while the demand for building sites continues to grow. As land is developed, prime agricultural land is lost and environmental quality decreases. Recreational demand may begin to decrease if environmental quality deteriorates further. The study examined a number of control scenarios that might be used, but concluded that the ecological implications could not be clarified because of the inadequacy of the ecological data base. Nevertheless it was suggested that present recreational use may already be more than the sensitive meadows can tolerate, yet doubling of the recreational use is not unlikely, and may be disastrous (Holling, 1978). Recent studies suggest that these alpine systems will be further stressed by global climate change and that the alpine tourism industry should take this into account in its strategic planning (Nilsson and Pitt, 1991).

The introduction of mass tourism to places such as Antarctica or to the more remote islands of the Pacific and Indian Oceans could bring with it new pressures on wildlife and, where there are local human communities, place severe stress on them. Roszak (1988) noted that at least 7,000 tourists visited Antarctica in the summer of 1988. IUCN's Strategy for Antarctic Conservation recognizes that there can be benefits as well as threats from Antarctic tourism, and that while there are no grounds for opposing people's desire to visit the Far South, careful revision of tourism regulations under the Antarctic Treaty is urgently required. Management guidelines are needed which should aim to encourage responsible and safe tourism practices, avoid conflict between tourist and other uses of the region and minimize harmful environmental impacts (IUCN, 1991). The social and ecological consequences of introducing tourism to new areas are not always easy to foresee, and cost-benefit analyses are not only extremely difficult to do but are unlikely to reveal such possibilities in any case. The need for careful environmental management of any tourism development in these areas is obvious.

In the broader environmental context, tourism can at times adversely affect the quality of life of the community in the place where it occurs, as a result of social and economic disruptions. For example, tourism may lead to changes in social structure as a result of the introduction of foreign values and higher wages for some, or to prostitution, crime and health problems for the local people as well as the tourists (Pasini, 1988). Moreover, competition for resources can have disruptive effects on the structure of the local economy through reducing the supply of labour available to other sectors, and can lead to excessive dependence on this one form of economic activity (European Community, 1990).

A recent study of the impact of tourism on the island of Bali, Indonesia (which increased enormously following the construction of an international airport in the late 1960s) concluded that 'the conflicts within the host society are pervasive and largely tied to the swarming in and out of tourists of the 'mass tourism' sort. The momentum of mass tourism at present is such that little can be done to amend, or improve, the situation, let alone put the brakes on it' (Francillon, 1991). The study noted that some land owners are engaged in building small-scale accommodation for a less affluent and structured tourism, but observed that this trend towards small-scale operations better adapted to local conditions amounts to very little in terms of tourist carrying capacity since Bali's limits have probably already been passed.

In many areas, National Park management excludes local people from grazing and hunting lands to which they once had free access, and penalizes those who poach wild animals for meat or other products. Furthermore, because such local people often derive little benefit from tourism development (most of the profits of which go to tour operators and hotel owners) they feel alienated from it and may even become antagonistic and in the case of National Parks, tempted to break the law.

The health aspects of tourism can be complex. In the search for new and unusual environments, tourists are likely to enter areas where diseases such as malaria and leishmaniasis are endemic and the risk of infection is high. In some destinations, tourists may be exposed to sewage-contaminated waters leading to increased rates of diarrhoea and other intestinal diseases. Because of the speed of modern-day transport, these diseases can spread rapidly to other countries: for example the intestinal parasite *Giardia* appears recently to have reached the back-country of New Zealand (New Zealand Department of Conservation, 1991).

Conversely, tourism can be a very positive environmental force. It can provide a commercial rationale for conserving buildings and environments which otherwise might be destroyed. For example, the protection of monuments and natural areas, the establishment of National Parks, the provision of attractive pedestrian areas and the conservation of historic buildings can be, and often are, justified on the basis that the long-term interests of the tourist economy demand that these resources be maintained (European Community, 1990). Tourism can also lead to increased knowledge and appreciation of other cultures and thus to greater understanding between peoples.

Responses

In various parts of the world there have been successful responses to the challenge of managing tourism on a sustainable basis. One of the best examples is to be found in Australia's Great Barrier Reef Marine Park, which was established by federal legislation in 1975 and now covers virtually the entire Great Barrier Reef region, an area of some 344,000 square kilometres. A single management authority, involving the governments of both the State of Queensland and the Commonwealth of Australia, is responsible for planning the region to allow reasonable use while conserving the reef and its environment. Tourism is the major industry in the region and works in close co-operation with the Great Barrier Reef Marine Park Authority (Box 3).

A major achievement of the last 20 years is that many countries now require that proposed major new developments are subject to some form of environmental impact assessment which can also extend to consideration of impacts on the

---

BOX 3

Environmentally sound management of tourism on Australia's Great Barrier Reef.

Economically, the most important activity in the Marine Park and on the adjacent mainland of Queensland, is tourism, much of it focused on the natural environment. The value of Great Barrier Reef tourism is high. An estimated $US200 million was spent in 1987-88 at island resorts, for commercial boat trips and on private recreational boating.

Because of their potential for negative environmental impacts, tourism projects need careful management by the Great Barrier Reef Marine Park Authority (the government agency established by legislation in 1975 to establish and manage the Marine Park). The primary strategy used in zoning is to provide both for protection of the Reef and for various uses. Conflicting uses are separated as far as possible. Zones range from those in which there is virtually no restriction on activities to those in which all uses (other than scientific research) are prohibited. A recent innovation is the introduction of *No Structures* sub-zones in the Cairns Section of the Marine Park (see map). The purpose is to ensure that a proportion of reef sites that are near centres of human population and open for tourism (and therefore subject to heavy use) are not all taken up with permanent or moored structures.

Tourist operations and structures are allowed by permit in most zones. All tourist operations are assessed for potential impacts before a permit is issued. Any deleterious impacts must, as far as possible, be omitted from the operation or reduced to insignificant levels. The remaining potential impacts are carefully checked through a monitoring programme managed by the Marine Park Authority and paid for by the tourist operator. For construction projects such as marinas, the Authority appoints on-site environmental supervisors with the power to stop construction if predetermined environmental parameters are exceeded. Other control measures used are monetary bonds or bank guarantees and insurance to provide funds in the event that the project fails and the Authority is left with the task of completing it or restoring the site.

social and cultural environment. Combined with appropriate economic analysis, the information obtained should enable wise decisions to be made on such projects, and experience suggests that well done environmental impact studies do result in the avoidance or minimization of environmental damage, and in some cases even to the abandonment or disallowance of projects.

Mitigation policies and other management practices have met with some success. In 1973 the Amboseli National Park in Kenya was expected to reach its carrying capacity at 70,000 to 80,000 visitors a year. However, under

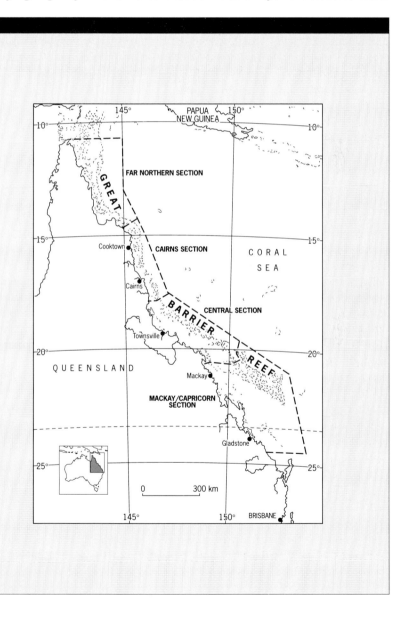

improved management practices, it was estimated that the park's annual capacity could exceed 250,000 visitors with no greater social or ecological impact than would have been caused by a smaller number of unregulated tourists (Western, 1986). Similarly, to combat deforestation in Nepal, which has been aggravated by the fuel needs of trekkers, the Annapurna Conservation Area Project sets guidelines for fuel use. Fuel-efficient water heaters have been introduced, and trekking groups now cook with kerosene. This switch to kerosene alone is expected to save over 1,600 kilograms of wood per day (Lindberg, 1991).

New plans for protected area management are increasingly being worked out in dialogue with local people so that they can be involved in the management of the area, gain employment, and receive a share of the economic benefits from the associated tourism. In Nepal's Royal Chitwan National Park, for example, the local people are now allowed into the Park for two weeks each year to harvest grass for thatching, which is worth about $US1 million a year to the 59,000 villagers who take part. At Khao Yai Park in Thailand, which had suffered severely from poaching and the encroachment of cultivators, villagers have been enlisted as guides and porters for groups of tourists hiking in the mountains; the wages were some three times those of normal village labour and demonstrated to the villagers that tourism brought economic benefits (McNeely, 1990).

Tourist areas in the industrialized countries are not immune from the need for action. According to a study that considered the potential environmental impacts of the forthcoming Single Market within the European Community (European Community, 1990), tourism has been identified as an economic activity of considerable significance. In particular, the situation for Greece was analysed. The study indicated that there would not be a large increase in tourist numbers, but that the type of tourism product supplied was likely to shift in the direction of large-scale, relatively self-contained complexes, located in areas of striking beauty. Such developments will require very stringent controls if they are to be relatively benign environmentally. Conversely, if the environmental aspects are not well planned and managed, the overall effects will almost certainly be negative.

Some guidelines have been developed, based on work carried out by several organizations including OECD, WTO and UNEP which set out the principles and identify actions that would help to improve the tourism-environment relationship (Box 4).

Concluding remarks

Clearly, the concept of sustainable development can and should be applied to tourism throughout the world. It will help to protect the often sensitive environments on which the tourism industry depends and without which it cannot thrive. Involvement of all the stakeholders - governments, the industry

BOX 4

Principles and actions that would improve tourism-environment relationships.

- Tourism development plans should be fully integrated with regional land-use and development plans; they should pay particular attention to environmental considerations, especially with respect to the quality of air, water (both for human consumption and for recreation), soil conservation, the protection of natural and cultural heritage and the quality of life in associated human settlements.
- Environmental impact assessments should be undertaken for all major tourism developments, to evaluate the potential damage to the environment in the light of forecast tourism growth and peak demand. Alternative sites for development should be considered, taking into account local constraints and the limits of environmental carrying capacity. This capacity includes physical, ecological, social, cultural and psychological factors.
- Planning authorities should seek out and take into account the views of their communities on the environmental and social impact of tourism projects.
- Decisions should be based on the fullest available information concerning the environmental implications of development proposals. Where essential information is lacking, decisions should be deferred until it becomes available.
- Adequate environmental measures at all levels of planning should be defined and implemented. Particular attention should be paid to peak demand and its consequences for sewerage, solid waste disposal, noise pollution, and to building and traffic density control. Developments should be as energy efficient as possible, minimizing their contribution to energy consumption through appropriate choice of equipment and the encouragement of access by public transport.
- In the most endangered zones, comprehensive improvement programmes should be formulated and implemented. Powers should be used to limit developments in sensitive areas and secure legislation should protect rare, endangered and sensitive environments.
- The principle that 'pollution prevention pays' is applicable to tourism, as is the 'polluter pays principle'. However it should be remembered that payment does not help if the polluter has destroyed the resource.
- Major incentive actions should be taken in both the public and the private sectors to spread tourism demand over time and space in order to use accommodation and other tourism facilities efficiently.
- All components of the tourism industry - host communities, tourists, travel agents, tourist operators, developers, owners and planning authorities - need to educate themselves on the mechanisms and benefits of an environmental perspective. Government and industry should share the responsibility for providing the necessary information programmes.

Source: UNEP (based on an IEO/PAC presentation to the Parliamentary Conference on Tourism, The Hague, April 1989.)

at large, the local communities affected by it, and the travelling public - in safeguarding and ensuring the continued use of the tourism resource is essential if sustainable tourism is to be achieved.

Managing the environmental threats likely to be caused by increasing tourism will be possible only with adequate planning and co-ordination. For countries that already have tourist industries or the potential to develop them, special governmental tourism boards or similar agencies may be required. Although the primary planning initiative must rest with national governments, because tourism is a world-wide industry and involves ecosystems of global concern, collaboration on a regional and even a global scale is generally necessary if the appropriate relationship between tourism and environmental protection is to be developed and maintained.

Apart from good planning and management of the tourist resorts themselves, the primary means of limiting the impact of tourism on the environmental quality of an area is by limiting the number of visitors, but good management should also involve the use of strategies to mitigate potential damage. Any

---

BOX 5

Selected indicators of tourism impact.

Tourist pressure indicators

i. Number of tourists in a well-defined tourist area, and density of tourists.
ii. Tourist numbers relative to local population.
iii. Areas covered by structures in the tourists areas.
iv. Traffic flow in the area.
v. Amount of waste disposed.
vi. Amount of waste water and sewage disposed.

Ecological impact indicators

i Lake (and river) water ambient quality:
  • bacteria concentration;
  • horticultural spray concentration;
  • nitrate and phosphate concentration;
  • transparency of water.
ii. Impacts on mountain slopes:
  • land slides and avalanches due to construction or over-use;
  • floods due to tourist construction or erosion;
  • erosion in general;
  • impacts on plants and animal life and consequential destruction of biotopes.
iii. Impacts on lake shores and moorlands:
  • reduction of lake shore vegetation through the extension and over-use of beach areas;
  • loss of aquatic communities;
  • drainage of swamps and the impact of toxic sprays.

---

strategy will be highly site-specific but some generalizations can be made. Ecological damage caused by tourism infrastructure can be reduced if facilities are sited carefully and appropriate visitor management techniques are used. For example, by diversified siting of viewing trails and varying the timing of visits, natural area managers can reduce both visitor congestion and the disturbance of flora and fauna.

A key principle for dealing with many tourist-related environmental problems is achieving the right relationship between the type and scale of the tourist activity and the carrying capacity of the different ecosystems likely to be affected. Furthermore, the ecological carrying capacity has to be complemented by consideration of other factors such as the carrying capacity of the social community into which the tourists are being introduced. Assessment of these capacities and adjusting the level of tourist activity to stay within them has to be seen as crucial to the prevention of future environmental and societal damage.

The development of a standard set of environmental indicators for use in

Nuisance and aesthetic indicators
i. Noise impact of traffic.
ii. Disruptive settlement pattern in tourist villages.
iii. Disruptive construction in recreation areas (e.g. on mountain slopes and lake shores).

Indicators of diverse impacts
i. Impact on agriculture:
 • extension of agriculture to supply tourist demand for specific agricultural products (environmental impact diverse);
 • reduction of agricultural area to allow tourist development.
ii. Impact on areas adjacent to tourist region:
 • transient traffic flows;
 • extension of road work;
 • potential pollution of waters from the tourist regions.

Financial indicators
i. Revenues:
 • local taxes;
 • levies on merchandise sold;
 • entertainment tax;
 • tourist levy.
ii. Expenditures:
 • waste collection and disposal;
 • waste water and sewage control;
 • erosion, flood and land-slide control;
 • noise control.

Source: Juhasz, in Mercer (1991).

areas where tourism is important could be a useful measure in helping governments, tourism operators, the tourists themselves and local people to assess the impact of tourism on the particular environment affected, and could provide an early warning of system overload. Such indicators would need to be agreed on an international basis and methods for their objective measurement established. One proposed set of indicators is given in Box 5.

Education can also play an important role, by informing tourists of what is acceptable behaviour in relation to plants and animals and perhaps even by influencing their desire to see or get close to certain species. For example, if it is explained that human presence decreases the cheetah's hunting success, some tourists may be willing to forgo cheetah viewing. Similarly, educating visitors about the damage that can be done to fragile ecosystems by off-road driving may help to reduce its incidence.

Environmental authorities should regularly monitor the state of the environment in tourist areas and take steps, in co-operation with local officials and the tourism operators, to counter any adverse effects that may be emerging. There should also be close integration, at the national level, of tourism and environmental policies in order to pre-empt the possibility of conflict between tourism development and the maintenance of environmental quality.

REFERENCES

Ascher, F. (1985) *Tourism: transnational corporations and cultural identities.* UNESCO, Paris.

Breiling, M. (1991) Some remarks concerning sustainable development and landscape change in Austrian alpine regions. Paper presented to International Symposium on Advances in Landscape Synthesis Research, May 1991, Bratislava, CSFR.

Briand, F., Dubost, M. Pitt, D. and Rambaud, D. (1989) *The Alps: A System under Pressure* IUCN/ICALPE, Chambery, France.

El-Hinnawi, E. and Hashmi, M. H. (eds.) (1982) *Global Environmental Issues.* Tycooly International, Dublin.

European Community (1990) *1992 - The Environmental Dimension.* Task Force Report, Environment and the Internal Market Economics. Verlag, Bonn.

Francillon, G. (1991) The dilemma of tourism in Bali. *In: Sustainable Development and Environmental Management of Small Islands.* Beller, W. d'Ayala, P. and Hein, P. (eds.). UNESCO and Parthenon Publishing, Paris.

Gately, D. (1988) Taking off: The U.S demand for air travel and jet fuel. *The Energy Journal*, **9**, No 4.

Grenon, M. and Batisse, M. (eds.) (1989) *Futures for the Mediterranean Basin: The Blue Plan.* Oxford University Press, Oxford.

Holling, C.S. (ed.) (1978) Obergurgl: Development in High Mountain Regions of Austria. *In: Adaptive Environmental Assessment and Management.* (Int. Ser. in Applied Systems Analysis, Vol 3). John Wiley, Chichester.

IUCN (1991) *A Strategy for Antarctic Conservation.* IUCN, Gland, Switzerland.

Lindberg, K. (1991) *Policies For Maximizing Nature Tourism's Ecological and Economic Benefits.* World Resources Institute.

McNeely, J. (1990) Conservation must pay. *Zoogoer*, January-February 1990, pp. 4-8.

Mercer, D. (1991) *A Question of Balance: Natural Resource Conflict Issues in Australia.* Federation Press, Leichhardt, NSW.

New Zealand Department of Conservation (1991) *Giardia and Back-country Water Sources.* Department of Conservation, Wellington.

Nilsson, S. and Pitt, D. (1991) *Mountain World in Danger - Climate Change in the Mountains and Forests of Europe.* Earthscan Publications, London.

OECD (1980)*The Impacts of Tourism on the Environment; General Report.* OECD, Paris.

Pasini, W. (ed.) (1988) Tourist Health: A New Branch of Public Health. *Proceedings of the International Meeting on Prevention and Control of Infections in Tourists in the Mediterranean Area, Rimini 1988*, World Health Organization and World Tourism Organization.

Roszak, T. (1988) Leave the wilderness alone! *New Scientist*, June 2, pp. 63-64.

Western, D. (1986) Tourist capacity in East African parks. *Industry and Environment* **9**: 14-16

WTO (1980) *Manila Declaration.* World Tourism Organization, Manila.

WTO (1990) Economic Review of World Tourism, World Tourism Organization, Madrid.

CHAPTER 16

Population and resources

Introduction

Environmental resources

Environmental resources fall into two distinct classes: the global (e.g. atmosphere, ozone layer) and the essentially national. National resources can further be classified into the non-renewable (e.g. minerals, fossil fuels) and the renewable (biomass products of plants or animals, biodiversity). It is a historical fact that most industrialized countries have long ago depleted a variety of their renewable natural resources and even some in the developing regions (Chapters 7 and 8). The bulk of renewable natural resources are now in the developing regions, where they form an essential base for biomass-based economies in countries where populations have been increasing at high rates.

Apart from fossil hydrocarbon fuels, the depletion of non-renewable resources has not proved to be a serious constraint in development efforts in recent history. In fact, the rates of consumption of some such resources have been declining of late as synthetic substitutes have replaced many traditional materials and recycling is reducing the rates of growth of demand on non-renewable resources (Chapter 12). As a generalization that is borne out by many of the facts in this book, it could be stated - with at least some justification - that the current environmental problems in the developed regions are mainly problems of deterioration or depletion of global environmental resources at the national and/or trans-national levels. In contrast, those of the developing regions are mainly problems of depletion of renewable biomass resources and their productivity. Particularly in the rural areas of less developed countries, environmental degradation is due to the exploitation of their ecological systems beyond their reproductive capacities, thus causing severe stress on the national resource base.

The levels and patterns of natural resource use and of waste production and management determine the collective impact of people on natural resources and the overall environment. It is, therefore, necessary to consider all demographic factors including population size, rate of growth, movement, distribution, and age/sex structure in order to formulate realistic policies and plans for the achievement of sustainable use of natural resources. Rapid population growth, for instance, often results in increased numbers of people moving to marginal lands or to regions with different ecosystems in pursuit of a new basis for daily sustenance. In the process, some degradation of the natural resource base is bound to happen, especially if people apply practices developed in one region to the systems of a new and different environment. This is happening at present in many developing countries with rapidly increasing populations for whom agriculture and subsistence activities are the predominant economic base. In the industrial countries, which presently experience low, zero, or even negative population growth and very moderate levels of population movement while having high levels of consumption of resources, it is necessary to consider the cumulative impact of average per capita resource consumption as well as the collective use of resource-depleting practices and polluting technologies.

People, resources, environment and development

Examination of the manner in which population and environmental policies have been evolving over the last two decades has resulted in a growing consensus on the nature and scope of the interrelationships between people, resources, environment and development. These interrelationships have been the subject of study since the UN General Assembly Resolution 3345 (XXIX) called for such studies in 1974.

The International Conference on Population held in Mexico City in 1984 recognized that a major immediate challenge for population policy was the disequilibrium between rates of change in population and changes in resources, environment and development (UN, 1984). There is no simple correlation between population and the environment. Population, environment and development factors interact in different ways in different places (UNEP, 1985). Not only the pace of development, but also its content, location and the distribution of its benefits, determine, in good measure, the state of the environment. These factors also influence the growth and distribution of the population. Environmental resources provide the basis for development, just as environmental factors constitute part of the improvement in the quality of life that development is meant to bring about. Similarly, the size of population, the rate of its growth and the pattern of its distribution influence the state of the environment, just as they condition the pace and composition of development. The Mexico City Conference observed that 'in many countries the population has continued to grow rapidly, aggravating such environmental and natural resource problems as soil erosion and desertification, which affect food and agricultural production'. These problems, as well as others such as air pollution, fresh water shortages, degradation of coastal zones and loss of biological diversity, and their root causes in our lifestyles and development activities, both past and present, are dealt with in other chapters in this book. This chapter focuses mainly on the impact of population growth and other demographic changes over the last two decades on natural resources - both in their depletion and in the degradation of quality.

As populations grow, the task of providing for their needs and well-being through environmental management becomes more challenging. The concept of carrying capacity is relevant, in general terms, to consideration of the relationship of population growth to the natural resource base. The carrying capacity varies from one locality to another, and depends on the life-styles and consumption patterns of the population (UNEP, 1985). The carrying capacity does not relate only to the human population, but also to the animal and plant population. The conflicts between the needs of each complicate the issue, and the number of variables involved in the analysis can be quite large - particularly when we take into consideration the impact of factors such as trade and transfer of technology.

While it is apparent that more people require more food, fuel and clothing, which must come from the Earth's resources, it is still not common practice to address the complex relationships between population and environment in an integrated fashion. Population growth has generally been identified as the main villain of environmental degradation, but this disregards the complex relationships between environmental degradation and different consumption patterns as well as other demographic variables such as chaotic patterns of population movement and settlement and distorted age-sex structures in selected geographical areas. Although the latter phenomena are often related to population growth, they are a consequence of complex interlinked economic, sociological and cultural patterns as well. The Report of the Sixth Session of the United Nations Population Commission held in March 1991, recognizes the need to develop policies that would incorporate population concerns into environmental management programmes. The report states that:

'The failure to take fully into account the possible effects of other factors that might contribute to environmental degradation characterizes many analyses of population-environment interrelationships at the national and global levels and thus limits their value in assessing the impact of demographic variables.' (p.3 E/CN.9/1991).

This limited understanding of the causes and consequences of demographic phenomena and the role they play in managing the environment accounts in part for the slow progress towards the formulation and implementation of adequate policies and programmes. Considerable attention has, however, been given to human population dynamics and resource use in the new World Conservation Strategy (IUCN/UNEP/WWF, 1991). Furthermore, in recent years an increasing number of countries have been gradually moving towards the formulation and implementation of national conservation strategies. The next logical step is to weave the relevant demographic variables into the different natural resource sectors addressed in these strategies and, more important still, into the relevant socio-economic development plans.

Natural resources policy, if it is to be realistic, must be based, among other things, on the most likely future size, movement and distribution of the population, and explicit targets need to be set for these variables, since - as discussed later on - for most countries significant changes are to be expected in coming years, particularly since progress in fertility decline has been slow.

Population trends 1972-92 and future projections

Continuing increases in world population

Since the Stockholm Conference in 1972, the world's population has grown from more than 3.6 billion in 1970, passing the 5 billion mark in mid-1987, to about 5.3 billion in 1991 and 5.48 billion in mid-1992 (Table 1; UNFPA,

1992). World population will reach 6 billion in 1998. While population growth rates are declining in many countries, large absolute increases in the world's population will continue for at least two decades, due to the young age structure of many populations, especially in developing countries, and the built-in 'demographic momentum'.

The first long-range (up to 2150) UN population projections since 1980 give the following estimates (UN Population Division, 1991):

- According to the *medium projection*, considered the most probable, world population will increase from 5.48 billion in 1992 to 10 billion in 2050 and level off at just over 11.6 billion in 2150. This last figure is 1.4 billion higher than the 1980 projection. Additions will average 97 million a year until the end of the century and 90 million a year until 2025, dropping between 2025 and 2050 to almost 61 million. It is only after 2050 that a significant slowdown will occur, by which time the world population would be almost double today's. Some 97 per cent of these increases will be in developing countries, with 34 per cent of world population growth in Africa and 18 per cent in South Asia. By 2050, Africa's population will be three and half times the present level, and almost five times by 2150.

- The *low projection*, based on the assumption that fertility will drop world-wide below the replacement level (an assumption that is highly dubious), forecasts that world population will peak in 2050 and fall thereafter.

- The *high projection* predicts that, at an average rate of 2.2 children per woman, world population will continue to rise indefinitely (12.5 billion by 2050 and 20.8 billion by 2150). At the higher rate of 2.5 children for each woman, world population will reach the staggering figure of 28 billion by 2150. In both cases, population will still continue to grow (Figure 1).

Thus, the 1990s may well prove to be a critical demographic turning point. If fertility decline can be sustained, it may be possible to achieve a stable population by the middle of the next century. On the other hand, a continued rapid increase of the world's population will seriously affect the possibility of achieving widespread improvements in the quality of life and a balanced, well-managed use of the Earth's natural resources. Indeed the very large numbers indicated by the United Nations' highest estimate of future population growth may not materialize because the resource base might prove inadequate to support such large numbers, and consequently the levels of mortality might be higher than estimated. It is obvious that such a phenomenon would signify a failure of human civilization. Therefore it must be prevented by concerted, efficient action.

Faster population growth in developing countries

As mentioned earlier, more than 95 per cent of the growth in population is taking place in developing countries. Between now and the turn of the century,

Table 1: *Total populations (in millions) and average annual growth rates (percentages) by decade, 1960-2000.*

Country groups	1960	1960-1970	1970	1970-1980	1980	1980-1990	1990	1990-2000	2000
Developing countries									
North Africa	54	2.53	69	2.55	89	2.80	117	2.32	148
Sub-Saharan Africa	210	2.60	271	2.98	364	3.13	495	3.25	681
South and East Asia	794	2.41	1,008	2.27	1,262	2.23	1,573	2.05	1,928
West Asia	46	3.23	63	3.41	88	3.67	126	3.15	171
Mediterranean	47	1.92	57	1.81	68	1.75	81	1.43	93
Western hemisphere	217	2.74	284	2.41	361	2.16	447	1.88	538
Subtotal, developing countries	1,367	2.51	1,752	2.45	2,230	2.44	2,838	2.29	3,558
China and Asian planned economies	704	2.36	889	1.87	1,070	1.39	1,229	1.32	1,400
Developed market economies	633	1.10	705	0.85	768	0.60	816	0.49	857
Eastern Europe and former USSR	313	1.08	348	0.81	378	0.70	495	0.57	429
Total, 151 countries	3,016	2.05	3,694	1.87	4,445	1.75	5,286	1.68	6,244
World total	**3,019**	-	**3,698**	-	**4,450**	-	**5,292**	-	**6,251**
Least developed countries	200	2.43	254	2.58	328	2.59	424	2.90	564

Source: UN (1988).

the population of developing countries will grow by over 900 million, nearly 25 per cent, while that of the industrialized countries will increase by only 58 million or five per cent (Table 2). The average annual rate of population growth in the developed countries decreased from 0.86 per cent per year in the period 1970-75 to 0.53 per cent per year in the period 1985-90. In contrast, the annual rate of population growth in the developing countries as a whole

Figure 1: *Population projections to the year 2150 (high, medium and low scenarios).*

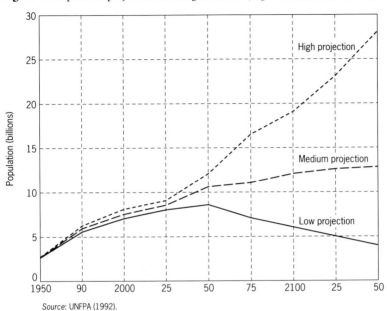

Source: UNFPA (1992).

Figure 2: *Population projections to the year 2150 by region (medium variant).)*

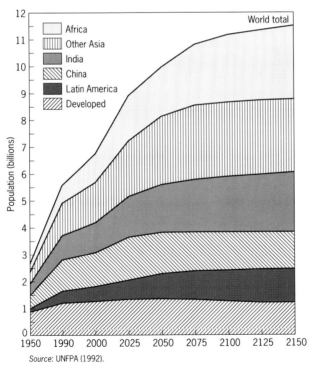

Source: UNFPA (1992).

479

decreased from 2.38 per cent per year in the period 1970-75 to 2.10 per cent per year in the period 1985-90, and has since remained more or less constant (Figure 3). However, there are significant regional differences. In East Asia, Southeast Asia, Central America and the Caribbean there were marked declines in population growth rates in the 1980s. In Africa, by contrast, the growth rate has increased over the last decade, to reach an estimated 3 per cent per year. Even within Asia, growth rates differ from one sub-region to another. China, which has almost a fifth of the world population, has reduced its population growth rate from 2.20 per cent per year in the period 1970-75 to 1.23 per cent per year in the period 1980-85, increasing to 1.39 per cent per year in 1985-90.

Infant mortality rates fell from 94 per 1,000 births in 1970-75 to 71 per 1,000 births. Industrialized countries have the lowest infant mortality rate (9 per 1,000 births), while the rate is more than 100 per 1,000 births in 34 developing countries (UNICEF, 1990). Death rates have also fallen world-wide and life expectancy rose from an average of 56.7 years in 1970-75 to an average of 61.5 years in 1985 90. In the developed countries, life expectancy now exceeds 73 years, but is 60 years in developing countries. In Africa it is only 52 years, in South Asia 57 years, and in Latin America 66 years.

Table 2: *Long-term world population projections by region, stationary population, and year when net reproductive rate (NRR) reaches one, 1990 to 2100.*

	Population (millions)				
	1990	**2050**	**2100**	**Stat.**	**year when NRR=1**[1]
Developing regions	4,086	8,716	10,200	10,020	2060
Africa	642	2,275	2,962	3,049	2060
Asia/Developing Oceania[2]	2,996	5,728[2]	6,194	6,374[2]	2055
Latin America	448	1,146	1,192	1,201	2030
Developed regions	1,206	1,319	1,310	1,314	2030[3]
Europe, USSR, Japan, N. Zealand, Australia	930	885[4]	982	891[4]	2030[3]
North America	276	332	328	329	2030[3]
World	**5,292**	**10,035**	**11,330**	**11,514**	**2060**

Notes:

1. The net reproductive rate (NRR) is the average number of daughters that would be born to a woman (or group of women) if she passed through her lifetime from birth conforming to the age-specific fertility and mortality rates of a given year. An NRR of 1.0 means each generation of mothers is having enough daughters to replace itself in the population.

2. Developing Oceania includes Melanesia, Micronesia and Polynesia; the World Bank projections include Australia and New Zealand under the broad category of Oceania.

3. Most countries in the developed regions have reached replacement fertility (of 2.1 births per fertile woman) much earlier than 2030.

4. These figures from the World Bank include all of Oceania and exclude Japan.

Source: Bulatao *et al.* (1990).

Figure 3: *Average annual population growth rates (percentage) in five-year periods, 1965–2010.*

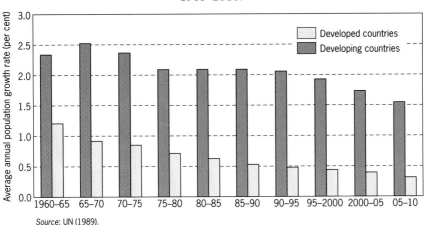

Source: UN (1989).

BOX 1

The eleven most populous countries.

Eleven countries are expected to exceed 100 million by 2020. Their relative positions change greatly. Japan starts in 1950 larger than six of the others and ends as number eleven; Nigeria starts three-fifths as large as Brazil and ends 25 per cent greater. The USA and former USSR increase in about the same proportions; so do Brazil and Mexico. Among the less developed countries of Asia, China increases less than threefold over the seventy years; Indonesia a little more than threefold; Bangladesh and Pakistan nearly fivefold.

Countries expected to exceed 100 million by 2020 (millions of persons). Number in parenthesis is the ratio of population in 2020 to that in 1950.

Year	Japan	USA	USSR[1]	Brazil	Mexico	Nigeria
1950	84	152	180	53	27	33
1980	117	228	265	121	69	81
2000	130	268	315	179	109	162
2020	133 (1.6)	304 (2.0)	358 (1.9)	234 (4.4)	146 (5.4)	302 (9.2)

Year	Bangladesh	China	India	Indonesia	Pakistan
1950	42	555	358	80	40
1980	88	996	689	151	86
2000	146	1256	964	211	141
2020	206 (4.9)	1436 (2.6)	1186 (3.3)	262 (3.3)	198 (5.0)

Note: 1. Now C.I.S., the Baltic States and Georgia.

Source: Keyfitz (1991), copyright 1991 by the American Assemby.

Changing age structures

The UNEP 1990 *State of the Environment Report* (UNEP, 1990) considers the issue of inter-generational equity as primarily a concern for the growing population of children and the deteriorating environment that hampers their development. Children under the age of 15 accounted for more than 1.7 billion or 33 per cent of the world's population in 1990. In developed countries, children accounted for only 21 per cent of total population compared with 36 per cent in developing countries. In Africa, the percentage is substantially higher at 45 per cent. Before the turn of the century, another 1.5 billion births will occur in the developing regions (60 per cent of these births in Asia), swelling the number of young dependents to unprecedented levels (Figure 4).

Figure 4: *(Upper graph) size of child population in millions and (lower graph) percentage of children (persons under age 15) in total world population, 1950–2025.*

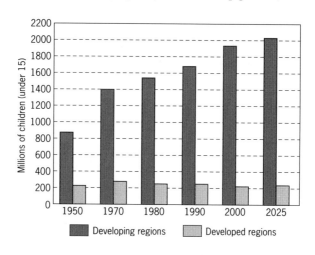

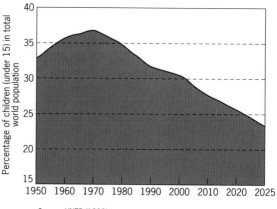

Source: UNEP (1990).

In 1970, there were about 77 persons under age 15 per 100 persons in the working ages of 15-64 in the developing regions as a whole, compared to only 44 in developed countries (Figure 5). By 2025 these 'child dependency ratios' will be less than 30 for developed countries but still close to 40 in the Third World.

Figure 5: *Child dependency by region, 1950–2025. Child dependency is the number of children (persons under age 15) per 100 people in the working age-range 15–64 years.*

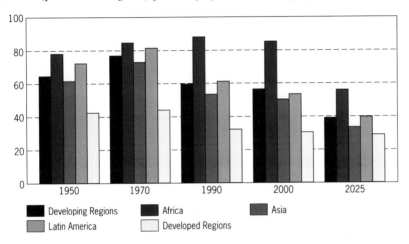

Source: UN World Population Prospects (1988).

The stimulus for population growth provided by the age structure among developed countries has disappeared, but the base of the age pyramid in developing countries continues to expand (Figure 6). The very young age structure in developing countries guarantees a continued large population growth even if fertility rates were to decline to replacement levels in the next few years, simply because there will be so many more couples in the child-bearing ages. In contrast, for the developed regions the age structure will flatten out over time because of low and falling birth and death rates that combine to produce an older population.

Increasing old-age dependency ratios

The changing population structure creates problems also for the growing number of elderly citizens. For the world as a whole, the old-age dependency ratio (those over 65 years old as a percentage of those between the age of 15 and 64) will rise from 10 per cent in 1970 to 12 per cent in 2000 (Figure 7). The world's population aged 60 and above was over 323 million in 1990 but is expected to swell to over 610 million in 2000 (UNFPA, 1973). Furthermore, the growth of the population aged 80 and above will be highest during the 1990s (UN, 1990).

Figure 6: *Population age pyramids for developed and devloping countries, 1985 and 2025.*

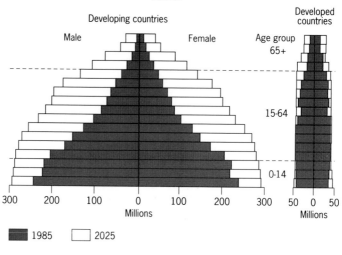

Source: UN Population Division (1990).

Figure 7: *Child and old age dependency rates (number of dependents per 100 persons in age-range 15–64).*

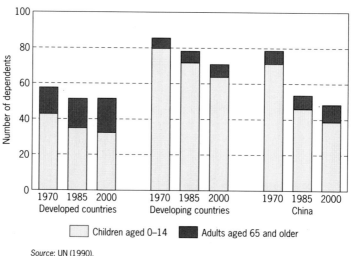

Source: UN (1990).

While most of the elderly in developed countries tend to reside in urban areas, in the developing regions the majority of the older people live in rural communities which have inadequate health care and other services. This change in the age structure of rural populations in developing countries has certain implications for sustainable use of natural resources, since older people tend to perpetuate the application of traditional practices to natural resource use. While many traditional practices may have been well adapted to slow-growth or no-growth populations, they could become inadequate when populations grow rapidly.

The job creation challenge

Changes in age structure over the past 40 years have produced a large working-age population. During the 1950s and 1960s, about 35 per cent of the world's population consisted of children under 15 years of age. These eventually moved into the main working age bracket of 25-59 years which began to increase rapidly during the eighties.

The average annual rate of growth in labour force participation rates is expected to decline from 2.1 per cent during 1970-80 to about 1.5 per cent in 1990-2000 (Figure 8). The average number of working-age people entering the labour force will fall from 41 million annually in the 1980s to 39 million annually in the 1990s but more than 35 million of these workers, or 90 per cent, will be added to the labour force in developing countries which already have serious problems of unemployment and underemployment.

Figure 8: *Labour force participation rates, 1990. (Labour force as a percentage of population aged 10 years and over.)*

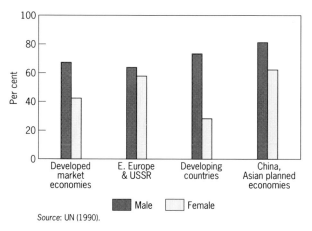

Source: UN (1990).

The crucial issue for the 1990s will be whether growth in employment will be able to absorb the increasing number of new entrants to the world's labour force. During the next decade the developing regions will have to generate over 30 million new jobs every year just to absorb into the work force children already born (Leonard 1990). This means that the natural resource base of these countries will have to provide an adequate existence for a rapidly increasing number of people. For instance, India alone has to absorb almost 10 million entrants a year in the coming decade. Since 73 per cent of the Indian population is presently living in rural areas, it follows that the natural environment of India will have to provide a living for around seven million of these new workers each year in coming years. This expansion will greatly increase the pressures on India's natural resources and will jeopardize their sustainable use in certain areas.

Another consequence of increased rural population, and a consequent severe stress on natural resources, is migration of the poor from rural communities to urban centres. However, if urbanization accelerates as fast as expected, demographic trends make virtually inevitable another doubling of rural labour force entrants in many developing countries, putting massive physical, economic and environmental strains on developing-country institutions. For instance, while China has more than 30 million workers in non-farm jobs, its rural labour force is ten times as big (IFAD, 1986).

Migration and urbanization

Historically, migration has been motivated by economic considerations. Rapid population growth, not accompanied by corresponding growth in the number of new workplaces outside agriculture or related occupations, has resulted in millions of people in developing countries occupying and exploiting marginal lands or migrating to urban centres.

Apart from rapid population growth, the pressure to move into remote and ecologically fragile rural environments, or to the fringes of urban centres, also comes from declining agricultural productivity, and from landlessness in the places of origin of the migrants as the carrying capacity of the local environment is exceeded. The physically precarious conditions typical of most remote areas have a negative effect on the natural environment, as well as on the migrants themselves. These aggravate the already declining agricultural productivity due to soil degradation/desertification or result in deforestation.

As for rural-urban migration, the number of urban residents in developing countries will be almost double that of industrialized countries at the turn of the century (Chapter 17), since the rate of population growth in urban agglomerations in developing countries is on the average three times the rate in more developed regions. However, despite continuing urbanization, in most developing countries the population remains mainly rural. During the 1990s about a third of total population increases in the developing regions will still occur in rural areas (Merrick, 1989; UN, 1990).

It is a matter of continuing debate whether this trend towards urbanization has a greater or lesser impact on the environment than would be the case if the same number of people were dispersed in villages or rural areas. While it often happens that urban centres lead to degradation of the areas immediately surrounding them through deforestation, alienation of arable lands, dumping of wastes etc., it may also be the case that the efficiencies and economies of scale generated by urban life lead to a lessened impact overall. The economic effects of urbanization have been extensively discussed without any unanimous conclusions being reached on the advantages or disadvantages, while the literature investigating the ecological effects is small and even less conclusive (Keyfitz, 1991).

Environmental refugees

Another way in which population redistribution occurs is through the movement of 'environmental refugees' from areas hit by natural disasters or suffering long-term ecological decline. Such environmental crises may be due to natural causes such as earthquakes, volcanic eruptions, extremes of weather causing landslides, floods and droughts, or to less direct factors such as human mismanagement of resources (for instance overgrazing of land, deforestation, overuse of soils, and pollution of air and water) or armed conflicts (Chapters 6, 9 and 19). Environmental degradation is generally both a cause and a consequence of population movement: a cause, because declining environmental carrying capacity forces people to seek another place to live; and a consequence, because increasing population pressure on the resources in the receiving areas exacerbates environmental stress, thus creating more environmental refugees. As an example, at least 10 million people in Africa were left homeless during the 1980s because of extended drought (UNDIES, 1990). UNEP estimated in 1988 that a total of 4.5 billion hectares world-wide (about 35 per cent of the world's land area) were in various stages of desertification and affecting more than 850 million people in arid, semi-arid and sub-humid areas. World-wide, more than 400 million people in rural areas are affected by dry climates; 64 per cent of these people cultivate highly erodible rain-fed croplands. In the Philippines migration to over-logged areas has resulted in the conversion of 175,000 hectares of forest land every year between 1975 and 1985 (Cruz *et al.*, 1988; Cruz and Cruz, 1990). In the Rondonia area in Brazil, the population of small-scale forest cultivators increased at over 15 per cent per year since 1975 - a rate that is many times larger than Brazil's annual population growth rate (Malingreau and Tucker, 1988). Similar mass migrations into tropical forests have occurred in Colombia, Ecuador, Peru, Bolivia, Côte d'Ivoire, Nigeria, Thailand and Indonesia (Myers, 1990; Southgate, 1990; Southgate and Runge, 1990).

As many as 70 million people, mostly from developing countries, are working (legally or illegally) in other countries. Every year, over one million people emigrate permanently to other countries and close to that number seek asylum. The number of refugees rose from 2.8 million in 1976 to 17.3 million in 1990 - a sixfold increase (UNFPA, 1992).

Implications of population pressure on resources and responses

Population pressure and resource degradation

People contribute to resource depletion and/or degradation in several ways. First, there is the direct impact of a growing population on the rate of resource consumption. Second, some of the environmental problems arising from rapid population growth are compounded by the differences in life-styles, and consequently consumer demand, within countries as well as between countries. The present demand pattern of natural resources world-wide calls for massive transfers of these resources from relatively poor regions of rapidly growing populations and declining natural resources to affluent regions of low population and resource-intensive ways of life. Most of the demand for natural resources in the developing and less developed countries originates from the industrialized countries (Figure 9, Box 2)

Figure 9: *Net Imports[1] of IPC commodities[2], 1970–89 (in $US millions).*

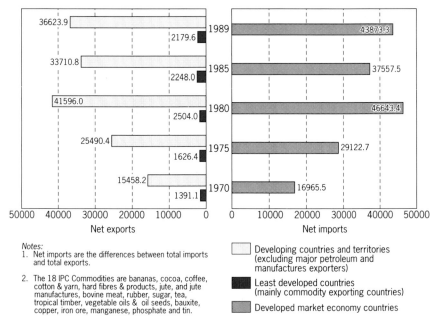

Notes:
1. Net imports are the differences between total imports and total exports.

2. The 18 IPC Commodities are bananas, cocoa, coffee, cotton & yarn, hard fibres & products, jute, and jute manufactures, bovine meat, rubber, sugar, tea, tropical timber, vegetable oils & oil seeds, bauxite, copper, iron ore, manganese, phosphate and tin.

☐ Developing countries and territories (excluding major petroleum and manufactures exporters)

■ Least developed countries (mainly commodity exporting countries)

▨ Developed market economy countries

Source: Based on data in UNCTAD (1991).

Third, rapid population growth has served to increase the number of poor households world-wide. As mentioned earlier, poor households often rely on marginal environments for survival, thus establishing a cycle of poverty and resource degradation. In many cases the cycle is perpetuated through a decline in the economy and consequently unstable national policies. This decline in the economy has been aggravated, as discussed in Chapter 24, by the unfavourable trade environment depressing the prices of most commodities, while those of the exports of industrial countries continued to increase (Table 3). It is estimated (UNDP, 1992) that some $US500 billion a year is lost to the poor countries because of restrictions in and unequal access to, international trade, financial and labour markets. This is nearly ten times what these countries receive in foreign aid.

	BOX 2

Developing countries amongst the ten principal exporters of some IPC commodities (ranked by 1989 values).

Commodity	Countries
Copper	Zambia, Zaire, Peru, Papua New Guinea (4)
Timber	Indonesia, Malaysia, Singapore, Brazil (4)
Cotton and cotton yarn	Pakistan, Egypt, Turkey, Brazil (4)
Sugar	Cuba, Thailand, Mauritius, Brazil (4)
Coffee	Brazil, Colombia, Indonesia, Côte d'Ivoire, Guatemala, Ethiopia, Costa Rica, Uganda (8)
Iron ore	Brazil, Liberia, Venezuela (3)
Natural rubber	Malaysia, Indonesia, Thailand, Liberia, Sri Lanka, Nigeria, Côte d'Ivoire, Viet Nam, Cameroon (9)
Tobacco (unmanufactured)	Brazil, Turkey, Zimbabwe, Malawi (4)
Cocoa beans	Côte d'Ivoire, Ghana, Malaysia, Nigeria, Brazil, Cameroon, Indonesia, Singapore, Ecuador (9)
Tea	India, Sri Lanka, Kenya, Indonesia, Turkey, Bangladesh, Malawi (7)
Bananas	Ecuador, Honduras, Costa Rica, Colombia, Philippines, Mauritius, Guatemala, Panama, Saint Lucia (9)
Phosphate rock	Morocco, Jordan, Togo, Senegal, Syrian Arab Republic, Nauru (6)
Tin	Malaysia, Brazil, Indonesia, Singapore, Bolivia, Thailand, Peru (7)
Bauxite	Guinea, Brazil, Jamaica, Guyana, Sierra Leone, Ghana (6)
Jute and products	Bangladesh, India, Thailand (3)
Manganese ore	Gabon, Brazil, Ghana, India, Morocco (5)
Hard fibres & manufactures	Brazil, Philippines, Sri Lanka, India, Kenya, Mexico, United Republic of Tanzania (7)

Source: UNCTAD (1991).

Finally, resource degradation occurs when the population exceeds the capacity of social institutions to cope with environmental problems. Village and household social systems (property rights and communal rules of access to resources and the increasing involvement of women in resource production, utilization and conservation) affect the outcome of excessive population growth. The breakdown of these systems under the pressure of growing populations, government actions, changes in market relations or the unequal distribution of holdings, undermines the effectiveness of these systems.

Table 3: *Changes in world commodity prices, 1975–89, in constant 1980 $US per unit of measure.*

	1975	1976	1977	1978	1979	198
Cocoa (kg), New York & London	1.98	3.21	5.41	4.23	3.61	2.6
Coffee (kg), Brazil	2.94	5.17	9.70	3.97	4.26	4.5
Tea (kg), World average	2.21	2.41	3.84	2.72	2.36	2.2
Rice (t), Thailand	578.2	399.5	388.9	456.5	363.3	433.9
Maize (t), US	190.5	176.5	136.1	125.1	126.6	125.3
Wheat (t), Canada	288.7	234.1	165.4	167.5	189.0	190.8
Sugar (kg), World	0.72	0.40	0.26	0.21	0.23	0.6
Beef (kg), US	2.11	2.48	2.15	2.66	3.16	2.7
Groundnut Oil (t), Nigeria	1,364.7	1,163.3	1,217.6	1,340.6	974.5	858.8
Cotton (kg), Index	1.85	2.66	2.22	1.95	1.85	2.0
Wool (kg), New Zealand	4.37	5.35	5.11	4.66	4.86	4.6
Natural rubber (kg), New York	10.49	13.71	13.10	13.75	15.61	16.2
Petroleum (barrel), OPEC	16.7	18.1	17.7	15.8	19.0	29.4
Aluminium (t), Europe	1,099.0	1,353.0	1,416.0	1,298.0	1,667.0	1,730.0
Copper (t), London	1,970.0	2,199.0	1,870.0	1,696.0	2,177.0	2,183.0
Iron Ore (t), Brazil	36.0	34.4	30.9	24.1	25.6	26.2
Nickel (t), Canada	7,277.0	7,808.0	7,433.0	5,729.0	6,563.0	6,519.0

Source: WRI (1992), based on World Bank data.

Unequal consumption of resources between developed and developing countries

Average figures for per capita GDP, GNP, income, consumption of different commodities, etc., are grossly inadequate for depicting a reliable picture of the situation under investigation. Such crude indicators conceal the true scale of discrepancies between nations and within them. With this in mind, such averages still reveal considerable disparities in per capita GDP between developed and developing regions and even within each group (Figure 10). The disparities between the rich and poor populations of the world have widened dramatically in recent years. In 1960, the richest 20 per cent of the world's population had incomes 30 times greater than the poorest 20 per cent. By 1990, the richest 20 per cent were getting 60 times more than the poorest 20 per cent (UNDP, 1992). The income gap between the richest and poorest in a country is also disturbingly wide. According to a rough estimate made for 41 countries, in which the industrial countries are over-represented, the inequality ratio is 65 to 1. For the whole world, the ratio may well be over 150 to 1 (UNDP, 1992).

Disparities in income are largely responsible for the marked differences in average consumption of resources between developed and developing nations for both renewable and non-renewable resources. The North, with about one-fourth of the world's population, consumes 70 per cent of the world's energy, 75 per cent of its metals, 85 per cent of its wood and 60 per cent of its food (Chapters 7, 11, 12, 13; UNDP, 1992). Nine of the ten countries with the lowest calorie intakes are least developed countries (Table 4). The results of a recent

1981	1982	1983	1984	1985	1986	1987	1988	1989
2.07	1.75	2.20	2.53	2.35	1.83	1.62	1.19	0.94
3.85	3.20	3.26	3.48	3.49	4.50	1.90	2.02	1.66
2.01	1.95	2.41	3.64	2.07	1.70	1.38	1.34	1.54
480.4	295.6	286.6	265.6	225.10	185.60	186.61	226.23	243.60
130.2	110.3	140.8	143.2	117.00	77.20	61.34	80.23	84.80
195.4	168.0	175.5	174.3	180.70	141.60	108.19	134.74	153.00
0.37	0.19	0.19	0.12	0.09	0.12	0.12	0.17	0.21
2.46	2.41	2.53	2.40	2.25	1.85	1.93	1.89	1.95
)37.8	590.2	735.9	1,071.3	943.7	501.8	405.2	442.8	589.5
1.84	1.61	1.92	1.88	1.37	0.93	1.34	1.05	1.27
4.25	3.96	3.77	3.87	3.71	2.92	3.66	4.35	4.07
12.46	10.11	12.82	11.55	9.64	8.33	9.05	9.66	8.50
33.0	34.3	30.5	30.6	29.3	12.1	13.9	10.3	12.4
331.0	1,071.0	1,548.0	1,445.0	1,160.0	1,112.0	1,303.0	1,910.0	1,552.0
733.0	1,493.0	1,648.0	1,453.0	1,478.0	1,212.0	1,420.0	1,953.0	2,166.0
24.2	26.1	24.8	24.3	23.7	19.4	18.0	17.4	20.1
)24.0	4,881.0	4,837.0	5,008.0	5,108.0	3,422.0	3,947.0	10,404.0	10,122.0

estimate (Parikh and Parikh, 1991) of the disparities in the consumption of some commodities are given in Table 5, as 'average disparity ratios' (Developed/ Developing, or ADR) and 'extreme disparity ratios' (USA/India, or EDR).

Table 4: *The ten countries with lowest potential calorie consumption per head.*

Country	Potential calorie consumption (per capita)
Mozambique	1,608
Ghana	1,733
Guinea	1,782
Sierra Leone	1,868
Rwanda	1,881
Haiti	1,902
Bangladesh	1,922
Central African Republic	1,940
Mali	2,020
Cameroon	2,040

Notes:
Potential calorie consumption is an estimate of average nutrients that reach the consumer, after deducting allowances for other uses, wastage and transport losses.
These figures overestimate actual consumption since they do not allow for wastage within the home or for food fed to pets.
The WHO recommends a minimum daily adult calorie consumption of 2,600 per head (with variations for age, occupation, etc.)

Source: The Economist Book of Vital World Statistics, Hutchinson (1990).

Table 5: *Consumption patterns for selected commodities: distribution among developed and developing countries.*

Category	Products	World total (mmt)	Percentage Share		Per Cap (Kg. or M²)		ADR Developed/ Developing	EDR USA/ India
			Developed	Developing	Developed	Developing		
a) Food	Cereals	1801	48	52	717	247	3	6
	Milk	533	72	28	320	39	8	4
	Meat	114	64	36	61	11	6	52
b) Forest	Round wood	2410	46	54	388	339	1	6
	Sawn wood	338	78	22	213	19	11	18
	Paper etc.	224	81	19	148	11	14	115
c) Industrial	Fertilizers	141	60	40	70	15	5	6
	Cement	1036	52	48	451	130	3	7
	Cotton & wool fabrics	30	47	53	15.6	5.8	3	6.4
d) Metals	Copper	10	86	14	7	0.4	19	245
	Iron & steel	699	80	20	469	36	13	22
	Aluminium	22	86	14	16	1	19	85
e) Chemicals	Inorganic chemicals	226	87	13	163	8	20	54
	Organic chemicals	391	85	15	274	16	17	28
f) Transport vehicles	Cars	370	92	8	0.283	0.012	24	320
	Commercial vehicles	105	85	15	0.075	0.0006	125	102

Notes: 1. • Cereals data 1987
      • Milk data include cow milk, buffalo milk and sheep milk (1987)
      • Meat data include cow milk, buffalo milk and sheep milk (1987)
      • Round wood includes fuel wood + charcoal and industrial round wood (1988)
      • Sawn wood includes that extracted from sawlogs and veneer logs (1988)
      • Paperboards include newsprint, printing and writing papers and other paper + paperboard (1988)
      Statistical Year Book 1987; Handbook of Industrial Statistics 1989; International Trade
      Statistics Yearbook 1987 and UN FAO book of production 1989
   2. • Fertilizer consumption data include nitrogen phosphate and potash fertilizers
      Statistical Year Book 1987; Handbook of Industrial Statistics 1989; International Trade
      Statistics Yearbook 1987; UN FAO Book of Production 1989
      • Cotton and Wool Fabric: Handbook of Industrial Statistics, 1988 UNIDO (Cotton fabrics + Woollen fabrics and excluded synthetics will alter these figures substantially; USSR & China excluded due to non-availability of data)
      • Cotton and Wool fibre total consumption figures in billion square metres and per capita consumption figures in kg. or square metres
      Statistical Year Book 1987
   3. • Per capita data are calculated

Source: Parikh and Parikh (1991).

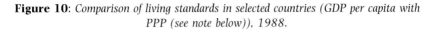

Figure 10: *Comparison of living standards in selected countries (GDP per capita with PPP (see note below)), 1988.*

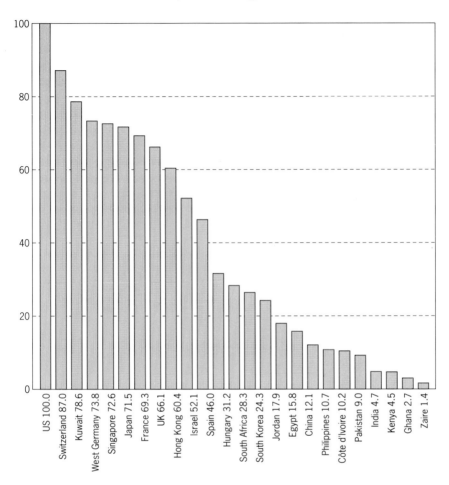

Note:
PPP is UN Purchasing Power Parity, which adjusts for differences in the cost of living.
Source: *The Economist Book of Vital World Statistics* (1990).

Annual consumption of cereals, averaging 716 kg/person in the developed regions, averages only 247 kg/person in the developing regions, with the lowest figure of 130 kg/person in Africa and the highest of 800 kg/person in Australia. Per capita meat and milk consumption, averaging 60.6 kg/person in developed countries, averages a mere 10.7 kg/person in the developing regions. The same pattern applies for some manufactured basic needs (fertilizers, textiles, cement). For chemicals, minerals and metals, with the exception of iron and steel (i.e. for copper, aluminium, inorganic and organic chemicals), 85 per cent is consumed in the developed regions.

High levels of per capita energy consumption are generally typical of the advanced economies of the industrialized countries (see Chapter 13). While coal, oil and gas (non-renewable resources) are the main sources of energy in these countries, fuelwood is the main source in the less developed countries (see Chapter 13). Under population pressure it was estimated that in 1980 over 1.2 billion people were meeting fuelwood needs by cutting wood faster than it is being replaced (FAO, 1987). Consumption of fuelwood exceeds supply by 30 per cent in Sahelian countries in Africa, 70 per cent in Sudan and India, 150 per cent in Ethiopia, and 200 per cent in Nigeria (Anderson and Fishwick, 1984). This has been one of the main causes of desertification in these countries (see Chapter 6).

Cycle of population growth, poverty, and resource degradation

As a consequence of rapid population growth in less developed countries, there will be sizcable increases in the number of the world's poor. The World Bank estimates that in 1985 more than 1.1 billion people were below the poverty line of $US370 per capita a year (World Bank, 1990). This is equivalent to about one-third of the entire population of the developing countries. Another estimate (UN, 1989) is that the number of the very poor rose from 944 million in 1970 to 1.156 billion in 1985, an increase of 212 million in the number of people living on the very edge of existence. The number is expected to rise to 1.3 billion by the year 2000 and 1.5 billion by 2025. Asia has 500 million poor people (the largest number at present). By the year 2000 it is estimated that Africa's share will increase from 30 per cent to 40 per cent, overtaking Asia (UNDP, 1991).

While the proportion of the world's population who are poor actually declined in the last decade, the total number of poor households increased as a consequence of population growth. The incidence of malnutrition in developing regions declined from 25 per cent in 1969-70 to 20 per cent in 1983-85, but the actual number of undernourished people increased from 460 to 512 million during the same years (World Bank, 1990) and is projected to increase further to over 532 million by the turn of the century (UN World Food Council, 1987).

Combined with rapid population growth, unequal distribution of landholdings rather than overall land shortage creates the most severe pressure on the poor to exploit marginal environments (Durning, 1989; Repetto, 1989). For example, in Latin America, more than 70 per cent of agricultural households are either landless or near-landless since 10 per cent of the population own 95 per cent of the total arable land (Thiesenhausen, 1989). As a result, millions of small farmers are forced to subdivide already small farms among their children until farm sizes become too small to provide subsistence. Combined with the failure of the rural economy to support increased productivity of agricultural lands, poor households have little alternative but to move into available marginal lands. Though not always totally unsuitable for farming, these lands are often highly susceptible to soil erosion, loss of soil fertility, and desertification. These environmental problems, in turn, undermine the livelihood of impoverished migrants, compounding the cycle of population growth, poverty and resource degradation (Durning, 1989; Leonard, (1990).

Focusing on women

Any effective population and environment policy aimed at changes in fertility rates, child mortality, population movement and structure, as related to changes in the use of natural resources, must focus on women. The rationale is simple; it is women who bear the children, and it is they who are in many instances the primary users and managers of a number of essential natural resources such as water, agricultural land, wood, and wild plants and fruits, to name just a few. Women account for about half of food production activities in most of the developing countries, producing as much as three-quarters of Africa's food (WHO, 1990). The essential tasks of collecting fuel and water, which poor households must provide for themselves because of the absence of public services, are done mostly by female household members. As such, they are in close contact with the natural resource base and are the main actors in a sustainable natural resources strategy.

The need to focus more explicitly on women is becoming more and more urgent since the number of poor households headed by women has expanded rapidly, especially in rural Africa, East Asia and Latin America (Leonard 1990). These households were found, on average, to have less access to productive resources because of cultural and economic restraints. In several countries, discriminatory practices, in terms of access to social services, credit and property ownership, prevent women from participating fully in the economy (Box 3).

A woman's education has a strong impact on the health and size of her family. In Kenya, for example, mothers with no education had a child mortality rate of 109 deaths of children under five years per 1,000 born, compared to the lower mortality rate of 72 among women with at least a primary education (Sadik, 1990). Children of mothers with no schooling in Indonesia were three times more likely to die before their fifth birthday than children of mothers with primary and secondary education (Sadik, 1990). Four to six years of education among women lowers fertility by as much as five per cent in Asia and Africa and 15 per cent in Latin America (Merrick, 1989). Women with at least seven years of schooling had on average 2.2 fewer children than those with no education at all (Sadik, 1990).

Important as family planning may be in a policy strategy, even more important is to provide the motivation for couples to have fewer children. The status of women in both economic and educational terms is a key factor in providing this motivation. Women do not have to be literate to know that they want fewer children, but they may well need the agreement, even the permission in some cases, of their partners to use family planning methods.

BOX 3

Economic empowerment of women.

Along with the sense of urgency brought about as a result of economic crises, population growth and environmental degradation, there has been the growing recognition that these and the other basic problems of development have a far better chance of being solved with greater involvement of women as active participants and agents of change. An important component of this investment will be the economic empowerment of women in both the formal and informal sectors.

A thriving subsistence sector, in which women play such a large role, can make an important contribution to development through the provision of food, clothing and other necessities of life. Most of the unpaid family workers in agriculture are women. Women are also in the majority among the urban small-scale self-employed, such as street vendors and home-based production or services. Yet they receive one-third to one-half of men's incomes in this sector, more because of lower returns than less time spent. When women engage in a multitude of subsistence activities, as they do, their labour is especially likely to be devalued.

These patterns can be traced in large part to the social stereotyping related to women's reproductive roles, and to women's responsibilities for child care and housework. The double burden of production and reproduction, well documented in recent studies on time-use, falls heavily on women in low-income households, who are least likely to have access to safe and effective health and family planning services and whose economic contributions are most crucial for the welfare of the family, especially the children. A survey of 74 developing countries indicates that 22% of households in Africa, 20% in the Caribbean, 18% in Asia, 16% in the Middle East and 15% in Latin America are headed by women.

The trend towards the feminization of poverty is exacerbated by migration to urban areas, often by husbands and sons leaving their homes to find employment. Often fewer than half the families remaining at home receive money from absent males. Environmental degradation and pollution have a particularly deleterious effect on women and their daughters, who must search for hours for a stretch of unpolluted water or for fuelwood.

Source: World Health Organization (1990).

Local level responses to demographic stress

Traditionally, communal management and the participation of women in agriculture and resource conservation have significantly contributed to equitable control over the use and distribution of resources. This is especially true in developing regions, where locally adapted systems of allocation tended to promote sustainable use of resources (Gibbs and Bromley, 1989; Repetto, 1988). In recent decades, however, population pressures, leading to over-exploitation of resources, have often contributed to the breakdown of these systems (Berkes, 1989).

Resource degradation occurs when the population is forced to convert resources into other forms of capital (Watson, 1989). Institutional responses break down as population growth leads to competition over scarce resources. Often, customary property rights are replaced or ignored, such as when ethnic mountain tribes are forced by lowland migrants to move into more remote, upper forests (Bromley, 1989; Grima and Berkes, 1989).

Rapidly increasing population pressures in most resource-poor areas, specifically along the margins of tropical forests, the edges of deserts, and coastal areas, may be so intense that households are unable to adopt new techniques or evolve social strategies to keep up with food requirements (Southgate and Runge, 1990). Under these circumstances, short-term approaches such as expanding cultivation into marginal lands, reducing fallow periods and over-fishing are frequently adopted (Grima and Berkes, 1989).

In some countries, population growth exacerbates the effects of government policies which have favoured the urban sector (Repetto, 1990). In most of sub-Saharan Africa, for example, migrant smallholders engage in soil-degrading intensive cultivation because of lack of appropriate economic and tenurial support (Durning, 1989). Regional differences in public infrastructure, urban-biased employment and wage policies, and programmes that promoted consolidation of agricultural landholdings contributed to mass migration into frontier environments.

The effects of macroeconomic policies on resource degradation are strong and varied. Rural labour supply, which affects frontier migration rates, depends not only on underlying population growth but also on the growth of urban labour demand and the general pattern of industrialization (Repetto, 1988).

In addition, insecure tenure discourages conservation of resources, whether the insecurity arises from the breakdown of communal management systems (Berkes, 1989), lack of government restrictions on property ownership, or irregularities in the distribution of use-rights to the resources (Repetto, 1989). Thus, resolving population pressures on marginal lands and ecologically sensitive areas requires a different set of approaches - one that incorporates activities with regard to population growth and family planning, population movement and settlement, health and sanitation, education, and the alleviation of poverty, as well as tackling food production, agricultural development and employment generation (Leonard, (1990).

Concluding remarks

Humanity and nature are interdependent. Changes in one are accompanied by changes in the other. It follows that significant progress in arresting environmental resource degradation can be achieved only by major institutional changes which would be the core of any environmentally sound development and conservation strategy. The most fundamental of these changes is the willingness of governments and bilateral and multilateral development institutions to adopt an integrated approach, addressing both the physical and biological aspects of the issue in tandem with the human aspects of sustainable use and conservation of the natural environment. Such an approach will need to be supported by the systematic application of environmentally benign practices and technologies as well as the reduction of over-consumption in the rich North and of the rich people in developing countries. It also requires the fastest possible reduction of population growth rates, the alleviation of poverty and the reversal of poverty-related environmental degradation.

Adjustment of population trends

The proposed United Nations Fourth Development Decade includes human resource development as a major goal, an integral part of which is promoting the availability of adequate health care and sanitation, as well as family planning and reproductive health services. Because of the key role of the population factor in natural resource use and environmental degradation, improving the quality of life will also require significant steps in resolving major global resource and environmental problems.

An efficient strategy for environmental conservation will, according to the arguments presented in the preceding pages, necessarily be accompanied by a population policy based upon present demographic trends and their likely future outcomes. Explicit targets for population growth, movement and distribution will have to be selected according to realistic prospects for the sustainable use of natural resources. A programme aimed at the adjustment of demographic trends and the minimization of their impact on natural resources would have three components:

a) the expansion of existing family planning and health care services in areas where the needs are greatest;

b) raising the socio-economic status of women, and

c) developing and implementing programmes that would reduce direct population pressures on natural resources.

Initiatives to achieve the first target must include improvements in women's education, widespread availability of family planning and reproductive health services, greater participation of males in family planning programmes, and the strengthening of community-based distribution networks by involving more non-governmental organizations and by using commercial and social marketing channels.

There should be increased spending on population and family planning programmes. The 1989 Amsterdam Forum on Population in the 21st Century proposed that for such financial support to be effective it must be at the level of about $US9 billion a year for core population activities. Women are the key agents in bringing about changes in fertility. Therefore all the world's women should have access to quality reproductive health services. Such services would enable women to plan their reproductive life, space their children, have safe pregnancies, and improve their own and their families' well-being.

Anti-poverty programmes should ensure the full participation of women in their planning and implementation. Women should have widespread access to credit. Agricultural extension programmes should be 'woman-friendly', especially since women are the main providers for over one-third of all households in less developed countries and produce about 60 per cent of the food grown. Programmes should promote and facilitate the recruitment and deployment of more female agricultural extension workers and the formulation and implementation of extension programmes that will adequately address women's needs.

There should be a basic reform of customs, policies and laws that are biased against women, specifically of those relating to women's access to land, credit, memberships in cooperatives, property, and government services. These should be integrated into national planning activities and made part of development assistance programmes.

Women's education should be given higher priority especially since most families in the developing world are reluctant to educate daughters for socio-cultural or economic reasons. As of 1990, only 55 per cent of women in developing countries were literate, compared with 75 per cent of men (UNESCO, 1991).

Breaking out of the cycle of population growth, poverty and resource degradation

As the previous pages demonstrate, an environmental policy that focuses on the conservation of resources, without due regard to the livelihood of the people who depend on these resources for living, results in poverty. By the same token, development policies that focus on economic growth to meet the needs of the people without concern for their impacts on the resources on which growth is based can only end up in economic decline and poverty. An effective strategy tackling simultaneously the three-cornered problem of population, poverty and resource degradation should begin by focusing on people. The objective of providing sustainable livelihoods for all should provide the integrating factor that ensures that policies address the issues of resource management, development and poverty eradication simultaneously (UNCED, 1992).

The ultimate objective is that all poor households are provided with the opportunity to earn a sustainable livelihood, while ecologically-vulnerable areas are handled in an integrated manner encompassing resource management, poverty alleviation and employment generation. This calls for a range of activities at national and regional levels, involving governments and citizen's groups and supported internationally, to achieve better environmental management practices, ensure the flow of reliable and up-to-date information on demographic and environmental changes, and to improve training (UNCED, 1992).

It should also be noted here once more that breaking away from this vicious cycle is closely linked to the current consumption patterns in countries other than those where most people live below the poverty line. A reduction in the rates of consumption of natural resources that come mainly from the developing and least developed countries, combined with effective family planning and a more equitable system of international trade, could pave the way to restoring the balance between population and resources.

Reducing population pressure on resources: changing consumption patterns

Pressure on natural resources is not caused only by poverty and increasing population in developing countries. Over-consumption created by wasteful life-styles of the rich, and the unsustainable production technologies used in both the North and the South, is equally disrupting to the environment and its natural resources. Policies and programmes must be developed to limit the squandering of resources by the rich. Profound changes should be promoted in cultural attitudes and behaviour regarding the natural environment, and basic reforms in the management of wastes, in consumer preferences, and in industrial development are necessary. These will require changes in education curricula, marketing and communication, public policies and laws, in conjunction with greater support for family planning and reproductive health information and action programmes.

The assessment of the impact of demographic variables on natural resources needs to be improved, first through more effective systems of population census collection and secondly though more reliable measurements of population impacts on resources. Such measurements should be devised initially for critical environmental zones experiencing demographic stress (such as tropical forests and coastal areas).

The analysis of the division of labour between different family members and between social groups is a necessary precursor of efforts to achieve sustainable development. Because of deepening poverty, certain sub-groups (such as the urban poor, landless farmers, subsistence fishers and environmental refugees) who are often marginalized by present development policies, are over-exploiting resources out of sheer necessity for survival. An important sub-group is composed of ethnic minorities who are often forced to abandon long-settled ancestral lands due to migrant encroachment or displacement by public development projects.

General economic and social policy reforms should be instituted to alter the circumstances that induce mass migration of the rural poor to marginal environments and to cities. Such reforms would include policies that generate employment, support agricultural development and agrarian reform, provide widespread access to government services and poverty alleviation programmes, and promote more equitable distribution of income and access to land and other resources. Because the context in which these reforms occur will vary by country and by region, local-level studies should be encouraged along with cross-country comparisons of patterns of interaction of population factors and environmental change.

The implications and consequences of current consumption patterns, technological practices, the world economic situation, resource and demographic distribution patterns in different walks of life, as well as approaches for fostering sustainable development in the variety of socio-economic sectors, have been the main concern of this book. Throughout, it was clear that the move away from wasteful environment-degrading production and consumption in no way results in depressed living standards, either in developed or developing regions. Not only is reduction of waste and the rational use of resources environmentally-desirable, but the immediate and long-term economic and social benefits are now much clearer and better defined.

REFERENCES

Anderson, D. and Fishwick, R. (1984) *Fuelwood Consumption and Deforestation in African Countries*. The World Bank, Washington, D.C.

Berkes, Fikret. (1989) Introduction. *In*: Fikret Berkes, (ed.) *Common property resource: ecology and community-based sustainable development*. Belhaven Press, London.

Bromley, Daniel W. (1989) Property relations and economic development: the other land reform. *In*: *World Development*, **17(6)**: 867-77.

Brown, Lester B. (1990) *State of the World 1990*. W.W. Norton and Company, New York.

Bulatao, Rodolfo A., Bos, Eduard. Stephens, Patience W. and Vu, My T. (1990) *World population projections* 1989-90 edition. The Johns Hopkins University Press for the World Bank, Baltimore.

Cruz, Maria Concepcion J., Zosa-Feranil, Imelda, and Goce, Cristela (1988) Population pressure and migration: implications for upland development in the Philippines, *In*: *Journal of Philippine Development*, **15(1)**: 15-46.

Cruz, Wilfrido D. and Cruz, Maria Concepcion J. (1990) Population pressure and deforestation in the Philippines. *In*: *Asean Economic Bulletin*, **7(2)**: 200-212 (November).

Durning, Alan B. (1989) *Poverty and environment: reversing the downward spiral*, Worldwatch Paper No. 92. Washington, D.C.

The Economist (1990) *The Economist Book of Vital World Statistics*, Hutchinson, London.

FAO (1987) *An Interim Report on the State of Forest Resources in Developing Countries*. FAO, Rome.

Gibbs, Christopher J.N. and Bromley, Daniel W. 1(989) Institutional arrangements for management of rural resources: common-property regimes. *In*: Fikret Berkes, (ed.) op. cit., p. 22-32.:

Grima, A., Lino, P. and Berkes, Fikret. (1989) Natural resources: access, rights-to-use and management. *In*: Fikret Berkes, (ed.) op. cit., p. 33-54.

IFAD (1986) *Annual Report*, International Fund for Agricultural Development, Rome.

IUCN/UNEP/WWF (1991) *Caring for the Earth: A Strategy for Sustainable Living*, IUCN, Gland.

Keyfitz, Nathan. (1991) Population Growth Can Prevent the Development That Would Slow Population Growth, *In*: Jessica T. Mathews (ed), *Preserving the Global Environment: The Challenge of Shared Leadership*, W.W. Norton, New York.

Leonard, H. Jeffrey, (ed) (1990) *Environment and the Poor: Development Strategies for a Common Agenda*. Transaction Books, New Brunswick.

Malingreau, J.P. and Tucker, C. J. (1988) Large-scale deforestation in the Southern Amazon basin of Brazil. *In*: *Ambio*. **17**:49-55.

Merrick, Thomas W. (with PRB staff) (1989) World population in transition. *In*: *Population Bulletin*. 41(2).

Myers, Norman (1990) The world's forests and human populations: the environmental interconnections. *In*: *Population and Development Review*.

Parikh, J. and Parikh, K. (1991) 'Role of Unsustainable Consumption Patterns and Population in Global Environment Stress, *In*: *Sustainable Development*, **Vol.1, No.1**, pp 108-18, New Delhi.

Repetto, Robert (1988) *The forest for the trees? Government policies and misuse of forest resources*. World Resources Institute, Washington, D.C.

Repetto, Robert (1989) Population, resources, environment: an uncertain future. *In*: *Population Bulletin*. 42(2).

Repetto, Robert (1990) Deforestation in the tropics. *In*: *Scientific American*. **262(4)**: 36-42.

Sadik, Nafis (1990) *State of world population 1990*. United Nations Fund for Population Activities, New York.

Southgate, D. (1990) The Causes of Land Degradation on 'Spontaneously' Expanding Agricultural Frontiers in the Third World, *In: Land Economics*, **66**: 93-101.

Southgate, D. and Runge, C.F. (1990) *The Institutional Origins of Deforestation in Latin America*, Department of Agricultural and Applied Economics, University of Minnesota, Minneapolis.

Thiesenhausen, William C. (ed.) (1989) *Searching for agrarian reform in Latin America*. Unwin Hyman, Boston.

UNCED (1992) Combating poverty, changing consumption patterns and demographic dynamics and sustainability, A/CONF.151/PC/100/ Add.2, discussed in the fourth session of the Preparatory Committee for the United Nations Conference on Environment and Development, New York, 2 March - 3 April, 1992.

UNCTAD (1991) *Commodity Yearbook*, United Nations Conference on Trade and Development, Geneva.

UNDP (1991) *Human Development Report, 1991*, United Nations Development Programmes, Oxford University Press, New York and Oxford.

UNDP (1992) *Human Development Report, 1992*, United Nations Development Programme, Oxford University Press, New York and Oxford.

UNEP (1985) *The State of the Environment, 1985*, UNEP, Nairobi.

UNEP (1990) *State of the Environment Report: Children and the Environment*, UNEP and UNICEF, Nairobi.

UNESCO (1991) *The World Education Report 1991*, United Nations Educational Scientific and Cultural Organization, Paris.

UNICEF (1990) *Children and Development in the 1990s*, UNICEF, New York.

United Nations, (1984) *Report of the International Conference on Population*, UN, New York.

United Nations (1989) *World Population Prospects, 1988*. ST/ESA/SER.A/106, United Nations, New York.

United Nations (1990) *1989 Report on the World Social Situation*, UN, New York.

United Nations Department of International Economic and Social Affairs (1990) *World population monitoring 1989. United Nations Population Studies No. 113*, New York:

UNFPA (1973) *The determinants and consequences of population trends*. United Nations Fund for Population Activities, New York.

UNFPA (1992) *A World in Balance*. State of the World Population, 1992, United Nations Fund for Population Activities, New York.

United Nations Population Division (1990) *1990 Revision of world population prospects: computerized data base and summary tables*. United Nations Department of International Economic and Social Affairs, New York.

United Nations Population Division (1991) *Long-Range Population Projections*, ST/ESA/ SER.A/125, UN, New York.

United Nations World Food Council (1987) *The global state of hunger and malnutrition and the impact of economic adjustment on food and hunger*. World Food Council Thirteenth Ministerial Session, Beijing, China.

Watson, Dwight J. (1989) The evolution of appropriate resource-management systems. *In*: Fikret Berkes. op cit.

WHO (1990) *Women, Health and Development*, Progress Report by the Director General of the World Health Organization, Executive Board 87th Session, Provisional Agenda (EB87/22).

World Bank (1990) *World development report 1990: poverty*. Oxford University Press, New York.

WRI (1992) *World Resources 1992 1993*, World Resources Institute, Oxford University Press, New York and Oxford.

CHAPTER 17

Human settlements

Introduction

A human settlement is a community - a group of people living in one place (Holdgate *et al.*, 1982). The development of such a community for productive purposes involves a transformation of the natural environment into a man-made environment that includes a variety of structures and institutions designed to meet the community's needs for work, recreation and other aspects of human life. It thus has a natural setting, a physical infrastructure of housing, transport, water, waste disposal and energy sources; and a social infrastructure of political, educational and cultural services.

Throughout the world, the single most common form of human settlement is the village. Cities and towns are far fewer in number than villages, isolated farmsteads or herding camps. In 1970, 62.9 per cent of the world population lived in rural areas; in 1990 this proportion declined to 57.4 per cent and is expected to decline further to about 40 per cent by the year 2025 (UN, 1989) as a result of rural-urban migration and transformation of rural areas into suburban ones.

Growth of urban areas

Urbanization dates back to the third millennium BC. Since that time, the process of urbanization has undergone many dynamic changes, and the Industrial Revolution has been the main factor in this change. Cities such as Cleveland, Detroit and Pittsburg in the United States; Liverpool and Glasgow in the United Kingdom; Lille and St. Etienne in France; Dortmund, Essen and Duisburg in Germany; and Turin in Italy first developed as industrial centres. High levels of urbanization were, therefore, reached at a relatively early stage in the industrialized countries. In 1950, more than half of the developed countries' population already lived in cities, and this percentage has grown steadily - 66.6 per cent in 1970, and 72.6 per cent in 1990. By the year 2025, eight out of ten residents of developed countries will live in cities (UN, 1989).

In the developing countries, urbanization has been growing at a much faster rate than in the developed nations (Figures 1 and 2). In 1950, the percentage of the total population living in urban areas in developing countries was only 17.0 per cent, in 1970 it reached 25.4 per cent, and in 1990 it was 33.6 per cent. By the year 2025, more than half of the population of developing countries (57 per cent) will live in cities. Some regions in the developing world are already highly urbanized. For example, in Latin America about 70 per cent of the population live in urban areas (Figure 3). High-income countries in the Middle East are already 80 per cent urban. By the year 2025, urbanization, at the current rate, is expected to reach as high as 85 per cent in Latin America, 58 per cent in Africa and 53 per cent in Asia. In general, the urban population in the developing countries is growing by 3.6 per cent a year, compared to industrialized regions where the urban population is growing by only 0.8 per cent a year (UNFPA, 1991). The rate of growth of urbanization in the low-income countries is much higher than that in the middle-income or high-income countries (Figure 4).

Figure 1: *The growth of urban populations in developed and developing regions, 1970–2000.*

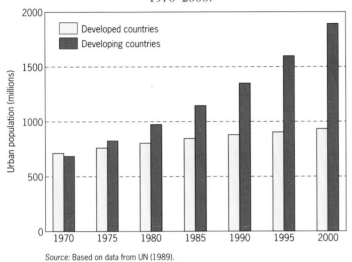

Source: Based on data from UN (1989).

Figure 2: *Urban population as a percentage of total population, by region, 1970 and 1990.*

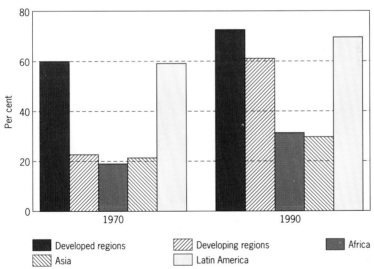

Source: United Nations, Department of International Economic and Social Affairs, *Prospects of World Urbanization*, (New York: United Nations Population Studies No. 112); Population Reference Bureau, *World Population Data Sheet 1990*, (Washington D.C.: Population Reference Bureau, Inc).

Figure 3: *Percentage urbanization in four major regions, 1950–2025.*

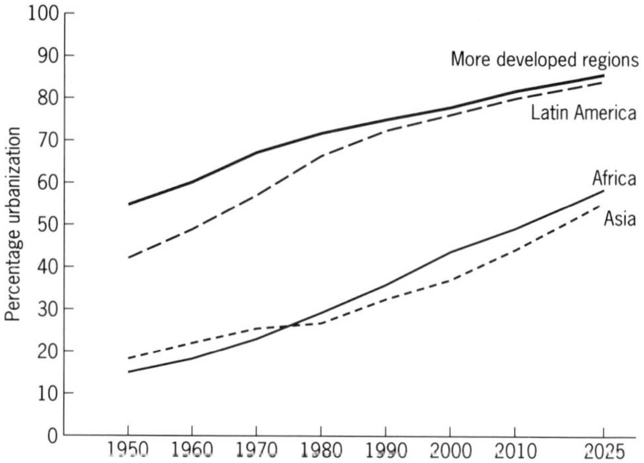

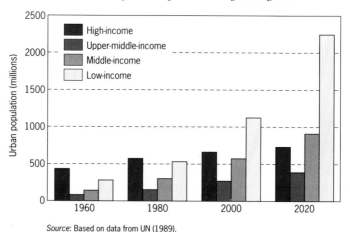

Source: Based on data from UN (1989).

Figure 4: *Urbanization and income level. Urbanization is proceeding most rapidly in the poorest countries; consequently a growing proportion of the new demand for shelter, infrastructure and services is occurring where competition for resources is greatest and where management capabilities are generally low.*

Source: Based on data from UN (1989).

A striking feature of the present urbanization trends in the developing countries is the increasing relative share of large and metropolitan cities. In 1960 there were 114 cities in the world containing a million or more residents: 62 of these cities were in developed countries and 52 were in developing nations. In 1980, the total number of million-plus cities increased to 222: 103 in developed countries and 119 in developing countries. By the year 2000 it is expected that there will be 408 such cities in the world, and 639 by 2025.

Of the latter, only 153 will be in developed countries, and the bulk - 486 cities - will be in developing regions (Kasarda and Rondinelli, 1990). Even more striking is the rise in the number and size of mega-cities, those with more than 5 million residents. In 1950, Buenos Aires was the only city in a developing country to have five million residents. By the year 2000 it is expected that at least 30 cities in developing countries will be housing more than five million people: of these, 25 cities will have a population of more than nine million each (Figure 5).

This runaway growth of urban areas is due to increases in the population within urban settlements and to migratory and other transfers of population from rural to urban places. The ratios between the two sources of increase vary

Figure 5: *Growth of the world's major cities. By the year 2000 it is estimated that 75% of Latin America's population, 42% of Africa's and 37% of Asia's will be urbanized, and 25 of the mega-cities will have populations of over nine million.*

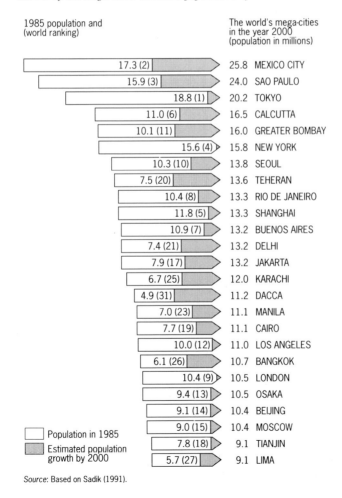

1985 population and (world ranking)	The world's mega-cities in the year 2000 (population in millions)	
17.3 (2)	25.8	MEXICO CITY
15.9 (3)	24.0	SAO PAULO
18.8 (1)	20.2	TOKYO
11.0 (6)	16.5	CALCUTTA
10.1 (11)	16.0	GREATER BOMBAY
15.6 (4)	15.8	NEW YORK
10.3 (10)	13.8	SEOUL
7.5 (20)	13.6	TEHERAN
10.4 (8)	13.3	RIO DE JANEIRO
11.8 (5)	13.3	SHANGHAI
10.9 (7)	13.2	BUENOS AIRES
7.4 (21)	13.2	DELHI
7.9 (17)	13.2	JAKARTA
6.7 (25)	12.0	KARACHI
4.9 (31)	11.2	DACCA
7.0 (23)	11.1	MANILA
7.7 (19)	11.1	CAIRO
10.0 (12)	11.0	LOS ANGELES
6.1 (26)	10.7	BANGKOK
10.4 (9)	10.5	LONDON
9.4 (13)	10.5	OSAKA
9.1 (14)	10.4	BEIJING
9.0 (15)	10.4	MOSCOW
7.8 (18)	9.1	TIANJIN
5.7 (27)	9.1	LIMA

☐ Population in 1985
▨ Estimated population growth by 2000

Source: Based on Sadik (1991).

from one country to another. It is believed that about a third of urban population increase in Africa and Asia, and 58 per cent in Latin America, has been due to migration or reclassification of rural settlements as urban (UNFPA, 1992). The rate of national population growth has a strong impact on urbanization. A study of 97 developing countries found that in the 48 countries with the most rapid national population growth, urban areas were growing at an average of 6.1 per cent a year; in the 49 countries with slower overall population growth, cities were growing at only 3.6 per cent a year (UNFPA, 1992). National population growth rates also seem to have some effect on the rate at which a country becomes more urban than rural. The urban share of total population increased by an average of 2.8 per cent a year in the 48 countries with faster population growth. In the 49 slower-growing countries the urbanization rate was only 1.8 per cent (UNFPA, 1992).

Migrants are pulled towards the city in anticipation of better access to education, health services and jobs, but they are also pushed off the land by rural poverty and environmental degradation (the so-called 'environmental refugees'; (El-Hinnawi, 1985)). In the 1980s, at least ten million Africans were forced off their land by extended drought to end up in urban areas or refugee camps. Such migration tends increasingly to cross national boundaries. Nearly one million Haitian 'boat people' - one-sixth of the entire populace - have fled the country in search of better lives elsewhere. This migration was fuelled, in part, by widespread environmental deterioration which has impoverished large areas of the island (Sadik, 1991). This intra-regional mobility (Chapter 16) has created various problems, both for the receiving countries and for the countries of origin. These problems are aggravated by the fact that most migrants end up in urban areas. For example, the increase in intra-regional mobility to the oil-exporting countries since the early 1970s has drawn increasing numbers of farmers from rural areas of the countries of origin. When the demand for unskilled labour slackened in the mid-1980s, return migration became a major issue for many countries of origin of migrant workers. Returnees tend to settle in urban areas and compete for the limited urban job opportunities rather than return to rural areas.

Most of the physical growth of cities takes place in illegal, unplanned squatter settlements. One billion people across the world are living in shanty towns at the present time, and 60 per cent of city dwellers in developing countries will be squatters by the end of the century (UNCHS, 1990a). Half the populations of Calcutta, Lusaka and Colombo are now living in shanties, as do two-thirds of the people of Bogotá and El Salvador. The proportion is even higher in Mexico City and Lima. The over-crowding of cities and lack of shelter (even illegal squatter settlements) have created millions of homeless people around the world. It is estimated that 100 million people world-wide have no form of shelter at all (UNCHS, 1990a). In India, half a million people sleep the night on the streets of Calcutta and a further 100,000 men, women and children spend the night on the pavements of Bombay. Such homeless people are now found in most of the cities of the developing world and are becoming conspicuous in the cities of the developed world as well. In the UK, over 400,000 people were officially classified as homeless in 1989 (including 196,000 children). About 120,000, many of them children, were living in 'bed and breakfast' hotels, half of which were considered unfit for human habitation. Many others live in unfit or overcrowded accommodation. The number of the 'hidden homeless', mainly single people between 16 and 18 years of age, was estimated at 180,000 in 1986. Thousands live rough in 'cardboard cities' in all the major towns of the UK. In the USA, some 40 per cent of those classified as poor live in depressed areas, mostly in inner cities (UNDP, 1991). It has been estimated that two million people are homeless in the USA (WHO, 1992). Families represent the majority of the homeless population in many large cities in the USA (e.g. in New York, Philadelphia, Portland). In Los Angeles, California, aproximately 35,000-50,000 are homeless persons. A survey of 196 homeless families showed that during the previous five years, over half had doubled up with others, three-fifths had lived in a hotel or motel, and one-third had lived on the street or in a car (Wood *et al.*, 1990). To house these people and the additions to the urban population world-wide until the end of the century, some 650 cities of one million each need to be built. Their annual cost by conventional construction methods, assuming modest standards, would roughly amount to some $US500 billion (Sachs, 1986).

While homelessness in developing countries is mainly caused by the massive influxes from rural areas to urban centres, homelessness in high-income countries has different causes. In countries such as Great Britain, Canada and the USA, much of the problem is at least partially attributable to economic policies. Amongst the principal causes of the problem are cuts in public spending on housing, private sector focusing on middle and upper class housing, decline in private rental and in housing conditions, loss of single room occupancy units and low-cost housing in general, basic economic changes resulting in unemployment and declining wage levels relative to housing costs, racial discrimination in employment and housing, demographic changes leading to smaller households, and the inflexibility of occupancy policies in government subsidized housing (Daly, 1990).

Intra-urban differentials

As cities increase in size, slums and squatter settlements proliferate. Although migration from rural to urban areas is a factor in the increase of population in these areas, natural increase within these areas is often a more important factor. In some squatter settlements virtually all the inhabitants are relatively long-term urban dwellers. This makes the view that the problem is one of the peasant unable to integrate as a town-dweller appear increasingly untenable (Harpham and Stephens).

The unplanned growth of urbanization has resulted in an acute shortage of housing in many countries. The gap between the demand and availability of housing has widened in the last two decades, particularly in the developing countries. Deteriorating economic conditions, the increasing cost of land and construction, and ineffective policies have caused this gap to widen. In the developing countries, the percentage of households unable to afford the normal-standard dwellings in selected cities (e.g. in Cairo, Manila, Bangkok and others) has increased over the last two decades from 35 to 75 per cent. The result has been increased overcrowding in older parts of the cities, and the proliferation of sub-standard housing and squatter settlements. The average rate of occupancy in the developing countries is now about 2.4 persons per habitable room, as compared to about 0.7 in the developed countries (UNDP, 1992).

The urban poor are not found only in developing countries. As mentioned earlier, even in the richest countries intra-urban differences are rising. Income disparity in the UK, for example, grew between 1979 and the late 1980s. The number of people below the poverty line increased from 8.2 million (15 per cent of the population) to 19.3 million (18 per cent). In the USA about 38 million people (13 per cent of the total population) are poor and a further 11 million are 'near poverty' (UNDP, 1991). There are many characteristics common to the lives of the dispossessed in poor areas of cities: sub-standard shelter, lack of - or inadequate - water and sanitary disposal services, inadequate health care, failing schools, overcrowding, premature adulthood, and a high rate of crime. They are sometimes referred to as 'socially marginal', 'spatially marginal', or 'politically marginal'. The sharing of equal misery in squatter settlements and poor areas has transformed some of these areas into almost closed societies within the urban conglomeration. They often make fertile ground for crime, drug trafficking and other illegal activities.

Redlining and other neglectful policies have resulted in a vicious cycle of deterioration in the poor areas of cities in developed and developing countries. The costs of such neglect are staggering. Not only does the quality of life of the poor suffer from such neglect, but the direct and indirect costs to society of tolerating an underclass of urban poor is also mounting. In the USA, for example, it is estimated that such costs amount to $US230 billion annually (El-Hinnawi, 1992). Harder to quantify, but far more important, are the long-term consequences of depriving both the underprivileged and society of their economic potential - the chance to become better educated, better skilled and more self-reliant.

The nature of the problems facing low-income groups are often poorly understood (WHO/UNEP, 1988). For example, location is often more important for them than quality of accommodation. A large family would choose to squeeze into one room in an inner city tenement because it is within walking distance of the main centres of employment, even when better accommodation is available on the outskirts of the city. Broadly speaking, each city has its characteristic mix of housing alternatives for the lower-income groups. Some are legal, e.g. rented rooms in tenements, and cheap boarding or rooming houses, or accommodation in peripheral locations. Others are illegal, such as spontaneous settlements on unoccupied sites, carefully-planned and organized 'invasions', or the illegal subdivision of purchased or rented dwellings. Each category has its distinguishing problems and each calls for a different approach.

Many children from poor families who do not find work end up on the streets. Some 100 million children live adrift on the streets of the world's cities today (Robilant, 1989) on the margins of the adult world, surviving by scavenging, stealing and finding transient jobs such as selling small items, shining shoes, guarding and washing cars or in unlawful and criminal activities. As street children have neither a voice nor a vote, their plight is often overlooked by politicians and city planners alike.

The socio-economic and environmental conditions of the slums are best illustrated by the intra-urban differentials in health. For example, in Manila the infant mortality rate for the whole city was 76 per 1,000 against 210 per 1,000 in Tondo, a squatter area. Neonatal mortality in Manila was 40 per 1,000, while it was 105 per 1,000 in Tondo. In Buenos Aires, mortality due to tuberculosis was three times higher in the peripheral areas than in the city as a whole. In Quito the infant mortality rate in upper-class districts was under 5 per 1,000, while for infants of manual workers in squatter settlements it was 129 per 1,000 (Harpham and Stephens, 1991; WHO, 1991a). Similar studies exist which point to intra-urban differentials in morbidity. For example, in Sao Paulo the incidence of diarrhoea in the lowest socio-economic stratum was 13.1 episodes per 100 children-months compared with 9.6 episodes in the next stratum and 3.6 episodes in the upper stratum. In Panama City, of 1,819 infants presenting at clinics with diarrhoeal diseases, 46 per cent came from slums, 23 per cent came from shanty towns while none came from the better housing areas. Studies in Durban, Singapore, Guatemala and Seoul have all found a higher prevalence of *Ascaris* and *Trichuris* in poorer parts of the city as compared to more wealthy parts (Harpham and Stephens, 1991). Such mortality and morbidity is associated with poor housing conditions, and lack of safe water supplies and sanitation facilities. In some countries, e.g. in Bangladesh, those living in slums and squatter settlements are also more prone to the impacts of natural disasters (Chapter 9).

The environment in and around sub-standard human dwellings offers an important habitat for a wide range of insects and rodents (fleas, cockroaches, bugs, mosquitoes, flies, rats and other insects and rodents). These animals transmit a number of diseases. Among the best known are malaria, typhoid, yellow fever, filariasis and dengue. Chagas disease is transmitted by bugs that flourish in cracks and crevices of poor-quality houses in Latin America. According to WHO (1991b), about 500,000 people become infected every year, 300,000 of them children. Between 10 and 15 per cent of infected people die during the fever that is typical of the acute phase of Chagas disease. The rest become chronically infected, and ultimately suffer heart and other chronic disorders. WHO estimates that between 16 and 18 million people in South America are infected while another 90 million are at risk.

The above-mentioned intra-urban differentials reflect the inappropriate urban development policies currently in force in most countries. Cities grow and deteriorate while little or nothing is being done to respond to the social and environmental costs which are part of urban change. Official statistics normally mention the number of houses, factories or offices built, the lengths of paved roads or water mains laid, and the sewers being constructed in the richer parts of urban areas. They totally ignore the poor areas, because nothing, or only very little, is being done in such areas. No wonder that the poorer areas of the cities have become a source of unrest and political instability in some countries. However, it should always be remembered that the slums and depressed areas are closely linked to the core city and that the two sides exchange goods and services at a scale the full extent of which is seldom realized.

Urbanization, environment and human health

Economists tend to look at cities as the site of many enterprises, whose concentration creates both positive and negative externalities but requires a costly infrastructure. From an ecological viewpoint, urban systems appear to provide very unnatural habitats. Human ecologists have been advocating the study of cities as ecological systems. However, most of the studies conducted within the UNESCO MAB Project on Ecological Approaches to Human Settlements, launched in the early 1970s, deal with the impact of the cities on the natural environment and its food-producing systems or describe the energy flows inside the city. Sachs (1986) attempted a more detailed approach and considered the city as a predominantly man-made ecosystem with paradigmatic analogies in relation to natural ecosystems. He pointed out that 'Such a perspective emphasizes the actual and potential interrelations and complementarities between different human activities conducted in the city. Whenever possible, loops must be closed and residues from one production transformed to inputs into some other. The urban ecosystem appears thus as a vast potential of physical and human resources to be identified and tapped in an attempt at arresting deterioration, and hopefully improving the quality of urban life'.

Despite the fact that rapid urbanization and the growth of mega-cities have brought serious physical and social problems, many cities have become, and are becoming, the crossroads of international business and culture. Cities contribute to economic growth and social transformation by providing economies of scale and proximity that allow industry and commerce to flourish and create much-needed jobs, by supporting modern education and health and social services, and by offering a wide variety of commercial and personal services essential to meeting human needs (Kasarda and Rondinelli, 1990). In most developing countries, modern productive activities are concentrated in large urban centres, often at a level much higher than their share of national population (Chapter 12). Abidjan, the capital of Côte d'Ivoire, with 15 per cent of the national population, accounts for more than 70 per cent of all economic and commercial transactions in the country. Bangkok, with about 10 per cent of the population of Thailand, accounts for 86 per cent of GNP in banking, insurance and real estate, and 74 per cent of manufacturing. Sao Paulo, with about 10 per cent of Brazil's population, contributes over 40 per cent of industrial value added and one-quarter of net national product (UNCHS, 1985). In general, cities in developing countries are contributing a disproportionately large percentage of GNP. This is no longer the case in the industrialized countries where zoning and the more developed infrastructure have moved most manufacturing and even some service industries away from the urban centres. It has also been claimed that accelerated urbanization consistently leads to lowered rates of population growth and might well be the preferred policy for almost every developing country in the twenty-first century (Ramachandran, 1992).

Cities have also contributed to economic growth in surrounding rural areas. Cities transform, distribute and consume large amounts of agricultural products and goods produced in rural areas, thus providing much-needed markets for the rural sector (Rondinelli, 1987). More than 9,000 tonnes of vegetables are brought to Shanghai daily from surrounding rural production areas in China for sale in the city's markets; fruit supplies to the city have increased from about 148,000 tonnes in 1980 to more than 360,000 tonnes in 1986 (MacKenzie, 1987). More than 8,000 tonnes of fruit and vegetables are sold every day in the principal wholesale market alone in Mexico City (Melendez, 1987). In some countries, increasing urban demand for agricultural products has led to considerable changes in agricultural systems. Because urban-consumer crops are increasingly deregulated in many developing countries, farmers near cities have shifted to growing such crops instead of conventional staple crops (Chapter 11). Increased inflow and outflow of people, materials and products to and from urban areas has created the need for more transport systems and their infrastructure, which has meant the use of more land areas, creation of more traffic congestion and air pollution in and around cities.

However, rapid and mostly unplanned urbanization has also created many economic, social and environmental problems. These include the increasing demand for various resources; the production of waste which enters the surrounding atmospheric, aquatic and terrestrial ecosystems; large-scale population movements; increasing population densities; and changing habitats for all species including mankind. On average, it has been roughly estimated that a city of 1 million inhabitants consumes every day about 625,000 tonnes of water, 2,000 tonnes of food and 9,500 tonnes of fuel, while at the same time generating 500,000 tonnes of waste water, 2,000 tonnes of solid wastes and 950 tonnes of air pollutants (Sadik, 1991). But there are great differences between cities in different parts of the world. New York City, for example, produces three times as much waste per person per day as Calcutta.

As cities grow, their boundaries creep outward, eating up farmland and forest. In several Western European countries, it is estimated that about two per cent of agricultural land is being lost per decade to urban growth. The figure for the United States is about 2.5 million hectares of prime farmland per decade. While figures are not available from developing countries, planners report that the best cropland is steadily being consumed by urban expansion (Brown *et al.*, 1985). The extent to which cities control the use of land varies widely. Japan and Western Europe have the world's most comprehensive urban land-use controls: control of land use in Eastern Europe, North America and Australia has not been as effective (Lowe, 1992). Developing countries have the loosest controls over how cities develop. Inappropriate land-use planning in urban areas has resulted in a wide array of air pollution, water pollution and waste management problems.

The overcrowding of urban areas in many countries has led to the increasing disappearance of green areas. Cities have become complexes of steel, concrete, bricks and stone structures with no life other than the masses of people moving around. This has largely contributed to the incidence of many disorders, sometimes referred to as urban diseases. High-rise buildings, built in order to make the best possible use of expensive land in densely populated urban areas, may present special hazards to physical and mental health. The difficulty of access to play areas deprives children of one of their rights and affects their learning and development. The elderly and the physically disabled suffer from restricted movement in such high-rise structures, and residents in upper storeys may be at special risk in case of fire and explosion (WHO, 1991a). In some developing countries, frequent breakdown of lifts imposes excessive stresses on residents, especially the aged and infirm. The trend in the last two decades to construct all-glass high-rise buildings not suited to local climates has added excessive pressures to the already strained water and energy supply systems. In addition, it has aggravated indoor air pollution problems (Chapter 1).

Rapid industrial development and population growth on ecologically fragile land also create serious physical problems. The Valley of Mexico, in which Mexico City is located, has undergone serious adverse transformations as the result of industrialization and over-development. The valley has lost almost all of its lakes, which have turned into large salt basins. More than 73 per cent of the valley's forests have been cut, and 71 per cent of the soil has been eroded, destroying vegetation and adversely modifying the microclimate (Schteingart, 1989).

Inappropriate location of industrial installations in urban areas and/or the lack of control of population densities around such installations has led to a dramatic increase in the number of victims of industrial accidents. The massive explosion at the liquefied petroleum gas storage facility in the crowded San Juanico neighbourhood of Mexico City in November 1984 killed 452 people, injured 4,248 and displaced 31,000. In 1992, an explosion in the sewers of Guadalajara caused considerable damage. Another example is the Bhopal accident in India (Chapter 9). These examples illustrate the precarious nature of a city where many millions of inhabitants live cheek by jowl with a variety of potentially dangerous installations.

Although the water supply to urban areas in the developing countries has improved over the last two decades, in 1990 there were about 244 million people - or 18 per cent of the urban population - without access to clean water supplies (Chapter 4). Such figures disguise differences between countries and within the same country. In some countries, such as Zimbabwe, Chile, Jordan and Malaysia, access to safe drinking water in urban areas is now 100 per cent (WHO, 1992a). In contrast, in countries such as Bangladesh only 25 per cent of the urban population has access to safe water; in the Central African Republic, 13 per cent; in Zaire, 59 per cent, and in India 79 per cent (WHO, 1992a). Access to clean water does not necessarily mean piped water to houses; it includes communal standpipes from which water could be fetched.

The proportion of the urban population served with sanitary facilities did not improve between 1970 and 1990. In 1970, 29 per cent of the urban population was not served by any sanitary facilities; in 1990, the proportion was 28 per cent. Considerable differences exist between developing countries and within individual countries. For example, while 100 per cent of the urban population in Zimbabwe, Chile and Malaysia are served with sanitary facilities (WHO, 1992), most urban centres in Africa and Asia have no sewerage system at all - including many cities with a million or more inhabitants (Hardoy and Satterthwaite, 1989). In many urban areas in the developing countries, sewage treatment plants are either lacking, not operating, or are restricted to primary treatment. The discharge of untreated sewage into inland waters and the sea has been a major source of water pollution in many areas (Chapters 4 and 5).

Untreated industrial wastes and domestic sewage are causing widespread pollution of inland water resources and coastal areas in many countries (Chapters 5 and 12). Large amounts of untreated discharges from Asian cities are wiping out local fisheries and seriously affecting water supplies from the Tjiliwung River in Jakarta, the Han River in Seoul, and the Pasig River in Manila. In the industrialized countries, the pollution of the Rhine and the Danube in Europe by discharges from cities has been well documented. The same is true of many rivers, coastal zones and open seas in the Eastern European countries and the former Soviet Union. Air pollution in urban centres is discussed at some length in Chapter 1, while Chapters 12 and 13 give information on the contribution of extractive and manufacturing industries, as well as the energy sector, to air pollution and hazardous wastes generation. Chapter 14 covers the impacts of urban transport on the environment. Air pollution, high-rise structures and local microclimatic conditions have transformed cities into 'heat islands', where air temperature is higher than in the surrounding suburban and rural areas. It has been estimated that for the industrialized countries as a whole, 42 kilograms of air pollutants are emitted annually per 100 people and nearly 10 tonnes of hazardous and special wastes are generated annually per square kilometre (UNDP, 1992).

With increasing urbanization, waste disposal is becoming a serious problem. The daily per capita domestic refuse generation in cities of the developed countries has been estimated at between 0.7 and 1.8 kilograms, whereas the figure is somewhere between 0.4 and 0.9 kilograms in the developing countries (Cointreau et al., 1984). On average, the amount of municipal solid wastes generated in the developed countries increased from 318 million tonnes in

1970 to 400 million tonnes in 1990; an increase of about 25 per cent. In the developing countries, the amount of refuse was about 160 million tonnes in 1970, and doubled to 322 million tonnes in 1990. Garbage collection services are inadequate or non-existent in most residential areas in Third World cities, and an estimated 30-50 per cent of the solid wastes generated within urban centres is left uncollected. It accumulates on streets, open spaces between houses and on wasteland. Such uncleaned refuse, particularly in hot climates, constitutes a breeding ground for all sorts of vectors and pathogenic organisms. Where municipal solid wastes are managed, hand picking of refuse is the most viable economic option. This often supports a large army of scavengers (mostly children), who extract various materials from the waste and sell them. It is paradoxical that the poorest countries are achieving a high level of recycling in this way, despite the small proportion of saleable matter in the waste. As the whole family is usually employed in scavenging, the young and the elderly are exposed to a wide variety of pollution effects, obnoxious odours and disease vectors. Most of the scavengers have chronic skin, eye and respiratory diseases and frequent intestinal problems.

The health of a city's people is largely determined by physical, social, economic, political and cultural factors in the urban environment, including the processes of social aggregation, migration, modernization and industrialization, and the circumstances of urban living, which may vary with climate, terrain, population density, housing stock, type of industrial base, income distribution and transport systems in use (Chapter 18). The impact of urban processes on health is not just the sum of the effects of the various factors taken individually, since they interact synergistically with one another (WHO, 1991b). In some countries rapid urbanization has resulted in levels of pollution that outstrip the natural absorptive capacity of the city's ecosystem, and adverse health effects have increased because controls are lacking or unreliable.

Urban dwellings may either increase the risk of, or protect against, the hazards of communicable and noncommunicable diseases and of injuries, depending on the design, standard and location of the dwellings. The siting of dwellings in relation to industry affects human exposures to toxins in air, water and soil. The siting of residential areas also affects traffic hazards and the burden of vehicular emissions. In cities in England and Germany, degeneration of the nervous systems of children living in areas close to main roads (resulting in problems in concentration and learning) has been associated with chronic ingestion of lead from vehicular emissions. Similar observations have been made in the USA (UNEP, 1990; WHO 1991b). Noise pollution from traffic, in manufacturing, and in overcrowded dwellings may lead to hearing loss and, more commonly, interfere with sleep. Local climatic change in urban areas may lead to increased heat stress caused by changes in radiation balance. This may magnify the impact of air pollution (Chapter 1) by helping to produce temperature inversions and by decreasing natural ventilation.

The increase in tourism in coastal cities has created increasing pressures on the infrastructure and services in some areas, and breakdown of water and electricity supplies is now common (Chapter 15). Because many coastal cities discharge their sewage into the sea without treatment, coastal bathing may become a health hazard. Studies carried out in Canada, Egypt, France, Hong Kong, Israel, Spain and the United States have shown the growing incidence of eye infections, skin complaints, gastro-intestinal symptoms, and ear, nose and throat infections due to exposure to polluted bathing water (Saliba and Helmer, 1990). Such pollution of coastal waters, resulting mainly from inappropriate urban planning and development, has affected tourism and recreation in several coastal cities (with resultant social and economic problems).

Rural settlements

Over the last two decades, most governments in the developing countries have seemed more preoccupied with mounting economic problems and increasing deterioration of urban areas than with rural development plans which have been almost completely neglected. Living conditions in rural areas are, in general, not better today than they were in 1970. In fact more agricultural land has deteriorated (Chapter 6) and more forests have been cleared (Chapter 7). Houses are still much below standard, made of mud bricks, bamboo, wood or other locally-available material. Electricity is still a rare commodity in most rural homes, and wood, agricultural residues and cow dung are still the main source of fuel in most rural areas. Most roads remain unpaved, and conditions of transportation and communication have not improved much.

Although the percentage of rural population with access to clean water increased markedly from 14 per cent in 1970 to 63 per cent in 1990, there were still many people without access to clean water supplies. The percentage of the rural population with some sort of sanitary facilities increased from 11 per cent in 1970 to 49 per cent in 1990, but there were still about 1,364 million people without any such facilities.

In rural areas, obtaining water and making it more readily available for domestic use has traditionally been women's work. In many developing countries, women (and children) still have to walk long distances to bring water home. Nearly 30 per cent of women in rural Egypt have to walk more than 60 minutes per day to meet their water needs (Shahin, 1984). Although pump water is used for cooking and drinking, canal or pond water is used for washing and bathing in rural areas. This is the main reason for the spread of schistosomiasis among rural children of 10-14 years of age in Egypt.

Women and children are also responsible for collecting branches, bushes, crop residues and cow dung for use as fuel at home. Women in northern Ghana may need a whole day to collect three days' supply of fuelwood. In rural Kenya, some women spend 20-24 hours per week at the task (Dankelman and Davidson, 1988; Fortmann, 1986; Cecelski, 1987). Burning this fuel in open or semi-contained fires results in several emissions that adversely affect women's and children's health, most importantly through chronic obstructive pulmonary disease (Chapter 1).

An important trend that has emerged in the last two decades is the increasing 'urbanization' of rural areas in the developing countries. The process has, however, been unplanned, taking place on an ad hoc basis according to prevailing economic conditions. For example, the increased demand in urban-consumer agricultural products (which are more profitable to farmers) has changed the pattern of agriculture in some rural areas. With more income, the inhabitants of these areas gradually began to emulate the life-styles of urban populations. Villages that were previously self-sufficient in food now 'import' their bread from urban areas. Increasing numbers of rural households are now using non-renewable energy sources (heavily subsidized by governments) mostly in an inefficient and wasteful manner, much as is happening in urban areas. Such transformation of life-styles in rural areas that previously respected nature and its resources could create serious long-term environmental impacts as well as socio-economic problems.

Responses

National responses

Over the last two decades, both developed and developing nations have been grappling with the problem of improving the condition of human settlements. This has covered a wide range of issues, from land-use planning, financing, rent control, construction technology, infrastructural services and rehabilitation of inner cities and slums, to attempts at regionalization, decentralization, public participation and the improvement of management practices. A global survey carried by the UN Centre for Human Settlements (Habitat) in 1986 concluded that the basic human settlements needs are largely met in developed countries; but that the situation in developing countries gives cause for considerable concern, with unsatisfied needs both in urban and rural settlements reaching disturbing dimensions in the cities and becoming even worse in rural areas. The survey found little evidence in most developing countries to suggest that the resulting negative impacts on the environment discussed in the previous sections are being checked or reversed. The consequences of the interactions between demographic and urbanization trends in the general climate of world recession and economic decline in many developing countries over the last decade threaten to become unmanageable (UNCHS, 1986). However, as has been indicated earlier, the developed countries still face some serious problems, particularly in inner cities.

As for the infrastructural services, and in spite of various approaches to solving or alleviating transport problems (such as tunnels, flyovers, ring roads, mass transit systems, underground railways, designation of traffic-free zones and regulation of private motor transport) congestion, considerable slowing down of traffic and even gridlock are still prevalent all over the world (see Chapter 14). The situation is deteriorating more rapidly in the mega-cities that are proliferating in the developing countries. Apart from the economic losses incurred, air pollution is becoming much higher in the metropolitan areas in developing countries and has reached dangerous levels in many cities, causing serious health hazards (Chapter 1). As has been discussed earlier, attempts at improving infrastructural services of drinking water and sanitation have not as yet managed to achieve the goals of the International Drinking Water and Sanitation Decade (see also Chapter 4).

In the field of construction methods and building materials technologies, the period has witnessed a variety of promising initiatives in devising construction methods appropriate to the prevailing situations in developing and least developed countries where the number of homeless and those living in almost sub-human conditions continues to increase. The quality of many building materials that are already well-known in many countries have proved to be capable of considerable improvement so as to become low-cost durable building materials at reasonable cost. However, these have not come into wide use mainly due to the prejudices of building authorities and professionals (UNCHS, undated).

Energy efficiency in human settlements - a crucial issue that covers the whole range of uses from the manufacture of building materials to construction techniques, building design and household energy efficiency - has also received attention. The manufacture of building materials (steel, cement and bricks in particular) is very energy-intensive and is estimated to account for up to 70 per cent of the total energy requirements in one type of building. Decentralized small- and medium-scale manufacture has proved to be advantageous, as has the use of low-cost, low-energy materials (UNCHS, 1991). In household energy efficiency, building design and construction has paid little attention to reducing the use of artificial light or to passive cooling and insulation techniques and has generally had little effect on stemming the proliferation of air conditioning world-wide. Apart from the energy consumption involved there are the impacts on the ozone layer (Chapter 2). Many energy authorities have devised combinations of regulatory measures, incentives and advisory services to rationalize the use of energy in the home. Both UNIDO and UNCHS have produced a series of useful publications on energy efficiency in housing construction as well as other technological aspects of construction (UN, 1986; UNCHS, 1991; UNIDO, 1987). The health hazards of asbestos and radon emissions in some buildings are examples of the new issues emerging in construction technologies over the last two decades.

International responses

Attempts at a holistic approach to human settlement problems can be traced back to the Stockholm Conference. One of its important decisions was to call for a United Nations Conference on Human Settlements (Habitat), to address the issue of the habitat of man in an increasingly populated and rapidly urbanizing planet. The Conference, held in 1976 in Vancouver, Canada, adopted a broad and ambitious plan of action and led to the establishment of the Commission on Human Settlements and the United Nations Centre for Human Settlements (Habitat). Since its inception, the Commission has devoted high-priority attention to meeting the basic human settlements needs of the poor in developing countries - shelter, infrastructure and services - in a sustainable manner. This concern gained momentum with the decision of the General Assembly to designate 1987 as the International Year of Shelter for the Homeless (IYSH). This initiative greatly contributed to developing world-wide awareness of the predicament of the homeless in both developed and developing countries, and of the growing difficulties in providing adequate shelter, infrastructure and services to all. Research promoted by the IYSH revealed that access to adequate shelter by the urban and rural poor was a development objective visibly losing ground, together with shelter related provisions such as water supply, sanitation, waste management, land and energy. The International Drinking Water Supply and Sanitation Decade (1980-1990) sought, with some success as indicated earlier, to meet two of the basic infrastructural needs of human settlements.

Ten years after the Habitat Conference, the need had become apparent for a new global assessment of human settlements trends, needs and priorities for action. As a result, in 1987, the Commission on Human Settlements discussed and endorsed a document called *A New Agenda for Human Settlements* (UNCHS, undated). One of the innovative features of the 'New Agenda' was the introduction of 'enabling strategies' as the approach needed to find a realistic solution to growing world-wide human settlements problems. The 'enabling approach' became the main focus of the 'Global Strategy for Shelter to the Year 2000' (Box 1), adopted by the General Assembly of the United Nations in 1988. Defined as the strategy's most important operational principle, the enabling approach is intended to mobilize the full potential and resources of all actors in the shelter improvement and production process. For the most part, governments' role will be to establish legislative, institutional and financial frameworks that will enable formal and informal business sectors, non-governmental organizations, community groups and households to make optimal contributions to national shelter-delivery systems. The strategy states that, ultimately, the enabling principle implies that the people concerned will be given the opportunity to improve their housing conditions according to the needs and priorities that they themselves define. This community-participation approach is intended to ensure that shelter development programmes connect to the social, physical and economic needs and potentials of targeted low-income communities, in order to mobilize these as powerful additional resources for a sustained and affordable shelter-development process.

Efforts to achieve sustainable management of settlements, particularly of rapidly growing cities and urbanizing regions, are receiving much more

---

BOX 1

The Global Strategy for Shelter to the Year 2000.

- Aims for *shelter for all* by the year 2000:
- Promotes an *enabling approach* whereby governments enable the people to improve their shelter situation themselves, by making it easier to:
  - obtain security of tenure and other legal requirements;
  - work with appropriate building codes;
  - have access to finance for housing; and
  - have access to low-cost building materials.
- Harnesses the energies and skills of the people themselves to improve their homes through the enabling approach, involving the participation of the community, especially women.

The main goal of the Global Strategy is to encourage governments to set up national shelter strategies based on an enabling approach. For example, in Nicaragua, where many people have been displaced by the internal situation, and others made homeless by Hurricane Joan in 1988, attention is now being focused on low-income groups. Uganda is currently devising a shelter strategy to integrate housing needs into its large-scale task of reconstruction, and Indonesia is focusing on suitable methods of housing finance to assist its enormous population.

Financial arrangements, in all developing countries, must now benefit low-income groups if the 'enabling approach' is to be successful. The demand for financial assistance is so great that new approaches to finance are needed, such as Turkey's special fund for housing which is supported by taxes on luxury items, and Sri Lanka's saving scheme which reaches the people through post offices and other outlets. In Brazil, Mexico and the Philippines, a certain percentage of each person's salary is saved for a compulsory saving scheme for housing.

Non-governmental organizations are also helpful in organizing finance for people of low income to purchase or to build. NGOs are working with the Government of Sri Lanka in implementing its Million Houses Programme and the Government of Kerala, India, with its Subsidized Aided Self-help Housing Project of which the first phase will be the construction of 25,000 housing units.

Most governments now appreciate the involvement of NGOs in national housing issues. These organizations, which are usually small, well-organized and work well with the people, can form important links between governments and the people. Apart from raising finance, NGOs have shown their worth in training, research, generating employment and lobbying to change laws.

Adequate housing for all by the year 2000 can only be achieved if people are able to help themselves in their bid for low-cost housing. Governments must enable people to improve their living conditions. In the long run, it is the energy and initiative of the shanty dwellers that will produce new cities from old.

Source: UNCHS (1990b).

attention and support than in the past. One example of this new orientation is co-operation between UNEP and UNCHS on the environmental aspects of urban development, regional development and management institutions, which culminated in the publication, in 1987, of *Environmental Guidelines for Settlement Planning and Management* (UNEP/UNCHS, 1987).

Another is the Urban Management Programme, jointly executed by the World Bank and UNCHS, with core funding from UNDP. The Urban Management Programme represents a joint approach by the United Nations family of agencies, together with external donor agencies, to strengthen the institutional capacity of municipalities, and also to strengthen the contribution that cities in developing countries can make toward economic growth, social development, alleviation of poverty and improvement of the urban environment. This latter component will provide action guidelines in five priority areas:

(i) coverage and efficiency of municipal waste collection services;

(ii) municipal waste treatment and disposal;

(iii) co-ordinated pollution control across levels of government and urban sub-sectors;

(iv) protection and sustainable utilization of vital resources; and

(v) incorporation of environmental planning and management techniques into city-wide strategic planning.

Of great importance in this context is the Technical Working Group Meeting on Urban Environmental Indicators, held in 1990 in Barcelona. This meeting was convened to agree on a core set of indicators for urban environmental policy analysis and decision-making. The Group's principal recommendations can be summarized in two proposals - one regarding the testing and application of a core set of urban environmental indicators and a second on establishing mechanisms for the management and exchange of urban data - including environmental information - on a global scale.

The co-operation between UNCHS and UNEP referred to above is expected to revolve largely around the UNCHS - initiated Sustainable Cities Programme. The principal objective of this programme is to improve the environmental planning and management capacity of municipal authorities, so as to strengthen their ability to define the most critical environmental issues, to identify the tools available to address these issues, and to involve all those whose co-operation is necessary to achieve set goals. Furthermore, as a global programme, 'Sustainable Cities' is designed to promote the sharing of expertise and experiences between cities in different regions of the world. By the end of 1990, the Programme was exploring possible participation of some twenty cities, five of which have already entered the project-formulation stage (UNCHS, 1990c).

Concluding remarks

In the OECD region, the most important change in the pattern of human settlements in recent decades has been the 'counter-urbanization' occurring outside major cities. The high quality infrastructure and services usually associated with large cities has become generally available to inhabitants of suburban areas, smaller urban centres and even rural areas. Much improved transport and communications systems have allowed many factories to locate outside major cities, and increased numbers of urban workers live now in suburban or rural areas. This process of 'counter-urbanization' has reduced the pressure on major cities and led at the same time to marked development of suburban and rural areas. Although this has greatly reduced the rural-urban differences in OECD countries, the intra-urban differentials have increased in many urban areas.

The sustainable development of human settlements in developing countries cannot be achieved by *ad hoc* or fragmented solutions. The urban situation cannot be dissociated from what is going on in the countryside. Part of the urban crisis is due to the stream of immigrants from rural areas where they are unable to earn even the most miserable living. The investment required to accommodate them in the cities is far greater than the outlays that would be necessary to provide them with agricultural and related jobs if access to land were only made possible by appropriate institutional measures. An overall national development strategy can only succeed if the urban *and* rural situations are tackled together. This echoes the postulate put forward by Abdalla in 1979 that 'Third World countries ought to industrialize without uprooting the peasants' (Abdalla, 1979). In fact the scale of inter-linkages between rural and urban areas is now so great that sustainable urban development and sustainable rural development cannot be separated. This poses many challenges for the future.

The UNCHS submission to UNCED ends by stating that 'Since poverty alleviation and human development have emerged as a prerequisite for sustainable development, the shelter and related infrastructure and services of the poor will require priority attention and action. Human settlements are where the poor of the world live, and it is in human settlements where opportunities for improved health conditions, income generation and access to the fundamental sources of livelihood can be translated into tangible results. This reality is made even more pressing by the fact that the living environments of the poor are increasingly located in urban contexts where access to adequate shelter, infrastructure and services are a fundamental prerequisite for survival' (UNCHS, 1992).

REFERENCES

Abdalla, I.S. (1979) *Depaysanisation ou developpement rural? Un choix lourd de conséquences.* IFDA Dossier 9, International Foundation for Development Alternatives, Nyon, Switzerland.

Brown, L.R. *et al.* (1985) *State of the World-1985.* N. W. Norton, New York.

Cecelski, E. (1987) Energy and rural women's work. *International Labour Review,* **vol. 126,** p. 41.

Cointreau, S.J. *et al.* (1984) *Recycling from municipal refuse.* World Bank Technical Paper No. 30. World Bank, Washington, D.C.

Daly, G. (1990) Health Implications of Homelessness: Reports from Three Countries. *J. Soc & Soc. Wel.* **XVII:** 111-26.

Dankelman, I. and Davidson, J. (1988) *Women and Environment: Alliance for a Sustainable Future.* Earthscan Publications, London.

El-Hinnawi, E. (1985) *Environmental Refugees.* UNEP, Nairobi.

El-Hinnawi, E. (1991): Sustainable agriculture and rural development in the Near East. Regional Doc. No.4 FAO/Netherlands Conference on Agriculture and Environment. FAO, Rome.

El-Hinnawi, E. (1992): Personal comunication

Fortmann, L.P. (1986): Women in subsistence forestry. J. *Forestry,* July 1986, p. 39.

Hardoy, J.E. and Satterthwaite, D. (1989) *Squatter Citizen.* Earthscan Publications, London.

Harpham, T. and Stephens, C. (1991) Urbanization and health in developing countries. *World Health Statistics* Quarterly, **44,** 62.

Kasarda, J. and Rondinelli, D.A. (1990) Mega-cities, the environment and private enterprise. *Environmental Impact Assessment Rev.,* **10,** 393.

Lowe, M.D. (1992) *In:* L.Brown (ed.) *State of the World-1992.* Earthscan Publications, London.

Mackenzie, C. (1987) The supply lines to Shanghai. CERES - *The FAO Review,* 20, 16.

Melendez, G.P. (1987) Markets for the Mexican megalopolis. CERES - *The FAO Review,* 20, 21.

Ramachandran, A. (1992) Statement of the Executive Director of Habitat to UNEP's Governing Council Special Session, Nairobi, February, 1992.

Robilant, A.D. (1989) Street Children. *In:* C. Moorehead, (ed.) *Betrayal.* Barrie and Jenkins, London.

Rondinelli, D.A. (1987) Cities as agricultural markets. *The Geographical Review,* **77,** 408.

Sachs, I. (1986) Work, Food and Energy in Urban Ecodevelopment. *Development: Seeds of Change,* **vol. 4,** p.2.

Sadik, N. (1991) *Safeguarding the Future.* UNFPA, New York.

Saliba, J.L. and Helmer, R. (1990) Health risks associated with pollution of coastal bathing water. *World Health Statistics Quarterly,* **43,** 177.

Schteingart, M. (1989) The environmental problems associated with urban development in Mexico City. *Environment and Urbanization,* **1,** 40.

Shahin, Z.M. (1984) Women, Water Supply and Sanitation: Socio-cultural and economic aspects. *Proc. INSTRAW Interregional Seminar,* Cairo, p. 277.

UN (1986) *Case Studies on Measures for Energy-Efficient Shelter and Infrastructure,* Nairobi.

UN (1989) *World Population Prospects.* United Nations, New York.

UNCHS (1985) *The role of small and intermediate settlements in national development.* UNCHS, Nairobi.

UNCHS (1986) *Global Report on Human Settlement 1986*, UNCHS, Nairobi.

UNCHS (1990a) *Shelter and Urbanization.* UNCHS, Nairobi.

UNCHS (1990b) *The Global Strategy for Shelter to the Year 2000*, UNCHS, Nairobi.

UNCHS (1990c) *Sustainable Cities Programme.* UNCHS, Nairobi.

UNCHS (1991) *Energy Efficiency in Housing Construction and Domestic Use in Developing Countries*, UNCHS, Nairobi.

UNCHS (1992) *Improving the Living Environment for a Sustainable Future*, UNCHS, Nairobi.

UNCHS (undated) *A Compendium of Information on Selected Low-cost Building Materials*, UNCHS, Nairobi.

UNDP (1991) *Human Development Report 1991*, Oxford University Press, New York and Oxford.

UNDP (1992) *Human Development Report 1992*, Oxford University Press, New York and Oxford.

UNEP (1990) *Children and Environment.* UNEP, Nairobi.

UNEP/UNCHS (1987) *Environmental Guidelines for Settlements Planning and Management*, UNEP/UNCHS, Nairobi.

UNFPA (1991) *Population Issues.* UNFPA, New York.

UNFPA (1992) *The State of World Population.* UNFPA, New York.

UNIDO (1987) The Building Materials Industry: its role in low-cost shelter programmes, *Sectoral Studies Series*, **No.39**, UNIDO, Vienna.

WHO (1991a) Environmental Health in Urban Development. *WHO Technical Report Series* **No. 807**. WHO, Geneva.

WHO (1991b) Control of Chagas Disease. *WHO Technical Report Series* **No. 811**. WHO, Geneva.

WHO (1992a) *Annual World Health Statistics - 1991.* WHO, Geneva.

WHO (1992b) *Our Planet, Our Health* (Report of the WHO Commission on Health and the Environment), WHO, Geneva.

WHO/UNEP (1988) *Urbanization and its implications for Child Health*, WHO, Geneva.

Wood, D., Valdez, R.B., Hayashi, T., and Shen, A. (1990) Homeless and Housed Families in Los Angeles: A Study Comparing Demographic, Economic and Family Function Characteristics, *AJPH*, **80**: 1049-52.

CHAPTER 18

Health

Introduction

The Constitution of the World Health Organization (WHO) defines health as 'a state of complete physical, mental and social well-being and not merely the absence of disease or infirmity'. Such a state is conditioned by a variety of factors ranging from the genetic, the social and the emotional, to the natural and man-made environment. Consequently, consideration of such a state of health will have to take into account factors such as development objectives and strategies, and economic relations (Figure 1) as well as the prevailing social structures, beliefs and value systems in the community.

Figure 1: *Influencing factors in (upper diagram) the downward spiral of inappropriate development and (lower diagram) the upward spiral of sustainable development.*

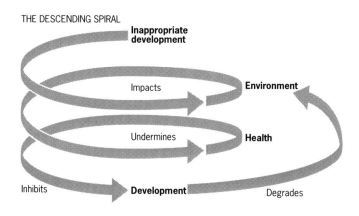

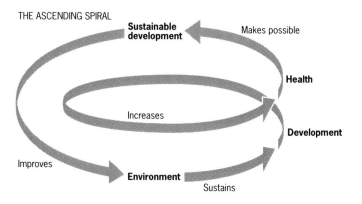

Source: UNEP (1986).

In this chapter we restrict the discussion to a review of the manner in which human health and well-being have been affected by changes in the human environment over the last two decades. The main elements of the environment involved here are the natural (the air we breathe, the water we drink, the food we eat, the radiations we are exposed to) and the man-made (the habitat, the place of work, the means of transport, the social and recreational facilities). These elements impact on human health in two ways: through physical and chemical agents (radiation, chemical compounds and emissions of gases, liquids or solids) and through pathogenic agents.

The generic impact of these agents on human health has been addressed in the preceding chapters and the responses in different fields and at different levels (scientific, social, national and transnational) are reviewed, and future trends summarized. We proceed here to show the overall results in the health field of these impacts in different parts of the world and over the last two decades.

Although infant and child mortality rates have declined steadily over this period (Figure 2), and life expectancy has increased from an average of 56.7 years in 1970-75 to an average of 61.5 years in 1985-90 (Chapter 16), the overall picture is still alarming in many respects (WHO, 1992).

Figure 2: *Estimates and projections for infant and child mortality in the least developed and other developing countries, 1950–2025.*

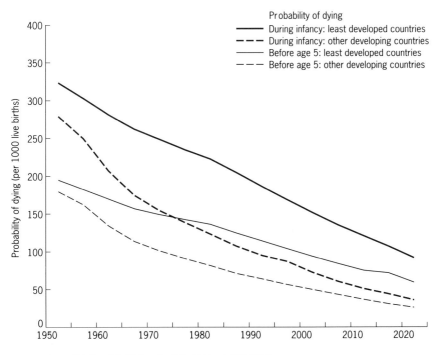

Source: Data from United Nations Population Division, in WHO (1989).

In the developing countries:

- Four million infants or children die every year from diarrhoeal diseases, largely as a result of contaminated food or water.

- Two million people die from malaria each year and 267 million are infected.

- Hundreds of millions suffer from debilitating intestinal parasitic infestations.

- 2,500 million people suffer from illnesses linked to insufficient or contaminated water and lack of sanitation.

- About 2,500 million people, mostly in rural areas, are at serious risk of contracting chronic respiratory disease and cancer as a result of indoor air pollution from open fires for cooking or heating.

On the global scale:

- Hundreds of millions suffer from respiratory and other diseases caused or exacerbated by biological and chemical agents.

- Hundreds of millions are exposed to unnecessary chemical and physical hazards at home, in the work-place or the wider environment (500,000 die and tens of millions are injured in road accidents every year).

- More than one thousand million city dwellers are regularly exposed to levels of air pollution that exceed WHO guidelines.

- The greenhouse effect could trigger new epidemics of tropical diseases in regions that are at present free of them (Box 1).

- Stratospheric ozone depletion is likely to lead to increases in skin cancer and cataract-related blindness and may damage the body's resistance to disease.

The world health situation 1972-92

The global figures for infant and child mortality and life expectancy quoted above conceal wide variations from country to country. Moreover, the causes of death differ dramatically between developed and developing countries. Infectious and parasitic diseases are still by far the leading cause of mortality in developing countries. Indeed, in those countries the risk of dying from one of these diseases is virtually identical to the risk in the developed countries of dying from cardiovascular diseases - the leading cause of death in the latter countries (Figure 3).

In developing countries, the commonest diseases are water-related. Eighty per cent of all illness is attributed to unsafe and inadequate water supplies, and

---

BOX 1

Climatic change: potential effects on disease.

Malaria.
Distribution: 100 countries. Prevalence of infection: 170 million. At risk of infection: 2,100 million. Climate change spread risk: Highly likely.

Schistosomiasis (Snail fever).
Distribution: 74 countries. Prevalence of infection: 200 million. At risk of infection: 600 million. Climate change spread risk: Very likely.

Lymphatic filariasis (Elephantiasis).
Distribution: Global, except for Europe. Prevalence of infection: 90.2 million. At risk of infection: 900 million. Climate change spread risk: Likely.

Leishmaniasis (Oriental sore).
Distribution: 80 countries. Prevalence of infection: 12 million. At risk of infection: 350 million. Climate change spread risk: Unpredictable.

Dracunculiasis (Guinea-worm disease).
Distribution: Sub-Saharan Africa, India, Pakistan. Prevalence of infection: Under three million. At risk of infection: 140 million. Climate change spread risk: Unlikely.

Onchocerciasis (River blindness).
Distribution: Africa and Latin America. Prevalence of infection: 17.5 million. At risk of infection: 85.5 million. Climate change spread risk: Likely.

African trypanosomiasis (Sleeping sickness).
Distribution: 36 countries in Sub-Saharan Africa. Prevalence of infection: 25,000 new cases a year. At risk of infection: 50 million. Climate change spread risk: Likely.

Source: WHO (1992).

---

half the hospital beds are occupied by people with water-related illnesses (US AID, 1987). As mentioned earlier, diarrhoeal disease is the leading cause of infant and childhood deaths. Cholera, malaria, dracunculiasis (guinea worm) and other parasitic infections remain prevalent, despite eradication campaigns that have achieved some improvement. Furthermore, the ratio of cardiovascular diseases to infectious and parasitic diseases has been increasing steadily of late in developing countries (Figure 4)

In the two decades under review the situation regarding some specific communicable diseases improved, while for others it deteriorated. Those diseases that are the target of the expanded programme of immunization (EPI)

Figure 3: *Causes of death in developed countries (left-hand bars and pie diagram) and developing countries (right-hand bars and pie diagram)1985.*

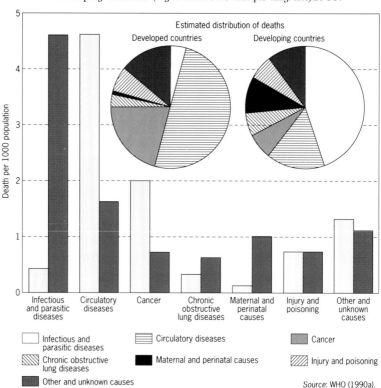

Estimated distribution of deaths

Developed countries   Developing countries

Death per 1000 population

Infectious and parasitic diseases	Circulatory diseases	Cancer
Chronic obstructive lung diseases	Maternal and perinatal causes	Injury and poisoning
Other and unknown causes		*Source*: WHO (1990a).

Figure 4: *The emergence of cardiovascular diseases in the cause-of-death structure in developing countries, 1985–2015.*

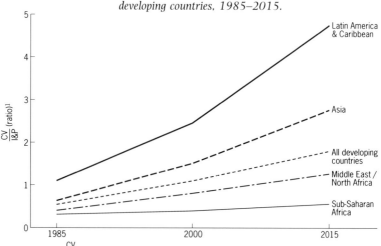

Latin America & Caribbean

Asia

All developing countries

Middle East / North Africa

Sub-Saharan Africa

$\frac{CV}{I\&P}$ (ratio)[1]

Note: 1. $\frac{CV}{I\&P}$ ratio = ratio of deaths from cardiovascular diseases to those due to infectious and parasitic diseases.

Source: WHO (1991c).

(viz. poliomyelitis, tetanus, measles, diphtheria, pertussis and tuberculosis) are largely declining as vaccination coverage increases (Figure 5). On the other hand, HIV infection and AIDS, unknown before 1980, are showing alarming increases. Sub-Saharan Africa is suffering the main brunt of this disease, but other regions, both in the developed and developing countries, face similar futures (Figure 6). Cholera has recently flared up in the Americas, with a resulting dramatic increase in the number of cases reported to WHO.

Aggregate global or regional figures tend to hide differences between and within countries. The graphic portrayal of schistosomiasis and malaria 'hot spots' illustrates this (Figures 7 and 8). These problems are largely local in nature, depending on a variety of features among which environmental factors are the most significant. Thus, while the global total of malaria cases has been relatively stable over the last decade, important epidemic outbreaks have contributed to the high incidence of illness and death in specific situations.

Similar variations can be found in the developed world. There is, for example, a striking contrast between Eastern and Western Europe. Figure 9 shows, *inter alia*, the difference in the incidence of cardiovascular disease in the two regions, while Figure 10 shows the regional trends in cardiovascular mortality rates between 1970 and 1988. The rise in cardiovascular disease in Eastern Europe contrasts with the trend in infant mortality, which has declined steadily in both Eastern and Western Europe.

Cancer rates provide similar variations in space and time (Figures 11, 12 and 13). Figure 11 also shows that the incidence of the various kinds of cancer in the developed and developing countries is markedly different. Both stomach and

Figure 5: *Global annual reported cases of pertussis, tuberculosis and measles, 1974–90 (provisional data only for 1990).*

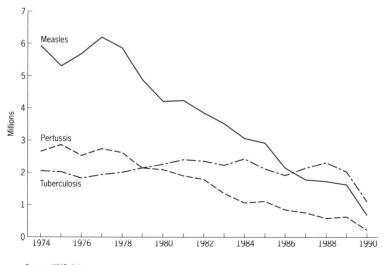

Source: WHO data.

Figure 6: *Estimated and projected incidence of AIDS by region, 1982–2000.*

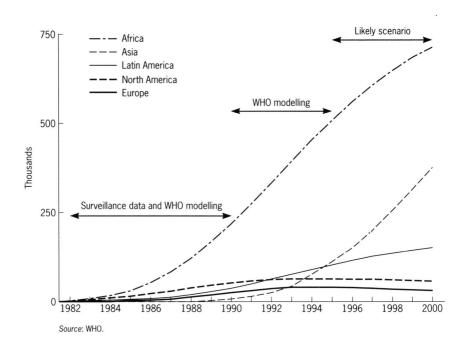

Source: WHO.

lung cancer show marked trends in Europe. The general decline in stomach cancer (Figure 12) is almost certainly related to dietary habits and changes in food preparation or preservation techniques. Tobacco smoking is the most important cause of lung cancer. In countries with a history of prolonged cigarette usage, 80-90 per cent of lung cancer and 30 per cent of all cancer deaths are attributed to tobacco smoking. The changing mortality from lung cancer in different European countries is shown in Figure 13.

In developed countries, concern over relationships between environment and health has often focused on exposure to pollution, both at work and in the wider environment. The past two decades have seen some improvement in air quality in urban environments of developed countries, although these gains have been partly offset by declining quality in the cities of the developing world and increased ozone levels in some rural areas (Chapter 1). Generally speaking, many of the pollutants that caused concern (sulphur dioxide, particulates, carbon monoxide, lead, mercury and organochlorine pesticides) have come under much closer control over the past two decades and their impacts have been reduced. Fertilizers and industrial effluents are still causes of serious concern. While pesticides, in particular, still cause many deaths in the developing world (Chapters 4 and 5), solid wastes, particularly hazardous wastes, have become a serious health hazard both in developed and developing countries (Chapter 10).

Figure 7: *Schistosomiasis: changes in incidence related to environmental and social changes (highlights from 76 countries surveyed).*

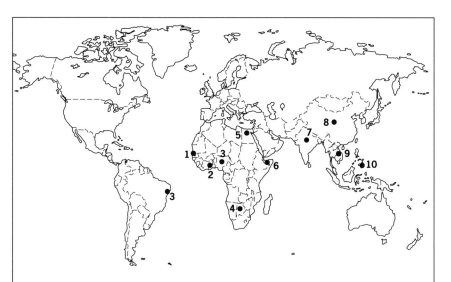

1 SENEGAL
An epidemic of intestinal schistosomiasis (1987–1990) has been occurring in Richard Toll since the construction of the Diama Dam.

2 GHANA
Since the construction of the Akosombo Dam (Volta Lake), intestinal schistosomiasis has been increasing in the Volta Delta.

3 URBAN SCHISTOSOMIASIS
Is now present in and around major cities in Northeast Brazil and Africa.

4 BOTSWANA
Urinary schistosomiasis was reported in the Okavango Delta in 1989.

5 EGYPT
Since the construction of the Aswan Dam (Aswan Lake), intestinal schistosomiasis has dominated the Nile Delta and has spread towards Upper Egypt.

6 SOMALIA
Intestinal schistosomiasis is now in the north due to war and refugee migration.

7 INDIA
Urinary schistosomiasis was first reported near Hyderabad in 1989.

8 CHINA
Acute infections reappeared in Hubei and Hunan Provinces after 1988.

9 MEKONG BASIN
S. mekongi is now reported in new areas.

10 PHILIPPINES
Schistosomiasis has been reported in the deforested areas of Mindanao.

Source: WHO.

Figure 8: *Malaria: the increasing threat of incidence in frontier areas of development.*

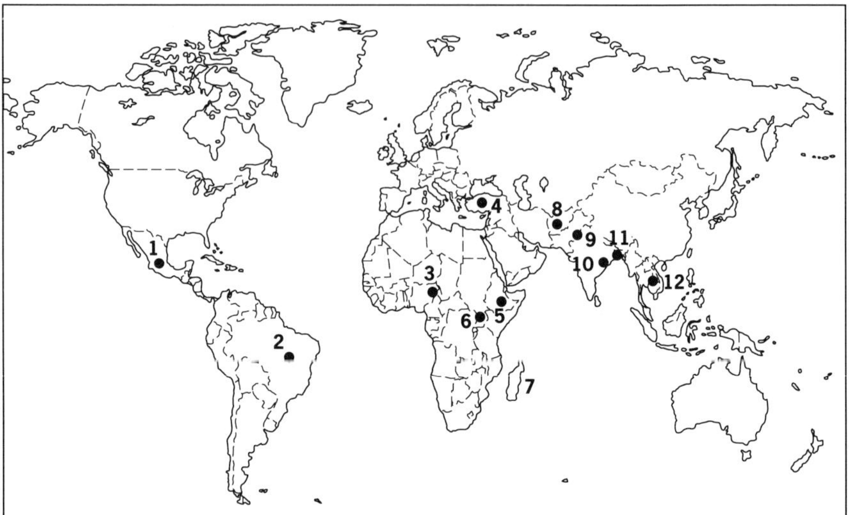

1 MEXICO
Second largest source of malaria in the Americas.

2 BRAZIL–AMAZONIA
50% of all malaria in the Americas, due to deforestation, mining and overcrowded settlements.

3 AFRICAN CITIES
Severe drug-resistance, inadequate sanitation, overburdened services.

4 TURKEY
1976–1977, 100 000 cases in Anatolia due to poor drainage and environmental management.

5 ETHIOPIA
150,000 persons died in 1985 epidemic. Repeated epidemics in highlands due to degraded environment, drought & famine.

6 EAST AFRICAN HIGHLANDS
(Rwanda, Burundi, Zaire) Epidemics in the highlands due to irrigation and possibly increasing temperatures (1989–1991).

7 MADAGASCAR
Dramatic malaria epidemic in 1988 with more than 25,000 deaths in central plateau.

8 AFGHANISTAN
War, interruption of control programme and displaced populations, about 300,000 cases per year.

9 PAKISTAN AND NW INDIA
Deterioration of irrigated environment. Increased desertification.Increasing epidemic risk.

10 ORISSA
Degradation of environment in tribal areas.Population pressure.

11 BANGLADESH
The Chittagong Hill Tract similar to Indochina. Increasing trend: around 10,000 cases in early 70s to about 45,000 now with epidemics reaching up to 70,000.

12 INDOCHINA PENINSULA
Rapidly increasing risk due to deforestation, gem mining in forest areas, civil unrest.

Source: WHO.

Figure 9: *Differences in cause-specific mortality between Central-Eastern Europe and the rest of Europe around 1988/89.*

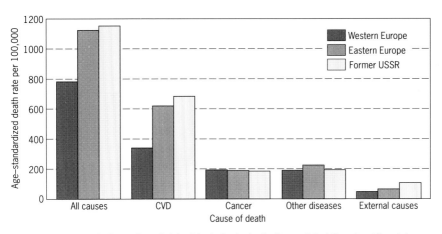

Note: Average for Eastern Europe includes Bulgaria, Czechoslovakia, Hungary, Poland, Romania and Yugoslavia.
Source: WHO.

Figure 10: *Subregional trends in cardiovascular mortality in age range 0–64 years, 1970–90.*

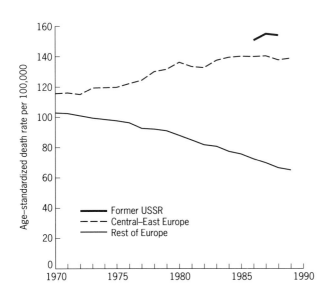

Note: Average for Central-Eastern Europe includes Bulgaria, Czechoslovakia, the former DDR, Hungary, Poland, Romania and Yugoslavia.
Source: WHO.

Figure 11: *Number of new cancer cases by type and ranking in developed and developing countries, 1980.*

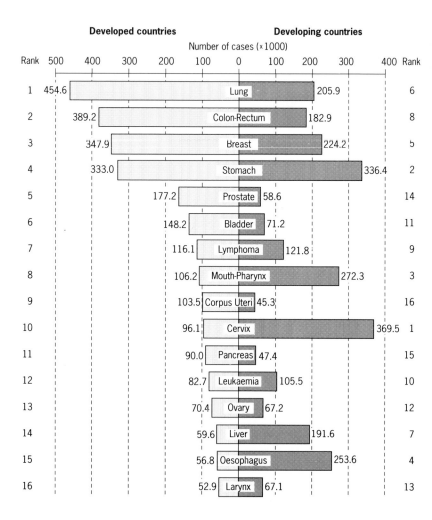

Source: Parkin *et al.* (1988).

Figure 12: *Trends in mortality from stomach cancer in the European region, 1950–85.*

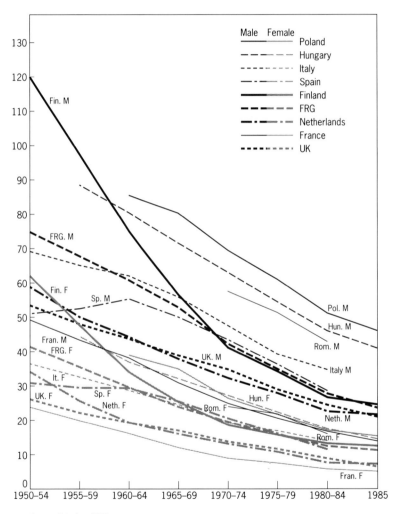

Source: Data from WHO.

Figure 13: *Trends in mortality from lung cancer in the European Region, 1950–85.*

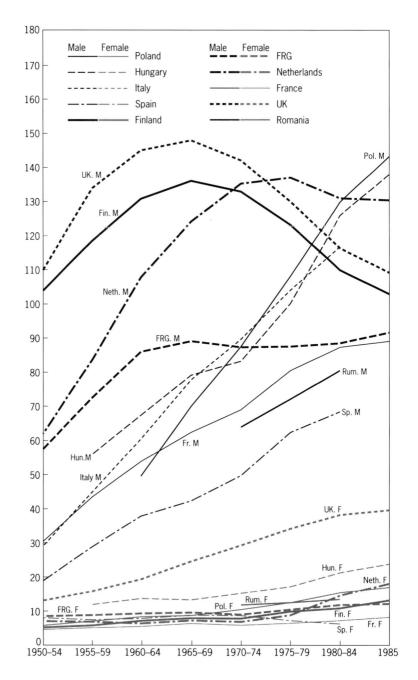

Source: Data from WHO.

In all countries, nutrition and health are closely related. In poorer regions of developing countries under-nutrition is a major factor increasing vulnerability to infectious and parasitic diseases (see later). By contrast, in developed countries, obesity and high intakes of saturated animal fats are believed to be linked to the high incidence of cardiovascular diseases. Understanding of such relationships has advanced considerably over the past two decades.

Finally, mental health is no less important than physical health. Bad housing and a degraded physical and social environment result in psychosocial health disorders, such as depression, drug and alcohol abuse, suicide, child and spouse abuse, delinquency and personal violence. It is now recognized that the environment plays an important role in violent behaviour. Apart from the genetic and societal causes of mental diseases, evidence has been accumulating over the last two decades of the role played by biochemical factors in mental ill-health. Certain biochemical abnormalities are environmentally-induced and some could be inherited. For example, exposure to some heavy metals and synthetic compounds has been linked to brain tumours and abnormal behaviour. Exposure to low doses of lead in childhood has been associated with long-term impairment in the functioning of the central nervous system (Needham *et al.*, 1990).

The environment and health

Rural life styles and health

A key advance in the past twenty years has come through the appreciation of the health significance of land-use changes in the developing world. A large proportion of the world's poor are rural people who work on farms, or do non-farm work that depends in part on agriculture. The relationship between their poverty and their health thus largely depends on agriculture and associated environmental issues. Their health depends particularly on the availability and productivity of farmland, forest resources (especially fuelwood), and water resources, as well as on risks from toxic chemicals, especially pesticides. Some problems of the same kind confront rural populations in high-income countries, but the impacts are slight compared with those in the low-income non-industrialized countries, which are more dependent on their natural resources for economic and social growth (Myers, 1989).

Land availability, land use and nutrition

The amount and quality of farmland available is a key environmental asset whose use determines food and cash income for those who own the land and those who work on it. These factors are critical for health. In extreme situations of environmental degradation, where soil depletion, erosion and water scarcity are all present, a whole region's welfare is often jeopardized by the scarcity of food, leading to world-wide calls for emergency action. In less extreme situations, the process of environmental degradation and increasing health vulnerability is more subtle and difficult to characterize. It has often been neglected until disaster strikes - yet it is in just these situations that soil and water conservation measures can be most effective.

Various studies during the past two decades have demonstrated the relationship between land availability and use and health status and vulnerability. The links between access to land and malnutrition are particularly well established (Norse, 1985). Table 1 illustrates the relationship between the absolute amount of land available for agricultural exploitation and childhood nutritional status in Nepal in 1979 (Nabarro, 1981). Similar findings were reported from an area in the Punjab where 54 per cent of the children of landless labourers were moderately or severely malnourished compared to less than 39 per cent of the children of the landowners (Levinson, 1974). In Bangladesh both food consumption and nutrient intake appear to be directly related to landholding, with the landless consuming only around 80 per cent of the calories and protein consumed by those possessing more than 1.2 hectares of land (FAO, 1982).

Table 1: *Percentage of children with muscle wasting (weight-for-height less than 80% of standard) in Nepal in 1979, in relation to availability of cultivated land per household.*

Area of land cultivated by the household (ha)	Age 12–35 months	Age 36–59 months	Age 60–95 months
0.0 – 0.5	22	11	2
0.5 – 1.0	23	4	1
over 1.0	2	1	1

Source: Nabarro (1981).

Figure 14 illustrates the precarious state of cereal food stocks among landless households in Bangladesh (Chowdhury *et al.*, 1981). Among these families, stocks were entirely used up during the critical lean month of October, when rice is most expensive and employment most difficult to obtain. Land-owning families, however, generally had sufficient stocks to maintain adequate family consumption throughout the year without resort to market purchases.

Several studies have illustrated the complexity of the relationship between health status and land usage. Health vulnerability appears greatest among those dependent on paid employment in agriculture (especially when growing cash crops) and less among land-owners and those growing food crops. Generally speaking, cash crops exacerbate maldistribution because they accelerate the processes of social differentiation: cash-crop adopters prosper faster, sometimes at the expense of non-adopters (UN/ACC Sub-committee on Nutrition, 1989).

Figure 14: *Comparison of average cereal stocks amongst land-owners and landless families in Bangladesh.*

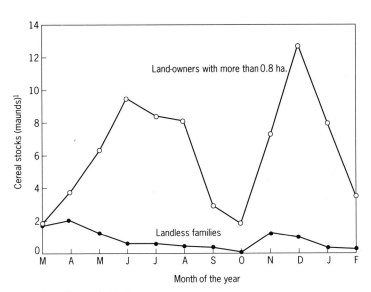

Note: One maund = 41 kilograms.
Source: Chowdhury *et al.* (1981).

For example, in Rio Grande do Sul, Brazil, infant mortality rates (IMRs) varied in the late 1970s from 70 deaths per thousand in the south to just over 20 in the northern districts. The pattern was correlated with land holdings and land use. The IMR was highest in areas with large land holdings used for cattle raising, which required much waged labour. These areas also had a higher proportion of low birth-weight babies, and lower mean intake of protein and energy. Areas with small properties, used for crop agriculture with a higher level of self employment had lower IMRs, and better child health and nutrition (Figure 15). Children of land-owners were less likely to be malnourished than children of labourers (Victora *et al.*, 1986). Other studies have demonstrated that where complementary advantage is taken of both cash and food crops, people are less vulnerable from a health point of view (as measured by food security) (Maxwell, 1988).

Health risks often coincide with the seasonality that is so characteristic of rural life. The World Bank notes that incomes in rural households vary substantially according to the season (World Bank, 1990). For example, wage work is readily available only at certain times of the crop year, and it often depends on the weather. In many African countries the dry season puts an extra burden on women, who may have to walk several kilometres to find water or fuelwood. In some busy seasons heavy agricultural work coincides with depleted food stocks and higher prices. Undernutrition and illnesses are more common at certain times of the year. The rains typically increase water

Figure 15: *Comparison of stunting and wasting in children from small properties (land-owners) and large properties (labourers) in Brazil.*

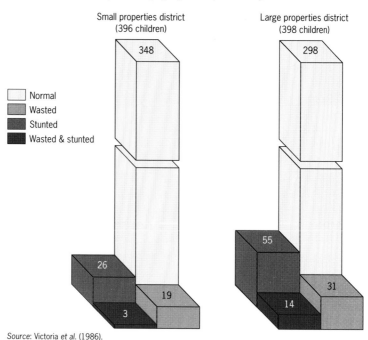

Source: Victoria *et al.* (1986).

contamination and the incidence of waterborne diseases. Acute weight loss during the 'hungry season' has been documented among farmers in The Gambia: adult weight fluctuated as much as 4.5 kilograms within one year. In northeast Ghana, losses of six per cent of body weight were recorded. Among women farmers in Lesotho the figure was seven per cent, and for pastoralists in Niger it was five per cent. For vulnerable groups such as children, the aged, and others whose biological defences are already weakened, seasonal weight change can be extremely damaging (World Bank, 1990).

The higher figure of seasonal undernutrition and illness among women reported by the World Bank is consistent with many studies that have shown that they are often more vulnerable in rural areas. In most rural areas, their labour is already over-stretched (in Africa, women produce an estimated 60 per cent of the food for household consumption and provide 80 per cent of agricultural labour (ILO, 1989a). Any further demands mean lower health and nutritional status, not only for women but also for their children since there is less time for breast feeding, food preparation and child care. These increased demands could be catastrophic within the context of diminishing access to and degradation of environmental resources (land, energy and water), and the all too common constraints of public expenditure, especially for social services (health, education, child care, etc.).

Cooking fuels and health

The health vulnerability of rural farm-women is aggravated by the fuelwood crisis. Wood, charcoal and other biomass fuels are traditional in lower-income countries and situations. Women are usually responsible for fuel collection, preparation and use, and children in almost all developing societies must help their mothers with these tasks. Girls, in particular, take part in fuel preparation, cooking and tending the fire (Dankelman and Davidson, 1988). The time and energy spent by women and children in collecting firewood means that less time and energy are available for household hygiene, income generation, education and schooling. Furthermore, foods that require longer cooking times may be omitted from the household diet as a fuel-conserving measure, thus further restricting an already limited choice of foods and increasing the risk of disease.

The burning of both biomass and fossil fuels are major sources of air pollution within houses, especially those with open hearths and inadequate ventilation (Chapter 1). Women and children are exposed to the greatest risk because they spend long hours inside, preparing and cooking food and taking care of the house. The most important health effects of air pollution are respiratory, and range from acute infections (particularly in children) to development of chronic obstructive pulmonary disease in the girls and women who tend the fires. It is estimated that 700 million women and their children are at risk of developing such serious respiratory diseases.

Forests and health

Forests in developing countries often serve as food banks, especially for poor members of local communities. Numerous types of fruits, nuts, leaves, roots and shoots are periodically collected. Forests also harbour many types of mammals, reptiles, birds and insects which can be hunted and consumed, while lakes and rivers in forest areas contain fish and other aquatic animals. These are important sources of protein for many households. By selling them, many households earn significant amounts of much-needed supplementary cash income. Forests are the dominant source of household energy for cooking, heating, construction materials, animal fodder and traditional medicines. They are thus an important source of income and employment within rural areas (Barraclough and Ghimire, 1990).

Forests also provide valuable and sometimes critical resources during hard times. Such resources serve to fill in seasonal shortfalls of food and income as well as providing seasonally crucial agricultural inputs that help in reducing risks and lessening the impacts of drought and other emergencies (FAO, 1989). For example, during the two-month hungry season, the poor forest people of Madhya Pradesh in India rely almost entirely on green leaves gathered in the forests as their main source of food (Cecelski, 1985). Recent studies have shown that the consumption of a variety of indigenous plants was crucial to the survival of many Sudanese during the 1985-86 famine. In particular, a protein-rich dish called 'kawal', made by fermentation of the leaves of the weed Cassia *obtusifolia*, was frequently the sole support of people whom food shipments failed to reach.

A high proportion of pharmaceutical products originates directly or indirectly from wild species. Apart from the increasing use of plants in the pharmaceutical industry, a much wider range of materials is used world-wide as traditional herbal remedies from wild habitats (Farnsworth *et al.*, 1976).

Water resources development and health

Water is essential for human health, well-being and development. Fresh and marine water yield food, while irrigation is the key to agriculture in many regions (Chapters 4 and 11). Access to safe drinking water and to sanitation are major contributors to general community health, and most water resources development projects have been directed towards these ends. But water also acts as a transport agent for many diseases (Figure 16), and is the habitat for a number of disease vectors.

Of all the diseases associated with irrigation, schistosomiasis is perhaps the best-documented. Apart from the fact that the very nature of irrigation systems creates favourable conditions for the survival of the snails that host the parasite during one phase of its life cycle, the problem is aggravated by social behaviour and poor sanitation practices. Other diseases aggravated by irrigation schemes are onchocerciasis (river blindness), which is endemic in some parts of Africa, Latin America and Yemen. The worm causing the disease is transmitted by black flies of the genus *Simulium*, which require swift-flowing, well oxygenated fresh water for the survival of their larvae (UNEP, 1986).

Development of water resources has often involved altering water courses, and especially the construction of dams to permit the extension of irrigated agriculture. The dams lead to the inundation of large areas of land and often this results in the displacement of many thousands of people (Goldsmith and Hildyard, 1984) with consequent settlement and health implications. The extensive areas of aquatic habitat created can also lead to an increase in health risks. A good example of what can happen - one of many - is provided by the Diama dam on the Senegal river. This, with a second dam at Manantali in Mali, is intended to allow the irrigation of some 300,000 hectares (Diallo *et al.*, 1990). The Diama dam was completed in August 1986. Disease surveillance began in May 1987. It has since become clear that one of the most important outbreaks of intestinal schistosomiasis in the history of modern West Africa is occurring in this area. The first cases of *Schistosoma mansoni* infection were reported in early 1988. In the last quarter of 1989, 71.5 per cent of 2,086 passive stools examinations were positive (Talla *et al.*, 1990). Before the construction of the dam, *Biomphalaria pfeifferi*, the snail vector for *S. mansoni*, was sparsely present and often not found in the Senegal river basin when looked for. The environmental changes caused by the dam, and believed to be responsible for the change in transmission patterns in the river basin, were changes in river water salinity favouring growth of the snails and stabilization of water level leading to more favourable living conditions for them.

Figure 16: *Diseases associated with water.*

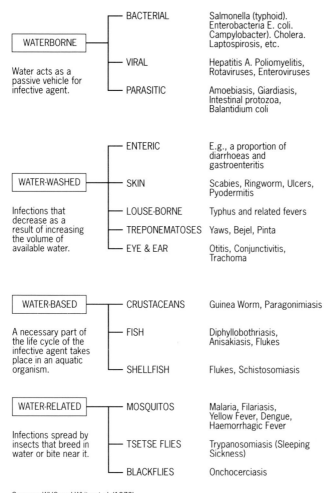

WATERBORNE Water acts as a passive vehicle for infective agent.	BACTERIAL	Salmonella (typhoid). Enterobacteria E. coli. Campylobacter). Cholera. Laptospirosis, etc.
	VIRAL	Hepatitis A. Poliomyelitis, Rotaviruses, Enteroviruses
	PARASITIC	Amoebiasis, Giardiasis, Intestinal protozoa, Balantidium coli
WATER-WASHED Infections that decrease as a result of increasing the volume of available water.	ENTERIC	E.g., a proportion of diarrhoeas and gastroenteritis
	SKIN	Scabies, Ringworm, Ulcers, Pyodermitis
	LOUSE-BORNE	Typhus and related fevers
	TREPONEMATOSES	Yaws, Bejel, Pinta
	EYE & EAR	Otitis, Conjunctivitis, Trachoma
WATER-BASED A necessary part of the life cycle of the infective agent takes place in an aquatic organism.	CRUSTACEANS	Guinea Worm, Paragonimiasis
	FISH	Diphyllobothriasis, Anisakiasis, Flukes
	SHELLFISH	Flukes, Schistosomiasis
WATER-RELATED Infections spread by insects that breed in water or bite near it.	MOSQUITOS	Malaria, Filariasis, Yellow Fever, Dengue, Haemorrhagic Fever
	TSETSE FLIES	Trypanosomiasis (Sleeping Sickness)
	BLACKFLIES	Onchocerciasis

Scource: WHO and White *et al.* (1972).

Irrigation schemes increase the extension of land under water, thus providing more breeding sites for vectors and reducing the diversity of habitats that could favour, for example, one species of mosquito rather than another. In a terrain with a moderately fast-flowing river and a wet and a dry season in West Africa, the favoured mosquito would be *Anopheles gambiae* and malaria transmission would occur mainly during the wet season. Were the river to be dammed, the streams draining into the man-made lake upstream of the dam would be slowed down and the surrounding land would be swampy, thus providing a suitable habitat for *A. funestus* that would replace the previous species. This new species breeds all year round, and malaria, which was formerly a seasonal disease, becomes permanent. Similar changes in the incidence of diseases may occur in areas where different species of insect prevail due to environmental changes (UNEP, 1986).

Other effects included dramatic increases in the rodent population of irrigated areas. This is of particular importance for several viral and parasitic infections, relapsing fever, leptospirosis and leishmaniasis. Apart from losses in agricultural stocks, the increased rodent population has also caused an increase in bites, especially among children under five years of age.

Agrochemicals and health

Another aspect of rural development is the increasing use of agrochemicals (Chapter 11). While these chemicals have played an important part in increasing food production and thus have contributed positively to human health, many have also proved to be harmful if not used properly. While not as visible as urban chemical contamination, the impact of these chemicals on the health of agricultural workers is of growing importance.

Pesticides cause most health concern, the main hazard being acute poisoning. WHO estimates that some three million people world-wide suffer annually from single, short-term exposure (including that resulting from suicide or attempted suicide) with 220,000 deaths. In 1986, in Sri Lanka as a whole, 57 per cent of admissions of cases of poisoning and 66 per cent of deaths by poisoning, were due to pesticide poisoning. In 1986, pesticide poisoning was the sixth leading cause of death in government hospitals. According to one estimate, in some countries about seven per cent of all agricultural workers involved with intensive pesticide use are likely to experience symptoms of poisoning each year where effective training programmes are not in place (Jeyaratnam et al., 1990). In one survey of acute pesticide poisoning it was found that the proportion of agricultural workers handling pesticides was 29.8 per cent in Indonesia, 91.9 per cent in Malaysia, 38.3 per cent in Sri Lanka, and 41.4 per cent in Thailand (Jeyaratnam, 1987). Table 2 demonstrates how deaths due to pesticides have increased in Sri Lanka between 1977 and 1981 (Perera and Gunatilleke, 1990). The fourteen-fold increase in Kurunegala may reflect the fact that this district, in addition to vegetable

Table 2: *Deaths due to pesticides in some districts of Mahaweli, Sri Lanka.*

District	1977	1978	1979	1980	1981
Kandy	24	13	17	39	25
Matale	20	45	73	33	53
Kurunegala	8	4	26	122	111
Anuradhapura	11	12	33	24	55
Nuwara Eliya	-	45	22	62	63
Badulla	7	9	16	28	42
SRI LANKA	236	217	463	641	690

Source: WHO, derived from Perera and Gunatilleke (1990).

growing, is one of the largest paddy growing areas with very high and expanding use of agro-chemicals.

Pesticides and other agro-chemicals can also affect human health because of their dispersion in the environment. Women and children are particularly vulnerable since they are the main food producers in many developing countries. The contamination of potable water supplies with pesticides and fertilizers may poison humans and livestock. Persistent chemicals like the organochlorines linger in soil and water for many years and may become concentrated in animals.

Food poisoning

Food crops are often directly contaminated by pesticides. It has been estimated that 90 per cent of human pesticide intake has occurred through the food chain (WRI, 1988). UNEP/GEMS has been systematically collecting data on pesticide contamination in a number of foodstuffs, mainly from developed countries. For example, DDT complex levels in the fat of cow's milk showed a marked decline in Japan and the Netherlands in the 1970s. A similar trend was also observed in finfish in Japan and the USA (UNEP, 1986).

Aflatoxin contamination in food is a major hazard in warm humid countries where food storage facilities are inadequate and crops are often left in moist conditions before harvesting. Mycotoxins, in general, have been known to cause serious outbreaks of poisoning, but their main significance may well be in long-term exposure to them. Combined with hepatitis B, they may be an important factor in primary cancer of the liver - one of the most common cancers in Asia, and Africa south of the Sahara. Aflatoxins produced by *Aspergillus flavus* in cereals, peanuts and soya beans are by far the most serious threats to human health (UNEP, 1986).

Urban life-styles and health

Although each person in a city breathes air, drinks water, eats food, sleeps, works and moves about, each does so in a unique familial, socio-cultural and communal environment. That is why different people encounter different environmental hazards. In spite of this, much is known about the general impact of the urban environment on people's health. Epidemiological studies have clearly shown that the odds against child survival and longevity are greater for those city dwellers who are severely exposed to malnutrition, inadequate shelter, poor sanitation, pollution, poor transportation, and the psychological and social stresses resulting from socio-economic deprivation (WHO, 1991a).

Urban poverty is growing in absolute, if not relative, magnitude (Chapter 16). Hundreds of millions of urban dwellers in the Third World now live in what might be termed life and health threatening homes and neighbourhoods (WHO, 1991a). While the picture may not be as dramatic in most cities of the developed world, and the trends are less clear, there are still millions who live in miserable situations in the cities of the North with a quality of life far below that of most of their fellow citizens (Chapter 17).

Health vulnerability in urban slums

One of the advances during the past twenty years has been in understanding the complexity of the causes of health vulnerability in particular situations. All risk groups and factors need to be studied and their changes over time related to changes in health and the environment. Furthermore, factors such as income, employment, assets and educational status, and the social dynamics and behavioral patterns of urban life which lead to exposures to particular risks (sexually-transmitted disease, child abuse, and stress-related illness) need to be taken into account (Cooper Weil *et al.*, 1990). The multi-factoral nature of poverty is well illustrated in a 1986-87 study of an urban slum in Bangladesh (Pryer, 1989, 1990).

The sample area was an established inner-city slum with a population of 2,200 where environmental conditions were appalling, being overcrowded and unsanitary. A cross-sectional survey of 208 households was conducted. Despite an outward impression of homogeneity it was found that a high degree of inequality existed in the slum, which could be subdivided into three groups. The richest families (34 households), mostly comprising slum landlords and traders, had average monthly incomes more than three times the local food poverty line. A second group (125 households) was in an intermediate position, while the poorest labouring households (49 households) were below this food poverty line. As Figure 17 indicates, the study showed that the poorest households are particularly disadvantaged relative to their better-off neighbours, and even more so with respect to the 'rich' families.

Nutritional and environmental diseases abound in these conditions. Cross-sectional surveys indicate that the prevalence of diarrhoea amongst children under five was 25 per 100 child days. Relatively high rates of ill-health and malnutrition were not, however, exclusively confined to young children - 43 per cent of mothers and 42 per cent of fathers had a Body Mass Index (ratio of weight in kilograms to height in meters squared) below 18, which is considered indicative of adult malnutrition.

Care is needed in interpreting statistics of health trends in urban areas. City health statistics often tend to look better than rural ones. An important reason is that there are usually better health care facilities in urban areas and it is therefore possible for city-dwellers to get treatment for serious illnesses. Another reason is that these statistics usually do not include the inhabitants of slum areas, as these areas are not officially recognized (Basta, 1977). A third reason is that these statistics aggregate all areas of the city, with no separation of groups living in slums and squatter areas from those living in better housing and under more favourable environmental conditions. Changing city politics, alternating between acceptance of squatters and their physical repression, clearly interfere with the gathering of meaningful comparative statistics,

Figure 17: *Health and employment in three social groups in an urban slum in Khulna, Bangladesh. C1 is the highest-income group, mainly slum landlords and traders: C2 is an intermediate group; C3 is the lowest-income group. BMI is an index of leanness. It is calculated as weight in kilograms divided by height in metres squared, and a BMI of less than 18 is considered indicative of undernourishment.*

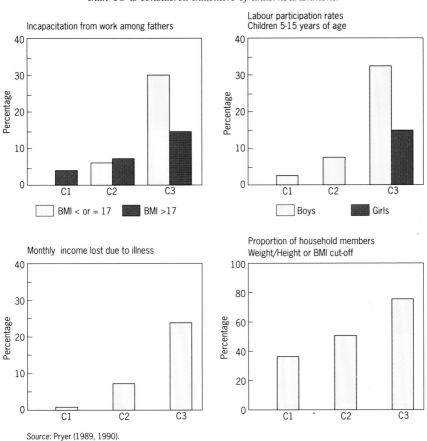

Source: Pryer (1989, 1990).

especially over time. Nevertheless, there have been studies that have been able to look at the city through a finer lens (Wray, 1985; Guimaraes and Fischmann, 1985). The picture that emerges from the first study is that at least in one sample of urban slums in Bangkok, Thailand, malnutrition is far worse than in the rural communities chosen for comparison. The second study revealed that the levels of infant mortality in a poor shanty town are much higher than in the most affluent part of Porto Alegre, Brazil. The same study demonstrated that infant deaths were especially high near a watercourse called the Moniho Arroyo, where 52.9 per cent of the deaths were attributed to intestinal diseases as compared to 15.9 per cent in the more distant sector. A specific environmental feature, a contaminated watercourse, clearly had a significant impact on mortality patterns.

The health of the homeless

The homeless state, whether in developed or in developing countries (Chapter 17), is clearly characterized by an increased risk to health: little or no protection from the elements; lack of access to basic sanitation or water sources; inadequate nutrition due to the absence of cooking facilities and of money to buy prepared foods; and the spread of disease as a result of overcrowding in temporary accommodation (Cooper Weil *et al.*, 1990). Very often, homeless people are not recognized as official residents and their births and deaths are therefore ignored in official records, which of course leads to serious underestimates of homelessness and the accompanying degree of ill health and premature mortality, and also helps city authorities to continue to ignore the problem (Acheson, 1990).

In the South, the plight of the 'street children' is an extreme case of homelessness, and illustrates the health risks attendant upon this condition. Such children are often a product of massive migration from rural areas and the resultant breakdown of family life, or of the death or separation of their parents. Unsupervised by adults, some children spend their days on the street but are able to return home at night; others have no home to return to and sleep wherever they can find shelter. These abandoned children inevitably suffer the consequences of lack of sanitation and clean water, occupational accidents, sexually transmitted disease, drug abuse, crime, and all the other effects of striving to cope alone, resulting in a deep sense of insecurity and emotional conflict (Tabibzadeh *et al.*, 1989).

Health problems in urban settlements of developed countries

One of the main health risks in urban settlements in developed countries is due to the use of certain materials in buildings. The use of asbestos is now completely banned; but some organic materials and metal compounds widely used in paints, pipes and furnishings, as well as others in daily use (such as cleaning agents and solvents) can cause poisoning. Radon emissions pose a serious health hazard, particularly in air-tight buildings (Chapter 1). In one study, the concentration of radon indoors was found to be six times higher than its concentration outdoors (Nazaroff and Teichman, 1990). Smokers are at a much higher risk of developing cancer when exposed to radon. Outdoor pollutants are also known to penetrate indoors. Recently, people started complaining of what came to be known as the 'sick building syndrome' which has been shown to be epidemiologically related to sealed buildings that develop high temperatures and levels of dust as well as passive smoking (Chapter 1).

Noise from a variety of sources has become a serious problem in modern urban settlements. Apart from being a nuisance, it can give rise to serious health problems, such as loss of hearing or increased reaction times. Complex tasks are disrupted at relatively low noise levels and higher anxiety levels and risk of hostile reactions have been known to result. High frequency and impulsive noise are usually more disruptive.

Health in unsettled situations

Migration from rural to urban areas is often driven by landlessness, poverty and homelessness. Such unsettled people often live in situations totally lacking the amenities expected in human settlements. Not only are these people exposed to above average health risks, but they can also contribute to the establishment of new disease foci as they move from place to place, a phenomenon no doubt linked with the persistence of malaria in the newly-developed areas of Brazil and the rapid spread of cholera in South America.

Colonization settlements

Colonization projects, which are initiatives to develop rural areas, have led many hundreds of thousands of people to move. They are being used by many countries to 'redistribute population from densely populated, resource-scarce areas to sparsely populated, resource-rich areas' (Findley, 1984). These projects are fraught with problems, well illustrated by recent colonization experiences in Brazil. This has led to a quadrupling of population in some rural areas over a 20-year period, from under four million in 1970 to over 17 million in 1990. An estimated nine million people have settled in the rural areas of the Amazon along a network of new roads extending over 45,000 kilometres (Wilson and Alicbusan, 1990).

This massive colonization has been accompanied by the spread of malaria (Figure 18), which has, in turn, had a major impact on the colonization process in a number of different ways. Ill workers cannot carry out heavy farm work.

Figure 18: *Rise in the incidence of reported malaria cases in Amazonia, 1970–89.*

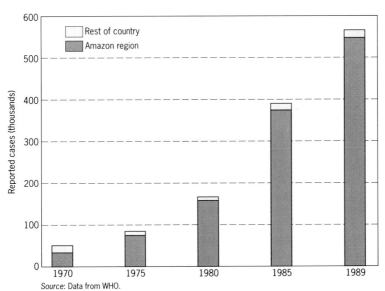

Source: Data from WHO.

The very threat of infection is enough to convince families that they must minimize unnecessary exposure and so, instead of the whole family moving to the settlement area, only the young adults work on the farm, and only when farm work is required. After a number of serious episodes with the disease, many decide to abandon their farms and seek work elsewhere (Sawyer, 1987).

Serious illness requires treatment, which in spite of the presence of extensive anti-malarial services in the area, is often not easily accessible. Transportation costs can be significant, and where private medical services are sought, cost can be prohibitive. An analysis of settlements in the 1970s indicates that many first owners abandoned their farms due to sickness (Henriques, 1988). More recent data from Rondonia show that the cost of malaria treatment represents a major expenditure, especially for low-income households. In one project area, Machadinho, malaria strikes, on average, more than three times per year and one-eighth of family income is used to pay for treatment (Sawyer, 1987). The actual burden on low-income families is likely to be significantly greater. They are less able to afford personal protection (mosquito nets cost $US5 each), to improve their housing, or to clear local breeding sites. In Machadinho, only ten per cent of new settlement plots are in the hands of their original owners.

The malaria epidemic in the Amazon now accounts for more than 60 per cent of all reported malaria cases in the Americas. Within Brazil itself, 97 per cent of the cases come from Amazonia, which has 15 per cent of the population of the country. Two states (Rondonia and Para) report 70 per cent of the cases, and four municipalities of Rondonia and four of Para have more than 60 per cent of the cases of these states.

Malaria, while posing the most serious threat to health in the Amazon, is not the only one. Other diseases - leishmaniasis, schistosomiasis, tuberculosis and leprosy - are also present, and accidental injury, especially during the clearing and burning season, can equally interrupt family settlement.

The story of malaria in the Amazon is not a unique one: many other areas of the world are suffering comparable difficulties. The absence of health care in newly settled areas has serious consequences because the combination of poor sanitation and exposure to an unfamiliar environment results in increased morbidity and mortality rates. In the Selva Lacandona community of Frontera Echevarria, Chiapas, Mexico, the crude death rate was found to be as high as 50 per 1,000 persons. Several surveys also report that colonists have experienced much higher incidence of parasitic and respiratory infections (Findley, 1984). In Thailand, about one million farm households are occupying forest lands as squatters. Lacking secure land ownership, they have less access to credit. Consequently, they cannot borrow money to buy equipment, pesticides or improved seed strains, so their land is less productive. 'Forest malaria' in Thailand is an important health problem and deforestation is an important component of environmental degradation.

The drama of cholera in South America illustrates the threat posed by poor, infected people moving from one poor environment to another, especially when they are carrying the micro-organisms with the potential of starting epidemics in previously disease-free areas. At the end of January 1991, an epidemic of cholera broke out on a continent which had witnessed no outbreak this century. Striking first in the coastal cities of Chancay and Chimbote, north of Lima, Peru, the epidemic then spread with unexpected speed and intensity (WHO, 1991b). Its movement to other countries in South America and the world is shown in Figure 19.

Figure 19: *Spread of the cholera epidemic through South America and beyond during 1991. Dates given are those of the first notified cases (as at 27 August 1991).*

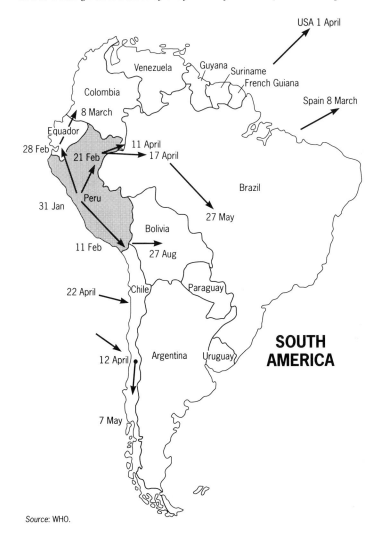

Source: WHO.

This diagram illustrates the rapidity with which a disease, which is fundamentally linked with a poor environment, especially unsafe water and inadequate sanitation, can spread. Clean water and hygienic practices can break the chain, but these are hardest to come by in unsettled situations, which no doubt has played an important role in this epidemic. In fact, the epidemic in Peru has had a very low mortality rate because of a successful campaign to educate women in the use of oral rehydration therapy.

Industrialization and health

These examples, while capturing many of the features of a variety of human settlements, do not shed much light on two further aspects of development activities which, independently of other deleterious features of urban life, have dramatic effects on human health and the environment, namely industrialization and waste accumulation. Industrial activities, while contributing in many ways to human well-being, also have adverse effects on health through the release of harmful chemical, physical and biological agents. The main pathways of transmission to people are air, water, wastes and contaminated food. Cities, particularly in developing countries concentrate populations and industrial sites (Chapter 12). This combination inevitably leads to higher levels of exposure of a greater number of people.

Most knowledge of the adverse effects of pollutants on health has been obtained in workplaces in developed countries, and relates to single agents (Chapter 10). There is little doubt, however, that the combined effects of multiple exposures, which happens most frequently in highly industrialized areas in both the developed and the developing world, can produce adverse health consequences which are more than simply additive. For example, the effects of sulphur dioxide on health increase in the presence of particulates and, as mentioned earlier, exposure to radon increases the risk of cancer for smokers. In 1984, WHO/UNEP established the 'Human Exposure Assessment Locations' (HEALs) programme (see Chapter 20) in order to monitor total human exposure to pollutants and assess the combined risk from air, food and water pollutants.

In recent years, a number of industrial accidents have alerted the public to the threat posed by the release of large quantities of toxic chemicals. Chapters 9 and 12 deal with natural and industrial hazards and accidents and the various approaches taken to identify hazards, prevent accidents or mitigate their adverse impacts as much as possible through sophisticated risk assessment techniques and adequate preparedness to contain these impacts when accidents do happen.

Urban waste and health

As the main centres of production and consumption, urban areas naturally generate large quantities of waste. Many major cities world-wide are facing a solid waste crisis as accessible landfills are filling up and rising transportation and processing costs make it harder and harder for services to keep up with the rapid growth of waste. Only between 30 and 70 per cent of solid wastes are currently collected in cities in developing countries. Disease vectors proliferate on waste and in situations where human excreta are added to garbage, increasing health risks considerably.

Particularly dangerous situations are created when hazardous and/or contaminated material, such as clinical wastes and wastes from small chemical and metal processing factories within the city, are mixed with ordinary municipal solid waste, since the latter is normally collected and dumped at a disposal site without much attention being paid to its immediate and long-term environmental consequences. Immediate dangers are faced by waste pickers or scavengers who search through dump sites for wastes of potential use, either directly or after recycling. These people, who are generally poor, living at or near the dump, are of sufficient importance in some cities as to 'play an important role in recovering resources from the waste materials, and thus reducing the waste volume to be ultimately disposed of (Gotoh, 1989). In 1987, 241 people were injured and two died when an abandoned radio-active cancer-treatment device was discovered among rubbish in Goiania, Brazil, and the materials re-utilized (Chapter 9).

The numbers of people engaged in this informal resource recovery are not exactly known (Furedy, 1989). There are estimates, however: 25,000 obtain wastes from Manila's Smokey Mountain dump, with perhaps 60,000 more depending upon these wastes for their basic needs. The waste found is used in diverse ways. It serves as building material for the shelter of the poor. Plastics, tin cans, bottles, bones, feathers, intestines, hair, leather and textile scraps find their way into industries. Families engaged in waste-picking may earn as much as, or more than, an unskilled worker, but this is only achieved by most family members, including young children, working for very long hours.

Few studies of the health status of waste-pickers have been carried out. However, a recent project in Calcutta examined the prevalence of respiratory diseases, diarrhoea, viral hepatitis, intestinal parasites, skin disease, immunization status and nutritional status in this group. A control group was taken from a population with similar socio-economic backgrounds but with different occupations (cultivation and fishing). Figure 20 summarizes the results obtained. While both the control group and the waste-pickers were found to be suffering high levels of ill health, the waste-pickers had greater prevalence of respiratory disease, diarrhoea and intestinal parasites (Nath et al., 1991).

The working environment

The working environment is that part of the human environment in which people spend their working hours every day. Its quality is conditioned by a combination of physical, psychological and socio-economic factors that fall into two main categories. Occupational safety deals with the physical, chemical and biological hazards and mental stresses present in the working environment, and their prevention and control. Working conditions include the organizational framework, length and distribution of working time, contents of work and human relations - all factors that have significant repercussions on the health and well-being of workers. In 1975, the Director General of ILO presented a report entitled *Making work more human*. This resulted in the 'International

Figure 20: *Prevalence of major diseases in a Calcutta slum.*

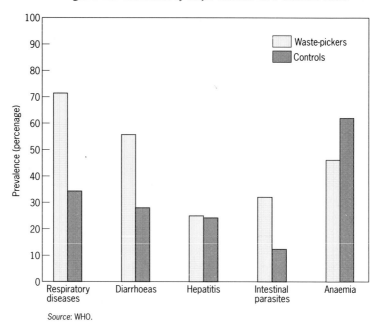

Source: WHO.

Programme for the Improvement of Working Conditions and Environment' (PIACT) (Chapter 12). ILO has since published a series of manuals on occupational health, safety and the working environment in several industries (e.g. ILO 1985, 1989a and 1991).

Occupational hazards depend on the processes involved, the safety precautions taken and the ability to respond effectively to accidents. It has been estimated that there are about 32.7 million occupational injuries and 146,000 deaths at the work-place every year (WHO, 1990a). The main problem in developing countries is the almost complete absence of basic safety measures in the many small workshops that predominate in those countries.

An example is the case of lead poisoning among household members who were exposed to lead-acid battery repair shops in Kingston, Jamaica (Matte *et al.*, 1989; Box 2). The Jamaica study illustrates two important points:

i.   The potential of environmental health hazards in the informal work sector, and

ii. the particular vulnerability of children to toxic chemical poisoning.

The first point is of increasing importance given the fact that more and more governments are looking to the informal sector as a source of employment for the growing numbers of unemployed. One major difficulty posed by this sector is the virtual absence of any environmental control measures or standards. The vulnerability of children is due to several factors such as their small ratio of body

Twenty-two children from Kingston, Jamaica, were hospitalized for lead poisoning between January 1986 and March 1987. The effects vary with the quantities of lead present, but include damage to the kidney, liver, nervous system and reproductive system. As well, growth is impaired and blood synthesis interfered with. An epidemiological investigation revealed that the most likely source of exposure was through ingestion of contaminated soil. (Matte *et al.* 1989). Soil contamination resulted from lead fumes generated during the repair process of smelting of scrap lead along with lead dust being blown from scrap piles, from scrap being dragged or carted through yards, or from lead dust being tracked on the shoes of workers.

Blood samples were taken from children living near repair shops as well as in the neighbourhood and hospital, the latter two serving as an epidemiological control. In all age groups, blood lead levels were significantly higher among subjects who lived in households where repair shops existed. Blood lead levels were higher in children less than 12 years of age than in those 12 or older. Children living in households where workers lived, but where no repairs were actually carried out, also had higher lead levels.

Source: Matte et al (1989).

volume to body surface area, their high metabolic and oxygen consumption rates, their different body composition and the effect of progressive maturation and differentiation of body systems (Belsey, 1987).

The Kingston example is but one of many. Studies in the UK demonstrated elevated levels of lead in people living near to lead smelters, especially the families of workers. Blood lead levels (BLLs) as low as 10 µg/dl, once considered safe, are now known to adversely affect cognitive development and behaviour in children, with potentially long-term consequences. In 1984, an estimated three to four million children in the USA had BLLs ≥15 µg/dl (CDC, 1991). It is estimated that a further 675,000 young children in the USA had high concentrations of lead in their blood, and millions world-wide have been exposed to 'potentially toxic amounts of lead' due to the increased use of leaded petrol over the last few decades (Postel, 1986). In those countries that have introduced unleaded gasoline, BLLs have dropped markedly (Chapter 1).

Even though no accurate statistics on occupational diseases are available, there is no doubt that the proliferation of new chemicals whose toxicity is still unknown (Chapters 10 and 12) means that it will be some time before research would reveal the full impact of exposure for long periods of time to what may at first have been considered inoffensive substances. Furthermore, the increasing use of radioactive materials and X-ray machines in industry increases the risks of exposures of increasing numbers of workers to ionizing radiation.

Although noise is not restricted to the workplace, noise levels in factories are generally higher than elsewhere, because of the large number of moving parts and their concentration in a limited space. A great number of developed countries have now adopted noise-limiting standards that reduce both the physical and the psychological ill effects of excessive noise.

Conditions of work have come to receive more attention of late. In almost all industrialized countries, working time has been reduced, either by legislation or by collective agreements. Perhaps the most important development has been in the developed countries where problems of work organization and job content have received a good deal of in-depth analysis. With the increasing recourse to better-educated workers doing more sophisticated jobs and the decline in dependence on motor skills and dexterity, job satisfaction now becomes a major concern of the new generations of workers in knowledge-intensive fields. This is quite different from previous emphasis on the discomfort, fatigue or stress of less skilled workers in the older industries.

Responses

In the previous chapters, whether those dealing with the major environmental issues (Part One of this volume) or the impacts of developments in the different sectors of socio-economic activity (Part Two), the responses to environmental problems, including those impacting on human health, have been addressed. These responses have ranged from better monitoring of impacts of various activities on the environment and better understanding of the manner in which the environment is affected, to measures of mitigating undesirable impacts through combinations of actions on both the technological and social fronts. On the technological front, action has involved minimization or treatment of emissions, effluents and wastes; reduced material and energy inputs; cleaner technologies and recycling. On the social front, new regulations and legislation to reduce pollution and wasteful use of resources have been enacted at the national, regional and international levels (Chapters 22 and 23); economic incentives and penalties have been applied, and technical and advisory services and public awareness campaigns have been initiated.

At the higher level of integrating development, health and the environment, a start was made with the 'Primary Health Care' approach for achieving health for all, adopted at the international Conference on Primary Health Care, held in Alma-Ata, USSR, in 1978. This has shaped and defined present-day thinking regarding the place of health development in overall national development. The meeting concluded that any distinction between economic and social development was no longer tenable:

'Economic development is necessary to achieve most social goals and social development is necessary to achieve more economic goals. Indeed, social factors are the real driving force behind development. The purpose of development is to permit people to lead economically productive and socially satisfying lives. Only when they (people) have an acceptable level of health can individuals, families and communities enjoy the other benefits of life. Health development is therefore essential for social and economic development, and the means for attaining them are intimately linked' (WHO, 1978).

Most health care systems have found it very difficult to move in the direction outlined at Alma-Ata because they have been overwhelmed by the basic problem of running their curative services. Few, if any, low-income countries have been able to establish an adequate public health capacity, even of the most rudimentary kind. Most still lack the necessary full-time staff with public health and related skills, or the logistical and communication capacity to collect and analyse data of the kind required. For many of them the austerity measures imposed by structural adjustment requirements have made the situation even more difficult, leading to lowered staff morale and motivation.

Following a recent review by WHO of the world situation, the WHO Director-General called for a *New Paradigm for Health*, embracing a world view, in which health is seen as central to development and to the quality of human life. As such, health should become a major political issue at cabinet level of government. Health, as a human right, cannot be left entirely to market forces, because the most disadvantaged will become even worse off. Furthermore, the new paradigm must be people-oriented:

'We must humanize our approach to health development - a process that begins and ends with the people themselves. This means paying due attention to the individual, the family and the community, but especially to the underprivileged and those who are at risk, such as women and children and the elderly. It means building a sustainable health infrastructure, 'bottom-up', in communities where people live and work, with appropriate response and support from district and more central levels, thus closing the existing gap between programme delivery and community initiatives' (Nakajima, 1991).

A 'post Alma-Ata' period of reflection has started. The desired features of the health sector's contribution to national development are now being examined critically, publicly and sectorally. A contribution is called for which is more in tune with today's political, economic and social realities. Central to this will be the prevention and control of health problems deriving from a rapid and widespread degradation of the human environment.

Concluding remarks

The health sector will have considerable difficulty in contributing effectively to the process of 'post Alma-Ata' reflection. Few national health sectors are organized to address development issues from a truly holistic perspective that goes beyond the delivery of curative services. It will take considerable political adroitness within the health sector to mobilize the human and financial resources required to address the issues identified above. The necessary action programme covers a range of policy issues that still lack the information and analysis needed for discussion and decisions.

At the level of the health sector itself, certain strategic lines can be suggested. These include mapping health-environment linkages so as to relate health vulnerability to specific economic development programmes through the intermediary factors of environmental degradation (e.g. better understanding of the interrelationship between rural, urban and unsettled contexts). This means that the capacity of ministries of health has to be strengthened so that action can be initiated to prevent or mitigate the adverse health consequences of the impacts of development activities.

Effective handling of health-environment interactions can only be achieved, particularly in most developing countries, if the issue of poverty is addressed seriously. As has been argued in several parts of this book, poverty is not only the result of scarcity of resources and population growth - important as these are. Social organization, national goals and development strategies, as well as international economic relations, are also important factors in environmental degradation and the resulting poverty and deteriorating health standards. This is related directly to consumption patterns and life-styles, both in the developed regions and amongst those groups in developing regions that adopt such consumption patterns and life-styles. It is these that impact on the economies and the environment of the developing countries, particularly those that rely mainly on the export of commodities for economic survival (Chapters 11 and 16).

It is also these life-styles that are mainly responsible for the emergence of a whole range of new health problems associated with, for example, eating and drinking habits, leading a more sedentary life and the hazards brought about by chemical or radioactive emissions, effluents and solid wastes, whether in the home, or the work and leisure environments. This, in turn, relates to the quality and quantity of material and energy inputs and the technologies used in satisfying the social demand for goods and services in different walks of life.

REFERENCES

Acheson, E.D. (1990) Edwin Chadwick and the world we live in. *The Lancet.* **336**: 1482-85.

Barraclough, S. and Ghimire, K. (1990) Social Dynamics of Deforestation in Developing Countries: Principal Issues and Research Priorities. *UNRISD Discussion Paper* 16. November 1990.

Basta S.S. (1977) Nutrition and Health in Low Income Urban Areas of the Third World: *Ecology of Food and Nutrition,* **6**: 113-24.

Belsey, M.A. (1987) Toxic Disasters with Crude Chemicals: An Approach to Identifying Risks in Infants and Children. In: *Attitudes to Toxicology in the European Economic Community* (ed P.L. Chambers). John Wiley & Sons Ltd.

CDC (1991) *Centres for Disease Control Morbidity and Mortality Weekly Report.* **40 (12)**: 194 (March 29).

Cecelski, E. (1985) *The Rural Energy Crisis, Women's Work and Basic Needs: Perspectives and Approaches to Action.* ILO, Geneva.

Chowdhury, A.K.M., Huffman, S.L. and Chen, L.C. (1981) Agriculture and Nutrition in Matlab Thana, Bangladesh. In: *Seasonal dimensions to rural poverty* (ed. R. Chambers *et al.*). Frances Printer, London.

Cooper Weil, D.E., Alicbusan, A.P., Wilson J.F., Reich, M.R. and Bradley, D.J. (1990) *The impact of development policies on health.* WHO, Geneva.

Dankelman, I. and Davidson, J. (1988) *Women and Environment in the Third World - Alliance for the Future.* Earthscan Publications Ltd., London.

Diallo, S., Ndir, O., Square, D., Gaye, O. and Th. Dieng (1990) Prévalence des Bilharzioses et des Autres Parasitoses Intestinales Dans le Basin du Fleuve Sénégal. *Programme 'Eau, Santé et Développement' Rapport annuel 1990.*

FAO (1982) *The State of Food and Agriculture,* 1981. FAO, Rome.

FAO (1989) *Forestry and food security. FAO Forestry Paper,* 90. FAO, Rome.

Farnsworth, N.R. and Morris, R. W. (1976) Higher Plants - the sleeping giant of drug development. *Amer. J. of Pharmacology,* **148**: 46-52.

Feder, G., Onchan, .T. and Chalamwung, Y. (1988) Land Policies and Farm Performance in Thailand's Forest Reserve Areas. *Economic Development and Cultural Change.* **Vol 36, No. 3**, 483-501.

Findley, S.E. (1984) Colonist Constraints, Strategies and Mobility: Recent Trends in Latin American Frontier Zones. *ILO World Employment Programme Research Working Paper WEP 2-21/WP.145.*

Furedy, C. (1989) Social Considerations in Solid Waste Management in Asian Cities. *Regional Development Dialogue,* **10 (3)**: 13-38.

Goldsmith, E. and Hildyard, N. (1984) *The Social And Environmental Effects of Large Dams.* Wadebridge Ecological Centre, Wadebridge, U.K.

Gotoh, S. (1989) Issues and Factors to be Considered for Improvement of Solid Waste Management in Asian Metropolises. *Regional Development Dialogue,* **10 (3)**: 1-10.

Guimaraes J., de Lima, J. and Fishmann, A. (1985) Inequalities in 1980 Infant Mortality among Shantytown Residents and Non-shantytown Residents in the Municipality of Port Alegre, Rio Grande do Sul, Brazil. *PAHO Bulletin,* **19 (3)**: 235-51.

Henriques, M.H.F.T. (1988) The Colonization Experience in Brazil. In: *Land Settlement Policies and Population Redistribution in Developing Countries* (ed. A.S. Oberai) 317-54.

ILO (1985) *Introduction to Working Conditions and Environment.* ILO, Geneva.

ILO (1989a) *Women and Land. Report on the Regional African Workshop on Women's Access to Land as a Strategy for Employment Promotion, Poverty alleviation and Household Food Security.* ILO, Geneva.

ILO (1989b) *Major Hazard Control*. ILO, Geneva.

ILO (1991) *Preventing Major Industrial Accidents*. ILO, Geneva.

Jeyaratnam, J., Lun, K.C. and Phoon, W.O. (1987) Survey of acute pesticide poisoning among agricultural workers in four Asian countries. *Bull. WHO*, **65 (4)**: 521-27.

Jeyaratnam, J. (1990) Acute pesticide poisoning: a major global health problem. *World Health Statistics Quarterly*, **43**: 139-144.

Levinson, M. (1974) An economic analysis of malnutrition among young children in rural India. *Cornell/MIT International Nutritional Policy Series*.

Matte, T.D., Figureueroa, J.P., Ostrowski, S., Burr, G., Jackson-Hunt, L., Keenlyside, R.A. and Baker, E.L. (1989) Lead Poisoning Among Household Members Exposed to Lead-Acid Battery Repair Shops in Kingston, Jamaica. *Int. J. Epid.* **Vol 18**, No. 4: 874-81.

Maxwell, S. (1988) Editorial. *IDS Bulletin*, **19 (2)**: 1-4.

Myers, N. (1989) The Environmental Basis of Sustainable Development. *In: Environmental Management and Economic Development* (Schramm, G. and Warford, J. J. eds.) World Bank, Washington, D.C.

Needham, H.L., *et al.* (1990) The long term effects of exposure to low doses of lead in childhood. *The New England Journal of Medicine*, **Vol. 322**: 32.

Nabarro, D. (1981) Social, Economic, Health and Environmental Determinants of Nutritional Status. *Food and Nutrition Bulletin*, **Vol. 6**, No. 1.

Nakajima, H. (1991) Statements of Dr. Hiroshi Nakajima, Director-General, to the Executive Board and the World Health Assembly. A44/DIV/4. WHO, Geneva.

Nath, K.J., Chakravarty, A.K., Chakravarty, I. and Kahali, S.D. (1991) Socio-Economic and Health Aspects of Recycling of Solid Waste Through Scavenging. (Unpublished final report of a WHO sponsored Study).

Nazaroff, W.W. and Teichman, K. (1990) Indoor radon. *Environmental Science and Technology*, **Vol. 24**: 774.

Norse, D. (1985) Nutritional implications of resource policies and technological change. *In: Nutrition and Development* (eds. Biswas, M. and Pinstrup-Andersen, P.). Oxford University Press, Oxford.

Perera, P.D.A. and Gunatilleke, G. (1989): The Mahaweli Project. Case study included in *Health Implications of Public Policy* (1990). Originally prepared as background by the Indian Institute of Management, Bangalore, for the WHO symposium 'The Implication of Public Policy on Health Status and Quality of Life', Bangalore, October 1989.

Postel, S. (1986) Altering the Earth's Chemistry: Assessing the Risks. *Worldwatch Paper, 71*.

Pryer, J. (1989) When breadwinners fall ill. *IDS Bulletin*, **20**: 49-57.

Pryer, J. (1990) *Socio-economic aspects of Undernutrition on Ill-health in an Urban Slum in Bangladesh*. Report for Save the Children Fund (UK). June 1990.

Rajeandran and Reich, M. R. (1981) Environmental health in Malaysia. *Bull. Atom. Sci.* April 1981: 30-35.

Sawyer, D. (1987) *Economic and Social Consequences of Malaria in New Colonization Projects in Brazil*. Paper presented at special conference on malaria in association with Xth International Conference on the Social Sciences and Medicine. Sitges, Spain. 26-30 October, 1987.

Tabibzadeh, I., Rossi-Espagnet, A. and Maxwell, R. (1989) *Spotlight on the Cities: Improving Urban Health in Developing Countries*. WHO, Geneva.

Talla, I., Kongs, A., Verle, P., Belot, J., Sarr, S. and Coll, A.M. (1990) Outbreak of Intestinal Schistosomiasis in the Senegal River Basin. *Ann. Soc. Belge Med. Trop.* **70**: 173-80.

UN/ACC Sub-Committee on Nutrition (1989) Does Cash-cropping Affect Nutrition? *SCN News*, a United nations ACC/SCN publication. No. 3, 1989.

UNCHS (1987) *Global Report on Human Settlements 1986*. Oxford University Press, Oxford.

UNEP (1986) *The State of the Environment: Environmental Health*. UNEP, Nairobi

UNEP (1991) *Environmental Data Report, Third Ed. 1991-92*. Basil Blackwell, Oxford.

US AID (1987) Accent for Health: Living proof of progress. *Horizons*, Spring, 1987.

Victora, C.G., Vaughan, P., Kirkwood, B., Martines, J.C. and Barcelos, L.B. (1986) Child Malnutrition and Land Ownership in Southern Brazil. *Ecology of Food and Nutrition*, **18**: 265-75.

Warford, J.J. (1991: Environment and Development - Saitama Paper (Draft), February 23, 1991.

White, G.F., Bradley, D.J. and White, A.U. (1972) *Drawers of Water*. Chicago University Press, Chicago.

WHO (1978) *Primary Heath Care*. Report of the International Conference on Primary Health Care, Alma-Ata, USSR, 6-12 September 1978. WHO, Geneva.

WHO (1986) *Intersectoral Action for Health. Technical discussions, Geneva, May 1986*. Background document A39/Technical Discussions/1.

WHO (1989) *World Health Statistics*. WHO, Geneva.

WHO (1990a) *World Health Statistics*. WHO, Geneva.

WHO (1990b) Global Estimates of Health Situation Assessment and Projections 1990. Unpublished report.

WHO (1991a) Environmental Health in Urban Development. Report of a WHO Expert Committee. *WHO Technical Report Series 807*. WHO, Geneva.

WHO (1991b) *Wkly. Epidem. Rec.*, **No. 20**, 1991: 141-45.

WHO (1991c) *World Health Statistics*, WHO, Geneva.

WHO (1992) *Our Planet, Our Health: Report of the WHO Commission on Health and the Environment*. WHO, Geneva.

Wilson, J.F. and Alicbusan A.P. (1990) Development Policies and Health: Farmers, Goldminers and Slums in the Brazilian Amazon. Unpublished World Bank draft, February 1990.

World Bank (1990) *World Development Report 1990*. Oxford University Press, New York and Oxford.

WRI (1988) *World Resources 1988*. Basic Books, New York.

Wray, J.D. (1985) *Nutrition and health in urban slums: an overview*. Paper presented at the workshop on Community Health and the Urban Poor, organized by LSHTM, Oxfam and UNICEF. Oxford, 1985.

CHAPTER 19

Peace and security

Introduction

The right to live in a peaceful and secure environment is fundamental to human well-being. Most nations have strived for this over the centuries, and people everywhere want peace and security above all else. Without them, the development of just, equitable and healthy societies cannot take place, yet nations continue to resort to violence as a means of settling their differences.

Violence is a prehistoric means of solving disputes which time and culture have endowed with endless sophistication, but otherwise have left unchanged. Centuries of enlightenment and science have merely enabled people to kill more of one another, more quickly and efficiently, than their ancestors were able to do. War and the preparations for war are inimical to development because they squander scarce resources and damage or destroy the international confidence that is essential to sustainable development and the improvement of the environment at regional and global levels.

Recognition of the hazards posed by modern weapons systems, and the need for peaceful international relations as the only satisfactory foundation for environmentally-sound development and human prosperity, increased during the last two decades. The United Nations Charter takes as its starting point the need to maintain peace and security through the suppression of acts of aggression and the settlement of international disputes in conformity with the principles of justice and international law. But the United Nations Charter also envisaged more positive action. It has been recognized that peace and security require the fulfilment of the other great principles enunciated in the preamble to the Charter: social progress, better standards of living, freedom, fundamental human rights, recognition of the dignity and worth of human individuals and the equal rights of men and women and of large and small nations. Yet the record of the last two decades is, to say the least, disappointing.

The world economic situation has been erratic. The gap between rich and poor has widened, and malnutrition, lack of access to safe water and sanitation and primary health care, and ignorance and poverty, are still rampant in many developing countries. The last two decades have seen escalating external debt, falling real prices for commodities, and adjustment policies that have exacted a severe toll on the poor, for many of whom the 1970s and 1980s were 'lost decades'. As a result, civil strife and conflicts prevailed in many regions, especially in developing ones.

In spite of the great destruction and suffering caused by World War II, which claimed the lives of at least 50 million people, armed conflicts continue to erupt, destroying life and the means of life in many areas. Since the end of World War II, about 200 armed conflicts have taken place, mostly in developing countries. Such wars have killed more than 20 million people and displaced several millions more while most have caused considerable economic and environmental damage. Ironically, the five permanent members of the UN Security Council were directly or indirectly involved in many of these conflicts.

Trends in military activity and consumption, 1972-92

Military expenditure

In the last two decades (1970-1990), the world spent about $US17 trillion, at 1988 prices and exchange rates, on military activities. In other words, the global military expenditure was an average of $US850 billion per year ($US2.33 billion per day = $97 million per hour = $1.6 million per minute). In current dollars, the annual global military expenditure has reached more than $US1000 billion (SIPRI, 1986, 1988, 1990). According to data of the US Arms Control and Disarmament Agency (ACDA), the spending peaked in 1987 and has been declining since. During the 1970s, military expenditure increased in real terms at an average annual rate of 2.5 per cent. Since 1980, however, the average annual real increase has been 3.5 per cent, higher than the annual growth rate of gross domestic product (GDP) of about 2.8 per cent. Global arms

Figure 1: *World military expenditure, 1970–90, in $US billions at 1988 prices, and percentage distribution of world military expenditure.*

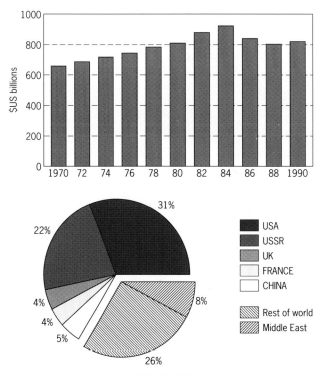

Source: Based on data from SIPRI (1986, 1990).

spending, as a percentage of GNP, peaked in 1983. Five countries (USA, former USSR, China, UK and France) account for about two-thirds of global military expenditure (Figure 1)

The world's military spending dwarfs any spending on development assistance. For example, in 1988 the total military expenditure was about $US850 billion. In the same year, the total official development assistance was only about $US35 billion (World Bank, 1990), i.e. about four per cent of global military expenditure.

Although military spending, as a share of gross national product (GNP), has decreased slightly on a global basis and in industrialized countries, it has increased in developing countries and especially in the Least Developed ones (Figure 2). On a regional basis, Latin America devotes the smallest share of its GNP - about 1.5 per cent - to military spending. The Middle East and North Africa, on the other hand, spend the highest share of GNP (about 12.6 per cent) on military activities (World Bank, 1988; UNDP, 1991).

Figure 2: *World military expenditure, 1960 and 1986, expressed as percentage of GNP.*

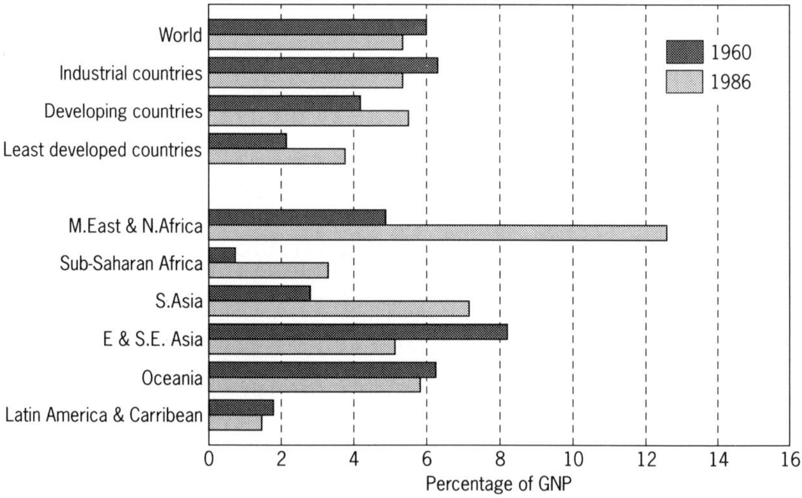

Source: based on data from UNDP (1991).

Associated with the increase in military expenditure is an increase in the cost of military research and development (R&D), which totals more than $US80 billion per year. Military R&D expenditure is concentrated in a few countries (the USA and former USSR dominate, followed by UK, France, China and Germany). In these countries the military sector spends ten to twenty times as much as the civilian sector on R&D input per unit of output. Globally, military R&D accounts for about one-third of the total expenditure on all scientific research and development. The result has been an increase in the rate of replacement of older weapons with new systems and 'families' of more destructive ones.

The arms supermarket

While it seems likely that arms of one kind or another have been traded since antiquity, it was not until the mid-nineteenth century that the modern armaments industry began to take shape. Today it is a global industry, controlled and manipulated by large producers and influenced by a cluster of political, military and economic factors. The arms trade flourishes in times of war. Historically, the arms suppliers and the middlemen have fuelled conflicts and even created new ones - directly or indirectly - to keep the industry thriving.

In the last two decades, the cumulative value of global arms sales reached $US410 billion (at constant 1985 prices) - about $US20 billion per year. During the period 1971-85, five countries (the former USSR, USA, France, UK and Italy) accounted for 88.5 per cent of arms exports; however, during the period 1980-85 the proportion attributable to these countries declined to 83.4 per cent with other countries such as the Federal Republic of Germany, China, Spain, Israel, Czechoslovakia, the Netherlands and Brazil increasing their share. The Middle East is the most significant arms importing region in the developing world, accounting for about 46 per cent of all major weapons system transfers between 1971 and 1985, and still some 43 per cent in 1990 (Figures 3 and 4) (Brzoska and Ohlson, 1989; SIPRI, 1990). Furthermore, the cost of individual items of military equipment continues to rise, so that the arms 'race' leads to even greater levels of expenditure in order to maintain equivalence.

The expansion of arms sales has largely been financed by credits from supplier countries. It has been estimated that about 50 per cent of all arms imports into developing countries have been financed by export credits (UN, 1989). The costs of such military credits amount to 30 per cent of all inflow of real debt to the developing countries.

Besides selling more and more arms of ever-increasing sophistication, the large producers are also exporting the technology to manufacture arms. For most of the post-World War II period, such arrangements were limited to the allies in the two major military blocks, NATO and the Warsaw Pact countries. It has now been extended to include other countries (e.g. South Korea, the Philippines, Egypt, Israel, Turkey, Pakistan, Argentina).

The stepped-up acquisition of modern arms and of military manufacturing capabilities by developing countries has also generated an increased demand for another type of non-hardware transfer - the provision of military technical skills and other specialized services. Because many developing countries lack the trained personnel to operate and maintain high-technology equipment, they increasingly look to defence contractors to provide such services as part of major arms agreements.

The impact of arms acquisitions on external debt varies considerably from country to country. For those with abundant foreign exchange, military expenditure is not necessarily a major constraint on civilian public spending and economic growth. On the other hand, in countries with scarce foreign exchange the cost of arms purchases and the establishment and maintenance

Figure 3: *The global arms race: (top left) percentage share of total arms exports, (top right) percentage share of total arms imports, and (bottom) global arms sales, 1970–90, in $US billions at 1985 prices.*

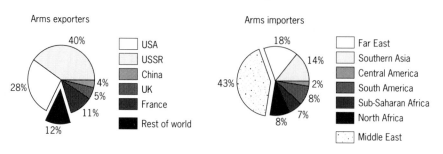

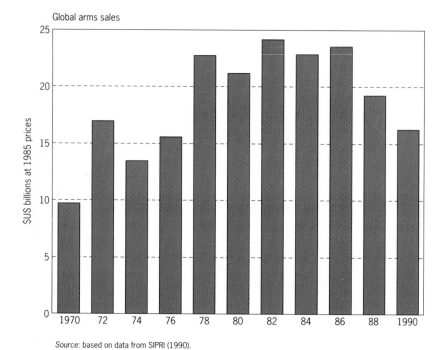

Source: based on data from SIPRI (1990).

of a viable arms industry are formidable, and can be managed only by increased borrowing. Therefore, high military spending contributes significantly to fiscal and debt crises, complicating stabilization and adjustment and negatively affecting economic growth and development.

Human resources

Military-related employment is of the order of 50 million people world-wide (UN, 1989). Among them are three million scientists and engineers and some

Figure 4: *(Left) the ten leading arms importers and (right) the ten leading arms exporters, 1985–89, at 1985 prices.*

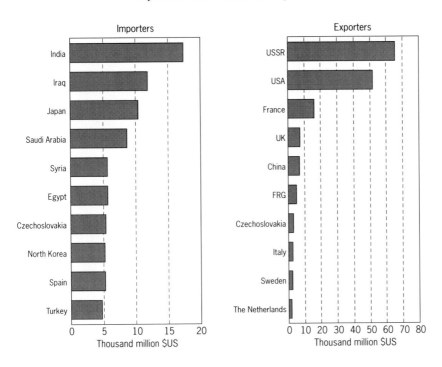

Source: based on data from SIPRI (1990).

eight to ten million production workers. Also included in this figure are the 29 million people that constitute the world's regular armed forces (11 million in developed countries and 18 million in developing ones.) In the last two decades the developed countries have remained much more militarized than the developing countries, in terms of both absolute expenditure and man-power.

Natural resources

An analysis of the utilization of natural resources by the military shows clearly that it competes for resources that might otherwise be available for social welfare and economic development. Considerable land areas are set aside for military training and weapons testing. Prime land is used in several countries, especially in developing countries, for the construction of military installations and service buildings, often without due consideration being given to their potential value for other uses more important for national development (such as agriculture, housing, tourism, etc.).

The relationship between mineral resources and the military aspects of their consumption is complex at both the national and international level. In part, this is because strategic minerals, including some key ones, are unevenly distributed. This leads simultaneously to requirements for external supplies and to possibilities for monopolistic behaviour among suppliers. It has been estimated that between 2 and 11 per cent of 14 minerals is consumed for military purposes: aluminium, chromium, copper, fluorspar, iron ore, lead, manganese, mercury, nickel, platinum group, silver, tin, tungsten and zinc (Kim, 1984). For aluminium, copper, nickel and platinum, estimated global military consumption was greater than that of all Latin American countries combined. The military demand for some metals has pushed their prices up during the last two decades. The military consumption of petroleum is about six per cent of the total world consumption, close to one-half of the entire consumption of all the developing countries.

Recent developments

During the latter part of the 1980s there were a number of profound changes in the world political situation that are bound to lead to significant changes in the military activity and expenditure trends that have been apparent for the greater part of the last two decades. A series of negotiations and agreements between the two major military power blocks on arms limitation and reduction were followed by the dissolution of the Warsaw Pact and the break-up of the Soviet Union (Box 1). While many uncertainties remain about the future disposition and control of the military forces of the former USSR, there seems every likelihood that there will eventually be major reductions in military expenditures and a general reduction in the levels of military preparedness on the part of both blocks during the coming years.

The handling and safe disposal of mass destruction weapons (chemical, biological or nuclear) is proving quite complicated and hazardous, particularly where drastic political, economic and social upheavals are experienced by the countries concerned. The implementation of the most recent agreements between the former USSR and the USA is running into difficulties in the independent states of the former USSR. It has been reported in the Russian media that many arms specialists of the former USSR are leaving their jobs and that their substitutes lack the expert knowledge for handling a large number of relatively old and unsafe nuclear warheads which need careful monitoring and upgraded safety measures.

These uncertainties are likely to continue for some time until the position is clearer. Some countries may choose to remain at a relatively high level of military preparedness. Furthermore, there is a possibility that some of the military hardware and technical expertise no longer required by the major power blocks may be diverted through arms sales - legal or otherwise - to other countries, especially to those regions where tensions remain high. Until there is a world-wide and sustained trend towards reduction in military expenditures, the environment will remain under threat.

BOX 1

Major developments in arms limitation and control.

1971   Treaty on the prohibition of the emplacement of nuclear weapons and other weapons of mass destruction on the sea-bed and on the ocean floor and in the soil thereof (Sea-Bed Treaty).

1972   First Strategic Arms Limitation Treaty (SALT I), restricting the number of ICBMs and long-range strategic bombers.

1972   Convention on the prohibition of the development, production and stockpiling of bacteriological (biological) and toxin weapons and on their destruction (BW Convention).

1977   Protocols I and II to the Geneva Convention of 1949 relating to the protection of victims of armed conflicts.

1977   Convention on the prohibition of military or any other hostile use of environmental modification techniques (Enmod Convention)

1979   Second Strategic Arms Limitation Treaty (SALT II), further restricting strategic weapons (not ratified).

1981   Convention on the prohibition or restriction on the use of certain conventional weapons which may be deemed to be excessively injurious or to have indiscriminate effects (Inhumane Weapons Convention).

1985   South Pacific nuclear free zone treaty (Treaty of Rarotonga).

1986   Agreement to regulate the exchange of information on R&D activities in biological and chemical weapons.

1987   Agreement to curtail the number of intermediate/short range nuclear forces (INF).

1989   Preliminary agreement to sign an international non-proliferation treaty on chemical weapons, destroying stocks within 10 years and approving inspections.

1990   Joint declaration on the non-proliferation of chemical weapons.

1990   Signature by 16 NATO and 6 Warsaw Pact countries of the Treaty on Conventional Armed Forces in Europe (CFE).

1991   Signature of Strategic Arms Reduction Treaty (START), reducing USA and USSR nuclear warheads and bombs and allowing on-site inspection for the first time.

Source: SIPRI (1990).

Environmental impacts of military activity

Conventional warfare

Threatening the survival of a population by destroying the ecological conditions for its existence has been used as a military strategy on many occasions. George Washington, for example, drew up such a strategy to defeat Indian tribes assisting the British during the American Revolution (Lumsden, 1975). After World War I, the technique of substituting air power for ground forces evolved. In the exercise of colonial power the British used the 'air method' to 'interrupt the normal life of tribespeople' in Western Asia. Their strategy was to deprive the offending tribes of their normal means of livelihood by forcing them to abandon their grazing grounds and water wells and to make life intolerable for them. The use of air power during World War II led to the area bombing (referred to as carpet bombing) of cities - a strategy explicitly designed to destroy the urban environment and thus bring pressure to bear on the enemy through the non-combatant population.

The Second Indochina (or Vietnam) War of 1961-75 is noted for the widespread and severe environmental damage caused by massive rural area bombing, resulting in extensive chemical and mechanical forest and crop destruction, wide-ranging anti-personnel harassment and area denial, and enormous forced population displacements. In short, this strategy represented the intentional disruption of both the natural and human ecologies of the region. As a result, the mass exodus from rural to urban areas led to the creation of vast refugee camps and urban slums in cities which lacked the economic base to support large increases in population. This in turn led to a rapid deterioration of the urban environment, and loss of adequate potable water, sanitation and public transport services.

Recent wars have demonstrated that although it is relatively easy to reconstruct the built environment, there are two types of war damage that are much more difficult to repair: damage to the natural environment and damage to the social fabric. There can be little doubt that conventional warfare in the Third World would exacerbate all the major problems already existing in these areas: environmental degradation, malnutrition, disease, unemployment, inflation and debt, rendering both physical and social reconstruction after a war extremely difficult.

There have been remarkable technological advances in conventional weapons in terms of their military efficacy. In particular, a whole array of guided weapons has emerged which can hit specified targets with a very high degree of reliability, and be very destructive while minimizing unwanted collateral damage. Such weapons systems are complex, very expensive and highly demanding in terms of training and maintenance. As the world has become more developed and industrialized, the tens of thousands of industrial installations (from nuclear and conventional power plants to refineries and chemical factories) scattered through them constitute military targets and could be destroyed during war directly or by collateral damage. Releases from such installations (radioactivity, toxic gases, etc.) could have potential local, regional and/or international environmental impacts (PRIO/UNEP, 1990).

The destructive potential of modern, high-technology conventional warfare has been recently demonstrated by the 1991 war over Kuwait. Using an updated scenario of the strategy of 'air power', 'area denial' and 'life-supporting systems denial', a significant proportion of the infrastructure of Iraq was destroyed in a period of about 30 days. The war left almost the total Iraqi population (some 18 million) without basic services, causing many deaths and a great deal of suffering through malnutrition and the spread of diseases. The war over Kuwait led to the destruction of various industrial installations releasing massive emissions into the air (especially those from the burning oil wells of Kuwait) and into the sea (the oil spills into the Gulf, during the war). Such environmental vandalism, although not new (see PRIO/UNEP, 1990, for historical examples of deliberate destruction of dams and industrial facilities), has been unparalleled in the history of conventional warfare.

Chemical weapons

The use of poisons in war dates back through recorded history, but the scientific and technological advances of the twentieth century have helped such weapons become much more lethal. Many nations now have chemical war-fighting technologies - offensive, defensive, or both - but the principal chemical arsenals are in the hands of the USA and the USSR.

The first modern chemical weapons were essentially toxic industrial by-products. These maimed or killed by blistering the skin, injuring the lungs, or disrupting cell respiration. More lethal chemical weapons known as nerve agents were developed during the 1930s. These impair the body's ability to regulate muscle action, leading to death through respiratory failure.

The introduction of chemical warfare in modern times dates from World War I, with the use by both France and Germany of tear-gas at about the same time in 1914. This was followed by the use of chlorine, phosgene and diphosgene, and mustard gas by both sides, mainly in projectiles. The Italians first used aerial spraying to distribute mustard gas in Ethiopia in 1936-37 while herbicides were first used in a defoliation programme by the British in Malaysia in the late 1940s - early 1950s (SIPRI, 1971; Holmberg, 1975). The most extensive use of herbicides in war was in the Second Indochina War. More than 72 million litres (91 million kg) of fluids containing about 55 million kilograms of active herbicidal ingredients were used to spray vast areas, especially in South Vietnam (SIPRI/UNEP, 1984a).

The major anti-plant agents that were employed in Vietnam were colour-coded 'Orange', 'White' and 'Blue'. Agents Orange and White consist of mixtures of plant-hormone-mimicking compounds which kill by interfering with the normal metabolism of treated plants. Agent Blue, on the other hand, consists of a desiccating compound, which kills by preventing a plant from retaining its moisture. About 44 million litres of Agent Orange, 20 million litres of Agent White and 8 million litres of Agent Blue were sprayed in the period from 1961 to 1971. The total area sprayed was 1.7 million hectares (1.1 million ha of dense forest, 0.4 million ha of other woody vegetation and 0.2 million ha of field crops and paddy rice). Many parts of that area were sprayed several times (SIPRI/UNEP, 1984a).

The use of these agents resulted in large-scale devastation of crops, widespread death of trees and their ultimate replacement by grasses, damage to coastal forest systems, widespread site debilitation via soil erosion and loss of nutrients in solution, decimation of terrestrial wildlife primarily via destruction of their habitat, losses in freshwater fish and a decline in offshore fisheries. The impact on the human population has included long-lasting neurotoxic effects as well as the possibility of increased incidence of hepatitis, liver cancer, chromosomal damage, spontaneous abortions and congenital malformations.

Although the Geneva Protocol on Chemical and Bacteriological Warfare was signed in 1925 and entered into force in 1928, the protocol is essentially a commitment by the signatories not to use chemical weapons first, but it does not forbid the production or stockpiling of such weapons. Efforts to improve the chemical arms control regime are ongoing.

Biological weapons

Biological warfare is the use of living organisms - generally pathogenic micro-organisms - for hostile purposes. Such biological agents, or biological weapons, can be bacterial, fungal, viral, rickettsial or protozoan. They must be alive, although they can be in spore or other dormant forms. Biological weapons are conceived as suitable for attacking personnel, livestock or crops; some may even work against non-living materials.

Known instances of biological warfare are rare. The smallpox virus is strongly suspected of having once been used as a biological weapon in the mid-eighteenth century (Westing, 1985). The only verified instance of biological warfare in modern times occurred during the Second Sino-Japanese War of 1937-45. Between 1940 and 1944 Japan attacked some 11 Chinese cities with various strains of bacteria. It was reported that at least 700 Chinese died from plague alone. A number of claims have been made by one nation or another that they have been attacked with biological weapons, but so far no concrete evidence has been found to support such allegations.

Although the armed forces of a nation attacked with biological weapons might have at least some level of prophylactic or, more likely, therapeutic protection from an anti-personnel agent, the civilian population almost certainly would not; infants, the aged and ill people would be most vulnerable. Virulent human disease organisms introduced into a region could cause an epidemic that would linger and spread, perhaps even to neighbouring countries and beyond. Some potential anti-personnel agents might also damage livestock and other animals; conversely, some potential anti-animal agents might harm humans. Attacks with anti-plant agents could lead to widespread civilian food shortages and starvation as well as to soil erosion and other forms of land degradation.

Advances in biotechnology and genetic engineering could add new dimensions to biological warfare. The effectiveness of existing biological agents could be enhanced and new ones could be created. Agent pathogenicity could be manipulated, including infectivity or virulence. An agent could be made more or less resistant to prophylactic or therapeutic treatments; its resistance to adverse light, temperature, moisture, and other conditions could be enhanced, thereby enabling it to be disseminated more readily; and techniques for mass production could be improved.

The use in war of biological weapons is forbidden by the Geneva Protocol of 1925. The Bacteriological and Toxin Weapon Convention of 1972 forbids the very possession of biological weapons and associated hardware. The deliberate use of harmful micro-organisms to modify the biotic composition and dynamics of ecosystems would also be prohibited under the Environmental Modification Convention of 1977 (see below). The major weaknesses of these conventions are, on the one hand, the limited number of governments that ratify them (two members of the Security Council are not parties to the 1977 Convention), and the absence of concrete provisions for verification of implementation. Adherence is expected to be maintained by shared attitudes, mutual self-interest, and military disutility.

Nuclear war

The introduction of nuclear weapons added an entirely new dimension to warfare. Quantitatively it has brought an enormous increase in explosive power over that of the most modern conventional weapons. The two bombs dropped on Hiroshima and Nagasaki in 1945 had a yield of 12.5 kiloton TNT and 22 kiloton TNT respectively, and their devastating effects are well documented. Nuclear weapons developed later represent a dramatic increase in destructive power (from kilotons to megatons). It is estimated that the number of nuclear warheads in the world stands between 37,000 and 50,000, with a total explosive power of between 11,000 and 20,000 megatons (equivalent to between 846,000 and 1,540,000 Hiroshima bombs). Such power of destruction has prompted many calls for the total elimination of nuclear weapons (e.g. the Stockholm Conference of 1972: Box 2).

Despite widespread condemnation of nuclear weapons, their production and testing continue. The total number of known nuclear tests from 1945 to 1990 was 1,818, of which 489 were in the atmosphere and 1,329 were underground (Figure 5). The Partial Test Ban Treaty signed in 1963 has banned nuclear weapon tests in the atmosphere, in outer space and under water. It has helped

---

BOX 2

'Man and his environment must be spared the effects of nuclear weapons and all other means of mass destruction. States must strive to reach prompt agreement, in the relevant international organs, on the elimination and complete destruction of such weapons.'

Principle 26, Declaration of the UN Conference on the Human Environment, Stockholm, 1972.

---

to curb the radioactive pollution by nuclear explosions. However, some countries have not so far become parties to the Treaty, and since 1963, 64 atmospheric tests have been carried out (41 by France and 23 by China). Underground testing of nuclear weapons has continued, with 870 such tests in the period from 1970 to 1990 (SIPRI, 1990).

In the 1980s, several studies were carried out to predict the impacts of a large-scale nuclear war (Ehrlich *et al.*, 1983; Turco *et al.*, 1983, 1984; Ehrlich, 1984; Grover, 1984; Covey *et al.*, 1984; UN, 1985a; NRC, 1985; Svirezhev, 1985; SCOPE, 1985, 1986; Dotto, 1986 and Peterson, 1986). In spite of several uncertainties, different scenarios of nuclear war (involving exchanges of 5,000 to 10,000 megaton yield) lead to an estimated casualty rate of 30 to 50 per cent of the total human population, the vast majority of whom would be in the Northern Hemisphere. Despite this devastation, perhaps 50 to 70 per cent of the human population in both the northern and southern hemispheres might survive the direct effects of a large-scale nuclear war. But they would likely be affected by what has become known as the 'nuclear winter' (see Chapter 20).

Figure 5: *Numbers of atmospheric and underground nuclear explosions, 1945–90. (The Test Ban Treaty was signed on 5 August 1963.)*

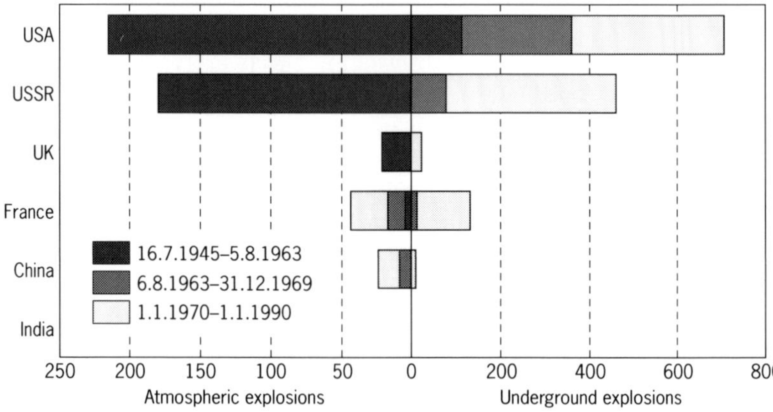

Source: based on data from SIPRI (1990).

In the aftermath of a large nuclear war, darkened skies might cover large areas of the Earth, perhaps for several weeks or even months as large, thick clouds of smoke from widespread fires blocked sunlight (SCOPE, 1985, 1986). The impact would be greatest over the continents of the Northern Hemisphere, where most of the smoke would probably be produced by nuclear ignited fires and where the average temperatures in some areas might drop some tens of degrees Celsius to below freezing for several weeks to months after the war. Climatic disturbances might persist for several years, even in countries not directly involved in the war. Rainfall in many regions of the world might be greatly reduced. Temperature and precipitation changes could also occur in the tropics and the Southern Hemisphere - less extreme than those in the Northern Hemisphere but still significant. Tropical and subtropical regions could experience unprecedented cooling and severe cold spells, accompanied by significant disturbances in precipitation patterns.

According to the SCOPE study, world agriculture and major terrestrial and marine ecosystems could be severely disturbed and their plant and animal populations stressed by rapid, dramatic changes in the normal climatic regime. Crop losses, caused not only by climate disturbances but also by the post-war disruption in supplies of essential inputs such as energy, machinery, fertilizers and pesticides, could create widespread food crises in both combatant and non-combatant nations. The failure of major food production and distribution systems, and the inability of natural ecosystems to support large numbers of people, could reduce the human population of the Earth even further than the war itself.

In addition to the potential climatic effects, a large-scale exchange of nuclear weapons would cause considerable devastation from the direct effects of fire, blast and local fall-out of radioactivity. Other impacts could include severe disruption of communications and power systems; further damage to the ozone layer in the upper atmosphere which protects life on Earth from the sun's biologically damaging ultraviolet radiation (see Chapter 2); intense local and long-term global radio-active fall-out; and severe regional episodes of air and water pollution caused by the release of large amounts of toxic chemicals and gases.

Environmental warfare

In the 1970s there was speculation about the possibility of environmental modification being used to cause economic or other damage to the population of an enemy (Box 3) (Goldblat, 1975; Barnaby, 1976; SIPRI, 1977 SIPRI/UNEP, 1984b). Environmental warfare could, at least in principle, involve damage caused by manipulations of celestial bodies (Sullivan, 1983) or space, the atmosphere, the land, the oceans or the biota. With respect to hostile manipulations of the upper atmosphere, it has been suggested that techniques might be developed in the future which would make it possible to alter the electrical properties of the ionosphere in such a way as to disrupt enemy communications.

BOX 3

'Nature shall be secured against degradation caused by warfare or other hostile activities.'

Article V, World Charter for Nature, UN General Assembly, New York, 1982.

Successful manipulation of the land for hostile purposes would depend for the most part on the ability to recognize and take advantage of local Earth instabilities or available pent-up energies. Landslides can in some instances be triggered in mountainous areas. The activation of volcanoes that are on the verge of eruption is another possibility. The destruction of dams and natural water impoundments for hostile action has been practised in some wars (SIPRI/UNEP, 1984b).

The biosphere can be manipulated for hostile purposes by applying chemicals, by introducing exotic living organisms, or by mechanical or incendiary means. Forests can be devastated for hostile purposes over huge areas by spraying them with herbicides, as was demonstrated during the Second Indochina War. At certain times and places self-propagating wildfires could be initiated which could damage or obliterate large tracts of forest as well as prairie grassland. Marine ecosystems could be locally disrupted with hostile intent by destroying offshore oil wells, near-shore oil loading facilities, or loaded oil tankers. Recovery from such disruption would be likely to take several years. The oil spill created as a result of the war over Kuwait in 1991 is an example. Figure 6 shows the scale of this spill in comparison with other major oil spills at sea.

The Convention on the Prohibition of Military or any other Hostile Use of Environmental Modification Techniques (the Environmental Modification Convention) was signed in 1977 and entered into force in 1978. As of 1 January 1990, 55 countries have become Parties to the Convention. The States Party undertake not to engage in military or other hostile uses of environmental modification techniques having widespread, long-lasting or severe effects as the means of destruction, damage or injury to any other State Party (SIPRI/UNEP, 1988).

Militarization of outer space

Outer space is becoming more and more militarized. Earth-orbiting satellites are used by the military to enhance the performance of Earth-based armed forces and weapons. Between 1970 and 1990 about 2,800 satellites were launched (an average of 140 per year), 75 per cent of which served military purposes (military communications, photo-reconnaissance, electronic intelligence, early warning, navigation, geodetic purposes, meteorology, etc). According to SIPRI (1990), on 31 December 1989 there were 227 operational military satellites

Figure 6: *The oil spill created as a result of the war over Kuwait, compared with other major oil spills at sea.*

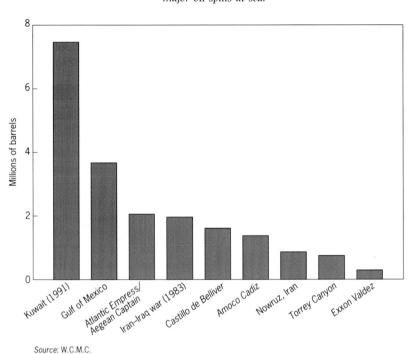

Source: W.C.M.C.

in orbit (USA, 118; USSR, 100; China, 4; France, 2; UK, 2 and Japan, 1). In addition, the USA and the USSR have been actively pursuing military programmes in space (e.g. the US Strategic Defense Initiative (the SDI, or 'Star Wars' programme), the anti-satellite weapons programmes (ASATs), etc).

Remnants of war

The hazards of war do not end with the coming of peace. The remnants of war constitute a variety of problems in areas where military operations took place. The word 'remnants' here refers to a variety of relics, residuals or devices not used or left behind at the cessation of active hostilities. They include non-exploding devices; unexploded land mines, sea mines and booby traps; unexploded munitions; materials such as barbed wire and sharp metal fragments; wreckage of tanks, vehicles and other military equipment; and sunken ships and downed aircraft (Box 4).

The assessment of the magnitude of the remnants of war is difficult, especially in relation to unexploded mines. Often there is no exact information on the location of emplaced mines. Sometimes they have been delivered by means of artillery or aircraft, in which case the ability to record their location is very limited, if not impossible. Unavailability of accurate maps, or specific geographical

---

BOX 4

Remnants of war.

- In Poland, 14,894,000 land mines and 73,563,000 bombs, shells and grenades have been recovered since 1945.

- In Finland, over 6,000 bombs, 805,000 shells, 66,000 mines and 370,000 other high-explosive munitions have been cleared since the end of World War II.

- In Indochina, about 2 million bombs, 23 million artillery shells and tens of millions of other high-explosive munitions were left unexploded after the war.

- In Egypt, following the 1973 Arab-Israel War, about 8,500 unexploded items were removed from the Suez Canal and more than 700,000 land mines were cleared from the terrain near the Canal. Yet hundreds of thousands of land mines and unexploded shells are still scattered around the Gulf of Suez and in Sinai.

Source: SIPRI/UNEP (1985).

---

and meteorological circumstances, may also limit minefield records. The information on unexploded dud munitions is extremely vague; normally there are no records and their location and magnitude can be estimated only very roughly.

Material remnants of war can affect ecological balances by disturbing the soil, destroying vegetation, killing fauna and introducing poisonous substances into the environment (SIPRI/UNEP, 1985). The economic implications of the material remnants of war are no less serious than the environmental ones. While some of the material is useful and can be exploited for non-military use, often by the poor, this is not without risk. Unexploded remnants of war endanger people, livestock and wildlife; impede the development of an economic infrastructure (roads, power and telephone lines, etc); make land unsafe to farm; and hamper the development of tourism and the discovery and development of mineral resources. Unexploded remnants of war at sea or in rivers interfere with navigation and fishing.

Areas that have been mined or otherwise contain unexploded remnants of war become unavailable for economic development or other social pursuits. The alternative to the abandonment of such lands, often large areas, is to undertake clearing activities, which include localization, identification, and neutralization of remnants of war. This overall process requires highly specialized personnel and equipment. It is a time-consuming process and extremely risky, as is shown by the casualties registered during clearing operations (SIPRI/UNEP, 1985). Similar problems can arise with peacetime military training areas.

Refugees

Wars and lesser armed conflicts have generated millions of refugees in the world. Their exact number is not known, partly because of the lack of an internationally-accepted definition of who is a refugee and who is not (El-Hinnawi, 1985). While the official estimate of the United Nations High Commissioner for Refugees is about 14 million, other estimates range as high as 27 million.

These refugees have not only suffered economic losses, but their whole social fabric has been disrupted. In most cases refugees live in camps in border areas where living conditions are harsh and adequate infrastructure is lacking. Malnutrition, infectious diseases and social disruptions are common. In some cases the repatriation of refugees becomes virtually impossible, and they continue to live in misery for decades. In spite of various international efforts to alleviate the problem (in particular by several non-governmental organizations), the number of refugees continues to increase with each increase in international tension and military activity.

Responses

The impact of military expenditure on development

> ### BOX 5
>
> 'The building-up of arms in large parts of the Third World itself causes growing instability and undermines development'.
>
> 'MORE ARMS DO NOT MAKE MANKIND SAFER, ONLY POORER'
>
> Independent Commission on International Development Issues,
>
> The Brandt Commission, 1980.

In the last two decades, the course of the world economy has been erratic, and it remains turbulent and uncertain in the early 1990s. After a sharp recession at the beginning of the 1980s there was a gradual recovery, but in both the developing and developed countries economic growth was much slower than in the 1970s. At the beginning of the 1990s, the world was again in widespread recession.

Despite the implementation of austerity measures in some cases and determined efforts to achieve economic growth, many developing countries seem to be falling even further behind. Money cannot be found to support new development projects because any current account surpluses are devoted to debt servicing and in many cases to military expenditures.

The social consequences of these austerity plans and the lack of economic progress has given rise to widespread resentment, manifested in protests and rioting in a number of countries. Some governments have fallen in the face of it, while others have been elected by promising to control the foreign debt burden. Yet, it is in the developing countries, where these problems are most acute, that military expenditure, including expenditure on arms imports, is growing most rapidly.

Several studies of the relationship between disarmament and development (see for example, UN, 1982, 1983, 1985b, 1989) have stressed the fact that the arms race and development compete for the world's finite resources. Furthermore, the arms race constitutes a major threat to international security; by hindering development - through the diversion of limited resources - it is an important source of national and regional insecurity.

The negative consequences of the arms race are succinctly stated in the Final Document of the 1987 International Conference on the Relationship between Disarmament and Development (Mårtenson, et al., 1987):

'The continuing arms race is absorbing far too great a proportion of the world's human, financial, natural and technological resources, placing a heavy burden on the economies of all countries and affecting the international flow of trade, finance and technology, in addition to hindering the process of confidence-building among States. The global military expenditures are in dramatic contrast to economic and social underdevelopment and to the misery and poverty afflicting more than two-thirds of mankind. Thus, there is a commonality of interests in seeking security at lower levels of armaments and finding ways of reducing these expenditures.'

There is no doubt that disarmament would have positive economic and social consequences as it would release added resources for civilian uses (the so-called 'peace dividend'). The resources released could be used immediately in the efforts to eradicate hunger and poverty, eliminate illiteracy and protect children from the double dangers of physical violence and underdevelopment. However, it is obvious that such a 'peace dividend' will not be available soon.

The major political changes in Eastern and Central Europe and the former USSR do offer a great opportunity for realizing this 'peace dividend'. It has been calculated that the range of actions proposed for stabilizing global population, reducing deforestation, conserving biological diversity, starting a reafforestation programme, conserving energy, protecting topsoil on croplands and reducing Third World debt to manageable proportions could be funded with an immediate five per cent reduction in present levels of military expenditure (about $US900 billion per annum) rising to an 18 per cent reduction in the year 2000. Even then, military expenditure would remain as high as $US739 billion per annum (IUCN/UNEP/WWF, 1991).

Indications are that disarmament has become a reality in the last few years. Weapons of various sorts are being destroyed in compliance with treaties and inspection is permitted. Furthermore, the world has recently witnessed unilateral steps that go beyond the obligations of treaties. Yet, it should be remembered that not all weapons are destroyed. Some of the remaining nuclear weapons are being stored and could be redeployed. Recently, there have been reports of attempts to market nuclear materials, technology and equipment, as well as to recruit scientists and engineers, particularly in Third World countries (PRIO, 1991).

Evolving concepts of security

Since World War II, the concept of national security has acquired an overwhelmingly military character, rooted in the assumption that the principal threat to security comes from other nations, arising from various ethnic, economic or resource-use disagreements. Commonly veiled in secrecy, considerations of military threats have become so dominant that new threats to the security of nations, threats with which military forces cannot cope, are being ignored.

During the last two decades it has become increasingly clear that military means alone are no longer adequate to provide tangible security benefits. It has been realized that the security of nations depends at least as much on economic well-being, social justice and ecological stability. Environmental problems - climate change, ozone depletion, transboundary air pollution, land degradation, deforestation and soil erosion - threaten to destroy the human habitat and undermine economies everywhere. Pursuing military security at the cost of social, economic and environmental well-being is akin to dismantling a house to salvage materials in order to erect a fence around it (Renner, 1989).

The world's increasing economic interdependence and the fragility of the biosphere make challenges to national security more and more complex. They also make it more urgent that these challenges be dealt with effectively. This fact of modern society calls not only for novel solutions to the economic and ecological problems themselves, but also for new ways of defining national security. Neither in developing nor in industrialized countries can security be dissociated from the economic and social reality, both internal and external. In other words, economic and politico-military aspects of security are intertwined in all groups of States.

This thinking led to the evolution of new concepts of security over the last two decades. Examples of such concepts are the 'balance of power', 'deterrence', 'peaceful coexistence' 'collective security', 'common security', etc. (UN, 1986). The all-embracing concept of common security, advanced by the Independent Commission on Disarmament and Security Issues (Palme Commission) in 1982, is based on the belief that genuine security can be achieved only on the basis of co-operation and co-ordination among all States, including those considered adversaries. There is a distinct and important difference between the concept of 'common security' and the traditional doctrine of 'collective security'. Economic, social and ecological vulnerability constitute a challenge to achieve common security, which extends in scope to include non-military fields.

The complex relationship between disarmament, development and security has been explored in the Final Document of the 1987 International Conference on the Relationship between Disarmament and Development (Mårtenson, et al., 1987). The argument was summed up as follows:

> 'Security is an overriding priority for all nations. It is also fundamental for both disarmament and development. Security consists of not only military, but also political, economic, social, humanitarian, human rights and ecological aspects. Enhanced security can, on the one hand, create conditions conducive to disarmament and, on the other, provide the environment and confidence for the successful pursuit of development. The development process, by overcoming non-military threats to security and contributing to a more stable and sustainable international system, can enhance security and thereby promote arms reduction and disarmament. Disarmament would enhance security both directly and indirectly. A process of disarmament that provides for undiminished security at progressively lower levels of armaments could allow additional resources to be devoted to addressing non-military challenges to security, and thus result in enhanced overall security.'

This broader and all-embracing concept of security, involves two major interlocking elements: the societal and the environmental. In its turn, societal security comprises political security (freedom of speech, participatory democracy, an independent judiciary, etc.); military security (on a purely defensive and non-provocative stance); economic security (good quality of life for all citizens); and personal security (equity, equality, respect, safeguarding of life and property) (Westing, 1989a, 1989b).

Environmental security is an inseparable component of comprehensive international security, the upholding of which is a shared responsibility of the entire international community (PRIO/UNEP, 1989). Environmental stress is both a cause and an effect of political tension and military conflict. Nations have often fought to assert or resist control over raw materials, energy supplies, land, river basins, sea passages and other key environmental resources (see SIPRI/UNEP, 1986 for a detailed overview on global resources and international conflicts, and Homer-Dixon, 1991 for an analysis of the relationship between environmental change and acute conflict).

Such conflicts are likely to increase as these resources become scarcer and competition for them becomes more acute. This scarcity is not so much natural as man-made; it is a consequence of unsustainable forms of development. According to the World Commission on Environment and Development (WCED, 1987) unsustainable development pushes individual countries up against environmental limits; thus major differences in environmental endowments or variations in usable land and raw materials may precipitate and exacerbate international tensions and conflicts.

Environmentally-based conflicts are already happening. In developing countries they may have a territorial aspect in that the direct dependence of people on land and water, and their harvests, have pushed national or sub-national collectives into rivalries with one another. In many civil wars and border wars, resource scarcities have been a catalyst of hostilities. Industrialized countries do not generally have similar mutual tensions over resources to the same degree. Yet the resource factor can often be discerned as a motivation for their external interventions (direct and indirect) and even their military strategies. The security of oil supplies was without doubt one of the factors - if not the main one - in the recent war over Kuwait. However, such interventions might jeopardize, rather than secure, oil supplies. Stability of supplies of raw materials seems more likely to be assured by the development of satisfactory long-term relations and the building of mutual confidence rather than military coercion and intervention (UN, 1989).

Environmental degradation has historically been a cause of tension and armed conflict. In times of drought, when the land became degraded, or when the water wells dried up, nomads moved to other areas to gain access to more fertile land and to water, to grow their subsistence crops and sustain their herds. Sometimes tribes had to resort to armed conflict to gain access to such resources. In the last two decades, the movement of millions of African environmental refugees has created both internal and interstate tensions (El-Hinnawi, 1985). The consequences of global climate change (particularly sea-level rise) are likely to greatly increase the number of environmental refugees in the second half of the next century.

Border-transcending environmental degradation most immediately affects neighbouring countries, as illustrated by disputes over water resources. At least 214 river basins are multinational. Before the creation of the Commonwealth of Independent States, 155 of these were shared between two countries; 36 among three countries; and the remaining 23 among four to twelve countries. About 50 countries have 75 per cent or more of their total area falling within international river basins, and an estimated 35 to 40 per cent of the world's population lives in these basins (SIPRI/UNEP, 1986). Disputes have revolved around water diversion or reduced water flow, industrial pollution, the salinization or siltation of streams, and floods aggravated by soil erosion.

The predicament of the Least Developed Countries merits special attention (UN, 1991). After World War II, we have witnessed more growth and more inequalities, both at the national level and between regions and sub-regions. It is more likely that in the changing world scene, with the opening up towards the East and the emergence of large regional economic blocs, inequalities will increase. The last two decades have provided abundant evidence that inequalities in access to natural resources or to an adequate environment lead to regional insecurities that occur first in or around the least developed countries.

It must be admitted, however, that the various attempts at analysing the relationship between environment and armed conflict (e.g. Robinson, 1979; Galtung, 1982) - valuable as they are - still need systematic empirical studies, particularly in clarifying the indirect effects and the causal chains involved. Work on these conceptual issues is currently in progress in peace research institutes and UN organizations.

It is abundantly clear that many nations, regions and the world as a whole are faced with serious environmental problems, many of which are becoming increasingly severe. Actions that come to grips with these problems will serve to reduce the threat to national and international security in the expanded sense. Concerted international actions are often called for because many of the problems are not confined to single countries, or because the problems have progressed too far or are too complex to be dealt with by the intellectual or material resources of any one nation. Exclusive national policies are ill-suited for a world that faces border-transcending environmental destruction on an unprecedented scale. Environmental security depends critically on international co-ordination and co-operation.

International action and future outlook

There are some obvious contradictions in the attitude of the world community to the whole question of military activity (Box 6). On the one hand, the numerous conventions, treaties and agreements provide clear evidence of a widespread desire to prevent the more devastating forms of warfare. On the other hand, the evidence of continuing military expenditure around the world implies a lack of conviction in the practicability of disarmament or even of holding forces and arsenals at constant size. And there are further conflicts between the demands for agricultural, social and economic development so vital to the future of the world, in particular to the developing countries, and the increasing allocation of limited resources for military purposes.

BOX 6

Contradictions

* The United Nations Environment Programme, the organization responsible for safeguarding the global environment, spent over the last *ten years* $US450 million; i.e. less than *five hours* of global military spending.

* The total annual Official Development Assistance (ODA) extended to the developing countries is $US35 billion; i.e. 15 days of global military spending.

Clearly the questions of disarmament, development and environmental security are closely linked and represent some of the most important issues before the international community today. Development can hardly proceed at the required pace and a healthy environment cannot be guaranteed amid a widening and constantly escalating arms race. Moreover, development and environmental efforts are threatened by the armaments, especially nuclear weapons, already stockpiled, the use of which either by intent or in error or through sheer madness would severely jeopardize humankind's very existence.

One of the most urgent tasks, therefore, is to arrest the technological development spiral at the centre of the international arms race and, through substantial and substantive disarmament measures, pave the way for major reductions in world military expenditure. A major breakthrough in the disarmament field would release - both in the medium term and in the long run - vast financial, technological and human resources for more productive uses in both developed and developing countries in an international political climate of reduced tension.

The key to future spending cuts will be the evolving situation in Europe and the Middle East. The legacy of confrontation in Europe absorbed $US600 billion of the global military spending of $US950 billion in 1989. It has been suggested that the industrialized countries could reduce their spending by two to four per cent a year during the 1990s if the present understandings between the Eastern and Western powers come to fruition and lasting peace could be achieved in the Middle East. This would translate into savings of $US200-300 billion a year by 2000.

Not all these savings would release funds for other areas however. Substantial costs will be incurred in dismantling the vast military facilities and in disposing of some of the more dangerous weapons systems, especially chemical, biological and nuclear weapons. There will be significant costs for retraining military personnel and probably some unemployment benefits, while the defence industries would have considerable capital investment to write off (UNDP, 1991). Nevertheless, after these costs are met there would still be huge savings available for use for other purposes. The most immediate prospect is that the 'peace dividend' would be used to balance national budgets and reduce or prevent deficit spending (UNDP, 1991). However, determined efforts should be made to ensure that part of the dividend is used to address some of the major social and environmental problems outlined in this volume.

In order to re-channel resources from the military to the civilian economy and to provide sufficient resources to reverse environmental degradation, a planned conversion process must be set in motion in parallel with a disarmament process. Yet conversion has political, economic and technical dimensions (UN, 1989). Its implementation depends on the political will of states and their readiness to take concrete measures on arms reductions and disarmament. Unilateral measures to curtail the military burden, and hence to initiate a conversion process, can be taken by any state, but in the real global political sense, disarmament has to be started by the major powers on the basis of verifiable agreements to reduce armaments and eliminate particular military capabilities. Progress in this context would not only lead to conversion of the economies of those powers from a military to a civilian nature, but may also encourage medium and small states to start reducing and converting their own military capabilities.

Conversion is more than just a theory. Following World War II, some 30 per cent of the United States GNP was transferred from the war industry into civilian uses. Today China stands as an example. In 1985, the country decided to utilize part of its military industrial capacity to manufacture civilian goods. Civilian production now accounts for 20 per cent of the output of China's military factories, and that share is projected to reach 50 per cent by the year 2000 (Renner, 1989, 1990). Conversion produces more jobs and helps to meet the growing socio-economic needs of people, and is of vital importance for environmental protection. In the United States, for example, spending $US1 billion on guided missile production creates about 9,000 jobs. Spending the same amount on educational services creates 63,000 jobs, and on air, water and solid waste pollution control, 16,500 jobs. A $40 billion conversion programme could bring a new gain of more than 650,000 jobs (Renner, 1990) (Box 7).

BOX 7

Trade-offs between military and social and environmental priorities.

- 6-7 hours of global military spending ($US700 million)
Eradication of malaria - the killer disease that claims the lives of one million children every year.

- 5 days of global military spending ($US11.5 billion)
Annual cost of implementation of the UN Action Plan to Combat Desertification over 20 years.

- 3 days of global military spending ($US7 billion)
Funding of Tropical Forest Action Plan over 5 years.

- 1 Apache helicopter ($US12 million)
Installation of 80,000 hand pumps to give Third World villages access to safe water.

- 1 Patriot missile system ($US123 million, without missiles)
Establishment of 5,000 low-cost housing units to free 5,000 families from life in slums.

- 1 day of the 1991 war over Kuwait ($US1.5 billion)
Global 5-year child immunization programme against six deadly diseases, preventing thereby the annual death of one million children.

The lost economic opportunities that military spending entails are echoed in investment trends, productivity and inflation. Undoubtedly, society would benefit if the resources now absorbed by defence were used in the civilian realm. Conversion is an economic opportunity for the future. And it is one of the most important prerequisites for achieving environmentally-sound and sustainable development.

BOX 8

Signs of hope.

*The Conference on Security and Co-operation in Europe (CSCE) is an example of a regional security undertaking that covers not only political and military, but also economic, environmental and humanitarian issues.

*The EMINWA programme (Environmentally Sound Management of Inland Waters) launched by UNEP in 1986 is designed to assist governments to integrate environmental considerations into the management and development of inland water resources, with a view to reconciling conflicting interests and ensuring the regional development of water resources in harmony with the water-related environment throughout entire water systems.

REFERENCES

Barnaby, F. (1976) Towards environmental warfare. *New Scientist*, **69**, 6.

Brzoska, M. and Ohlson, T. (1989) *Arms transfers to the third world, 1971-1985.* SIPRI, Stockolm and Oxford University Press, Oxford.

Covey, C. *et al.* (1984) Global atmospheric effects of massive smoke injections from a nuclear war. *Nature*, **308**, 21.

Dotto, L. (1986) *Planet Earth in Jeopardy.* J. Wiley, Chichester.

Ehrlich, P.R. *et al.* (1983) Long-term biological consequences of nuclear war. *Science*, **222**, 1293.

Ehrlich, A. (1984) Nuclear Winter. *Bull. Atomic Scientists*, April 1984, **40(4)** 1S-15S.

El-Hinnawi, E. (1985) *Environmental Refugees.* United Nations Environment Programme, Nairobi.

Galtung, J. (1982) *Environment, Development and Military Activity*, Universitetsforlaget, Oslo.

Goldblat, J. (1975) The prohibition of environmental warfare. *Ambio*, **4**, 187.

Grover, H.D. (1984) The climatic and biological consequences of nuclear war. *Environment*, **26**,(4),6.

Holmberg, B. (1975) Biological aspects of chemical and biological weapons. *Ambio*, **4**, 211.

Homer-Dixon, T.F. (1992) On the threshold - Environmental changes as causes of acute conflict. *International Security*, **16**,(2),76

IUCN/UNEP/WWF (1991) *Caring for the Earth. A Strategy for Sustainable Living.* IUCN, Gland.

Kim, S. S. (1984) *The Quest for a Just World Order.* Westview Press, Boulder, Colorado.

Lumsden, M. (1975) 'Conventional' war and human ecology. *Ambio*, **4**, 223.

Mårtenson, J. *et al.* (1987) Report of the international conference on the relationship between disarmament and development, UN General Assembly, Publication No. A/CONF.130/39, UN, New York.

NRC (1985) *The effects on the atmosphere of a major nuclear exchange.* National Research Council, National Academy Press, Washington, D.C.

Peterson, T. (1986) Scientific studies of the unthinkable - the physical and biological effects of nuclear war. *Ambio*, **15**, 60.

PRIO (1991) Communication from the International Peace Research Institute, Oslo, dated 30 October, 1991.

PRIO/UNEP (1989) *Environmental Security*, International Peace Research Institute, Oslo.

PRIO/UNEP (1990) *Environmental Hazards of War.* Sage Publication, London.

Robinson, J.P. (1979) *The Effects of weapons on ecosystems*, UNEP/Pergamon Press, Oxford.

Renner, M. (1989) Enhancing global security. *In*: Brown, L. (ed.) *State of the World - 1989*, W.W. Norton & Co., New York.

Renner, M. (1990) Swords into plowshares: converting to a peace economy. *Worldwatch Paper*, 96., Worldwatch Institute, Washington, DC

SCOPE (1985) Environmental consequences of nuclear war. Vol. II Ecological and agricultural effects. *SCOPE Report*, **No 28.** J. Wiley, Chichester.

SCOPE (1986): Environmental consequences of nuclear war. Vol.I Physical and atmospheric effects. SCOPE Report, **No. 28**. J. Wiley, Chichester.

SIPRI (1971) *The problem of chemical and biological warfare. Vol. 1. The rise of CB weapons.* Armqvist and Wiksell, Stockolm and Humanities Press, New York.

SIPRI (1977) *Weapons of Mass Destruction and Environment.* Taylor and Francis, London.

SIPRI (1986) *World Armaments and Disarmament*. SIPRI Yearbook 1986, Oxford University Press, Oxford.

SIPRI (1988) *World Armaments and Disarmament*. SIPRI Yearbook 1988, Oxford University Press.

SIPRI (1990) *World Armaments and Disarmament*. SIPRI Yearbook 1990, Oxford University Press. Oxford.

SIPRI/UNEP (1984a) *Herbicides in War*. Taylor and Francis, London.

SIPRI/UNEP (1984b) *Environmental Warfare*. Taylor and Francis, London.

SIPRI/UNEP (1985) *Explosive Remnants of War*. Taylor and Francis, London,

SIPRI/UNEP (1986) *Global Resources and International Conflict*. Oxford University Press, Oxford.

SIPRI/UNEP (1988) *Cultural Norms, War and the Environment*. Oxford University Press, Oxford.

Sullivan, W. (1983) *In*: SIPRI/UNEP *Environmental Warfare*. Taylor and Francis, London.

Svirezhev, Y.M. (1985) *Ecological and demographic consequences of nuclear war*. USSR Acad. Sci. Computer Centre, Moscow.

Turco, R.P. *et al.* (1983) Nuclear Winter. *Science*, **222**, 1283.

Turco, R.P. *et al.* (1984) *The climatic effects of nuclear war*. Scientific American, **251 (2)**, 33.

UN (1982) *The relationship between disarmament and development*. E82.IX., United Nations, New York.

UN (1983) *Economic and social consequences of the arms race and of military expenditures*. E83.IX.2., United Nations, New York.

UN (1985a) *Climatic effects of nuclear war, including nuclear winter*. A/40/440, Report of Sec. General, United Nations, New York.

UN (1985b) *Study on conventional disarmament*. E85.IX.1., United Nations, New York.

UN (1986) Concepts of security. *Disarmament Study Series*, 14, United Nations, New York.

UN (1989) Study on the economic and social consequences of the arms race and military expenditures. *Disarmament Study Series*, 19, United Nations, New York.

UN (1991) Disarmament, Environment and Development and their relevance to the Least Developed countries, a joint UNIDIR/ UNEP project, UN GV.E.91.0.19, New York.

UNDP (1991) *Human Development Report 1991*. Oxford University Press, Oxford.

WCED (1987) *Our Common Future*. Oxford University Press, Oxford.

Westing, A.H. (1985) The threat of biological warfare. *Bioscience*, **35**, 627.

Westing, A.H. (1989a) Comprehensive human security and ecological realities. *Environmental Conservation*, Geneva, **16**:295.

Westing, A.H. (1989b) Environmental component of comprehensive security. *Bulletin of Peace Proposals*, Oslo, **20**:129-134

World Bank (1988) *World Development Report*. Oxford University Press, Oxford.

World Bank (1990): World Development Report. Oxford University Press, Oxford.

PART THREE
THE RESPONSE

An introductory overview

The world has been changed in many ways as a consequence of human actions during the past two decades. Although it has become increasingly evident that the world community is at risk from some of the changes it has caused, many of the pressures continue. As the earlier chapters have shown, pollution and the misuse of land and water continue to threaten the sustainability of development and to undermine national economies. There have been responses, in the form of international conventions and agreements, national laws, world environment reports, a mass of scientific and popular literature, and a swelling tide of public concern. Environmental politics, even though it began a century ago, is essentially a phenomenon of the second half of the twentieth century. 'Green consumerism' is a recent product of concern in Europe and North America.

But for all this, the rate of response lags behind the rate of change, and the risk of collision between humanity and the world environment is mounting. The risk varies in nature and severity from region to region, and is closely related to the state of development. In the developed countries, where human numbers are more or less stable, the chief problems are the over-consumption of resources and the emission of far more pollutants - especially greenhouse gases - than the biosphere can safely absorb. For these countries the challenge is to reduce resource consumption and waste while maintaining standards of living and increasing the quality of life.

In the developing countries, millions still lack anything approaching a tolerable quality of life. The 'pollution of poverty' is pervasive. Population growth remains rapid, and imposes added burdens on the already over-stretched health care, education and other infrastructures of those countries. The debt burden and adverse terms of trade add yet greater pressures: the flow of resources from the developing world to the developed world, estimated at ~US35 billion in 1989, erodes essential development.

The future demands a new vision and a new approach. Both have to be established at all levels, from the individual to the global. The central message is one of responsibility – of one individual to another, of community to community, of state to state, and of today's generation to those that will come afterwards. While individual circumstances and priorities differ, the ethic of care for people and the Earth has to be universal.

In this section of the book, four key elements in the human response to environmental problems are considered:

- the evolution of public perceptions and attitudes;

- the development of scientific, economic and regulatory tools for environmental understanding and management;

- the strengthening of national policies and institutions, and

- the development of international policies, institutions and actions.

These four chapters draw on all that has gone before, and prepare the way for the final chapter which examines the challenges ahead and suggests priorities for action.

The approach to the future must be designed in hope, but it must also be built on realism. It is necessary to learn from failures as well as from successes. Managing the environment is done by managing people - or rather by persuading and leading people. In the end, what people do depends on what they believe. An ethic or a religious code is often more powerful than a government edict. While governments have a duty to lead, they succeed only when they lead a majority of people who accept the basic concepts and share the objectives of the leadership.

That is why we have to start with individuals: with what makes them choose the actions they take. We must see the environment as they do. This is why knowledge and understanding are so crucial. The first two chapters in this section set the scene and review how people have changed their outlook in response to the new knowledge that has become available to them during the past two decades. The two following chapters review the rapid growth in national, regional and global institutions that has been such a dramatic feature of the past 50 years, and of the last 20 in particular. Taken together, these chapters show how action is determined - and constrained - by information and attitudes. National and international policies follow from the way people perceive their world, and from the goals they seek. That is the basic lesson of this section of the book.

CHAPTER 20

Understanding the environment

Introduction

More than two million papers are published every year in science (including medicine). This is a twentyfold increase since 1940, and the number of papers is increasing (Bussard, 1990). Assessing such a massive output over a twenty-year period is quite impossible. However, there is no doubt that understanding in all branches of science deepened tremendously between 1972 and 1992. This is particularly true in the environmental field.

A general framework for the environmental sciences is shown in Box 1. Over the last several decades there have been major developments in all four of the categories listed: surveys and monitoring of environmental systems have increased dramatically, and a vast amount of data has become available through remote sensing, especially from space platforms. There have also been advances in sensor sensitivity, telecommunications and computer capabilities. Development of these tools has stimulated many of the major advances in understanding and predicting environmental processes.

BOX 1

A general framework for environmental studies.

1. Description (field surveys and monitoring)

2. Explanation (analysis and modelling)

3. Prediction (modelling)

4. Management (environmental engineering; environmental policy-making)

Steps 1 and 2 are iterative. Better models lead to improved monitoring systems.

Step 3 requires explicit assumptions about externalities, e.g. future emission rates of greenhouse gases in the case of climate models.

Management strategies (Step 4) may be designed to reduce the predicted environmental impacts or to protect society from these impacts.

One consequence of these developments has been the increasing attention given to long-term futures (50 or even 100 years ahead). However, expanding the time horizon increases uncertainty, as well as requiring that a broad range of socio-economic and environmental factors are taken into account. The complexity of models and analyses has therefore increased greatly.

Advances in the environmental sciences

Advances in the natural sciences

The science of ecology began with Haeckel in 1866, and from the outset the social implications of its teachings about the holistic nature of the environment were recognized - by Haeckel himself and by social thinkers such as Nietsche and Bolsch (Bramwell, 1989). Ecology developed rapidly in the following decades with the introduction of such concepts as the community and the ecosystem, and recognition of processes such as succession. But the modern view of interacting socio-economic/ecological systems has its roots in the work of W. I. Vernadsky (Vernadsky, 1945; Yanshin, 1988), whose book *The Biosphere* was published in Russian in 1926. Vernadsky warned of the increasing rates and scales of environmental transformations taking place, and he dreamed of a time when the biosphere and the technosphere would be in harmony: this happy state he called the *noosphere*. Modern concepts of sustainability and sustainable development demand such harmony.

During the 1970s and 1980s, much has been learned about the global nature of environmental systems. Particularly impressive are the insights gained into the biogeochemical cycling of elements essential for life, notably carbon, nitrogen, oxygen, phosphorus and sulphur. This cycling has been called the environmental life-support system by Tolba and White (1979), who have emphasized that many large-scale environmental issues - climate warming, stratospheric ozone depletion, acid rain and soil degradation - are inter-related through disruptions in the global cycles of carbon, sulphur, nitrogen and phosphorus. Current understanding of these cycles has been synthesized, and new research directions proposed, in several of the monographs produced by SCOPE (the Scientific Committee on Problems of the Environment), including those on the global carbon cycle (SCOPE 13, 1979), the sulphur cycle (SCOPE 19, 1984) and the interactions amongst the cycles (SCOPE 21, 1983). The importance of the biosphere in modulating the geochemical cycles of these substances is now fully recognized. This view is expressed, for example, in the idea of a global metabolism and in the Gaia hypothesis (Lovelock, 1979, 1988) which states that the biosphere has been in homeostatic equilibrium with its global environment over many millions of years (the hypothesis is named for Gaia, the Greek goddess of the Earth). Only in recent decades has the system begun to unravel. Ecologists are deeply divided on the validity of the Gaia concept: according to Grinevald, the Swiss historian of science, Gaia 'is the major cultural and scientific revolution of our time' but some ecologists feel that Gaia is an untestable hypothesis and is 'an unscientific attempt to deify the biosphere' (quoted in Mann, 1991).

The last two decades have seen considerable advances in understanding the behaviour of geophysical and ecological systems, and of plant and animal populations, and in the development of a capacity to model them mathematically. If changes within a system are sufficiently small for it to be expected to remain close to the range of previous experience, it is possible to develop rational environmental management plans by applying the ideas of *carrying capacity of a region, sustainable yield of a renewable resource,* and *assimilative capacity of a watershed* or *airshed*. These tools have developed considerably in the last 20 years although there has been increasing recognition that some systems are inherently unstable, and can suddenly switch from one mode to another, the timing being quite unpredictable. Holling (1973) has introduced the idea of *resilience* of an ecosystem, which means that although most ecosystems can withstand environmental stresses within certain limits, they may not survive a major shock. In this case the system may shift to a new equilibrium state. Mathematical models can also display such sudden shifts. For example, May (1974) developed a model of a very simple biological system comprising only one species, which showed that the population could remain in equilibrium, oscillate between two or more values, or vary chaotically. These ideas suggest that if environmental conditions are changing, human attempts to maintain an ecosystem in its current state will inevitably lead to a loss of resilience.

Such shifts to new environmental equilibria may take place on a local or regional scale, and sudden changes may occur after a considerable period of apparent stability during which underlying physico-chemical factors are being modified. For example, the pH of an acidifying lake may not change much for a number of years, but once the buffering capacity of the lake is used up it may drop suddenly to a new level that is dangerous for fish populations (NAS, 1986). Soils that are naturally acid, but have been limed for agriculture, may release toxic heavy metals if liming ceases when the land is afforested. Stigliani (1991) has elaborated the concept of chemical 'time bombs' of this kind in soils, sediments, coastal waters and wetlands. On the regional scale, deep-sea sediments south of Greenland, as well as measurements in the Greenland icecap, show several climatic flips between two apparently stable modes in the geological past. Broecker (1987) has hypothesized that ocean currents are driven by salinity gradients, and that in the area just south of Greenland a mechanism may exist for shutting down the North Atlantic oceanic conveyor belt, cooling the sea and adjacent land areas.

A related idea that has been much popularized in recent years is that of *chaos*, which is a measure of the sensitivity of a system to initial conditions (Gleick, 1988). It was first studied in the context of weather forecasts by Lorenz (1963), who demonstrated that the atmosphere quickly loses its predictability. One of the objectives of GARP (the Global Atmospheric Research Programme) undertaken by the WMO and ICSU in the 1970s was to obtain sufficiently detailed weather observations to test Lorenz's ideas. The GARP programme demonstrated that the atmosphere contained only very modest predictability.

A third field of study is that of *fractals* (Mandelbrot, 1977), which introduced the idea of similarity of phenomena occurring on different scales. As Mandelbrot pointed out, the length of a coastline increases in an inverse logarithmic way with the measurement scale. Although not yet fully exploited, Mandelbrot's ideas have relevance for scaling up from local to global environmental experiments (Rosswall *et al.*, 1988).

These and other concepts have been drawn together in the development of new approaches to environmental management. The ecological insights into the dynamics of ecosystems and the critical conditions governing stability or change, have led in turn to analyses of how far it is practicable to use certain systems sustainably. The concept of sustainable development embraces economic and ethical elements as well as ecological ones, but it is clear that development is sustainable only if it is ecologically sound (see Clark and Munn, 1986 and IUCN/UNEP/WWF, 1991, for example). The intellectual challenge for ecologists is that sustainability must be achieved not simply under conditions of slow change, but also under the rapid environmental and socio-economic transformations expected in coming decades. Studies that attempt to evaluate these changes and to manage the resulting impacts are therefore becoming increasingly important. In this connection, the scientific problem of how to achieve *re-development* of degraded ecosystems has been given increased attention in the last decade; see for example, Regier and Baskerville (1986) and the collection of papers in issue no. 2, volume 17 (1988) of *Ambio*.

There have, of course, been major advances in many other branches of both basic and applied science during the past 20 years which have contributed in important ways to the intellectual context within which environmental understanding has evolved. Physics, chemistry, mathematics and geophysics have all contributed much. Within the engineering sciences, the idea of *industrial metabolism* and *materials flows* has been pursued by Ayres (1989), Stigliani (1990) and others (see also Chapter 12). Major advances in medical science have contributed considerably to human quality of life. Smallpox has been eradicated, and there is now a much better understanding of other major infections and about the causes of cancer. On the other hand, the appearance of many new kinds of industrial chemicals, and rapid advances in measurement techniques, have led to the appearance of a suite of new health issues relating to toxic substances such as PCBs and pesticides. In fact, a new scientific discipline, *environmental toxicology* (or *ecotoxicology*), has been born (Butler, 1978). At the international level, research and synthesis in this important field are being promoted by SGOMSEC (Scientific Group on Methodologies for the Safety Evaluation of Chemicals - SCOPE/ICSU/IPCS/WHO/UNEP/ILO) and IPCS (International Programme on Chemical Safety - WHO/ILO/UNEP).

These various advances have involved world-wide co-operation. This has always been a feature of science, but it has strengthened in the past two decades, a substantial portion of the growth having taken place in the developing world. Although led mainly by scientists from developed countries, the International Biological Programme (IBP) (1964-74) had national adhering organizations in 26 developing countries and undertook a number of field studies there (Worthington, 1975). The UNESCO Man and the Biosphere (MAB) programme had 22 projects in the humid and subhumid tropics between 1973 and 1980 and involved 60 countries in these zones . The work of the Scientific Committee on Problems of the Environment (SCOPE) of the International Council of Scientific Unions (ICSU) is also very widely based.

An important feature of the past 20 years has been the increasing scientific co-operation between various international agencies, especially within the UN system, and the nongovernmental professional community, especially within ICSU. Such co-operation is well illustrated by the Global Atmospheric Research Programme (GARP) involving WMO and ICSU, and the World Climate Research Programme (WMO, UNEP and ICSU).

Box 2 illustrates the structure of a major contemporary programme, the International Geosphere-Biosphere Programme (IGBP). The IGBP had its origin in the early 1980s as a response to mounting concerns about the increasing

BOX 2

The IGBP Core Projects

Project	Related question	Status
IGAC (International Global Atmospheric Chemistry)	1	established
STIB (Stratosphere-Troposphere Interactions and the Biosphere)	1	proposed
JGOFS (Joint Global Ocean Flux Study)	2	established
GOEZS (Global Ocean Euphotic Zone Study)	2	potential
LOICZ (Land-Ocean Interactions in the Coastal Zone)	3	proposed
BAHC (Biospheric Aspects of the Hydrological Cycle)	4	established
GCTE (Global Change and Terrestrial Ecosystems)	5	established
GCEC (Global Change and Ecological Complexity)	5	potential
PAGES (Past Global Changes)	6	established
GAIM (Global Analysis, Interpretation and Modelling)	7	proposed
DIS (Data and Information System)	7	established
START (System for Analysis, Research and Training)	7	established

Source: IGBP (1990).

rapidity of social, technological and environmental changes that were expected to transform the Earth in the twenty-first century. Its objective is:

To describe and understand the interactive physical, chemical and biological processes that regulate the Earth system, the unique environment that it provides for life, the changes that are occurring in this system, and the manner in which they are influenced by human activities (IGBP, 1990). In this connection, it should be emphasized that IGBP complements the WCRP (World Climate Research Programme).

IGBP has developed a number of first-priority Core Projects, focused on seven questions (IGBP, 1990):

1. How is the chemistry of the global atmosphere regulated, and what is the role of biological processes in producing and consuming trace gases?

2. How do ocean biogeochemical processes influence and respond to climate change?

3. How do changes in land use affect the resources of the coastal zone, and how will changes in sea level and climate alter coastal ecosystems?

4. How does vegetation interact with the physical processes of the hydrological cycle?

5. How will global change affect terrestrial ecosystems?

6. What significant climatic and environmental changes have occurred in the past, and what were their causes?

7. How can our knowledge of components of the Earth system be integrated and synthesized in a numerical framework that provides predictive capability?

The Core Projects relating to these questions are listed in Box 2. They provide a very full menu of studies for the next decade at least, involving many hundreds of scientists.

Advances in the social sciences

Although social scientists are making increasingly important contributions to our understanding of the environment, progress has not been as spectacular as in the natural sciences. There are several reasons for this, including the following:

- The funds available to social scientists for environmental research, and for collaboration in international programmes, are generally far smaller than those available to natural scientists.

- In many international environmental programmes, the social science content is either non-existent or an add-on. Contributions from specialists in economics, politics and the humanities often have to fit into research agendas set by biophysical scientists.

- Natural scientists are well organized through ICSU and adhering bodies. The equivalent international organization for social scientists is the ISSC (International Social Science Council), which has few national connections with Academies of Science and ICSU committees. However, good relations exist between the ICSU and ISSC Secretariats and a social scientist has recently been added to the Secretariat of the IGBP.

- Measurements of social and environmental phenomena have been pursued in isolation and there has been little attempt to ascertain relationships between patterns of change in environmental and socio-economic variables.

This imbalance is gradually being corrected as the roles of human population growth, resource consumption and life-styles are increasingly recognized as major factors transforming the biosphere. It is now appreciated that it is impossible to draw up plans and strategies for environmental management without understanding the sociology and economics of the people involved. Resource planning requires models of demographic trends and resource consumption patterns. The roles of special groups - women, children and the aged - are being closely analysed. The traditional wisdom of indigenous people is also being evaluated for the insights it can provide into sustainable resource use. Developments in social science research have led in turn to the development of new techniques to promote environmental management. These range from evaluations of the environmental impacts of new policies and industrial developments, examined through environmental impact assessment (which increasingly considers economic and social factors) to combinations of regulatory and economic instruments to promote optimal resource use and minimal pollution. Such principles as the Precautionary Principle, the Polluter Pays Principle and the User Pays Principle have originated at least as much from the social sciences as from the world of pollution-control technologies.

International environmental programmes in the social sciences began in the 1970s with UNESCO/MAB and some elements of ICSU/SCOPE (Price, 1990). With the launching of IGBP, a consortium of international bodies, including the ISSC, the International Federation of Institutes for Advanced Study (IFIAS), the United Nations University (UNU) and UNESCO began work on a complementary programme, The Human Dimensions of Global Change Programme (HDGCP) (Burton and Timmerman, 1989), which has in turn fostered the creation of related programmes, including the ISSC Human Dimensions of Global Environmental Change Programme (HDGECP) (Jacobson and Price, 1990).

The fact that environmental change has important human dimensions has been recognized for a long time. However, the report of the World Commission on Environment and Development has provided a fresh research agenda for social scientists, particularly in the fields of *economy-environment interactions* and *sustainable development*. A great deal of critical effort has gone into defining just what 'sustainable development' really means (or should mean). The process of definition must embrace the natural and social sciences, linking ecology, economics, sociology, demography, medicine and ethics. Research on human interactions with the environment needs to embrace three basic aspects: how human activities may alter the environment; how societies and individuals are affected by environmental change; and how they may adapt to it or intervene to modify its rate or direction (WMO, 1991). These three areas will provide social scientists with major research agendas over the coming decades. The difficulties of this research should not be underestimated: the challenges may indeed be greater than those facing natural scientists. (How does one develop quantitative models of human behaviour, for example?).

Advances in environmental monitoring systems and data inventories

Since the Stockholm Conference, there has been an enormous growth in public demand for information on the state of the environment. In addition, there has been an increasing requirement by modellers, resource managers and other specialists for particular kinds of data sets. This has led to considerable interest in the scientific aspects of monitoring, stimulated in part by UNEP through its Global Environmental Monitoring System (GEMS). There have been important national initiatives, such as those of the USSR State Committee for Hydrometeorology and Control of the Natural Environment, which helped in the development of conceptual frameworks through a series of symposia on environmental monitoring (see, for example, Izrael, 1983). One of the current issues, of course, is how to manage the increasing flow of environmental data that users demand. Climate data from US satellites and weather stations alone are accumulating at the rate of more than 1,000 magnetic tapes a day, a flow that will increase the current archives one hundredfold by the year 2000 (Anderson, 1991). Fortunately, modern technology in the form of laser-read disc storage provides a more durable, compact and easily-read system which is keeping up with increasing demands for data handling.

As emphasized by Izrael and Munn (1986), there are many reasons for monitoring the environment, and data collected for one purpose may not be suitable for another. During the last two decades there has been increasing recognition of the need to define the explicit objectives for monitoring, and several studies have been designed to compare the cost-effectiveness of alternative monitoring systems. In this connection it is important that a monitoring system should have a conceptual framework, based, for example, on the biogeochemical cycling of elements, or the movement of toxic substances through food chains. Such an *integrated* approach to monitoring has been strongly advocated by Izrael (1980) and associates. In the radiological health field, international bodies such as UNSCEAR have for some time designed their monitoring systems in close relation to their models of environmental pathways and bioaccumulation of radionuclides (see UNSCEAR, 1988).

Examples of environmental monitoring systems

The great diversity of monitoring purposes is evident when examples of the various monitoring systems now extant are considered. Three important examples, described in the following paragraphs, are:

- GOS, the Global Observing System;
- GEMS, the Global Environmental Monitoring System;
- LTER, Long-term ecological field research programmes.

GOS is co-ordinated by the WMO and operated by Member States. It has its roots in weather observations dating well back into the last century. Over many years, such questions as units of measurement, instrument intercalibration, siting criteria and network density have been resolved by international agreement. Paralleling GOS, the WMO has recently established the Global Atmospheric Watch (GAW) (Figure 1) which monitors the chemical composition of the atmosphere and of precipitation. This network originated conceptually in the 1950s with the programmes of the Scripps Institution to monitor CO_2 at Mauna Loa, Hawaii, and of the Meteorological Institute of the University of Stockholm to analyse the chemical constituents of precipitation.

GEMS was one of the first programmes to be established by UNEP in the early 1970s. Activities within GEMS now span 142 countries, and include monitoring in three main areas (atmosphere and climate, environmental pollutants, and renewable resources) as well as data exchange and assessments. GEMS has several components, and works in collaboration with various partner organizations. These are illustrated in Box 3.

In recent years, a whole new science dealing with Geographical Information Systems has arisen. In a GIS the characteristics of a particular area are collated in a computer program which permits rapid integration, intercomparison and manipulation of data. The use of GIS will permit much more profound study of the behaviour of societies and their interactions with ecosystems. Box 4 illustrates how a Geographical Information System can be applied to model a vector-borne disease.

In addition to GEMS, there are a number of other global data banks and information systems. They include :

- the ICSU World Data Centres for selected solar, geophysical and environmental data;

- the Carbon Dioxide Information Analysis Centre (CDIAC) at Oak Ridge, USA;

Figure 1: *The WMO Global Atmosphere Watch.*

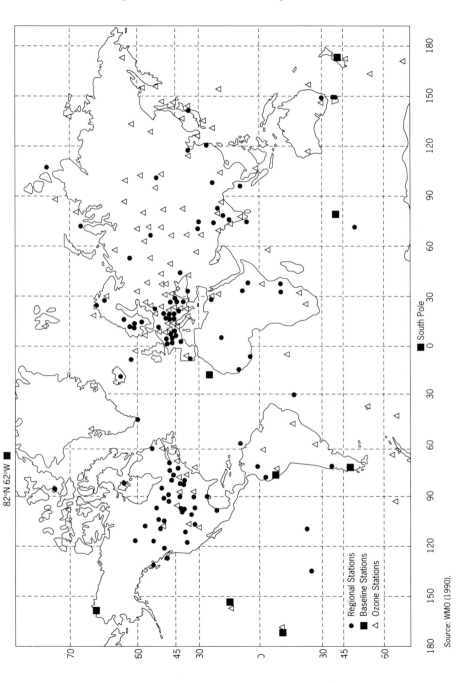

Source: WMO (1990).

BOX 3

Programme elements and collaborating institutions within GEMS.

STRUCTURE: GEMS is a UN-wide global system to which many organizations contribute. Since 1975 it has been co-ordinated by a small Secretariat within UNEP. The main collaborating organizations are :

WMO (whose Background Air Pollution Monitoring Network, BAPMoN, has been operational since 1969);

WHO (which has recently established with UNEP a Human Exposure Assessment Locations Programme (HEALS) within GEMS (see below);

FAO (with which UNEP has collaborated in global assessments of forests and land degradation);

IUCN and WWF (with which UNEP has established the World Conservation Monitoring Centre - see below).

GEMS networks monitor changes in atmospheric composition and the climate system, freshwater and coastal pollution, air pollution, food contamination, deforestation, ozone depletion, the build-up of greenhouse gases, acid rain, the extent of global ice cover, and many issues related to biological diversity. The detailed working of GEMS is illustrated by three following examples : HEALS (Human Exposure Assessment Locations), GRID (Global Resource Information Database), and the links with the World Conservation Monitoring Centre (WCMC) and the Monitoring and Assessment Research Centre (MARC, London).

HEALS. The HEALS programme was established in 1984 by WHO and UNEP to monitor the total human exposure to particular pollutants reaching the body by various pathways (e.g. through breathing, food and water) (Ozolins, 1990; WHO, 1990). Seven countries currently participate (Brazil, China, India, Japan, Sweden, USA and Yugoslavia), and the pollutants currently monitored are mercury, fluorides, volatile organic chemicals and respirable particulate matter. This long-term programme will provide comparable assessments of human exposures to pollutants around the world, and ultimately may permit the identification of trends. In fact the collaboration achieved by doctors, chemists and engineers in the field of human exposure assessment (through personal monitors, for example) is one of the success stories of the last two decades.

GRID. Established in 1985, GRID provides environmental data in readily accessible form (UNEP, 1990). GRID data are stored within a geographic information system (GIS), which uses a common geographical co-ordinate system, permitting the manipulation and display of data sets from disparate sources. Thus, for example, spatial patterns of topography, soil, vegetation, temperature and rainfall can be overlaid. Supported by regional centres in Nairobi, Geneva and Bangkok, the activities of GRID are of three main types: database acquisition and management; GIS applications; and GIS technology transfer. Some of the applications undertaken in the pilot phase of GRID indicate the potential benefits to be derived:

- an estimate has been provided of the number of elephants in Africa;

- an evaluation has been made of the suitability of coastal sites in Costa Rica for the development of aquaculture;

- identification has been made of high-risk areas for East Coast Fever, a major killer of cattle in Africa*. See Box 4 (UNEP, 1990).

UNEP (GEMS) provides support to two important research centres:

1. The World Conservation Monitoring Centre (Cambridge, UK). This Centre, supported through a partnership between UNEP, IUCN and the World Wide Fund for Nature (WWF), provides key information on biological diversity, including data on threatened plant and animal species and their habitats. The Centre also maintains a data bank on the world's protected areas, and on international trade in wild species and derivative products.

2. The Monitoring and Assessment Research Centre (London, UK). In partnership with the WHO, this Centre provides assessments of data from GEMS and other sources; organizes training workshops on environmental monitoring; and undertakes research on specific aspects of monitoring.

---

* In similar kinds of assessment, Linthicum et al., (1987) used NOAA satellite data to infer precursor conditions for outbreaks of Rift Valley Fever in Kenya while Steele (1991) has described a US System called FEWS (Famine Early Warning System) based on the amount of chlorophyll inferred from infrared scans of Earth from satellites.

BOX 4

Modelling vector-borne disease with a GIS.

In 1987, the International Laboratory for Research on Animal Diseases used GRID to carry out an investigation into the environmental factors that limit the range of East Coast Fever, a major cattle killer in Africa, which is transmitted by a vector – the brown ear tick.

A GIS is ideal for this type of study as it can identify the potential range of a disease by overlaying maps created from data on the climatic and ecological conditions essential for the vector's survival.

The brown ear tick can live only within a certain temperature range and at specific humidity levels. In order to create GIS maps of the tick's possible occurrence, data on temperature, rainfall and evaporation had to be obtained for the whole of Africa. Vegetation is also a critical factor in the tick's survival, so vegetation type and distribution were taken into account. Buffalo are not susceptible to the disease, but often carry the tick, so data on their distribution were also used.

The data on climate and vegetation were processed by a GIS to create an environmental suitability map which showed all the areas in which the tick could survive. Data on cattle ranges were overlaid by the GIS onto the environmental map in order to identify the areas where all factors conducive to the tick's survival coincided with cattle distribution. In such areas there is a high risk of the disease occurring.

East Coast Fever is not, however, present in the high-risk areas which GIS identified in Ethiopia and Central West Africa. This is probably because infected cattle cannot range over the entire area due to natural barriers such as desert and tropical forest.

Should infected cattle be transported into high-risk regions, there is a potential danger of a severe outbreak of East Coast Fever.

The map below left shows the current distribution of East Coast Fever, and the map below right shows the areas identified by the GIS as being environmentally suited to the disease.

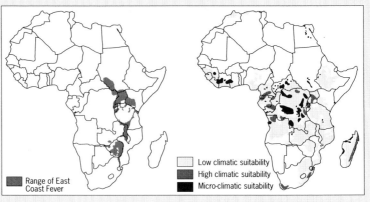

Range of East Coast Fever

Low climatic suitability
High climatic suitability
Micro-climatic suitability

- the IOC General Format System for handling geo-referenced oceanographic data;

- the Global Run-off Data Centre (GRDC) at Koblenz, Germany;

- the Global Resources Information Database (UNEP).

Another feature of the past 20 years has been the publication of State of the Environment Reports, and the establishment of information centres through which people can gain access to the results of monitoring. Among these:

- UNEP INFOTERRA provides a referral system guiding enquirers to information sources;

- the World Resources Reports (see e.g., WRI, 1990), the UNEP Environmental Data Reports (see e.g., UNEP, 1991), the OECD Reports (see e.g., OECD, 1989, 1991) and publications of several other bodies provide international overviews, while many countries publish their own national series of environmental data and accounts of the state of the national environment.

There are several long-established ecological field research programmes, including those at Rothamsted, UK, which date from the 1850s (Johnson, 1991) and at Hubbard Brook, USA, dating from 1963 (Likens, et al., 1977). In the last two decades, however, interest in this kind of activity has increased tremendously, stimulated by the recognition that slow and sometimes subtle changes are taking place in many of the world's ecosystems. Studies of phenomena such as forest dieback in Europe and desertification in Africa require the continuation of measurement series for decades, especially since there is evidence that some such processes are cyclical.

A summary of much of the long-term field research initiated in recent decades is given by Risser (1991). In the United States, many ecologists have been involved for more than a decade in a programme called LTER (Long-Term Ecological Research). In the former USSR, ecological monitoring is a component of the national integrated monitoring system, the main emphasis being on monitoring pathways of trace substances through the biosphere.

Optimal design of environmental monitoring systems

In the last two decades there has been considerable research into the optimization of monitoring systems, particularly with respect to the acid deposition issue in North America, and in cases where system operating costs are expected to be high. A review of the methods used to monitor air pollution has been given by Munn (1981), and the principles are widely applicable. If sufficient historical data banks are available, as well as a good model of the system being studied, the best approach is to compute root-mean-square (rms) differences between model-predicted and observed values, and to locate monitoring stations where the rms errors are greatest. Other factors are of course important, such as the objectives for monitoring and the value of long time-series of observations. If only excedences of environmental standards are of interest, for example, monitoring stations would be located where observed or predicted concentrations are likely to be high.

Advances in interpretation

Computer models

Computer models in the natural sciences

Mathematical models have long been used to predict weather, tides, urban air pollution and many other geophysical phenomena, as well as some ecological processes. In such applications, model performance can easily be evaluated by comparing observations and predictions. But models may also be used to simulate conditions that have never occurred; in fact, comparisons may be made of several alternative 'futures' based on a range of possible human interventions in coming decades.

As a result of increasing understanding of complex environmental systems, and of advances in computer technology, model performance has improved greatly in the last 20 years. What is surprising, however, is the increased degree of public acceptance of model-derived 'futures', as will be illustrated in the following four examples.

Climate change models. Box 5 gives a simplified description of the construction and use of a climate change model (IPCC, 1990). Assuming a doubling of CO_2 concentrations, these models yield an estimated rise of from 1.9 to 5.2 degrees C in globally-averaged surface air temperature. Climate modellers are relatively confident that warming of this magnitude will occur unless the current emissions of greenhouse gases are stabilized very quickly (IPCC, 1990; WMO, 1991). At the time of the Stockholm Conference there was already evidence that CO_2 concentrations were increasing and that global warming would follow (SMIC, 1971). What is most remarkable, however, is that in the intervening years, most world leaders have been persuaded to take these projections seriously and to examine the policy consequences, despite the fact that the global temperature trend in recent decades has not been sufficient to require any explanation other than natural variability.

Acid rain models. One of the contributions to the Stockholm Conference was the Swedish Case Study on the acidification of Scandinavian lakes and forests. Evidence was presented that acidification was due in part to the long-range transport of pollutants (LRTP), particularly the oxides of sulphur and nitrogen. Subsequently the OECD and then the ECE sponsored major LRTP monitoring and modelling programmes in Europe. Acidification also became a major public issue in North America, and there have been several major monitoring and modelling programmes in the last 15 years.

BOX 5

Climate change models.

The most highly developed climate change model is the *general circulation model* (GCM), which is based on the laws of physics together with descriptions in simplified physical terms (called parameterizations) of smaller-scale processes such as those due to clouds and deep mixing in the ocean. In a climate model, an atmospheric component, essentially the same as a weather prediction model, is coupled to a model of the ocean, which can be equally complex.

To make a climate forecast, the model is first run for a few (simulated) decades. The statistics of the model's output provide a description of the model's simulated climate which, if the model is a good one, will bear a close resemblance to the climate of the real atmosphere and oceans. The above exercise is then repeated with increasing concentrations of greenhouse gases. The differences between the two sets of statistics provide an estimate of climate change. The long-term change in surface air temperature following a doubling of CO_2 is generally used as a benchmark to compare models. The range of results from model studies is 1.9 to 5.2 degrees C. Most results are close to 4.0 degrees C but recent studies using a more detailed but not necessarily more accurate representation of cloud processes give results in the lower half of this range. Hence, the model results do not justify altering the previously accepted range of 1.5 to 4.5 degrees C.

Source: After IPCC (1990).

Because the biogeochemical processes are complex, with important feedbacks, acidic deposition models necessarily contain many components. Nevertheless, through careful peer reviews considerable confidence in model outputs has been built up among both scientists and those concerned with public policy. An example is the European RAINS model, which provides a range of country-by-country projections of acidic deposition for given European sulphur emission strategies (Alcamo *et al.*, 1987, 1990). In its optimization mode, the model provides estimates of the emission reductions required across Europe if, for example, a target loading in Scandinavian lakes is not to be exceeded. Moreover, the model computes how this may be achieved at either minimum cost or minimum amount of sulphur removal.

Nuclear winter models. Nuclear winter simulations postulate a large injection of smoke and dust into the atmosphere following a nuclear war in the Northern Hemisphere (Pittock *et al.*,1989). The predictions for this scenario vary with the season, but in general are characterized by major cooling at the Earth's surface, beginning in the Northern Hemisphere but spreading to the Southern Hemisphere within several months. As in the earlier examples, the model predictions cannot be compared with real data; nevertheless, a number of studies including the 1985 international synthesis by SCOPE had a major influence on public awareness of the dangers of nuclear war, and contributed significantly to intergovernmental debates on nuclear arms reductions. In a similar type of application, the climatic effects of smoke from the burning oil wells in Kuwait were modelled, the results being that although a 4 degree C drop in surface temperature might occur in the Gulf region, no appreciable impact on global climate was expected (Bakan *et al.*, 1991). This prediction was broadly confirmed by observation, although the temperature drop in the headwaters of the Gulf was in fact only some 2°C relative to the seasonal average (Pellew, 1991).

Models of the impacts of environmental change on the biosphere. Given a range of environmental futures, there remains the problem of estimating the impacts of these futures on the biosphere (and on society; see below). This is often the most uncertain part of the assessment.

A recent and interesting study by Parry (1990) deals with the impacts of climate change on world agriculture. Parry's conceptual framework consists of:

(a) A hierarchy of models relating to: climate change; direct relationships between climatic variables and plant growth, crop yield and rangeland carrying capacity; second-order relationships between yields and production, employment and profitability at the farm level; higher-order relationships between regional/national agricultural yields and agricultural employment, activity rates in non-agricultural sectors, etc.;

(b) Identification of changes caused by ecological changes, e.g. changes in rates of soil erosion, pest and disease outbreaks, groundwater depletion and acid deposition patterns;

(c) Identification of technical adjustments by the farmer, and regional/national policy responses.

The weakest link in this conceptual chain is the very first one: climate change models are inadequate at the regional level. The analysis is therefore supplemented by a sensitivity study designed to identify the climate elements to which crops are especially sensitive, and to quantify threshold changes in the climatic elements that would seriously impact crop yields. This helps to focus the analysis.

Computer models in the social sciences

The Limits to Growth project. For many social scientists the basing of public policy on the simulation of future conditions using input values that are unprecedented causes concern. This reaction is nowhere better illustrated than in the history of the Limits to Growth project, whose main results were published in paperback form (Meadows *et al.*, 1972) prior to completion of the normal scientific review process. Evolving from Forrester's systems dynamics models (Forrester, 1972), the Limits to Growth project was carried out at MIT under Jay Forrester and supported by the Club of Rome. Later, another book was published to explain the assumptions of the world model and the smaller experimental models from which it was developed (Meadows and Meadows, 1973). However, most people – both critics and supporters – read only the paperback, which sold more than a million copies in more than 20 languages.

The model extrapolated population and resource utilization trends, both of which had been increasing exponentially, in a world of finite resources and in which 'pollution' had negative impacts on life expectancy. The model was run under a variety of conditions, with different rates of population growth, resource availability and human response to food shortages and pollution. The model was so structured that projections of exponential growth in a finite world led inevitably to a system crash. Unfortunately, the project's title closed off some of the possibilities for debate and perhaps was responsible for the criticism that the model evoked. (See, for example, Cole *et al.*, 1973.)

Criticism came from three main sources – the business community, the modelling community and proponents of Third World concerns. The first group stated that resource availability always responds to scarcity through the price mechanism and technological innovation, and that the model ignored this. Some members of the modelling community were alarmed at the crudity of the model, and pointed out that there were no usable data for human mortality response to 'pollution'. From the Third World perspective, the suggestion that world economic growth should be limited while two-thirds of the world's people lived in poverty was a callous and unrealistic conclusion based on a purely technical analysis. None the less, the study provoked a tremendous debate and undoubtedly stimulated research, including studies such as *Mankind at the Turning Point* (Mesarovic and Pestel,1974), and the OECD *Interfutures Study* (OECD, 1979). Limits to Growth had a major impact on public perception, and was one reason why, by the mid-1970s, many people had come to support the idea of a conserver society.

Regional disaggregations of the world model. Many researchers replaced the aggregated nature of the model (whereby all the world's people, resources and pollution were totalled) with regional sub-systems. (See Bruckmann (1978) and Hughes (1980), for example). Although this produced more varied results, showing that some regions would fare better than others, it was difficult to escape the 'Limits' hypothesis. (For a summary of the output of ten such studies, see Wils (1982)). The problem remained of how people would respond to the growing pressures of population growth and environmental degradation.

Input-output models. The regional models were attempts to *disaggregate* the world model and they inherited many of its limitations. An alternative is to *build* on what is known, such as the structure of an advanced industrial economy. A good example of this approach is provided by the economic input-output models pioneered by Leontief (1970). These proved to be adaptable to environmental concerns by adding on the physical resource inputs used by the various economic sectors (including 'environmental goods' such as air and water) and the unwanted outputs, or 'pollution'. This approach had been adapted to predict the environmental impacts of land-use changes (Isard, 1972), and was used in the broad field of resource management (Braat and van Lierop, 1984; Whitney *et al.*, 1987). Indeed, one of the major shifts in attitude that this approach produced was not simply to say, 'What is the environmental cost of economic growth?', but to ask 'How do we manage the production of residuals on a regional scale?' (Bower, 1977; Basta and Bower, 1984). The term 'residuals' has replaced 'pollution' to emphasize that waste is an integral part of our economic activities, and that such materials have an uneven impact on the environment.

Resource management on a regional scale. The virtue of the input-output model is the clarity of its structure and the relative simplicity of its data requirements, such as the linear relationships between economic sectors. However, because of the degree of aggregation and the difficulty of quantification of many environmental variables, and the necessary simplification of relationships among variables, input-output models have not so far helped answer practical questions of environmental management policy.

Several regional models for resource management have been developed at the International Institute for Applied Systems Analysis (Laxenburg, Austria), some of which make intensive use of the interactive capacity provided by today's computers, allowing decision makers to pose a series of questions as events unfold, thereby allowing a response to the inherent uncertainty of complex systems. (See Fedra (1984, 1985) and Fedra and Loucks (1985), for example). In Senegal, a groundwater model has been developed by the Department of Geology (University of Dakar) and the Ministry of Water Resources (Gaye *et al.*, 1988; Niang, 1988). The system includes a model of drawdown and replenishment of the major eastern aquifer, and a management data base for rural water supply throughout Senegal.

The relationship between agricultural productivity and population growth is closer to parabolic than to linear. Increasing population provides labour to derive more output, until population exceeds the carrying capacity of the land and the system begins to lose its productivity. One might argue that mechanization and chemical inputs have broken this Malthusian mould, but in low-income subsistence farming regions on poor soils, the model may still hold. This relationship has been incorporated into a model of inter-regional migration in Niger and Senegal (Tellier, 1983, 1985). The relationship is also significantly influenced by social and institutional factors such as the land tenure system, as is evident in Japan and the Republic of Korea.

In complex systems that contain many non-linear relationships and a great variety of feedback mechanisms, we may introduce the concept of evolution – with new forms emerging that did not exist before. This is the assumption behind the 'evolutionary models' developed at the Université Libre de Bruxelles under the guidance of Ilya Prigogine who encouraged the application of the concept of irreversibility (which he developed in physical chemistry) to a great variety of biological and social phenomena (Prigogine and Stengers, 1984). Such models have been used as tools to study the evolution of urban systems, a regional economy and the management of an ocean fishery (Sanglier and Allen, 1989; Allen and McGlade, 1987).

Even at the project level, simulation models are being used increasingly to resolve conflicts among objectives and group interests, and to achieve satisfactory distribution of the costs and benefits of environmental changes entailed by proposed projects. The methodology of adaptive environmental assessment and modelling (Holling, 1978) is relevant in this connection.

Reducing uncertainty

Uncertainty has been a major issue for environmental scientists over the last two decades. The main developments have been along three lines: efforts to reduce uncertainty through research and observations; efforts to quantify uncertainty, largely through statistical methods; and efforts to manage the consequences of uncertainty. One of the obstacles, of course, is the great variability in the environment and in populations of humans and other species.

BOX 6

Some recent outdoor ecosystem experiments

Class 1 — Long-term natural experiments
- The ecological effects of El Niño in the Galapagos Islands.
- Effects of fire on stream water chemistry.

Class 2 — Experiments having both natural and anthropogenic components
- Desertification.
- Wetlands (effects of drainage, wastewater releases, etc.).

Class 3 — Experiments having mainly anthropogenic components (Observer status only for the investigator)
- Recent deforestation in the Amazon Basin.
- Historical cases of deforestation .
- Acid rain (both decreases and increases in loadings).
- Pollutant releases over decades from agricultural, domestic and industrial sources.

Class 4 — Manipulation of ecosystems
- Clear-cutting of forested watersheds.
- Whole lake manipulation of the Experimental Lakes Area in Northern Ontario (eutrophication, acidification, radionuclides, etc.).
- Acidification of whole catchments in Norway.
- Liming of acidified lakes.

Source: Mooney et al., (1991)

This has led to much recent study of two of the causes of uncertainty in ecological predictions: patchiness and rare events, e.g. volcanic eruptions, forest fires, floods and droughts. Some recent work is summarized in Box 6. Another major area of recent effort has been in the field of model performance testing (Federov, 1989; Munn *et al.*, 1988; Alcamo, 1986).

Broadening the basis of environmental assessments

Economic aspects of environmental assessment

Resource economics and environmental economics

The current economic paradigm, based largely on economic growth and more loosely on the tenets of capitalism, enjoys continued popularity in spite of recurring bouts of recession and growing pessimism about the future. At the same time, the World Commission on Environment and Development has called for the integration of economics and the environment, using the concept of sustainable development. In this connection, Daly (1987) makes a useful but often overlooked distinction between growth and development. *Growth* is an increase in the flows of matter and energy through the economy, whereas *development* is an improvement in non-physical characteristics. Munn (1990) equates *development* with *evolution*, suggesting that sustainable development is achievable in a time of increasing global change only through strategies that promote the ability of society to adjust to change. The 1991 Strategy for Sustainable Living (IUCN/UNEP/WWF, 1991) defines *sustainable development* as 'improving the quality of human life while living within the carrying capacity of supporting ecosystems'. Many writers have pointed out that *sustainable growth* is a contradiction in terms, because nothing physical can grow indefinitely.

Resource economics focuses primarily on methods to broaden conventional neoclassical economics to include the true costs of natural resources, particularly common goods such as air and water, in the determination of optimal allocation of resources. Non-market valuation methods have been developed which enable economists to determine the cost or benefit attributable to non-priced natural resources. However, the fundamental precepts of neoclassical economics are incorporated. One of these is the circular flow of goods and services in the economy, i.e., *production factors* to *firms* to *markets* to *consumers* to *production factors*, and so on. 'Value added' at each stage ensures that continuous economic growth is possible. Economic activity is neat, circular and independent of any other system. Environmental economics, similarly following essentially neo-classical tenets, addresses issues of pollution control, standard-setting, waste management and recycling, externalities of private enterprise action, conservation and use of common property resources and so on, from the standpoints of providing guidance for efficient allocation of resources and sound environmental policy.

As the concept of sustainable development is drawn into mainstream politics, the emphasis may again be placed on economic growth as a long-term solution to the global problematique. Literature on resource economics and environmental economics is growing, as the debate on sustainable development centres on ways of internalizing the 'externalities' of production and economic growth. These externalities range from global warming to local problems such as soil erosion. The emphasis remains on mobilizing the market mechanism through adjustment of price signals to influence the behaviour of households and enterprises and so achieve environmental objectives in tandem with social and economic ones. Reconciliation of the local with the national and the global, and the short-term with the long-term, remains an elusive goal.

Cost-effectiveness analysis (CEA)

Resource economics employs a number of tools, critically reviewed by Lave and Gruenspecht (1991). One of the easiest (in principle) to use is Cost-Effectiveness Analysis (CEA), where the concern is to minimize the cost of achieving a specified environmental standard or objective set through a scientific-cum-political process. In the acid deposition field, for example, the objective might be to meet target loadings of sulphur at minimum cost over a large region, taking into account that control costs vary from industry to industry, and that the cost of control increases with increasing severity of control. An analysis of this kind has been made by Amann (1990) within the European RAINS model.

Benefit-cost analysis (BCA)

BCA was originally designed to evaluate the net benefits of project proposals, but in the last three decades serious attempts have been made to extend it into the social, environmental and natural resource fields. BCA compares the monetary value of streams of benefits and costs of a proposed policy, programme or project over defined space and time boundaries. This comparison can be straightforward for project construction costs but can be complex when considering environmental and social costs and benefits such as those relating to air quality or aesthetics, particularly when the issue involved is long-term and large-scale. For a recent BCA application to urban air pollution control programme in the United States, see Krupnick and Portney (1991).

Despite the growing use of BCA in the environmental field, critics contend that the methods are flawed, or that the approach (using monetary values for environmental goods and services which have inherent value) is unethical. Although current methods are in principle capable of including all benefits and costs, the controversy over the appropriateness of BCA as an evaluation tool continues. Five particular issues arise

- How to calculate both *direct and indirect costs*;

- How to incorporate a *societal perspective*. An activity that is beneficial overall may still disadvantage some members of society. BCA has no mechanism to deal with equity: while those who gain can compensate those who lose, they often do not.

- How to measure intangible *extramarket values* such as air quality. (Economists have developed some means of dealing with this problem, for example by using 'shadow prices');

- How to relate the benefits and costs calculated for each year into the future to a common scale. This is done by discounting the value to a base year using present values, but the discount rate chosen can greatly affect the results. A high discount rate reflects a bias towards present-day concerns. A low discount rate is more in keeping with the tenets of sustainable development with respect to intergenerational equity. Many consider that when dealing with natural resources that are effectively irreplaceable, a zero discount rate has to be adopted.

- How to deal with uncertainty, both in financial and biological forecasts (Swartzman, 1982).

These issues are discussed by Bland (1986), Freeman (1979), Hyman (1981), Pearce *et al.*, (1989), Pearce and Markandya (1989).

Ecological economics

Increasing realization in recent years that *resource economics* has left out vital economic/environmental linkages has led to the emergence of *ecological economics* as an alternative (Daly, 1990). The assumption of circular flow in the market economy as found in resource economics, is not grounded in ecological principles, nor does it recognize ecological inputs or outputs. Essential to an understanding of ecological economics are three concepts: *throughput, carrying capacity* and *entropy*. Our economy, and indeed our very survival, relies on *ecological throughputs*. We need water for production and transportation, minerals and soil for agriculture, and so on. At the other end, we rely on nature to act as a sink for our wastes.

The second concept is *carrying capacity*. Under steady-state conditions, the carrying capacity of an ecosystem can be modelled with reasonable confidence. However, for an ecosystem including people, the real carrying capacity will depend on human consumption patterns (Arizpe *et al.*, 1992) and will vary geographically and over decades. Thus it is difficult to estimate the population that the world can sustain. Erring on the side of caution is therefore desirable; as Daly (1977) observes: 'minimizing future regret is wiser than maximizing present benefit'. Another kind of problem arises because while the carrying capacity of a region can be estimated for cattle or elephants, the estimates for people depend on assumptions about their life-styles.

While natural inputs can be easily identified, and carrying capacity is intuitively understandable, *entropy* is neither easily identified nor understood. However, it is a very useful 'trump card' for ecological economics. 'In entropy terms, the cost of any biological or economic enterprise is always greater than the product. In entropy terms, any such activity necessarily results in a deficit' (Georgescu-Roegen, 1973). Regardless of how efficient production processes become in terms of minimizing externalities or reducing waste, production will always contribute to the ever-increasing state of entropy in the universe. Economic activity should therefore strive to maintain the desired level of goods in society, while minimizing entropy contribution.

The goal of ecological economics might best be described as finding the best ways of living lightly on the planet, and striving for a 'frugal society' (i.e. one that is based on a definition of thrift in terms of economic efficiency, and on achieving environmentally sound economic development (Goodland *et al.*, 1991). This will not be achieved through adherence to a single dogma such as neoclassical economics: rather, 'methodological pluralism' (Norgaard, 1989) is to be encouraged. This approach is reflected in the activities of the newly formed International Society for Ecological Economics, which has begun publication of the journal *Ecological Economics*. (See also *Ambio*, 1991 and Constanza, 1991).

Demographic aspects of environmental assessment

With increasing public concern about the growth in world populations, demographers have been called upon more and more to make projections and to collaborate in interdisciplinary studies of the environmental impacts of rapid population growth. Their tasks have been made easier by the continued development of computer systems and new frames of geographical reference such as postal codes.

Given the age structure of a population at a given time, as well as assumptions about future fertility, mortality and immigration/emigration, a population forecast can readily be made. However, Lutz *et al.*, (1990) note that because of demographic 'surprises' (baby booms, epidemics, immigration waves, etc.), demographers can provide only a range of 'futures' for a series of explicit assumptions rather than a specific prediction. Some illustrative scenarios are given in Figure 2 (Lutz *et al.*, 1990), and in Chapter 16.

As Chapter 16 emphasizes, there are important interactions between *population, resources, development* and *environment*. Demographers, in co-operation with representatives of many other disciplines, are making an important contribution to the search for pathways leading towards sustainable development (Repetto and Holmes (1983) Shaw; (1989)). However, several controversial questions remain unanswered, such as how an increased

Figure 2: *Total population for Europe and North America 1990-2050 according to ten scenarios.*

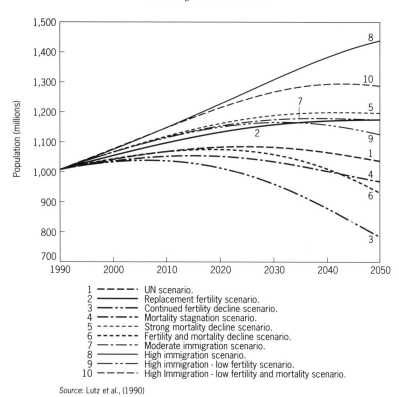

1	— — — ·	UN scenario.
2	————	Replacement fertility scenario.
3	—·· —	Continued fertility decline scenario.
4	—·—··	Mortality stagnation scenario.
5	- - - - - -	Strong mortality decline scenario.
6	· · · · · ·	Fertility and mortality decline scenario.
7	—·—··	Moderate immigration scenario.
8	————	High immigration scenario.
9	—·· —	High immigration - low fertility scenario.
10	— — — ·	High Immigration - low fertility and mortality scenario.

Source: Lutz et al., (1990)

standard of living affects birth rates, or what effect rapid population growth has on welfare. These uncertainties need to be explored in projections based on different assumptions about birth control and population growth (including changes due to migration) and the implications for GNP per caput, food production and other measures of welfare (Arizpe *et al.*, 1992; Keyfitz, 1992).

Environmental ethics

'Environmental ethics' is a term that has recently come into use to describe the principles that should guide humanity's relationship to the environment. The history of environmental ethics under other names is long, encompassing the whole range of humanity's experience and use of nature (Passmore, 1980); but the current global environmental situation has given fresh impetus to the exploration of the values underpinning modern society. Some of this exploration is being carried out in the philosophical community, some in religious groups, and some simply among ordinary people who want to know what they ought to be doing to 'think globally, and act locally' (Rolston, 1988; Callicott, 1989).

There are two different kinds of ethical approach to the environment. The less radical kind suggests that we can apply standard ethical understandings derived from philosophy or theology. This is the approach used by IUCN, UNEP and WWF in their new publication, *Caring for the Earth : A Strategy for Sustainable Living* (1991). The second kind is more radical, and argues that a complete rethinking of our relationship to the environment is required, and that by doing that we shall inevitably be forced to change many things we take for granted, including what we mean by 'ethics'. These two kinds of 'environmental ethics' have been referred to as 'shallow ecology' and 'deep ecology' (Naess and Rothenberg, 1989), although both use the term 'ecology' in a sense somewhat removed from its scientific meaning.

Recent debates over environmental ethics have focused particularly on the validity of extrapolating to the natural world ideas and practices that are currently applied to humanity, and on assertions of the intrinsic value of Nature, no matter what might be the claims of humankind. The first has considered whether higher animals, plants or other natural beings could be said to have 'rights' which must be respected by human beings (Singer, 1977; Regan, 1983) and granted legal standing (Stone, 1975). The second kind of argument may be held to have gained recognition in the World Charter for Nature, adopted by the United Nations General Assembly in 1982, which states that 'all forms of life merit respect, regardless of their usefulness to humanity'. Another area of debate has centred on the need for 'bills of environmental rights' for people, including rights to clean air, water and a healthy environment - tempered by a matching obligation not to impair other people's enjoyment of these benefits.

Other elements of the debate encompass some of the following issues:

- the preservation of wilderness areas and biosphere reserves as 'fiduciary trusts' (Brown-Weiss, 1988);

- setting out responsibilities for future generations (Partridge, 1980);

- the development of communal rules for the use of common property areas (Hardin, 1968; McKay and Acheson, 1987).

Recent years have also witnessed a growing number of initiatives among the world's religious traditions to reconsider their role in environmental issues. This was given a substantial boost by the holding of the 25th anniversary celebrations of the World Wide Fund for Nature in Assisi, Italy, in 1986, which served as the focal point for a conference and celebration by representatives of the world's major religions. Subsequent to that event, a global network on religion and ecology has been established through WWF International. IUCN has established an Ethics Working Group (Engel and Engel, 1990), while the Commission of the European Communities has held a debate on 'environmental ethics: man's relationship with nature'. There has also grown up in many parts of the world a new series of ritual events, some associated with religious traditions, including environmental sabbaths, local environmental faith events and celebrations centring on Earth Day.

These developments in the broad field of 'environmental ethics' have been a particularly important feature of the past 20 years. An ethical imperative linking the actions of the current generation to future generations (i.e. the debate about intergenerational equity) has become a central element in the definition of 'sustainable development'. The new Strategy for Sustainable Living (IUCN/UNEP/WWF, 1991) begins with a call for a new ethic that expresses the respect and care that people owe to one another and to the Earth. Questions about the kind of world we wish to sustain into the foreseeable (and unforeseeable) future are now central to the debate over environmental policy. The new interest in environmental ethics is, in part, a reflection of concern that placing an economic value on the environment is insufficient to capture its significance to human well-being, and provides insufficient motivation for the action that needs to be taken. There are calls for a Declaration and Covenant that commit states to the world ethic for living sustainably, and define their rights and responsibilities accordingly. Clarification of these ethical issues may well be high on the international agenda in the years to come.

Tools for environmental management

Introduction

Human behaviour is central to environmental and resource management, and if it is to be understood there must be collaboration between natural and social scientists. It is therefore fortunate that the institutional and educational barriers that have prevented this for a long time are now being removed. Nevertheless, in many governmental and intergovernmental bodies a separation of responsibilities, funding and employee skills still exists, and sectoral divisions still hamper integrated environmental management (Chapter 22). Fortunately, a number of new policy tools have been developed in the past two decades, as described below.

The ecosystem approach; a framework for environmental assessment

At the time of the Stockholm Conference, interdisciplinary researchers were turning from a preoccupation with the natural components of ecosystems to a recognition of the importance of both natural and human cultural attributes. ICSU's International Biological Programme (IBP) had emphasized the biological parts of ecosystems but had treated human cultural involvement largely as an external force. Contemporaneously, UNESCO's International Hydrological Decade (IHD) had focused mostly on non-living aspects of large-scale watersheds. In 1973, UNESCO launched its more comprehensive programme, MAB, and soon after that, IHD was replaced by the International Hydrological Programme (IHP).

In recent years, the concept of an *integrated ('ecosystem') approach* has gained momentum, particularly in the context of environmental management at the regional level, for example in the Great Lakes Water Quality Agreement of 1978 between Canada and the United States. The approach is also central to the Convention on the Conservation of Antarctic Marine Living Resources (CCAMLR) adopted in 1980. Adopting an ecosystem approach means that the entire ecological–jurisdictional system within a region must be integrated within a single conceptual framework. Managing a river, for example, includes management of the human uses of the surrounding drainage basin in the context of changing institutional and jurisdictional arrangements.

Restoration of degraded systems must be based on ecological and social understanding (Rapport *et al.*, 1985; Regier and Baskerville, 1986). The aim is to achieve sustainable re-development: a concept closely linked to that of *ecosystem integrity*, which is concerned with fostering the health of ecosystems as self-organizing systems, and avoiding their degradation (Edwards and Regier, 1990).

Environmental resource accounting

Gross National Product, Gross Domestic Product and similar indicators were designed at a time when the Keynesian economic model was gaining acceptance. The possibility of natural resource scarcity seemed remote. In recent decades, however, there has been widespread recognition that purely economic indicators are inappropriate for judging the success or failure of development, or for guiding policy and planning (Repetto, 1989). More recently, several researchers have developed methods which include natural resource assets in the national accounts. This activity is called *resource accounting* or *environmental accounting*. Towards the end of the 1980s, several governments (Norway, France, Netherlands) began to consider how these might be applied to give a more complete measure of human well-being than that provided by national income accounts, and in the process a more informed basis for environment and development policies and planning.

Although there is increasing consensus that national accounts must be revised, incorporating evaluations of damage to the natural resource base, treating government, private sector and household expenditures consistently, and regarding changes in stocks of natural resources as part of changing national income (Repetto, 1989b), there is little agreement on how this should be done. Peskin (1990), after reviewing proposed approaches, observes that none of the systems is able to address the full range of concerns. In Norway, a system of national resource accounts has been developed; these are satellite accounts, not directly linked to the national income accounts. In France, the approach taken is to look at environmental changes resulting from human activity. These accounts include material and energy flows between economic activities and the natural environment, as well as the natural resource accounts used in Norway. Again, the accounts are expressed in physical units.

The establishment of physical accounts is a necessary step towards incorporation of environmental goods and services in national accounts. However, this on its own is not sufficient. These physical measures must be converted to economic values despite the difficulties encountered in attempting economic valuation.

One method is to use economic rent in the valuation of natural resource stocks (Repetto, 1989b). Landefeld and Hines (1985) discuss three methods for estimating the value of such stocks:

(1) the present value of future net revenues;

(2) the transaction value of market purchases and sales of the resource *in situ*; and

(3) the net price, or unit rent, of the resource multiplied by the relevant quantity of the reserve.

Hueting (1989, 1990) argues that construction of shadow prices for environmental services that are comparable with market prices of human-produced goods and services is impossible (the valuation problem). However, he believes that the GNP can be corrected through deducting expenditures on measures that compensate for, mitigate, or prevent losses of environmental functions. Standards of sustainable economic development, including indicators of health, must first be set. Then the costs of achieving these standards would be used to measure the extent to which a society does not have a sustainable GNP.

Resource accounting is an active field of research. Most current studies are in the context of developed countries, and analyses of the types described above would be of little value in regions where people lived mainly at the subsistence level or where they did not have land-tenure security. That a resource accounting system is a valuable tool for long-range economic planning in developing countries has, however, been demonstrated by Repetto *et al.* (1989) for Indonesia. Over the years 1971 to 1984, the Gross Domestic Product increased at an average annual rate of 7.1 per cent but if resource depletion were given a dollar value and subtracted from the conventionally calculated GDP, the average annual increase would be only 4 per cent, with decreases in 1979 and 1980.

Policy instruments for environmental management

Careless technology coupled with inadequate standards, controls and institutional arrangements has often allowed producers and consumers to disregard the costs of environmental degradation. These costs, which are known as 'technological externalities' (e.g. the emission of pollutants into air and water bodies) are the focus of many studies in environmental economics.

Box 7 lists twelve policies that have been used. The choice of instruments would depend on local circumstances and on the particular type of pollutant

BOX 7

Some institutional instruments for water, land and air quality management. (Upper items are oriented towards 'administrative' fiat, while lower items are market oriented.)

1. Prohibition, e.g. harvesting of endangered species, dumping of some toxics or contaminants into an ecosystem.
2. Regulation, e.g. of phosphorus concentrations in sewage effluents to control eutrophication; land-use zoning.
3. Direct government intervention to modify some ecosystemic feature, e.g. building dams to create lakes.
4. Grants and tax incentives, e.g. to industry for installing pollution control equipment, or to a municipality to accept solid wastes.
5. Buy-back programmes, e.g. government purchase of flood-prone land.
6. Liability for compensation, e.g. USA Superfund provisions.
7. Compulsory insurance, e.g. to compensate the victims of pollution for damages.
8. Effluent charges, e.g. fees for waste disposal scaled according to the direct cost of treatment or to the indirect cost associated with deleterious impacts on a receiving ecosystem; effluent charges may be incorporated into 'delayed pollution control charges'.
9. Resource rent, e.g. tax or charge on harvesters of a resource in order to recover a fair return for the owners (all the people) of the resource, and also to foster efficient use of the resource by discouraging over-capitalization.
10. Management of the demand, e.g. through rate structures involving marginal cost pricing and/or peak responsibility pricing to improve overall efficiency of use and to foster conservation.
11. Transferable development rights, e.g. incentives to industry to develop in designated areas.
12. Transferable individual quotas, e.g. rights to emit specific quantities of pollutants or to harvest specific quantities of fish or wildlife.

Source: Adapted from Regier and Grima (1985).

to be controlled or the resource to be conserved. For example, pricing of water to reflect its scarcity is likely to reduce water consumption (and help manage demand) but it may not work in areas where water shortages exist. For a pollutant that is perceived to be noxious and to have serious human health effects (e.g. leaded gasoline), outright prohibition is more desirable than imposing a tax. Generally speaking, the policy mechanisms that are market-based (i.e. those towards the bottom of the box) work better when a *reduction* in the loading of a pollutant is required rather than a complete ban. Such pollutants include SO_2 (which causes acid rain), CO_2 (climate warming) and nitrates and phosphates (eutrophication of aquatic ecosystems).

Criteria that should guide the choice of economic policy instruments include effectiveness, efficiency, ease of monitoring and administration, provision of incentives for the producer (or consumer) to reduce stress on the ecosystem (i.e. the polluter-pays principle) and political acceptability (Baumol and Oates, 1979; Pearce *et al.*, 1989; OECD, 1975). Here only regulation/prohibition and transferable emission permits will be discussed as examples.

Regulation and prohibition. By and large, protecting environmental quality has relied on technology-related command-and-control regulation and prohibition. The most common form is an effluent standard but others include specification of best available technology, an ambient environmental quality standard or outright prohibition. Other examples include land-use zoning and setting limits on harvesting wildlife, fishing and logging. The main disadvantage of the regulatory approach is that uniform standards do not distinguish between industries that could reduce their loadings at a low cost and those that could not. Other disadvantages are that there is no incentive to reduce the pollutant loading further once an industry achieves the standard, and that the enforcement of regulations requires expensive monitoring and policing and is subject to legal delays. Regulation is the most common mechanism for controlling pollution and resource use because it has an easy-to-understand rationale and, at least in theory, it should work. It gives citizens, politicians and administrators a perception of control over the problem. It has wide political appeal; even industry is prepared - albeit reluctantly - to consider pollution control as one of the costs of doing business. However, this policy instrument clearly reduces the choice for consumers and producers and is not consistent with achieving efficient allocation of resources in a market economy.

Transferable emission permits. This mechanism allows for property rights to be attached to emission or discharge permits at a specific site. In effect, the assimilative capacity of ecosystems is allocated to those who bid the highest (i.e. show the highest willingness-to-pay). The initial level of pollution that the community is prepared to accept is set through the political process (Dales, 1968). For example, a country may decide to halve current pollution levels. Then it would be up to those industries who most need permits to bid for them; other polluters would have an incentive to adopt abatement technology, or change the input mix, production process or output. Environmental groups could also bid for permits and not use them, thus enhancing air or water quality.

Transferable permits allow choice to industry *and* provide a constant incentive to reduce the level of pollution and therefore to sell some of their transferable discharge permits. Such a market in transferable permits would accommodate an increase in economic activities and also encourage technological innovation and economic development. In addition, the community would be able to set total pollutant loads directly rather than relying on the effectiveness of best available technology required by regulation of each polluter. An example of a limited form of emission rights in lieu of regulations is that of the Ontario Hydroelectric Power Commission, which operates coal-fired power stations, and which is permitted an annual province-wide SO_2 emission limit rather than a specific limit for each station.

Environmental impact assessment

The modern *environmental impact assessment* (EIA) process began with the passage of the National Environmental Policy Act (NEPA) in the United States on 1 January 1970, even though large engineering works (hydroelectric dams, nuclear power stations, etc.) had been subject to environmental scrutiny many years earlier. NEPA required that all federal agencies should prepare an *Environmental Impact Statement* (EIS) prior to taking any major action or introducing legislation that would significantly affect the environment. NEPA was a pace-setter which, amongst other things, opened the development process to public scrutiny. In subsequent years an EIA-type procedure has been introduced in many other countries and jurisdictions, and has been accepted by the World Bank and other inter-governmental bodies.

An EIA is 'an activity designed to identify and predict the impact of an *action* on the biogeophysical environment and mankind's health and well-being, and to interpret and communicate information about the impacts' (Munn, 1979), where an action could be a proposal to construct a large engineering work, or to implement new legislation, policies or operational procedures with environmental implications. In the 1970s, EIAs generally did not include assessments of socio-economic impacts. Today, some jurisdictions include these factors within EIAs while in other cases, separate *socio-economic impact assessments* (SIAs) are required.

Over the last 20 years there has been a marked improvement in the quality of EIAs and EISs, and in their usefulness. Many guidelines and analyses have been prepared for training purposes. There has also been a good deal of attention paid to the regulatory, procedural and institutional aspects of the EIA process. Much of the credit must go to bodies such as the ECE and CEARC (Canadian Environmental Assessment Research Council), who have provided a focus (and are continuing to do so) for developing effective EIA methods and institutional procedures. On the non-governmental side, the International Association for Impact Assessment, which was launched in 1981, has played an increasingly important role.

A synthesis of the state-of-the-art in the late 1970s is given in SCOPE 5 (Munn, 1979). Of the many advances that have been made subsequently, the most important ones are as follows:

Valued ecosystem components (VECs). Following Beanlands and Duinker (1983), under this approach the first step in an EIA is the identification of a list of environmental components/indicators that are of importance to the various groups concerned with the proposed actions. This tends to sharpen the scope of the assessment.

Cumulative impacts. EIAs are generally not concerned with the long-term cumulative effects of an action, yet these may ultimately lead to the collapse of an ecosystem. A serious examination of this problem took place at a joint US-Canada Workshop (CEARC-NRC, 1986), and a research agenda has been defined (CEARC, 1988).

Social impact assessment (SIA). Whenever large engineering works are proposed, many of the concerns raised by the public relate to quality of life - jobs, disruption of cultural values, relocation of populations etc. - rather than to the quality of the environment. For this reason there has been a trend towards requiring parallel *social impact assessments* (SIAs) (Finsterbusch and Wolf, 1977; CEARC, 1985). Such assessments may be particularly important in developing countries: in Indonesia, for example, it has been suggested that SIAs may be more important than EIAs (Suprapto, 1990).

Environmental risk assessment. In most proposals for development, there is the possibility that a rare event could cause serious harm. For example, a tank-truck loaded with toxic chemicals might overturn and catch fire on a proposed new highway or in a tunnel. To mention this eventuality in an impact statement raises the alarm; however, not to mention it may subsequently lead to a charge of concealment by local citizens. The problem is compounded by the fact that the public cannot distinguish amongst differences in very low probabilities, particularly when the consequences are widely disparate. Grima *et al.*, (1989) believe that risk assessment methodologies, including ways of presenting risk information to the public, need to be codified, but in order for this to be successful all parties concerned should be involved in the process. As pointed out by Grima *et al.*, EIAs, SIAs and risk assessments are complementary tools in environmental management.

Large-scale international EIAs. International EIAs are of two types: (1) those relating to proposals for specific projects that may cause environmental impacts in neighbouring countries; and (2) those concerned with already existing environmental degradation, such as acidified lakes and stratospheric ozone depletion, requiring concerted regional or global action. In both cases, there has been increasing recognition of the need to provide a policy focus for the assessments, and to try to obtain some uniformity in legislative procedures and EIA methodologies. European countries have, under the auspices of the ECE, adopted a regional convention on EIAs in a transboundary context (ECE, 1991). This includes not only a scientific framework for EIAs but also an administrative procedure, and a listing of the types of enterprises to be included under the convention (thermal power stations, large waste disposal installations, large dams and reservoirs, etc.). In North America, the US National Air Pollution Assessment Program (NAPAP) has conducted an integrated assessment of acid precipitation (NAPAP, 1990). The recent assessment by the Intergovernmental Panel on Climate Change (IPCC, 1990) which was an important input into the Second World Climate Conference and into subsequent ministerial meetings dealing with the preparation of a draft convention on greenhouse gases, was also an international EIA.

Technology assessments (TAs). Technology moves forward at a remarkably rapid pace, but the resulting environmental changes are rarely assessed, or even monitored once the technology enters the market-place. This is an area of study that needs much greater attention, particularly with respect to cumulative effects that might cause severe impacts over the long term. Procedures have been established in several countries to screen emerging and imported technologies from the standpoint of their environmental significance. However, no examples of model TAs can yet be cited. See Heaton *et al.*, (1991) and CSTD (1991).

EIAs in the context of sustainable development. Recent world-wide discussions of sustainable development have influenced the content of EIAs, which today are being broadened to include consideration of: (1) the long-term and large-scale effects of the proposed action, and (2) long-term changes in other factors (e.g. in technology, population and climate) and their impacts on the proposed action. A useful checklist of sustainability criteria that could help define the initial scope of EIAs and SIAs is given in Box 8 (Jacobs and Sadler, 1990). The Commonwealth of Australia is proposing to develop guidelines for inclusion of climate change in EIAs.

Management of risk and uncertainty

There has been a marked rise in public concern about hazards over the last 20 years, paralleled by an improvement in theoretical and empirical capabilities to make better decisions in situations involving risk. Numerous conferences (e.g. Fowle *et al.*, 1988; Miller *et al.*, 1986), new journals such as *Risk Analysis* and *Risk Abstracts* and new programmes by government agencies (e.g. EPA, 1984; Ruckelshaus, 1985; CEARC, 1986) are evidence of this.

Risk assessment has emerged as a significant tool for environmental decision-making (Miller *et al.*, 1986). It has also helped to focus public attention on health and environmental issues and has been instrumental in clarifying the role of public participation and conflict resolution in the environmental field.

A significant step has been the recognition that risks need to be assessed in an integrative framework rather than in isolation. For example, the threat of climate warming has acted as a catalyst that has led to consideration of forest depletion, fossil fuel consumption and CFC use as an interrelated set of concerns. More comprehensive analytical frameworks and improved technical instruments for characterizing risks augur well for future risk assessments (e.g. Whyte and Burton, 1980; Covello *et al.*, 1986). Equally encouraging is the interest that the various stakeholders have shown in the public debate on risk decision and mitigation. The need to involve social scientists and the public in emergency planning and accident mitigation has also been recognized (see Bourdeau and Green, 1989).

Significant progress has also been made in promoting the role of public involvement in environmental decision making. In particular:

• the citizens most closely affected by environmental risk decisions are increasingly insistent that they be involved;

BOX 8

A preliminary checklist of sustainability criteria.

Economic sustainability

Why is the proposed development needed? What is the economic justification? How is the development expected to meet human needs, improve net social welfare or community well-being? Does the project require a financial or environmental subsidy?

Ecological sustainability

What potentially significant or irreversible cumulative effects are anticipated? To what extent might the development deplete renewable resources or impair ecological integrity locally, regionally or globally? How will the development affect nutrient cycling, soil capability, biomass, water quality, etc.? What compensatory measures can offset deterioration in resource productivity or ecological capacity?

Social/community sustainability

What is the social/community rationale for the development? To what extent, for example, does the proposal promote fair and equitable distribution of benefits and costs? How does it maintain choice of life-styles, take into account minority rights, and meet community aspirations. Who will share in the benefits, or receive compensation for unavoidable impacts?

Policy and institutional integration

What are the key interdependencies among economic activity, natural processes and social/cultural values? How have these changed in the past, and will they be likely to change in the future? Where are the spatial boundaries of these interactions best drawn; do they have local, regional or global impacts or implications? To what extent does the proposed planning/assessment process identify the substantive issues and their policy and institutional implications, suggest alternative actions for resolving problems, elaborate decisional criteria, and establish conditions for monitoring, auditing and evaluating progress in each of these areas? How will this process foster an adaptive approach to coping with scientific policy and technological uncertainty, changing values and intra- and inter-generational equity?

Source: After Jacobs and Sadler (1990).

- planners increasingly wish to incorporate stakeholders' views and objectives into their plans;

- politicians sometimes use public hearings to clarify group interests involved in decision-taking, and to demonstrate the technical and political complexities of reaching consensus on an issue (e.g., the disposal of high level radioactive wastes).

The significance of public participation is reinforced by the increasing openness of government in many countries, and the more widespread opportunities that now exist for people to appeal to the media and the courts (Sewell and O'Riordan, 1976). Recent progress up the 'ladder of participation' (Arnstein, 1969) has involved a number of elements, some of which are shown in Box 9. A variety of procedural mechanisms for enquiry, appeal and arbitration have been developed (O'Riordan *et al.*, 1985; Grima, 1989; Bingham, 1984).

This dovetailing of the need for scientific understanding of environmental issues and that for public acceptance and support has been one of the most significant developments in many industrialized countries over the last two decades. It has resulted in vigorous debate and heightened awareness about ozone depletion, climate warming, acidic precipitation, deforestation, landfill site selection, and the transport and management of toxic materials, to mention only a few of the important issues. In a very real sense the forum for discussion has been broadened to include not only scientists, governments and professionals but also stakeholder groups, interested citizens, the mass media and educators, a truly remarkable revolution.

Environmental audits

Environmental audits have become an integral part of the EIA process (Munro *et al.*, 1986). For large engineering works, these audits should cover the construction, post-construction and decommissioning phases, i.e. a time period of at least 40 to 50 years. For assessments that have a bearing on international conventions/treaties (e.g. on acid rain, stratospheric ozone depletion or shared water resources), environmental audits may need to continue for a century or more, merging into programmes of general monitoring.

When some of the early EIAs were later evaluated it was found that many of their predictions were vague and not testable. This difficulty has now been largely overcome, although it must be emphasized that: (1) in EIAs where several alternatives (in siting, safety features, road networks, etc.) are assessed but only one of the alternatives is ultimately selected, it will never be possible to decide whether the optimal choice was made; and (2) it is very difficult to specify the environmental changes caused by an action as distinct from those that would have occurred anyway.

An environmental audit 'compares measured impacts of the project with pre-project conditions and with the predicted effects of the project. The audit

---

BOX 9

Mechanisms for public involvement in risk management.

A. Formal appeals from administrative decisions
- political elections
- court action
- boards consisting of citizens
- ombudsman

B. Institutional mechanisms for participation
- statutory public hearings
- committees of legislative bodies
- meetings of local government councils and committees
- Royal commissions/Congressional committees
- citizen representation on boards
- public meetings
- referenda

C. Public information
- brochures, pamphlets, newsletters
- slide, film and TV shows
- radio and talk shows/open-line shows
- press releases, feature articles, letters to editors

D. Attitude surveys
- opinion polls
- questionnaires
- advertisements with reply coupons
- content analysis of mass media
- workshops, seminars
- Delphi exercises

E. Stakeholders' initiatives
- non-government, citizens' organizations
- professional organizations
- citizens' task forces

F. Mechanisms for sharing power and benefits
- assistance of third-party mediation
- arbitration agreements
- compensation agreements

Source: After Regier et al.,(1980) revised (1990) .

---

measures the relative accuracy of the prediction of impacts and their management through mitigation and compensation. On the basis of the scientific evidence, the audit analyses the causes of departures of actual from predicted results as objectively as possible (Krawetz et al., 1987)'. Environmental audits provide early warning that the environment is not behaving as predicted and that corrective measures may be necessary. Audits also help improve the next generation of EIAs for that type of action, providing a better information base with respect to cumulative impacts, ecosystem behaviour and risk assessment. They need to be broad enough to encompass the social context as well as the long-term ecological consequences of projects (Walker and Sinclair, 1990).

Environmental audits are also being developed to monitor the environmental performance of corporations, alongside the customary audit of financial performance. Such audits are being introduced as a response to the demand by the public and shareholders for corporate responsibility towards the environment.

Long term policy formulation; achieving consensus under uncertainty

Adaptive management strategies

The essence of adaptive management is to keep policy options open as long as possible while at the same time improving the knowledge base to reduce uncertainty. The technique has been elaborated by Holling (1978) and Walters (1986), who *inter alia* have used AEAM (Adaptive Environmental Assessment and Management) Workshops as a training ground for policy analysts, research managers and others. An integrating mechanism used at these workshops is a block diagram or a very simple simulation model which can be useful in the formative stages of an assessment even before the issues are well formulated and the quantitative relationships amongst key variables are known. The technique is particularly helpful in assessments that involve socio-economic as well as ecological factors (e.g. water quality in a lake which receives pollution from multiple sources; large-scale pesticide spraying; management of an off-shore fishery). Assisted by programmers working in the background, a simulation model of a system can be constructed in two to three days, based on specialist knowledge garnered from the participants. Such models help to define priority research and monitoring requirements, and they lead to better communication amongst specialists. They may also highlight the unexpected: in one widely quoted assessment of a proposed development in the Canadian sub-arctic, model predictions suggested that one of the main impacts would be a serious depletion of fish and game by construction workers during their days off (Munn, 1979).

Decision support systems (DSS)

A DSS is a computer-based tool that provides policy analysts and decision-makers with useful information in easily accessible form. The DSS is designed specifically to answer 'what if' questions. An example is the European RAINS model for acid deposition, mentioned earlier, which is capable of comparing the effectiveness of various sulphur-reduction options for the whole of Europe. Many environmental and ecological models are converted to DSSs, but they must be designed carefully if they are to be useful to policy analysts who, for example, are interested in the effects of acid rain on fish populations, not on the water chemistry of the lakes in which the fish swim.

Another kind of DSS is the *expert system*, in which a computer program is used to answer various 'what-if' questions in situations containing major data and knowledge gaps. Information and models from a wide variety of sources are introduced to overcome these deficiencies, the models being selected according to knowledge rules provided by experts. As an example, Lam *et al.*'s (1988) model RAISON has been used to obtain regional soil acidification scenarios for southern Quebec, based on water chemistry data collected in only a few parts of the region, and in watersheds at differing stages of acidification. RAISON searches for data-rich areas with soil characteristics similar to those where water chemistry data are lacking, and applies well-tested models to obtain synthetic data sets for subsequent analysis.

Many other kinds of DSSs exist, including *decision trees* and various mathematical techniques for optimizing decisions.

Gaming, policy exercises and backcasting

For more than a decade, environmental scientists have been dissatisfied with traditional approaches to long-term policy formulation. While the advice of academics or consulting engineers is commonly gathered as a basis for decisions by ministers, the advice is generally filtered and presented by their staff without face-to-face contact and discussion. In other cases, panels and commissions appointed by governments or Academies of Science are given the task of achieving consensus on some controversial issue referred to them. These panels are appropriate for analysing complex but well-bounded scientific questions, but members may lack knowledge of the wider economic or social background to policy choices. If the responsible Cabinet Minister is presented only with a short summary of the panel report, without the opportunity to discuss it with the authors, he/she may not appreciate the fine distinctions and shades of uncertainty in the original.

One way of overcoming these difficulties is through policy discussions, which resemble the approach of *gaming* used for many years by military analysts, but applied only recently to the environment-resource field (Brewer, 1986; Toth, 1988). These *policy exercises* ideally last at least a full day, and bring together five to ten senior people from government or (as appropriate) business and advocate organizations, together with a roughly equal number of environmental scientists. Such an approach is most suited to the presentation and analysis of issues that involve significant uncertainty and have major socio-economic implications. Such issues could include climate warming, forest decline, urbanization, energy futures and land degradation.

Examples of policy exercises are described by Stigliani *et al.*, (1989) in connection with a study of possible future environments for Europe up to the year 2030. The point of view taken in that study was that because of long-term uncertainty, the best approach was to construct a range of plausible futures, including one or two 'not-impossible' ones, and to try to design robust environmental policies that would be sustainable, no matter what surprises might ensue. The policy-exercise approach has also been used in the context of climate change (Jager *et al.*, 1991). Although some of the participants in that exercise found their endpoints (year 2050) improbable, they agreed that if a policy exercise had been held in 1930, some recent events (the stratospheric ozone hole, personal computers, fax machines) would have seemed equally improbable.

The policy exercise has been criticized on the grounds that it is elitist (only a few people are involved in the workshops) and that the participants never explicitly discuss the kinds of socially desirable futures that they would want (Munn, 1991). Thus an approach called *backcasting* has been advocated (Robinson, 1988) in which workshop participants begin by defining the set of futures that they would want - or preferably, a much larger group of people is canvassed for their views well in advance of the workshop.

As the two approaches are further tested and refined, a common modus operandi will probably emerge. Thus the questions, 'How do we get there from here?' and, 'How did we get here from there?' will no doubt be found to be complementary. Policy exercises are valuable learning experiences, and could provide a means of improving the quality of analysis available to senior ministers, and of removing some of the concerns over present perceived inadequacies in their decision process.

Achieving public consensus

Over the last 15 years one of the major changes has been the increasing call for greater public participation in the development and implementation of environmental policies. In response, governments and the academic community have been testing a variety of approaches including :

- **Information sharing**: Agencies that gather data are under increasing pressure to share their data with the public.

- **Data interpretation**: Data interpretation (in reports, policy recommendations, etc.) is coming under greater public scrutiny, in addition to scientific peer review.

- **Risk assessment**: There is increasing demand to formalize risk assessment as a process open to public involvement.

- **Enforcement**: There is a demand for stricter enforcement of existing laws, and for greater penalties. Enforcement actions and records are coming under greater scrutiny.

- **Compliance**: More environmental activities which have been traditionally considered the purview of governments are being undertaken by the public at large, thus placing greater emphasis on public education and self-reliance. As the costs of enforcement increase, this approach might gain in popularity.

Establishing priorities amongst long-term environmental threats

The ranking in importance of a number of long-term environmental threats is an essential but quite difficult task. Issues such as acid rain, stratospheric ozone depletion, climate change and sea-level rise operate on quite different time and space scales; the nature of the potential impacts varies greatly; and the persons/ species involved can vary from people in flooded coastal areas to ecosystems in acidified lakes. Clearly some kind of weighting system is required for the characteristics of these various threats, as two recent studies illustrate.

In the European Futures Study (Stigliani *et al.*, 1989), ten environmental threats (called *dilemmas*) were identified, and the factors considered in their ranking were: intensity; extent; preventability (possibility of early warning; uncertainty involved; technology available to mitigate the threat; costs; social acceptability); environmental response time; and adaptability of society. For each of these factors, a degree of seriousness was used, viz., low, medium and high. As another example of an attempt to establish priorities, Norberg-Bohm *et al.*, (1990) have looked at 28 environmental hazards, ranking them according to 18 attributes. The resulting comparative analysis for the United States and India is given in Box 10, which shows some important differences between the two countries.

BOX 10

Preliminary comparative analysis of the most important environmental hazards in the United States and India. The hazards given are those ranked in the upper quartile of an original list of 28 hazards.

	USA	India	USA and India
Current issues	Acid rain, smog, indoor air quality	Fresh water quality, soil degradation, pest epidemics	No priority issues
Future issues	Stratospheric ozone	Climate change	Animal habitat, radiation exposure, wildlife stock, floods
Current & future issues	Toxic air pollutants	Soil productivity	Fresh water quality, droughts

Source: Norberg Bohm *et al.*, (1990).

Challenges for the future

Sustainable development: the scientific challenge

The ecological-economic partnership

The World Commission on Environment and Development identified two kinds of problems facing society: those that are immediate (for example the need to meet the basic needs of people living today, particularly in developing countries); and those that are long-term (for example environmental hazards such as climate warming and sea-level rise). These two very different types of crises are tightly coupled, the conceptual linkage being provided by the idea of *sustainable development.*

Paraphrasing Clark (1990), the long-term crisis has four main characteristics:

• It is simultaneously local and global, and it is as much a problem for developed as for developing countries.

• The environmental changes expected will take place over decades and centuries.

• The causes and effects of these changes are intimately linked to energy, agriculture, population, life-style and other factors.

• Uncertainty is a dominant characteristic of the crisis.

There are analytical techniques for quantifying uncertainty, provided that the environment is in a steady or slowly changing state, and that an historical data base is available (flood and earthquake frequencies, etc.). However, there is no historical experience with stratospheric ozone depletion, for example, so even though Clark (1985) is right that society's responses to long-term hazards ought to be viewed as an extension of classical risk management, it is not easy to decide how to proceed. The best approach may well be that adopted by the Intergovernmental Panel on Climate Change (IPCC) which brought together the most knowledgeable people in the world to interpret the results of the various models available, and make best estimates, high estimates and low estimates of likely change (Chapter 3). However, these estimates do not define confidence limits, and it is not easy to see how this can be done: Clark (1985), using data from Dickinson (1985), does make some initial attempts in this direction.

Quantitative indicators of sustainable development

How can one tell whether a proposed action or policy will lead towards or away from a condition of sustainable development? As has been noted by Pezzey (1989), several pathways may lead in the right direction; sustainability criteria are therefore to be regarded as pointers rather than measures of optimality. In this connection, four questions must be asked (Munn, 1990): Sustainable for whom? Sustainable for what purposes? Sustainable at the subsistence or luxury level? Sustainable under what conditions? The selection of appropriate sustainability indicators is therefore a most difficult research task. Box 11 provides one list of indicators (Munn, 1992) but undoubtedly there are other possibilities. As a global measure, for example, Ehrlich and Virousek (see Holden, 1990) have suggested monitoring the changes in the photosynthetic output used or preempted by people (currently 25 per cent globally, or 40 per cent for the terrestrial component).

BOX 11

Sustainability indicators.

A: General indicators
    1. State of health of natural ecosystems (primary productivity, efficiency of nutrient recycling, species diversity, population fluctuations, prevalence of pests, etc.) (Rapport, 1989).
    2. Resilience indicators (biodiversity; spatial patchiness, etc.).

B: Indicators of threats to ecosystem integrity
    1. Increasing population (especially urbanization).
    2. Increasing energy consumption (type, per capita and total amounts, energy required to produce a unit of manufactured goods, etc.).
    3. Increasing consumption of water (per capita and total).
    4. Rates of depletion of renewable and non-renewable resources (forests, prime agricultural land, shoreline, wetlands, mineral reserves, etc.).
    5. Increasing amounts of wastes.
    6. Transportation indicators (number of cars, kilometres of road, number of airline passengers, etc.).

C: Indicators of reduced threats to ecosystem integrity
    1. Increasing output of production per unit of natural resources used (both renewable and non-renewable).
    2. Increasing extent of recycling efforts.
    3. Conservation of scarce or highly valued resources (endangered species, wilderness, historical monuments, etc.).
    4. Increasing afforestation and restoration of degraded land.
    5. Declining sales of pesticides and herbicides.
    6. Declining economic and energy subsidies given to the natural resource sectors.
    7. The degree of citizen involvement in 'environmentally friendly' actions.

Models of sustainable development

A credible set of sustainability indicators presupposes a credible model of the interactions between population, development, resources and environment. But ecological and socio-economic systems interact at several levels, in both time and space; for example, local ecosystems may react, sometimes years later, to changes in the regional or global socio-economic systems. One conceptual model of this population/development/resources/environment system has been developed by IIASA (Shaw *et al.*, 1991).

New information systems

Curiously, the rapid growth in the environmental sciences in the last 20 years has further disadvantaged many scientists in developing countries and Eastern Europe, where libraries and research institutes have had severe budget and blocked currency problems. At the State University of Brno, CSSR, for example, the latest issue of *Science* (as of May 1990) was dated 1963 (Lollar, 1990), and similar situations exist in Africa, Latin America and parts of Asia. Although most of the South American participants at a 1990 SCOPE Workshop in Chile were aware of the Report of the World Commission on Environment and Development, few had ever seen a copy.

The problem is immense and there are no quick solutions. Two of the major difficulties are rising costs (of both publication and shipping), and the enormous range of publications available, which means that there must be continuing dialogue between donors and users. Mention should also be made of the fact that because current practices in the publishing world cannot easily be changed, 'emergency relief' plans to help developing countries must operate under the present system. The flow of paper from the printing houses cannot be slowed down, and is in fact increasing. (The number of scholarly journals in all fields has increased in the last 20 years from 70,000 to 108,590 (Hamilton, 1990)). Environmental scientists in developing countries cannot possibly keep pace, which means that the preparation and distribution of review papers and syntheses should be given priority.

Some of the efforts being made to solve this problem are as follows:

- Many UN and non-governmental bodies sponsor the publication of valuable technical reviews and workshop reports. See, for example, the IUBS-UNESCO/MAB series in *Biology International* and two recent popular books: on climate change and agriculture (Parry, 1990) and on nuclear winter (Dotto, 1986). Nevertheless, these reports rarely achieve their readership potentials, and the organizations involved should be encouraged to increase the number of free copies sent to developing countries.

- The Third World Academy of Sciences in Trieste has consistently provided scientific publications to the Third World. During the period 1986 to October 1989, for example, 195 libraries received journals and 451 institutions received books (Salam, 1991).

- The American Association for the Advancement of Science's (AAAS) Sub-Saharan Africa Journal Distribution Program provides more than 200 journals to 175 libraries in 35 countries.

- ICSU is developing strategies to help alleviate publication shortages in the Third World, working with UNESCO, AAAS, the Third World Academy and other bodies. One of the proposals is that scientific societies and commercial publishers print a few extra copies of their journals and books for direct mailing *at their own expense* to Third World libraries and institutions.

- Teleconferencing is an emerging way of exchanging information. In August 1990, for example, 3000 agricultural scientists were linked world-wide by television at a symposium on plant biotechnology (Anderson, 1990). Total cost was about $US300,000, which was much less than the cost of holding an international conference of the same magnitude in one city.

- Electronic information systems will some day provide collections of journal articles instantaneously to users anywhere in the world. However, as pointed out by Buckingham (1990): 'the immediate need is not for more information, but for selection, evaluation and analysis'. At the moment 'information overload', is a serious problem.

- A North American paediatric hospital has 'adopted' a Chinese hospital, sending it duplicate material from its library collections. This 'one-on-one' approach can be particularly fruitful.

These efforts are helpful but insufficient. Other measures, such as the establishment of regional environmental clearing houses should be examined. These would be located in existing institutes in developing countries, and their main task would be to provide usable knowledge (Ravetz, 1986) to scientists in surrounding areas.

Concluding remarks

Progress in the environmental sciences has been remarkable in the last two decades, and effective technologies and management measures have been discovered for quite a few of the troubles that beset the biosphere. However, this new knowledge is not being applied in policies sufficiently quickly. Unfortunately too, many of the environmental prescriptions that worked 20 years ago are no longer applicable - because of the unprecedented changes that are just beginning and will accelerate in the twenty-first century. The Earth is indeed being transformed, environmentally, technologically, politically and socially. Climate warming is only one of the host of changes to be expected in coming decades, and the challenge is to develop and implement joint environmental-economic policies that will promote long-term sustainable development. The priorities for science, as a base for environmentally sound development, have been evaluated in the reports of two recent Conferences (NAVF, 1990; ICSU, 1992). Some principal conclusions of ICSU's International Conference on an Agenda for Science for Environment and Development in the twenty-first Century (ASCEND 21) are summarized in Box 12.

BOX 12

International research priorities in the field of environment and development.

(1) The main development problems to which science could make a contribution are:
- population and per capita resource consumption
- depletion of agricultural/land resources
- inequity and poverty
- climate change
- loss of biological diversity
- industralization and waste
- water scarcity
- energy consumption

(2) The recommended research priorities in the field of environment and development are:
- intensified research into natural and anthropogenic forces and their inter-relations, including the carrying capacity of the Earth and ways to slow population growth and reduce over-consumption
- strengthened support for international global research and observations of the total Earth system
- research and studies at the local and regional scale on: the hydrological cycle; impacts of climate change; coastal zones; loss of biodiversity; vulnerability of fragile ecosystems; impacts of changing land use, of waste and of human attitudes and behaviour
- research on transition to a more efficient energy supply and use of materials and natural resources
- special efforts in education and building up of scientific institutions as well as involvement of a wide segment of the population in environment and development policy problem-solving
- regular appraisals of the most urgent problems of environment and development and communication with policy-makers, the media and the public
- establishment of a forum to link scientists and development agencies along with a strengthened partnership with organizations charged with addressing problems of environment and development
- a wide review of environmental ethics.

Source: ASCEND 21 (1991).

REFERENCES

Alcamo, J. (1986) A framework for assessing atmospheric model uncertainty, In RR-86-5, IIASA, Laxenburg, Austria, pp. 5-11.

Alcamo, J., Amann, M., Hettelingh, J.-P., Holmberg, M., Hordijk, L., Kamari, J., Kauppi, L., Kauppi, P., Kornai, G., and Makkela, A. (1987) Acidification in Europe: a simulation model for evaluating control strategies, *Ambio*, **16**, 232-45.

Alcamo, J., Shaw, R. and Hordijk, L. (eds.) (1990) *The RAINS Model of Acidification*, Kluwer Academic Pub., Dordrecht, The Netherlands, 402 pp.

Allen, P.M. and McGlade, J.M. (1987) Modelling complex human systems: a fisheries example, *European J. Operations Res.*, **30**, 147-67.

Amann, M. (1990) Energy use, emissions and abatement costs, in *The RAINS Model of Acidification* (J. Alcamo, R.Shaw and L.Hordijk, eds.) Kluwer Academic Pub., Dordrecht, The Netherlands, pp. 61-113.

Anderson, C. (1990) Ready for the future? *Nature*, **346**, 687.

Anderson, C. (1991) Too much of a good thing, *Nature*, **553**.

Arizpe, L., Constanza, R. and Lutz, W. (1992) Primary factors affecting population and natural resource use, In ASCEND, Proc. Int. Conf. on an Agenda of Science for Env. and Develop. into the 21st Century, ICSU, Paris, France (in press).

Arnstein, S.R. (1969) Eight rungs on the ladder of citizen participation, *J. Amer. Inst. Planners*, **35**, 216-24.

ASCEND 21 (1991) Conference Statement, Int. Conf. on an Agenda of Science for Environment and Development into the 21st Century, ICSU, Paris, France.

Ayers, R.U. (1989) Industrial metabolism, in *Technology and Environment* (J.H. Ausubel and H.E. Sladovich, eds.), National Academy Press, Washington, D.C., USA.

Bakan, S., Chlond, A., Cubasch, U., Feichter, J., Graf, H., Grassl, H., Hasselman, K., Kirchner, I., Latif, M., Roeckner, E., Sausen, R., Schlese, U., Schriever, D., Schult, I., Schurmann, U., Sielmann, F. and Welke, W. (1991) Climate response to smoke from the burning oil wells in Kuwait, *Nature*, **351**, 367-71.

Basta, D.J. and Bower, B.T. (eds.) (1984) *Analyzing Natural Systems*. Analysis for Regional Residues - Environmental Quality Management, Resources for the Future, Washington, D.C.

Baumaol, W.J. and Oates, W.E. (1979) *Economics, Environmental Policy and the Quality of Life*, Prentice Hall, Englewood Cliffs, N.J.

Beanlands, G.E. and Duinker, P.N. (1983) *An Ecological Framework for Environmental Impact Assessment in Canada*, FEARO, Env. Canada, Ottawa, 132 pp.

Bingham, C. (1984) *Resolving environmental disputes; a decade of experience*, The Conservation Foundation, Washington, D.C.

Bland, P.F. (1986) Problems of price and transportation: two proposals to encourage competition from alternative energy rescources, *Harvard Env. Law Rev.*, **10**, 345-416.

Bourdeau, P. and Green, G. (1989) *Methods for Assessing and Reducing Injury from Chemical Accidents*, SCOPE 40, John Wiley, Chichester, U.K., 330 pp.

Bower, B.T. (ed.) (1977) Regional Residuals Environmental Quality Management Modelling, Res. Paper R-7, Resources for the Future, Washington, D.C.

Braat, L.C. and van Lierop, W.F.J. (eds.) (1986) *Economic-Ecological Modeling for Environmental and Resource Management*, North Holland Pub. Co., Amsterdam, The Netherlands.

Bramwell, A. (1989) *Ecology in the Twentieth Century: A History*, Yale University Press, New Haven and London.

Brewer, G.D. (1986) Methods for policy exercises, In *Sustainable Development of the Biosphere* (W.C. Clark and R.E. Munn, eds.), Cambridge University Press, Cambridge, U.K., pp 445-473.

Broecker, W.S. (1987) Unpleasant surprises in the greenhouse? *Nature*, **328**, 123-6.

Brown-Weiss, E. (1988) *In Fairness to Future Generations: International Law, Common Patrimony and Intergenerational Equity*, Transnational Publishers, Inc., Dobbs Ferry, N.Y.

Bruckmann, G. (ed.) (1978) SARUM and MRI: Description and Comparison of a World Model and a National Model, Proc. 4th IIASA Symp. on Global Modelling, Pergamon Press, UK.

Buckingham, M. (1990) Communication problems, *Nature*, **347**, 581.

Burton, I. and Timmerman, P. (1989) Human dimensions of global change: a review of responsibilities and opportunities, *Int. Social Science J.*, **XLI**, No. 3 (August).

Bussard, A. (1990) Unp. lecture, ICSU General Assembly, Sofia, Bulgaria.

Butler, G.C.(ed.) (1978) *Principles of Ecotoxicology*, SCOPE 12, John Wiley and Sons Ltd, Chichester, U.K., 350 pp.

Callicott, J.B. (1989) *In Defence of the Land Ethic: Essays in Environmental Philosophy*, State University of New York Press, Albany, N.Y.

CEARC (1985) *Social impact assessment: a research prospectus*, CEARC/FEARO, Env. Canada, Ottawa, Canada, 16 pp.

CEARC (1986) *Philosophy and themes for research*, CEARC/FEARO, Env. Canada, Ottawa, Canada.

CEARC (1988) *The assessment of cumulative effects: a research prospectus*, CEARC/FEARO, Env. Canada, Ottawa, Canada, 11 pp.

CEARC-NRC (1986) *Cumulative Environmental Effects: A Binational Perspective*, CEARC/FEARO, Env. Canada, Ottawa, Canada, 175 pp.

Clark, W.C. (1985) On the practical implications of the carbon dioxide question, WP-85-43, IIASA, Laxenburg, Austria, 84 pp.

Clark, W.C. (1990) Towards useful assessments of global environmental risks, *In: Understanding Global Environmental Change* (Kasperson, R., Dow, K., Golding, D. and Kasperson, eds.) Earth Transformed Project, Clark University, Worcester, Ma., pp. 5-19.

Clark, W.C. and Munn, R.E. (eds.) (1986) *Sustainable Development of the Biosphere*, Cambridge Univ. Press, Cambridge, U.K., 491 pp.

Cole, H.S.D., Freeman, C., Jahoda, M. and Pavitt, K.L.R. (eds.) (1973) *Thinking about the Future. A Critique of The Limits to Growth*, Chatto and Windus, London, U.K.

Constanza, R. (1991) *Ecological Economics: The Science and Management of Sustainability*, Columbia University Press, New York, 435 pp.

Covello, V.T., von Winterfeldt, D. and Slovic, P. (1986) Risk communication: a review of the literature, *Risk Abstracts*, **3 and 4**, 171-182.

CSTD (1991) *Environmentally Sound Technology Assessment*, Centre for Sci. and Technol. for Develop., New York, USA

Dales, J. (1968) *Pollution, Property and Price*, University of Toronto Press, Toronto, Canada.

Daly, H. (1977) *Steady-State Economics*, W.H. Freedom and Co., San Francisco, Ca.

Daly, H. (1987) The economic growth debate: what some economists have learned but many have not. *J. Envir. Econ. Manage.* **14**, 323-36.

Daly, H. (1990) Sustainable development: from concept and theory towards operational principles, *Population and Development Review*, Proc. from a Hoover Institution Conference

Dickinson, R.E. (1986) Impact of human activities on climate - a framework, in *Sustainable Development of the Biosphere* (W.C. Clark and R.E. Munn, eds.) Cambridge University Press, Cambridge, U.K., pp 252-289.

Dotto, L. (1986) *Planet Earth in Jeopardy*, John Wiley and Sons Ltd., Chichester, U.K., 142 pp. .

ECE (1991) Draft convention on enviromental impact assessment in a transboundary context, ENVWA/R.36, ECE, Geneva, 22 pp.

Edwards, C.J. and Regier, H.A. (eds.) *An Ecosystem Approach to the Integrity of the Great Lakes Basin in Turbulent Times*, Int. Joint Comm., Windsor, Ont. and Great Lakes Fisheries Comm., Ann Arbor, Mich., Pub. 90-4, 299 pp.

EPA (1984) *Risk Assessment and Management: Framework for Decision-Making*, EPA, Washington, D.C.

Engel, J.R. and Engel, J.G. (1990) *Ethics of Environment and Development*, Belhaven Press, London.

Evernden, N. (1985) *The Natural Alien*, Univ. of Toronto Press, Toronto, Canada.

Federov, V.V. (1989) Kriging and other estimators of spatial field characteristics with special reference to environmental studies, *Atmos. Envir.*, **23**, 175-84.

Fedra, K. (1984) Interactive water quality simulation in a regional context: a management-oriented approach to lake and watershed modeling, *Ecol. Model.*, **21**, 209-32.

Fedra, K. (1985) A modular interactive simulation system for eutrophication and regional development, *Water. Res. Research*, **21**, 143-52.

Fedra, K. and Loucks, D.P. (1985) Interactive computer technology for planning and policy modeling, *Water Res. Research*, **21**, 114-22.

Finsterbusch, K. and Wolf, C.P. (eds.) (1977) *Methodology of Social Impact Assessment*, 2nd ed., Dowden, Hutchinson and Ross, Inc., Stroudsberg, Pa.

Forrester, J.W. (1972) *World Dynamics*, Wright-Allen Press, Cambridge, Mass.

Fowle, C.D., Grima, A.P. and Munn, R.E. (eds.) (1986) *Information Needs for Risk Management*, Inst. for Env., Studies, University of Toronto, Toronto, Canada.

Freeman, A.M. (1979) *The Benefits of Environmental Improvement*, John Hopkins Press, Baltimore, Md.

Gaye, C.B., Gelinas, P.J., Faye, A., Oullet, M. and Therrien, P. (1988) Exploitation et gestion des eaux aquifers cotiers: le cas de la nappe infrabasaltique à Dakar, Sénégal, Preprint, Conf. Eaux et Développement, Dakar, Déc.

Georgescu-Roegen, N. (1973) The entropy law and the economic problem, in *Toward a Steady-State Economy* (H.Daly, ed.), W.H. Freedom and Co., San Francisco.

Gleick, J. (1988) *Chaos: Making a New Science*, Penguin, New York, 354 pp.

Goodland, R., Daly, H., El Serafy, S. and von Droste, B. (eds.) (1991) *Environmentally Sustainable Economic Development: Building on Brundtland*, UNESCO, Paris, 100 pp.

Grima, A.P. (1989) Environmental risk assessment and community impact mitigation, in *Risk Perspectives on Environmental Impact Assessment* (A.P. Grima, C.D. Fowle and R.E. Munn, eds.), Env. Monog. No. 9), Inst. for Env. Studies, U. of Toronto, pp. 99-112.

Grima, A.P., Fowle, C.D. and Munn, R.E. (1989) *Risk Perspectives on Environmental Impact Assessment*, Env. Mon. No. 9, Inst. for Env. Studies, U. of Toronto, Toronto, Canada, 164 pp.

Hamilton, D.P. (1990) Publishing by - and for - the numbers, *Science*, **250**, 1331-32

Hardin, G. (1968) The tragedy of the commons, *Science*, **162**, 1243-48.

Heaton, G., Repetto, R. and Sobin, R. (1991) *Transforming Technology: An Agenda for Environmentally Sustainable Growth in the 21st Century*, World Resources Institute, Washington, D.C., 39 pp.

Holden, C. (1990) Multidisciplinary look at a finite world, *Science*, **249**, 18-19.

Holling, C.S. (1973) Resilience and stability of ecological systems, *Ann. Rev. Ecology and Systematics*, **4**, 1-23.

Holling, C.S. (1978) *Adaptive Environmental Assessment and Management*, IIASA, John Wiley and Sons, Chichester, U.K., 377 pp.

Hueting, R. (1989) Correcting national income for environmental lossses: toward a practical solution, in *Environmental Accounting for Sustainable Development*, (Ahmad, Y.J., Salah, El Serafy and Lutz, E., eds.) The World Bank, Washington, D.C.

Hueting, R. (1990) Should national income be corrected for environmental losses? A theoretical dilemma, but a practical solution, in *The Ecological Economics of Sustainability* (R. Constanza, ed.) Working Paper 32, The World Bank, Washington, D.C.

Hughes, B.B. (1980) *World Modelling*. The Mesarovic-Pestel World Model in the Context of its Contemporaries, Lexington Books, Lexington, Mass.

Hyman, E.L. (1981) The valuation of extramarket benefits and costs in environmental impact assessment, *Env. Impact Assess. Rev.* **2**, 227-58

ICSU (1992) Proc. Int. Conf. on an Agenda of Science for Environment and Development into the 21st Century, ICSU, Paris.

IGBP (1990) *The Initial Core Projects*, IGBP Rep. No. 12, Swedish Academy of Sci., 250 pp.

IPCC (1990) *Climate Change; the IPCC Scientific Assessment*, Cambridge Univ. Press, Cambridge, U.K., 403 pp.

Isard, W. (1972) *Ecological-Economic Analysis for Regional Development*, Tree Press, New York.

IUCN/UNEP/WWF (1991) *Caring for the World: A Strategy for Sustainability*, IUCN, Gland, Switzerland, 133 pp.

Izrael, Yu.A. (1980) Main principles of the monitoring of natural environmental pollution, In, *Proc. Symp. on the Development of Multi-Media Monitoring of Env. Pollution*, Special Env. Rep. No. 15, WMO, Geneva, Switzerland.

Izrael, Yu.A. (ed.) (1983) *Integrated Global Monitoring of Environmental Pollution*, Gidrometeoizdat, Leningrad, USSR, 371 pp.

Izrael, Yu.A. and Munn, R.E.(1986) Monitoring the environment and renewable resources, In, *Sustainable Development of the Biosphere* (W. Clark and R.E.Munn, eds.), Cambridge Univ. Press, Cambridge, U.K., pp. 360-375.

Jacobs, P. and Sadler, B. (1990) *Sustainable Development and Environmental Assessment: Perspectives on Planning for a Common Future*, CEARC-FEARO, Env. Canada, Ottawa, Canada, 182 pp.

Jacobson, H.K. and Price, M.F. (1990) *Framework for Research on the Human Dimensions of Global Environmental Change*, Int. Social Sci. Council, Paris, France.

Jäger, J., Sonntag, N., Bernard, D. and Kurz, W. (1991) *The challenge of sustainable development in a greenhouse world: some visions of the future*, Stockholm Env. Inst., Stockholm, Sweden, 85 pp.

Johnson, A.E. (1991) Benefits from long-term ecosystem research: some examples from Rothamsted, In *Long-Term Ecological Research; an International Perspective* (P. Risser, ed.), SCOPE 47, John Wiley and Sons Ltd., Chichester, U.K., pp 89-113.

Keyfitz, N. (1992) Completing the worldwide demographic transition: the relevance of past experience, *Ambio*, **21**, **1**, 26-30.

Krawetz, N.M., MacDonald, W.R. and Nichols, P. (1987) *A framework for effective monitoring*, CEARC/FEARO, Env. Canada, Ottawa, Canada, 92 pp

Krupnick, A.J. and Portney, P.R. (1991) Controlling urban air pollution: a benefit-cost assessment, *Science*, **252**, 522-8.

Lam, D.C.L., Fraser, A.S., Swayne, D.A., Storey, J. and Wong, I. (1988) Regional analysis of watershed acidification using the expert system approach, *Env. Software* **3**, 127-34.

Landefeld, J.S. and Hines, J.R. (1985) National accounting for non-renewable resources in the mining industries, *Rev. of Income and Wealth*, **31**, 1-20.

Lave, L. and Gruenspecht, H. (1991) Increasing the efficiency and effectiveness of environmental decisions: benefit-cost analysis and effluent fees: a critical review, *J. Air Wast Manag. Assoc.*, **41**, 680-93.

Leontief, W. (1970) Environmental repercussions and the economic structure: an input-output approach, *Rev. Economics and Statistics*, **52**, 403-22.

Likens, G.E., Bormann, F.H., Pierce, R.S., Eaton, J.S., and Johnson, N.M. (1977) *Biogeochemistry of a Forested Ecosystem*, Springer-Verlag, New York, 146 pp.

Linthicum, K.J., Bailey, C.L., Davies, F.G. and Tucker, C.J. (1987) Detection of Rift Valley Fever viral activity in Kenya by satellite remote sensing imagery, *Science*, **235**, 1656-59.

Lollar, C. (1990) Eastern Europe awakens to new worldwide scientific opportunities, *Science*, **250**, 1155.

Lorenz E. (1963) Deterministic nonperiodic flow, *J. Atm. Sci.*, **20**, 448-64.

Lovelock, J.E. (1979) Gaia: *A New Look at Life on Earth*, Oxford Un. Press, Oxford, U.K., 157 pp.

Lovelock, J.E. (1988) *The Ages of Gaia: A Biography of Our Living Earth*, W.W. Norton and Co., New York, N.Y., 252 pp.

Lutz, W., Prinz, C., Wils, A.B., Buttner, T. and Heilig, G. (1990) Alternative demographic scenarios for Europe and North America, *In: Future Demographic Trends in Europe and North America* (W. Lutz, ed.), pp.523-69.

Mandelbrot, B. (1977) *The Fractal Geometry of Nature*, W.H. Freeman, New York, 468 pp.

Mann, C. (1991) Lynn Margulis: science's unruly earth mother, *Science*, **252**, 378-81.

May, R,M. (1974) Biological populations with nonoverlapping generations: stable points, stable cycles and chaos, *Science*, **186**, 645-7.

McKay, B.J. and Acheson J.M. (eds.) (1987) *The Question of the Commons*, University of Arizona Press, Tucson, Ariz.

Meadows, D.H., Meadows, D.L., Randers, J. and Behrens III, W.W. (1972) *The Limits to Growth*, Potomac Associates, Washington, D.C.

Meadows, Denis and Meadows, Donella H. (1973) *Towards Global Equilibrium: Collected Papers*, Wright-Allen Press, Cambridge, Mass.

Mesarovic, M. and Pestel, E. (1974) *Mankind at the Turning Point*, Hutchinson, London, U.K.

Miller, C.T., Kleindorfer, P.R. and Munn, R.E. (eds.) (1986) *Conceptual Trends and Implications for Risk Research*, IIASA, Laxenberg, Austria.

Mooney, H.A., Medina, E., Schindler, D., Schulze, E.-D., and Walker, B.H. (1991) *Ecosystem Experiments*, SCOPE 45, John Wiley and Sons Ltd, Chichester, U.K.

Munn, R.E. (ed.) (1979) *Environmental Impact Assessment*, SCOPE 5, 2nd edition, John Wiley and Sons, Chichester, U.K., 190 pp.

Munn, R.E. (1981) *The Design of Air Quality Monitoring Networks*, MacMillan Pub. Ltd., London, U.K., 109 pp.

Munn, R.E. (1990) Towards sustainable development, In *Proc. of a Symp. on Managing Env. Stress*, University of New South Wales, Australia, pp 11-19.

Munn, R.E. (1991) A new approach to environmental policy making: the European futures study, *Sci. Total Env.*, **108**, 163-72.

Munn, R.E. (1992) Monitoring for ecosystem integrity, In *Proc. Workshop on Ecological Integrity and the Management of Ecosystems*, University of Waterloo, Canada (in press).

Munn, R.E., Alcamo, J. and Federov, V. (1988) Evaluating the performance of air quality models in a policy framework, In, *Air Pollution Modeling and its Application 6* (H. van Dop, ed.), Plenum Press, New York, pp. 237-56.

Munro, D.A., Bryant, T.J. and Matte-Baker, A. (1986) *Learning from experience: a state-of-the-art review and evaluation of environmental impact assessment audits*, CEARC-FEARO, Env. Canada, Ottawa, 48pp.

Naess, A. and Rothenberg, D. (1989) *Ecology, Community and Lifestyle*, Cambridge University Press, Cambridge, U.K.

NAPAP (1990) *Integrated Assessment, Questions 1 to 5*, 3 volumes, External Review Draft, NAPAP, Wasington, D.C., 600 pp.

NAS (1986) *Acid Deposition: Long-Term Trends*, National Academy Press, Washington, D.C., 506 pp.

NAVF (1990) *Sustainable Development, Science and Policy: The Conference Report*, Norwegian Res. Council for Science and Humanities, Oslo, 579 pp.

Niang, A. (1988) Présentation de la cellule de gestion des eaux souterraines du Sénégal, Preprint, Conf. "Eaux et Développement", Dakar, Déc.

Norberg-Bohm, V., Clark, W.C., Koehler, M. and Marrs, J. (1990) Comparing environmental hazards: the development and evaluation of a method based on a causal taxonomy of environmental hazards, In, *Usable Knowledge for Managing Global Climatic Change* (W.C. Clark, ed.), The Stockholm Environmental Institute, Swedish Academy of Science, pp. 18-67.

Norgaard, R. (1989) Linkages between environmental and national income accounts, in *Environmental Accounting for Sustainable Development*, (Ahmad, Y.J., El Serafy, Salah and Lutz, E. eds.), UNEP-World Bank Symposium, The World Bank, Washington, D.C.

OECD (1975) *The polluter pays principle: definition, analysis, implication*, OECD, Paris.

OECD (1979) *Interfutures: Facing the Future, Mastering the Probable and Managing the Unpredictable*, OECD, Paris.

OECD (1989) *OECD Environmental Data Compendium 1989*, OECD, Paris.

OECD (1991) *State of the Environment 1991*, OECD, Paris, 297 pp.

O'Riordan, T., Kemp, R., and Purdue (1985) How the Sizewell B inquiry is grappling with the concept of acceptable risk. *J. Env. Psych.* **5**, 69-85.

Ozalins, G. (1990) Global efforts to produce accurate human exposure data, *J. Air Waste Manag. Assoc.* **40**, 962-4.

Parry, M. (1990) *Climate Change and World Agriculture*, Earthscan Pub. Ltd, London, U.K., 157 pp.

Partridge, E. (ed.) (1980) *Responsibilities for Future Generations*, Prometheus Books, Buffalo, N.Y.

Passmore, J. (1980) *Man's Responsibility for Nature*, 2nd edition, Charles Scribner's Sons, New York, N.Y.

Pearce, D. and Markandya, A. (1989) *Environmental policy benefits: monetary valuation*, OECD, Paris, France

Pearce, D., Markandya, A. and Barbier, E.B. (1989) *Blueprint for a Green Economy*, Earthscan, London, U.K.

Pellew, R. (1991) Disaster in the gulf, *IUCN Bulletin*, **22**, 17-18.

Peskin, H.M. (1990) in *The Ecological Economics of Sustainability*, (R. Constanza ed.), Env. Working Paper 32, World Bank, Washington, D.C.

Pezzey, J. (1989) *Economic analysis of sustainable growth and sustainable development*, Env. Dpt. Working Paper No. 15, World Bank, Washington, D.C., 88 pp.

Pittock, A.B., Ackerman, T.P., Crutzen, P.J., MacCracken, M.C., Shapiro, C.S. and Turco, R.P.(1989) *Environmental Consequences of Nuclear War*, Vol. 1, 2nd edition, John Wiley and Sons Ltd, Chichester, U.K., 359 pp.

Price, M.F. (1990) Humankind in the biosphere, *Global Env. Change*, **1**. 3-13.

Prigogine, I. and Stengers, I. (1984) *Order out of Chaos*, Bantam Books, New York.

Rapport, D.J., Regier, H.A. and Hutchinson, T.C. (1985) Ecosystem behaviour under stress, *Amer. Naturalist*, **125**, 617-40.

Ravetz, J.R. (1986) Usable knowledge, usable ignorance: incomplete science with policy implications, In *Sustainable Development of the Biosphere* (W.C. Clark and R.E. Munn, eds.) Cambridge University Press, Cambridge, U.K., pp. 415-32.

Regan, T. (1983) *The Case for Animal Rights*, University of California Press, Berkeley, Ca.

Regier, H.A., Whillans, T. and Grima, A.P. (1980) Rehabilitation of the Long Point ecosystem, *J. Urban and Env. Affairs*, Univ. Waterloo, Waterloo, Ont.

Regier, H.A. and Grima, A.P. (1984) The nature of Great Lakes ecosystems, *Int. Bus. Lawyer*, June issue, pp. 261-9.

Regier, H.A. and Baskerville, G.L. (1986) Sustainable re-development of regional ecosystems degraded by exploitive development, In, *Sustainable Development of the Biosphere* (W. Clark and R.E. Munn, eds.) Cambridge University Press, Cambridge, U.K., pp. 75-101.

Repetto, R. (1989) A new means of calculating wealth can save the environment, *Washington Post*, May 28,

Repetto, R. and Holmes, T. (1983) The role of population in resource depletion in developing countries *Population and Development Rev.* **9**.

Repetto, R., Magrath, W., Wells, M., Beer, C. and Rossini, F. (1989) *Wasting Assets; Natural Resources in the National Income Accounts*, World Resource Institute, Washington, D.C.

Risser, P. (ed.) (1991) *Long-Term Ecological Research; an International Perspective*, SCOPE 47, John Wiley, Chichester, U.K., 294 pp.

Robinson, J. (1988) Unlearning and backcasting: rethinking some of the questions we ask about the future, *Technol. Forecast. Soc. Change*, **33**, 325-38.

Rodhe, H. and Herrera, R. (1988) *Acidification in Tropical Countries*, SCOPE 36, John Wiley and Sons Ltd., Chichester, U.K., 424 pp.

Rolston, H. (1988) *Environmental Ethics: Duties to and Values in the Natural World*, Temple University Press, Philadelphia, Pa.

Rosswall, T., Woodmansee, R.G., and Risser, P.G. (eds.) (1988) *Scales and Global Change*, John Wiley and Sons Ltd., Chichester, UK, 355 pp.

Ruckelshaus, W.D. (1985) Risk, science and democracy, *Issues in Sci. and Technol.*, **1**, 19-38.

Salam, A. (1991) Spreading the word, *Nature*, **353**, 457-8.

Sanglier, M. and Allen, P. (1989) Evolutionary models of urban systems: an application to the Belgian provinces, *Env. and Planning,A.*, **21**, 477-98.

SCOPE 13 (1979) *The Global Carbon Cycle* (Bolin, B., Degens, E.T., Duvineaud, P. and Kempe, S., eds.) John Wiley and Sons Ltd, Chichester, U.K., 491 pp.

SCOPE 19 (1984) *The Global Biogeochemical Sulphur Cycle* (Ivanov, M.V. and Freney, J.R., eds.) John Wiley and Sons Ltd, Chichester, U.K., 470 pp.

SCOPE 21 (1983) *The Major Biogeochemical Cycles and their Interactions* (Bolin, B. and Cook, R.B., eds.) John Wiley and Sons Ltd, Chichester, U.K., 554 pp.

Sewell, W.R.D. and O'Riordan, T. (1976) The culture of participation in environmental decision-making, *Natural Res. J.*, **16**, 1-22.

Shaw, R., Gallopin, G., Weaver, P. and Öberg, S. (1991) Sustainable development: a systems approach, Contract report to UNCED and Canada External Affairs, IIASA, Laxenburg, Austria, 122 pp.

Shaw, R.P. (1989) Rapid population growth and environmental degradation: ultimate versus proximate factors, *Env. Conservation*, **16**, 199-208.

Singer, P. (1977) *Animal Liberation: A New Ethics for our Treatment of Animals*, Avon Books, New York, N.Y.

SMIC (1971) *Inadvertent Climate Modification*, MIT Press, Cambridge, Mass., 308 pp.

Stigliani, W.M. (1990) Chemical emissions from the processing and use of materials: the need for an integrated emissions accounting system, *Ecolog. Econom.*, **2**, 325-41.

Stigliani, W. (ed.) (1991) *Chemical time bombs: definition, concepts and examples*, IIASA Exec. Rep. 16, IIASA, Laxenburg, Austria, 23 pp.

Stigliani, W., Brouwer, F.M., Munn, R.E., Shaw, R.W. and Antonofsky, M. (1989) Future environments for Europe: some implications of alternative development paths, *Sci. Total Env.*, **80**, 1-102.

Steele, D. (1991) Spotting famine from space, *Nature*, **350**, 545.

Stone, C. (1974) *Should Trees Have Standing? Towards Legal Rights for Natural Objects*, Avon Books, New York, N.Y.

Suprapto, R.A. (1990) Social impact assessment and planning: the Indonesian experience, *Impact Assess. Bull.* **8**, 25-9.

Swartzman, D. (ed.) (1982) *Cost-benefit analysis and environmental regulations*, The Conservation Foundation, Washington, D.C.

Tellier, L.-N. (1983) Prospective économique et équilibre alimentaire dans le cadre du tiers monde: un modèle non-linéaire, I.N.R.S. - Urbanisation, Etudes et Documents no. 37, I.N.R.S., Montreal.

Tellier, L.-N. (1985) Demographic growth and food production in developing countries: a non-linear model, *Ekistics*, **313**, 383-5.

Tolba, M.K. and White, G.F. (1979) Global Life Support Systems, Joint Press Release of UNEP and SCOPE, 2 pp.

Toth, F. (1988) Policy exercises, *Sim. and Games*, **19**, 235-76.

UNEP (1990) *GEMS*, UNEP, Nairobi, 32 pp.

UNEP (1991) *Environmental Data Report*, Third Ed., UNEP/Blackwell, Oxford, U.K., 408 pp.

UNSCEAR (1988) *Sources, Effects and Risks of Ionizing Radiation*, United Nations, New York, 647 pp.

Vernadsky, W.I. (1945) The biosphere and the noosphere, *Amer. Scientist*, **33**, 1-12.

Walker, B.H. and Sinclair, A.R.E. (1990) Problems of development aid, *Nature*, **343**, 587.

Walters, C. (1986) *Adaptive Management of Renewable Resources*, MacMillan, London, U.K., 374 pp.

Whyte, A.V. and Burton, I. (1980) *Environmental Risk Assessment*, SCOPE 15, John Wiley and Sons Ltd., Chichester, U.K.

Whitney, J.B., Dufournaud, C.M. and Murck, B.W. (1987) An examination of alternatives to traditional fuelwood use in the Sudan, *J. Env. Manag.*, **25**, 319-46.

WHO (1990) *The Human Exposure Assessment Location*, WHO, Geneva, 7 pp.

Wils, W. (1982) A decade since *Limits to Growth: a review of ten global studies*, ABC Programme Office, Delft Hydraulics Lab., Delft, The Netherlands.

WMO (1990) *WMO and global warming*, Pub. no. 741, WMO, Geneva, 24 pp.

WMO (1991) *Proc. Second World Climate Conf.*, WMO, Geneva, 575 pp.

Worthington, E.B. (1975) *The Evolution of IBP*, Cambridge University Press, Cambridge, U.K.

WRI (1990) *World Resources 1990-91*, Oxford University Press, Oxford, U.K., 383 pp.

Yanshin, A.L. (1988) Reviving Vernadsky's legacy: ecological advances in the Soviet Union, *Environment*, **30**, 7-9, 26-7.

CHAPTER 21

Perceptions and attitudes

Introduction

Human perceptions of the environment, and attitudes towards it, have evolved as an integral part of the long history of human interactions with the rest of nature. Hunter/gatherer cultures revealed some of their perceptions in dramatic cave and rock-shelter paintings and carvings in regions as far apart as Australia and Western Europe. The development of cultivation and livestock husbandry in various parts of the world about 10,000 years ago allowed people to exchange the uncertainties of hunting and wandering for the routines of settlement, but it also brought new kinds of impact on the environment, and greatly changed human relationships with it. Settled peoples developed skills in pottery, building and in the mining of ores and the smelting of metals. Historical records indicate that such activities took place some 7,000 years ago in Egypt, Iran and Thailand.

The recognition that people can damage or deplete the natural resources on which they depend is also ancient. Plato, in *Critias*, described deforestation and soil erosion as the negative side of power (McCraken, 1987). Some civilizations declined because their resource management, especially of water and soil, was not good enough to maintain agricultural productivity. Conservation measures began with religious sanctions that protected some species of animals, and certain plants and forest groves, and with practices to prevent soil erosion and to maintain fertility. Some early civilizations created reserves to protect wildlife or natural areas (O'Riordan, 1981; Goudie, 1986; McCraken, 1987; Nicholson, 1987).

The evolution of increasingly self-reliant urbanized cultures, especially in Western Europe, seems to have been accompanied by an alienation from nature, and even a hostility towards it. The world-wide concern about the environment that is so evident today is a relatively recent phenomenon, stimulated by the equally recent recognition of the pervasive nature of human impact, and the threats that this could pose. This concern has led, in turn, to demands for changes in approach, at international, national, community and individual levels.

What people do depends on what they believe: on their religion, ethics or codes of conduct. It also depends on what they know, and how free they are to act. Everyone's priorities are influenced by their social circumstances, and the deprived poor have more immediate preoccupations than the environment of other groups far away in space or time. Contemporary public attitudes reflect today's environment, today's information, and today's economic, social and political circumstances. Changing human behaviour towards the environment similarly demands the alteration of perceptions and attitudes, and especially the creation of circumstances under which the new behaviour is seen to be both rational and possible. People will accept a new ethic for sustainable living when they are persuaded that it is right and necessary to do so, when they have sufficient information, and when they are enabled to obtain the required knowledge and skills (IUCN/UNEP/WWF, 1991).

This chapter describes how perceptions and attitudes have changed since 1972, the factors governing those changes, and the environmental consequences. These changes have been closely interlinked with the growth and dissemination of scientific knowledge, the development and application of new economic and technical tools for environmental management, described in Chapter 20, and the growth of national and international institutions, described in Chapters 22 and 23.

Over-generalization is dangerous. The world is environmentally and culturally diverse, and global generalizations often obscure the critical detail of 5.3 billion human situations and attitudes. Too much 'environmental' literature is written from the narrow perspective of a few thousand authors from the developed countries. They tend to emphasize problems such as pollution, deforestation and the loss of biological diversity. But this is only one perspective - and for several billion people it is a distorted one. Poor people have a perspective of the environment dominated by insufficient food and water, inadequate shelter and fuelwood, no safe waste disposal, and a lack of health care, education and employment. These attitudes are, in the end, more powerful because they are more prevalent.

Public perceptions and their development

The historical background

The cultural heritage of humanity is one of our greatest prizes. One of the many strands in the evolution of attitudes and perceptions in recent decades has been the acceptance that the life-styles and traditions of many indigenous peoples are based on a real understanding of the environment, and that other peoples can learn much from them.

The individual's perception of environment is moulded (Figure 1) by tradition (especially as conveyed through parental influence); personal observation and experience; education; and non-formal information from a diversity of sources. As individuals age, the proportional emphasis of these sources commonly changes. Cultural pressures, education and non-formal information often reduce the perceived importance of tradition. The tragedy is that this process often inculcates values that reduce rather than enhance sustainable living - that is, living that is in enduring harmony with other components of the world of nature.

Figure 1: *Influences on the individual's perception of environmental issues.*

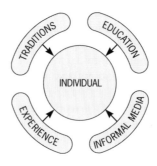

Table 1: *Foundation of selected private environmental organizations, 1865—1914.*

Year	Organization	Country
1865	Commons, Open Spaces and Footpaths Preservation Society	Britain
1867	East Riding Association for the Protection of Sea Birds	Britain
1870	Association for the Protection of British Birds	Britain
1880	Fog and Smoke Committee (National Smoke Abatement Institution from 1882)	Britain
1883	American Ornithologists Union Natal Game Protection Association	United States South Africa
1885	Selborne Society	Britain
1886	Audubon Society (lapsed 1889, revived 1905)	United States
1891	Society for the Protection of Birds	Britain
1892	Sierra Club	United States
1895	National Trust	Britain
1896	Massachusetts Audubon Society	United States
1898	Coal Smoke Abatement Society (now National Society for Clean Air)	Britain
1903	Society for the Preservation of the Wild Fauna of the Empire	Britain
1909	Swiss League for the Protection of Nature Swedish Society for the Protection of Nature Wildlife Preservation Society	Switzerland Sweden Australia
1912	Society for the Promotion of Nature Reserves	Britain
1913	British Ecological Society	Britain

In the developed countries of North America and Western Europe, modern environmental attitudes and perceptions grew from concern over the damage caused by unconstrained industrialization. The environmental pioneers of the nineteenth century, such as Alexander von Humboldt and George Perkins Marsh, expressed concern about the impacts of human transformation of the landscape (Marsh, 1864). These were followed by demands for action to curb industrial pollution (Ashby & Anderson, 1981) and by the growth of a 'Conservation Movement'. In 1864 the first National Park was established in Yosemite, and conservation societies blossomed rapidly in North America and Europe (Table 1).

It is a curious paradox that the development of ecological science had only a belated impact on the modern environmental movement. One reason was the desire of many academic ecologists to study 'natural' systems unconfounded by the variables of human impact, but during the 1960s the move to a more human-related ecology gained ground, propelled in part by Soviet scientists who demanded that the proposed International Biological Programme should focus on 'the study of the biological basis of man's welfare' (Waddington, 1975). The IBP led on to the Man and Biosphere programme (MAB) of UNESCO. In parallel, the late 1960s and early 1970s saw the publication of some major scientific reviews, including the *Study of Critical Environmental Problems* (SCEP, 1970), the *Study of Man's Impact on Climate* (SMIC, 1971), and the modelling studies promoted by the Club of Rome, whose *Limits to Growth* (Meadows *et al.*, 1972) had an immense public impact. The tide of publications on the environment by articulate and informed scientists soon swelled to a flood, and some books, for example Rachel Carson's *Silent Spring* (1962), caught the attention of the media.

The period up to 1972 saw the growing together of these various strands. The wildlife conservation movement spread through the developed countries and became increasingly demanding of better protection for nature and its habitats. Nature conservation organizations were founded in a number of countries. The national conservation bodies became more vociferous and broadly based, and the new politics of environment took root in the shape of "Green" parties that increasingly stressed the need for a more caring and holistic approach to the environment and its resources, which they perceived as the fundamental base for the human future (Bramwell, 1989). The IBP and its successor, MAB, and the activities of international environmental organizations such as UNESCO, ICSU, IUCN, WWF and other non-governmental bodies, made ecology relevant as the base for new policies. The philosophy that George Perkins Marsh had enunciated in the 1860s at last became mainstream thinking in the environmental movement, at least in Europe and North America.

There are few records of attitudes in the developing countries during this period. However, respect for nature is an integral element in many great religions, including Islam and Buddhism, and in cultures from Polynesia to the Amazon. There are few more eloquent statements of the need to respect and care for the Earth than the letter written by Chief Seathl (Seattle) of the Dwamish and allied tribes of Puget Sound in the Pacific Northwest region to the President of the United States in 1855 (Royston, 1979). Undoubtedly their beliefs, coupled with tradition and their closeness to the land, made many rural people naturally careful of their environment. Their impact was in any event trivial compared to the swelling impact of the new northern industrialization which had already begun to promote consumerism, waste natural resources, and draw resources from south to north. But the tide of economic, cultural and political dominance was such that the principles of environmental management that many 'southern' rural communities had developed awaited rediscovery by the dominant world powers for many decades.

In the period immediately before the 1972 Stockholm Conference, the environment was not an issue of public concern in the majority of the developing countries. Many of the calls for nature protection in Africa or for the protection of the Amazonian rain forests, for example, came from outside the countries concerned. At the time of independence, most of the African countries had more pressing political agendas than the protection of the environment, although their dependence on agriculture was near absolute, and the safeguarding of the productivity of the land and the provision of adequate water in irrigated areas were major concerns. The demands for conservation by the environmental movements of the developed countries seemed to be at best a diversion, and at worst an obstruction to the development process in the countries with greatest human needs.

The period 1972 - 1992

During the 1970s, environmental perceptions broadened enormously and environmental issues became established as a permanent feature of national and international policy. The United Nations Conference on the Human Environment, in Stockholm (1972), was the most important event in the growth of environmental awareness. It came about because public pressure, backed by scientific findings about the impacts of industrial emissions, persistent pesticides and other pollutants in the late 1960s, stimulated the necessary political will, at least in developed countries. Developing countries were initially suspicious, but a seminal meeting at Founex, Switzerland, in 1971 allowed their central concerns over the environmental problems caused by lack of development to be voiced, and opened the way for their active participation in the Stockholm Conference itself (Tolba, 1982; WCED, 1987). In 1972 the Club of Rome, in its first report *The Limits to Growth* (Meadows *et al.*, 1972), called attention to the possibility that development might be halted, or even reversed, as a result of the interactions between human population growth, pollution and resource constraints and, although many of its projections and assumptions came under detailed criticism, it added another strand to the concept of sustainability. Finally, the Action Plan for the Human Environment adopted by the Stockholm Conference, the establishment of the

United Nations Environment Programme in the same year, and the enthusiasm of non-governmental organizations, before, during and after the Conference, gave the environmental movement further impetus and effective expression in the international community.

After Stockholm, a further meeting at Cocoyoc, Mexico, in 1974 reinforced understanding of the imperative for action to deal with the acute problems of poverty in the developing countries, and laid the foundations of the now-familiar concept of sustainable development. The widely-publicized environmental debate at Stockholm, and the momentum it gave to world action, influenced millions of people around the world. A number of major studies and reports appeared during the 1970s, including that of the UN Conference on Desertification, held in Nairobi in 1977 (UNCOD, 1977), the OECD Interfutures study (OECD, 1979), and the report of the Brandt Commission (1980). A list of key documents published between 1970 and 1991 appears in Box 2. During the 1980s, there were perhaps three major strands in the evolution of public awareness and attitude:

a) alarm over widely-publicized tragedies and threats (see Box 1) which had a direct impact on public attitudes, particularly to the acceptability of nuclear power (Figure 2);

b) the increasing recognition that development is essential, and that conservation and development have to be part of the same process. The Stockholm Conference established this principle for the first time, and by doing so made conservation relevant to those whose chief concern was to eradicate what had come to be called 'the pollution of poverty';

c) the recognition that concern for the environment was not some peripheral green fancy or additional sectoral element among the many preoccupations of governments, but a demand that the real wealth of nations be cherished.

These ideas were reflected in the growth of national laws and departments concerned with the environment (Chapter 22), and of international legal instruments (Chapter 23).

The role of the media

The media were reactive rather than innovative during much of the period between 1970 and 1990. Coverage rose and fell in response to disasters as much as to advances in scientific understanding. All the events listed in sections A and B of Box 1 received extensive coverage by the press, radio and television. The Bhopal accident was ranked as the second biggest news item in 1984 by the Associated Press editors, and the Ethiopian drought ranked third (Wilkins, 1987).

The number of people reached by the media also increased dramatically during the period. Between 1975 and 1985 alone, the number of radios world-wide increased by 58 per cent, while the number of television receivers nearly

---

BOX 1

Major environmental incidents prompting public concern.

A. Large-scale industrial accidents
Seveso (1976) - chemical escape and hazardous waste.
Three Mile Island (1979) - radioactivity.
Bhopal (1984) - chemical escape.
Basel (1986) - chemical escape.
Chernobyl (1986) - radioactivity.

B. Oil pollution of the sea
Amoco Cadiz (1978).
Exxon Valdez (1989).
Iran-Iraq War (1980s).
War over Kuwait (1991).

C. Environmental degradation
Desertification, deforestation and drought.
Famine, floods and storms.
Soil erosion.
Extinction of species and loss of biological diversity.
Depletion of non-renewable resources.
Pollution, national and trans-boundary.
Ozone depletion.
Climate change.
Proliferation of high-risk technologies, toxic chemicals and hazardous wastes.
Worsening spiral of poverty and environmental damage, especially in developing countries.

---

doubled. The biggest expansion was in developing countries where the number of radios increased 116 per cent and that of television sets increased threefold (Stevenson, 1986) (Table 2). The UNESCO target set in 1961 of one television set per 50 people has been achieved for the Third World countries as a whole. India had one million television sets in 1980, 7 million in 1985 and was expected to have 35 million sets in 1990; transmission would then have reached some 300 million people.

The principal medium (or media) disseminating environmental information varies from country to country, and depends on education and socio-economic levels. A study in the United States of America showed that newspapers and television, in that order, were the major sources of environmental information (Ostman and Parker, 1986/87), and that as adult education increased, newspapers were increasingly preferred to television. Other print media,

---

<table>
<tr><td colspan="3" align="center">**BOX 2**</td></tr>
</table>

Major reports on environment and development.

1980	*World Conservation Strategy*	IUCN/UNEP/WWF
1982	Report of Session of Special Character of Governing Council	UNEP
	The World Environment, 1972 - 1982	UNEP
1987	*Our Common Future*	World Commission on Environment and Development
1987	*Environmental Perspective to The Year 2000 and Beyond*	Governing Council, UNEP
1991	*Caring for the Earth: A Strategy for Sustainable Living* (Successor to the *World Conservation Strategy*)	IUCN/UNEP/WWF

especially books, were perceived as more believable sources of scientific information on the environment. Another study in the United Kingdom showed that 52 per cent of a sample of the population trusted television news programmes about the environmental impacts of nuclear power, while 33 per cent expressed more trust in newspaper and magazine articles (Market and Opinion Research International, 1987). A third study showed that television is the most important source of environmental information among children and the elderly (Fortner and Lyon, 1985). In developing countries, where illiteracy rates are high, radio and television predominate, but there are also wide variations, even in the same country. For example, in Egypt, television and newspapers are the main source of information in major urban centres, but radio and television predominate in semi-urban and rural areas, while radio is the main source of information in remote areas (El-Hinnawi, 1986).

The power of the mass media to influence attitudes is generally weakest in rural areas of developing countries, where environmental pressures are strongest. Traditional interpersonal forms of communication appear to be far more effective in such areas (for example in Southeast Asia). The mass media generally appears alien, almost always projecting the values and priorities of elites in the cities, and rarely communicating in local dialects. Experiments and studies demonstrate that poor people respond most readily when communicators relate to their own local circumstances and cultures, and when they interact with their audiences.

Figure 2: *Public opposition in selected countries to the building of additional nuclear power-generation facilities: before and after Chernobyl.*

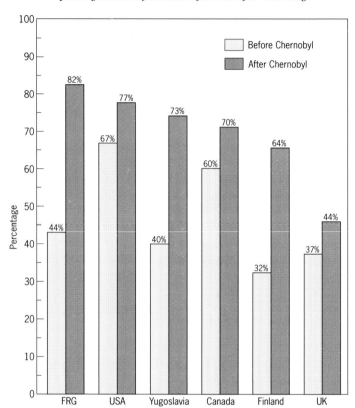

Source: UNEP, *The State of the Environment* (1988), based on data given in Flavin, Reassessing nuclear power: the fallout from Chernobyl, *Worldwatch*, Paper, No. 75 (1987), and Riordan, T.O. and others, Nuclear accidents and emergency planning: Sizewell B. Inquiry in the light of the Chernobyl disaster. *Project Appraisal*, vol. 1 (1986).

Public attitudes to media reports (and, indeed, to statements by scientists and public figures) have become much more critical. An opinion survey in the United States (Ostman and Parker, 1986/87) found that only about 55 per cent of respondents perceived media personnel as 'telling the truth' about environmental topics; 58 per cent perceived a political leaning; and about 81 per cent felt the media sensationalized and were selective in order to maximize their audience. A survey of Indian students (Sekar, 1981) found that 74 per cent felt that newspapers could play only a limited role in creating mass interest in current environmental problems: inadequacy and superficiality of coverage were two of the reasons for this belief, along with mass illiteracy and indifference among readers. On the other hand, a British survey showed that all media are trusted more than government ministers and officials, or the nuclear industry, for information about the environmental impact of nuclear

Table 2: *The numbers of radio and television receivers in developed and developing countries, 1965/75/85.*

Radio receivers

Area	Number of radio receivers (in millions)			Number per 1,000 inhabitants		
	1965	1975	1985	1965	1975	1985
World	573	1032	1776	170	255	362
Africa	6	17	62	26	58	142
Arab States	6	17	48	56	121	247
Asia	51	132	452	27	58	162
Europe	222	348	523	272	478	675
Latin America/Caribbean	34	81	134	137	251	327
North America	251	424	532	1173	1797	1992
Oceania	3	13	25	171	819	1000
Developed countries	498	841	1182	488	762	988
Developing countries	75	191	594	32	55	150

Television receivers

Area	Number of TV receivers (in millions)			Number per 1,000 inhabitants		
	1965	1975	1985	1965	1975	1985
World	192	414	710	57	102	145
Africa	0.1	0.6	6	0.4	2	13
Arab States	0.9	3.4	17	8	24	65
Asia	24	58	130	13	25	45
Europe	81	189	280	120	260	362
Latin America/Caribbean	8	27	59	32	84	145
North America	78	133	209	355	584	783
Oceania	2	5	9	137	262	360
Developed countries	181	373	564	177	325	472
Developing countries	11	41	146	4.7	14	39

Source: Childers, *Development (2/90).*

power. Those trusted most of all were people living close to nuclear power plants (MORI, 1987). Two surveys carried out around the Three Mile Island plant in the year after the accident had similar findings (Goldsteen and Schorr,1982; Mitchell, 1982), showing that public trust of information from Federal and State officials was very low, while information given by local officials had greater credibility (see Figures 3 and 4). It seems that there is somehow a public mistrust of information originating from government officials in almost all countries.

Figure 3: *Trust and credibility. Public responses in the USA, following the Three Mile Island accident, to the question, 'Do you feel the information was truthful from...'*

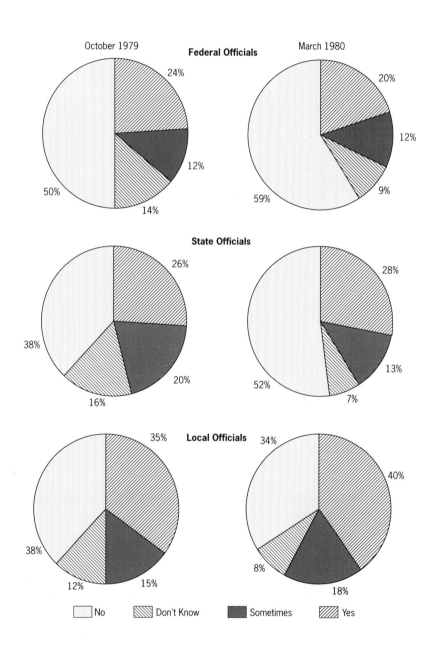

Source: UNEP, *State of the Environment* (1988), based on data given in R. Goldsteen and J. K. Schon, The long-term impact of a man-made disaster, an examination of a small town in the aftermath of the Three Mile Island accident, *Disasters*, **vol. 6** (1982).

Figure 4: *Trust and credibility. Percentage of public in the UK expressing a 'great deal'
or 'fair amount' of trust in the truth of official information about the environmental
impacts of nuclear power in the UK.*

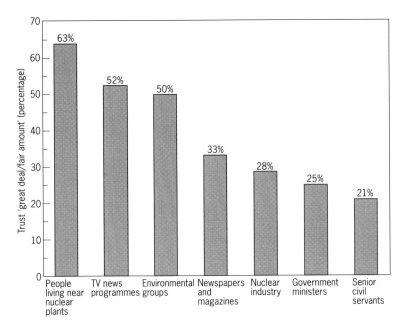

Source: *Nuclear Power, UK*, after Market Opinion Research International,
Public Attitudes to Nuclear Power (MORI, London, 1987).

Public opinion polls as a guide to attitudes

Despite various limitations (Ashby, 1987) linked to the size, structure and
characteristics of the population samples surveyed, public opinion polls still
provide the most useful measure of changing public attitudes. While public
opinion polls carried out in the late 1960s and early 1970s concentrated
mainly on local environmental issues, those conducted more recently have
often been expanded to include national, regional and global issues, as well as
issues related to socio-economics, politics, development and quality of life
(UNEP, 1988a).

High levels of public concern and consciousness about environmental issues
have been recorded in several polls, including the Omnibus Survey for the
European Economic Communities in 1986; surveys in the US by Louis Harris
and Associates in 1986 and 1989, and by Cambridge Reports in 1989; a survey
in Canada by Environics Research Group and Angus Reid Associates in 1988;
and the Gallup Survey for the *Daily Telegraph* in the UK in 1988. Repeated
surveys in the United States of America have revealed both consistently high
and broadly increasing levels of concern during the 1980s (Figure 5).

Figure 5: *Percentage of people who agree that protecting the environment in the United States is so important, 'that requirements and standards cannot be too high and continuing environmental improvements must be made regardless of cost'.*

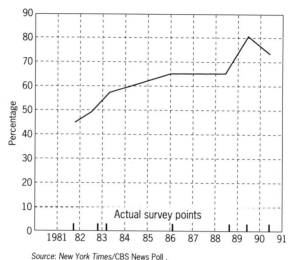

Source: New York Times/CBS News Poll .

These polls (UNEP, 1989) indicate that the majority of the public in Europe and North America:

a) were chiefly concerned with air and water pollution and waste disposal, followed by species extinction and natural resource depletion;

b) did not believe that the relevant authorities were doing enough to protect the environment; and

c) favoured increasing government regulation and spending to control environmental degradation, even if it meant higher taxes or prices.

Most of those polled in the United Kingdom and Canada believed that preservation of the environment should take precedence over economic growth. A sizeable proportion of the people sampled in Scandinavia and in the European Economic Community and the US were already engaged in activities to protect the environment.

A survey of member states of the European Economic Community in 1986 (CEC, 1986), revealed that the public were in general not particularly conscious of pollution in their own local area, found little to complain about regarding their local environment, and considered that the infrastructure services in place were being well managed. People did, however, express concern about new developments in their neighbourhoods, especially if they had environmental disadvantages. There was evidence, for instance, that even where the public was not opposed to nuclear power stations or toxic waste facilities in general, they were vigorously opposed to having them sited in their neighbourhoods. This 'not-in-my-back-yard' (NIMBY) attitude has also been recorded in other industrialized countries (Johnson, 1987; Environment Agency of Japan, 1982).

There have been far fewer surveys of opinion in developing countries, and even the most extensive, the Louis Harris/UNEP survey of 1989 (Box 3), interviewed only small samples, all of them in urban areas. However, such evidence as exists indicates that public concern about local environmental problems varies widely from country to country, and even from place to place within the same country (UNEP, 1988a). For instance, while people in urban centres may be more concerned about air or noise pollution, people in rural areas may be more concerned about lack of safe drinking water, adequate sanitation or proper garbage disposal. An interestingly common phenomenon in Third World countries is that while many people are deeply unhappy about the sordid physical environment that surrounds them, they are often inclined to accept it as their 'fate' (Johnson, 1987) or become 'acclimatized' to such conditions (El-Hinnawi, in prep.).

There is also considerable variation from country to country in the public judgement of the priorities for action on national environmental problems. For example, a 1984 survey in the US (OECD, 1987) indicated that the public was very concerned about nuclear waste disposal, followed by industrial waste disposal, damage to the marine environment, water pollution and air pollution. A 1982 survey in Japan (Environment Agency, 1982) revealed that air pollution was the top public concern, followed by water pollution, damage to natural scenery, noise pollution and garbage disposal. A 1986 survey in the European Economic Community (CEC, 1986) indicated that, on the whole, the public was concerned about damage to the marine environment, pollution of rivers and lakes, disposal of industrial wastes and air pollution, in decreasing order of priority. A 1986 survey in Australia (Department of Arts, Heritage and Environment, 1987) revealed that the public was most concerned about pollution, followed by conservation of flora and fauna and deforestation. A 1986 survey in Canada (SOE, Canada, 1986) indicated that the primary public concerns were water pollution and acid rain. Other environmental issues that have gained public attention more recently in Canada include energy, nuclear power, municipal wastes, toxic substances, rare and endangered species and deforestation. Meanwhile, in many Third World countries, water pollution and sanitation head the list of public concerns along with other major issues such as deforestation (India, Kenya, Indonesia and Brazil), desertification (Ethiopia and other countries of the Sahel), groundwater deterioration (Qatar and Bahrain) and air/water/noise pollution and garbage disposal (Egypt) (UNEP, 1988a).

Concern is one thing: willingness to invest in an appropriate response can be another. A 1990 survey by Yankelovich Clancy Shulman for *Time International* in the USA (McDowell, Schoenthal and Thompson, 1990), revealed that 94 per cent of the public polled considered protecting the environment a very important issue, while 63 per cent supported stronger laws and regulations to get the job done. However, when it came to financing measures to preserve the environment, the public polled were sharply divided:

BOX 3

The Louis-Harris UNEP poll.

The New York-based firm of Louis Harris and Associates conducted a poll for UNEP between February 1988 and June 1989. The countries covered were:

Africa: Kenya, Nigeria, Senegal, Zimbabwe.
Asia and the Pacific: China, India, Japan, Saudi Arabia.
Latin America and the Caribbean: Argentina, Brazil, Jamaica, Mexico.
Europe: Hungary, Germany (Federal Republic), Norway.
North America: USA.

The sample in industrialized countries was designed to be representative of all persons aged 16 and above. In developing countries the sample was confined to major urban centres. Leaders were polled as a separate group except in the USA.

The principal conclusions were:

1. There was widespread concern about the environment. In more detail:
   (a) most people in 15 of the 16 countries rated their national environmental quality as only fair or poor. In Saudi Arabia the majority of the public considered it excellent or pretty good, as did leaders in Zimbabwe, Saudi Arabia and Norway;
   b) all groups in all countries except Saudi Arabia considered that the environment had become worse in the past ten years;
   (c) large majorities in all countries believed that there was a direct link between the quality of the environment and public health;
   (d) in all countries, large majorities of the public believed that more should be done by Government to protect the environment. Except for Brazil, the leaders agreed.
   (e) there was grave concern over the pollution of drinking water, rivers

while 48 per cent were willing to 'go full speed ahead' in 'spending money to clean up the environment', 47 per cent believed that, given other national problems, it would be better to 'go slow'. Despite their general desire for a cleaner environment, 64 per cent of the public polled admitted that they personally 'should be doing more' to achieve this goal, while 80 per cent contended that 'there are so many contradictory things being said about the environment that it is sometimes confusing to know what to do'. With regard and lakes, air and land. There was more limited concern over loss of agricultural land, deforestation, desertification, radioactivity, toxic wastes and acid rain. There was less awareness and concern over climate change, global warming and ozone layer depletion.

2. Human action was accepted as the cause of environmental degradation, and industrial activity and governmental inaction were seen as the two most serious causes of pollution.

3. Most people, including leaders, were pessimistic about both the 5- and 50-year outlooks for the environment. Younger people and women were more concerned than older people and men.

4. None the less, most people did not accept that environmental degradation was inevitable. They believed that if the environment was given a higher priority, and if their governments gave more information to the public, steps would be taken.

5. Most people believed that protecting the environment should be a broadly-based activity involving their government, international organizations, business, farmers, voluntary organizations and individual men and women. In most countries (except China) business and farmers were seen as having a major role.

6. There was wide support for stronger action from governments and international organizations. In most countries, strict environmental laws were accepted. In all the countries surveyed except Brazil, China, India and Saudi Arabia both the public and leaders believed that international co-operation to protect the environment was essential.

7. In most countries, most people were willing to make personal sacrifices to protect the environment. Large majorities were willing to join in voluntary action. In most countries, higher taxes dedicated to environmental improvement were acceptable. Younger people were more willing to make sacrifices than other people. Women, in particular, preferred to see health risks reduced even if standards of living suffered because of the costs incurred.

to environmentally friendly activities, either on a regular or occasional basis, 83 per cent of the public polled said that they returned cans and bottles to a store or recycling centre, 84 per cent shopped for ecologically safe products, 64 per cent saved newspapers for recycling, 77 per cent purchased products made from recycled materials, 54 per cent avoided the products of companies with poor environmental records, and 48 per cent contributed money to environmental groups.

It seems clear that by the end of the 1980s people in many parts of the world were informed, concerned and committed to the cause of environmental protection. Public opinion polls, asking the same questions about environmental issues over a period of years, indicate little or no waning of support for stronger measures, despite adverse economic circumstances (UNEP, 1988a). Such public support has in fact increased in the US since 1980 (OECD, 1985; Table 3). Likewise, public perception that environmental protection was an important and urgent issue in member states of the European Economic Community did not change between 1982 and 1986 (CEC, 1986).

Some of the surveys imply both a growing desire for more, rather than less, government regulation and an increasing awareness of the long-term nature and broad scope of environmental issues (UNEP, 1988a). A notable shift appears to have taken place recently in public perception in several industrialized countries based on a recognition that the state of the environment and economy are interlinked and interdependent, and that international co-operation to solve the world's environmental problems is essential.

The change in perception of the relationship between environment and development has been particularly important. At the time of the Stockholm Conference, and indeed for some years after it, the way environmental concerns were voiced by leaders of developed countries fuelled concern among the leaders of the developing world that the campaign was essentially negative and restrictive: that in the process of halting environmental changes the action demanded would constrain the freedom of developing countries to deal with essential social tasks.

Although traces of this attitude linger, it is now generally evident to both the public and politicians that environment and development are not only mutually inclusive, but also mutually interdependent. The challenge thus lies in achieving development that is both people-centred and conservation-based, and that uses natural resources in a sustainable way (IUCN/UNEP/WWF, 1991).

This poses major challenges to both developed and developing countries, neither of whom (by and large) achieve sustainable resource use. The former need to curb their present excessive consumption of natural resources, their waste of materials and energy, and their pollution of the biosphere with 'greenhouse gases' and other substances, while the latter need to value environmental resources as their 'natural capital' and to avoid eroding the wealth on which future generations will have to live. Global co-operation is increasingly recognized to be necessary, and to need expression in action that will reverse the present net resource flows from the 'south' to the 'north', which in 1989 totalled $US35 billion (see Chapter 16).

An increasing proportion of people in many countries now accept this need for development strategies that enable people to live off 'nature's interest', rather than 'nature's capital' (WCED, 1987; UNEP 1988b; IUCN/UNEP/WWF, 1991). They accept the simple proposition of intergenerational equity: that future generations should not inherit less environmental capital than the present generation inherited. This leads in turn to a recognition of the following priority goals :

- A political system that secures the right of the citizen to know about his or her environmental situation, and to participate effectively in decisions about it.

- Local and national administrative systems that are flexible and can be adapted and adjusted on the basis of the monitoring of progress towards social and environmental targets.

- An economic system that favours the conservation and sustainable use of environmental resources, and contributes to an enhanced quality of life throughout the community.

- A social system that avoids the tensions arising from disharmonious development, or solves them when they do arise.

- A production system that respects the obligation to preserve the environmental base for development.

- A technological system that can provide and install new and sustainable processes.

- An international system that addresses the substantial obstacles to equitable world trade, transfer of technology, and world financial flows that will advance sustainability.

Responses

The response to this quickening of public concern is taking many forms:

- People are seeking individual ways towards a more secure future for themselves and their descendants, through life-styles that use natural resources sustainably. The rise of 'green consumerism' and the emergence of a new ethic of environmental care are both elements in this response.

- People are joining together to promote action at community level. Citizens' groups and environmental non-governmental organizations have become more numerous and stronger since 1972.

- People are pressing governments for action. Indeed, governments are often following where the non-governmental sector has led.

Individual actions: consumer choices

One manifestation of increased public concern for the environment, particularly in developed countries, is the individual's demand for information that will allow him or her to choose products 'friendly' to the environment. The driving force for such changes in consumer attitudes and behaviour is the growing public awareness of the overall costs and consequences of unsustainable life-styles and consumption patterns.

The result (reflected in such documents as the Declaration of the Asia-Pacific Consumer Conference held in Omiya in 1989), is the emergence of a new consumer society wherein people's consumption of products and services is based not only on criteria such as price, quality, durability, performance and after-sale-service, but also on three other important considerations:

- Is it ethical? (Does the producer engage in business practices that exploit lenient legal, social and environmental regulations in other countries, as compared to regulations in the home country of the producer? Is the producer involved in malpractices such as bribery and corruption?).

- Is it ecologically sound? (Does the producer care for the environment, and is the product or service environmentally harmful?).

- Is it equitable? (Does the producer take into account the traditions and economics of local communities and vulnerable sectors of the society?).

As a consequence, environmental considerations have become an integral part of the international consumer movement. Care for the environment is enshrined in the manifesto of consumer rights and responsibilities accepted and advocated by the International Organization of Consumers Unions.

In those countries where green consumerism has become established (Secrett, 1988), its advocates deliberately avoid products and services that:

- endanger the quality of the environment or the safety of other consumers;

- cause environmental degradation during the extraction of natural resources (which serve as raw or source materials), or during their manufacture, use or disposal;

- consume a disproportionately large amount of energy during their manufacture, use or disposal;

- cause unnecessary waste, either because of overpackaging or because of an unduly short 'lifespan';

- utilize materials derived from threatened species or threatened environments; or

- adversely affect other countries and communities, especially those in the developing world.

Over the past decade, many guides and catalogues have been published to aid green consumers to select 'environmentally friendly' products and producers. There have also been several well-documented successes to demonstrate the influence of green consumers, for example in changing the marketing practices of corporations and supermarket chains. Increasing public awareness about environmental issues, and escalating pressure from green consumers, have resulted in a growing demand for products such as phosphate-free detergents, lead-free petrol and ozone-friendly aerosols. 'Green' firms such as Body Shop - the international cosmetic retailer and manufacturer - have been phenomenally successful, while the business of consultancy agencies such as Brand New Product Development and Sustainability - established to advise companies on green strategies - are booming (George, 1989).

Increasingly, green consumers are not confining their attention to the environmental characteristics of end products. They are also considering the environmental implications of all the steps in the process by which a product moves 'from cradle to grave', including exploration, extraction and exploitation of raw materials, energy efficiency, recycling and re-use traits, life-span and utility, final disposal, marketing practices, and the environmental record of the producers. In response, industry has embarked on a new wave of marketing and sales promotion strategies, including 'green advertising' and 'eco-labelling' for products and services projected as being 'green', 'natural' or 'environmentally friendly'.

Green consumerism none the less has its share of shortcomings and misgivings. As the logic of industry is to expand consumption, even the most green-tinted of corporations are unlikely to promote less consumption. The green consumer movement, on the other hand, is moving beyond educating consumers to purchase green products, by urging them to consider whether they need a particular product at all (Kellner, 1990). This creates one area of potential conflict. Another arises because some firms keen to take advantage of the fast growth of the 'green sector' of business have made dubious claims for their products, or engaged in misleading advertising (Kellner, 1990).

There is evidence of the contribution green consumerism can make to the achievement of overall sustainability. It is not just that it is open to being exploited by established interests in business and politics. Rather, it is that the construction of a better world requires not only individual actions, but also collective actions and political and institutional restructuring based on a sound understanding of the limits of growth and the concept of ecological sustainability (Irvine, 1989).

The growth of environmental groups

The membership of environmental groups has grown enormously during the last two decades, and the concerns of citizens' groups have broadened (Table 3). In the earlier part of this century, environmentalism was essentially synonymous with wildlife conservation and had its main supporters in the upper-income groups, but since the late 1960s it has gained widespread popular support and become concerned with all aspects of the natural environment. It has also become concerned with the interrelationships between the natural environment and the human situation, and with poverty and environmental degradation.

Table 3: *Growth in membership of (A) British and (B) American environmental organizations, 1970-90.*

A: Membership of British organizations (in thousands)

	1971	1975	1981	1985	1989	1990
National Trust	278	539	1,047	1,323	1,900	2,000
Royal Society for the Protection of Birds	121	274	421	475	680	960
Greenpeace	...	...	8	50	300	387[a]
World Wide Fund for Nature	12	...	60	91	213	231
Royal Society for Nature Conservation	64	107	143	166	205	250
Friends of the Earth	...	5	20	30	180	200
Green Party	...	...	0.5	4	17	16[a]

Note: (a) 1 March 1991
Sources: Organizations.

B: Membership of American organizations (in thousands)

	1970	1975	1980	1985	1990
National Wildlife Federation	2,600	...	4,600	4,500	5,800
Greenpeace[a]	...	6	80	450	2,000
Sierra Club	114	153	182	363	566
National Audubon Society	105	255	310	425	515
Wilderness Society	66	...	63	97	363
Environmental Defence Fund	10	40	45	50	150
Natural Resources Defence Council[b]	...	15	35	65	140

Notes: (a) Founded in 1971   (b) Founded in 1970
Sources: Organizations.

There are many kinds of environmental groups. Some are organized locally to fight local problems such as a threat of pollution or some apparently inappropriate form of development. Others deal with a special issue, but on a national scale. Yet others are primarily concerned with the use of the environment and who should benefit from it. Some have been described as 'sustainable development' or 'appropriate technology' groups. Women have played a particularly important part in many of them, and the Chipko Movement in India is one important example. Throughout the 1970s, local people - largely, but not exclusively, women - stopped the felling of their forests by outside contractors, often by hugging the trees ('chipko' means 'to embrace'). The Movement insists that producing 'soil, water and oxygen' rather than timber or resin is the main purpose of the forests. However, it has shown that under proper popular control they can also provide fuel, fodder, small timber and fertilizer for local people while being preserved (Bandyopadhyay and Shiva, 1987).

There are many other examples of effective environmental groups that reflect public concern in their particular circumstances. They include the Green Belt Movement in Kenya, similar movements in Mexico, Sri Lanka, Indonesia and the Philippines, and action groups opposed to nuclear power and chemical waste dumps in many developed countries. These bodies have augmented and complemented the work of long-established conservation groups such as the Sierra Club or National Audubon Society in the USA, or the National Trust and the Royal Society for the Protection of Birds in the UK (see Table 3). Many cities and regions have citizens' groups concerned with the improvement of their local environment. Chapter 22 contains further information on the growth of environment-related NGOs in developing countries.

The environmental NGO movement has become increasingly international, with the emergence of powerful bodies such as Friends of the Earth, Greenpeace and the World Wide Fund for Nature (WWF). A unique link between the non-governmental and governmental sector has been provided since 1948 by IUCN, The World Conservation Union, which links in membership some 55 states, 100 government agencies and 450 NGOs. The range of conservation, development and humanitarian NGOs, and of industry groups concerned with the environment, has expanded steadily during the 1970s and 1980s, and the contacts between the NGO and government sectors have also strengthened. Through environmental groups, therefore, individuals are increasingly able to influence national and world policy.

Collective actions: business, industry and commerce

The environmental movement, both at individual and group level, has had a profound influence on world-wide industry. Whereas in the 1960s and 1970s industry tended to regard environmental concern as a peripheral nuisance, to be evaded where possible, in the 1980s many companies have themselves become active in developing environmental policies. Corporate managers are beginning to see that improving the environment, while improving the bottom-line, is the smart way of conducting business. Based on the premise that profit-making opportunities of the 1990s will be in manufacturing and marketing 'environmentally sound' products and services, initiatives such as developing cleaner production processes, offering products that generate less waste, devising safer pest control strategies and cleaning up past damage are fast becoming top-priority investment areas (Shea, 1989). Spurred on by the phenomenal growth and impact of the environmental awareness movement sweeping across the globe, there has been a remarkable 'greening of the corporate boardroom' in several countries, with an almost astonishing turn-around in the attitudes of many major companies toward environmental issues (Collison, 1989).

A major initiative was taken in the 1980s by the Chemical Manufacturer's Association of the United States, which launched an initiative on Community Awareness and Emergency Response (CAER). The European chemical industry followed a similar path, and in 1986 the Canadian Chemical Producers Association (CCPA) commissioned a survey to evaluate public opinion towards the industry. This revealed that chemical manufacturers were just about the least trusted business sector in Canada, and stimulated the CCPA to develop and promote a policy called 'Responsible Care' (Collison, 1989). Responsible Care is a 'cradle-to-grave' stewardship programme for chemicals, based on the principle that manufacturers have a moral, if not legal, responsibility for their products long after they have sold them. Soon after the Union Carbide gas leak tragedy in Bhopal in 1984, CCPA made Responsible Care a condition for membership. It has now been adopted widely in the chemical manufacturing industry world-wide. As its centrepiece, Reponsible Care comprises six codes of practice covering R&D, manufacturing, transportation, distribution, waste management, and community awareness and emergency response. Likewise, contending that 'candor was essential to gain any credibility with the public', Dow Canada embarked on an 'open-door policy' - inviting the public, the press and environmental groups to inspect its chemical plants, as well as reporting to the public all spills, even those it was not obliged to report (Collison, 1989).

In another example, the US-based 3M corporation, as part of its 1989 Earth Day activities, launched a recycling awareness programme aimed at 'establishing a recycling mind-set' among employees at work and at home (Conservation Exchange, 1989). As far back as 1975, 3M initiated its now-famous '3P' (Pollution Prevention Pays) programme, focusing on pollution reduction at its source and on turning wastes into resources. By 1989, 3M had claimed to have saved $US408 million through implementing its 3Ps programme, and to have prevented 111,000 tons of air pollutants, 15,000 tons of water pollutants and 380,000 tons of sludge and solid waste from being released into the environment.

In some industries, the pressure to be green is subtly changing the links among companies as well as those between companies and their customers. ICI, for instance, offers to take back contaminated sulphuric acid from its customers in the oil industry and clean it up: similarly, the problem of disposing of heavy metals in batteries has led Philips to join other companies in setting up a body designed to recycle 80 per cent of nickel cadmium batteries by 1991.

The financial community is also paying close attention to the increasing environmental awareness world-wide and to the burgeoning success of environmental entrepreneurs. According to the US Environmental Protection Agency (EPA), the environmental market - which is expected to grow from $US50 billion in 1989 to $US200 billion by the year 2000 - will be 'one of the most important venture capital arenas of the 1990s' (Conservation Exchange,

1989). It is thus not surprising that Hambrecht & Quist, a leading investment banking and venture capital firm, recently announced the creation of a new venture capital fund focusing on investments in environmental technology areas, including alternative clean-up technologies, resource recovery, product recycling, process control technologies etc.

Japan's biggest stockbroking company, Nomura Securities Co., has introduced an 'Earth Environment Fund'. Meanwhile, the investment house James Capel scored a notable first in the UK with its 'Green Index'. Companies listed in the index include Swallowfield Plc, which produces 'ozone-friendly' aerosols, and Will Shaw Plc, which makes hardened metal powders used in lead-free car engines. In the US, a survey undertaken by Calvert Social Investment Fund - one of the largest funds of its kind - on the investment priorities of its 40,000 participants, indicated that the environment is set to become the dominant factor in socially responsible investment (Shea, 1990). The fund is not only already avoiding companies producing CFCs and pesticides which are toxic and in violation of US EPA, state and local environmental regulations, but is also stepping up its efforts to find companies that market innovative environmental products or that at least have well-defined environmental policies.

In an attempt to better monitor and evaluate the environmental performance of corporations, the US-based Coalition for Environmentally Responsible Economies (CERES) is lobbying corporations to accept and adopt an environmental code of conduct, dubbed the 'Valdez Principles' (after the *Exxon Valdez* oil spill), aimed at committing corporations to a policy of protecting the biosphere, using natural resources sustainably, reducing and disposing of wastes responsibly, using energy wisely, reducing risks, marketing safe products and services, compensating for damage, disclosing hazards and harms, appointing environmental directors and managers, and conducting and publicizing annual assessments and audits of environmental performance. To encourage this trend, the US Centre on Transnational Co-operations (UNCTC) has developed a set of criteria for sustainable development management.

The overall approach can be summarized quite simply. 'Companies know that good environmental performance is no longer an option for industry; it is a precondition for success. While this outlook is not universally accepted, it is gaining among both blue chip corporations and the thousands of new businesses that are spurring economic growth in North America and Europe' (Shea, 1989). Meanwhile, others contend that, 'The green wave is bringing in a multi-billion dollar tide of business. Many companies are going green purely to capture a bigger share of the market. If a company cannot increase profits by turnover of sales, it has to differentiate its products to garner a bigger market share, and green is a good way of value-distinguishing one's products' (George, 1989). Another major growth area has been in environmental consultancy firms, set up to help both public and private sector organizations avoid undue environmental impact, and optimize their use of environmental resources. Figure 6 shows how rapidly these consultancies have grown in the United Kingdom between 1950 and 1990.

Figure 6: *Growth in number of British environmental consultancies, 1950–90.*

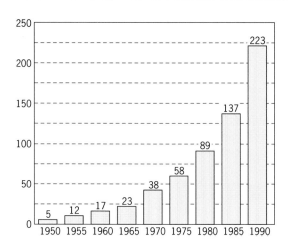

Source: *ENDS* Report. Environmental Data Services, London, June 1990.

Collective actions: education and information

Public attitudes are moulded and developed by both formal education and non-formal information via the mass media, and both have grown in response to rising concern during the 1970s and 1980s. In many countries non-formal information channels have been more persuasive than formal education, and there is no doubt that a powerful reinforcement to public concern has been provided by media 'personalities' who have themselves become concerned and begun to convey the environmental message. In 1989 The Environmental Media Association (EMA) was launched, to promote public awareness of environmental issues. Programmes on nature and wildlife - such as the British Broadcasting Corporation's natural history productions - became established in the 1950s as television began to reach a mass audience, and they proved extremely popular. Today, wildlife series frequently achieve higher audience ratings in the United Kingdom than sports programmes, and an hour-long Saturday afternoon environment programme regularly reaches about a third of Belgian homes (Elkington *et al.*, 1988). Regular environmental series have also been established for more than two decades on British commercial television, the United States Public Broadcasting Service, Australian national television, and Canadian radio (among others). Two environmental film festivals, Ecovision and Wildscreen, are regularly held in Europe, and a recent survey showed that 95 per cent of European television stations were willing to participate in environmental co-productions. The Television Trust for the Environment recently reviewed environmental films from developed and developing countries throughout the world (TVE, 1976) and has stimulated production in 23 countries.

BOX 4

Principles of the Belgrade Charter on environmental education.

Environmental education should:

- consider the environment in its totality - natural and man-made, ecological, political, economic, technological, social, legislative, cultural and aesthetic;
- be a continuous life-long process, both in school and out of school;
- be interdisciplinary in its approach;
- emphasize active participation in preventing and solving environmental problems;
- examine major environmental issues from a world-wide viewpoint, while paying due regard to regional differences;
- focus on current and future environmental situations;
- examine all development and growth from an environmental perspective;
- promote the value and necessity of local, national and international co-operation in the solution of environmental problems (UNESCO/UNEP, 1975).

In addition, environmental education should include among its primary objectives measures for assisting individuals and groups to:

- acquire awareness of and sensitivity to the total environment and its allied problems;
- acquire basic understanding of the total environment, its associated problems, and humanity's critically responsible presence and role in it;
- acquire social values, strong feelings of concern for the environment, and motivation for actively participating in its protection and improvement;
- acquire skills necessary for solving environmental problems;
- evaluate environmental measures and education programmes in terms of ecological, political, economic, social, aesthetic and cultural criteria;
- develop a sense of responsibility and urgency regarding environmental problems, and to ensure appropriate action to solve these problems.

Formal education has, of course, played a major part, stimulated by both intergovernmental and NGO activities. Following the recommendations of Stockholm 1972, UNEP and UNESCO launched an 'International Environmental Education Programme' (IEEP) aimed at fostering the exchange of information and experience in the field of environmental education. Under the auspices of the IEEP, an 'International Workshop on Environmental Education' was held in Belgrade in 1975, which led to the adoption of 'The Belgrade Charter - A Global Framework for Environmental Education' (Box 4) (UNESCO/UNEP, 1975).

The UNESCO/UNEP 'Intergovernmental Conference on Environmental Education' held in Tblisi in 1977 underlined the need for concerted environmental education at the global level. One of the resolutions of the Tblisi Conference stresses that environmental education should strive to enable individuals and communities to understand the complexities of the natural and man-made environments arising from the interaction of their biological, physical, chemical, social, economic, cultural, ethical and political aspects, besides acquiring the knowledge, values, attitudes and practical skills needed for participating in a responsible and effective manner in the anticipation and solving of environmental problems and in the management of environmental quality. It was further resolved that environmental education should ideally demonstrate the economic, political and ecological interdependence of the modern world, wherein environmentally detrimental decisions and actions of individual countries can have international repercussions (UNESCO, 1977).

From 1977 to 1987 the International Environmental Education Programme (IEEP) has been associated with the world-wide effort to incorporate an environmental dimension into the educational systems and practices of states. This has been widely achieved through the production and widespread dissemination of resource material and a continuous series of international, regional and local workshops and seminars for teacher trainers and curriculum developers as well as educational administrators.

In 1987, ten years after Tblisi, UNESCO and UNEP organized the Moscow Congress on Environmental Education and Training which produced a global strategy (now printed and disseminated in eight languages). This 'International Strategy for Action in the field of Environmental Education and Training for the 1990s' is providing member states and institutions with a framework and guidelines for the next decade, and a basis for the preparation of their own national strategies for Environmental Education and Environmental Training for the 1990s (UNESCO/UNEP, 1988).

The IEEP entered its sixth phase in 1990-91. It has become a major force in the development of environmental education world-wide, involves over 25,000 educators, and has been active in more than 140 countries since 1975. Environmental education lays the foundation for longer-term sustainable development. Recent trends in its evolution include a strong emphasis on environmental ethics, and, on the practical front, on the development of innovative, often multi-media teaching-learning material adapted to each region or locality. The IEEP is pursuing these goals in 1991 and 1992.

The period between 1972 and 1992 has seen a considerable increase in the provision of general environmental training, which attempts to build awareness of the environment as a whole and to develop individual capacities to deal with environmental concerns. The target groups for such general training include policy-makers, planners, engineers, architects, industrialists, agriculturalists, and town planners. Training of people on the ground, including women who in many countries are the actual managers of the environment, is also of great importance. In many developing countries there is a great shortage of extension workers, especially those able to teach a wide environmental view to farmers and land managers. None the less, there is evidence that when campaigns of

Figure 7: *Changes in farmers' attitudes towards various conservation issues between the first and fifth years of a conservation awareness campaign in Rwanda.*

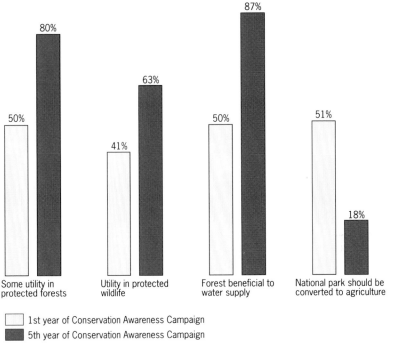

1st year of Conservation Awareness Campaign
5th year of Conservation Awareness Campaign

Source: UNEP, *State of the Environment* (1988), based on data in Harcourt *et al.*, Public attitudes to wildlife and conservation in the Third World, *Oryx*, **20**, 1986.

explanation are mounted, there is a marked positive response (Figure 7).

Specialized environmental training at the higher educational and post-graduate levels seeks to develop the problem-solving capabilities of professionals so that they can deal with specific environmental problems and sectors. The target groups for such training include ecologists and other biologists, economists, health scientists, hydrologists, soil scientists, foresters, oceanographers, engineers and those engaged in pollution monitoring and control. As the 1980s have proceeded, increasing emphasis has been placed on training for integrated rather than sectoral environmental management. An important contribution is being made by business and management schools, and by specialized courses within industry. These are, in a sense, linked to the formulation by various industrial sectors of codes of professional ethics relating to the environment, and to the increasing emphasis on environmental targets that industrial managers must meet (Box 5).

All UN agencies have been involved to some extent in environmental training, but UNEP, ILO, UNESCO, UNIDO and UNU, together with non-UN organizations such as IUCN, have played a major part. UNEP has to a considerable extent provided the necessary coordination.

BOX 5

Code of Environmental Ethics for Engineers

The WFEO Committee on Engineering and Environment, with a strong and clear belief that man's enjoyment and permanence on this planet will depend on the care and protection he provides to the environment, states the following principles.

TO ALL ENGINEERS

When you develop any professional activity:

1. Try with the best of your ability, courage, enthusiasm and dedication to obtain a superior achievement, which will technically contribute to and promote a healthy and agreeable surrounding for all men, in open spaces as well as indoors.
2. Strive to accomplish the beneficial objectives of your work with the lowest possible consumption of raw materials and energy and the lowest production of wastes and any kind of pollution.
3. Discuss in particular the consequences of your proposals and actions, direct or indirect, immediate or long term, upon the health of people, social equity and the local system of values.
4. Study thoroughly the environment that will be affected, assess all the impacts that might arise in the state, dynamics and aesthetics of the ecosystems involved, urbanized or natural, as well as in the pertinent socio-economic systems, and select the best alternative for an environmentally sound and sustainable development.
5. Promote a clear understanding of the actions required to restore and, if possible, to improve the environment that may be disturbed, and include them in your proposals.
6. Reject any kind of commitment that involves unfair damages to human surroundings and nature, and manage the best possible social and political solution.
7. Be aware that the principles of ecosystemic inter-dependence, diversity maintenance, resource recovery and interrelational harmony form the basis of our continued existence and that each of those bases poses a threshold of sustainability that should not be exceeded.

Always remember that war, greed, misery and ignorance, plus natural disasters and human induced pollution and destruction of resources, are the main causes of the progressive impairment of the environment, and that you, as an active member of the engineering profession, deeply involved in the promotion of development, must use your talent, knowledge and imagination to assist society in removing those evils and improving the quality of life for all people.

Approved by the Committee on Engineering and Environment of the World Federation of Engineering Organizations, in the 6th Annual Plenary Session, New Delhi, 5 November 1985.

The political dimension

One of the manifestations of changes in public perception of the environment is the emergence and consolidation of 'green politics' and 'green parties' in several countries, particularly in Europe.

The building blocks of green politics - rooted in the 'four pillars' of ecology, social responsibility, grass-roots democracy and nonviolence - are carved from an integrated philosophical and practical approach to address the prevailing ecological, economic and political crises, all of which are seen to be interrelated and global in nature. The primary stimulus for green politics can thus be attributed to a recognition that we are confronted with a multifaceted, world-wide crisis which touches every aspect of our lives - our health and welfare, the quality of our environment, our social relationships, our economy, technology and politics, and our very survival on planet Earth (Spretnak and Capra, 1985).

Proponents of green politics contend that the more people perceive the linkages between the fundamental principles of ecological wisdom, a truly secure state of peace, a sustainable economy and participatory democracy - with power being generated directly from the grass-roots - the more they realize the absence of such ideals and values among conventional political parties. In this context, the rise of green politics can be said to have been triggered by a people-centred revision of existing political ideals and values - a revision that has led to a new vision of reality among growing numbers of people which not only transcends conventional political frameworks but also emphasizes the 'interconnectedness and interdependence of all phenomena', as well as the 'embeddedness of individuals and societies in the cyclical processes of nature' (Spretnak and Capra, 1985).

Barely noticed by the mainstream media and established political parties when they first emerged a decade ago, the green parties of Western Europe and Australia became, during the 1970s and 1980s, an electoral force to be reckoned with. Notable election victories in 1988 and 1989 placed representatives of green parties in city halls and national parliaments (Renner, 1990). The influence has spilled over to the longer-established traditional parties, who have themselves given increasing prominence to the environment. Outside the formal green parties, there are many 'green politicians' who regularly champion 'green issues' at the level of political decision-making. Many governments have issued environmental policy statements during the period under review, and have strengthened national institutions dealing with the environment (Chapter 22).

Concluding remarks

Public concern for environmental issues is much greater in 1992 than it was in 1972. Even though there is limited information from developing countries (and especially rural communities), the world-wide spread of information is being reflected in responses at the individual, citizens' group and political levels. It is provoking changes in the approach of the media, the formal education system and governments.

In many countries it is becoming evident that the key to a sustainable future lies with individuals, informed by science, monitoring and open information policies, and strengthened by their capacity to join environmental movements and take an increasingly direct role in the management of the local environment. The dramatic advances in scientific knowledge described in Chapter 21 provide power to those citizens' groups and to governments. In turn, the expansion of national institutions, described in Chapter 22, allows a co-ordinated and effective response to the problems facing countries today.

The *Strategy for Sustainable Living* published by IUCN, UNEP and WWF in 1991 emphasizes the need for more and better information, education and training. It calls for the world-wide adoption of an ethic for sustainable living, under which every person takes responsibility for his or her impacts on nature, and each generation undertakes to leave to the future a world as diverse and prosperous as the one it inherited. Universal primary education, incorporation of environmental education at all levels of the formal system, stronger training services (including extension services), action by governments to make people aware of the need to stabilize resource consumption and population, encouragement to 'green consumer' movements, and a range of actions to strengthen the capacity of local communities to care for their own environments, are all highlighted. Looking to the future, local communities and citizens' groups are likely to play an increasingly crucial role in environmental care and sustainable management, and information, education and enhanced communications will be essential tools. The whole field covered in this chapter can be predicted as a growth area for the coming decades.

REFERENCES

Ashby, E. (1987) *Reconciling man with environment*. Stanford University Press, Stanford.

Ashby, E. and Anderson, M. (1981) *The Politics of Clean Air*. Clarendon Press, Oxford.

Bramwell, A. (1989) *Ecology in the 20th Century*. Yale University Press, New Haven and London.

Brandt Commission (1980) *North-South : A Program for Survival*. MIT Press, Cambridge, Mass.

Bandyopadhyay, J. and Shiva, V. (1987) Chipko : rekindling India's forest culture. *Ecologist*, **17**, p.26.

Carson, R. (1962) *Silent Spring*. Houghton Mifflin, Boston.

CEC (1986) *The Europeans and their Environment*. Commission of the European Communities, Brussels.

Collison, R. (1989) The greening of the boardroom. *Business Magazine*, July 1989, Toronto.

Conservation Exchange (1989) *Growing greener - corporations come round to the environmental cause*. **vol. 7**, no 3, Washington, D.C.

Department of Arts, Heritage and Environment (1987) *State of the Environment in Australia*. Australian Government Publishing Service, Canberra.

El-Hinnawi, E. (1986) Environmental awareness in Egypt. Proc. of the Egyptian-German seminar on Environmental Awareness, Cairo.

El-Hinnawi, E. (in prep.) *Third World and Environment*. Cassell-Tycooly, London.

Elkington, J. *et al.* (1988) *Green Pages - the business of saving the world*. Routledge, London.

Environment Agency of Japan (1982) Public opinion poll on environmental pollution. *Japan Environment Summary*, **vol.10**, Tokyo.

Fortner, R.W. and Lyon, A.E. (1985) Effects of a Cousteau television special on viewers' knowledge and attitudes. *J. of Environmental Education*, **vol. 16**, p.12.

George, M.K. (1989) *Seeing the green light*. South, London.

Goldsteen, R. and Schorr, J.K. (1982) The long-term impact of a man-made disaster: an examination of a small town in the aftermath of the Three Mile Island accident. Disasters, vol.6.

Goudie, A. (1986) *The human impact on the natural environment*. Blackwell, Oxford.

Irvine, S. (1989) Consuming Fashions? The limits of green consumerism. *Ecologist*, **19**, no.3.

IUCN/UNEP/WWF (1991) *Caring for the Earth : A Strategy for Sustainable Living*. IUCN, Gland, Switzerland.

Johnson, B.B. (1987) Public concerns and the public role in siting nuclear and chemical waste facilities. *Environmental Management*, **vol. 11**.

Kellner, J. (1990) Beware the Greencon. *New Internationalist*, **No. 203**, January 1990, Oxford.

Marsh, G.P. (1864) *Man and Nature*. Harvard University Press, Cambridge, Mass.

McCraken, R.J. (1987) Soils, Soil scientists and Civilization. *Soil Science Society of America Journal*, **vol. 51**, p. 1395.

McDowell, J., Schoenthal, R. and Thompson, D. (1990) Is the planet on the back burner? *Time International*, 24 December 1990, New York.

Meadows, D.H., *et al.* (1972) *The Limits to Growth*. Potomac Associates, Washington, DC.

Mitchell, R.C. (1982) Public response to a major failure of a controversial technology. In: D. L. Sills *et al. Accident at Three Mile Island : the Human Dimension*. Westview Press, Boulder, Colorado.

MORI (1987) *Public Attitudes to Nuclear Power*. Market and Opinion Research International, London.

Nicholson, E.M. (1987) *The New Environmental Age.* Cambridge, University Press, Cambridge.

OECD (1979) *Interfutures.* Final Report of the Research Project on the Future Development of Advanced Industrial Societies in harmony with that of Developing Countries. OECD, Paris.

OECD (1985) *The State of the Environment in 1985.* OECD, Paris.

OECD (1987) *OECD Environmental Data Compendium.* OECD, Paris.

O'Riordan, T. (1981) *Environmentalism.* Pion, London.

Ostman, R.E. and Parker, J.L. (1986/87) A public's environmental information sources, and evaluation of mass media. *J. of Environmental Education,* **vol. 18**, p.9.

Renner, M.G. (1990) Europe's Green Tide. *Worldwatch,* **vol 3**., No.l, Washington D.C.

Royston, M.G. (1979) *Pollution Prevention Pays.* Pergamon Press, Oxford.

SCEP (1970) *Study of Critical Environmental Problems. Man's Impact on the Global Environment.* MIT Press, Cambridge, Mass.

SMIC (1971) *Study of Man's Impact on Climate. Inadvertant Climate Modification.* MIT Press, Cambridge, Mass.

Secrett, C. (1988) Green Consumerism. *New Economics,* **No.7**, London.

Sekar, T. (1981) Role of newspapers in creating mass concern with environmental issues in India. *Int. Journal of Environmental Studies,* **vol. 17**, p. 115.

Shea, C.P. (1989) Doing well by doing good. *Worldwatch,* November-December 1989. Washington DC.

Shea, C.P. (1990) Environmental Investing. *Worldwatch,* **vol. 3**, No.1, Washington DC.

SOE Canada (1986) *State of the Environment Report for Canada.* Environment Canada, Ottawa.

Spretnak, C and Capra, F. (1985) *Green Politics - The Global Promise.* Paladin-Crafton Books, Collins Publishing Group, London.

Stevenson, R.l. (1986) Radio and television growth in the Third World, 1960-1985. *Int. Journal of Mass Communication Studies,* *vol. 38,* 1986.

Tolba, M..K. (1982) *Development without Destruction.* Tycooly International, Dublin.

TVE (1976) *Switching on the Environment.* Television Trust for the Environment, London.

UNCOD (1977) *Desertification : Its causes and Consequences.* UN Conference on Desertification, Pergamon Press, Oxford.

UNEP (1988a) *The Public and the Environment. The State of the Environment, 1988.* UNEP, Nairobi.

UNEP (1988b) *Environmental Perspective for the Year 2000 and Beyond,* UNEP, Nairobi.

UNEP (1989) World-wide Concern about the Environment. *Our Planet,* **vol. 1**, Nos. 2-3, UN Environment Programme, Nairobi.

UNESCO/UNEP (1975) International Workshop on Environmental Education. Belgrade, 13-22 October, 1975. Document ED-76 / WS / 95.

UNESCO/UNEP (1988) *International Strategy for Action in the field of Environmental Education and Training for the 1990s.* Proceedings of the UNESCO-UNEP Conference held in Moscow, 1987. UNEP and UNESCO, Nairobi and Paris.

UNESCO (1977) *Education and the Challenge of Environmental Problems.* Report of the Intergovernmental Conference in Environmental Education, Tblisi, 14-26 October 1977. UNESCO Doc. ED-77/CONF/ .203/Col.3.

Waddington, C.H. (1975) The Origin. *In:* Worthington, E.B. *The Evolution of IBP.* Cambridge University Press, Cambridge.

WCED (1987) *Our Common Future.* Report of the World Commission on Environment and Development. Oxford University Press, Oxford.

Wilkins, L. (1987) *Shared Vulnerability. Media Coverage and Public Memory of the Bhopal Disaster.* Greenwood Press, Westport Conn.

CHAPTER 22

National responses

Introduction

In the early development of human societies, interactions with the environment were predominantly at local, and especially village, level. This is still the case in most of the world. In 1970, 75 per cent of the population of developing countries was rural, although by 1980 this figure had fallen to 71 per cent, and to 63 per cent by 1990 (UN, 1991). Agriculture continues to provide over two-thirds of total employment in many of the lower-income countries, and to provide over three-quarters of their exports (Table 1: MacNeill, Winsemius and Yakushiji, 1991). But the development process has, in most countries, been accompanied by the progressive expansion of the industrial sector, and urbanization has brought the growth of municipal utilities and aggregated many human impacts on the environment into 'point sources'.

Table 1: *Resource-dependence of selected developing countries.*

Economies	Agricultural production as a percentage of GDP		Employment in agriculture as a percentage of total employment		Exports of primary products as a percentage of total exports	
	1965	1986	1965	1980	1965	1986
LOW INCOME						
Burma	35	48	64	53	99	87
China	39	31	81	74	54	36
India	47	32	73	70	51	38
Indonesia	56	26	71	57	96	79
Sri Lanka	28	26	56	53	99	59
Ethiopia	58	48	86	80	99	99
Ghana	44	45	61	56	98	98
Kenya	35	30	86	81	94	84
Nigeria	53	41	72	68	97	98
Tanzania	46	59	92	86	87	83
MIDDLE INCOME						
Bolivia	23	24	54	46	95	98
Colombia	30	20	45	34	96	82
Costa Rica	24	21	47	31	84	65
Thailand	35	17	82	71	95	58
Senegal	25	22	83	81	97	71
Zimbabwe	18	11	79	73	71	64
INDUSTRIAL MARKET						
Canada	6	3	10	5	63	36
Japan	9	3	26	11	9	2
Spain	15	6	34	17	60	28
United States	3	2	5	4	35	24

Note: The agricultural sector comprises agriculture, forestry, fishing and hunting. Primary products, in addition to agriculture, include fuels, minerals and metals.

Source: MacNeill, Winsemius and Yakushiji (1991).

Local policies governing human interactions with the environment are as ancient as history. Traditional ways of apportioning hunting rights, and controlling how land and water are used, or livestock managed, unquestionably began long before we have written records. National policies for the apportionment of land, protection of forests and control of hunting and fishery are also ancient. There is nothing surprising about this, for there is nothing more fundamental to a nation than its land and water, and rights to the use of such resources must inevitably have been among the first to be codified and guarded by the central organization of a state.

Every country is therefore likely to have had policies and institutions for the environment long before that particular label was applied to them. This chapter looks at how sub-national and national policies, laws and institutions have evolved, especially over the past 20 years.

Environmental policies span many sectors, not all of them traditionally labelled 'environmental'. Besides land tenure, agriculture, forestry and fishery management, they include development control, measures to prevent pollution, and measures to regulate industry and transport. In recent years, recognition that virtually every activity has some impact on the environment has meant that every field of policy has had its environmental component, even if it is not dominated by environmental concerns.

Environmental policy has become increasingly cross-sectoral and integrated. While it is necessary to have specific environmental policy objectives for each of the many sectors of a modern and complex state, it is also necessary to have a broad over-view and system of guidance, designed to optimize the use of the 'natural capital' which the environment represents. This emergence of an increasingly integrated environmental policy has been one of the dominant features of the last 20 years.

Environmental policy is developed by institutions at many levels, including local authorities, citizens' groups, environment and development NGOs, religious groups and industry. All have their place and importance alongside national governmental institutions. Another trend in the past 20 years has been the recognition of the diversity of human groups that need to work in mutual awareness, if not in co-operation, in order to ensure coherent and effective policy for the environment (IUCN/UNEP/WWF 1991).

A third trend of the past 20 years has been the recognition of the interdependence of nations and the need for this to be expressed in regional policies and institutions that span broad areas of environment, and for whose management a number of sovereign states need to come together. The most obvious examples are large river basins (40 per cent of the world's population live in the catchment area of a river shared between several states), and areas of sea whose ecosystems are interlinked, and which need managing as units (IUCN/NOAA, 1991).

The diversity observed in policies and institutions is both inevitable and adaptive. The world's environment is diverse: peoples have adapted to it, culturally and have adopted environmental policies that are an expression of sustainable living under local conditions. The hunting lore of Inuit or Amerindian nations; the water-use practices of the mountain peoples of Papua New Guinea, or the former Inca Empire in the high Andes; the irrigation systems of ancient Egypt or Mesopotamia, and their modern counterpart in the *falaj* system in Oman, are all expressions of this kind of adaptive pattern. One important feature of recent decades has been the rediscovery of such systems, and the recognition that they have much to offer of relevance to the present day, even though they have been undermined in many parts of the world by the imposition of external cultures and technologies and by the pressure of mounting human numbers.

This chapter considers:

- the history of adaptive environmental policies, in all continents; and

- the pattern of evolution of policies and institutions at national and sub-national level between 1970 and 1990;

  - in industrialized countries;

  - in developing countries;

Boxes illustrate particular case histories. Unfortunately, relatively few time series and statistics are available, and these relate largely to the few parameters in this field that are quantifiable - such as the number of institutions, the number of members in particular institutions, and the number of laws that have been enacted. It is much more difficult to evaluate the effectiveness with which particular kinds of institution have implemented particular kinds of policy, yet such an evaluation is clearly necessary if the world community is to learn from experience and adopt the patterns that are most likely to be adaptive to a changing future.

History of environmental policies

Within historical times there have been two evident trends in policy, relating respectively to the management of resources that continued wild and were subject to various kinds of hunting, and to the management of resources that passed into ownership but required rules to prevent mis-use. As agriculture expanded in Europe, western Asia and China, areas of land began to be set aside as hunting preserves, and their use became increasingly restricted to kings and rulers. Since many preserved hunting areas were also important for their timber, such measures may well have had a dual protective value in countries where agriculture was increasingly encroaching on to the woodland resource. Today, complex systems for closed seasons and licensing of hunting still characterize North America and Europe, where hunting wild species is an important recreational and even economic activity.

In the North Atlantic, seabird-hunting cultures established by the Viking people in Iceland, Faroe and northwest Scotland developed very strict rules about cropping seasons and the size of the permissible harvest. In contrast, fisheries and marine mammals have traditionally been 'open access' resources. In other words, there were originally no ownership rights to restrict access, and as a result there was a strong incentive for exploiters to obtain as much of the resource as they could use or market. This competitive process often led to the destruction of the underlying resource, as well illustrated by the history of exploitation of fur seal populations in the Southern Hemisphere between around 1775 and 1890 and by the more modern history of pelagic whaling (Bonner, 1982; Gulland, 1976, 1987).

The depletion of such resources led belatedly to the adoption of international agreements and regulations in the shape of regional fisheries agreements and the International Whaling Convention (Chapter 23). The other parallel trend has been the extension of controls by coastal states outward to sea so that strict national jurisdictional rules could be applied, thus restricting the openness of access to the resource. This has been a classic pattern for fisheries in most parts of the world, with the extension of territorial waters from the original 3 mile limit to 12 miles, and the more recent establishment of EEZs extending 200 miles from land. In parallel, there have also been developments of controls in the shape of net sizes and other regulations on fishing gear, together with seasonal regulation and the establishment of sanctuary areas.

On land, resource utilization has largely been regulated by grants of land ownership, either to individuals or to groups. Where there was communal ownership, elaborate policies backed by enforceable rules evolved early in history. Water management in arid regions has been subject to tight codes of practice from very early times, for reasons that are obvious given the limiting role of water in agriculture in such locations (Box 1). In parts of Northwest Europe, grazing rights in common pasturage have often been apportioned in terms of a right to particular farms within a village to graze a specified number of animals on the shared pasture, a particular unit of grazing carrying the right (for example) to graze one horse or one cattle beast or three sheep. In Africa, pastoral practices have also been subject to unwritten, but elaborate, traditions (Box 2).

These traditional resource-use policies were both adaptive and sustainable as long as the number of people seeking to exploit the resource did not rise dramatically. Many, however, have broken down under the combined pressures of population growth, the development of technologies that have permitted more intensive use of the resource, or the superimposition of alien systems, often in conjunction with colonial expansion.

The situation in 1970

By 1970, the only remaining international 'open access' resources were offshore fisheries, marine mammals and the minerals of the international deep

BOX 1

Water management in the Omani *falaj* system.

Falaj systems have three components: water collection from one or more wells, water transportation through channels which may be surface or underground and may extend to distances of up to around ten kilometres, and water distribution sections to the irrigated cropland.

The underground distribution canal is usually open to daylight at the head of the village which benefits from it, and this is the site for drinking water collection. Houses therefore tend to cluster at this point. From the drinking area the water flows on to washing areas and then into the gardens via a network of open, lined channels. Crops are planted in basins varying in size, but commonly rectangular in shape.

Falaj oasis agriculture is based on a three-crop system, with the date palm as the mainstay of farming. Other trees, grapevines, beans, sorghum and vegetables are also cultivated, and cereals tend to be grown at the end of the *falaj* system where water flow is less assured.

Water allocations are cyclic, and are based on a time unit of half an hour which may be sub-divided. Water is distributed on a recurring cycle which may vary in length from approximately six days or less, and there may be an allowance for days on which water rights may be sold at auction for one day per week. This gives individuals a temporary chance to gain access to water and increase their existing allocation.

The control and management of each settlement's *falaj* is delegated to an individual manager (*Wakil*) selected from among the garden owners. This individual adjudicates on disputes, sells short-term tenancies when water is available, holds the income from these sales and keeps the system in good repair from the mother well as far as the village main channel.

The whole system is an elaborate social structure, which depends heavily on traditional and hereditary rights and practices, and the acceptance by the whole community of the inherent fairness of the arrangement. In a number of villages it is being undermined because individuals have augmented their water flow by sinking wells within their property, and these compete with one another for the water table.

Source: Gabriel (1988).

BOX 2

Traditional pastoral practices in Africa.

The resource exploitation strategy of pastoralists on African rangelands involves a dozen or so major forms of behaviour, which combine to optimize resource use. These behavioural elements include:

- *Movement.* This, the most obvious element of any pastoral strategy, involves a continuing search for food among patchy, fluctuating and low-density resources, but allows the exploitation of uneven distributions of moisture and plant nutrients, which follow no precise pattern from season to season.
- *Resource reservoirs.* Certain areas within low-density resource rangelands do have higher production (e.g. highlands, swamps and permanent rivers). These are used as fallback areas which allow temporary heavy occupation by stock and people during extreme stress periods.
- *Division of the human population into dispersed sub-populations.* These are often large groups (10,000 or more people) with large home ranges over which they move.
- *Small animal management groups.* Within the larger units, people own bunches of animals whose watering and grazing can be closely controlled.
- *Individuation.* Each pastoral population consists of a large number of herd owners. These show considerable flexibility in movement within the groups.
- *Local groups.* The two organizational levels - the small animal management groups and the large dispersed populations - are tied together by temporary clusters of herding groups. This facilitates information exchange and increases physical security.
- *Competitive exclusion.* This has been accomplished not so much by perimeter defence as by pre-emptive claims on resources, but has in the past included aggressive exclusion, by raiding and localized warfare.
- *Range management.* Burning has been used traditionally to induce new shoot growth, while sub-surface water tapped through wells dug in dry river beds can increase the capacity of an area to support livestock.
- *Management of herd structure and species composition.* There is an element of control of both these parameters so as to maximize livestock populations in a given area. Multi-species composition of livestock has the value of using both browse and grass species in the plant community, and of turning the different periodicities of reproduction and growth to the task of providing continuous supplies of human food. Ratios of cattle to small stock (sheep and goats) vary according to rangeland type, with smaller stock being favoured in the more arid rangelands. Goats replace sheep in the harsher environments. Within the cattle herds, management maintains a high percentage of adult females, as sources of both milk and blood.

Source: Dyson-Hudson (1984)

sea bed. Even these were subject to actual or proposed international regulation under Fisheries Conventions, the International Whaling Convention or the UN Convention on the Law of the Sea. Although some traditional rights within 200-mile (or-12 mile) limits had been conceded, and in Northwest Europe an elaborate common policy had been worked out for fisheries management among the states of the European Community, coastal states were increasingly managing offshore fisheries resources out to a 200-mile limit. This extension has not been achieved without friction, of which the most striking example was the so-called 'Cod War' in the North Atlantic over rights of access to Icelandic offshore fisheries, in which fleets of foreign fishing vessels were for a period protected by naval vessels. The resolution of this conflict in favour of the coastal state was an important step toward the general policy of extending coastal state rights over the resources of their adjacent seas.

On land, tenure had been codified in most countries long before 1970, but often remained a socially divisive issue. In some countries, indigenous peoples were dispossessed in favour of settlers or absentee landlords from other nations or regions, and the re-assertion of traditional rights has been an important cause of policy review in recent decades in many countries, including the United States of America, Canada, Australia and New Zealand. In areas where land was originally divided into large units there has been increasing pressure for re-allocation to empower small farmers, many of whom had been tenants with very limited rights on large estates.

Redistribution of land has been an important political issue in a number of South American and Asian countries, and in those parts of the former colonial territories in Africa that were occupied by expatriate settlers. There have been problems of land-use policy because the larger holdings are often efficient producers of cash crops for export, but indigenous population growth has led to an increasing demand for reallocation of land for food production to meet local needs. In other regions, encroachment on forest (even with indigenous occupants) has been encouraged by the policies of some governments to promote road building and provide tax credits and subsidies for the clearance of forests for cropping and cattle ranching. Forest dwellers frequently had no secure rights to the land they occupied, whereas outsiders could readily establish rights through clearance (in many countries, clearance of land has been a pre-requisite for claiming land rights). Such policies often led to unsustainable land use: recently they have been reviewed in several countries (see Pearce, 1991).

By 1970, most countries had established forest policies, even though these were largely restricted to government-owned forests, were not proof against encroachment, commonly dealt with exploitation rather than sustainable management, and often transferred forest management concepts from northern temperate regions to areas of tropical land to which they were less suited. In several developed countries, economic policies were encouraging reafforestation of land formerly used for grazing, and a number of these countries had established national forestry agencies. However, the tendency of these agencies and private landowners to plant monocultures of non-indigenous species, especially conifers and eucalypts, was in turn a cause of environmental controversy.

Pollution control laws evolved slowly in the industrialized world, usually as a reaction to situations that had become socially intolerable. The first modern laws to deal with air pollution from industry were the United Kingdom Alkali etc Works Regulation Acts of 1863 and 1872, enacted only following massive public complaint about the environmental devastation and threats to health caused by the discharge to the atmosphere of hydrochloric acid fumes from the manufacture of caustic soda (Ashby & Anderson, 1981). Following the 1952/53 London smogs which caused 3,000 and 4,000 deaths respectively, a government committee was appointed in the UK to investigate the problem. The reports of the committee (1953 and 1954) were the basis for the Clean Air Act of 1956. Another Clean Air Act (1968) extended and improved the 1956 Act. Similarly, laws to control water pollution have tended to be enacted only after intolerable pollution of river systems, public complaint, public enquiry, and the recognition by central government that action was essential. As a result, piecemeal pollution control legislation expanded in Europe and North America from the 1860s onward, and as late as 1970 control systems were commonly fragmented, dealing sectorally with air and water pollution, and with responsibility for action vested in a multiplicity of different central and local government agencies.

The evolution of environmental policies up to 1970 was sectoral, reactive and uneven. In many countries it was intimately linked to the granting of rights to certain sections of the population, often to the exclusion of others. Sometimes the rights of one sector to damage the environment in the pursuit of personal gain had pre-eminence over the general interest of the community to be provided with clean air, safe water supplies and a productive environment. In many countries, industrial development and urban growth assumed priority over the protection of air, water and land and the safeguarding of agricultural interests. Many cities were permitted to expand over prime farmland, and the cash value of land approved for urban and industrial development was commonly much greater than that for agricultural use, reflecting the increased potential earnings from the land under the new pattern of use rather than its intrinsic worth to the community. The need for systems of economic evaluation that truly reflected the service to the community provided by wild land and agricultural land had not been recognized, and economic cost-benefit equations and the judgemental process behind permission for land conversion tended to favour the urban and industrial and the short-term over the long-term.

Institutions for environmental management also grew sectorally and sporadically. By 1970, many developed countries had departments of government that were concerned with aspects of environmental management, notably agriculture, fisheries, forestry, town and country planning, industrial development, transport policy, and central government's responsibilities for local and municipal affairs. At local level, there was commonly devolved responsibility for land management and development control within the area of the authority's interests, and for the control of local sources of pollution. In developing countries, a number of Ministries of Health had responsibilities that included environmental matters. However, coherent environmental policies based on monitoring and with guiding strategies generally did not exist in any country.

There was a burst of activity just prior to the Stockholm Conference. Sweden established a National Environmental Protection Board in 1969, while the United States established its Council on Environmental Quality and its Environmental Protection Agency (EPA) in 1970, under a National Environment Policy Act (NEPA). The United Kingdom established a Royal Commission on Environmental Pollution and a Department of the Environment in 1970, and Canada established its Department of Environment in the same period. In 1971 Japan established its Environmental Agency and France established a Ministry for the Environment and the Protection of Nature. It was not, however, until after the Stockholm Conference that Departments of Environment and cross-sectoral coordinating machinery for environmental affairs were established in many countries. The same pattern also characterized national law. Among OECD nations only four major national environmental laws were passed in the five years 1956-60, 10 in 1961-65 and 18 in 1966-70, but between 1971 and 1975, 31 such measures were adopted (OECD, 1979; Holdgate, Kassas & White, 1982).

The period up to 1970 was also characterized by a widening influence of what can broadly be termed 'ecological thinking' on national policies, laws and institutions and a growth in citizens' groups concerned with environmental issues (see Chapter 21). These private organizations - the ancestors of today's world-wide non-governmental environmental movement - were founded to focus on very specific issues such as the protection of birds and the safeguarding of sites of outstanding importance for their natural beauty and wildlife. The vast proliferation of environmental NGOs, many of them with wide cross-sectoral interests, and their expansion to cover international affairs, has been a post-1970 phenomenon.

These trends were undoubtedly stimulated by increasing concern in developed countries (reviewed in Chapter 21) over the adequacy of the world's environmental systems to support mounting human populations, and the recognition in the 1960s of the unforeseen environmental impacts of modern industry, and especially pesticides. (Carson, 1962; McCormick, 1989). But at the time of the Stockholm Conference, these concerns were only beginning to reach developing countries, which had their own preoccupations, especially with 'the pollution of poverty'.

In 1972, therefore, the national situation around the world was one of extreme unevenness in policy, law and institution. While there was world-wide recognition of the need for productive agriculture and forestry and the protection of fisheries, in most countries these had not been given an environmental label. Environmental protection and pollution control were largely a concern of northern developed countries. A few of the latter had established Departments of the Environment and national environment protection agencies, backed by special councils to establish national targets across the whole environmental field, but they were a minority. The scientific base for environmental policy was fragmented, and there were few national reports on the state of the environment and few national monitoring systems. Indeed the first real flurry of such national reports was provoked by the Stockholm Conference itself, which received 80 such documents (McCormick, 1989).

Trends in the industrialized countries, 1972-92.

Any division between 'industrialized' and 'non-industrialized' or between 'higher income' and 'lower income' countries is inevitably arbitrary. In this chapter, 'industrialized countries' will be taken to include the OECD and former CMEA states even though the CMEA itself ceased to have formal existence towards the end of the period under consideration and included some non-industrialized countries such as Mongolia and Cuba.

At the beginning of the period under review, all these countries had a compendium of national laws relating in one way or another to the environment (usually by sectors, including agriculture, forestry, development control and the protection of outstanding landscape and wildlife habitat). The administrative units responsible for implementing national laws and policies were generally similarly sectoralized, but there was an emerging recognition that environmental policies and administrative units needed to be drawn together in some more comprehensive way. There was also a recognition that the intensifying problems of environmental impact required measures that would anticipate and prevent damage rather than consume resources in cleaning up devastation created by unwise industrial expansion. These trends found their expression in the rapid development of policies, laws and institutions in virtually all industrialized countries, with increasing co-operation through the Organization for Economic Cooperation and Development (OECD), which established an Environment Committee in 1970 (see Chapter 23).

The data base for these trends in the industrialized countries is none the less uneven. There is good information for the OECD countries, provided through a number of reports published by that organization (1979, 1980, 1985, 1991). In addition, detail of the emergence of environmental policies in the European Economic Community has been recorded in a series of European environmental year-books (e.g. DocTer, 1987, 1991). The record for the central and eastern European states is less complete, but is partly provided in a series of CMEA publications (e.g. CMEA, 1979). More recently, surveys of the condition of the environment in central and eastern European countries have been provided in a series of publications by IUCN (1990, 1991).

Policies

The emergence of environmental policy in the industrialized countries over the past 20 years has been characterized by the following main features.

- A recognition of the need for sectoral policies to take more account of environmental impacts and constraints. Agricultural policies, for example, have been broadened to include much tighter controls over pesticides, fertilizers and farm wastes. Energy, industry and transport policies have all devoted increasing attention to reductions in emissions to air and water, and to the handling of hazardous wastes. Environmental impact assessment has been a feature of development control in most developed countries.

- An increasing recognition of the need for cross-sectoral policies. Links between environment, energy, transport and technology policies have been strengthened. Pollution control policies have become cross-sectoral, seeking the 'best practicable environmental option'. This requires 'a national integrated policy for waste disposal, which determines that waste shall be put where it will do least harm, not just where it is under least control' (Royal Commission, 1976, 1984).

- Replacement of a formerly largely reactive approach to pollution control by an anticipatory and preventive one. During the two decades there has been an increasing focus on pollution abatement at source. The precautionary principle and polluter pays principle are more and more widely recognized. Screening of potentially hazardous substances before their introduction to the general market has become more stringent, partly as a reaction to the unforeseen impacts of persistent organochlorine pesticides and chlorofluorocarbons. The concept of 'critical loads' (defined as the maximum amount of pollutant an ecosystem can withstand without stress) has increasingly come to guide policy.

- Increasing interest in economic instruments as an incentive to energy and materials efficiency and pollution control. This development recognizes that while strict regulation remains necessary, within a market economy such instruments are often the most efficient way of bringing about the necessary environmental improvement.

- Increasing promotion of energy efficiency, energy conservation, and environmentally sound processes in industry, transport and domestic environments.

- Recognition of the international, and often regional, nature of many environmental problems, with an increasing broadening of the response.

- Increased attention to public information and participation. It has become evident that the most effective way of ensuring the success of environmental policies is to involve all sectors of the community in the discussion of their objectives, and to inform the public about the state of their environment and the success of the actions being taken to protect it.

- Better environmental science and monitoring. These are the foundations of sound policy and public understanding. Through them, the loop between policy, implementation and effect can be closed. Because of the international dimension of many environmental problems, monitoring is increasingly subject to international co-ordination, and linked into global programmes (Chapter 20).

During the 1970s and 1980s, as in the preceding decades, unforeseen events continued to force sudden adjustments in policy (see Chapter 21, Box 1). These were one cause of the action early in the 1970s to prevent the pollution of the sea from the dumping of toxic materials and through accidents involving large oil tankers. In the 1960s and 1970s there had been a prevailing view of the ocean as an effectively limitless disposal system, but it was clear by the time of the Stockholm Conference that this was not a resource that should be relied on, and since that date a whole series of national and international measures curbing marine pollution have been set in hand. Many of the latter are regional in scope, and are described in Chapter 23.

Disasters at Flixborough, Bhopal, Seveso, Basel and Chernobyl led in turn to the demand for much more stringent action on the safety of industrial plant, better protection and higher health standards for workers in industry, more certain containment of hazardous chemicals, close controls over the movement of hazardous waste and improvements in the safety of nuclear reactors, all of which were reflected in both national and international responses, the latter being described in Chapter 23 (see also Chapters 10, 12, 13 and 21). These events were a salutary reminder of the fact that risk is inherent in any complex industrial operation, and there was therefore a strand of commitment to improving risk assessment and to the judgement of new policy on the basis of its results (Kates, 1978; Royal Society, 1981, 1983).

These trends in policy were codified very clearly at a meeting of OECD Environment Ministers held in May 1979 (OECD 1980). This was the first statement of the modern form of environmental protection policy to emerge from the developed countries as a whole, and it resulted from a critical analysis of the trends in various sectors as they affected the environment. The meeting also noted the positive value in developing industrial techniques that were designed from the outset to avoid environmental problems. The key components of the OECD Ministerial statement are set out in Box 3.

Within the OECD member countries, numerous guiding principles have been evolved since 1970. The first of these was the 'polluter pays principle' which essentially states that the costs of pollution should not be externalized. An industry or municipality should itself bear the costs, without subsidy, of the actions needed to meet environmental standards and avoid environmental

---

BOX 3

Environmental policies as stated in a declaration by OECD Environment Ministers, 1979.

- The environmental consequences should be considered at an early stage of the decision process.

- More effective instruments should be developed, including both economic and other means, in order to integrate environmental policy with policy in other sectors (primarily land-use planning, energy and the chemical industry).

- Both economic and fiscal instruments should be used in combination with regulation in order to induce public and private enterprises and individuals to anticipate the environmental consequences of their actions.

- The design, development and use of processes that conserve resources and energy and protect and enhance the environment should be encouraged.

- Laws and regulations should avoid conflicting requirements and consequent delays in decisions affecting the environment.

- A system should be developed to account for changes in environmental quality and related resource stocks.

- Public participation should be encouraged where decisions with significant environmental consequences are to be taken. The public should be provided with information on the risks, costs and benefits associated with the decisions.

- Environmental education should be promoted in order to increase awareness.

- International co-operation within OECD should be strengthened, in order to achieve harmonization of national environmental policies.

- The OECD states should cooperate through appropriate international organizations with all countries, and particularly developing countries, in order to help to prevent environmental deterioration.

Source: OECD (1980).

---

damage. As a consequence, market prices should reflect the full costs of environmental damage arising from pollution - or, more appropriately, of the costs of preventing such damage. Similarly, the *'user pays principle'*, which is a development of it, requires that prices reflect the full social cost of use or depletion of a resource.

In parallel with these demands there has been the recognition that the standards set to curb environmental damage need to be higher than those in vogue at the beginning of the 1970s, and that a *precautionary principle* should be adopted. This broadly demands that if an activity or substance clearly carries a significant risk of environmental damage it should either not proceed or be used, or should be adopted only at the minimum essential level and with maximum practicable safeguards. The precautionary principle has been applied to curbing discharges to the waters of the European coastal seas, notably at the Ministerial Conferences on the North Sea in 1987 and 1990 (Chapter 23).

To put these principles into practice requires a capacity to evaluate the risks posed by activities and make certain that the decision process prevents an activity being undertaken until a proper assessment has been made. Hence the practice of environmental impact assessment (EIA) as a necessary preliminary to the granting of permission for a development has gained universal acceptance within the OECD countries, and has been accepted in principle at least in the former CMEA countries. EIA unavoidably requires a capacity to assess risk, and as Chapter 20 indicates, there are reasonably well developed techniques for doing this for short-term and medium-term site-specific factors. The requirement to internalize costs and to take account of the full social cost of an activity has led to a major search for a whole new process of environmental accounting, in which the economic value of environmental resources and the functioning of environmental systems is more adequately reflected (see Chapter 20; WRI, 1990).

As a result of these evolutions in policy, there has been increased investment in the developed countries in three particular spheres: environmental protection itself (Table 2), governmental investment in environmental research (Figure 1), and public information.

The volume of public information has increased enormously in the period between 1970 and 1990. Most developed countries have published government policy statements about the state of the nation's environment, and the intentions of the national authorities in this field. Compendia of environmental data have been published in 17 OECD member countries (Canada, the United States of America, Japan, Australia, Austria, Denmark, Finland, France, Germany, Ireland, Italy, The Netherlands, Norway, Portugal, Sweden, the United Kingdom and Yugoslavia) as well as for the European Community as a whole (OECD, 1991). National reports on the state of the environment have been published yearly since 1970 in the United States; since 1969 in Japan, and on a less frequent basis in most OECD countries and by the European Community (Table 3).

Table 2: *Investment in environmental protection in OECD member countries, 1970-85.*

	As percentage of GNP	As percentage of all investment
Public sector	0.13 – 1.17	0.24 – 2.87
Private sector	0.08 – 0.86	0.28 – 1.62
Total	0.82 – 1.52	1.28 – 3.15

Source: OECD (1991).

Figure 1: *Government expenditure on environmental research and development in selected countries, 1975–83.*

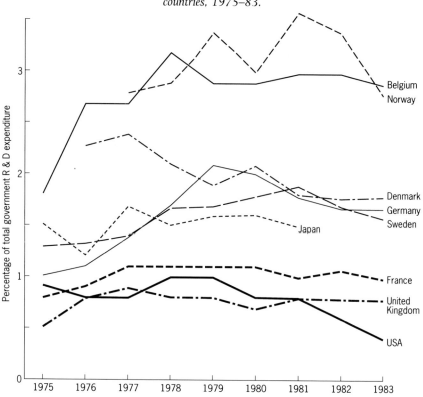

Note: These expenditures refer to R & D expenditures on pollution abatement, but do not include expenditures concerning pollution abatement through changes in production processes themselves.

Source: OECD (1985).

A number of countries have established independent bodies to act as evaluators of the state of the environment or of particular environmental problems, publishing their reports for the information of the community at large. Such bodies include the Council on Environmental Quality (USA) (CEQ, 1970-), The Royal Commission on Environmental Pollution in the United Kingdom (Royal Commission, 1971-), and the Environment Agency in Japan (Environment Agency, Japan, 1979-). Other significant public information documents have appeared from time to time in Norway, Sweden, Finland, Denmark, The Netherlands, France, Germany, Canada, and most other OECD states. However, in central and eastern Europe there has been much less effective information flowing to the public on environmental progress and conditions, not least because of a failure in most of those countries to achieve the stated goals of environmental policy or even to conform strictly with the

Table 3: *"State of Environment" reports available in 1990.*

Continent	Country	Date
Africa	Kenya	1987
	Madagascar	1987
	Mauritius	1985
	Morocco	In preparation
	Mozambique	1985
	Zimbabwe	1990
North America	Canada	1986, 1986
	Costa Rica	1988
	Mexico	1986
	Panama	1985
	USA	1988, 1988
South America	Brazil	1984, 1985
	Chile	1984, 1985
Asia	Bahrain	1988
	China	1988
	Cyprus	1989
	Hong Kong	1988
	India	1985
	Indonesia	1990
	Israel	1988
	Japan	1988
	Kuwait	1987
	Malaysia	1987
	Nepal	In press
	Qatar	1987
	Saudi Arabia	1989
	Thailand	In press
Europe	Austria	1989
	Belgium	1989
	Finland	1988
	France	1988
	Germany	1989 (FR); 1990 (DR)
	Greece	1983
	Hungary	1990
	Ireland	1987
	Italy	1987, 1989
	Liechtenstein	1984
	Luxembourg	1984, 1988
	Netherlands	1987, 1989, 1989, 1989
	Norway	1988
	Poland	1989
	Portugal	1989
	Sweden	1990
	Switzerland	1989
	UK	1978-90, 1990
	Yugoslavia	1987
Oceania	Australia	1986

Source: UNEP (1991).

standards established by law (IUCN, 1990, 1991). This situation is now being remedied. In the former USSR, for example, a general statement of environmental policy was published in 1990 (translated into English in IUCN, 1991).

A final important development in policy has been the requirement in many OECD countries that there should be public access to information on the environment held by public authorities, including especially the results of monitoring of the performance of industry in achieving the standards and consent conditions imposed on it by government. Legislated freedom of access now exists in the United States, Canada, New Zealand and the European Community (under Directive 90/313/EEC of 1990), and it is clearly fast becoming the norm. The difficulty has lain in striking an appropriate balance between the public interest and the need for industry to maintain commercial confidentiality over technical advances that secure market advantages. The main categories of information released into the public domain have been the quantities and nature of hazardous materials held by particular industrial plants, and the quantities of discharge to the environment of specified substances.

Public concern, expressed through the 'Green Consumer Movement' (Chapter 21), regarding the environmental 'friendliness' of products, has led to a further policy of requiring the labelling of consumer goods to include information that will allow a prospective purchaser to judge whether or not they meet criteria of environmental acceptability. Labelling schemes, pioneered in Germany and extended in Canada and Japan, are now being adopted in the Nordic countries, Australia, New Zealand and in the European Community and elsewhere (Sand, 1990).

Laws and Instruments

There has been a major evolution of environmental law in the industrialized countries between 1970 and 1990. Table 4 gives some general statistics, which are further extended by the detailed analysis of European Community legislation in Chapter 23. Both in the EEC and in OECD generally, the surge in environmental legislation between 1950 and 1970 appears to have been succeeded by a more measured process from 1970 onwards, in which the character of the legislation has changed to bring about a more integrated and cross-sectoral series of policies, and increasingly to apply within the territory of individual countries international obligations entered into on a regional or global basis.

Much national environmental law has been concerned with regulating activities that have the potential to cause environmental hazard. Such regulations concern, for example, the containment of toxic substances, in storage, in use and in transportation; the authorization of discharges to the environment (which normally require specific permits, industrial plant by industrial plant, from an appropriate control authority); and the setting of standards for emissions, which must be met either by point sources of emission or by motor vehicles, aircraft and other emitters. Another whole dimension of

Table 4: *Environmental laws adopted by OECD member countries, 1950-84.*

Type of legislation	1950-59	1960-69	1970-79	1980-84
General	0	2	8	7
Water	3	6	16	4
Wastes	0	1	13	2
Air	1	6	8	4
EIA	0	2	8	1
Other	0	3	11	5

Source: OECD (1979, 1985).

law and regulation is concerned with standards for manufactured products, ranging from vehicles and aircraft through to consumer goods. Yet another dimension is concerned with the procedures that must be adopted before new developments are sanctioned (most clearly illustrated by the laws requiring environmental impact assessment before development proceeds, in virtually all industrialized countries), or prior to the approval for use of pharmaceuticals, pesticides and other particular classes of product. A final area of recent legislation concerns public access to information, and the rights of the public to sue companies or others that cause damage or create risk by their activities.

The emergence of an increasing body of law and regulation has been paralleled by changes in the way in which legal instruments are interpreted and enforced in a number of countries. The public has acquired an increasing right to challenge the way in which departments of government and official agencies carry out their functions, and this move to make public authorities accountable to the public at large through the courts for the achievement of national policy goals in the environmental field is a development that has clear implications for the future. In the United States of America, for example, most environmental statutes give private citizens the right to sue polluters when the US Environmental Protection Agency or a state government has not done so. After the Bhopal disaster in India, Congress enacted the Emergency Planning and Community Right-to-Know Act, aimed at reducing such risk in the United States. This Act requires the reporting of releases of toxic substances so as to help citizens and local communities plan for and respond to such incidents (see Chapter 12 for details).

During the 1960s and 1970s an increasing number of countries have recognized that economic instruments can, within market economies, be an effective means of improving the environment and then maintaining high environmental quality. By 1988, 153 different economic instruments were reported to be in use in the OECD countries. Of these, 81 involved charges of various kinds, 41 were subsidies, and 31 were of other kinds. User and administrative charges were commonest, and included fees for waste collection and sewerage services, while administrative charges were usually for licences to release substances to the environment. Not all of these charges covered the full economic cost of the activity in question, some simply recovering the costs of the administrative activities of the licensing authority. Among the 'other' economic instruments in use were charges added to the cost of projects in order to encourage recycling, for example of oil, batteries and containers (OECD 1991). Table 5 illustrates the range of charges and types of charge systems being used in OECD countries. Differential taxes, though not a direct charge, have been found useful in influencing consumer behaviour at an individual and household level. For example, Denmark, Finland, Germany, The Netherlands, New Zealand, Sweden, Switzerland and the United Kingdom impose different levels of taxation on leaded and unleaded gasoline, while The Netherlands, Sweden, Japan and Germany use taxation as an instrument to promote low-pollution vehicles.

The evolution of natural resource accounting as a base for economic instruments that will in turn promote sound use of natural resources was an active phenomenon in many industrialized countries in the late 1980s. This new development has been a reflection of the fact that gross national product

Table 5: *Types of environmental charge systems in use or under consideration in selected countries.*

	Effluent				User	Product	Administrative	Tax differentiation
	Air	Water	Waste	Noise				
Canada					X	X		X
United States				X	X	X	X	
Australia		X	X		X		X	
Japan	X			X				
Austria		X			X			X
Belgium		X	X		X		X	X
Denmark			X		X	X	X	X
Finland					X	X	X	X
France	X	X		X	X	X		
Germany		X	(X)	X	X	X	X	(X)
Greece	X				X		X	X
Italy		X			X	X		
Netherlands	X	X		X	X	X	X	X
Norway					X	X	X	X
Portugal		X					X	
Spain			X		X		X	
Sweden	X				X	X	X	X
Switzerland	(X)			X	X	(X)		X
Turkey			X					
United Kingdom		X		X	X		X	X

Notes: X = applied; (X) = under consideration.

Source: OECD (1989, 1991).

and related aggregate economic accounts are not adequate as a means of measuring the depletion and degradation of natural resources, even where these resources are a primary source of national income. Norway, France, Canada and The Netherlands have all now proposed or established systems of environmental accounts, and the methodology has been extended more broadly through the work of the United Nations Statistical Office in modifying the UN System of National Accounts (Chapter 20; Weiller, 1983; United Nations 1986; Repetto *et al.* 1989; WRI, 1990).

Institutions

Institutional responsibility for the environment in industrialized countries is commonly divided between the national and the state/district/municipal/local authority levels. Moreover, there are commonly divisions, at both these levels, between departments with overall responsibility for policy and agencies that have operational responsibility for setting technical standards, granting operational consents, levying charges, monitoring performance and prosecuting infringements.

The governmental sector has no monopoly of institutions concerned with the national environment. During the last 20 years there has been an increasing emergence of self-regulation institutions within industry, linked to the development of new and environmentally compatible technologies in which sectors of industry have commonly co-operated. A third institutional development has been the massive growth in non-governmental organizations, usually in the form of groups whose business it is to demand higher standards of environmental performance, protect particular regions or interests, or draw attention to inadequacies of performance.

At national level, the early 1970s were characterized in many developed countries by a sudden initiative to incorporate what were then perceived as new environmental concerns within the mandates of existing sectoral ministries. This was done in Sweden (Ministry of Agriculture), The Netherlands (Ministry of Health) and Germany and Finland (Ministry of the Interior) (OECD, 1979). In other countries, including the United Kingdom and France, new ministries were formed by the bringing together of pre-existing sectoral departments. In yet other countries, including the United States, Japan, Norway, Canada, France, Australia and New Zealand, more specialized environment ministries were created. In the United States, the Environment Protection Agency, established in 1970, has responsibility for the implementation of much legislation and policy for the protection of the environment, while the Council on Environmental Quality exists to advise the President on the need for further action in the environmental field (a task which in the United Kingdom is devolved especially on the Royal Commission on Environmental Pollution).

Toward the end of the 1980s there was increasing concern within governments over the inadequacy of having sectoral Departments of Environment established alongside others, when it was apparent that the environment was actually the natural capital of nations, the foundation of quality of life and prosperity, and influenced by all the various sectoral activities (WCED, 1987). Many governments have therefore established new machinery to better co-ordinate the activity of the sectoral departments. In Norway, the Ministry of Environment co-ordinates and examines annually the budget proposals for environmental spending in all ministries. The total governmental spending on environmental issues is presented as Norway's 'Green Budget' to the Parliament (Miljoverndepartementet Norway, pers. com., 1992; Norway 1992). In New Zealand, the Resource Management Act of 1991 integrates resource management across all sectors except energy production, while in 1986 the Ministry for the Environment was given a 'reporting function' which requires a report from the Ministry to accompany any proposal with significant environmental implications submitted to Cabinet or its Committees (MfE, 1987). In the United Kingdom, the Prime Minister has assumed chairmanship of a committee to oversee environmental policies, while the Secretary of State for the Environment presides over a Standing Committee of Ministers which examines the implementation of those policies in greater detail. A Minister in each Department of State has been nominated to have responsibility for considering the environmental implications of all of that department's policies and spending programmes, and for following up the relevant policies set out in a major government statement in 1990 (UK, 1990). The Netherlands and a number of other countries are taking a similar line. An unpublished survey of 101 countries by UNEP shows that by 1991/2 nearly all industrialized countries surveyed had established a ministry or department concerned solely or principally with environmental protection and natural resource management (UNEP, 1992).

In many industrialized countries (notably the United States, Australia, Switzerland, Germany, Canada, and to a lesser extent the United Kingdom, France, Austria and the Nordic countries), provinces, cantons and states within national federations, or districts, municipalities or other major local units within the unitary state, have substantial responsibility for setting goals for environmental protection, and for enforcing standards and regulations. In a number of countries there has been a debate about the proper balance between the central or federal government and the state or provincial administrations. The broad conclusion seems to be that broad strategy has to be national, but worked out through a dialogue involving the components of the state, and that while standards and policies must be applied everywhere, the enforcement mechanism is often most effectively decentralized. Making this machinery work means establishing effective consultation within the countries, often down to local community and village level, and the need to empower local communities to take more direct care for their own environments is being increasingly stressed (e.g. IUCN/UNEP/WWF 1991).

In the industrial field, Chambers of Commerce, national business councils and specialized industrial organizations are playing an increasing part in environmental policy. Some such organizations are concerned with promoting products perceived to be superior in energy use or environmental impact, while others seek to represent industrial sectors in dialogue with government. In Canada, for example, the Canadian Manufacturers' Association has produced a policy paper which describes sustainable development as both a challenge and an opportunity for industry, while the Canadian Chamber of Commerce has established a Task Force for the Environment and machinery designed to stimulate and guide national industry (Canada, 1991). The need for technological transformation is appreciated in many countries and within sectors of industry, and this will be a growth area during the 1990s (Heaton, Repetto and Sobin, 1991).

A dramatic feature of the last 20 years has been the growth in non-governmental organizations concerned with environmental affairs. For example, the aggregate membership of the major environmental organizations in the USA increased from under three million in 1971 to over nine million in 1990. The figures for the UK are even more startling - a combined membership of under half a million in 1971 grew to over 3.6 million in 1990 (Chapter 21). Another recent trend has been the recognition by governments of the need to involve all these institutions - governmental and non-governmental alike - in the discussion of national policy. In Canada, round tables that involve government, industry and the non-governmental environmental movement have been established at both national and provincial levels (Canada, 1991). A similar mechanism has been established in Australia for a national study of ecologically sustainable development. It is widely accepted that discussion between the governmental sector as the regulator of action and the establisher of overall policy for the environment, industry as the source of the invention and technological advance that will be needed in order to maintain quality of life while minimizing environmental impact and the waste of resources, and the environmental non-governmental movement with its extensive scientific knowledge and commitment, and ability to mobilize public opinion, is the best way of forging an integrated approach to a sustainable future (IUCN/UNEP/WWF, 1991).

The former CMEA countries

The evolution of environmental policy, law and institutions in the former Soviet Union and the countries of Eastern Europe has largely paralleled that described above for the OECD countries. In both cases policies and legislation have historically been based on the concept that the environment is a public good, and that value can only be attributed to objects or commodities produced with the help of human labour. The approach has therefore been regulatory rather than market-oriented, and administered by centralized agencies in both East and West (Jancar, 1987). The trends in the former CMEA countries have also been similar to those in the OECD for most of the period under review. In 1987, Jancar referred to 'similar bottlenecks in environmental administration' and suggested that 'both [East and West] are seeking solutions along somewhat similar lines'.

The trend during the last two decades towards sectional policies that take account of environmental inputs and constraints, and the growing recognition of the need for cross-sectoral policies and over-arching legislation, of the international nature of environmental problems, of the need for better environmental science and for increased public participation, have all been evident in the former CMEA countries, as in the OECD. These policies have had their legislative parallels; for example in the 1977 constitution of the USSR, which introduced articles devoted to environmental protection, and in the 1976 amendment to the Polish constitution, by which measures aimed at protecting both the environment and the public's right to enjoy it in an unspoilt state were incorporated. During the same period, over-arching laws on resource use and environmental protection were enacted in East Germany (1970), Romania (1973), Hungary (1976) and Poland (1980) (Kolbasov, 1990). Bulgaria had already enacted such legislation in 1967 (IUCN, 1991). The principal difference between trends in the CMEA and OECD countries is that the former did not, at least until very recently, share the trends towards the use of market forces and economic instruments, and the promotion of increased energy and resource use efficiency, to the same extent as did most of the OECD countries.

Another difference has been in the effectiveness with which environmental measures have been implemented, leading Kolbasov to write in 1990, 'A low level of implementation in all the socialist countries is causing grave concern'. One reason given for this is the fragmentation of administrative responsibility which existed in many of the CMEA countries despite their centralized planning systems. Table 6 illustrates this fragmentation, which was exacerbated in countries such as the Soviet Union by a federal system of government (Jancar, 1987). In 1988 the USSR established a State Committee for Environmental Protection, with the corresponding state committees in the Union and Autonomous Republics, Territories, regions, districts and cities coming within its system. However, the fragmentation persisted, with responsibility divided between this new State Committee and the State Committees for forestry, hydrometeorology and industrial safety, and the ministries of water supply construction, geology, fisheries, the interior and public health (Kolbasov, 1990). The drawbacks of such approaches are summarized in a recent report on the state of the environment in the Czech Republic (Czech Ministry of the Environment, 1990). In individual industries, for example, the protection of individual components of the environment was scattered, downgraded and even suppressed because of the priority of economic interests. National committees were often unable to reconcile their responsibilities for socio-economic development with those for environmental protection. Direct investors at the local level either waited for the allocation of funds from the centre, or sometimes even diverted those earmarked for environmental protection to the solution of production problems.

The massive economic, social and political changes of the last few years in the countries of the former Soviet Union and Eastern Europe make it rather more difficult to describe ongoing trends in those countries. Recent re-evaluation of government institutions and legislative machinery suggest there could be further, and quite rapid, evolution of environmental law and institutions in the short and mid terms. There have also been far-reaching reviews of the state of the physical environment in several of the former CMEA countries, for example in Hungary, Czechoslovakia, the Soviet Union (and, since the establishment of the Commonwealth of Independent States, in Russia) and Poland, which suggest that a re-appraisal of the mechanisms for environmental assessment and regulation is taking place in those countries (e.g. Hinrichsen and Enyedi, 1990; Vavrousek, 1990; Izrael and Rovinsky, 1991; Environmental Resources Ltd, 1990; WRI, 1992). The same changes have resulted in a rapid growth in citizens' participation and NGO activity in these countries. While the 1970s saw a considerable increase in published material on, and media coverage of, environmental issues in the USSR, for example, and a corresponding increase in the independence and influence of environmental experts, direct citizens' input was until recently limited to the local level of government. Environmental NGOs such as student environmental groups and nature protection associations existed, but mass citizens' input tended to be organized from above rather than consisting of a spontaneous grass-roots activity (Jancar, 1987). For example, in Bulgaria, the National Committee for Nature Protection organized mass participation in environmental projects and policies as well as public education on environmental issues, as did trade and professional organizations, The Young Communist League and other, more specialized, social organizations. In Hungary, East Germany and Yugoslavia public participation in environmental issues was similarly organized from above, with the emphasis on public education and voluntary input to environmental projects rather than critical input to public policy-making (Enyedi, Gijswijt and Rhode, 1987). In contrast to these 'politically safe' movements, independent environmentalism has spearheaded political change in many Eastern European countries. As early as 1983, the Charter 77 movement in Czechoslovakia published clandestinely a report of the Czechoslovak Academy of Sciences detailing the extent of the country's environmental problems. In September 1980, the Polish Ecological Club - the first independent environmental group in Eastern Europe - was founded in Krakow (Brown *et al.*, 1991). At present, independent environmental groups exist in all Eastern European countries. In 1990 Kolbasov observed that *perestroika* had brought about a broader public participation in the monitoring and control of resource allocation and environmental protection in the CMEA countries. The formation of a large number of autonomous, grass-roots NGOs involved in monitoring activities impacting on the environment has led, in his terms, to 'a particularly strict public control ... over fulfilment of ecological requirements' in relation to industrial, agricultural and municipal projects (Kolbasov, 1990).

Table 6: *Fragmentation of responsibilities for environmental measures in the former USSR.*

	USSR State committees						Branch ministries		
	GP	GS	GKNT	GKG	GL	GKAP	MIu	MZ	MG
Air									
Planning	X	X	X	X	X	X		X	
Use									
Control		X		X		X	X	X	
Expertise		X		X	X			X	
Water									
Planning	X	X	X	X	X	X		X	
Use				X	X	X		X	
Control				X	X	X	X	X	
Expertise		X	X	X	X			X	
Land and soil									
Planning	X					X			
Use						X			
Control					X	X	X	X	
Expertise						X		X	
Underground									
Planning	X		X		X			X	X
Use									
Control			X		X		X	X	X
Expertise			X		X			X	X
Forests									
Planning	X	X			X	X			
Use					X				
Control					X		X		
Expertise		X		X					
Protected land									
Planning		X	X		X	X			
Use									
Control						X	X		
Expertise						X			
Wildlife and plants									
Planning		X			X	X			
Use						X			
Control					X	X	X		
Expertise					X	X			
Human environment									
Planning	X	X	X	X				X	
Use									
Control		X	X	X			X	X	
Expertise		X	X	X				X	

Descripton of column heads: GP = Gosplan; GS = Gosstroi; GKNT = State Committee on Science and Technology; GKG = Goskomgidromet; GL = Gosleskhoz; GKAP = State Agro-Industrial Committee; MIu = Ministry of Justice; MZ = Ministry of Health; MG = Ministry of Geology; MNNP = Ministry of Oil Refining and Petrochemical Industry; MXM = Ministry of Chemical and Petroleum

Source: Jancar (1987).

| and organizations | | | | | | | | | Territorial administration | | |
MNNP	MXM	MEE	MCoal	MRyb	MLDTs	MVD	MMVKH	AN	Republic	Region	local
							X	X	X	X	X
	X	X		X	X						
X	X			X			X	X			
X		X	X	X	X		X	X	X	X	X
X		X	X	X	X		X		X	X	X
		X	X	X	X	X	X				
X		X	X	X	X		X	X	X		
X							X		X	X	X
X		X		X							
						X					
							X		X	X	X
X	X	X					X		X	X	X
X	X	X					X		X	X	X
		X	X				X				
X	X	X	X				X	X	X	X	
				X					X	X	X
				X	X					X	X
					X	X		X	X		
								X	X	X	
						X				X	
								X	X	X	
			X						X	X	X
			X							X	
			X			X		X		X	
									X	X	X
										X	
						X			X	X	

Machine Building; MEE = Ministry of Power and Electrification; MCoal = Ministry of Coal; MRyb = Ministry of Fish Industry; MLDTs = Ministry of Timber, Woodworking and Pulp and Paper Industry; MVD = Ministry of Internal Affairs MMVKH = Ministry of Land Reclamation and Water Resources; AN = Academy of Science.

Trends in developing countries

The data base on national action within the developing world is far less extensive than for the developed countries. However, as a broad generalization, the developing world has tended to follow a pattern similar to that of the developed countries, but with a different range of concerns and on a different time-scale.

The primary environmental concern in the developing world is obviously the management of land and fresh waters in order to maintain the production of food and other essential resources, and the exports which are the mainstay of these economies (Table 1). Development is clearly essential in order to improve the quality of life, eliminate poverty and support the infrastructure needed in order to deliver the health care, education and other institutions essential to the national future. It is, however, increasingly recognized that development processes must be based on care for the environment, and this is well illustrated by the large number of developing countries that have prepared substantial national conservation strategies (in the form of national environmental plans for the sustainable use of their resources). Many such countries have also provided reports on the state of the environment, statements of environmental policy, and in some cases considerable volumes of statistics (see Table 3 and CSE, 1982; NEPA, China, 1990).

It is more convenient to summarize the development of policies, laws and instruments and institutions within the developing world regionally, because there are important regional differences, and this is the sequence of the following paragraphs. However, the overall trend reflects an almost universal change in expressed attitudes to the environment, from limited concern about policies for its conservation at the time of the Stockholm Conference to the broad recognition by 1990 that the conservation of natural resources is an essential part of development.

Asia-Pacific region

The primary thrust of environmental policy among most of the developing countries in the region is the maintenance of agricultural and forest productivity, often in the face of pressures from mounting population growth and a conflict between the requirement for maximum revenue in the short term and the protection of the interests of the future. However, a number of countries have enunciated much broader statements of environmental policy and several are becoming increasingly preoccupied with managing industrial growth. For example, the integrated national planning of the economy, including environmental planning, its development through national land-use planning and the associated planning of environmental pollution control, have all been stressed as policy goals in China (UNEP, 1988).

Similarly, Indonesia has stated its commitment to development based on the sustainable use of the country's resources, with improved planning and management of the environment, better use of human resources and natural resources such as land and water, and more attention to the environmental dimension in land and resource-use practices (Soerjani, 1989). Similar broad policy statements are evident in, for example, national reports from the region to the UN Conference on Environment and Development.

Within this broad policy context, the predominant emphasis is on planning, and on regulation, including some regulations to prevent environmental pollution in areas such as China where air pollution, particularly from industry, is already an acute problem (Qu Geping & Woyen Lee, 1984).

In the forest sector, many developing countries in the Asia-Pacific region have provided economic incentives for forest exploitation without insisting on investment in sustainable use, including replanting. However, a special reforestation fee in Suba, Malaysia, accounted for 6–11 per cent of forest revenue between 1973 and 1983, while a reforestation deposit fee in Indonesia, introduced in late 1980, is refundable to loggers upon presentation of evidence of adequate replanting programmes. In practice this tax has not encouraged replanting because the cost is estimated to be two to three times the refundable amount. None the less, Indonesia has plans to tap the sums available and itself invest in reforestation (WRI, 1988).

The environmental laws in the region are diverse. Some countries have relatively wide-ranging and sophisticated series of laws (e.g. India, Malaysia, the Philippines and Sri Lanka) while others (e.g. Brunei, Maldives and the Pacific Island nations) have rather less developed environmental legislation. Papua New Guinea has environmental care built into its constitution. While the primary thrust of such legislation has, like that of policy, been to regulate the use of natural resources in agriculture and forestry, pollution control has also been covered and a number of countries have also required environmental impact assessment and have established specialized agencies to whom environmental protection tasks have been entrusted. EIA is specifically part of the planning process in Papua New Guinea and the Philippines, while China, Indonesia, Iran, Korea, Maldives, Malaysia, Pakistan, Sri Lanka and Thailand have included it in general legislation. Those countries that introduced EIA legislation in the 1980s (Indonesia, Malaysia and the Philippines) have revised their regulations and guidelines a number of times since then. However, in many countries the limiting factor is not the adequacy of the written law but the efficiency with which it can be enforced.

Institutions in the region are similarly diverse. Under its Environmental Protection Ordinance of 1983, Pakistan has established a Pakistan Environment Protection Council whose responsibilities include the establishment of comprehensive national environmental policy, the inclusion of environmental considerations in national development plans and policies, the enforcement of national environment quality standards, and the power to direct government agencies, bodies or persons to take measures to control pollution. A Pakistan Environmental Protection Agency has also been established in order to administer the ordinance, prepare policy for approval by the Council, publish an annual report on the state of the environment, establish appropriate national environmental quality standards, coordinate environmental policies and undertake a host of detailed actions (Mumtaz, 1989).

India has also established a National Committee on Environmental Planning, and a Department of the Environment in central government, while a number of the component states have also set up their own Departments of Environment. An Environmental Information Service has been proposed for the collection, processing and dissemination of environmental information (CSE, 1982). However, the size of such institutions can limit their functions: the Department of the Environment in India is very small, despite its breadth of responsibilities.

China established environmental planning in the late 1960s, and in 1974 established an Environmental Protection Leading Group under the State Council. In 1982, a Ministry of Urban and Rural Construction and Environmental Protection was established, under which there was an Environmental Protection Bureau in charge of this work throughout the whole of the People's Republic. Further changes in 1984 led to the establishment of the Environmental Protection Commission of the State Council, and the National Environmental Protection Agency which is both a standing office of the Environmental Protection Commission and has a reporting line through the Ministry of Urban and Rural Construction and Environmental Protection (UNEP, 1988). In 1991 China announced the establishment of a national Advisory Council on the Environment, with very senior membership from within the People's Republic together with a group of senior international experts.

In Saudi Arabia, as an example of a western Asian nation, the Monitoring and Environment Protection Agency has had prime responsibility for environmental management since 1980, while wildlife conservation in the Kingdom of Saudi Arabia is the responsibility of the National Commission for Wildlife Conservation and Development.

As a general summary, ministerial and departmental authority for environmental interests now exists in most of the states of the region. A considerable number now have national environmental protection agencies which are increasingly concerned with the overall planning of the management of natural resources. Few, if any, of these existed in 1972. The cross-sectoral integration of these interests is, however, far from easy, especially given the complexities of the environmental situation in the region. The rural level, and especially the village communities, are often the key entities in environmental management, and yet it is not easy to involve them in overall policy. Strategies to develop land and water use systems that are built on local resource availability, people's own knowledge and traditions, and local skills, needs and ecological circumstances are however now being developed, for example in India (Agarwal & Narain, 1989).

A corresponding increase can be seen in the number, membership and effectiveness of NGOs concerned with the environment in this region during the period under consideration. Asia, and South Asia in particular, has the largest number of NGOs in the developing world. It also has a more active NGO involvement in policy making than other regions. India has an estimated

12,000 independent development organizations and tens or even hundreds of thousands of local groups dealing with environment and development issues. Bangladesh has more than 10,000 environment-related NGOs (most are small) and in the Philippines (where the role of NGOs is recognized in the constitution) the Caucus of Development NGOs has more than 1,300 affiliated members (WRI, 1992). In Indonesia, the number of NGOs in the 1980 directory of the Indonesian Forum for Environment was 450 (WALHI, 1992). However, comparison of figures between countries can be misleading, as definitions of 'environmental NGOs' can differ markedly. In some countries, national statistics are difficult to obtain and umbrella organizations list only those NGOs relating to their field of specialization. For example, in India the most recent list of environmental organizations compiled by the Ministry for the Environment (1983) contains only 273 names. In a 1992 edition of a directory prepared by the World-wide Fund For Nature, India has approximately 1,000 entries. Neither list purports to be exhaustive (WWF-India, pers. com., 1992). Similarly, the 1991 yearbook of the China Association of Science and Technology lists only eight NGOs registered as 'environmental' with the Ministry of Civil Affairs (Chinese Society of Environmental Sciences, pers. com., 1992) and the Thai Environmental Engineers' Association in 1992 listed 30 environmental NGOs (pers. comm., 1992)

Africa

Environmental policy in the African countries has been especially concerned with agricultural development, the management of the publicly-owned forest estate, the management of wildlife resources, especially for tourism, and above all the acceleration of the development process.

Africa is an immensely diverse continent, with a wide range of human traditions, climates and vegetation types. Some countries such as Ethiopia, the coastal countries of West Africa, Uganda, Kenya, Tanzania, Zambia, Zimbabwe and South Africa have substantial tracts suitable for rain-fed agriculture, and policy has been directed especially to the development of cash crops for export. Even in the Sahel, often thought of only as a region afflicted by drought, agricultural exports have risen steadily in the past two decades (IUCN, 1989), although not as fast as population growth. In other countries, the balance between food crops and cash crops for export has been difficult, and policies have needed substantial adjustment over the period (FAO, 1986a, 1986b).

A part of this policy adjustment has been necessary because of the difficulties in the transportation of food from the rural areas to the rapidly growing urban centres, together with the fact that urban populations have developed more of a taste for imported food (such as wheat and rice) than the produce easily grown and marketed in many African countries. A further problem has arisen because of the importation of increasing volumes of food in a period when world markets are depressed, which has in turn depressed agricultural production within the continent (FAO 1986b). Another policy dilemma lies in the reconciliation of rapidly growing populations, which are taking more land to cultivate and graze, and the surviving areas of wildlife habitat which may be extremely important bases for tourism, notably in eastern and southern Africa.

A number of African countries have developed national conservation strategies or national environmental plans. They include Zimbabwe, Zambia and Botswana, with studies in progress or planned in Tanzania, Kenya, Uganda, Ethiopia and a number of western African States including Chad. Nigeria has published its national policy on the environment (Nigeria, 1989). This latter policy, like the national conservation strategies, emphasizes that the goal of government is to achieve a sustainable environment, in part by conserving and using the environmental and natural resources for the benefit of present and future generations, restoring maintaining and enhancing ecosystems and ecological processes, preventing disasters such as flood, drought and desertification, integrating population and environmental factors in national development planning, preventing the depletion of forests, and solving public health and other important problems. Thus the policy base in a number of countries is clear, but immense difficulties remain in converting it into the practical and effective action that secures development, improves the national economy, and creates the circumstance in which populations will come into stable balance with an already stressed environment.

Overall national conservation strategies and environmental plans are relatively few and far between in Africa. Those that do exist display a strongly sectoral structure within government, with very little cross-sectoral machinery yet established, despite the call for integration in many conservation strategies and environmental plans. This is a challenge for the future, which a number of governments are beginning to address.

Environmental law in Africa generally follows the same sectoral pattern. In a number of countries, however, separate laws put in place by sectoral ministries have been linked together in over-arching legislation, and some new laws have been put into place. The need to update and integrate legislation dealing with environment and resource management has received increasing attention in recent years, with agencies such as the Economic Commission for Africa and UNEP (through its Regional Office for Africa) offering assistance to governments. The following areas have been suggested as requiring particular efforts over the next few years: identification of gaps in existing environmental legislation; the harmonization and strengthening of legislation to make it more comprehensive and responsive to local realities and specifications; putting in place umbrella national environmental legislation covering key sectors; establishment of appropriate frameworks for the effective implementation of environmental legislation; building capacity for modifying, revising or strengthening national environmental legislation in response to emerging and unforeseen issues; and the training of national legal experts in the formulation of environmental legislation (UNEP-ROA, 1991)

In many African countries environmental institutions also take the form of sectoral departments, though this too is changing. By 1992, 37 out of 51 countries on the continent had established Ministries of Environment, Natural Resources or Nature Conservation/Protection. In the remainder, responsibility for environment and natural resource management remained with other, usually sectorally based, ministries or agencies. Most of these are concerned with agriculture, forestry, lands and natural resources, tourism and wildlife or variants on these themes (UNEP-ROA, 1991). Kenya, Tanzania, Zaire and several other countries have government agencies responsible for wildlife conservation and the management of National Parks. Nigeria, however, has established a National Environment Protection Agency, at federal level, to harmonize policy formulation and the implementation of environmental law and the effective operation of linkages within and among the various levels of government. In addition to these administrative institutions, some African countries have established high-level institutional arrangements for formulating environmental policy. Senegal was the first country in the region to establish an inter-ministerial council for the environment (and, in Senegal's case, for urban planning) chaired by the Head of State. Guinea-Bissau, which like Senegal has a cross-sectoral ministry dealing with environmental issues, has recently established a similar council of ministers, also presided over by the Head of State.

The same period has also seen a rapid growth in NGOs and citizens' groups dealing with environmental issues. An umbrella organization for African NGOs working in the environment, 'The African NGOs Environment Network (ANEN)' was formed in 1982 with a membership of 21 organizations. This number increased more than ten-fold in the first six years of its existence, and more than doubled again in the next two years. In 1990, the membership was 530 organizations in 45 countries. The ANEN data base of NGOs working in the field of environment and development in Africa in 1990 listed over 3,700 NGOs (ANEN, 1990). Although African NGOs are involved in the same range of activities - from advocacy and policy development to grass-roots conservation and development efforts - a higher proportion is involved in practical environment and development work than among their counterparts in other regions (WRI, 1992). A 1989 list produced by ANEN with the assistance of UNEP names 39 nationally organized indigenous NGOs undertaking conservation and rehabilitation work in respect of one or more of the following areas: deserts and arid lands; rivers and lake basins; forests and woodlands (ANEN, 1989). In Morocco, six major nationally organized NGOs can be readily identified as dealing solely with environmental protection or some aspect of it, but many more exist at a sub-national level, or deal with a broad range of issues, including the environment. In Egypt, the number of environmental NGOs rose from around 19 in 1972 to 80 in 1992 (Arab Office of Youth and Environment, 1992).

Latin America and the Caribbean

In 1972, a meeting organized by the Economic Commission for Latin America and the Caribbean identified four key problem areas within the region - human settlements, natural resource management, environmental pollution and international problems (ECLAC, 1991).

These areas of policy have been developed further during the past two decades. A number of conferences have laid down guidelines for development in the region as a whole (see below; IUCN, 1975). A policy statement on the preservation and management of the Caribbean environment was issued by the Caribbean Community in 1989 as the Port of Spain Accord. An overall policy for development in the Caribbean region is being formulated by the West Indian Commission (1990). The continued problems which policies in Central America must confront have been well surveyed by reviews including that by Leonard (1987).

Many countries in the region have published national development strategies during the past two decades, although few (apart from Costa Rica) have produced national conservation strategies and environmental plans. Environmental Profiles have been produced by five English-speaking countries as a first step in resource assessment, and as a base on which full environmental strategies might be developed. The process of establishing cross-sectoral policies, in countries where environmental damage has occasionally been caused by the pursuit of sectoral goals, remains an urgent need. These reviews and reappraisals have already led to some extremely striking changes in national policy, well illustrated in Brazil by the reversal of the economic incentives and federally subsidized highway developments that were opening up large areas of forest in Amazonas to settlement and to the conversion of forests into pastures which proved economically unsustainable (Pearce, 1991). Recent developments have also included the association of a number of countries in the Amazon region to forge a common policy for the forests of that river's catchment area, and the establishment of a Central American Commission on Environment and Development.

There is a considerable diversity of legal instruments bearing on environmental resources in the Latin American and Caribbean region, but this tends to be fragmented and sectoral, and there are few overall national bodies of environmental law which come together to provide an integrated framework for the protection of the environment and the sustainable management of its resources. Specific legislation has, however, dealt with the protection of forests, the safeguarding of National Parks and other protected areas, the regulation of settlements, and the protection of air and water from pollution. A baseline survey of environment-related law has recently been undertaken in the English-speaking Caribbean by the Center for Environmental Law of the University of West Indies as a first step in assessing the need for strengthening, and ultimately rationalizing, the system within the framework of the Caribbean Community (University of West Indies, 1991). A review of environmental law and institutions in borrower members of the Inter-American Development Bank (IDB) (that is most countries in the region with the exception of Cuba and a number of English-speaking island countries of the Caribbean) was published by IDB in 1991 (Brañes, 1991). Table 7 summarizes its findings on the balance between environmental and sectoral legislation affecting the environment.

There is a great diversity of institutions for environmental management in Latin America and the Caribbean, and the division of responsibility between central and provincial or state governments further complicates the situation. A few countries have set up Ministries for the Environment, but it is not yet clear that this has facilitated inter-sectoral co-ordination in the planning and management of resource use. Of the seven environmental ministries in the region, only one (Venezuela) is exclusively dedicated to the environment. A number of countries have established environmental committees or commissions, some of which share responsibility for environmental management with sectoral agencies (Brañes, 1991). Table 8 summarizes the distribution of institutional arrangements for environmental management in the IDB member states of the region.

Table 7: *Legislation governing environmental protection in IDB member countries of Latin America and the Caribbean.*

Countries	True environmental legislation	Sectoral legislation with environmental relevance							
		Water	Forests	Wildlife	Soils	Marine and coastal ecosystems	Non-renewable natural resources	Human settlements	Environmental sanitation
Argentina	•	X	X	X	X	X	X	X	X
Bahamas		X	X	X	X	X	N/A	X	X
Barbados		X	X	X	X	X	N/A	X	X
Bolivia	•	X	X	X	X		X	X	X
Brazil	X	X	X	X	X	X	X	X	X
Colombia	X	X	X	X	X	X	X	X	X
Costa Rica	•	X	X	X	X	X	X	X	X
Chile		X	X	X	X	X	X	X	X
Dominican Rep.		X	X	X	X	X	X	X	X
Ecuador	X	X	X	X	X	X	X	X	X
El Salvador		X	X	X	X	X	X	X	X
Guatemala	X		X	X	X	X	X	X	X
Guyana	•	X	X	X	X	X	X	X	X
Haiti		X	X	X	X	X	X	X	X
Honduras	•	X	X	X	X	X	X	X	X
Jamaica	•	X	X	X	X	X	X	X	X
Mexico	X	X	X	X	X	X	X	X	X
Nicaragua		X	X	X	X	X	X	X	X
Panama		X	X	X	X	X	X	X	X
Paraguay		X	X	X	X		X	X	X
Peru	•	X	X	X	X	X	X	X	X
Surinam		X	X	X	X	X	X	X	X
Trinidad & Tobago		X	X	X	X	X	X	X	X
Uruguay		X	X	X	X	X	X	X	X
Venezuela	X	X	X	X	X	X	X	X	X

Note: • These countries have proposed or preliminarily proposed true environmental legislation.

Source: Adapted from Brañes (1991).

Table 8: *Environmental management institutions of IDB member countries of Latin America and the Caribbean.*

Borrowing member countries of the IDB	Environmental agencies				Non-environmental agencies				
	Ministries	Systems	Committees	Other	Agriculture	Health	Planning	Natural resources	Labour
Argentina			X			X	X		
Bahamas									
Barbados									X
Bolivia			X						
Brazil	X	X							
Colombia				X	X	X			
Costa Rica			X					X	
Chile			X						
Dominican Rep.			X	X					
Ecuador							X		
El Salvador			X						
Guatemala			X						
Guyana				X					
Haiti							X		
Honduras			X						
Jamaica						X	X		
Mexico	X		X						
Nicaragua			X	X					
Panama			X						
Paraguay							X		
Peru							X		
Surinam			X				X		
Trinidad & Tob.	X								
Uruguay	X								
Venezuela	X								

Source: Brañes (1991).

Many countries in Latin America have a decentralized system, like that in Colombia which has five planning regions, each with a Regional Council on Economic and Social Planning (CORPES). Each of these bodies formulates regional plans, prepares investment programmes, and co-ordinates and supervises the activity of various sectoral agencies. This structure was developed following an extensive discussion of a national environmental profile, which recognized the need for co-ordination of national activities, with an appropriate balance between central and regional administrations, and with the creation of a national environmental system in which the Administrative Department of Natural Resources and Environment (DANAR) would have a leading role in the formulation of overall strategy.

During the period under consideration, the number of NGOs and citizens' groups in the region dealing with the environment has also grown rapidly. It is estimated that there are currently more than 6,000 NGOs in the Latin American and Caribbean region that have goals related to environmental protection, most of them having been formed within the last 15 years. As in other regions this number includes organizations with broad objectives which include, but are not limited to, environmental goals, as well as those whose objectives are specific and limited to environmental matters (Brañes, 1991). A recent survey of 1,000 NGOs in Brazil found that 90 per cent of them were started since 1970, the majority (55%) dating from the 1980s. Just over half of those surveyed had church affiliations (37%) or were linked to other institutions such as universities (16%) (WRI, 1992)

The number and functions of environmental NGOs varies somewhat between countries, but the trend of increasing numbers and influence appears to be the same throughout the region. For example, Argentina, Mexico and Brazil all have a large number and wide variety of well-established NGOs. In Brazil, where they must be authorized and are subject to government supervision, there are close to 600 NGOs devoted particularly to ecological militancy and public awareness. Colombia's many environmental NGOs include the 'Green Councils', created in 1985 under the National Code on Renewable Natural Resources and Environmental Protection. Venezuela has approximately 200 environment-related NGOs with objectives varying from species conservation activities to environmental education and policy formulation (Brañes, 1991). In Jamaica the National Environmental Societies Trust (NEST), formed in 1989, lists twenty-five environment and conservation related organizations among its members (NEST, pers com, 1991). In some countries of the region environmental legislation makes specific provision for either NGO participation (e.g. Brazil, Columbia, Chile) or public participation (e.g. Mexico, Venezuela, Peru) in determining policy on environmental issues (Branes, 1991).

Concluding remarks

It is clear from this review that environmental policy is now a concern of virtually all nations. However, the programmes, laws and institutions that have been created have grown haphazardly and are largely sectoral. In most countries, different institutions are responsible for agriculture, forestry, fisheries, wildlife conservation, mineral resources, occupational and environmental health, development control, human settlements, industry, transport and tourism.

Recognition of the inter-sectoral nature of many environmental concerns is causing an increasing number of governments to develop cross-cutting policies, laws and institutions, and many are also seeking to bridge the gaps between central and local government, and between government, industry, and non-governmental organizations. Within government, the machinery commonly takes the form of interdepartmental committees, while the policies are often expressed in national environmental strategies, developed in consultation between the sectoral departments. Even then, the policies are generally implemented through the traditional sectoral machinery. Only a few governments have created high-level, cross-cutting procedures under the direct control of the Head of Government or a very senior minister, and assigned to all ministers responsibility for accounting for the environmental impacts of their departmental policies. Still fewer have required their Ministers of Finance to adopt new thinking about environmental valuation, or to scrutinize how the natural capital of the environment is used.

It is obviously necessary for governments to develop their environmental policy through a dialogue with business, industry, commerce, environmental and development NGOs, and citizens' groups, and to secure consensus through a wide process of consultation. Some nations (such as Canada) have developed model consultative machinery and more are moving tentatively towards wider processes of consultation.

The need for a national integrated environmental policy is reflected in national conservation strategies (as in Nepal (1988) and Pakistan (1991)). However, in many countries, especially in the developing world, the weakness is less in organizational structure, or even in consultative machinery, than in the resources available for the implementation of agreed policy or the enforcement of law. An authoritative review in India (Khoshoo, 1986) has suggested 12 areas where priority action is needed :

- population stabilization;

- integrated land-use planning;

- maintenance of the health of cropland and grassland;

- forests and woodland conservation;

- conservation of biological diversity;

- control of pollution;

- development of non-polluting renewable energy systems;

- recycling of wastes and residues;

- development of ecologically compatible human settlements;

- promotion of environmental education and awareness;

- updating environmental law; and

- the establishment of a new dimension of national security which includes the recognition that this depends on safeguarding the environment.

However, environmental departments in many developing countries have tiny staffs and budgets in relation to the demands made on them. Finance for such activities as nature conservation or environment protection is often far below the calculated minimum for efficiency (Leader Williams and Albon, 1988). In addition, revenue from natural resource use or tourism often bypasses local communities so that they have little incentive to conserve the resources on which these activities depend.

Another problem is the lack of information and data. Despite the expansion of global monitoring schemes, information about the state of the environment and on trends in it is inadequate in many developing countries. Communications with local communities are often poor, so that central authorities are out of touch with the situation at local level and with the needs of local groups. Formal and informal education, training, and other supportive measures that would help increase environmental awareness fall far short of meeting the needs of sustainable environmental management. Uncertainties over land tenure and environmental rights are another obstacle in many countries.

The result is that most developing countries lack the scientific base and the human resources to develop either the policies or the legislation for effective management of natural resources and the physical environment. A serious effort must be made by developing and developed countries working in partnership to build up a sufficient body of trained personnel - scientists, environmental economists, environmental lawyers and resource managers - in each developing country for the creation of effective government institutions dealing with environmental management as well as with sectoral issues of resource management. Only then will developing countries be able to undertake environmental monitoring and assessment, formulate environmental standards appropriate to local conditions, develop the tools of environmental accounting and enact appropriate environmental law; all of which are fundamental to their capacity for environmental management and sustainable development. Also very important is the role of non-governmental institutions and organizations and citizens' groups, both those that make a substantive contribution to environmental management through research and monitoring of environmental data and those that contribute to the policy-making process by raising public awareness and stimulating public participation in policy debates.

National, regional and international actions are clearly closely linked. The growth of regional programmes and institutions, backed by regional conventions and other legal instruments, has been a striking feature of the period between 1970 and 1990, and seems set to continue as the extent of the interest shared between nations becomes more and more apparent. The increasing recognition of global problems is in turn leading to the rapid growth in international legal instruments concerned with the environment (Chapter 23). Again, this growth in international obligations poses serious problems for many governments in the developing world, because the obligations they incur are far from cost- or resource-free. International scrutiny of the compliance with treaty commitments is becoming closer, with understandable criticism of non-compliance. Governments thus find themselves under dual pressure: to join in international agreements and to commit scarce resources to their implementation.

It is pointless to create ideal frameworks for environmental management unless governments have both the resources and the will to apply them. The current impetus for more and stronger environmental action will provide the necessary results only if the development process itself is well-conducted, and if human and financial resources can be committed to effective action on the ground.

REFERENCES

Agarwal, A. and Narain, S. (1989) *Towards Green Villages: A Strategy for Environmentally Sound and Participatory Rural Development.* Centre for Science and Environment, New Delhi.

ANEN (1989) *Directory of African Non-Governmental Organisations Involved in the Conservation and Rehabilitation Activities Related to Deserts and Arid Lands; Rivers and Lake Basins; Forests and Woodlands; and Seas.* African NGOs Environment Network/United Nations Environment Programme, Nairobi.

ANEN (1990) *Building the African Environment Movement from the Grassroots.* African NGOs Environment Network, Nairobi.

Arab Office of Youth and Environment (1992) Personal communication, held in file at UNEP.

Ashby, E. and Anderson, M. (1981) *The Politics of Clean Air.* Clarendon Press, Oxford.

Brañes, R. (1991) *Institutional and Legal Aspects of The Environment in Latin America, Including the Participation of Non-governmental Organisations in Environmental Management.* IDB, Washington.

Bonner, W.N. (1982) *Seals and Man: A Study of Interactions.* Washington University Press, Seattle.

Brown L. *et al.* (1991) *State of the World 1991.* Norton, New York and London.

Carson, R. (1962) *Silent Spring,* Houghton Mifflin, Boston.

Canada (1991) Canada's National Report. United Nations Conference on Environment and Development, Brazil, June 1992. Environment Canada, Ottawa.

CEQ (1970-) Environmental Quality. Annual Reports of the Council on Environmental Quality - annually from 1970. USA Government Printing Office, Washington D.C.

Chinese Society of Environmental Sciences (1992) Personal communication, held on file at UNEP.

CMEA (1979) Reports of Various Activities of the Bodies of the CMEA in 1978. Council for Mutual Economic Assistance, Moscow.

CSE (1982) *The State of India's Environment, 1982: A Citizen's Report.* Centre for Science and Environment, New Delhi.

Czech Ministry of Environment (1990) *Environment in the Czech Republic,* Academia, Prague.

DocTer (1987) *European Environmental Year Book.* DocTer International, London, UK.

DocTer (1991) *European Environmental Year Book.* Second Edition. DocTer International, London, UK.

Dyson-Hudson, N. (1984) Adaptive Resource Use Strategies of African Pastoralists. *In*: DiCastri, F., Baker, F.G.W. and Hadley, M. (eds), *Ecology in Practice,* Part I , Ecosystem Management. Tycooly International, Dublin, for Unesco, Paris.

ECLAC (1991) *Sustainable Development: Changing Production Patterns Social Equity and the Environment.* Economic Commission for Latin America and the Caribbean, Santiago de Chile.

Environment Agency, Japan (1979-) *The Quality of the Environment in Japan. Overviews of the state of Environment in Japan,* annually from 1979. Japanese Environment Agency, Tokyo.

Environmental Resources Ltd (1990) *Eastern Europe: Environmental Briefing,* ERL, London.

Enyedi G, Gijswijt, A.J. and Rhode, B (eds) (1987) *Environmental Policies in East and West.* Taylor Graham, London.

FAO (1986a) *African Agriculture: The Next 25 Years.* Food and Agriculture Organization of the UN, Rome.

FAO (1986b) *Atlas of African Agriculture.* Food and Agricultural Organization of the UN, Rome.

Gabriel, T. (1988) Agriculture and Human Resources, *Journal of Oman Studies Special Report* No. 3, 473-84.

Gulland, J.A. (1976) Antarctic Baleen Whales: History and Prospects. *Polar Record,* **18**:5-13.

Gulland, J.A. (1987) The Antarctic Treaty System as a resource management mechanism. *In*: Triggs, G.D. (ed) *The Antarctic Treaty Regime.* Cambridge University Press, Cambridge.

Heaton, G., Repetto, R. and Sobin, R. (1991) *Transforming technology: an agenda for environmentally sustainable growth in the 21st century.* World Resources Institute, Washington, D.C.

Hinrichsen, D. and Enyedi, G. (eds) (1990) *State of the Hungarian Environment.* Statistical Publishing House, Budapest.

Holdgate, M.W., Kassas, M. and White, G.F. (1982) *The World Environment, 1972-1982.* Tycooly International, Dublin.

IUCN (1975) *The Use of Ecological Guidelines for Development in the American Humid Tropics.* International Union for Conservation of Nature and Natural Resources, Gland, Switzerland.

IUCN (1989) *IUCN Sahel Studies, Volume 1.* IUCN, Gland, Switzerland.

IUCN (1990) Environmental Status Reports: 1988/89, IUCN Eastern European Programme, Gland, Switzerland

IUCN (1991) Environmental Status Reports: 1990. IUCN Eastern European Programme, Gland, Switzerland.

IUCN/NOAA (1991) *Large Marine Ecosystems.* Proceedings of an International Conference held at Monaco, 1990.

IUCN/UNEP/WWF (1991) *Caring for the Earth: A Strategy for Sustainable Living.* International Union for Conservation of Nature and Natural Resources, Gland, Switzerland.

Izrael, Y.A. and Rovinsky F.Y. (eds) (1991) *Review of the State of the Natural Environment in the USSR.* Department of Gidrometeoizdat, Moscow.

Jancar, B. (1987) *Environmental Management in the Soviet Union and Yugoslavia.* Duke University Press, Durham.

Kates, R.W. (1978) *Risk Assessment of Environmental Hazard.* SCOPE 8. Chichester, John Wiley & Sons, New York, Brisbane & Toronto: .

Khoshoo, T.N. (1986) *Environmental Priorities in India and Sustainable Development.* Indian Science Congress Association, New Delhi.

Kolbasov, O. (1990) Introduction to an unpublished report on environmental legislation of the CMEA Member Countries. Original held by United Nations Environment Programme.

Leader Williams, N. and Albon, S.D. (1988) Allocation of resources for conservation. *Nature,* **336**, 533-35.

Leonard, H.J. (1987) *Natural Resources and Economic Development in Central America.* Transaction Books, New Brunswick and Oxford.

MacNeill, J., Winsemius, P. & Yakushiji, T. (1991) *Beyond Interdependence. The Meshing of the World's Economy and the Earth's Ecology.* Oxford University Press, New York & Oxford.

McCormick, J. (1989) *The Global Environmental Movement.* Belhaven Press, London.

MfE (1987) *Ministry for the Environment Reporting Function: Guidelines for Government Departments.* Ministry for the Environment (New Zealand), Wellington.

Miljoverndepartementet (May 1992) Personal communication, held on file at UNEP.

Mumtaz, K (1989) *Pakistan's Environment: A Historical Perspective and selected biography with annotations.* JRC-IUCN, Karachi.

NEPA (National Environmental Protection Agency), China (1990) *The Conservation Atlas of China.* Science Press, Beijing, China.

Nepal (1988) *The National Conservation Strategy for Nepal.* National Planning Commission, HM Government of Nepal, Kathmandu.

NEST (1991) List of National Environmental Societies Trust member NGOs, personal communication held on file at UNEP.

Nigeria (1989) *National Policy on the Environment.* Federal Environmental Protection Agency, Federal Republic of Nigeria.

Norway (1992) Gronn Bok 1992, Miljotiltak I Statsbudsjettet. Oslo: Miljoverndepartementet.

OECD (1979) *The State of the Environment in OECD Member Countries.* Organization for Economic Cooperation and Development, Paris.

OECD (1980) Environmental Policies for the 1980s. Organization for Economic Cooperation and Development, Paris.

OECD (1985) *The State of the Environment in OECD Member Countries.* Organization for Economic Cooperation and Development, Paris.

OECD (1989) *Economic Instruments for Environmental Protection.* Organization for Economic Cooperation and Development, Paris.

OECD (1991). The State of the Environment in OECD Member Countries. Organization for Economic Cooperation and Development, Paris.

Pakistan (1991) *National Conservation Strategy.*

Pearce, D. (1991) Deforesting the Amazon: Towards an Economic Solution. *Ecodecision*, **Volume 1, No. 1**, Pages 40-49.

Qu Geping & Woyen Lee (1984) (eds) *Managing the Environment in China.* Tycooly International, Dublin.

Repetto, R., William Magrath, Michael Wells, Christine Beer and Fabrizio Rossini (1989) *Wasting Assets: Natural Resources in the National Income Accounts.* World Resources Institute, Washington D.C.

Royal Commission on Environmental Pollution (1971-) Reports of the Royal Commission, commencing with first report, CMND 4585. HMSO, London.

Royal Commission on Environmental Pollution (1976) *Air Pollution Control: An Integrated Approach.* 5th Report, CMND 6371, HMSO, London.

Royal Commission on Environmental Pollution (1984) *Tackling Pollution - Experience and Prospects.* 10th Report: CMND 9149. HMSO, London.

Royal Society (1981) *The Assessment of Perception of Risk. A Royal Society Discussion.* The Royal Society, London.

Royal Society (1983) *Risk Assessment.* Report of a Royal Society Study Group. The Royal Society, London.

Sand P.H. (1990) *Lessons Learned in Global Environmental Governance.* World Resources Institute, Washington D.C.

SCEP (1970) *Man's Impact on the Living Environment.* MIT Press, Cambridge, Mass.

Soerjani, M. (1989) *Promoting Environmental Study Centres in Indonesia in Support of Sustainable Development.* UNDP/ World Bank/Government of Indonesia INS 82/009.

Thai Environmental Engineers Association (1992) List of 1992 environmental NGOs, held by Thai Environmental Engineers Association. Personal communication, held on file at UNEP.

UK (1990) *This Common Inheritance.* Britain's Environmental Strategy. Cm 1200. HMSO, London.

UN (1991) *World Urbanization Prospects 1990.* ST/ESA/SER.A/121. United Nations, New York.

UNEP (1988) *Environmental Management and Planning in the People's Republic of China.* United Nations Environment Programme, Regional Office for Asia and the Pacific, Bangkok.

UNEP (1991) *Environmental Data Report 1991/92.* Blackwell, Oxford, Massachusetts.

UNEP (1992) *Saving Our Planet, Challenges and Hopes*. UNEP/GCSS.III/2. United Nations Environment Programme, Nairobi.

UNEP-ROA (1991) *Brief Review of the Development of Environmental Policies, Institutions and Initiatives in Africa*. United Nations Environment Programme Regional Office for Africa, Nairobi.

United Nations (1986) *A System of National Accounts*. Department of Economic and Social Affairs, Statistical Papers, Series F, No. 2, Rev.3. United Nations, New York.

University of West Indies (1991) *Environmental Laws of the Commonwealth Caribbean*. Caribbean Law Institute of the University of West Indies, Barbados.

Vavrousek, J. (1990) *The Environment in Czechoslovakia*. State Commission for Science, Technology and Investments, Prague.

WALHI (Indonesian Forum for the Environment) (1992) Personal communication, held on file in UNEP.

Weiller, E. (1983) The Use of Environmental Accounting for Development Planning. Report to the United Nations Environment Programme. United Nations, New York.

West Indian Commission (1990) *Let All Ideas Contend. Preparing the West Indies for the Future*. West Indian Commission Secretariat, Barbados.

WCED (1987) *Our Common Future*, Report of the World Commission on Environment and Development. Oxford University Press, Oxford and New York.

WRI (1988) *World Resources, 1988-1989*. Basic Books, New York.

WRI (1990) *World Resources, 1990-1991*. Oxford University Press.

WRI (1992) *World Resources, 1992-1993*. Oxford University Press, New York and Oxford.

WWF - India (1992) Personal communication from World Wide Fund for Nature - India, held on file at UNEP.

CHAPTER 23
International responses

Introduction

Before World War II, few international environmental agreements existed. There were also few institutions, with the distinguished exception of the World (originally International) Meteorological Organization, founded in 1873 to co-ordinate world weather observations and promote the understanding of climate, and the International Council of Scientific Unions which, founded in 1919 as the International Research Council, began to promote and co-ordinate international science. Most of the environmental agreements concentrated on protecting migratory birds and animals, managing transfrontier river basins and boundary waters, and conserving marine fish and mammals (Box 1). However, from 1944 onwards there was a surge of institution building. The Food and Agriculture Organization, World Health Organization and UNESCO were all established between 1945 and 1948 (Appendix I). Within a few years, new agreements were concluded protecting the oceans and living marine species, governing Antarctica, and banning nuclear tests under water, in the atmosphere and outer space, and in Antarctica. In addition, the Treaty on Principles Governing the Activities of States in the Exploration and Use of Outer Space, including the Moon and Other Celestial Bodies, signed on 27 January 1967, obliged states engaged in studying or exploring outer space to avoid harmful contamination and adverse changes in the environment of the Earth.

World War II highlighted the importance of consultation and dialogue to identify potential sources of friction early on, and so avoid future conflicts. The UN Charter not only established the United Nations itself in 1945, but contemplated that specialized agencies in economic, social, cultural, educational, health and related fields would be linked to it 'with a view to the conditions of stability and well-being which are necessary for peaceful and friendly relations among nations'. The Charter also called for the establishment of regional arrangements to further these ends (UN, 1945).

As this network of international institutions evolved, it was inevitable that specialized organizations and programmes would be founded to promote the co-operation that is essential if transboundary resources and the global commons are to be preserved. The International Union for Protection (later Conservation) of Nature and Natural Resources (IUCN), founded in 1948 with both state and non-governmental members, was charged with protecting natural areas and species. Others, including the International Maritime Organization (IMO), the International Labour Organization (ILO), and the World Health Organization (WHO) stemmed from efforts to protect the human environment by ensuring human health and safety.

These institutional developments have proceeded in parallel and close association with the evolution of international policy and philosophy on environmental issues. One of the most striking aspects of this evolution has been the way in which environmental concepts and terms have entered the language

---

BOX 1

Major international environmental agreements.

Early initiatives

1906 Convention concerning the Equitable Distribution of the Waters of the Rio Grande for Irrigation (US-Mexico water treaty).

1909 Boundary Waters Treaty concluded between Great Britain (on behalf of Canada) and the United States.

1911 Convention for the Protection and Preservation of Fur Seals.

1923 Convention for the Preservation of the Halibut Fishery of the Northern Pacific Ocean and the Bering Sea.

1931 Convention on the Regulation of Whaling.

1940 Convention on Nature Protection and Wildlife Preservation in the Western Hemisphere.

1946 International Convention for the Regulation of Whaling.

1951 International Plant Protection Convention.

1954 Convention for the Prevention of Pollution of the Sea by Oil.

1958 Convention on Fishing and Conservation of the Living Resources of the High Seas.

1959 Antarctic Treaty.

1964 Agreed Measures on the Conservation of Antarctic Fauna and Flora.

1963 Treaty Banning Nuclear Weapon Tests in the Atmosphere, in Outer Space and Under Water.

1967 Treaty on Principles Governing the Activities of States in the Exploration and Use of Outer Space, including the Moon and other Celestial Bodies.

Later international agreements

Species and habitat:

1971 Convention on Wetlands of International Importance Especially as Waterfowl Habitat (Ramsar Convention) IUCN/IWRB.

1972 Convention Concerning the Protection of the World Cultural and Natural Heritage (World Heritage Convention) UNESCO.

1973 Convention on International Trade in Endangered Species (CITES) - UNEP.

1979 Convention on the Conservation of Migratory Species of Wild Animals (Bonn Convention) UNEP.

BOX 1 CONTINUED

1980   Convention on the Conservation of Antarctic Marine Living Resources (CCAMLR) - CCAMLR Secretariat.

1991   Protocol to the Antarctic Treaty on conservation.

Marine pollution

1972   Convention on the Prevention of Marine Pollution by Dumping of Wastes and Other Matter (London Dumping Convention) - IMO.

1973   International Convention for the Prevention of Pollution from Ships and the Protocol of 1978 Relating Thereto with Annexes (MARPOL 73/78) - IMO.

1982   UN Convention on the Law of the Sea (LOS Convention) - United Nations Office of Ocean Affairs and Law of the Sea - OALOS.

1990   International Convention on Oil Pollution Preparedness, Response and Co-operation - IMO.

Hazardous substances

1986   Convention on Early Notification of a Nuclear Accident - IAEA.

1986   Convention on Assistance in the Case of a Nuclear Accident or Radiological Emergency - IAEA.

1989   Convention on the Control of Transboundary Movements of Hazardous Wastes (Basel Convention) - UNEP.

Pollution of the atmosphere

1979   Convention on Long-Range Transboundary Air Pollution (LRTAP) - UN/ECE.

1985   Helsinki Protocol on the Reduction of Sulphur Emissions or Their Transboundary Fluxes by at Least 30 Per Cent.

1988   Sofia Protocol Concerning the Control of Emissions of Nitrogen Oxides or Their Transboundary Fluxes.

1985   Vienna Convention for the Protection of the Ozone Layer (Vienna Convention) - UNEP.

1987   Montreal Protocol on Substances that Deplete the Ozone Layer (Montreal Protocol).

---

of international politics and diplomacy. One possible reason for this growth may have been the perceived political neutrality of the environment, allowing environmental actions to be promoted as a bridge across doctrinal divides (as in Europe, where the Economic Commission for Europe and the programme under the Conference on Security and Co-operation in Europe brought western and eastern nations together).

International organizations depend for their effectiveness on the commitment of governments (and in a few cases a wider membership) to support them. All have governing bodies representative of their members, in which policies are laid down. Many have national committees or institutions with which they have forged close links. The international and national institutional structures have thus evolved together.

This chapter concentrates on the evolution of international thinking about the environment, and the parallel development of international institutions (at both global and regional levels). It covers both intergovernmental and non-governmental bodies, as well as those few organizations that span both sectors.

The development of international institutions

The United Nations system: 1944 - 1972

Between 1944 and 1972 the United Nations system developed a substantial range of activities that bear on the environment. Like the counterpart national institutions they adopted an essentially sectoral approach. FAO was concerned with improving world food security and the conditions of rural people and hence with the development of agriculture and fisheries, actions extended by the International Fund for Agricultural Development (IFAD), established in 1976. FAO also took forestry into its mandate because of the importance of forests to rural populations. WHO focused on achieving the highest possible standards of human health — and hence, inevitably, became concerned with the environment because of human living conditions and their role in disease transmission. UNESCO was the parent body for the Intergovernmental Oceanographic Commission (IOC), and also supported major programmes of scientific study through ICSU which mounted the International Biological Programme (IBP) between 1964 and 1974 (Worthington, 1975). WMO's efforts to develop world-wide meteorological and hydrological observations and provide understanding of environmental processes continued campaigns that already had decades of history.

The UN Development Programme (1966) had as a key mission the promotion of economic and social development, much of which clearly depends on sound management of the environment. The World Food Programme (1961), UN Population Commission (1946) and other specialized bodies, and the various Regional Economic Commissions and other regional institutions (Appendix II) all became involved with environmental questions.

The UN Conference on the Human Environment, held at Stockholm in 1972, is commonly thought of as the event where international debate on the environment began, but it was, in reality, the culmination of a considerable process of discussion. It was also a presentation of rather different points of view. The industrialized world came to discuss international solutions to pollution problems which resulted from the growth of industrial activity. Their emphasis was on regulation and pollution control. They were also concerned with nature conservation, and especially with threats to particular species (such as whales) and habitats (such as rain forests). The developing countries expressed little interest in these problems as compared with 'the pollution of poverty' and the inefficiency of resource use caused by underdevelopment.

This divergence became a serious issue in the months leading up to the Stockholm Conference, and at one point seemed to endanger it. An important meeting that began to bridge the gap in perception was held at Founex, Switzerland, in June 1971 (UNEP, 1981a). The report of this meeting clearly and cogently defined many of the issues that were to confront governments from both developed and developing countries. At Founex it became clear that 'development' and 'environment' were two sides of the same coin.

The breadth of activity within the UN system was familiar to the governmental delegations at Stockholm, many of whom were also members of non-UN institutions such as the Organization for Economic Co-operation and Development (OECD) and IUCN, who were playing an active part in evaluating and responding to emerging environmental concerns. There were three main thrusts behind the Stockholm agenda:

(a) the recognition that the developing nations faced massive environmental problems - especially linked to poverty - that needed to be overcome by development, while the developed countries had problems because their development had followed the wrong course;

(b) the growing scientific understanding of the inter-relatedness of natural systems, largely a product of the International Geophysical Year (1956-57) and International Biological Programme (reviewed in 1968 at an Intergovernmental Conference of Experts on the Scientific Basis for the Rational Use and Conservation of the Resources of the Biosphere, convened by UNESCO together with FAO, WHO and IUCN);

(c) growing public concern over the cumulative impacts of human activities on global environments (Caldwell, 1990). This had been stimulated by disasters such as the 1967 wreck of the *Torrey Canyon* (the world's first supertanker accident) and reports of damage to wildlife through persistent contamination by DDT, mercury and other substances.

In addition, there was the increasing recognition that neither individual nations, nor the North or the South acting alone, could adequately protect the global environment.

The UN family was hardly in a position to respond coherently to the Stockholm agenda. Those agencies with programmes directly relevant to the environment, such as FAO, UNESCO and WHO, presented points of view that were naturally linked closely to their formal mandates. Forest depletion was seen through the eyes of foresters, soil loss by agronomists and soil scientists, and so on. Stockholm came before the period when all applicable natural sciences became grouped as environmental science. The holistic, or systems, approach was still something for a few academics. Moreover, the knowledge of the social scientists was hardly ever presented in the debate. Economic questions were discussed largely in relation to development and natural resource depletion. Government organization was similarly sectoral (and there were, as yet, hardly any Departments of Environment). The overall approach was compartmentalized, and the diplomatic community did not regard environment as a central concern at all.

The Stockholm Conference was the culmination of a two-year preparatory process, and the beginning of a new and intensified programme. In the preparatory process, Intergovernmental Working Groups drew up plans, later endorsed by the conference and put into operation by the United Nations Environment Programme (UNEP), for a global environmental monitoring system (GEMS), an International Register of Potentially Toxic Chemicals (IRPTC) and an International Referral System for Sources of Environmental Information (INFOTERRA). Concern over marine pollution led to the negotiation of the Convention on the Prevention of the Pollution of the Sea by the Dumping of Wastes and Other Matter, finalized in London in October 1972. In parallel, conventions on wetland conservation (Ramsar), world heritage sites (Paris), and the control of trade in endangered species (CITES) (Washington) were stimulated and pressed forward.

Given the background, it is remarkable that the Stockholm Conference in fact decided on a system of environmental co-ordination for the UN that took 15 years for many governments to adopt for themselves. Although the resolutions were not systematically organized, and few were truly global or even regional in scope, they none the less covered a wide spectrum of issues. The need for global information (as manifest through the establishment of Earthwatch) led towards global standard setting. The establishment of the UN Environment Programme, with its voluntary fund, was a major advance, and a spur both to the development of the UN system and to the adoption of a more integrated approach by governments, slowly though this proceeded. The limiting factor in the success of the UNEP machinery may well, in fact, have been the slowness in the adoption of similarly inter-sectoral approaches at national level.

The Stockholm conference and UNEP

The Stockholm Conference was also a success because it created a tremendous public interest in the environment (Chapter 20), and provided directives for international and national action. It began a process that linked environment inseparably with economic development: an issue explored further at a UNEP/UNCTAD Symposium held in October 1974 at Cocoyoc in Mexico (Pearson and Pryor, 1978; UNEP, 1981a). This meeting identified the economic and social factors inherent in environmental degradation, examined the increasing scarcity of resources and the rising pressures upon them, and crystallized the concept of 'development without destruction'. The Stockholm Declaration on the Human Environment and the Declaration of Principles (UNEP, 1981a) constituted a solid foundation for future work, and the Stockholm Action Plan was embodied in the United Nations Environment Programme. Principles approved at Stockholm on assessment and control of marine pollution provided guidance for the on-going development of the first comprehensive framework agreement dealing with the global commons — the 1982 UN Convention on the Law of the Sea, and its sections on protection and preservation of the marine environment.

The institutional and financial arrangements set out in the Stockholm Conference report provided the basis for the establishment of UNEP by the UN General Assembly. Governments at Stockholm had made it clear that they did not wish to establish a new UN Agency for the environment. The goal was a small coordinating secretariat, and an integrated programme. As established, UNEP had four principal components (UN, 1972):

(a) *The Governing Council* (GC): composed of 58 member governments elected on a rotating basis for four years. The GC initially met annually but later decided to meet every two years with additional special sessions every six years and others if necessary. It reports through ECOSOC to the General Assembly.

(b) *The Environment Secretariat*: headed by the Executive Director, who is elected by the UN General Assembly on the nomination of the UN Secretary-General, and headquartered in Nairobi.

(c) *The Environment Co-ordination Board* (ECB): established to ensure co-operation among all UN bodies having a mandate for environmental programmes. The ECB was to function 'under the auspices and within the framework' of the UN's Administrative Committee on Coordination (ACC), and under the chairmanship of UNEP's Executive Director, reporting annually to the UNEP Governing Council.

(d) *The Environment Fund*: established 'to enable the Governing Council . . . to fulfil its policy-guidance role for the direction and co-ordination of environmental activities' by financing 'wholly or partly the cost of the new environmental initiatives undertaken within the UN system', with particular attention to 'integrated projects'. The fund was also to be used for 'assistance for national, regional and global environmental institutions' with due account being taken of the needs of developing countries, and 'to ensure that the development priorities of developing countries [are] not adversely affected'. It was contemplated that a $US100 million, five-year fund would be created.

As originally conceived, the United Nations Environment Programme was just what its name implies: the complete agenda of the United Nations system for environmental matters. The Secretariat, guided by the Governing Council, was to be a reference point for environmental matters throughout the UN system. UNEP was not to be 'an executing agency and does not bear the prime responsibility in the UN system for executing environmental projects' (UNEP, 1985).

The ECB and the Secretariat were the agents responsible for 'coordinating, catalysing and stimulating environmental action primarily - but not exclusively - within the UN system' (UNEP, 1987a). Under the guidance of the Governing Council, the Secretariat was to review and assess the effectiveness of environmental programmes within the UN system. The General Assembly charged the Governing Council with providing general policy guidance for direction and co-ordination of environmental programmes within the UN system; reviewing the world environmental situation to ensure that emerging problems of international significance were adequately considered; promoting contributions by scientific and other professional communities to greater environmental understanding and to technical aspects of programme formulation and execution within the UN system; and reviewing the impacts of national and international environmental policies on developing countries and the additional costs to them of implementing environmental programmes and projects, as well as ensuring that such programmes and projects were compatible with national development plans and priorities (UN, 1972; Thacher, 1990a).

UNEP has come to play a leading role in developing environmental policies and promoting their implementation. Through the collection and dissemination of environmental information, the development of policy guidance, and efforts to mobilize support for environmentally-sound development projects - the UNEP triangle of environmental assessment, environmental management, and support measures - the programme has converted its limited authority and resources into a strong and highly visible environmental activism.

The UNEP programme

Environmental assessment and global awareness

The Earthwatch Programme, UNEP's environmental assessment arm, collects and disseminates environmental information. Its main component is the Global Environmental Monitoring System (GEMS), created in 1975 to monitor the global environment and undertake periodic assessments of its health (Chapter 20). GEMS links hundreds of national and international organizations, of which the most important are FAO, WHO, WMO, UNESCO, IUCN and the World Conservation Monitoring Centre (WCMC), established by UNEP, IUCN and WWF in 1988. In 1985, the Global Resource Information Database (GRID) was established to promote use of geographic information systems in support of environmental studies and applications.

Other components of UNEP's Earthwatch Programme include the International Register of Potentially Toxic Chemicals (IRPTC) and the International Environmental Information System (INFOTERRA). INFOTERRA was established in 1977 as a network of national reference points for queries requiring environmental expertise. Its directory lists over 6,500 institutions as sources, and the national offices have access to some 600 commercial data banks (UNEP, 1990a).

Prominent among public outreach efforts is the International Environmental Education Programme (IEEP), established by UNEP and UNESCO in 1975 pursuant to a Stockholm Conference recommendation. The IEEP 'has become the major international vehicle for the promotion of environmental education at the local, national, regional and global levels' (UNEP, 1987a; Chapter 21). Its aims, reaffirmed by the UNESCO-UNEP International Congress on Environmental Education and Training in Moscow in 1987 (UNESCO/UNEP, 1989; Chapter 21), are to develop educational materials and methods to train educators from pre-school through university level, in school and out of school.

UNEP has also taken great strides in reaching the public through information campaigns and programmes oriented toward the mass media, through sponsorship and involvement in a variety of international conferences and seminars, and with the establishment of 5 June as World Environment Day by the UNGA in 1972. Its involvement in the development and dissemination of major publications such as the *World Conservation Strategy* (IUCN/UNEP/WWF, 1980), *The World Environment 1972-1982* (Holdgate, Kassas, White, 1982), *Environmental Perspective to the Year 2000 and Beyond* (UNEP, 1988b), and *Caring for the Earth* (IUCN/UNEP/WWF, 1991) and its Executive Director's annual *State of the Environment* reports has helped bring environmental issues to the attention of policymakers and individuals around the world.

Co-operation with NGOs is another important element in UNEP's public awareness initiatives. The Environmental Liaison Centre International (ELCI) was established in 1974 in Nairobi as a direct result of the Stockholm Conference. It is a global coalition of NGOs whose aim is to strengthen NGOs working in environment and development, particularly in the developing nations. ELCI provides information, financial assistance and training to its network of more than 6,000 NGOs world-wide.

Through its Outreach Programme, established in 1983 (UNEP, 1988a), UNEP seeks to cultivate new audiences for environmental issues. The Programme involves prominent representatives of youth and women's groups, eminent religious figures, captains of industry, parliamentarians, and others well-established in their fields. In this way, UNEP builds a broad base of support in international, regional and national forums.

UNEP's Industry and Environment Office (IEO), now the Industry and Environment Programme Activity Centre (IE/PAC) in Paris was established in 1975 to bring industry, governments and non-governmental organizations together to work towards environmentally sound forms of industrial development. Through publication of technical guides, information transfer, training and technical co-operation programmes, IE/PAC seeks to:

- define environmental criteria and encourage their incorporation in industrial development;

- formulate principles and procedures to protect the environment and facilitate their implementation;

- promote the use of safe low- and non-waste technologies; and

- stimulate the exchange of information and experience on environmentally sound forms of industrial development throughout the world.

In recent years IE/PAC has placed major emphasis on the promotion of cleaner technologies through its 'Cleaner Production' programme and the International Cleaner Production Information Clearinghouse (ICPIC), and on development of its programme on Awareness and Preparedness for Emergencies at Local Level (APELL) (UNEP, 1990b).

Environmental policy and law

UNEP's annual *State of the Environment* report provides a means for bringing emerging environmental issues and problems of international significance to the attention of the world community. But it is also necessary to ensure that policymakers determine an appropriate response and act on it. In this respect, UNEP's responsibility for catalysing an effective policy response, and for co-ordinating follow-through within the UN system, is critical. It does this through the Governing Council deliberation and annual reports to the UN Economic and Social Council and General Assembly, and its role as secretariat for both the Committee of International Development Institutions on the Environment (CIDIE) and the UN Designated Officials on Environmental Matters (DOEM). UNEP has often used seed money from the Environment Fund to launch initiatives leading to policy guidance for states. It has used three particular techniques to promote international law:

- the 'action plan' approach - as a means of mustering both scientific consensus and political support to secure agreement on legally binding instruments (Thacher, 1990b);

- 'soft law' instruments - non-binding guidelines and principles adopted as guidance at the regional and global level - which foster more uniform standards and practices among nations and which may ultimately be incorporated into either binding international legal agreements or national law; and

- recourse to conventions to command agreement on general principles and institutional mechanisms among states parties, supplemented by protocols specifying more detailed rights and obligations as governments become willing to address them.

The action plan device evolved out of preparations for the Stockholm Conference and has been creatively combined by UNEP with the development of soft law and framework conventions. As a preliminary step, UNEP has commonly convened workshops at which scientific experts define the specific nature of environmental problems and needs. These meetings help identify and build political support for the concrete actions required to achieve environmental protection, and serve as a basis for conferences at which diplomats and legal experts begin considering principles and approaches that may ultimately lead to legally binding instruments. This approach was used first in the development of the 1975 Mediterranean Action Plan, which led to a framework convention in 1976 and subsequent protocols and the 1985 Vienna Convention on the Protection of the Ozone Layer and its 1987 Montreal Protocol (Thacher, 1990c; UNEP, 1981b; Petsonk, 1990). Similarly, the meetings held at Villach, Austria, and Bellagio, Italy, in 1987 laid the initial scientific foundation for the Intergovernmental Panel on Climate Change (IPCC), whose work led directly through the Second World Climate Conference to the process of drafting an international Climate Convention.

A comparable approach was adopted but with less success in response to mounting concern over desertification. As the Stockholm Conference closed, the 1968-73 Sahel drought was at its height. Widespread crop failure and livestock deaths brought great hardship to people living in arid areas in five African countries. A UN Conference on Desertification, held in Nairobi in 1977, adopted a Plan of Action with 28 recommendations (Chapter 6). Unfortunately, the subsequent response was slow and limited. Some training programmes were initiated, and a limited number of national action plans were prepared (although few were implemented). The call for contributions to a United Nations Special Account to support action against desertification was also disappointing. The 1977 Conference calculated that $US2.4 billion should be spent annually. However, because of opposition by a number of developed countries to the principle of special funds, only eight governments actually contributed to the Account, whose total balance stood at $US333,294 by 31 October 1991 (including over $US140,000 in accrued interest). The Special Account was ultimately closed, as decided by UN General Assembly on 19 December 1989 at the recommendation of UNEP's Governing Council.

UNEP's environmental law programme was outlined in 1975, and that autumn the UN General Assembly requested that UNEP report to it annually on the status of environmental conventions and protocols (UNEP, 1975; UN, 1975). This programme has evolved to cover both the development of soft law instruments and major international conventions. In 1978 UNEP's Governing Council commissioned an in-depth review of environmental law activities carried out by international organizations within and outside the UN system. This led to a report considered by senior government officials in environmental law who met in Montevideo in 1981 (UNEP, 1982a).

The Montevideo programme adopted by the UNEP Governing Council in 1982, laid out a comprehensive programme for the progressive development of international environmental law. As a priority, UNEP was to develop guidelines, principles or agreements on marine pollution from land-based sources, protection of the stratospheric ozone layer, and transport, handling, and disposal of toxic and dangerous wastes. Eight other action areas were noted: environmental emergencies, coastal zone management, soil conservation, transboundary air pollution, international trade in potentially harmful chemicals, protection of rivers and other inland waters against pollution, legal and administrative mechanisms for the prevention and redress of pollution damage, and environmental impact assessment (UNEP, 1982a).

In at least two of the three priority areas, UNEP has enjoyed remarkable success.

- Within the Regional Seas Programme nine regions have adopted action plans (Mediterranean, Kuwait, Red Sea and the Gulf of Aden, Wider Caribbean, West and Central Africa, East Africa, South-East Pacific, South Pacific and East Asia). An action plan has been drafted for the South Asia region, and the development of action plans has recently been initiated for the Black Sea, North-West Pacific and South-West Atlantic. Conventions and several Protocols concerning pollution in cases of emergency have been adopted by eight regions. UNEP has developed guidelines and principles concerning the protection of the marine environment against pollution from land-based sources; however, agreements on land-based sources of pollution have only been adopted for the Mediterranean, Kuwait and the South-East Pacific (Chapter 5).

- On the other hand, the Vienna Convention for the Protection of the Ozone Layer, its extremely successful Montreal Protocol, and the remarkable London Amendment and adjustments to the Montreal Protocol, set the stage for negotiations in 1991 and 1992 for a global framework convention on climate change.

- The 1987 Cairo Guidelines on management of hazardous wastes led ultimately to agreement on the 1989 Basel Convention on the Control of Transboundary Movements of Hazardous Wastes and Their Disposal (the Basel Convention).

Additional non-binding agreements cover such topics as offshore mining and drilling (UNEP, 1981b), shared natural resources, environmental impact assessment (UN, 1973; UNEP, 1981b) and the London Guidelines for the Exchange of Information on Chemicals in International Trade. These have been adopted by governments and are being used in the development of international and national legal regimes. Furthermore, during this period, regional agreements for the Zambezi River System and Lake Chad were adopted (Chapter 4). A new area that was not included in the Montevideo Programme on the Development and Periodic Review of Environmental Law but which has gained significance since then is the conservation and rational use of biological resources. The Governing Council of UNEP decided in 1989 to take action towards agreeing on a convention for the conservation of biodiversity. In the late 1980s UNEP started work, jointly with WMO, on the development of a convention on climate change. UNEP also played a major role in developing the CITES and CMS conventions providing for the conservation of wild fauna and flora, which were adopted and came into force during the 1970s and early 1980s (Chapter 8). UNEP's programme of assistance to developing countries in the formulation and implementation of national environmental laws and institutions has been carried out in 48 developing countries. UNEP also publishes several important legal publications on environmental law including the *Multilateral Treaties in the Field of Environment* and the *Register of Environmental Treaties* (UNEP, 1991c).

Mobilizing support for development co-operation

Although UNEP's mandate does not entail responsibility for the direct implementation of development projects, UNEP does influence these activities through its management of the Environment Fund, its Clearinghouse programme, its administration of various 'trust funds', and its secretariat function for CIDIE and DOEM.

In creating UNEP in 1972, the General Assembly directed that the Environment Fund finance wholly or partly new environmental initiatives undertaken within the UN system, with particular attention to integrated projects. A portion of the Environment Fund is used to cover support and management costs for UNEP, after which sixty per cent is devoted to global programmes such as GEMS and INFOTERRA (UNEP, 1990b). The remainder has been employed to catalyse the implementation of projects in developing countries (UNEP, 1987b).

By providing seed money from the Fund to launch projects carried out by other UN agencies, governments and NGOs, and by linking the needs of developing countries to the resources and expertise of the more affluent nations through the Clearinghouse programme established in 1982 (UNEP, 1988a, 1989a), UNEP has been able to stimulate environmental action well beyond the bounds of the UN system. Contributions made to projects initially financed by the Fund have been estimated to be four times the contribution made by the Fund itself (UNEP, 1988a). The Clearinghouse programme has assisted in channelling more than $US17 million worth of technical co-operation to developing countries. The programme 'works at both ends - helping developing countries to formulate projects and encouraging donors to finance them' (UNEP, 1987b).

The environment programmes of the UN agencies

Over seventy UN programmes and agencies are listed in directories of the UN system. The vast majority carry out functions relevant to national economic and social development and to managing the global commons (see Appendices I and II). These functions may be grouped roughly into data collection and research, policy formulation (including international treaties and 'soft law' codes and standards), and development co-operation activities, such as training and education, technical assistance and project funding. Virtually all of these agencies and programmes also contribute to global awareness by making available to the public information describing the issues and objectives of their programmes.

As the interconnections between avoiding environmental degradation, alleviating poverty and developing the human potential are more clearly understood, it has become increasingly apparent that virtually every organization in the UN system has some bearing on the environment and sustainable development. The Stockholm Conference prompted many agencies and programmes to examine their activities and highlight environmental components. Many of the major global conferences following Stockholm (Appendix III) which explored in-depth issues related to food and fresh water, science and technology, renewable energy sources, desertification and other topics, led to the establishment of additional inter-governmental and secretariat organs within the UN system. Yet, in a perverse manner, the proliferation of 'environmental' components and narrow, issue-oriented organs has increased the isolation of each instead of enhancing an integrated, multisectoral approach to problem-solving.

The Report of the World Commission on Environment and Development (WCED, 1987) and UNEP's report: *The Environmental Perspective to the Year 2000 and Beyond* (UNEP, 1988b) were a major spur to new thinking, and especially to a recognition that an integrated approach was essential. They have been complemented by several significant regional responses, including *From Our Own Agenda*, for Latin America and the Caribbean, *Economic Policies for Sustainable Development*, prepared under the auspices of the Asian Development Bank (1990), and a long-term perspective study by the World Bank, *From Crisis to Sustainable Growth*, dealing with sub-Saharan Africa. While these three latter documents are not specifically about the environment, they demonstrate how environmental considerations are now deeply integrated into development planning.

The United Nations system

Data and information services

The World Meteorological Organization (WMO) and the Intergovernmental Oceanographic Commission (IOC) are examples of organizations that specialize in global data collection and support major interdisciplinary scientific investigations - in world climate and weather and in the nature and resources of the oceans respectively. The IOC, in collaboration with its 'parent', UNESCO, has increasingly taken on the role of assisting developing nations to acquire a marine scientific research capability of their own. WMO is becoming more involved in the policy implications of climate and weather research with its sponsorship, together with UNEP, of the Intergovernmental Panel on Climate Change (IPCC).

Global statistics are maintained by a number of UN agencies. WHO has extensive information on various aspects of human health, and trends in factors that affect health. UNFPA maintains global and national population data and predicts trends. UNESCO and UNICEF collect world-wide data relevant to education and the well-being of children. FAO collects information on agricultural and food production and consumption (including fisheries), forestry and forest products, and related economic activities. The former UN Centre for Transnational Corporations (UNCTC), whose functions are now incorporated in the United Nations Secretariat's Department of Economic Development, synthesized a wealth of information on corporate activities as well as data on product bans and hazardous technologies and their alternatives. The UN Disaster Relief Organization (UNDRO) has begun to compile data on the possible effects of environmental degradation on natural disasters (Chapter 9). Each of the UN regional economic commissions also carries out various environmental data collection and dissemination functions.

Environmental policy and law

Many of the UN agencies provide guidance on environmentally-sound development, and prepare action plans, strategies (Box 2), principles, codes and guidelines (Box 3) as well as legally binding international treaties (Box 1). Some of these state objectives and principles, while others provide operational guidelines and practices. The more general are designed for review every five to ten years, while many of the more specific ones provide institutional mechanisms to evaluate and update their specifications. The latter include:

- FAO's 1983 **International Undertaking on Plant Genetic Resources**, which provides guidelines for conservation and exchange of genetic resources;

- FAO's 1985 **International Code of Conduct on the Distribution and Use of Pesticides**, which identifies potential hazards, establishes standards of conduct, and defines responsibilities for those engaged in pesticide distribution, regulation and use;

- the **Codex Alimentarius**, a programme jointly established by WHO and FAO in 1962, which has produced a large volume of food safety standards; and

- the 1985 **Tropical Forestry Action Plan**, sponsored by FAO, UNDP, IBRD and the World Resources Institute, a process created to re-orient forestry policies and review national implementation plans.

In addition, a large volume of 'soft law' guidance in the form of handbooks and codes is produced by agencies such as IMO and ILO, to assist states in designing specific national laws, standards and procedures to give effect to international treaties governing labour health and safety and ship safety and pollution control. In 1989, the IAEA Board of Governors approved international criteria for the safe disposal of high-level radioactive wastes to protect present and future generations against radiation.

BOX 2

Major Action Plans, Programmes and Strategies since 1972.

1972   Stockholm Action Plan.

1974   World Population Plan of Action.

1974   Universal Declaration on the Eradication of Hunger and Malnutrition (FAO).

1974   Cocoyoc Declaration on Patterns of Resource Use, Environment and Development Strategies.

1975   Regional Seas Programmes; Nine regional action plans adopted by 1991 (UNEP).

1975   International Environmental Education Programme (UNEP/ UNESCO).

1976   Vancouver Action Plan for Human Settlements.

1977   Mar del Plata Action Plan for Water Resources Management.

1977   Plan of Action to Combat Desertification.

1977   Tbilisi Declaration on Environmental Education (UNEP/ UNESCO).

1978   Buenos Aires Plan for Promoting and Implementing Technical Co-operation Among Developing Countries.

1979   Vienna Programme of Action on Science and Technology.

1979   World Climate Impact Programme.

1980   World Conservation Strategy (UNEP/IUCN/WWF).

1980   Declaration of Environmental Policies and Procedures Relating to Economic Development.

1981   Global Strategy for Health for All by the Year 2000 (WHO).

1981   Nairobi Programme of Action for the Development and Utilization of New and Renewable Sources of Energy.

1981   Agriculture: Toward 2000 (FAO).

1981   World Soils Charter (FAO).

1981-
1990   International Drinking Water Supply and Sanitation Decade.

1982   Montevideo Programme for Environmental Law (UNEP).

1982   World Charter for Nature (UNGA).

1982   World Soils Policy (UNEP).   *continued*

BOX 2 CONTINUED

1983 Strategy for the Protection of the Marine Environment (IMO).

1984 Programme of Action on Agrarian Reform and Rural Development (FAO).

1984 Global Action Plan for the Conservation, Management and Utilization of Marine Mammals (UNEP/FAO).

1984 Declaration of World Industry Conference on Environmental Management (WICEM).

1984 Strategy for Fisheries Management and Development (FAO).

1985 Action Plan for Biosphere Reserves (UNESCO).

1985 Tropical Forestry Action Plan (FAO, WRI, World Bank, UNDP).

1987 Environmental Perspective to the Year 2000 and Beyond (UNEP).

1987 International Strategy for Action in the Field of Environmental Education and Training for the 1990s (UNESCO/UNEP).

1988 Global Strategy for Shelter to the Year 2000 (UNCHS).

1990-
1999 International Decade for Natural Disaster Reduction.

1991 Strategy and Agenda for Action for Sustainable Agriculture and Rural Development (FAO).

1991 World Conservation Strategy for the 1990s (IUCN, UNEP, WWF).

1991 Declaration of the Second World Industry Conference on Environmental Management (WICEM II).

Increasingly, these programmes build on related initiatives. For example, in 1989, the FAO Conference incorporated into the pesticide code the 'prior informed consent' procedure adopted in UNEP guidelines related to chemicals in international trade (UNEP, 1989b). UNEP, ILO and WHO operate a joint International Programme on Chemical Safety, established in 1980, and ILO and IAEA a programme on protection of workers exposed to radiation. In a joint policy initiative of a slightly different nature, UNESCO's Man and the Biosphere Programme has since 1971, in collaboration with national committees throughout the world, sought to 'develop within the natural and social sciences a basis for the rational use and conservation' of natural resources and the environment (UNESCO, 1989).

Several UN agencies have played a significant role in the development of international environmental law. The IMO has produced an impressive record

BOX 3

Selected examples of UN organizations and their contribution to 'soft law'.

'Soft Law' is used here to refer to international standards, codes and guidelines that are not legally binding. Even though such standards do not hold the weight of legally binding international agreements, they have often been very effective means for furthering environmental protection and sustainable development. Furthermore, in some instances 'soft law' has served as the basis for negotiating legally binding agreements, e.g. UNEP's 1987 Cairo Guidelines for the Environmentally Sound Management of Hazardous Wastes served as the basis for the Basel Convention on that subject in 1989. While numerous UN organizations and other institutions have long been involved in the development of 'soft law', the following is a short list of some of the major contributions of several UN organizations in this area.

- WHO: Since 1951 the WHO expert committee on biological standards has met yearly to formulate recommendations for international standards for biological and pharmaceutical products; WHO has long been involved in the development of Environmental Health Criteria - particularly in relation to specific substances; WHO has also formulated guidelines for Drinking Water Quality; and, since 1962, WHO has been involved in the Joint FAO/WHO Food Standards Programme, administered by the Codex Alimentarius Commission.

- FAO: Joint FAO/WHO Food Standards Programme (1962); International Undertaking on Plant Genetic Resources (1983); International Code of Conduct on the Distribution and Use of Pesticides (1985).

- ILO: Since its inception ILO has been involved in developing labour-related international standards and guidelines such as those for the working environment and safety and health in construction.

- IMO: IMO has produced numerous guidelines, standards and codes related to maritime safety and the protection of the marine environment. Examples include: the International Maritime Dangerous Goods Code; Guidelines and Standards for the Removal of Offshore Installations and Structures on the Continental Shelf and in the Exclusive Economic Zone; and Guidelines for the Surveillance of Cleaning Operations Carried Out at Sea on Board Incineration Vessels.

continued

BOX 3 CONTINUED

- IAEA: Much of IAEA's work concerns the development of international codes, standards and guidelines for nuclear energy. Among them are; the five Nuclear Safety Standards Codes (1978); the Principles for Limiting Releases of Radioactive Effluents into the Environment (1986); and the International Criteria for the Safe Disposal of High-Level Radioactive Wastes (1989).

- UNEP: Principles of Conduct for the Guidance of States in the Conservation and Harmonious Utilization of Natural Resources Shared by Two or More States (1978); Guidelines on International Co-operation on Weather Modification (1980); Conclusions of the Study of Legal Aspects Concerning the Environment Related to Offshore Mining and Drilling within the Limits of National Jurisdiction (1982); Montreal Guidelines for the Protection of the Marine Environment Against Pollution from Land-Based Sources (1985); Cairo Guidelines for the Environmentally Sound Management of Hazardous Wastes (1987); Goals and Principles of Environmental Impact Assessment (1987); London Guidelines for the Exchange of Information on Chemicals in International Trade (1987 - amended 1989).

- UNCTC: Criteria for Sustainable Development Management of Transnational Corporations (1989).

---

of international agreements to prevent and control marine pollution from ships. UNESCO administers the 1972 World Heritage Convention. The IAEA has long been involved in developing codes of practice for safe installation and operation of nuclear power plant facilities. In the wake of the Chernobyl accident, two 1986 treaties were concluded under IAEA auspices: The Convention on Early Notification of a Nuclear Accident, and the Convention on Assistance in the Case of a Nuclear Accident or Radiological Emergency. ILO has developed several conventions on safety in the working environment, including a 1990 convention concerning safety in the use of chemicals at work. UNEP has played a major role in the development and implementation of international environmental law during the last two decades, as described earlier in this chapter.

Environmentally sound development policies

In 1979 an evaluation was carried out of the environmental policies and programmes of nine development assistance agencies. The publication *Banking on the Biosphere?* concluded that greater attention to environmental concerns was necessary at all levels of the development assistance process and among all of the institutions surveyed (Stein and Johnson, 1979). In response, UNEP together with nine development assistance agencies, including the World Bank and the United Nations Development Programme (UNDP), signed the *Declaration of Environmental Policies and Procedures Relating to Economic Development* in 1980. They pledged to create systematic environmental assessment and evaluation procedures for all development activities and to support projects enhancing the environment and natural resource base of developing nations. In addition, they formed the Committee of International Development Institutions on the Environment (CIDIE) to review regularly the implementation of the Declaration (CIDIE, 1988).

Today CIDIE is composed of 17 members. Apart from member institutions, annual CIDIE meetings are also attended by observers from bilateral assistance agencies, interested inter-governmental organizations and NGOs. Although CIDIE has made substantial progress in incorporating environmental considerations into the policies and activities of its member institutions, its full potential in terms of the effectiveness of such policies and activities in bringing about sustainable development has yet to be realized.

In the late 1980s, and following the report of the World Commission on Environment and Development, these agencies have given more attention to environmental issues. A number of UN agencies have considered introducing environmental impact assessment of proposed activities and the integration of environmental considerations into all stages of project planning, implementation and evaluation (Box 4). UNIDO in 1988 launched a five-year programme to strengthen the capacity of developing countries to incorporate environmental considerations into the design and implementation of industrial operations. ILO and UNEP together have initiated environmental information and training programmes on environmentally sound management practices for both employers' and workers' organizations (ILO, 1989). And in 1990 the UN Centre on Transnational Corporations developed Criteria for Sustainable Development Management to strengthen the participation of large industrial enterprises in efforts to preserve the environment.

The development community has grown increasingly aware of the links between poverty, education, culture, human health and shelter, and their impacts on sustainable development. In 1990, UNDP produced the *Human Development Report*, which contained a new human development index based, *inter alia*, on life expectancy at birth, adult literacy rates, and purchasing power. UNDP believes that donor priorities should be reoriented toward accelerating human development and correcting the imbalances among nations illustrated by the index, which it expects to update annually (UNDP, 1990a, 1990b). The donor community is also under considerable pressure to involve local communities and affected 'stakeholders' - bankers, entrepreneurs, politicians, unions, scientists, engineers, the organized public and indigenous communities - in project planning and review. The view that development assistance programmes should be driven primarily by the country and constituencies involved, should draw on local skills and expertise, and must command widespread national support, is gaining ground rapidly.

BOX 4

Selected examples of environmental assessment procedures.

The following is a brief summary of current environmental assessment procedures at the World Bank, the Inter-American Development Bank and UNDP. In this respect it should be noted that numerous other international and regional organizations such as FAO, UNIDO and the Asian Development Bank have also developed environmental assessment procedures specific to their own activities.

- World Bank: In 1989 the Bank issued Operational Directive 4.0 on environmental assessment (EA) guidelines in order to facilitate the systematic incorporation of environmental considerations into project and sector work - including projects of the IDA. While EA preparation is considered to be the responsibility of the borrower, the Task Manager for the activity monitors the EA process, with support from the appropriate Regional Environment Division. EAs are supposed to utilize the findings of the National Environmental Action Plan (if the country has such a plan) in order to ensure that the development options under consideration are environmentally sound. The Bank expects the borrower to take the views of affected groups and local NGOs fully into account in the preparation of EAs as well as in the design and implementation of the activity. Since 1989 the Bank has been preparing technical environmental guidelines to address particular sectors. These are contained in the *Environmental Assessment Sourcebook* of February 1991.

- IDB: In 1990 the IDB approved 'Procedures for Classifying and Evaluating Environmental Impacts of Bank Operations'. These procedures require that all prospective loans, including structural adjustments loans, be examined for their potential impacts. As at the World Bank, the IDB considers EA preparation to be the responsibility of the borrower. Furthermore, the IDB states that affected peoples and local groups must participate in a formal and verifiable way in the preparation of the terms of reference for the EA. Since the approval of the Procedures, the IDB has

---

Support for sustainable development and the protection of global environments

The Bretton Woods Institutions - the World Bank (IBRD) and the International Monetary Fund (IMF) - were established in order, respectively, to help rebuild countries devastated by World War II and to provide short-term financing for temporary balance-of-payments deficits. During the 1960s, they re-oriented their programmes to focus primarily on the developing nations. In 1990 the World Bank lent out approximately $US15.2 billion for projects including agriculture, forestry, rural development, industrial development, infrastructure and energy.

been developing detailed environmental guidelines for individual sectors - transportation, mining, sanitation and urban development, agriculture, industry and energy. These guidelines have been prepared for IDB staff responsible for identifying, analysing and negotiating loans. Their focus is on procedural aspects as opposed to technical environmental concerns. This approach was taken because the IDB felt it unnecessary to duplicate World Bank efforts to elaborate technical guidelines - given the similarities between the environmental policies of the World Bank and the IDB and the fact that the IDB members are also members of the World Bank.

- UNDP: UNDP's three volume *Policies and Procedures Manual* (PPM) details the procedures used in the project cycle by both UNDP and agencies that execute UNDP-funded projects. Although the PPM contains no EA guidelines or checklists for development assistance activities, UNDP's new *Environmental Management Guidelines* (EMG) present both a general policy overview and an operational strategy by which UNDP staff and others may introduce environmental management and sustainable development considerations into technical assistance schemes, country programming, and project cycles. The EMG offer operational guidelines for introducing 'environmental checkpoints' into selected steps of UNDP planning and executing procedures (outlined in the PPM). The EMG also call for the preparation of short Environmental Overviews (EO) to be part of project documents and country programmes. EOs are to provide UNDP and the country, institution or organization that requests UNDP assistance with basic information describing the characteristics of the environment where the proposed activity is to take place as well as potential environmental impacts. After reviewing the EO, the preparation of an in-depth Environmental Management Strategy (EMS) can be recommended for the proposed activity. The EMS is an on-going effort that must be closely monitored throughout the life of the activity. It is intended to provide alternatives for sound environmental management and is to be built on broad participation by all relevant 'actors'.

The United Nations Development Programme (UNDP) was established as a subsidiary organ of the UN General Assembly in 1966 as the principal UN mechanism for financing technical co-operation in developing nations. Total UNDP resources in 1989 were approximately $US1.3 billion. In addition, UNDP administers several trust funds, for example Science and Technology, Women, and the Population Fund. The original idea that UNDP would serve as the central funding mechanism for technical co-operation has long since been overtaken by events. UNDP continues to provide somewhat more than half the funding for technical co-operation projects executed by agencies such as WHO, WMO and UNIDO, but these funds are often supplemented by substantial voluntary contributions made directly by bilateral donors to individual agencies or to trust funds administered by them. UNEP, as noted earlier, administers several such trust funds. UNIDO established its own Industrial Development Fund in 1976 to undertake 'special projects'. Less than ten per cent of IAEA's technical assistance funds come from UNDP, while the UNICEF Children's Fund is completely independent of UNDP. At the other end of the spectrum, the value of the World Food Programme's assistance projects - many of which support irrigation, soil conservation and other development activities meant to increase food security - exceeds total UNDP resources.

The establishment of trust funds for environmental purposes, a practice later adopted by UNEP in the regional seas programme, was pioneered by UNESCO, for example in the World Heritage Fund established under the World Heritage Convention. Trust funds are based on voluntary contributions from governments and are used to help finance the implementation of an agreement. The 1990 meeting of the parties to the Montreal Protocol on Substances that Deplete the Ozone Layer carried this idea further by adopting an interim financial mechanism which required states parties, on the basis of the UN scale of assessments, to help fund assistance programmes and, explicitly, the transfer of technology, to enable developing nations to comply with control measures established by the protocol.

The UN structures for reporting and co-ordinating environmental activities

General

The General Assembly and the Economic and Social Council (ECOSOC) review environmental matters. The Security Council has not generally addressed these issues, although liability for environmental damage and the depletion of natural resources is reaffirmed in the Security Council resolution on the termination of hostilities between Iraq and Kuwait. The General Assembly meets every northern autumn in regular session in New York, while ECOSOC generally meets three times yearly in Geneva and New York, with one organizational session followed by two substantive sessions. It reviews reports from specialized agencies and United Nations organs, including UNEP, and forwards them to the General Assembly together with any recommendations. They are then considered in the Second Committee of the General Assembly, which deals with economic matters.

ECOSOC, in the context of its 1946 Resolution on negotiations on the relationship between the UN and the specialized agencies, requested the Secretary-General to establish a standing committee of administrative officers with himself as chairman. The heads of the specialized agencies were members of this committee, whose purpose was to ensure the fullest and most effective implementation of the agreements entered into between the United Nations and the specialized agencies. Membership of this committee, now known as the Administrative Committee on Coordination (ACC), has expanded over the years to include heads of major United Nations programmes such as UNDP, UNICEF, UNEP and UNCTAD, as well as the heads of the World Bank, the IMF and the IAEA. The ACC meets twice a year under the chairmanship of the Secretary-General and its reports and recommendations are submitted to ECOSOC and the General Assembly. In some instances, special ACC reports are submitted to the inter-governmental body concerned.

The Environment Co-ordination Board and subsequent co-ordination on environmental matters

As the world has increasingly realized the pervasive nature of environmental problems and the integral connection between environment and development, the necessity for a co-ordinated, multisectoral approach to sustainable development has become very evident. As originally conceived, the Environment Co-ordination Board (ECB), established as part of UNEP by the UN General Assembly in 1972, was to serve as the mechanism for fostering co-operation among UN bodies having a mandate for environmental programmes. The ECB was to function under the auspices and within the framework of the ACC. In 1977, at the time of the restructuring of the UN bodies dealing with economic aspects, the ECB functions were assumed directly by the ACC (UNEP, 1982b).

Once the ACC took over the ECB's function of co-ordinating environmental matters, the UNEP Governing Council urged the Executive Director to ensure that designated officials of the UN system met to prepare for ACC meetings on environmental issues. Hence, the term 'Designated Officials on Environmental Matters' (DOEM) was born (UNEP, 1982b, 1988c). DOEM meetings do not have a fixed membership, but nearly every UN system organization has designated one of its senior officials to attend this group, which is chaired by the Deputy Executive Director of UNEP and supported by a UNEP secretariat. Traditionally, the DOEM hold three meetings a year. In recent years, the DOEM have been formally designated as the mechanism to facilitate the preparation, monitoring and evaluation of the system-wide medium-term environment programme (SWMTEP), considered below (UNEP, 1982b, 1983, 1984, 1986).

In 1975, UNEP initiated the process of 'joint programming' in the form of 'bilateral' meetings between UNEP representatives and representatives of individual co-operating agencies. It soon became evident, however, that because of the wide range of activities of the various UN organizations involving the environment, it would be more productive to convene discussions with all the concerned agencies around a programme theme, such as fresh water or biodiversity. A series of thematic joint programming sessions on environmental matters was initiated by UNEP in 1977 (UNEP, 1982c). These have provided a fruitful opportunity for UNEP to develop effective co-operative relationships with other UN agencies and among them (Box 5).

Although thematic joint programming is organized by the DOEM, the exercises themselves rely on experts from relevant UN agencies, and may depend on institutional mechanisms that have been in place for some time. For example, the Inter-Agency Working Group on Desertification (IAWGD), which was established by the United Nations General Assembly on the recommendation of the 1977 UN Conference on Desertification, co-ordinates desertification activities throughout the UN system. Likewise, the Ecosystem Conservation Group (ECG), which is composed of UNEP, UNESCO, FAO and IUCN, in co-operation with WWF and the International Board for Plant Genetic Resources (IBPGR), is the mechanism that has been utilized for joint programming on biodiversity (UNEP, 1988c).

Although system-wide medium-term planning in the UN system was conceived in 1977 (UN, 1977), its implementation was delayed until the beginning of the 1980s. The first such medium-term plans in the UN were the environment plan SWMTEP I, which was approved in 1982, covering the period 1984-89, and SWMTEP II, adopted in 1988 to cover the period 1990 to 1995. Both cut across the entire UN system, laying out a co-ordinated plan for agencies in implementing their various environmental activities (UNEP, 1988c, 1988d). The process of formulation gave UNEP the opportunity to exchange views with all relevant members of the UN system on how to streamline or expand existing programmes and on how to fill gaps.

SWMTEP II is divided into subject areas focusing on environmental components such as the atmosphere, oceans and terrestrial ecosystems. In addition, it has cross-cutting sections focusing on social issues such as human health and welfare and economic sectors such as energy and industry. It is to be reviewed mid-term in 1993, and this evaluation, together with the final evaluation of SWMTEP I, will serve as the basis for proceeding with SWMTEP III.

The bases for ensuring the co-ordinated development of environmental programmes within the UN system exist. What is still missing is the process by which the implications of development policies and programmes are fully linked with those in the environmental area. This process is needed to strengthen the capacity of the UN system in its support for sustainable development.

Other global institutions

One of the dramatic features of the past two decades has been the massive growth of international institutions outside the United Nations system (but commonly linked to it by shared objectives and practical co-operation). In the environmental field, mention has already been made of IUCN, now known as The World Conservation Union, founded in 1948 with states, governmental agencies and non-governmental scientific and conservation organizations among its 650 members. It has played a significant role in the development of international environmental law and policy, in monitoring environmental

BOX 5

Examples of interagency programmes.

- World Conservation Monitoring Centre (IUCN/UNEP/WWF): to support conservation and sustainable development through the provision of information on the world's biological diversity.

- Intergovernmental Panel on Climate Change (UNEP/WMO): formed in 1988 to complete assessments on the scientific aspects of climate change; the ecosystem, economic and social impacts of climate change; and the appropriate policy responses. The assessments were presented to the Second World Climate Conference of 1990. Since that time, the IPCC has been advising the international negotiating forum on a climate change convention.

- World Climate Programme (WCP) (WMO/UNEP/UNESCO/IOC/ICSU): under overall co-ordination of WMO, the WCP comprises programmes concerned with the collection of climate data, the application of climate information in all economic sectors, the improvement of prediction capability, and the assessment of the socio-economic impacts of climate change and the policy implications.

- Joint Group of Experts on Scientific Aspects of Marine Pollution (GESAMP) (UNEP/IMO/IOC/FAO/UNESCO/WHO/IAEA/United Nations): expert members appointed in their individual capacities to provide: assessment of potential effects of marine pollutants; scientific bases for research and monitoring programmes; international exchange of scientific information relevant to assessment and control of marine pollution; scientific principles for the control and management of marine pollution sources; and scientific bases and criteria relating to legal instruments and other measures for the prevention, control or abatement of marine pollution.

- International Programme on Chemical Safety (IPCS) (UNEP/ILO/WHO): est. 1980 to assess the risks that specific chemicals pose to human health and the environment.

- Joint FAO/WHO Food Standards Programme: established to protect the health of consumers and ensure fair practices in the food trade by initiating and guiding the preparation, finalizing, publication and revision of international food standards and by promoting the co-ordination of all food standards work undertaken by international organizations. In 1962 the FAO/WHO Codex Alimentarius Commission was established to implement the programme.

- Ecosystem Conservation Group (UNEP/UNESCO/FAO/IUCN in co-operation with WWF and IBPGR): involved in policy and programme development as well as co-ordination efforts on matters related to the conservation of biological diversity.

continued

BOX 5 CONTINUED

- Interagency Working Group on Desertification (UNEP/UNESCO/UNDP-UNSO/UNIDO/WMO/WFP/World Bank/UNDRO/ILO/IFAD/FAO/DTCD/ECA/ESCWA): to review issues of co-ordination as they relate to the efforts to implement the Plan of Action to Combat Desertification - a forum for exchanging information on UN agency activities related to desertification control and for thematic joint programming on desertification.

- Joint Consultative Group on Policy (UNDP/WFP/UNFPA/UNICEF/IFAD): to co-ordinate assistance policies and programmes among the five major donor programmes of the UN. Increasingly, members are meeting in the field (at country level) to programme activities. In 1989, the organizations formally decided to synchronize their programme cycles with the respective governments' planning cycles.

- Tropical Forestry Action Plan (TFAP) (WRI/FAO/World Bank/UNDP): provides the framework for efforts of the international development assistance community to address tropical deforestation. TFAP serves to identify development constraints in tropical forestry and to set common priorities for countries and donors. WRI withdrew as a sponsor in 1991.

- United Nations Sudano-Sahelian Region (UNDP/UNEP): established in 1975 as a special entity within UNDP concerning drought and desertification in the Sudano-Sahelian region. In 1978, UNSO was given the mandate (by joint agreement between UNDP and UNEP) to act on behalf of UNEP in implementing the Plan of Action to Combat Desertification in the Sudano-Sahelian region. To this effect, UNSO is funded by UNDP and UNEP and provides information, advice and technical assistance to some 22 countries of the region.

- International Trade Centre (ITC) (UNCTAD/GATT): works with developing countries to set up effective national trade promotion programmes for expanding their exports and improving their import operations through the provision of information, advice, training and technical co-operation.

- Energy Sector Management Assistance Programme (World Bank/UNDP): to provide assistance to energy-related activities in developing countries. It is also supported by other UN agencies and numerous bilateral aid agencies.

conditions and trends, in technical co-operation projects in developing nations, and in raising public awareness world-wide. It engages in joint ventures with UN agencies and programmes, for example action in partnership with UNEP and WWF to support the World Conservation Monitoring Centre (WCMC), acting as a member of the Ecosystem Conservation Group (ECG), and producing two World Conservation Strategies (in 1980 and 1991) in partnership with

UNEP and WWF and with support from UNESCO and FAO.

Another world-wide institution that has had a major influence on environmental affairs is ICSU, The International Council of Scientific Unions, which maintains permanent Unions in major disciplines, operates specialist Scientific and Special Committees (such as SCOPE, The Scientific Committee on Problems of the Environment) and mounts major programmes such as the International Geophysical Year, International Biological Programme and International Geosphere/Biosphere Programme (IGBP). ICSU has often worked very closely with WMO and IOC. One of its great strengths is that its national member organizations are mostly Academies of Science, and ICSU activities thus draw on many of the world's most outstanding scientists.

In the 1970s and 1980s many environmental NGOs were established to provide independent advice (as the International Institute for Environment and Development, IIED, and the World Resources Institute, WRI, do) or to act as advocates on general or particular issues. The World Wildlife Fund (now World Wide Fund for Nature, WWF) was established to mobilize support for conservation, but as it has grown it has also broadened and now deploys funds substantially greater than those at the disposal of UNEP on a wide range of projects and programmes for the conservation and sustainable use of environmental resources. Greenpeace and Friends of the Earth are two other powerful global NGOs with a more active campaigning approach. In total the resources at the disposal of these major global NGOs are quite large.

The past two decades have also seen the strengthening of many 'humanitarian' or 'development assistance' NGOs such as Oxfam, Save the Children Fund, and Christian Aid. These bodies have become more and more involved with environmental issues because it has become evident to them that humanitarian aims such as famine relief are best extended and consolidated by measures to provide an environment within which people can live in a sustainable way.

Regional institutions and actions

The number of regional political and economic organizations that in one way or another are concerned with environment and development questions has also increased greatly since 1972 (Appendix II). They include the original regional security arrangements that came into being after World War II as contemplated in Chapter VII of the UN Charter; the five UN regional economic commissions established directly under ECOSOC; numerous regional offices of global UN agencies and programmes, such as those of UNEP, FAO and WHO; the regional development banks, and several other organizations and conferences that bear no relationship to the UN system at all, such as the OECD and the South Pacific Forum. This list is further swelled by the increasing decentralization of bodies such as IUCN, WWF and the growth of NGOs on the regional scale.

A primary reason for the growth in regional environmental activities has been the recognition that many severe environmental questions - such as control of transboundary air pollution, shared river basin management, and the protection and use of regional seas and their coastal areas - span national frontiers, and cannot be dealt with unless the states that share them act together for mutual benefit. Many species of animal, and especially birds, migrate or move across frontiers. Animal-borne diseases, ranging from malaria to rabies, are carried by their vectors from country to country and demand co-ordinated preventative measures. Trade carries cargoes of potentially hazardous substances through regional seas, necessitating co-operative arrangements to guard against, and deal with, accidents.

At regional level, policy, law and institutions have commonly evolved together. A problem, once admitted, leads to bilateral or multilateral regional discussion. Agreements then reached define the policies it has been agreed to pursue, and these require translation into national legal instruments and administrative practices, which in turn affect the nature of national institutions. International institutions are commonly set up as a framework for the onward development of policy, for monitoring its effectiveness, and for sharing expertise and resources.

Regional activities of UNEP

The Stockholm Conference called for a legal agreement to control ocean dumping and, within its framework, for regional agreements 'in particular for enclosed and semi-enclosed seas, which are more at risk from pollution'. In response, UNEP, as mentioned earlier, developed its Regional Seas Programme, widely acknowledged as one of its most successful ventures. The Mediterranean was tackled first, and an Action Plan was adopted in 1975. A Convention and two Protocols covering dumping by ships and aircraft and pollution by oil and other harmful substances in cases of emergency were signed in the following year. In 1980 the Protocol for the Protection of the Mediterranean Sea Against Pollution from Land-based Sources was signed by 12 States and the European Community. In 1982 the Protocol Concerning Mediterranean Specially Protected Areas was signed. This Mediterranean programme became a model for UNEP's other Regional Seas programmes, now covering more than 140 countries. The steps in each case have been:

a) a survey of the problems involved;

b) development of a five-component Action Plan which includes environmental assessment, environmental management, environmental legislation, and institutional and financial arrangements.

These regional action plans and their Conventions are described in more detail below.

Regional UN security and economic arrangements

The original post-war regional arrangements for international peace and security have evolved into fora concerned with a broader range of issues. For example, the Southeast Asian Treaty Organization (SEATO) evolved into the Association of Southeast Asian Nations (ASEAN). Its mandate has expanded from promoting regional peace and security to accelerating economic growth and co-operation among its members. Co-operation in the field of environment commenced in 1977, and is today reflected in the third ASEAN Environment Programme (1988-1992), adopted by the Third ASEAN Ministerial Meeting on the Environment (UNEP/ASEAN, 1988).

The UN regional economic commissions in recent years have all emphasized the need to orient economic development to take account of environmental issues. For the West Asian region, this means water resources management and conservation, whereas the European region has stressed atmospheric pollution and hazardous wastes (UNEP, 1989c). ECE established a group of Senior Advisers on Environment in 1971, and in 1987 merged it with its Committee on Water Problems. Under ECE auspices, major international environmental agreements have been concluded on long-range transboundary air pollution and, in 1991, on environmental impact assessment in a transboundary context. A Code of Conduct on Accidental Pollution of Transboundary Inland Waters, prepared with financial support from UNEP, was adopted in 1990. The ECE in 1991 finalized two conventions, one on the Protection and Use of Transboundary Watercourses and International Lakes, and another on Transboundary Effects of Industrial Accidents.

The ECA has established an intergovernmental Regional Committee for Human Settlements and Environment, and in 1984 convened a Conference of Ministers which produced the Kilimanjaro Declaration, the first high-level recognition of the role played by environmental degradation in the African crisis. In 1985, the first session of the African Ministerial Conference on the Environment (AMCEN) was held in Cairo, and a Regional Programme of Action for Co-operation on the Environment was adopted. The Ministerial Conference meets every two years to review progress in the implementation of the Cairo Programme. In 1991 with the support of the African Development Bank, OAU, the UNCED Secretariat, UNEP and UNSO, two ministerial-level regional conferences were held as part of the preparation for the UN Conference on Environment and Development in Cairo and Abidjan. They both benefited from the conclusions of the first African Regional Conference on Environment and Sustainable Development, which took place in Kampala in June 1989.

In Southeast Asia, joint action by the South Pacific Commission, the South Pacific Bureau for Economic Co-operation, ESCAP and UNEP led to the establishment of the South Pacific Regional Environmental Programme (SPREP) in 1982. The Ministerial-Level Conference on Environment and Development in Asia and the Pacific, held in Bangkok in October 1990, recognized that sustainability of the Earth could not be achieved in parts isolated from one another. Two documents presented at the Conference (*State of the Environment in Asia and the Pacific 1990*, and the *Regional Strategy for Environmentally Sound and Sustainable Development*) were included in the Regional Report to UNCED, prepared by ESCAP. The seven ESCAP intergovernmental committees consider such issues as rural development and environment, natural resources and energy, development planning and industry and technology, and are supported by an Environmental Co-ordinating Unit within the Secretariat.

The Economic Commission for Latin America and the Caribbean programme in Latin America has been oriented more toward policy development than development projects, and since 1978 it has turned its attention to identifying the environmental ramifications of development patterns prevailing during recent decades and to providing technical assistance to encourage the incorporation of the environmental dimension into development planning. The Ministers responsible for the environment in Latin America and the Caribbean have organized biennial meetings. The Caribbean Community, CARICOM, has also organized periodic ministerial conferences on the environment. The purpose of these meetings has been to establish an inter-governmental forum for consultation on the environment, and to provide a framework for launching regional cooperative programmes to address priority common problems of resource management.

Regional development banks

The regional development and investment banks for Africa, the Americas and Asia, the European Investment Bank, the European Bank for Reconstruction and Development and the Nordic Investment Bank seek to promote economic and social development within the regions they serve. The Inter-American Development Bank (IDB) and the Asian Development Bank (AsDB) are developing substantial environmental policy guidance and have set up specific internal structures to ensure that environmental considerations are taken into account in project planning. The IDB has also issued a document on 'Strategies and Procedures on Socio-Cultural Issues as Related to the Environment', in order to ensure consideration of the effects of its operations on indigenous tribal groups and local communities. The AsDB is seeking to develop indices of the social, economic and environmental aspects of development to use in national planning approaches. The recently-established European Bank for Reconstruction and Development is the first such body specifically to include environmental concerns in its Charter.

Other regional inter-governmental fora on the environment

The Organization of American States (OAS), founded in 1948 to provide a general forum for economic, political and social co-operation among the nations of North and South America, has evolved a variety of programmes to strengthen local capabilities in managing and conserving natural resources. The Caribbean Environmental Health Institute has been established as a facility within the Caribbean Community institutional framework to focus on marine pollution and related aspects of environmental health. The South Pacific Forum

(SPF) has been particularly active in fisheries management issues and in developing a regional information network and a centralized fisheries data base to monitor fishing activities. Within the Organization of African Unity (OAU), environmental considerations have received more attention in recent years as its connections with meeting basic human needs have grown more apparent.

UNEP, OAU and ECA sponsored the First African Ministerial Conference on Environment (AMCEN) in 1985. AMCEN is now a permanent institution with a Secretariat located in Nairobi. Its mandate is to address environmental problems, with special emphasis on natural resource conservation (soil, forests and water; marine environment and integrated rural development with special emphasis on rural energy requirements). Furthermore, the League of Arab States in co-operation with UNEP convened in 1987 the First Arab Ministerial Conference on Environment. The League of Arab States decided in 1988 to establish the Council of Arab Ministers Responsible for Environment (CAMRE) as a permanent body in the context of the League to deal with environment at ministerial level. CAMRE is entrusted with promotion of Inter-Arab co-operation with respect to major shared environmental problems with special emphasis on desertification, water resources, industrial pollution and public awareness.

Among the committees established by the OECD, the Environment Committee, which dates from 1970, has met several times at ministerial level. The OECD Development Advisory Committee is an important forum for donors to exchange information on development policy issues and their respective development programmes.

The European Community (EEC), whose environment programme was launched in 1973, supports numerous environmental activities and has issued a number of directives to govern member states' environmental practices. These EEC measures constitute the framework with which national legislation must conform (Haigh, 1989). The Single Act of 1986 strengthened the European Community Treaties by establishing the need for environmental policy firmly within the competence of the Community. Another regional organization in Europe that is evolving into a significant forum for discussion of environmental issues is the Conference on Security and Co-operation in Europe (CSCE), which grew out of the 1975 Helsinki accords. The CSCE held a meeting on the Protection of the Environment in Sofia, Bulgaria, in 1989 which decided to focus on the transboundary effects of international accidents, the pollution of transboundary watercourses and links, and the management of hazardous chemicals. In June 1991, European Ministers for the Environment met in Czechoslovakia and agreed to work together on common environmental problems. ECE is playing a considerable part in the follow-up to both Conferences, and is working with the EC Commission and others to prepare for Ministers an Environment Programme for Europe and a report on the State of the European Environment.

The Commonwealth Secretariat and the Lomé Conventions are not regional organizations but both play a role in supporting sustainable development. The

Commonwealth is composed of the 49 members of the association of Commonwealth countries. It issued the Langkawi Declaration on the Environment (UN, 1989), which sets a comprehensive long-term agenda for environmental activities. Through the Commonwealth Fund for Technical Co-operation, the Commonwealth finances cooperative endeavours. The Commonwealth Secretariat has been particularly involved in integrated coastal and marine management, and has recently established expert groups on the implications of global climate change for low-lying islands and to monitor regional climate change data. The Lomé agreements provide a basis for co-operation in economic development between the European Community and 69 Asia-Caribbean-Pacific (ACP) nations. The fourth agreement, signed in December 1989, includes a full section on the environment.

Responses in OECD

OECD has also been an important forum for international policy development during the past 20 years. It established an Environment Committee in 1970, and under that Committee has developed a number of significant principles and agreements. The main driving force behind these developments has been the growth of evidence of change and damage in OECD countries, some of which is transboundary in origin, and demands for regulations and controls that are similar across the economic region, avoiding non-tariff barriers to trade and creating the 'level playing field' favoured by industry. In 1974 this process led to the agreement of the Polluter Pays Principle, and there have also been agreements on specific controls and standards for particular industrial sectors and problems (OECD, 1980). These negotiations have inevitably been complex because they bring changes to an immensely complex economic system, and because the instruments range from soft law to somewhat harder treaty law.

The OECD region has been in the vanguard of much environmental regulation because it is the seat of many of the worst problems. Vehicle emissions, persistent pesticides, the control of parts of the nuclear cycle, and air and water pollution from stationary sources have all posed severe problems, often as the consequence of years of inaction. The questions now arising concern how far the standards and approaches developed in OECD can and should be extended to other parts of the world. There are now regular meetings between Environment Ministers from Western and Central and Eastern Europe at which these questions are discussed. In December 1991, OECD Environment and Development Ministers adopted a new Policy Statement (Box 6).

Regional scientific fora

Accurate data and information on environmental conditions and trends is essential for environmentally-sound development, as is sound scientific and technical advice and analysis. Several regional bodies have been established for these purposes. Some of the oldest have focused on marine fisheries management and marine pollution control, such as the International Council for the Exploration of the Sea (ICES), established in 1901 to co-ordinate marine

BOX 6

OECD policy on the environment.

OECD Ministers of Environment and Development Co-operation, Heads of UN Agencies and a Commissioner of the European Community met at OECD Paris on 2-3 December 1991 and adopted a new policy statement. The principal elements of this statement are:

1. Contributing to sustainable development world-wide is one of OECD's central objectives. It requires the concurrent pursuit of economic growth, higher living standards and a healthy environment.

2. The vicious circle that at present leads to poverty and environmental degradation must be broken through sound national economic and social policies, supported by international development co-operation.

3. A strengthened partnership for sustainable development is needed between OECD members and non-member countries, and the UN Conference on Environment and Development in 1992 should be used as a catalyst to this end.

4. OECD countries have a responsibility to use energy and raw materials more efficiently, and to reduce their impact on the global environment. They will need to re-evaluate their patterns of both consumption and production, and promote new technologies.

5. Non-OECD countries should be encouraged to join OECD countries in a commitment to sustainable development, developing their social and economic policies to that end and slowing population growth where it is too high to permit sustainable development. Additional financial resources must be made available to these countries as a part of a strengthened partnership.

6. Conditions should be established that encourage the private sector to create, market and use technologies that contribute to sustainable development. Technical assistance and co-operation between developed and developing countries should be strengthened, and barriers to technology transfer reduced. OECD knowledge and experience should be shared.

7. The international trading system should be improved and trade and environmental policies should be mutually supportive. Developments in GATT through the Uruguay Round and the Working Group on Trade and Environment must be brought to a successful conclusion.

continued

THE WORLD ENVIRONMENT 1972-1992

BOX 6 CONTINUED

8. The private sector is a driving force for economic growth in market-oriented economies, and at the same time creates resources needed to protect the environment. Co-operative efforts with firms in developing countries are to be welcomed.

9. Sustainable development can be achieved only through the active involvement of all sectors of society. More decision making should be devolved to local level, and the different roles of men and women in environmentally-related activities should be recognized. The increasing involvement of non-governmental organizations in promoting sustainable development is welcome.

10. UNCED should formulate agreed strategies for sustainable development and start a process of strengthening partnerships to attack environmental problems at national, regional and global levels. The Conference should be a decisive stage in a new process of multilateral co-operation on environment and development.

Source: OECD (1991).

---

scientific research in the North Atlantic region, the International Council for the Scientific Exploration of the Mediterranean (ICSEM) and, most recently, the North Pacific Marine Science Organization (commonly known as PICES) created in 1990.

Regional conventions and legal instruments relating to the environment

Regional agreements on air pollution

The recognition of the international dimension of air pollution led to some of the earliest regional action in North America and Western Europe. In 1972, Sweden identified transboundary air pollution as a major problem affecting southern Scandinavia, and reported accordingly to the Stockholm Conference (Sweden, 1972). In the same year, the OECD began a co-operative programme to measure long-range transport of air pollution (OECD, 1977). In 1977 the ECE set up a continent-wide co-operative programme for monitoring and evaluation of the long-range transmission of air pollutants in Europe, spanning both the Western, OECD, member countries and the Central and Eastern European nations that were then members of CMEA. This programme led directly to the signing of the ECE Convention on Long-Range Transboundary Air Pollution in 1979.

The evolution of this Convention is an interesting example of how regional measures have developed during the period under review. Initially, the primary thrust of the Convention was environmental monitoring and research, and the exchange of the results. Monitoring arrangements were formalized under a Protocol on Long-term Financing of the Co-operative Programme for Monitoring and Evaluation of Long-range Transmission of Air Pollutants in Europe in 1980, and by August 1991 all except two of the 32 States Party to the Convention had also ratified this Protocol (UNEP, 1991). Although the Nordic states emphasized from the outset that action had to go further and include measures to reduce emissions of sulphur oxides which were clearly implicated as a major source of 'acid rain', the agent of biological impoverishment of thousands of lakes and many hundreds of kilometres of rivers in the southern areas of Scandinavia, resistance from major industrial nations (especially at this stage the Federal Republic of Germany, the United Kingdom and the United States) blocked any action. A Protocol on the reduction of Sulphur Emissions or their Transboundary Fluxes by at least 30 per cent was none the less adopted in 1985, with Germany as a signatory, and by August 1991 this had 20 ratifications, including the Federal Republic of Germany but excluding the USA, the UK, the then German Democratic Republic, Hungary, Poland and Romania among major industrial states. A Protocol Concerning the Control of Emissions of Nitrogen Oxides or their Transboundary Fluxes, agreed in 1988, had 18 ratifications by August 1991, including the USA, Germany and the UK. It entered into force on 14 February 1991.

Canada and the United States are party to the ECE Convention, but events in North America have followed a parallel path rather than one identical to the situation in Europe. It has been clear for many years that many lakes in north-eastern North America have also sustained damage attributable to acid deposition, and in 1976 the Canadian Government identified acid rain as a high-priority environmental problem. A United States-Canada Research Consultation Group was established and produced its first report in 1979, and in 1980 a US-Canada Memorandum of Intent was adopted which covered research, monitoring, exchange of scientific information and gradual reduction in emissions (US-Canada, 1980). However, despite major research efforts including a National Atmospheric Deposition Project (NADP) set up by the US Environment Protection Agency, international action in North America has lagged behind that in Europe.

While action to curb emissions of sulphur and nitrogen oxides in both Europe and North America have moved slowly, and are a long way from reducing depositions to below the 'critical loads' which the most sensitive ecosystems can tolerate, the regional action on these two continents does provide something of a model of how transboundary air pollution problems can be addressed. There is no doubt that similar agreements, pressed ahead by more vigorous action than has characterized the two areas that have pioneered the regional approach, will be found desirable in other parts of the world as these industrialize and begin to face risks of transboundary air pollution.

Regional action to safeguard fresh water resources

In 1970, the United Nations General Assembly recommended that the International Law Commission (ILC) study the law on non-navigational uses of international watercourses, with a view to codifying and developing it (Holdgate, Kassas and White, 1982). Such analysis ran into difficulties of definition, and at the time of the Stockholm Conference in 1972 the potential problems posed by the impact of the actions of one state on other states through the creation of changes in shared rivers was already evident, and a cause of some friction. In the developed regions, the chief concern was over pollution, progressively increasing as a river flowed seawards, and posing particular problems for the nations in whose territory the estuary of the river lay. In the developing world the chief concern was over the impact on downstream states of water abstractions upstream, for agriculture and industry. The management of the Parana, Brahmaputra and Ganga all raised issues of the latter kind.

These problems were returned to at the United Nations Water Conference in 1977, which emphasized the need for co-operation. Since then, the trend in many regions has been towards joint discussion of the needs the various nations have for use of rivers that flow through several territories, the establishment of some kind of agreement setting out principles, the development of some kind of joint plan including agreements on abstractions and discharges, and the creating of a Commission or other body to monitor trends and provide a forum for discussion.

In Europe, by 1977, there were at least 40 bilateral, 6 trilateral, and one 5-member agreement affecting boundary waters, and international commissions had been established for the Danube, Rhine, Mosel and Saar rivers (UNWC, 1978). In North America a similar Commission had been established for the Great Lakes and other waters at the boundary between Canada and the United States. These Commissions have continued to work since then, and there is no doubt that water quality has improved in many parts of the systems concerned, although progress cannot be described as rapid. Also in North America, a treaty on the management of water resources at the frontier between the United States and Canada (British Columbia) was signed in October 1984. It excluded the possibility of flooding land in Canada through the construction of a dam on the Skagit River in the USA. This treaty is an excellent example of an international agreement on a trans-frontier water resource issue.

In Asia, the most significant new development relates to the Mekong Basin. Viet Nam, the Lao People's Democratic Republic and Thailand have agreed to participate in a Lower Mekong Basin Development Environment Programme (LMBDEP), and an Interim Committee has been established. In 1980, this established guidelines for environmental planning and assessment. In Africa, agreements had been reached before 1970 for the Senegal basin (1953), Niger (1964), and Lake Chad (1964). These have been extended by agreements for the Mano (1973) and Kagera (1977) and a series of cooperative arrangements between Egypt and the Sudan for the Nile, for which a Nile Basin Authority has been established. Discussions have been held on the establishment of similar measures for the Congo-Zaire system. In 1987 an agreement was signed by Botswana, Mozambique, Tanzania, Zambia and Zimbabwe for the Environmentally Sound Management of the Common Zambezi River System. This action plan was endorsed by the Council of Ministers of the Southern African Development Coordination Conference (SADCC) as a concerted action programme of SADCC. The implementation of the action plan is under way, and UNEP has given financial support to SADCC for initiation of its implementation. This was followed by a similar action plan for the Lake Chad Basin (Chapter 4). In South America, the Rio de la Plata was the subject of an agreement established before 1980. In fact, as discussed in Chapter 4, UNEP has pioneered the promotion of the concept and practice of Environmentally-Sound Management of Inland Waters (EMINWA).

Water management is of crucial importance in Africa, and many regional agreements have been put in place to respond to drought and desertification. The Permanent Inter-State Committee on Drought Control (CILSS), the Institut du Sahel and the United Nations Sudano-Sahelian Office (UNSO) were established in response to these concerns. Collaboration between Algeria, Egypt, the Libyan Arab Jamahirya, Mauritania, Morocco and Tunisia is working towards the establishment of a green belt across North Africa. A regional project has been established between the Libyan Arab Jamahirya, Chad, Egypt and the Sudan to manage the northeastern Saharan aquifer. An Intergovernmental Authority for Drought and Development (IGADD) has been established by Djibouti, Ethiopia, Kenya, Somalia, the Sudan and Uganda, working in their region much as CILSS has endeavoured to work in the western Sahel.

The trend to regional co-operation is thus well established. The need now is for action to follow agreement, and in many instances this will require resources. Meanwhile another trend is apparent. Because the condition of inshore seas is especially dependant on water-borne discharges from the land via rivers and outfalls, the need to manage both areas of coastline and the catchments draining to them is increasingly appreciated (IUCN/UNEP/WWF, 1991).

Regional agreements on the protection of the sea

The need for regional co-operation to safeguard and manage areas of sea began, in one sense, with various regional Fisheries Conventions as early as the nineteenth century. However, in recent years impetus has been given by concern over marine pollution from ships, by the dumping and offshore incineration of wastes, and from land-based sources. An Intergovernmental Working Group on Marine Pollution was established as a part of the preparatory process for the Stockholm Conference, and in parallel with its efforts regional agreements were drawn up to prevent marine pollution by the dumping of wastes in the northeast Atlantic (the Oslo Convention, agreed in 1972), and to protect the Baltic (the Helsinki Convention, 1974). In 1974 the

Paris Convention, also covering the northeast Atlantic, regulated pollution from land-based sources.

As noted above, soon after its creation, UNEP identified the protection of the sea as a priority concern, and established a regional seas programme. Its geographical coverage is shown in Figure 1, while Box 7 lists the regional agreements that have been adopted as a direct result of this programme and indicates their scope.

In a number of regions, action has been carried forward by regular ministerial meetings. In Europe there have been three such meetings to co-ordinate action to safeguard the North Sea, and these have produced declarations of policy, for example adopting the 'precautionary approach' and setting stringent targets for pollution reduction at their 1987 gathering (UK, 1987). In the South-East Pacific, a Convention on the Protection of the Natural

Figure 1: *The geographical coverage of the UNEP Regional Seas Programme.*

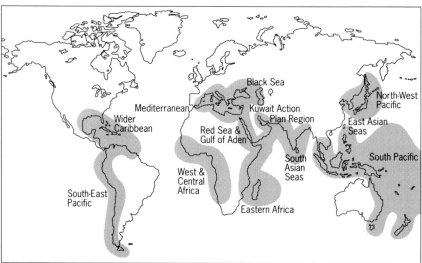

Source: UNEP

Resources and Environment of the South Pacific, adopted in 1986, has Protocols on the Prevention of the Pollution of the South Pacific Region by Dumping, and on Co-operation in Combating Pollution Emergencies in the South Pacific Region. The South Pacific Forum Fisheries Agency (SFFA) assists in the conservation and management of fisheries in the exclusive economic zones (EEZs) of the South Pacific island states.

West Asia is another region with particular marine problems and established frameworks for co-operation. The Red Sea and the Persian Gulf are both areas of land-locked water, traversed by many oil tankers, and the Gulf in particular, because it is shallow, is much at risk from pollution. The Kuwait Action Plan

---

BOX 7

Regional seas conventions (so far adopted).

- Convention for the Protection of the Mediterranean Sea against Pollution (Barcelona, 16 February 1976).

- Kuwait Regional Convention for Co-operation on the Protection of the Marine Environment from Pollution (Kuwait, 23 April 1978).

- Convention for Co-operation in the Protection and Development of the Marine and Coastal Environment of the West and Central African Region (Abidjan, 23 March 1981).

- Convention for the Protection of the Marine Environment and Coastal Area of the South-East Pacific (Lima, 12 November 1981).

- Regional Convention for the Conservation of the Red Sea and Gulf of Aden Environment (Jeddah, 14 February 1982).

- Convention for the Protection and Development of the Marine Environment of the Wider Caribbean Region (Cartagena de Indias, 24 March 1983).

- Convention for the Protection, Management and Development of the Marine and Coastal Environment of the Eastern African Region (Nairobi, 21 June 1985)

- Convention for the Protection of the Natural Resources and Environment of the South Pacific Region (Noumea, 25 November 1986).

Source: UNEP (1991b)

---

has eight parties, seven being Arab states, and the Red Sea and Gulf of Aden Plan has eighteen members, again seven of them being Arab states.

The Antarctic region

The Antarctic region, broadly defined, includes one-fifth of the Earth's surface, lying south of the limits of permanent settlement in South America, Africa and Australasia. Since 1959 there has been agreement under the Antarctic Treaty among all the nations active in the region to use Antarctica for peaceful purposes only, and to co-operate in scientific research. The compendium of legal instruments affecting the region has increased, with the adoption of Agreed Measures for the Conservation of Antarctic Fauna and Flora in 1964, a Convention on the Conservation of Antarctic Seals in 1972, and a Convention on the Conservation of Antarctic Marine Living Resources in 1980. A Protocol providing for the comprehensive conservation of the Antarctic environment and the exclusion of minerals exploration and exploitation for at least 50 years was signed on 4 October 1991.

The need for comprehensive international action to safeguard the Antarctic region, and to regulate the fisheries that have developed in its surrounding seas, is widely recognized. Detailed measures to define and protect areas important as habitat or because of outstanding landscape quality, to ensure that all developments are subject to environmental impact assessment, to ensure that wastes are removed from the region or destroyed without impact, and to regulate tourism, are all either in place or in process of negotiation. While the restriction of consultative status under the Antarctic Treaty to nations with a major presence or activity in the region has excited criticism in the United Nations General Assembly, the region as a whole remains the subject of more comprehensive international agreements for environmental protection than any other part of the Earth's surface. A comprehensive conservation strategy for the region was published in 1991 (IUCN, 1991).

Integrated regional policies and actions

It is increasingly recognized that a sectoral approach to environmental management is of limited effectiveness. Pollutants spread from air to rivers, land and the sea. About 98 per cent of the lead ultimately dissolved in sea water, 80-90 per cent of the organochlorine pesticides found in open water, and significant quantities of nitrate, reach the sea through the atmosphere (GESAMP, 1990). Coastal development, land-use practices, waste disposal methods, the management of the hydrological cycle and the control of transport all affect the quality of coastal seas. The need for an integrated approach to environmental protection is evident.

Over the past two decades, regional policies have become increasingly integrated. In Europe, the European Economic Community has developed environmental programmes spanning all 12 member states. The Community's first Environmental Action Programme, adopted in 1972, was primarily concerned with preventing pollution: later Programmes have become progressively wider in scope. Table 1 summarizes the legal instruments and policy decisions adopted by the Community up to 1991.

The European Community has also adopted enforcement procedures. An important aspect of European Community procedures is the right of individuals or citizens' groups to complain to the Commission over the non-observance by a member state of a Community instrument. A single complaint by a resident in one of two areas in the United Kingdom that failed to comply with the 1980 Directive on Air Quality Limit Values and Guide Values for Sulphur Dioxide triggered a Commission investigation that led to infringement procedures against seven member states.

The number of member states that were the subject of complaint by the Commission on grounds of non-compliance with Community directives rose

Table 1: *European Community laws and decisions, 1970-91.*

	Up to 1970	1971-75	1976-80	1981-85	1986-89	1990-1991
Air	2	4	8	9	12	4
Noise	2	1	4	7	4	-
Hazardous and toxic substances	1	5	17	10	7	15
Fresh waters and seas	1	7	16	20	5	7
Wastes	0	3	9	7	3	5
Flora and fauna	0	2	4	19	6	4

Source: data for 1970–89 from DocTer (1991).

from 10 in 1982 to 190 in 1985 and 460 in 1989. Unless the state concerned satisfies the Commission that action has already been taken to resolve the problem, the complaints are referred to the European Court of Justice, whose rulings have considerable political weight with member states. Not all rulings of the Court have been implemented, but most states have complied with its judgements. The spotlight of public opinion is increasingly turned upon offenders, and governments are increasingly recognizing the political advantage of a good record in this field.

The EEC is unique, but institutions and agreements for close regional co-operation on a wide range of environmental issues are gaining strength in most regions. In Asia and the Pacific, ASEAN co-operation on the environment began in 1977 when a draft Sub-Regional Environmental Programme (ASEP I) was prepared. The ASEAN Experts Group on the Environment (AEGE) has met annually since 1977: in 1990 it was replaced by Ministers with a group of ASEAN Senior Officials on the Environment (ASOEN) which is co-ordinated by the member countries on a three-year rotation and is designed to ensure the co-ordinated planning and implementation of regional environmental programmes. ASEAN countries have adopted several official declarations and agreements including the Manila Declaration on Heritage Parks and Reserves, the Bangkok Declaration on the ASEAN Environment (1984), an Agreement on the Conservation of Nature and Natural Resources (1985), and a Jakarta Resolution on Sustainable Development (1987).

Also in the Asian and Pacific Region, a South Asia Cooperative Environmental Programme (SACEP), with headquarters in Colombo, Sri Lanka, and the South Pacific Regional Environmental Programme (SPREP) based on Noumea, New Caledonia, have been the focus for co-ordinated regional programmes. Afghanistan, Bangladesh, Bhutan, India, Iran, Maldives, Nepal, Pakistan and Sri Lanka participate in SACEP, whose Council has emphasized the need to integrate environmental protection with economic development. A Regional Seas Programme for South Asian Seas is being co-ordinated by SACEP, with support from UNEP. The seven countries of South Asia (Pakistan, India, Nepal, Bhutan, Bangladesh, Sri Lanka and the Maldives) also participate in the South Asia Association of Regional Co-operation, which is preparing a report on the state of the region's environment with a special focus on natural disasters, and this will include a programme for bilateral and multilateral co-operation. For its part, SPREP with its advisory Association of South Pacific Environmental Institutions, formed in 1986, is promoting country-specific national conservation strategies, coastal zone management, and an evaluation of traditional knowledge of the environment.

In Africa, a decision to establish five Sub-Regional Environment Groups (SREGs) to promote regional and sub-regional co-operation in the field of environment was taken in 1983 by UNEP in response to the need, expressed in the Lagos Plan of Action for the economic development of Africa, for regional co-operation in the field of the environment. These groups cover Central Africa (CASREG), Eastern Africa (EASREG), Northern Africa (NASREG), Southern Africa (SASREG) and Western Africa (WASREG). The SREGs conducted a series of meetings in 1983 and early 1984 ending with the formulation of a Programme of Action on the African environment. The first African Ministerial Conference on the Environment (AMCEN), held in 1985, received the reports of the SREGs together with a report of the Executive Director of UNEP on the state of the environment and natural resources in Africa. On the basis of these reports the Conference adopted a Regional Programme of Co-operation On the Environment. This programme identified four priorities: halting environmental degradation, enhancing food production capacity, achieving self-sufficiency in energy and correcting the imbalance between population and resources.

There are several regional agreements on environmental matters in Africa. They include a Convention on the African Migratory Locust (Kano, 1962), a Phyto-Sanitary Convention for Africa (Kinshasa, 1967) and an African Convention on the Conservation of Nature and Natural Resources (Algiers, 1968). Regional Seas Conventions covering West and Central Africa and Eastern Africa as well as a protocol concerning co-operation in combating pollution in cases of emergency were adopted in 1981 and 1985 respectively. For the Eastern African Region a protocol concerning Protected Areas and Wild Fauna and Flora was also adopted in Nairobi in 1985.

There have also been a number of regional agreements in Latin America and the Caribbean. One group focuses specifically on natural resources and the environment and includes the Convention on Nature Protection and Wildlife Preservation in the Western Hemisphere (1940), the Convention for the Conservation of the Vicuna (1979), the Treaty for Amazonian Co-operation (1978) and a Recommendation concerning the formulation of an American Declaration on the Environment by the Organization of American States (OAS). The second group of instruments is of a more general character and includes the Central American Commission on Environment and Development, the

Andean Pact, the Caribbean Community (CARICOM), and the Acapulco Commitment to Peace Development and Democracy (1987), with subsequent declarations by the Group of Rio (ECLAC, 1991).

Institutional growth and environmental need: an analysis

The central challenge

The complexity and magnitude of the environment/development 'problematique' was recognized during the preparatory process for the Stockholm Conference. Twenty years later, the level of political interest in the environment is very different. Yet, ironically, many of the essential points recognized by the participants in the Founex and Cocoyoc meetings are still being raised in debate. The major problem of underdevelopment is back on the agenda of the developing countries in a major way. The debate over the over-consumptive, wasteful life-styles in the developed countries has, if anything, intensified. Many people are again asking the kinds of questions posed in *Limits to Growth*: for example concerning the possibility, let alone the desirability, of unabated growth in energy demand and the private transport sector. In the 1970s these questions generally related to energy resources running out: today they focus on the limits of the environment as a 'sink' or 'provider of services', but the central issue of nature's capacity to tolerate human impact remains.

The massive expansion of national and international institutions that has occurred during the past 20 years has produced as much sectoralization and fragmentation as synthesis. Only now is the world community really beginning to come to grips with the underlying economic and demographic forces that impede sustainable development. It is now clear that development strategies need substantial adjustment, and that this must go far beyond the technology of environmental management to incorporate trade, debt, social infrastructure and population policies. It is also evident that environmental costs and benefits must be incorporated into the technologies and processes of development from the initial planning stages. There is now much talk of a greater reliance on market mechanisms as a mode of environmental control. The use of economic instruments and full environmental cost accounting, largely unexplored until the mid-1980s, is the new fashion in environmental governance. Many governments are insisting that the costs of natural resource depletion must be accounted for and that those who generate pollution and waste must pay the price. Many of the 1972 Stockholm Conference recommendations were about reducing uncertainty through new research and monitoring, and waiting for the resulting clarification. Now, even when there are far more answers, the 'precautionary principle' is gaining general acceptance.

Despite twenty years of progress since Stockholm, much improved scientific knowledge and economic methodology (Chapter 20) and enormously expanded institutional frameworks, many essential tools are still not in place. Without better international collaboration, closer integration between international and national agencies, and much stronger partnership between the intergovernmental and non-governmental sectors - including the world of business and commerce - it will be impossible to master the knowledge, skills and resources to meet the global challenges that confront humanity.

What has been particularly lacking is an appreciation of the complex linkages between sectoral activities, macro-economic policies and sustainable development, and how to internalize them in policies and development strategies. This next step will require far greater collaboration among all concerned agencies, and a far greater effort to ensure that best use is made of all available resources for the promotion of sustainable development. It will also require better procedures to review and evaluate the effectiveness of on-going programmes.

Environmental diplomacy between nations has developed alongside national political shifts. Today the topic is a major focus of foreign affairs, alongside matters of security and trade. In the industrial region, the drive to see transnational environmental regulations has been impelled by the recognition that it is easier to negotiate transnational agreements between partners when economic disparities are least marked, and trade-offs are most possible. It is also increasingly accepted that the United Nations system, other international bodies, regional organizations and national bodies need to work in partnership, and become inter-sectoral, because environmental issues pervade all components of social and economic policy. Decision-makers in the environmental field are now using international fora to:

- identify emerging environment/development issues;

- design initiatives and programmes that address them in an integrated and co-ordinated way;

- ensure, through regular review, that these initiatives and programmes are implemented in a timely fashion; and;

- identify and deal with any impediments.

The world's institutional machinery is responding to these needs in two main ways. First, it is developing an increasing number of linkages, both within the UN system (e.g. CIDIE, DOEM, SWMTEP) and between the UN bodies and other governmental and non-governmental groups (e.g. ECG, joint programme activities). Second, the individual institutions are adopting a more integrated and trans-sectoral approach. It seems clear that both trends will continue.

Ingredients of the response

This broadening and interlinkage has dangers. It can erode the special expertise of a body, and hence the authority of its contribution. The response is, essentially, to link expert bodies each of which has its own distinct standing and speciality in a cross-sectoral way, and to marshal the groups best able to contribute to a response regardless of their affiliation. It means openness, awareness and information flow, but not rivalry or take-over bids.

The vehicle of a response to an environmental issue varies with circumstance and need. It can be expressed as an international declaration, action plan, strategy, guidelines, codes of practice or other forms of 'soft law', or binding international law. The choice of instrument must be made according to circumstance, following analysis and debate.

This in turn implies that the institutions need to intensify their capacity to collect and synthesize information, build data bases, monitor trends, and establish frameworks within which meetings to undertake the necessary evaluations can occur. The past two decades have seen an extraordinary proliferation of environmental conferences and meetings, with massive duplication and almost certainly a considerable waste of time and resources. A more co-ordinated approach would clearly be cost-effective.

Implementation of decisions also requires effective co-ordination, and procedures to ensure that the delay between agreements and their entry into force is minimized. In the 1970s, IMO concluded several agreements with a 'tacit acceptance procedure' and this is the quickest way of bringing instruments into force. Under this kind of system (first used in the 1972 London Convention on the Prevention of Marine Pollution by Dumping of Wastes and Other Matter) technical amendments adopted by majority vote enter into force for all parties within a given period unless a specified proportion (usually one-third or one-quarter) of the Parties object (Sand, 1990; ICAO, 1984; Kimball, 1989). Where agreements need implementation by states (as is commonly the case), financial assistance is often necessary, and recent agreements have included provisions for funds to support information exchange, technology transfer, training and technical assistance. Technical co-operation to help developing countries implement 'soft law' guidelines and standards, and also international legal agreements, has long been funded by WHO, IMO, IAEA, UNEP and UNESCO. As a general policy, international obligations should not be accepted by governments until institutional and financial mechanisms for implementation have been studied and provided for.

Responses within the UN system

In responding to the overall challenges of integration and effectiveness, the UN system has emphasized the need for a number of actions that apply fully in the response to environmental problems. These are: (a) a co-ordinated medium-term planning process and (b) programme and budget review. CPC, ACC, ECOSOC and the General Assembly are all concerned with reviewing the effectiveness of the activities supported by the UN system, and the Joint Inspection Unit (JIU) is a potential instrument for internal analysis of programme effectiveness as well as for internal audit of services and expenditure (UN, 1990a).

The system should also be able to review performance in or foster agreement on:

(a) *Implementation at regional level* (programmes of Regional Economic Commissions and of Regional Development Banks).

(b) *The implementation of international legal instruments* (a number of which now incorporate provisions for international inspection (e.g. 1959 Antarctic Treaty; 1968 Treaty on the Non-Proliferation of Nuclear Weapons; International Whaling Convention and various fisheries conservation agreements) and require reports from states parties).

(c) Development of new international legal instruments to address:

(i) the assessment of the environmental impact of activities that might have transboundary effects or influence the global commons; and

(ii) liability and compensation aspects under environmental legal instruments.

(d) Agreement on the establishment and funding of co-operative monitoring (the ECE co-operation initiative for monitoring and evaluation of air pollutants in Europe, EMEP, is a model here, and does include commitments for funding) (Sand, 1987).

(e) Agreement that each state party to an international instrument designates a management and scientific authority as a focal point, as done under the 1973 Convention in International Trade in Endangered Species of Flora and Fauna (CITES), and subsequently incorporated into international regimes for trade in chemicals, conservation of migratory species, transboundary movement of hazardous wastes and emergency response to marine pollution and nuclear accident.

(f) *Integration of environmental and development policies.* This action has been advanced by inter-agency discussions and conferences, and the establishment of the Committee of International Development Institutions on the Environment (CIDIE) in 1980 as described earlier in this chapter. At policy level, the establishment of the post of Director General for International Economic Co-operation, and of a Joint Consultative Group on Policy (JCGP) as a forum for co-ordinating the approaches of IFAD, UNDP, UNFPA, UNICEF and WFP, is an advance in this field. However, much needs to be done to strengthen national management capabilities - indeed this is perhaps the single greatest priority for development assistance (UN, 1990b: Box 5).

(g) *Examination of trade and debt issues.*

UNCTAD has been requested to review whether the environmental regulations applied by industrialized countries to particular products constitute non-tariff barriers to trade with developing countries

(UNCTAD, 1989). Another major issue that is repeatedly put forward is the relationship between debt and environmental deterioration. This issue must be addressed in the UN fora as a matter of urgency.

Concluding remarks and future prospects

The growth of the UN system, its specialized agencies, and other international bodies (including the non-governmental organizations), has been predominantly sectoral. In this respect the international system mirrors the national. Both international and national levels now face a cross-sectoral and cross-institution co-ordination challenge.

In the UN system, the response has been to establish a committee structure and to improve the machinery for information exchange. Real integration in the shape of joint programmes has been less evident. The ECB was an attempt to support inter-agency co-operation but was limited by the independent nature of the different agencies and programmes. The 'thematic joint programming' approach was extended in 1991 in the inter-agency co-operation with the Secretariat preparing the UN Conference on Environment and Development, held in Rio de Janeiro in June 1992. The issue in 1992 will be how to strengthen machinery like DOEM, and composite programmes such as SWMTEP, to produce really effective co-operation between the various organs and organizations of the UN system dealing with environment.

The past two decades have seen models for co-operation both within the UN and between UN agencies and other bodies, and these precedents seem likely to be copied in future. WMO and UNEP jointly established the Intergovernmental Panel on Climate Change (IPCC) in 1988, to review the scientific evidence, consider implications and evaluate what might be done to shield communities from the disruptive impact of such change. An expert group convened by IAEA developed the 'Code' of Practice for the Transboundary Movements of Nuclear Wastes', adopted at the General Conference of IAEA in October 1990. The code requires that transboundary movements of radioactive waste should only take place in accordance with internationally accepted safety standards, with prior notification and consent of the sending, receiving and transit states. This followed the adoption of the Basel Convention which dealt with hazardous wastes other than radioactive ones. FAO (fisheries), UNEP, and the contracting parties to the London Dumping Convention joined forces to review draft guidelines (prepared under the auspices of IMO in 1989) governing offshore installations and structures such as oil rigs, in order to implement article 60 of the 1982 UN Convention on the Law of the Sea.

Non-UN and non-governmental organizations have also been involved in an increasing number of cooperative programmes with the international agencies. In previous sections, reference has been made to the ECG and its role in developing and reviewing SWMTEP. ECG has also been the forum for the development of action, led by UNEP, to negotiate a new international convention on the conservation and rational use of biological diversity. Several other environmental agreements have been prepared in analogous ways. IUCN worked closely with UNEP to develop the 1979 Bonn Convention on Migratory Species, initially provided the secretariat for the CITES Convention (now serviced by UNEP), and continues to provide the secretariat for the 1971 Ramsar Convention. The collaboration between the Antarctic Treaty Consultative Parties and SCAR, the Scientific Committee on Antarctic Research of ICSU, has been close and effective for over 30 years. The World Conservation Monitoring Centre, (WCMC) established in 1988, is a co-operative venture between UNEP, IUCN and the World Wide Fund for Nature (WWF). Its primary function is the continuous collection, analysis and dissemination of information and data on the conservation of animal and plant species, wildlife trade, protected areas, and specific habitats such as tropical forests and coral reefs. Such collaboration offers great advantages to the UN system as well as to the partner non-UN organizations. It is likely to become an increasing feature of the future.

Such inter-agency and UN-non-UN co-operation is a reflection of a parallel process at regional and national levels. For example, in Europe national NGOs have played a leading part in pressing their governments to more forceful action, and have joined together on a European scale to move the environment programme of the European Community forward. Pressure from the NGO community was a leading factor in the achievement of the moratorium on commercial whaling under the International Whaling Convention. NGOs formed a special association, the Antarctica and Southern Ocean Coalition (ASOC), to press governments to adopt stronger conservation measures in Antarctica, ban minerals activities there, and do more to protect the ecosystems of the adjacent seas. Campaigns to remove lead from gasoline, curb motor vehicle pollution, stop acid rain and protect tropical forests have all had a major influence on state and intergovernmental action.

The need for dialogue between the governmental and non-governmental sectors has also been recognized in the structure of several recent conferences. UNEP has co-operated with the International Chamber of Commerce in two World Industry Conferences on Environmental Management (WICEM), held in Versailles, France, and Rotterdam, Netherlands, in 1984 and 1991. These meetings brought representatives of business and commerce together with environmental NGOs, and members of governments and the intergovernmental community. At Bergen in 1990 the Government of Norway and the UN Economic Commission for Europe hosted a European Ministerial Conference in preparation for UNCED. The Ministerial Conference was preceded by separate fora on the role of the scientific community, industry, labour organizations, youth and environmental NGOs, and the conclusions of these discussions were fed into the Ministerial session. This broadening of the sphere of discussion is likely to be an enduring feature of the future.

Inter-agency co-operation has recently extended into the area of financial support. It has become increasingly clear that a global action to deal with many pressing environmental problems will demand new finances, and adjustment in present financial patterns and flows. The establishment of a fund to assist developing countries to implement the Montreal Protocol and introduce substitutes for the chlorofluorocarbons that are the agents of destruction of stratospheric ozone is an important precedent. The Global Environment Facility (GEF) represents a further significant new departure. It is administered through a tri-partite arrangement between UNEP, UNDP and the World Bank. It is to provide concessional financing for global environmental programmes in four areas: protection of the ozone layer, reduction of greenhouse gas emissions and improvement in energy efficiency, protection of international marine and freshwater resources, and the conservation of biodiversity. The GEF, which is now experimental, is supported by donor governments and is not meant to diminish current funding for development co-operation.

It is clear that there have been major changes in national attitudes and consequent international action since the Stockholm Conference. Mathews (1991) has suggested seven new understandings on which these changes rest.

- Because environmental conditions are the foundation of the human situation, environment must be a component of mainstream political and economic policy.

- The need for regional and global approaches is changing the nature of national sovereignty.

- Because national security (in both the military and economic sense) depends on global security, north-south relations are bound to change.

- Institutional reform at national, regional and global level is a high priority.

- New groups (the scientific community, environmental NGOs, and business and commerce) are becoming important actors in the development of world policy.

- The economic tool kit needs revision and strengthening.

- Much can be done at little cost by correcting current policy failures.

New trends and concepts are apparent in international policy. An inter-sectoral approach is becoming evident. For the first ten years after Stockholm, few agencies in the UN system seemed to grasp the point that UNEP stood for the environmental programme of the whole UN, and not just the Secretariat in Nairobi. But during the second decade the adoption of two System-Wide Medium-Term Programmes (SWMTEPs I and II) covering all agencies was an important step forward. Now there are signs of appreciation that the UN programme on environmental issues needs to be developed in awareness of, and partnership with, the programmes of non-UN and NGO bodies. A further significant step is towards the design of these programmes in consultation with national authorities, so that the international programmes respond to clearly defined national needs.

A driving force behind the action to deal with environmental problems, especially those of regional and global character, is the recognition of its importance in building global security. The links between peace, environment and development are now evident (Chapter 19). Environmental degradation creates tensions leading to social unrest, political instability and armed conflict. The definition of 'security' is now being broadened to take in a defence against some current environmental threats which are as real as armed threats: global warming, ozone depletion, sea level rise, desertification or conflicts over shared water resources or over transboundary air or water pollution. 'The ultimate aim should be to achieve international environmental security as a component of the system of comprehensive security.' (UNEP, 1988e).

It is clear that, despite manifest failings and inadequacies, there have been important advances in international co-operation during the past two decades. The trends toward closer co-operation in programme planning, specific activities, funding and regional action are a heartening sign that the world's institutions can function as a community, while retaining the distinctive strengths of each. We do now possess institutions and infrastructure for dealing with virtually every conceivable kind of environmental issue. There does not seem to be any need to create new institutions to achieve sustainable development and environmental security, provided that the existing ones are given the full opportunity to work effectively.

Looking to the future, it is clear that the UN Environment Programme must provide a central co-ordinating programme for the whole UN system. The Governing Council of UNEP should provide a forum for effective dialogue among member states, leading to a programme that responds to both collective and individual national needs. Such a programme must be closely co-ordinated with the work of non-UN bodies, in the intergovernmental and non-governmental sectors. It must be built on the recognition that conservation and development are part of one process: the UN Environment Programme and UN Development Programme have, accordingly, to be very closely linked. Governments, and their publics, have similarly to be led to see that environmentally sustainable development is a major - an essential - political goal. Governments need to provide the legal frameworks, economic incentives, and infrastructure to get the job done. The world community must build on what has been done recently in the establishment of new funding mechanisms, the rescheduling and forgiveness of debt and the reform of world trading patterns to facilitate these processes. All these must be ingredients in the action of the next decade, and if they are there is good hope of success.

Appendix I: Global institutions

Specialized agencies and other autonomous bodies

Institution	Established	Status	Mandate
Food and Agriculture Organization	1945	Specialized Agency	To promote action aimed at raising levels of nutrition and standards of living, improving production and distribution of all food and agricultural products, and bettering the conditions of rural populations.
General Agreement on Tariffs and Trade	1946	Specialized Agency	To expand international trade and to reduce or remove tariff and non-tariff barriers to trade.
International Atomic Energy Agency	1957	Independent under aegis of UN	To seek to accelerate and enlarge the contribution of atomic energy to peace, health and prosperity throughout the world.
International Fund for Agricultural Development	1976	Specialized Agency	To mobilize additional resources to be made available on concessional terms for agricultural development in developing member states with an emphasis on helping the poorest of the poor.
International Labour Organization	1919	Specialized Agency since 1946	To improve living and working conditions through the adoption of international labour conventions and recommendations as well as through research and technical co-operation activities.
International Maritime Organization	1958	Specialized Agency	To facilitate co-operation among governments on technical matters affecting international shipping. IMO has a special responsibility for safety of life at sea, and for the protection of the marine environment from pollution caused by ships and other craft.
UN Educational, Scientific and Cultural Organization	1945	Specialized Agency	To contribute to peace and security by promoting collaboration among nations through education, science and culture.
UN Industrial Development Organization	1966	Specialized Agency since 1986	To encourage and extend, as appropriate, assistance to the developing countries for the development, expansion and modernization of their industries.

Institution	Established	Status	Mandate
World Health Organization	1948	Specialized Agency	To promote the attainment by all peoples of the highest possible levels of health.
World Intellectual Property Organization	1970	Specialized Agency	To promote the protection of intellectual property through co-operation among states and, where appropriate, in collaboration with other organizations.
World Meteorological Organization	1873	Specialized Agency since 1950	To facilitate international co-operation in the collection, analysis, standardization and dissemination of meteorological, hydrological and other related environmental information.
World Tourism Organization	1975	Specialized Agency	To promote and develop tourism with a view to contributing to economic development, international understanding, peace, prosperity and universal respect for, and observance of, human rights and fundamental freedoms.
International Monetary Fund	1944	Specialized Agency	To restructure and reorganize the international monetary system to provide short-term financing for temporary balance-of-payments deficits. Beginning in the 1960s the IMF turned its attention toward short-term stabilization programmes for developing countries. In the 1980s, it introduced structural adjustments lending to support macro-economic adjustment programmes of longer duration.
International Bank for Reconstruction and Development	1944	Specialized Agency; World Bank Group	To promote the international flow of capital for productive purposes and to assist in financing the rebuilding of nations devastated by the Second World War. Lending to projects or to finance reform programmes that will lead to economic growth in its less developed member countries is now the Bank's main objective.
International Development Association	1960	Specialized Agency; World Bank Group	To promote economic development by providing finance to the less developed areas of the world on much more concessionary terms than those of conventional loans.
International Finance Corporation	1956	Specialized Agency; World Bank Group	To promote the growth of the private sector and to assist productive private enterprises in its developing member countries, where such enterprises can advance economic development.

UN organs, programmes, commissions, and other bodies

Institution	Established	Status	Mandate
UN Children's Fund	1946	UN Organ	To provide assistance, particularly to developing countries, in the development of permanent child health and welfare services.
UN Conference on Trade and Development	1964	UN Organ	To promote international trade, particularly between countries at different stages of development, in order to accelerate economic growth in developing countries.
UN Development Programme	1966	UN Programme	To assist developing countries to accelerate their economic and social development. UNDP is the central funding and co-ordinating mechanism for technical co-operation in the UN system.
World Food Programme	1961	Joint organ of FAO and the UN	To provide food aid primarily to low income, food deficit countries, to assist in the implementation of economic and social development projects, and to meet the relief needs of victims of disasters.
UN Environment Programme	1972	UN Programme	To co-ordinate, catalyse, and stimulate environmental action primarily, but not exclusively, within the UN system.
UN Population Fund	1967	Subsidiary organ of the GA; under administrative authority of UNDP	To provide assistance in the field of population and to promote and co-ordinate population programmes and projects in the UN system.
UN Population Commission	1946	UN Commission	To study and advise ECOSOC on population issues, policies, etc. In 1975, the commission was requested to examine on a biennial basis efforts to implement the World Population Plan of Action.
UN Commission on Human Settlements	1977	UN Commission	To promote international co-operation in the efforts to solve human settlements problems.
UN Commission on Transnational Corporations	1974	UN Commission	To act as a forum within the UN system for the comprehensive and in-depth consideration of issues relating to transnational corporations.
Committee on the Development and Utilization of New and Renewable Sources of Energy	1982	UN Committee	To promote, guide and review efforts to implement the 'Nairobi Programme of Action for the Development and Utilization of New and Renewable Sources of Energy'.
Office of the UN Disaster Relief Coordinator	1971	Reports to Secretary-General	To mobilize, direct, and co-ordinate the relief activities of the various UN organizations and NGOs in response to a request for assistance from a state that has been stricken by a disaster.

Appendix II: Regional institutions

Institution	Established	Status	Mandate
Economic Commission for Africa	1958	UN Commission	To initiate and participate in measures for facilitating concerted action for the economic development of Africa, including its social aspects.
Economic and Social Commission for Asia and the Pacific	1947	UN Commission	To further the economic and social development of the Asia-Pacific region.
Economic Commission for Europe	1947	UN Commission	To generate and improve economic relations among its members and with other countries and to strengthen intergovernmental co-operation.
Economic Commission for Latin America and the Caribbean	1948	UN Commission	To facilitate concerted action for dealing with urgent economic problems; for raising the level of economic activity in the region; and for maintaining economic relations of the region.
Economic and Social Commission for Western Asia	1973	UN Commission	To further the economic and social development of the West Asian region.
African Development Bank	1964	Regional Bank	To contribute to the economic development and social progress of its members - individually and jointly.
Inter-American Development Bank	1959	Regional Bank	To contribute to the economic and social development of the regional developing member countries.
Arab Bank for Economic Development in Africa	1973	Regional Bank	To contribute to the economic and social development of developing member countries.
Asian Development Bank	1965	Regional Bank	To assist the economic and social advancement of developing member countries.
Caribbean Development Bank	1970	Regional Bank	To contribute to the harmonious economic growth and development of the member countries in the Caribbean.
European Bank for Reconstruction and Development	1991	Regional Bank	To assist in the economic reconstruction and development of the countries of Eastern Europe.
Organization of American States	1948	Regional Political	To facilitate peace and justice among the American nations, promote their solidarity, strengthen their collaboration, and defend their sovereignty, territorial integrity, and independence.

Institution	Established	Status	Mandate
Caribbean Community	1973	Regional Economic	To promote economic co-operation and understanding among member states.
Organization of African Unity	1963	Regional Political	To promote unity and solidarity among members and defend their sovereignty, territorial integrity and independence.
Economic Community of West African States	1975	Regional Economic	To facilitate economic integration and co-operation among West African States.
Economic Community of Central African States	1983	Regional Economic	To promote economic co-operation among member states.
League of Arab States	1945	Regional Political	To work toward peace and security in the Arab region; settle regional disputes; and promote co-operation among members.
Association of Southeast Asian Nations regional	1967	Regional Political	To accelerate economic growth among member nations; develop intra-regional co-operation; encourage social and cultural progress, and promote peace and security.
South Pacific Forum	1972	Regional Political	To maintain co-operation, aid and development among member nations of the Pacific region.
Organization for Economic Cooperation and Development	1961	Regional Economic	To promote policies designed to achieve the highest sustainable economic growth and employment, and to contribute to world-wide economic expansion and the expansion of world trade on a non-discriminatory basis.
European Economic Community	1958	Regional Economic	Advocates continuous and balanced economic expansion within the European community. Promotes economic stability, improvement of living standards in participating countries, and closer socio-political ties among members.
Conference on Security and Cooperation in Europe	1975	Regional Political	Supports and promotes European security based on economic, cultural, environmental, and humanitarian co-operation.

Appendix III: Major United Nations conferences since 1972

Date	Conference	Product	Follow-up
1974	World Population Conference	World Population Plan of Action (WPPA)	1984 and 1994 conferences
1974	World Food Conference	Universal Declaration on the Eradication of Hunger and Malnutrition	
1976	UN Conference on Human Settlements	Vancouver Action Plan for Human Settlements	Regular review by UNCHS; 1988 Global Strategy for Shelter to the Year 2000
1977	UN Water Conference	Mar del Plata Action Plan	International Drinking Water Supply and Sanitation Decade (1981-1990); Global Consultation on Safe Water and Sanitation for the 1990s (1990); International Conference on Water and the Environment 1991.
1977	UN Conference on Desertification	Plan of Action to Combat Desert-ification (PACD)	UNEP 1984 assessment; second assessment completed in 1992.
1978	UN Conference on Technical Co-operation among Developing Countries	Buenos Aires Plan for Promoting and Implementing Technical Cooperation among Developing Countries	
1979	UN Conference on Science and Technology for Development	Vienna Programme of Action on Science and Technology for Development	Elaboration and implementation of the Programme by the UN Centre for Science and Technology for Development
1979	World Climate Conference		Second World Climate Conference (1990)
1981	UN Conference on New and Renewable Sources of Energy	Nairobi Programme of Action for the Development and Utilization of New and Renewable Sources of Energy	

Date	Conference	Product	Follow-up
1981	UN Conference on Least Developed Countries		1990 conference
1984	World Conference on Agrarian Reform and Rural Development	Programme of Action on Agrarian Reform and Rural Development	
1984	World Fisheries Conference	Strategy for Fisheries Management and Development	
1990	Second World Climate Conference	Scientific/technical reports; Ministerial Declaration	Intergovernmental negotiations on a Climate Convention
1991	International Conference on Agriculture and the Environment	Strategy and Agenda for Action for Sustainable Agriculture and Rural Development	
1992	International Conference on Water and the Environment		
1992	UN Conference on Environment and Development	Earth Charter; Agenda 21; Declaration of Principles	

REFERENCES

Caldwell, L.K. (1990) *International Environmental Policy: Emergence and Dimensions.* 2nd Edn. Duke University Press, Durham and London.

CIDIE (1988) *Action and Interaction. The role and potential of CIDIE.* CIDIE Secretariat, UN Environment Programme, Nairobi.

DocTer (1991) *European Environmental Year Book.* 2nd Edition. DocTer International, London, UK.

ECLAC (1991) *Sustainable Development: Changing Production Patterns, Social Equity and the Environment..* Economic Commission for Latin America and the Caribbean, Santiago de Chile.

GESAMP (1990) *The State of the Marine Environment.* UN Environment Program, Nairobi.

Holdgate, M.W., Kassas, M. and White, G.F. (eds) (1982) *The World Environment, 1972-1982,* A Report by UNEP. Tycooly International, Dublin.

Haigh, N. (1989) *The environmental policy of the European Community and 1992. International Environment Reporter,* 13 December 1989. The Bureau of National Affairs, Inc. Washington DC.

ICAO (1984) *The Convention on International Civil Aviation. The first 40 years.* ICAO Public Information Office, January 1984, Geneva.

ILO (1989) ILO Contribution to Environmentally Sound and Sustainable Development. Report of the Governing Body of the International Labour Office to the 44th Session of the General Assembly in response to General Assembly resolutions 42/186 and 42/187. ILO, Geneva, 1989.

IUCN (1991) *A Strategy for Antarctic Conservation.* Gland, Switzerland: IUCN - The World Conservation Union, Gland, Switzerland.

IUCN / UNEP / WWF (1980) *The World Conservation Strategy.* IUCN, Gland, Switzerland.

IUCN / UNEP / WWF (1991) *Caring for the Earth: A Strategy for Sustainable Living.* IUCN - The World Conservation Union, Gland, Switzerland.

Kimball, L. A. (1989) International Law and Institutions: the Oceans and Beyond. *Ocean Development and International Law,* **vol. 21(2)**, 1990, P. 147. Taylor and Francis Ltd., London.

Mathews, Jessica T. (1991) Introduction and Overview. *In: Preserving the Global Environment: The Challenge of Shared Leadership,* (ed) J.T. Mathews, W.W. Norton, New York and London.

OECD (1977) *Cooperative Technical Program to measure the Long Range Transport of Air Pollution.* OECD, Paris.

OECD (1980) *Environmental Policies for the 1980s.* Organization for Economic Co-operation and Development, Paris.

OECD (1991) Meeting of OECD Ministers on Environment and Development, Paris, 2-3 December 1991. Document SG/PRESS (91) 71. Organization for Economic Co-operation and Development, Paris.

Pearson, C. and Pryor, A. (1978) *Environment North and South: an Economic Interpretation..* John Wiley and Sons, New York.

Petsonk, C.A. (1990) The role of the United Nations Environment Programme (UNEP) in the Development of International Environmental Law. *American University Journal of International Law and Policy,* **vol.5**, No.2, winter 1990, p. 365-72.

Sand, P. (1987) Air Pollution in Europe: International Policy Responses. *Environment,* **Vol. 29**, No. 10, December 1987, p. 16.

Sand, P. (1990) *Lessons Learned in Global Environmental Governance..* World Resources Institute, Washington DC.

Stein and Johnson (1979) *Banking on the Biosphere?* Lexington Books, Lexington, Massachusetts.

Sweden (1972) Air Pollution across National Boundaries. The impact on the environment of sulphur in air and precipitation. A/Conf.48/CS/1.2.26. Sweden's Case Study for the UN Conference on the Human Environment. Royal Ministry for Foreign Affairs. Royal Ministry of Agriculture. Swedish Preparatory Committee for the Conference. Stockholm 1971.

Thacher, P. S. (1990a) Background to Institutional Options for Management of the Global Environment and Commons. A Preliminary Paper for the World Federation of United Nations Associations' Project on Global Security and Risk Management. World Resources Institute, Washington DC.

Thacher, P.S. (1990b) Sea Pollution: The Mediterranean - A New Approach to Marine Pollution. Paper prepared for International Institute for Applied Systems Analysis (IIASA), 29 September 1990. World Resources Institute, Washington DC.

Thacher, P.S. (1990c) Alternative legal and institutional approaches to global change. *Colorado Journal of International Environmental Law and Policy*, **vol.1**, no.1, summer 1990. University Press of Colorado, Niwot, Colorado.

UK (1987) Ministerial Declaration, Second International Conference on the Protection of the North Sea. Department of the Environment, London.

UN (1945) Charter of the United Nations, Articles 55 and 57. United Nations Office of Public Information, New York.

UN (1972) General Assembly Resolution 2997 (XXVII), 15 December 1972.

UN (1973) General Assembly Resolution 3129 (XXVIII), 13 December 1973.

UN (1975) General Assembly Resolution 3436 (XXX), 9 December 1975.

UN (1977) General Assembly Resolution 32/197 of 20 December 1977.

UN (1990a) Application of Evaluation Findings in Programme Design, Delivery and Policy Directives. Report of the Secretary General. Document A/45/204, 17 April 1990. United Nations, New York.

UN (1990b) Implementation of the Recommendations of the Joint Inspection Unit. Report of the Secretary General. UN Document A/45/441, 13 September 1990. United Nations, New York.

UNCTAD (1989) Environment and Development: UNCTAD's Contribution to the Follow-up of General Assembly Resolutions 42/186 and 42/187. TD/B/1199 4 January 1989. United Nations Conference on Trade and Development, New York.

UNDP (1990a) *Human Development Report.* UN Development Programme, New York.

UNDP (1990b) 1989 Annual Report of the Administrator: A Watershed for Technical Co-operation. DP/1990/17, p. 2-3. UN Development Programme, New York.

UNEP (1975) UNEP Governing Council Decision 35 (III), May 1975.

UNEP (1981a) *In Defence of the Earth: the basic texts on Environment*: Founex, Stockholm, Cocoyoc. UN Environment Programme, Nairobi.

UNEP (1981b) Environmental Law: *An In-depth Review. UNEP Report*, no.2, 1981. UN Environment Programme, Nairobi.

UNEP (1982a) *Montevideo Programme for the Development and Periodic Review of Environmental Law.* UN Environment Programme, Nairobi.

UNEP (1982b) *Annual Report of the Executive Director.* Nairobi: UN Environment Program.

UNEP (1982c) *A Review of the Major Achievements in the Implementation of the Stockholm Action Plan.* UN Environment Programme, Nairobi.

UNEP (1983) *Annual Report of the Executive Director.* UN Environment Programme, Nairobi.

UNEP (1984) *Annual Report of the Executive Director.* UN Environment Programme, Nairobi.

UNEP (1985) *Annual Report of the Executive Director.* UN Environment Programme, Nairobi.

UNEP (1986) *Annual Report of the Executive Director*. UN Environment Programme, Nairobi.

UNEP (1987a) *Annual Report of the Executive Director*. UN Environment Programme, Nairobi.

UNEP (1987b) *UNEP Profile*. UN Environment Programme, Nairobi.

UNEP (1988a) *Annual Report of the Executive Director*. UN Environment Programme, Nairobi.

UNEP (1988b) *Environmental Perspective to the Year 2000 and Beyond*. UN Environment Programme, Nairob.

UNEP (1988c) *The United Nations System-Wide Medium-Term Environmental Programme, 1990-1995*. UN Environment Programme, Nairobi.

UNEP (1988d) *UNEP News*, April 1988, p.3. UN Environment Programme, Nairobi.

UNEP (1988e) Report of the Ad Hoc Experts' Group Meeting on Expanded Concept of International Security, UNEP, Nairobi, 23-26 February 1988. UN Environment Programme, Nairobi.

UNEP (1989a) UNEP Governing Council paper UNEP/GC.15/L.24. *The Clearing-House Function*. UN Environment Programme, Nairobi.

UNEP (1989b) *Annual Report of the Executive Director*. UN Environment Programme, Nairobi.

UNEP (1989c) Implementation of General Assembly Resolutions 42/186 on the Environmental Perspective to the Year 2000 and Beyond and 42/187 on the Report of the World Commission on Environment and Development. UNEP/GC.15/6/ Add.2. UN Environment Programme, Nairobi.

UNEP (1990a) *GEMS: The Global Environmental Monitoring System*. UN Environment Programme, Nairobi.

UNEP (1990b) *UNEP Profile*. UN Environment Programme, Nairobi.

UNEP (1991a) *Environmental Data Report*, Third Edition. Basil Blackwell, Oxford.

UNEP (1991b) Status of regional agreements negotiated in the framework of the Regional Seas Programme, Rev.3, UNEP, Nairobi.

UNEP (1991c) *Register of international treaties and other agreements in the field of the environment*, UNEP, Nairobi.

UNEP/ASEAN (1988) ASEAN Environment Program III (ASEP III), 1988-1992. UNEP-ASEAN Environment Program Series, February 1988. UN Environment Programme, Nairobi.

UNESCO (1989) Report on Progress made by UNESCO towards the Objectives of Environmentally Sound and Sustainable Development. SC-89/WS.26. UNESCO, Paris.

UNESCO/UNEP (1989) Report of the Conference on Environmental Education and Training, Moscow, 1987. Environment Programme, Nairobi and UNESCO, Paris.

UNWC (1978) *Proceedings of the UN Water Conference, Mar del Plata*, Argentina, 1977. Pergamon Press, Oxford.

US-Canada (1980) Memorandum of Intent between the Government of Canada and the Government of the United States of America Concerning Trans-Boundary Air Pollution. State Department, External Affairs, Washington DC and Ottawa.

WCED (1987) *Our Common Future*. The report of the World Commission on Environment and Development. Oxford University Press, Oxford.

Worthington, E.B. (ed) (1975) *The evolution of IBP*. Cambridge University Press, Cambridge.

PART FOUR

LOOKING AHEAD

CHAPTER 24

Challenges and opportunities

Introduction

Ten years ago, UNEP's publication *The World Environment 1972-1982* concluded with the observation that:

'At the Stockholm conference it was generally assumed that the world's system of national governments, regional groupings and international agencies had the power to take effective action. ... By the early 1980s there was less confidence in the capacity of national and international managerial systems to apply known principles and techniques, or in the effectiveness with which international debates lead to action ... Restoration of confidence and consensus in these areas may be the greatest challenge for those seeking to improve the world environment in the 1980s' (Holdgate *et al.*, 1982).

It is disturbing that the same statement is still valid a decade later. Indeed, many of the concerns identified in the earlier report remain the same. There are still serious gaps in our understanding of the environment, our ability to estimate the cost of repairing the damage we have done to it, and our knowledge of the cost of failing to take rapid action to halt its degradation. Twenty years after Stockholm it is still not possible to describe the state of the world environment comprehensively or to say with confidence that the governments of the world have the knowledge or the political will to deal with the global problems we already know exist.

The most significant concerns remain the lack of many of the prerequisites for informed decision-making and good environmental management. In particular:

- The data base is still of variable quality, with a shortage of data from developing countries. As a result, comprehensive data on the major environmental problems cannot be compiled, and 'best estimates' are all that are available.

- Despite great advances in science, remote sensing and the technical ability to monitor the world environment, these have not been generally applied, mainly because of a lack of equipment and trained personnel in many countries.

- There has been no general agreement on the socio-economic indicators of a healthy relationship between people and their environment or on standards for a decent environment.

- Comprehensive assessments of the environmental situation and of the Earth's carrying capacity are, in consequence, still difficult to produce.

Despite these concerns there has been clear progress in a number of areas during the past decade. The scientific assessment of stratospheric ozone depletion and understanding of the processes involved have progressed very rapidly and have been matched by international and national actions to redress the situation. A strong scientific consensus is now emerging on climate change and loss of biodiversity, their causes and the need for a collective response. Some progress has been made in dealing with the problems of hazardous wastes and toxic chemicals, and there are more and better overall assessments of the environment backed by improved data compendia.

The first section of this volume outlines the ten major environmental issues that are of concern today; the second deals with the human actions that give rise to such problems, while the third deals with human responses to these issues, individually and collectively, at the global, national and community level. These responses hold the key to our future and that of our descendants. It is the level of awareness and the individual perceptions of people that will ultimately determine whether or not governments and businesses integrate environmental concerns into their development plans and economic activities. In recent years there has been a marked increase in public awareness of environmental issues, and rising demands for action on them. There has also been an enormous growth in the scientific knowledge relating to them. Instruments such as the 'Precautionary Principle', the 'Polluter Pays' and 'User Pays' principles, and techniques such as Environmental Impact Assessment, have been adopted as policy by some governments. The debate of the last two decades on environment and development, and the studies of the inter-relationships between people, resources, environment and development called for by the UN General Assembly in the mid-1970s (see for example UNGA resolution 3345 (XXIX)) have also led to a growing consensus on the key issues and goals - most notably that of sustainability.

The grand debate: understanding environment and development interrelationships and practising interdependence

Concern for the environment is as old as human civilization. History abounds with examples of the wide variations in human understanding of the environment and in our ability to maintain it in a healthy condition. Those societies that managed to provide their material, cultural and spiritual needs in a sustainable manner were those that succeeded in reconciling their needs and aspirations with the maintenance of a viable environment. Whenever the outer limits of the physical environment were exceeded, civilizations declined or even vanished.

As signs of irreparable damage to the environment have threatened both the inner limits of basic human needs and the outer limits of the planet's physical resources, we have witnessed a quantum jump in concern for the environment over the last two decades. The Stockholm Conference in 1972 turned a new page in the book of human concern for the environment. We have seen intensive discussions of interactions between environment and development. A few examples suffice to show how much environmental concerns influenced them.

• **The World Models**: The Club of Rome decided early on to initiate a project on 'The Predicament of Mankind'. Phase One of the project resulted in the well-known *Limits to Growth* report in 1972, followed in 1974 by the second Club of Rome report, *Mankind at the Turning Point*. This heralded a series of world models, most prominent among which are the 'Latin American World Model' (Bariloche Model), the OECD 'Interfutures', and the UN 'Future of the World Economy' (Leontief). Despite their different objectives, methodologies, assumptions and approaches (and notwithstanding any shortcomings they had) in each of them the rational use of natural resources and the dangers of environmental degradation figure prominently. Chapter 20 contains a fuller discussion of these and other models.

• **The UN Development Decades**: The United Nations International Development Strategies for the Third and Fourth Development Decades (UN 1980, 1990), in particular, emphasize the dependence of human health and well-being on the maintenance of the integrity and productivity of the environment and its resources.

• **UN General Assembly Resolutions**: As early as 1974, the UN General Assembly passed its first resolution on the 'Interrelationships between Population, Resources, Environment and Development'. Ever since then there has been a continuing review of UN efforts in co-ordinating multi-disciplinary research and pilot field experiments on these interrelationships. The GA Resolutions 42/186 & 187 in 1987, welcoming and endorsing respectively the report of the World Commission on Environment and Development, *Our Common Future* (WCED, 1987) and UNEP's *Environmental Perspective to the Year 2000 and Beyond* (UNEP, 1988), heralded a period of intensive review of development strategies guided by the recommendations of the two reports. Since that time the General Assembly has received regular reports from the Secretary General of the United Nations on how governments and UN organs are implementing these recommendations. The debate finally led the General Assembly to convene the UN Conference on Environment and Development held in Rio de Janeiro, Brazil, in June 1992.

The last two decades also gave rise to a number of commissions and studies to investigate the current and future condition of the world at large and to recommend appropriate actions for sustainable and equitable development on a global scale. Their reports include *North-South: A Programme for Survival* (Brandt, 1980), *The Global 2000 Report to the President of the United States* (Barney, 1980), *The World Conservation Strategy* (IUCN/UNEP/WWF, 1980), *A World at Peace: Common Security in the Twenty-first Century* (Palme, 1989) and *Caring for the Earth: a Strategy for Sustainable Living* (IUCN/UNEP/WWF 1991).

The report of the World Commission on Environment and Development, UNEP's *Environmental Perspective to the Year 2000 and Beyond,* (1987) and *The Challenge to the South* - the report of the South Commission (South, 1990) specifically addressed the question of sustainable development.

Although these models, strategies and reports tackled different issues, using different criteria, and with different objectives, each addressed the complex relations between the physical environment, development efforts and the human condition.

In addition, during the last two decades UNEP has commissioned, sponsored and itself undertaken work on many of the concepts that lie behind sound environmental management. The resulting reports and publications have dealt with issues as diverse as the inter-relationships between people, resources, environment and development, the effect of different patterns of development and life-styles on the environment, the integration of environment into development planning and the techniques of resource accounting and environmental economics. There have also been innumerable national, regional and international meetings and publications addressing various aspects of the relationship between environment and development and the concept of sustainable development.

Twenty years after Stockholm it is both useful and sobering to review some of the main ideas, principles and concepts that were enunciated during this grand debate. They covered new international development strategies, new life-styles, the strong ties between environment and economics, the need for additional financial resources and technology transfer, inter- and intra-generational equity, the precautionary principle, the rights and responsibilities of states and several others.

This review, which is by no means comprehensive, reveals two significant facts. The first is *the enduring relevance and validity of these ideas and concepts* (Boxes 1-5). The second is that *issues identified two decades ago remain with us today.* The ideas have continued to be reiterated ever since the preparatory meetings for the Stockholm Conference - particularly the Founex meeting almost one year before the conference - and in the Conference Declaration and Action Plan. Despite all the reflection and debate, our environment continues to deteriorate and living conditions in many parts of the world continue to decline.

BOX 1

Environment and development.

1972   The capacity of the Earth to produce vital renewable resources must be maintained and, wherever practicable, restored or improved ...

The protection and improvement of the human environment is a major issue which affects the well-being of peoples and economic development throughout the world.

Stockholm Declaration

1972   Taking no action to solve these problems is equivalent to taking strong action ... A decision to do nothing is a decision to increase the risk of collapse.

Club of Rome: *The Limits to Growth*

1975   Environmental management implies sustainable development.

UNEP Governing Council

1981   Human beings ... must come to terms with the reality of resource limitation and the carrying capacities of ecosystems and must take account of the needs of future generations.

IUCN/UNEP/WWF: *World Conservation Strategy*

1987   In essence, sustainable development is a process of change in which the exploitation of resources, the direction of investments, the orientation of technological development, and institutional change are all in harmony.

WCED: *Our Common Future*

1987   Anticipatory and preventive policies are the most effective and economical in achieving environmentally sound development.

UNEP: *Environmental Perspective to the Year 2000 and Beyond*

1989   No government can say with certitude that it has attained environmentally sound development or, indeed, has available methodology to do so.

UNEP Governing Council: *Decision 15/2*

1990   To achieve sustainable development, policies must be based on the precautionary principle.

Bergen Declaration, Bangkok Declaration

1991   Humanity must live within the carrying capacity of the Earth. There is no other rational option in the longer term. Unless we use the resources of the Earth sustainably and prudently, we deny people their future. We must adopt life-styles and development paths that respect and work within nature's limits.

IUCN/UNEP/WWF: *Caring for the Earth*

BOX 2

Environment and economics.

1971   For society as a whole, environment is a part of its real wealth and cannot be treated as a free resource.

Environmental issues may come to exercise a growing influence on international economic relations. They ... could influence the pattern of world trade, the international distribution of industry, the competitive position of different groups of countries, their comparative costs of production, etc.

Some environmental actions by developed countries ... are likely to have negative effects on developing countries' export possibilities and their terms of trade.

Founex Report

1972   GATT, among other international organizations, could be used for the examination of the problems [of trade and the environment], specifically through the recently established Group on Environmental Measures and International Trade.

Stockholm Plan of Action

1987   Incomplete accounting occurs ... especially in the case of resources that are not capitalized in enterprise or national accounts: air, water and soil.

WCED: *Our Common Future*

1987   Environmentally related regulations and standards should not be used for protectionist purposes.

UNEP: *Environmental Perspective to the Year 2000 and Beyond*

1991   New models that incorporate ethical, human and ecological factors as well as economic ones are being developed. They are badly needed as we face the challenge of sustainable human development.

IUCN/UNEP/WWF: *Caring for the Earth*

BOX 3

Environment and international relations.

1972   States have ... the sovereign right to exploit their own resources pursuant to their own environmental policies, and the responsibility to ensure that activities within their jurisdiction or control do not cause damage to the environment of other States or of areas beyond the limits of their national jurisdiction.

Stockholm Declaration

1984   It cannot be too much or too often emphasized how crucial the dialogue between and amongst the developed and developing countries is for the development of appropriate solutions to environmental problems.

UNEP: *State of the Environment 1984*

1987   Awareness of the environmental aspects of international economic relations has increased but it has not yet found adequate expression in institutional practices and national policies.

UNEP: *Environmental Perspective to the Year 2000 and Beyond*

1989   The responsibility for ensuring a better environment should be equitably shared and the ability of developing countries to respond be taken into account.

Langkawi Declaration

1990   If the multiple bonds that characterize interdependence are convincingly present in any field, it is in that encompassing development and the environment ... But the transition is not occurring smoothly and harmoniously; it is turbulent and beset with conflict.

South Commission: *South Report*

1991   The ethic of care applies at the international as well as the national and individual levels. All nations stand to gain from world-wide sustainability, and are threatened if we fail to attain it.

IUCN/UNEP/WWF: *Caring for the Earth*

BOX 4

Additional financial resources.

1971   Looking at the problem [of who pays for the higher cost arising out of environmental concerns] strictly from the point of view of the developing countries, it is quite clear that additional funds will be required.

Founex Report

1972   The General Assembly ... *decides* that the Governing Council [of UNEP] shall have the following main functions and responsibilities:
... (f) To maintain under continuing review ... the problem of additional costs that may be incurred by developing countries in the implementation of environmental programmes and projects .

UN General Assembly: *Resolution 2997*

1981   The General Assembly [recognizes] that environmental deficiencies generated by the conditions of under-development pose grave problems and can best be remedied by accelerated development through the transfer of substantial quantities of financial and technical assistance.

UN General Assembly: *Resolution 36/192*

1989   The Governing Council [of UNEP] believes that the international community must not only commit itself generally to the proposition that such additional resources are needed, but must specifically identify their possible sources.

UNEP Governing Council

BOX 5

Technology transfer.

1971   It is quite likely that future technological developments in the developed world will be influenced by their current preoccupation with non-pollutive technology ... It is also obvious that some of this non-pollutive technology would be quite costly for the developing countries ... its export to developing countries under tied credits will further reduce the real content of foreign assistance.

Founex Report

1972   It is recommended that the Secretary General [of the UN] be asked to ... find means by which environmental technologies may be made available for adoption by developing countries under terms and conditions that encourage their wide distribution without constituting an unacceptable burden to developing countries.

Stockholm Plan of Action: *Recommendation 108*

1987   Transfer of clean, low-waste and pollution control technologies should be promoted ... The scope to make available such technologies at concessional prices to the countries in need should be explored.

UNEP: *Environmental Perspective to the Year 2000 and Beyond*

1989   Governments and intergovernmental organizations [are] urged to review terms of trade in pollution control technology with the objective of identifying and minimizing trade barriers.

UNEP Governing Council: *Decision 15/6*

1991   The Governing Council [of UNEP] calls upon the Executive Director: ... to promote the identification of ways and means to facilitate access by and transfer of technology to developing countries in respect of cleaner production methods, techniques and technologies.

UNEP Governing Council: *Decision 16/33*

Familiar problems get worse, while new environmental threats are revealing themselves. The present generation of human beings carries a terrible responsibility for its lack of action on almost every front.

> I do not wish to seem over-dramatic; but I can only conclude ... that the Members of the United Nations have perhaps ten years left in which to subordinate their ancient quarrels and launch a global partnership to curb the arms race, to improve the human environment, to defuse the population explosion, and to supply the required momentum to development efforts.
>
> *U. Thant (UN Secretary General, 1969)*

How far did we move to heed this warning? Specifically, how much have we improved the human environment since it was given? These world statistics give the answer:-

> In 1969 the world's population was 3.7 billion. In 1990 it reached 5.3 billion, and it is still growing by 1.7 per cent each year.[1]
>
> In 1989, $US850 billion was spent on arms, while official aid to developing countries was just $US34.1 billion and the developing countries paid $US59.5 billion in interest on loans.[2]
>
> Today, it is estimated that *every second* over 200 tonnes of carbon dioxide are released and 750 tonnes of topsoil are lost; while *every day*, 47,000 hectares of forest are destroyed, over 16,000 hectares of land are turned to desert, between 100 and 300 species become extinct and 40,000 children die of disease and malnutrition. As this goes on, 1,116 million people in developing countries are living below the poverty line.[3]
>
> *Sources*: 1. UNFPA 2. Brown *et al.* (1991) 3. World Bank/UNDP data.

This book, and the boxes in this chapter, demonstrate the need to commit ourselves *now* to specific, quantifiable targets over specific periods of time, and to set ourselves in earnest to achieve them. In addition, we have to ensure that we have the mechanisms to monitor our progress towards these targets, as well as the courage to acknowledge when we are wrong and the wisdom to correct our path. If we do not, we will be left with more lip service, more destruction, more misery. Surely the international community does not want this and cannot afford it.

811

The changing world scene

The world has not been standing still while the debate on environment and development has gathered momentum and while ideas, concepts and issues have emerged, been clarified and reiterated. The two decades since 1972 have seen major political, economic and social changes. The global political and economic landscape has altered, not gradually but in a number of dramatic and unforeseeable upheavals. As a result, the ideological and economic world maps of 1972 are no longer accurate in 1992: the geo-political assumptions that accompanied them do not hold true today, and the predictions of social change that were based on them have been proved inaccurate.

The challenge

The most dramatic and obvious political changes have been the most recent. The movement to democratic pluralism and the drastic economic, social and political changes in the former Soviet Union and the countries of Central and Eastern Europe have gripped world attention and often dominated the news media since the introduction of *perestroika* in the USSR in the mid-1980s. However, the causes are to be found earlier, in more subtle changes to the prevailing philosophies in both East and West which had more profound consequences than was immediately obvious. The change from an essentially bipolar world in which two super-powers and their supporters faced each other across an ideological and political abyss has created both opportunities and uncertainties. It may be some time yet before a new geo-political map is finally drawn, but the nature of that map and the world it represents will owe more to the fundamental causes of those changes than to the changes themselves.

The radical optimism of the early 1970s gave way under the pressure of the global economic recession which followed the second oil shock in 1978. The belief, which under-pinned the Second UN Development Decade and the call for a 'New International Economic Order', that institutional solutions could be found to human and social problems, was replaced by a more individualistic, inward-looking, market-oriented philosophy. Paradoxically, the same improvements in mass communications that have liberated individuals and fuelled popular demands for political reform have also led to an increased sense of individual helplessness in the face of mounting environmental crises, and greater popular distrust of politically-generated solutions to social, economic and environmental problems.

The search for greater economic efficiency has led not only to the phasing out of the 'command economies' world-wide, and particularly in Eastern Europe, but to a parallel deregulation of the financial, manufacturing and service sectors in the West. One result has been a freer movement of capital between countries and between sectors. This globalization of production was also boosted by profound technological changes, particularly the rapid development and spread of information technologies. Knowledge suddenly became a business asset even more valuable than physical or financial resources. The traditional trans-national corporation was transformed from a collection of semi-autonomous local affiliates into a widespread network of production and operational facilities that could be monitored and controlled from the centre in 'real time'. At the same time 'modular' production developed, with specialized firms producing just one module of a total product or service. As a result, many large corporations have come to depend on the small knowledge-intensive enterprise which spends a high proportion of its income on research and development. These changes, among other things, have further reduced the ability of national governments to influence the activities of trans-national corporations in order to achieve social or environmental goals. While the flight of capital from already under-capitalized countries in the South has continued, there have also been waves of foreign investment in those economies that took advantage of the changed economic scene and the gradual dismantling of the rather shaky protectionist measures instituted in many developing countries to nurture their emerging industries. The emergence of global markets and economic interdependence has largely been fuelled by the extraordinary growth in trade since World War II. In the 40 years following 1950, world manufacturing output increased seven-fold. Between 1950 and 1973, export volumes from industrialized countries increased by 10 per cent per year. Since 1984, the volume of world merchandise trade has consistently risen at a higher rate than growth in world outputs. However, both the trade boom and the growth in the global economy have been highly selective, by-passing most developing countries. Although there has been a significant increase in developing country exports of manufactured goods during the last two decades, it is mainly accounted for by the relatively small number of newly industrialized economies, and despite this increase, developing country exports are still only 15 per cent of world manufactures trade (UNCTAD, 1990).

The few value-added products that are generated in developing countries are often blocked by lack of market access, as developing country commodity exports are affected by the 'new protectionism' which followed the recession of the early 1980s. Non-tariff barriers, voluntary export restraints, direct and indirect subsidies and other obstacles have made developing country access to northern markets extremely difficult. According to the World Bank, the percentage of OECD country imports covered by non-tariff barriers almost doubled between 1966 and 1986 (Figure 1). Moreover, the percentage of trade affected by highly restrictve non-tariff measures is greater for developing countries than for industrialized countries (Figure 2). Subsidies on agricultural produce within the OECD are in the vicinity of $US300 billion per annum. The cost (in 1990 dollars) to the global South in 1980 of trade protectionism in developed countries has been estimated at around $US55 billion (World Bank, 1991).

Changes in the global economy have been at the centre of discussions in intergovernmental trade forums, and produced some serious strains in the

Figure 1: *The share of imports affected by all non-tariff measures, 1966 and 1986.*

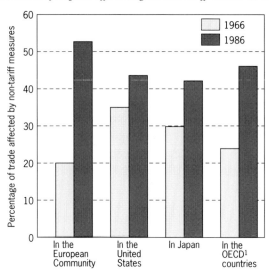

Note: 1. Excludes data for Australia, Austria, Canada, Iceland, New Zealand and Sweden.
Source: World Bank (1991).

Uruguay round of negotiations on the General Agreement on Tariffs and Trade (GATT). While there has been a considerable degree of agreement on the principle of lowering barriers to international trade, and in particular the removal of non-tariff barriers, agreement on concrete actions has been more difficult to achieve. The question of agricultural protectionism, which has significant implications for the world environment, has been one of the most intractable, producing significant disagreements between countries and trading blocs in the North as well as between North and South. While in some areas, notably primary production and commodity trade, a freer (but equitable) world market could reduce the pressure to exceed the carrying capacity of over-stressed ecosystems in developing countries, the current trend towards deregulation also poses a threat to the environment.

A totally deregulated world market suggests a possible conflict with international treaties and national legislation to protect the global commons, for example the Montreal Protocol or the CITES Convention (both of which contain restrictions on trade with non-parties) or measures that might be taken under conventions on climate change or biological diversity. Similar difficulties could arise in connection with any agreements that might be reached to ensure that developing countries have access to environmentally sound technology. The potential for such difficulties was clearly seen as early as the Founex meeting which preceded the Stockholm Conference (see Box 5). The objective, identified in 1971, of reconciling the need for an orderly, fair and efficient system of world trade with the need to preserve the ability of the Earth to sustain human life has now become pressing.

Figure 2: *Hard-core non-tariff measures applied against industrial and developing countries, 1986.*

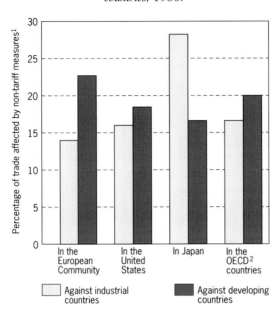

Notes: 1. Calculated using 1981 trade weights. Hard-core non-tariff measures include quotas, voluntary export restraints, the Multifibre Arrangement, and other highly restrictive measures.

2. Excludes data for Australia, Austria, Canada, Iceland, New Zealand and Sweden.

Source: World Bank (1991).

Over the past 20 years, both the World Bank and IMF have shifted development priorities from import substitution to export-led growth accompanied by severe structural adjustment programmes. For most developing countries with scant industrial capacities there is little to export but natural resources, making them almost totally reliant on commodity exports. However, commodity prices have fallen steadily since the early 1970s. By 1986, average real commodity prices were at their lowest recorded levels in this century (with the single exception of 1932, the trough of the Great Depression). Prices of two critical export crops – cocoa and coffee – fell even further between 1986 and 1989 (by 48 and 55 per cent respectively). The World Bank forecasts that commodity prices are unlikely to rise during this decade, with intensified South-South competition in saturated markets.

The effects on developing country economies of trade protectionism and commodity price collapses are compounded by external debt. It is now clear that the economies of the least developed countries will not be able to grow out of debt, as donor agencies had hoped. By 1989 the external debt of the countries rated by the World Bank as 'severely indebted' equalled almost three times their combined export earnings for that year. The total indebtedness of all developing countries was $US988 billion, approaching twice the value of their exports, 22 per cent of which went on debt servicing. The most indebted region (relative to its earning capacity) is Sub-Saharan Africa, with external debt almost equal to its combined GNP and approaching four times its total exports, a fourfold increase in just nine years (World Bank, 1991).

The combined effect of debt servicing and reduced aid is a net financial flow from the South to the North. In 1989, the developing countries paid $US59 billion in interest on their debts (World Bank, quoted in SIPRI, 1991), and received official development assistance of $34 billion (UNDP, 1991). In the same year the official debt of low- and middle-income countries grew by an average of four per cent (World Bank, 1991). Increasing interest payments on a spiralling debt burden can only be met by increasing exports.

For countries that are almost totally dependent on commodity exports in a hostile market this means placing greater pressure on the environment and a further reduction in living standards for their people. With a projected one billion additional people sharing scarce resources in the global South in the near future, the pace of environmental degradation seems certain to increase unless the debt crisis is resolved and greater equity introduced to the world's commodity markets.

The social and environmental costs of the economic changes of the last two decades have been severe. While the world economy has grown considerably during that period, much of the growth has been in countries that were already consuming an inordinate share of the world's resources. Many of the least developed countries had little economic growth and a substantial fall in per capita production during the 1980s. Furthermore, the growth was greater in the manufacturing and service sectors than in food production which (on a per capita basis) declined in the latter half of the 1980s (FAO, 1991). The fall in per capita food production has been most severe and most sustained in the least developed countries, while mountains of surplus food are stored in Western Europe and North America. Finally, the growth has been accompanied by a horrifying loss of the world's productive capacity. Between 1970 and 1990 the world population increased by about 1 billion. During the same period an estimated 480 billion tons of topsoil were lost and an additional 120 million hectares of desert were created (Brown et al., 1991).

These changes have been coupled with regional conflicts which have plagued much of Southeast Asia, Africa and the Middle East for the last two decades. As a result, glaring anomalies in global patterns of resource consumption have been made worse, the standard of living of the poorest half of the world's population has fallen even further and the incidence of famine and disease in the poorest countries has increased. Greater poverty in the least developed countries has led to increased environmental degradation and so to a vicious downward spiral of poverty and loss of productive capacity. Furthermore, the least developed countries now face the possible diversion of a substantial portion of existing aid flows and concessional development capital to help with the reconstruction of Eastern Europe.

Figure 3: *Net resource flows and net transfers to developing economies, 1980–89, (billions of dollars).*

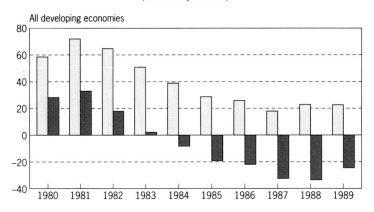

All developing economies

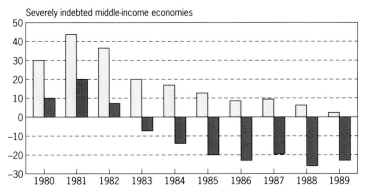

Severely indebted middle-income economies

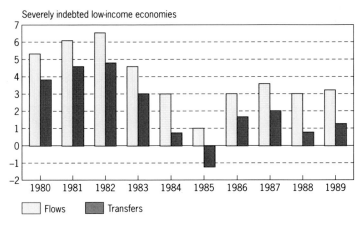

Severely indebted low-income economies

☐ Flows   ■ Transfers

Note: Data are for all economies reporting transactions to the World Bank and refer to long-term debt, excluding IMF.

Source: World Bank (1990).

The opportunity

To concentrate only on the negative statistics of the last two decades and ignore promising present trends and recent events would give a distorted and overly negative picture. While we must accept the reality that the two decades since the Stockholm Conference have seen a considerable degradation of the global environment and a further squandering of the world's stock of productive natural resources, there are also some grounds for optimism. A growing appreciation of the global nature of environmental problems and their implications - not just for the quality of life but for its very sustenance - has led to a new and more serious approach to environmental issues since the mid-1980s. Governments have displayed a greater willingness to act together to address environmental threats on a global basis, as was demonstrated by the successful negotiation between 1985 and 1987 of the Montreal Protocol on Substances that Deplete the Ozone Layer, its dramatic strengthening in 1990 and the large number of countries that have ratified it. This rapid and decisive action (at least in terms of international treaty negotiations) and the steps taken since then towards the negotiation of conventions on control and disposal of hazardous wastes, on climate change, and on the conservation of biodiversity, would have been hard to predict even a decade earlier. This willingness to act has been accompanied by an encouraging movement away from confrontation and towards a more co-operative approach by governments in forums dealing with environmental issues. It has thus been possible to develop new and innovative means (for example the funding mechanism established under the Montreal Protocol and the Global Environmental Facility) to address issues such as the transfer of environmentally sound technology to developing countries and to deal with major environmental problems.

Overall, the world community has accumulated a large number of legal instruments that commit it to co-operate in the environmental field. As Chapter 23 records, in the 50 years prior to 1972 fewer than 60 environmental treaties were agreed. During the 20 years since Stockholm almost 100 more have been concluded. There are now over 150 multilateral environmental treaties, yet the efficiency and effectiveness of these legal instruments leave much to be desired, mainly because of the limitations inherent in the traditional negotiating process, limited compliance with existing legislation, ineffective procedures for the enforcement and verification of implementation, and lack of mandatory jurisdiction in legal instruments for settling disputes. There is now a growing recognition among governments that the traditional means by which they negotiate and agree on collective action is unlikely to be able to keep pace with growing threats to the world's environmental health.

The basic responsibilities and rights of nations relating to the environment (Box 6) have been spelled out in a number of declarations and other statements of principle, including the 1972 Stockholm Declaration, the Nairobi Declaration of 1982, The World Charter for Nature (1982), the Report of the World Commission on Environment and Development (1987) and many others. All these have been adopted, endorsed or welcomed by the United Nations. In addition there have been many regional and other declarations of a similar nature.

BOX 6

Some principles governing the rights and obligations of people and states in relation to the environment.

Man has the fundamental right to freedom, equality and adequate conditions of life, in an environment of a quality that permits a life of dignity and well-being, and he bears a solemn responsibility to protect and improve the environment for present and future generations.

Stockholm Declaration, 1972 (Principle 1)

It is prohibited to employ methods or means of warfare which are intended, or may be expected, to cause widespread, long-term and severe damage to the natural environment.

Additional Protocol I to the Geneva Convention, 1977

[The General Assembly] proclaims the historical responsibility of States for the preservation of nature for present and future generations.

UNGA Resolution 35/8, 1980

States shall maintain ecosystems and ecological processes essential for the functioning of the biosphere, shall preserve biological diversity, and shall observe the principle of optimum sustainable yield in the use of living natural resources and ecosystems.

Report of the WCED, 1987 (Legal Principle 3)

States shall develop contingency plans regarding emergency situations likely to cause transboundary environmental interferences and shall promptly warn, provide relevant information to and co-operate with concerned States when emergencies occur.

Report of the WCED, 1987 (Legal Principle 19)

Since mass poverty is often at the root of environmental degradation, its elimination and ensuring equitable access of people to environmental resources are essential for sustainable development.

UNEP: Environmental Perspective to the Year 2000 and Beyond, 1987

Most of the declarations and legal instruments of the past two decades are of the nature of 'soft laws': that is to say, they do not have a legally binding effect but are statements of principle to which governments give general support or claim to subscribe. Their significance lies in the fact that, taken together, they clearly demonstrate the common, consistently held view of nations that their sovereign rights are circumscribed in the field of environment by a set of higher principles. Whether their espousal of these principles is enough to induce governments to agree on limitations to or exceptions from the rule of consensus, which is at the heart of the law on treaties, remains to be seen. However, the fact that almost all (if not all) governments have repeatedly subscribed to these declarations of principle without reservation for two decades encourages the belief that they may now be ready to streamline and strengthen the negotiating process in the interests of global preservation.

The last two decades have also witnessed the rise of non-governmental environmental groups and consumer organizations. These have gained great strength - particularly in the industrialized countries - and influenced, to a large extent, the attitudes of governments and industries towards the environment and its natural resources.

At the same time (and as described in Chapter 21), a combination of greater public awareness and growing consumer pressure have led to a 'greening' of many businesses. The creation of the Business Council for Sustainable Development, and the adoption, by the ICC-sponsored Second World Industry Conference on Environmental Management (WICEM II) of a 'Business Charter for Sustainable Development', are encouraging signs. The last few years have witnessed an emerging tendency among industry leaders in favour of adopting voluntary codes of practice, rather than waiting for (and resisting the imposition of) regulations to protect the environment. Several large-scale enterprises have openly and voluntarily committed themselves to substantial reductions in the release of pollutants over relatively short time periods. The growing realization that environmentally sound business practices are also good business practices augurs well for the future. In addition (also outlined in Chapter 21), a growing number of professional bodies (for example in the fields of medicine and engineering) have adopted codes of ethics or stated principles relating to professional obligations in regard to the environment.

These changes in perception and practice are obviously not by themselves going to solve the world's environmental problems. However, if they represent a genuine recognition of the seriousness of the situation the world faces, and a real willingness on the parts of industry, governments and individuals to act together to address it, then there is ground for optimism. The break-down of the political and ideological barriers dividing East and West presents an unequalled opportunity to divert a substantial part of the world's economy from producing weapons of destruction to repairing the environmental damage of the past. If the world is ready to take advantage of this opportunity then the next twenty years may tell a different story.

Epilogue

This review of environmental trends during the last two decades, and the interactions between human development needs and environmental resources, would be incomplete without reference to the lessons it holds for the future. The overriding message it conveys is that human beings and human institutions wait too long before taking action. On nearly every issue we have failed to apply the precautionary principle when it was appropriate - indeed prudent -to do so. We have had more than enough talk about environmental problems.

Throughout this book, and mainly in the concluding remarks of some chapters, we have given examples of achievable future actions which, with real international co-operation, could be carried out during the next two or three decades. They do not by themselves constitute a plan of action to address the environmental problems of the world, and should not be interpreted as such. They are, however, consistent with the goals set out in documents that have been welcomed or endorsed by the UN General Assembly[1] and by the Governing Council of UNEP[2]. More recent statements of goals have included the 'Priorities for Action' contained in the UNEP Director-General's *Saving our Planet* (Tolba, 1992), presented at the Special Session of the UNEP Governing Council in February 1992 and subsequently at UNCED; the 'International Research Priorities' identified by ASCEND 21 (November 1991), and UNCED's 'AGENDA 21', presented in June 1992.

The actions suggested in this report, as well as those in the documents cited above, cover a variety of fields. They range from a more comprehensive assessment of the state of the environment and the environmental impacts of human activities, to sharper tools of economic analysis to give proper estimates of the cost of environmental damage and rehabilitation. They will only be achieved on a global scale if a whole host of support measures can also be undertaken: development of national capabilities on a wide front of individual and institutional fields; closer international co-operation based on the spirit of true partnership, and a strong conviction that environmental issues cut across political, social and economic divides.

There is no priority, in this volume, between, for example, actions needed to combat land degradation and marine pollution or between the building in developing countries of an adequate capacity in environmental economics and legislation and the development of an effective, internationally agreed means of ensuring compliance with treaties on the environment. Each represents a

1. *Environmental Perspective to the Year 2000 and Beyond* (UNEP, 1988); *Our Common Future*, the report of the World Commission on Environment and Development (WCED, 1987).
2. *Goals and Targets: Future Orientation of UNEP's Programme* (UNEP 1987); *World Conservation Strategy* (IUCN/UNEP/WWF 1980); *Caring for the Earth: A Strategy for Sustainable Living* (IUCN/UNEP/WWF 1991); *Global Biodiversity Strategy* (WRI/IUCN/UNEP 1992).

priority in its own field. Despite the very real achievements and advances of recent years the state of the world environment and the living conditions of many of its people have continued to get worse. The message of this epilogue is the same as the burden of argument throughout this book; that we no longer have the luxury of picking and choosing what to do next: the state of the world environment demands that we take action simultaneously on a broad front, with no further delay. Unless the world community is prepared to act decisively and co-operatively to deal with the problems outlined in the first ten chapters of this book, and their basic causes described in the subsequent section, the rate of degradation of our living environment will increase rapidly. It is no exaggeration to say that the ability of the biosphere to continue to support human life is now in question. That fact should compel us to implement what we say we should do.

What are needed now are specific commitments to take specific actions, over specific periods of time, with the costs calculated, the sources of funding identified, and a clear indication of who will be doing what. Nothing less will suffice.

References

Barney G. (dir) (1980) *The Global 2000 Report to the President of the United States: Entering the 21st Century*, Pergamon Press, Oxford, UK.

Brandt (1980) Report of the Independent Commission on International Development Issues (Brandt Commission) *North-South: A Programme for Survival*, Pan, London.

Brown L. R., During, D., Flavin, C., French, H., Jacobson, J., Lenssen, N., Lowe, M., Postel, S., Renner, M., Ryan, J., Starke, L., Young, J. (1991) *State of the World 1991, A Worldwatch Institute Report on Progress Toward a Sustainable Society*, Norton, New York and London.

CIDIE (1991) *Action and Interaction: The Role and Potential of CIDIE*, CIDIE, Nairobi.

FAO (1991) *International Agricultural Adjustment - Seventh Progress Report*, FAO conference document.

Holdgate, W., Kassas, M., and White, G. (1982) *The World Environment 1972 - 1982*, Tycooly International, Dublin.

IUCN/UNEP/WWF (1980) *The World Conservation Strategy*, IUCN, Gland, Switzerland.

IUCN/UNEP/WWF (1991) *Caring for the Earth: a Strategy for Sustainable Living*, IUCN, Gland, Switzerland.

Palme (1989) *A World at Peace: Common Security in the Twenty-First Century*, Palme Commission on Disarmament and Security Issues Stockholm, Sweden.

SIPRI (1991) *The SIPRI Yearbook 1991: World Armaments and Disarmament*, Oxford University Press, Oxford, U.K.

South (Nyerere) Commission (1990) *The Challenge to the South Commission*, Oxford University Press, Oxford, U.K.

Tolba, M.K. (1992) *Saving Our Planet*, Chapman & Hall, London.

UN (1980) *Development and International Economic Co-operation*, A/35/592/Add.1, UNGA, New York.

UN (1990) *Development and International Economic Co-operation: International Development Strategy for the Fourth United Nations Development Decade (1991-2000)*, A/45/849/Add., UNGA, New York.

UNCTAD (1990) *Handbook of International Trade and Development Statistics 1990*, United Nations, New York.

UNEP (1987) *Environmental Perspective to the Year 2000 and Beyond*, UNEP, Nairobi.

UNEP (1992) *Status of Desertification and Implementation of the United Nations Plan of Action to Combat Desertification*, UNEP, Nairobi.

UNDP (1991) *Human Development Report 1991*, Oxford University Press, Oxford and New York.

WCED (1987) World Commission on Environment and Development *Our Common Future*, Oxford University Press, Oxford, UK.

World Bank (1991) *World Development Report 1991: The Challenge of Development*, OUP, New York.

WRI/IUCN/UNEP (1992) *Global Biodiversity Strategy*.

APPENDIX A

Contributors to the report

M. **Abraham**
Head, Information & Research
International Organization of Consumer
Unions
Penang
Malaysia

A. **Agarwal**
Director
Centre for Science and Environment
New Delhi
India

E.C.F. **Bird**
Geostudies
Victoria
Australia

A.K. **Biswas**
President
International Society for Ecological
Modelling (ISEM)
Oxford
England

Ms M. **Biswas**
International Society for Ecological
Modelling (ISEM)
Oxford
England

J. **Butlin**
Natural Resources & Environmental
Management
Cheshire
United Kingdom

M.T. **El-Ashry**
Director, Environment Department
The World Bank/IFC/MIGA
Washington, DC
USA

Ms M.C.J. **Cruz**
Visiting Scholar
Population Reference Bureau
Washington, DC
USA

E. **El-Hinnawi**
Professor
National Research Centre
Cairo
Egypt

O.A. **El-Kholy**
Emeritus Professor
University of Cairo
Egypt

K. von **Gehlen**
Professor
Institut für Geochemie und Petrologie
Frankfurt
Germany

M. **Hashimoto**
President
Overseas Environmental Co-operation
Center
Kanagawa Prefecture
Japan

M. **Holdgate**
Director-General
World Conservation Union (IUCN)
Gland
Switzerland

H.F. **Kaltenbrunner**
Director
Resources Planning Consultants b.v.
Delft
The Netherlands

Ms L.A. **Kimball**
Senior Associate
World Resources Institute
Washington, DC
USA

T. **Lambo**
Executive Director
Lambo Foundation
Lagos
Nigeria

T. **Mansson**
Senior Advisor/Expert
Ministry of Environment
Stockholm
Sweden

A.D. **McIntyre**
Emeritus Professor
University of Aberdeen
Scotland

J. **McNeely**
Chief Conservation Officer
World Conservation Union (IUCN)
Gland
Switzerland

B. **Moore**
Professor
University of New Hampshire
Durham
USA

R.E. **Munn**
Professor
University of Toronto
Canada

R. **Oldeman**
Head, Programmes and Projects,
International Soil References and
Information Centre (ISRIC)
Wageningen
The Netherlands

H. **Saddler**
Director
Energy Policy & Analysis Pty
Deakin
Australia

R. **Sandbrook**
Executive Director
International Institute for
Environment and Development (IIED)
London
United Kingdom

W. **Sombroek**
Director
International Soil References and
Information Centre (ISRIC)
Wageningen
The Netherlands

M.S. **Swaminathan**
Chairman
Swaminathan Research Foundation
Madras
India

P. **Timmerman**
Institute for Environmental Studies
University of Toronto
Canada

R. **Watson**
National Aeronautic and Space
Administration (NASA)
Washington, DC
USA

D.M. **Whelpdale**
Senior Research Scientist
Atmospheric Environment Service
Ontario
Canada

T. **Wigley**
Director
Climatic Research Unit
University of East Anglia
United Kingdom

T.M. **Wolters**
Ecoplan b.v.
Willemstad
The Netherlands

United Nations Centre for
Human Settlements (Habitat)
Nairobi
Kenya

World Health Organization (WHO)
Geneva
Switzerland

World Conservation Monitoring Centre
(WCMC)
Cambridge
United Kingdom

APPENDIX B

Reviewers, participants and commentators in workshops

A. **Al Gain**
Acting Director ROPME
Meteorology & Environmental Protection
Administration (MEPA)
Jeddah
Saudi Arabia

M. **Ardui**
First Secretary
Permanent Mission of Belgium to UNEP
Nairobi
Kenya

J.K. **Atchley**
Permanent Representative
Permanent Mission of the USA to UNEP
Nairobi
Kenya

D.N. **Axford**
Deputy Secretary-General
World Meteorological Organization (WMO)
Geneva
Switzerland

M.A. **Ayyad**
Professor
Plant Ecology, Faculty of Science
University of Alexandria
Egypt

M. **Batisse**
President Mediterranean Blue Plan
c/o UNESCO
Paris
France

M.C. **Baumer**
Senior Scientist
International Centre for Research on
Agriculture and Forestry (ICRAF)
Nairobi
Kenya

P. **Berthoud**
46 Chemin des Coudriers
Geneva
Switzerland

L.J. **Brinkhorst**
Director-General
Commission of the European Community
Brussels
Belgium

M.Z. **Cutajar**
Executive Secretary
Intergovernmental Negotiating Committee
for a Framework Convention on Climate
Change (INC/FCCC)
Geneva
Switzerland

D.F. **di Castri**
President
Scientific Committee on Problems of the
Environment (SCOPE)
Paris
France

Lord S. **Clinton-Davis**
S.J. Berwin & Co.
London
United Kingdom

H.B. **Dieterich**
Divonne-les-Baines
France

D.J. **Dixon**
Producer, Agricultural Programmes
BBC World Service
Dorset
United Kingdom

E. **El-Hinnawi**
Professor
National Research Centre
Cairo
Egypt

O.A. **El-Kholy**
Emeritus Professor
University of Cairo
Egypt

A. **Faiz**
Highways Adviser
The World Bank
Washington, DC
USA

I. **Fells**
Professor
University of Newcastle Upon Tyne
United Kingdom

R. **Frosch**
Vice President
General Motors Research Laboratories
Michigan
USA

P. **Garau**
Chief, Settlements Planning & Policies
Section
UNCHS (Habitat)
Nairobi
Kenya

G. **Garcia-Duran**
Permanent Representative
Permanent Mission of Colombia to UNEP
Nairobi
Kenya

T.H. **Gatara**
Project Development Advisor
International Planned Parenthood
Federation (IPPF)
Nairobi
Kenya

Q. **Geping**
Administrator
National Environment Protection Agency
Beijing
People's Republic of China

G. **Glaser**
Focal Point for Environmental Affairs
UNESCO
Paris
France

G.N. **Golubev**
Professor
Department of Geology
Moscow State University
Russian Federation

Mr. R. **Goodland**
The World Bank
Washington, DC
USA

Mr. M.J. **Grubb**
Research Fellow
Royal Institute of International Affairs
London
United Kingdom

Ms A. **Hammad**
Adviser to the Director-General on
Health & Development Policies
Word Health Organization (WHO)
Geneva
Switzerland

Sir P. **Harrop**
Chairman
UNEP-UK Committee
London
United Kingdom

B. **Hitchcock**
Permanent Mission of Australia to UNEP
Nairobi
Kenya

S. **Hoffman**
Assistant Secretary for Research
Smithsonian Institution
Washington, DC
USA

M. **Holdgate**
Director-General
World Conservation Union (IUCN)
Gland
Switzerland

W-K. **Hoogendoorn**
Permanent Representative, Permanent
Mission of the Royal Netherlands to UNEP
Nairobi
Kenya

J. **Houghton**
Chief Executive and
Director-General
Meteorological Office
Berkshire
United Kingdom

Ms A. **Hussein**
Cairo
Egypt

Ms. P. **Huston**
85 Carlton Mansions
London
England

R. **Jaffe**
Professor
Universidad Simon Bolivar
Caracas
Venezuela

M. **Kassas**
Emeritus Professor
University of Cairo
Egypt

S. **Keckes**
Rovinz
Yugoslavia

M. **Khalid**
Chairman, Centre for Our Common Future
Geneva
Switzerland

V.S. **Kitaev**
Permanent Representative
Permanent Mission of the Russian
Federation to UNEP
Nairobi
Kenya

W. **Kreisel**
Director, Division of Environmental Health
World Health Organization (WHO)
Geneva
Switzerland

G. **Kullenberg**
Secretary, Intergovernmental
Oceanographic Commission (IOC)
c/o UNESCO
Paris
France

I. **Lang**
Hungarian Academy of Sciences
Budapest
Hungary

G. **Lean**
The Observer
London
United Kingdom

C.A. **Liburd**
Permanent Representative
Permanent Mission of the Co-operative
Republic of Guyana to Kenya
Nairobi
Kenya

Dr. S. **Litsios**
World Health Organization (WHO)
Geneva
Switzerland

D.J. **Lisk**
Director
New York State College of Agriculture
and Life Sciences
Ithaca, New York
USA

S. **Lodgaard**
Director
International Peace Research Institute
(PRIO)
Oslo
Norway

W.L. **Long**
Director, Environment Directorate
Organization for Economic Co-operation
and Development (OECD)
Paris
France

R. **Lonngren**
Nasvagen 13
Broma
Sweden

R.A. **Luken**
United Nations Industrial Development
Organization (UNIDO)
Vienna
Austria

Y.A. **Mageed**
Associate Consultants & Partners
Khartoum
Sudan

M.S. **Mahdavi**
Permanent Representative
Permanent Mission of the Islamic Republic
of Iran to UNEP
Nairobi
Kenya

P.J. **Mahler**
Special Advisor to the Director-General
Food & Agriculture Organization of
the United Nations (FAO)
Rome
Italy

T.M. **McCarthy**
Managing Director
Haley, McCarthy & CIE S.A.
Trelex
Switzerland

D. **McMichael**
Environment and Heritage Consultants
Monaro Crescent
Australia

H.R.H. K.A. **Meesook**
President
International Council on Social Welfare
of Thailand under Royal Patronage
Bangkok
Thailand

K.E. **Mott**
Chief, Schistosomiasis Unit
World Health Organization (WHO)
Geneva
Switzerland

D. **Munro**
Project Director
World Conservation Union (IUCN)
Geneva
Switzerland

Mrs. W. **Mwagiru**
Environment and Development Officer
United Nations Conference on
Environment and Development (UNCED)
Nairobi

J.A. **Najera**
Director
Division of Control of Tropical Diseases
World Health Organization (WHO)
Geneva
Switzerland

P. **Ndegwa**
Chairman
First Chartered Securities Ltd.
Nairobi
Kenya

P. **Nijkamp**
Professor
Faculty of Economics
Free University of Amsterdam
The Netherlands

R. **Novick**
World Health Organization (WHO)
Geneva
Switzerland

P. **van der Oeven**
Head, Social Services Division
World Conservation Union (IUCN)
Gland
Switzerland

M. **Opelz**
International Atomic Energy Agency
(IAEA)
Vienna
Austria

T. **O'Riordan**
Professor
University of East Anglia
Norwich
England

M. **Osman**
Former Permanent Representative
Permanent Mission of the Arab Republic of
Egypt to UNEP
Nairobi
Kenya

F. **Penazka**
Permanent Representative
Permanent Mission of the Czech and Slovak
Federal Republic to UNEP
Nairobi
Kenya

K. **Piddington**
Former Director
Environment Department, The World Bank
Washington, DC
USA

G.T. **Prance**
Director
Royal Botanic Gardens
Surrey
United Kingdom

A. **Prost**
World Health Organization (WHO)
Geneva
Switzerland

X. **Qunbao**
National Environment Protection Agency
Beijing
People's Republic of China

A. **Ramachandran**
Executive Director
United Nations Centre for Human
Settlements (Habitat)
Nairobi

L. **Reijnders**
Professor
University of Amsterdam
The Netherlands

H. **Rodhe**
Professor
University of Stockholm
Sweden

L. **de Rosen**
Paris
France

G. **Rosenthal**
Executive Secretary
Economic Commission for Latin America
and the Caribbean (ECLAC)
Santiago
Chile

F. **De Rossi**
Permanent Mission of Argentina to UNEP
Nairobi
Kenya

R. **Schmidt**
Auf dem Acherchen 36
Wachtberg-Villip
Germany

I. **Schmitt-Tegge**
Umweltbundesamt
Berlin
Germany

B. **Shaib**
Maiduguri
Nigeria

D. **Skole**
Institute for the Study of Earth,
Oceans and Space
University of New Hampshire
Durham
USA

G. **Speth**
President
World Resources Institute
Washington
USA

B. **Stedman**
Hayward
Wisconsin
USA

R.W. **Stewart**
Global Change Science Officer
International Council of Scientific Unions
(ICSU)
Victoria, BC
Canada

M. **Strong**
Secretary-General
United Nations Conference for Environment
and Development (UNCED)
Conches, Geneva
Switzerland

E. **Tengstrom**
Professor
Centre for Interdisciplinary Studies of
Human Conditions
University of Goteborg
Sweden

R.S. **Timberlake**
Director, Transportation Studies
Dar Al-Handasah Consultants
Dubai
United Arab Emirates

S.C. **Trindade**
S E² T International, Ltd.
New York
USA

H. **Vainio**
International Agency for Research on
Cancer (IARC)
Lyon
France

National Co-ordinator of TFAP
Dept. d'Aff. Fonciers Environnement &
Conservation
Kinshasa
Zaire

Madame S. **Veil**
Chairman
WHO Commission on Health &
Environment
Geneva
Switzerland

C.C. **Wallen**
Consultant
World Meterological Organization (WMO)
Geneva
Switzerland

M.P. **Walsh**
Arlington
Virginia
USA

J. **Warford**
Environment Department
The World Bank
Washington, DC
USA

J.C. **Wheeler**
Director
Programme Integration, United Nations
Conference on Environment and
Development (UNCED)
Conches
Switzerland

G. **White**
University of Colorado
USA

Ms E. **Witoeler**
Vice-Chairman
DML (Friends of the Environment Fund)
Jakarta
Indonesia

B. **Zentilli**
Senior Programme Officer for Forestry
United Nations Conference on Environment
and Development (UNCED)
Conches
Switzerland

A.T.M. **Zahurul-Huq**
Professor
University of Dhaka
Bangladesh

Ze-Jiang **Zhou**
Director, Science & Technology Office
National Environmental Protection
Agency of China
Nanjing
People's Republic of China

Food and Agriculture Organization of the
United Nations (FAO)
Rome
Italy

United Nations Educational,
Scientific & Cultural Organization (UNESCO)
Paris
France

United Nations Children's Fund (UNICEF)
New York
USA

United Nations Population Fund (UNFPA)
New York
USA

The World Bank (IBRD)
Washington, DC
USA

World Health Organization (WHO)
Geneva
Switzerland

World Meteorological Organization (WMO)
Geneva
Switzerland

World Tourism Organization (WTO)
Madrid
Spain

International Fund for Agricultural
Development (IFAD)
Rome
Italy

International Maritime Organization (IMO)
London
United Kingdom

Intergovernmental Oceanographic
Commission (IOC)
c/o UNESCO
Paris
France

Economic Commission for Europe (ECE)
Geneva
Switszerland

Economic Commission for Latin America
and the Caribbean (ECLAC)
Santiago
Chile

Economic and Social Commission for Asia
and the Pacific (ESCAP)
Bangkok
Thailand

Commission of the European Economic
Community (EEC)
Brussels
Belgium

Greenpeace
Canonbury Villas
London
United Kingdom

International Council of Scientific Unions
(ICSU)
Paris
France

World Conservation Union (IUCN)
Gland
Switzerland

Organization for Economic Co-operation
and Development (OECD)
Paris
France

Population Council
Dokki
Egypt

World Resources Institute
Washington, DC
USA

Permanent Mission of Australia to UNEP
Nairobi
Kenya

Permanent Mission of Belgium to UNEP
Nairobi
Kenya

Permanent Mission of Canada to UNEP
Nairobi
Kenya

Permanent Mission of Finland to UNEP
Nairobi
Kenya

Permanent Mission of Japan to UNEP
Nairobi
Kenya

Permanent Mission of the United Kingdom
to UNEP
Nairobi
Kenya

Permanent Mission of the United States of
America to UNEP
Nairobi
Kenya

UNEP staff

Y.J. **Ahmad**
Special Adviser to the
Executive Director
Nairobi

Ms J. **Aloisi de Larderel**
Director
Industry & Environment (IE/PAC)
Paris
France

A. **Alusa**
Programme Officer
Global Environment Monitoring System
(GEMS/PAC)
Nairobi

Ms M. **Astralaga**
Programme Officer
Oceans & Coastal Areas (OCA/PAC)
Nairobi

M. **Atchia**
Chief
Environmental Education & Training Unit
Nairobi

Ms M. **Bjorklund**
Senior Programme Officer
Terrestrial Ecosystems Branch
Nairobi

A. **Buonajuti**
Chief
Follow-up & Evaluation Section
Nairobi

U. **Dabholkar**
Chief
Development Planning & Co-operation Unit
Nairobi

A. **Dahl**
Deputy Director
Oceans & Coastal Areas (OCA/PAC)
Nairobi

Ms. M. **de Amorin**
Director & Regional Representative
Regional Office for Africa (ROA)
Nairobi

H. **El-Habr**
Officer-in-Charge, Fresh Water Unit
Terrestrial Ecosystem Branch
Nairobi

S. **Evteev**
Special Adviser to the Executive Director
Nairobi

J.L. **Garcia**
Officer-in-Charge
State of the Environment Unit
Nairobi

N. **Gebremedhin**,
Chief
Technology & Environment Branch
Nairobi

Goh Kiam **Seng**
Director & Regional Representative
Regional Office for Asia and the Pacific
(ROAP)
Bangkok

M.D. **Gwynne**
Assistant Executive Director for Earthwatch
Co-ordination and Environmental
Assessment
Nairobi

T. **Hiraishi**
Co-ordinator
Support Measures
Nairobi

J. **Huismans**
Director
International Register of Potentially
Toxic Chemicals (IRPTC)
Geneva

J. **Illueca**
Chief
Terrestrial Ecosystems Branch
Nairobi

D. **Kaniaru**
Chief
Programme Co-ordination Unit
Nairobi

W.H. **Mansfield**
Special Adviser to the Executive Director
Nairobi

R. **Olembo**
Assistant Executive Director for
Environmental Management and Support
Measures
Nairobi

S. **Osman**
Director for Policy Development and
Inter-Organizational Affairs
Nairobi

B. **Rosanov**
Special Adviser to the Executive Director
Nairobi

Ms I. **Rummel-Bulska**
Co-ordinator
Interim Secretariat of the Basel Convention
(ISBC)
Geneva

P. **Schroder**
Acting Director
Oceans and Coastal Areas (OCA/PAC)
Nairobi

Bai-Mass Max **Taal**
Senior Programme Officer, Forests Unit
Terrestrial Ecosystems Branch
Nairobi

A. **Timoshenko**
Programme Officer
Environmental Law and Institutions Unit
Nairobi

M. **Uppenbrink**
Director & Regional Representative
Regional Office for Europe (ROE)
Geneva

P. **Usher**
Co-ordinator, Climate/Atmosphere
Global Environment Monitoring System
(GEMS/PAC)
Nairobi

B. **Waiyaki**
Programme Officer
Terrestrial Ecosystems Branch
Nairobi

P. **Woollaston**
Policy Advisor to the Executive Director
Nairobi

H. **Zedan**
Co-ordinator, Biological Diversity,
Microbial Resources and Related
Biotechnologies
Nairobi

Index

Note: index entries referring to Figures or Tables are indicated by *italic page numbers*; Boxes by **bold page numbers**.

A

AAAS (American Association for the Advancement of Science) 648
AASE (Airborne Arctic Stratospheric Expedition) 48
Abidjan city (Côte d'Ivoire) *439*, 514
ACC (UN Administrative Committee on Co-ordination) 761
accidents
  chemical 230—2
  energy activity related 395, 397
  media reaction to *666*
  nuclear 232—6, 395
  see also chemical...; nuclear accidents; oil tanker spills
acid rain
  causes 18, 19, 90, 428
  effects 7, 18—19, 28, 90—1, **229**
  history of international agreements **25**
  model simulations 20, 618—19
  research required 29
  see also nitrogen...; sulphur oxides
acidification damage 18—19, 28, 90—1, 168
  first reported 18, **25**
  regions with problems 19, *21*, 90
  remedial action 19, 91
  sensitive soils affecting 20, *21*
  soil degradation caused 150
action plan approach 747, 748
  major action plans listed **753—4**
adaptive environmental assessment/modelling 623, 641
adaptive management strategies 641
AEAM (Adaptive Environmental Assessment and Management) workshops 641
affluence, increase in environmental hazards due to 216
Afganistan, malaria in *538*
aflatoxin poisoning 551
Africa
  animals used in agriculture *286*
  arid lands *135*
  arms imports data *574*
  aviation fuel consumption data *418*
  car ownership data *412*
  cardiovascular diseases data *534*
  cereals consumption per capita *493*
  child population data *482, 483*
  clean-water supply data *95, 96*

climate-change effects **74**
commodities exported **489**
dams constructed *88*
deforestation *162, 163, 169*
desertification trends *137, 138, 139,* **142**
discharge-monitoring stations *88, 89*
drought affecting 224—5, *225,* 487
Economic Commission for 767, 792
effect of education on women 495
energy consumption data 165, *333, 334*
environmental policies 723—5
environmental reports published *709*
food production per capita *141, 144, 283, 284*
forest area *162, 169*
fuelwood consumption/deficit 165, *393,* 494
harvest of wild species 199
hazardous wastes imported *269, 349*
  Convention covering 273
homeless people 487
industrial output data *325, 326*
inland transportation of goods *423*
land use data *153, 288*
life expectancy 480
living standards *493*
malaria *538*
military expenditure *572*
nitrogen emissions data *13*
number of poor people *494*
population growth 477, *478,* 480
  future projections *479, 480*
precipitation-monitoring stations *88, 89*
public opinions on environmental issues **674—5**
radio/TV ownership *669*
regional agreements covering 780
sanitation facilities data *95, 96,* 517
schistosomiasis *537, 548*
soil degradation *137, 138, 152*
species loss *187*
Sub-Regional Environment Groups 780
sulphur emissions data/estimates *14, 22*
tourism data *454*
traditional pastoral practices **699**
urbanization *133,* 506, *507, 508*
vehicle emissions data *429, 430*
water availability *87*
water quality monitoring stations *89*
water resource agreements 100, 774—5
water use 85, *86*
wetlands
  area data *170*

R